Smart Civil Structures

Smart Civil Structures

by
You-Lin Xu
Jia He

CRC Press
Taylor & Francis Group
Boca Raton London New York

CRC Press is an imprint of the
Taylor & Francis Group, an **informa** business

MATLAB® is a trademark of The MathWorks, Inc. and is used with permission. The MathWorks does not warrant the accuracy of the text or exercises in this book. This book's use or discussion of MATLAB® software or related products does not constitute endorsement or sponsorship by The MathWorks of a particular pedagogical approach or particular use of the MATLAB® software.

CRC Press
Taylor & Francis Group
6000 Broken Sound Parkway NW, Suite 300
Boca Raton, FL 33487-2742

First issued in paperback 2019

© 2017 by Taylor & Francis Group, LLC
CRC Press is an imprint of Taylor & Francis Group, an Informa business

No claim to original U.S. Government works

ISBN-13: 978-1-4987-4398-3 (hbk)
ISBN-13: 978-0-367-87492-6 (pbk)

Library of Congress Cataloging-in-Publication Data

Names: Xu, You-Lin, 1952- author. | He, Jia (Civil engineer), author.
Title: Smart civil structures / You-Lin Xu, Jia He.
Description: Boca Raton : Taylor & Francis, CRC Press, 2017. | Includes
bibliographical references and index.
Identifiers: LCCN 2016034973| ISBN 9781498743983 (hardback : alk. paper) |
ISBN 9781498743990 (ebook)
Subjects: LCSH: Intelligent buildings. | Intelligent transportation systems.
| Intelligent control systems. | Public works.
Classification: LCC TA636 .X83 2017 | DDC 624.1--dc23
LC record available at https://lccn.loc.gov/2016034973

Visit the Taylor & Francis Web site at
http://www.taylorandfrancis.com

and the CRC Press Web site at
http://www.crcpress.com

In celebration of The Hong Kong Polytechnic University's 80th anniversary

To Wei-Jian, my wife

– Y.L. Xu

To Lei-Pin, my wife

– J. He

Contents

Foreword   xxi
Preface   xxiii
Acknowledgements   xxv
Authors   xxvii

PART I
Elements in smart civil structures

1 Introduction   3

   1.1   Civil structures 3
   1.2   Loading conditions and environments 6
       1.2.1   Dead loads 6
       1.2.2   Live loads 8
       1.2.3   Wind loads 8
       1.2.4   Earthquake loads 9
       1.2.5   Highway loads 10
       1.2.6   Railway loads 11
       1.2.7   Impact loads 11
       1.2.8   Temperature effects 12
       1.2.9   Corrosion 12
   1.3   Materials used in civil structures 12
   1.4   Design, construction and maintenance 13
   1.5   Necessity of smart civil structures 15
       1.5.1   Structural health monitoring technology 16
       1.5.2   Structural vibration control technology 17
       1.5.3   Definition of smart civil structures 19
   1.6   Historical developments of smart civil structures 20
   1.7   Organisation of this book 23
   References 24

2 Smart materials   31

   2.1   Preview 31
   2.2   Shape memory alloys 31
       2.2.1   Basic characteristics of shape memory alloys 31

      2.2.2   *Constitutive modelling of shape memory effect 34*
      2.2.3   *Applications of shape memory alloys in smart civil structures 35*
  2.3   *Piezoelectric materials 36*
      2.3.1   *Basic characteristics of piezoelectric materials 36*
      2.3.2   *Constitutive modelling of piezoelectric materials 37*
      2.3.3   *Applications of piezoelectric materials in smart civil structures 39*
  2.4   *Magnetostrictive materials 40*
      2.4.1   *Basic characteristics of magnetostrictive materials 40*
      2.4.2   *Constitutive modelling of magnetostrictive materials 41*
      2.4.3   *Applications of magnetostrictive materials in smart civil structures 42*
  2.5   *Electrorheological and magnetorheological fluids 43*
      2.5.1   *Basic characteristics of ER and MR fluids 44*
      2.5.2   *Constitutive modelling of ER and MR fluids 45*
      2.5.3   *Applications of ER and MR fluids in smart civil structures 48*
  2.6   *Optical fibres 49*
      2.6.1   *Basic characteristics of optical fibres 49*
      2.6.2   *Light propagation in optical fibre 50*
      2.6.3   *Fibre-optic sensors and their applications in smart civil structures 51*
  2.7   *Bio-inspired materials 52*
      2.7.1   *Self-healing materials 52*
      2.7.2   *Bio-inspired materials for sensing systems 53*
  2.8   *Nanomaterials 53*
  Notation 54
  References 55

3 Sensors and sensory systems   **61**

  3.1   *Preview 61*
  3.2   *Wind sensors 61*
      3.2.1   *Anemometers 61*
      3.2.2   *Pressure transducers 63*
      3.2.3   *Wind profile measurements 63*
  3.3   *Seismic sensors 64*
  3.4   *Load cells 65*
      3.4.1   *Load cells 65*
      3.4.2   *Weigh-in-motion 67*
  3.5   *Thermometers 68*
  3.6   *Strain gauges 69*
      3.6.1   *Foil strain gauge 69*
      3.6.2   *Vibrating wire strain gauge 69*
  3.7   *Displacement sensors 70*
      3.7.1   *Linear variable differential transformer 70*
      3.7.2   *Level sensing station 72*
      3.7.3   *Tilt beams 72*
      3.7.4   *Global navigation satellite system 73*
      3.7.5   *Camera 74*
  3.8   *Accelerometers 74*

3.9   *Fibre-optic sensors 76*

3.10   *Non-contact sensors 78*

3.11   *Weather stations 80*

3.12   *Chemical and corrosion sensors 80*

3.13   *Sensor performance and sensory systems 82*

Notation 83

References 83

4 Control devices and control systems   **87**

4.1   *Preview 87*

4.2   *Base isolation devices 87*

        4.2.1   *Elastomeric bearings 88*

        4.2.2   *Lead-plug bearings 89*

        4.2.3   *High-damping rubber bearings 90*

        4.2.4   *Friction pendulum bearings 90*

        4.2.5   *Other types of base isolation devices 91*

4.3   *Passive energy dissipation devices 92*

        4.3.1   *Metallic dampers 92*

        4.3.2   *Friction dampers 93*

        4.3.3   *Viscoelastic dampers 95*

        4.3.4   *Viscous fluid dampers 95*

        4.3.5   *Tuned mass dampers 97*

        4.3.6   *Tuned liquid dampers and tuned liquid column dampers 98*

4.4   *Active control devices 98*

        4.4.1   *Servovalve-controlled hydraulic actuators 99*

        4.4.2   *Pneumatic actuators 99*

        4.4.3   *Shape memory alloy actuators 99*

        4.4.4   *Magnetostrictive actuators 100*

        4.4.5   *Piezoelectric actuators 102*

        4.4.6   *Active mass damper devices 103*

        4.4.7   *Active tendon devices 103*

        4.4.8   *Active brace devices 104*

        4.4.9   *Pulse generation devices 105*

4.5   *Semi-active control devices 105*

        4.5.1   *Semi-active friction dampers 105*

        4.5.2   *Semi-active hydraulic dampers 106*

        4.5.3   *Semi-active tuned liquid dampers 107*

        4.5.4   *Semi-active stiffness control devices 108*

        4.5.5   *Electrorheological dampers 108*

        4.5.6   *Magnetorheological dampers 109*

4.6   *Hybrid control devices 110*

        4.6.1   *Hybrid mass dampers 110*

        4.6.2   *Hybrid base isolation devices 111*

        4.6.3   *Hybrid bracing control devices 112*

4.7   *Configuration of control systems and control performance 113*

References 113

5 Processors and processing systems — 121

  5.1 *Preview 121*
  5.2 *Configuration of a structural health monitoring system 121*
  5.3 *Configuration of a structural vibration control system 123*
  5.4 *Configuration of an integrated health monitoring and vibration control system 125*
  5.5 *Data acquisition and transmission system 126*
      5.5.1 *Configuration of DATS 126*
      5.5.2 *Hardware of DAUs 127*
      5.5.3 *Network and communication 128*
      5.5.4 *Operation of data acquisition and transmission 129*
  5.6 *Data processing systems for structural health monitoring 129*
      5.6.1 *Data acquisition control 130*
      5.6.2 *Signal pre-processing 130*
      5.6.3 *Signal post-processing and analysis 131*
  5.7 *Controller systems for structural vibration control 135*
  5.8 *Integrated data processing and controller systems 140*
  5.9 *Data management systems 141*
      5.9.1 *Components and functions of data management systems 141*
      5.9.2 *Maintenance of data management systems 142*
  5.10 *Structural health evaluation systems 142*
  5.11 *Wireless sensors and sensory systems 144*
      5.11.1 *Overview of wireless sensors 144*
      5.11.2 *Basic architectures and features of wireless sensors 144*
      5.11.3 *Challenges in wireless networks 146*
  5.12 *Power supply and energy harvesting 146*
  Notation 148
  References 149

PART II
Integration for smart civil structures

6 Multi-scale modelling of civil structures — 155

  6.1 *Preview 155*
  6.2 *Introduction to finite element modelling 155*
  6.3 *Review of multi-scale modelling of large civil structures 157*
  6.4 *Review of mixed-dimensional finite element coupling methods 158*
  6.5 *Linear constraint equations 160*
  6.6 *Numerical methods for generating constraint equations 161*
      6.6.1 *Substructure and nodal force model 161*
      6.6.2 *Application of unit force or moment 162*
      6.6.3 *Construction of coefficient matrix 162*
  6.7 *Nonlinear constraint equations 163*
      6.7.1 *Transformation of coordinate systems at interface 164*
      6.7.2 *Nonlinear constraint equations in a local coordinate system 164*
      6.7.3 *Nonlinear constraint equations in a global coordinate system 165*

6.8 *Verification of new mixed-dimensional finite element coupling methods* 166
    6.8.1 *Linear beam-to-plate coupling* 166
    6.8.2 *Frame structure* 169
    6.8.3 *Nonlinear analysis of beam–shell coupling* 170
6.9 *Concurrent multi-scale modelling of a transmission tower* 172
    6.9.1 *Background* 172
    6.9.2 *Physical model of a transmission tower structure* 173
    6.9.3 *Multi-scale modelling of the transmission tower* 174
    6.9.4 *Validation of multi-scale modelling* 175
Notation 181
References 182

7 System identification and model updating   **185**

7.1 *Preview* 185
7.2 *Mathematical description of a dynamic civil structure* 185
7.3 *Modal analysis and frequency response function* 186
7.4 *Modal identification in the frequency domain or the time domain* 189
7.5 *Modal identification in the frequency-time domain* 190
7.6 *Force identification* 192
7.7 *Model updating methods* 192
    7.7.1 *Objective functions and constraints* 194
    7.7.2 *Parameters for updating* 196
    7.7.3 *Optimisation algorithm* 197
7.8 *Multi-scale model updating* 199
    7.8.1 *Objective functions and updating parameters for multi-scale FE model* 199
    7.8.2 *Multi-objective optimisation algorithm* 200
    7.8.3 *Kriging meta-model for multi-objective optimisation* 202
    7.8.4 *Implementation procedure of multi-scale model updating method* 203
7.9 *Multi-scale model updating results of a transmission tower* 205
    7.9.1 *Multi-objective optimisation model for model updating* 205
    7.9.2 *Model updating results and discussions* 207
    7.9.3 *Comparison of Kriging method with QPSR method* 210
Notation 211
References 212

8 Multi-type sensor placement   **217**

8.1 *Preview* 217
8.2 *Review of sensor placement methods* 217
8.3 *Dual-type sensor placement method* 219
    8.3.1 *Strain–displacement relationship* 220
    8.3.2 *Theoretical formulations* 222
    8.3.3 *Numerical example* 225
8.4 *Experimental validation of the dual-type sensor placement method* 228

      8.4.1   *Overhanging beam and FE model 228*

      8.4.2   *Dual-type sensor placement 229*

      8.4.3   *Validation of reconstructed responses*
                  using dual-type sensing system 230

  8.5   *Multi-type sensor placement method 231*

      8.5.1   *State-space equation 232*

      8.5.2   *Theoretical formulations 232*

  8.6   *Multi-type sensor placement of a suspension bridge model 236*

      8.6.1   *Physical bridge model 236*

      8.6.2   *FE model of the Tsing Ma Bridge model 237*

      8.6.3   *Modal tests and model updating 239*

      8.6.4   *Framework of multi-type sensor placement*
                  and multi-scale response reconstruction 239

      8.6.5   *Numerical analysis and results 242*

      8.6.6   *Experimental validation 246*

Notation 248

References 250

9 Structural control theory   255

  9.1   *Preview 255*

  9.2   *Stability 255*

  9.3   *Controllability and observability 256*

  9.4   *Pole assignment 259*

      9.4.1   *Pole assignment by state feedback 260*

      9.4.2   *Pole assignment by output feedback 261*

  9.5   *Linear optimal control 261*

      9.5.1   *LQR control 262*

      9.5.2   *LQG control 263*

  9.6   *Independent modal space control 264*

  9.7   *Sliding mode control 266*

      9.7.1   *Design of sliding surface 266*

      9.7.2   *Design of controllers using Lyapunov direct method 268*

  9.8   *H_2 and H_∞ control 269*

      9.8.1   *Transfer function and its norms 270*

      9.8.2   *H_2 control algorithm 271*

      9.8.3   *H_∞ control algorithm 272*

  9.9   *Adaptive control 273*

  9.10 *Artificial intelligent control 275*

  9.11 *Semi-active control 276*

      9.11.1   *Simple bang-bang control law 277*

      9.11.2   *Optimal bang-bang control law 277*

      9.11.3   *Clipped optimal control law 278*

Notation 279

References 280

10 Control device placement   285

10.1   *Preview 285*
10.2   *Review of control device placement methods 285*
10.3   *Increment algorithms for control device placement 289*
10.4   *Suboptimal control gain and response 292*
10.5   *Equivalent optimal parameters of control devices 294*
10.6   *Numerical example for actuator placement 296*
10.7   *Numerical example for passive damper placement 302*
Notation 304
References 304

11 Collective placement of control devices and sensors   309

11.1   *Preview 309*
11.2   *Review of collective placement methods for sensors and control devices 309*
11.3   *Collective placement of control devices and sensors 312*
    11.3.1   *Increment-based approach for optimal placement of control devices 312*
    11.3.2   *Response reconstruction-based approach for optimal placement of sensors 313*
11.4   *Case study 318*
    11.4.1   *Determination of the configurations of the control system 319*
    11.4.2   *Investigation of the control performance with El-Centro ground excitation 323*
    11.4.3   *Investigation of the control performance with Kobe ground excitation 325*
Notation 328
References 329

PART III
Functions of smart civil structures

12 Structural damage detection   333

12.1   *Preview 333*
12.2   *Introduction to structural damage detection 333*
12.3   *Non-destructive testing methods 335*
    12.3.1   *Ultrasonic pulse velocity method 336*
    12.3.2   *Impact-echo/impulse-response methods 336*
    12.3.3   *Acoustic emission method 337*
    12.3.4   *Radiographic method 338*
    12.3.5   *Eddy current method 338*
    12.3.6   *Infrared thermographic method 339*
12.4   *Dynamic characteristics-based damage detection methods 339*
    12.4.1   *Natural frequency changes 340*
    12.4.2   *Mode shape changes 341*

12.4.3 Modal damping changes 342

12.4.4 FRF changes 342

12.4.5 Mode shape curvature changes 343

12.4.6 Modal strain energy changes 343

12.4.7 Flexibility changes 344

12.4.8 Comparison studies 344

12.4.9 Challenges in dynamic characteristics-
based damage detection methods 345

12.5 Dynamic response-based damage detection methods 346

12.5.1 Review of response-based damage detection
methods using WT and HHT 346

12.5.2 Experimental investigation of damage detection using EMD 347

12.5.3 Statistical moment-based damage detection method 354

12.6 Multi-scale damage detection method 360

12.6.1 RBF network for response reconstruction of damaged structure 360

12.6.2 Response sensitivity-based FE model
updating and damage detection 362

12.6.3 Experimental studies 365

12.7 Damage detection methods with consideration of uncertainties 369

12.7.1 Uncertainties in damage detection 369

12.7.2 Perturbation approach 370

12.7.3 Bayesian approach 372

12.7.4 Statistical pattern recognition 373

12.7.5 Monte Carlo simulation 373

12.7.6 Stochastic damage detection method
with parametric uncertainties 374

Notation 380

References 382

13 Structural vibration control 389

13.1 Preview 389

13.2 Introduction to full-scale implementations 389

13.3 Active control of adjacent buildings using hydraulic actuators 391

13.3.1 Equations of motion 391

13.3.2 LQG controller 393

13.3.3 Closed-form solution for dynamic characteristics 394

13.3.4 Closed-form solution for seismic response 395

13.3.5 Application of closed-form solutions 398

13.4 Semi-active control of a building complex using friction dampers 406

13.4.1 Modelling of building complex with variable friction dampers 408

13.4.2 Control strategy 410

13.4.3 Experimental investigation of building
complex with variable friction damper 412

13.5 Semi-active control of building complex using MR dampers 420

13.5.1 Experimental arrangement 420

13.5.2 *Multi-level logic control algorithm* 421

13.5.3 *Seismic control of building complex with MR damper* 423

13.6 *Multi-objective hybrid control of high-tech equipment in high-tech facility* 425

13.6.1 *Multi-objective hybrid control platform* 429

13.6.2 *Equation of motion of building with hybrid platform* 431

13.6.3 *Active control algorithm* 433

13.6.4 *Experimental investigation of multi-objective hybrid platform* 434

Notation 444

References 446

14 Synthesis of structural health monitoring and vibration control in the frequency domain   449

14.1 *Preview* 449

14.2 *Current research in the synthesis of structural health monitoring and vibration control* 450

14.3 *Integrated procedure using semi-active friction dampers* 451

14.4 *System identification and model updating* 453

14.4.1 *Equation of motion* 453

14.4.2 *FRF-based method with full excitation* 453

14.4.3 *FRF-based method with single excitation* 456

14.5 *Vibration control using semi-active friction dampers* 460

14.5.1 *Local feedback control strategy* 461

14.5.2 *Global feedback control strategy* 463

14.6 *FRF-based structural damage detection* 465

14.7 *Numerical investigation* 465

14.7.1 *Description of a numerical example building* 465

14.7.2 *Selection of additional stiffness* 467

14.7.3 *Effects of the number of natural frequencies included* 468

14.7.4 *Effects of the number of frequency points used* 471

14.7.5 *Comparison with a previous study* 471

14.7.6 *Damage detection using full excitation* 473

14.7.7 *Damage detection using single excitation* 473

14.7.8 *Seismic response control of the building structure* 475

14.8 *Experimental investigation* 480

14.8.1 *Experimental setup* 480

14.8.2 *Measured FRFs of the building complex* 483

14.8.3 *Stiffness identification of the building complex* 485

14.8.4 *Structural damage detection* 487

Notation 488

References 489

15 Synthesis of structural health monitoring and vibration control in the time domain   491

15.1 *Preview* 491

15.2 *Formulation of integrated system with time-invariant parameters* 492

15.2.1 *Identification of constant structural parameters and excitations* 493

15.2.2 Structural vibration control with identified
time-invariant parameters 497
15.2.3 Implementation procedure of integrated
system with time-invariant parameters 499
15.3 Numerical investigation of integrated system
with time-invariant parameters 500
15.4 Experimental investigation of integrated system
with time-invariant parameters 504
15.4.1 Experimental setup 504
15.4.2 Damage scenarios 506
15.4.3 Implementation of identification and control algorithms 507
15.4.4 Damage detection and vibration control under Kobe earthquake 509
15.4.5 Damage detection and vibration control
under Northridge earthquake 512
15.5 Formulation of integrated system with time-varying parameters 515
15.5.1 Identification of time-varying parameters and unknown excitation 515
15.5.2 Vibration control with identified time-varying parameters 520
15.5.3 Implementation procedures of integrated
system with time-varying parameters 523
15.6 Numerical investigation of integrated system with time-varying parameters 524
15.6.1 Description of example building structure 524
15.6.2 Accuracy of time-varying parameter and excitation identification 526
15.6.3 Performance of semi-active control with MR dampers 528
15.6.4 Comparisons 529
Notation 533
References 534

16 Energy harvesting for structural health monitoring and vibration control 535

16.1 Preview 535
16.2 Energy harvesting 535
16.2.1 Concept of energy harvesting 535
16.2.2 Wind energy harvesting 536
16.2.3 Solar energy harvesting 537
16.2.4 Vibration energy harvesting 538
16.2.5 Radio frequency energy harvesting 539
16.2.6 Thermal energy harvesting 540
16.3 Energy harvesting for structural health monitoring sensory systems 541
16.3.1 Sensors and structural health monitoring sensory systems 541
16.3.2 Power requirement and management 542
16.3.3 Energy harvesting methods for structural
health monitoring sensory systems 543
16.3.4 Applications of energy harvesting systems
to structural health monitoring 545
16.4 Integrated vibration control and energy harvesting systems 552
16.4.1 Electromagnetic dampers for vibration
damping and energy harvesting 554

16.4.2   *Energy harvesting circuits in electromagnetic dampers 555*
16.4.3   *Power flow in electromagnetic dampers 556*
16.4.4   *Energy harvesting efficiency of electromagnetic dampers 557*
16.4.5   *Testing electromagnetic dampers 558*
16.4.6   *Power flow in integrated systems 563*
16.4.7   *Analysis of structure-EMDEH systems 565*
16.4.8   *Application to stay cables 569*
Notation 577
References 578

17   Synthesis of energy harvesting, structural control and health monitoring   585

17.1   *Preview 585*
17.2   *Current research status on energy harvesting of vertical axis wind turbines 585*
         17.2.1   *Introduction to VAWTs 585*
         17.2.2   *Analytical methods for VAWTs 588*
         17.2.3   *Computational simulations of VAWTs 588*
         17.2.4   *Wind tunnel tests of VAWTs 589*
         17.2.5   *Field measurements of VAWTs 590*
17.3   *Control of blade pitch angles of vertical axis wind turbines 591*
         17.3.1   *Double disks multiple stream-tube model 592*
         17.3.2   *Startup control algorithm in Stage 1 596*
         17.3.3   *Power maximisation control algorithm in Stage 2 598*
         17.3.4   *Rated power control algorithm in Stage 3 598*
         17.3.5   *Parking control algorithm in Stage 4 602*
         17.3.6   *Pitch control system 603*
17.4   *Structural health monitoring of vertical axis wind turbines 607*
         17.4.1   *Wind loads on a VAWT 608*
         17.4.2   *Fatigue and strength analyses of laminated composite blades 609*
         17.4.3   *Fatigue and strength analyses of structural members 615*
         17.4.4   *Structural health monitoring system for the VAWT 619*
17.5   *Concept of smart vertical axis wind turbines 622*
Notation 624
References 624

18   Synthesis of structural self-repairing and health monitoring   627

18.1   *Preview 627*
18.2   *Current research status of structural self-rehabilitation 627*
         18.2.1   *Concept of structural self-rehabilitation 627*
         18.2.2   *Structural self-centring systems 628*
         18.2.3   *Self-healing materials 631*
         18.2.4   *Structural self-repairing materials and systems 633*
18.3   *Self-repairing concrete 636*
         18.3.1   *Epoxy-cement composites without hardener 636*
         18.3.2   *Method of investigation 637*
         18.3.3   *Method of evaluation of self-repairing ability of concrete 637*
         18.3.4   *Results and discussions 638*

18.4   Self-repairing concrete beams 639
    18.4.1   SR system using super-elastic SMAs and
            adhesive-filled brittle fibres 640
    18.4.2   Experimental design 640
    18.4.3   Experimental results and analysis 641
18.5   Self-repairing steel joints 642
    18.5.1   Impedance-based SHM system 642
    18.5.2   Self-repairing of loose bolted joints 643
    18.5.3   Integrated impedance-based SHM and SR system 643
    18.5.4   Experimental investigation and results 643
18.6   Self-diagnosis and self-repairing active tensegrity structures 644
    18.6.1   Description of an active tensegrity structure 645
    18.6.2   Self-diagnosis of active tensegrity structure 645
    18.6.3   Self-repairing of active tensegrity structure 648
    18.6.4   Experimental investigation and results 649
Notation 651
References 651

19   Synthesis of structural life-cycle management and health monitoring   657

19.1   Preview 657
19.2   Concept of SHM-based life-cycle management of civil structures 657
    19.2.1   Multi-scale modelling and model updating 659
    19.2.2   Multi-type sensor placements for response reconstruction 659
    19.2.3   Structural damage detection and health assessment 660
    19.2.4   Inspection, maintenance and repair 660
    19.2.5   Estimation of future loadings 661
    19.2.6   Damage prognosis and remaining life 661
    19.2.7   Life-cycle management strategies 662
19.3   Finite element model and model updating of Tsing Ma Bridge 663
    19.3.1   Tsing Ma Bridge 663
    19.3.2   Finite element model 663
    19.3.3   Model updating 667
19.4   SHM systems of Tsing Ma Bridge 669
    19.4.1   WASHMS in Tsing Ma Bridge 669
    19.4.2   GPS-OSIS in Tsing Ma Bridge 669
19.5   SHM-based loading assessment and models 669
    19.5.1   Highway loading 669
    19.5.2   Railway loading 673
    19.5.3   Wind characteristics and models 674
    19.5.4   Temperature loading 677
19.6   SHM-based stress analysis due to multiple dynamic loadings 679
    19.6.1   Stress analysis framework 679
    19.6.2   Verification of the framework 680
    19.6.3   An engineering approach 680

19.7 SHM-based fatigue damage prognosis and reliability 682
    19.7.1 Fatigue damage prognosis 682
    19.7.2 Fatigue reliability 683
19.8 SHM-based bridge rating system and inspection 685
    19.8.1 SHM-based long-span suspension bridge rating system 685
    19.8.2 Criticality and vulnerability factors 686
    19.8.3 Criticality and vulnerability analyses 689
    19.8.4 Inspection 690
Notation 692
References 692

20 Epilogue: Challenges and prospects   **695**

20.1 Challenges 695
    20.1.1 Multi-scale modelling of smart civil structures 695
    20.1.2 Multi-scale damage detection with substructure techniques 696
    20.1.3 Distributed sensor systems and networks 697
    20.1.4 Development of truly smart wind turbines 698
    20.1.5 Life-cycle management of smart civil structures 699
20.2 Prospects 699
References 700

Index   701

Foreword

A smart civil structure integrates smart materials, sensors, actuators, signal processors, communication networks, power sources, diagonal strategies, control strategies, repair strategies and life-cycle management strategies to perfectly perform any preset functions under normal environment and to confidently preserve the safety and integrity of the civil structure during extreme events. However, smart civil structures defined above have not been completely realised, and a good knowledge and understanding of smart civil structures have not been widespread among university students and practising engineers. One of the reasons is that there is no book, to my knowledge, that addresses this topic comprehensively and systematically, although a few books on either structural vibration control or structural health monitoring can be found in the market. In anticipation that the rapid development in smart materials, sensing technology, control technology, robotics technology, information technology, computation simulation and life-cycle management will eventually overcome the challenging issues in realisation of smart civil structures, a comprehensive book, like this, on the subject covering not only the fundamental knowledge but also the state-of-the-art developments will certainly facilitate learning and preparation of students and engineers to face the challenges posed by the smart civil structures of tomorrow.

Dr. You-Lin Xu, the first author of this book, has conducted teaching, research and consultancy work in the fields of structural vibration control and structural health monitoring for more than 30 years. He was engaged in wind-induced vibration control of tall buildings using tuned liquid column dampers in Australia from 1989 to 1995. Dr. Xu has also been involved in the structural health monitoring of the Tsing Ma suspension bridge in Hong Kong since 1995, the Stonecutters cable-stayed bridge in Hong Kong since 2003 and the Shanghai Centre, a tall building 632 m high in Shanghai, since 2010. Both the Stonecutters Bridge and the Shanghai Centre have been equipped with not only a structural health monitoring system but also a structural vibration control system, although the two systems are separately designed. Dr. Xu, in my opinion, is at the right stage of his career with the distinguished academic and professional background to synthesise his research into a comprehensive and exhaustive book. I am confident that this book will inspire researchers to pursue new methodologies and innovative technologies for the realisation of smart civil structures and it will be very well received both in academia and design practice.

Tsu Tech Soong
SUNY Distinguished Professor Emeritus
State University of New York at Buffalo
Buffalo, New York

Preface

Civil structures provide very fundamental means for a society, and the failure of civil structures could be catastrophic not only in terms of losses of life and economy, but also subsequent social and psychological impacts. The safety and serviceability of civil structures are therefore basic elements of a civilised society and a productive economy, and they are also the ultimate goals of engineering, academic and management communities. However, many civil structures in-service are under continuous deterioration and in fact defective owing to many factors such as man-made and natural hazards. In the past three or four decades, structural vibration control technology and structural health monitoring technology have been developed and incorporated into some important civil structures to enhance their functionality, safety and resilience. Structural self-repairing technology and structural energy harvesting technology have also been explored to improve their durability and adaptability. Nevertheless, these technologies have mostly been applied separately, even though they need similar hardware devices such as sensors, signal processors and communication networks.

With the recent rapid development in smart and bio-inspired materials, sensing technology, control technology, robotics technology, information technology, computation simulation and life-cycle management of infrastructure, it is time to integrate the aforementioned technologies together to create smart civil structures that can mimic biological systems with smart self-sensing, self-adaptive, self-diagnostic, self-repair and self-powered functions to perform any intended functions under the surrounding environment and to preserve the safety and resilience of the structures during strong winds, severe earthquakes and other extreme events. However, there is no book on the market suitable for graduate students and practising engineers, although a few books on either structural health monitoring or structural vibration control have been published in recent years. This is the original incentive of the authors in writing the present book to introduce the fundamentals to the state-of-the-art topics in smart civil structures.

This book is organised into three parts: elements in smart civil structures, integration for smart civil structures, and functions of smart civil structures. Part I contains five chapters covering the basic elements in a smart civil structure: civil structures, smart materials, sensors and sensory systems, control devices and control systems, and processors and processing systems. Part II includes Chapters 6 through 11 on the integration for smart civil structures. Several important issues are addressed: multi-scale modelling of civil structures, system identification and model updating, multi-type sensor placement, structural control theory, control device placement and collective placement of control devices and sensors. Part III of the book has nine chapters concerned with various functions of smart civil structures, which include structural damage detection, structural vibration control, the synthesis of structural health monitoring and vibration control, the synthesis of energy harvesting and structural vibration control, the synthesis of energy harvesting and structural control and health monitoring, the synthesis of structural self-repairing and health monitoring and the synthesis of structural life cycle management and health monitoring.

The first author has conducted teaching, research and consultancy works in the fields of structural vibration control and structural health monitoring for more than 30 years. He has fortunately been involved in long-term collaborative research and practice with the Hong Kong Highways Department since 1995 for the structural health monitoring of the Tsing Ma suspension bridge and the Stonecutters cable-stayed bridge in Hong Kong. He has also been favourably engaged in the structural performance monitoring of the Shanghai Center, a tall building of 632 m high in Shanghai, together with Tongji University. Both the Stonecutters Bridge and the Shanghai Center have been equipped with not only the structural health monitoring system but also the structural vibration control system, although the two systems are separately designed. The second author obtained his PhD degree from Hunan University, China, in 2012. He has then worked together with the first author as a postdoctoral fellow at the Hong Kong Polytechnic University on the synthesis of structural health monitoring and vibration control. Most importantly, both authors have been inspired by the work of many outstanding scholars and engineers in the past years, and we would like to dedicate this book to them.

In writing the book, the authors are always reminded that the book mainly serves as a textbook for graduate students and practising engineers to understand smart civil structures and straddle the gap between theoretical research and practical applications. The readers are assumed to have some background in structural analysis, structural dynamics, probability theory and random vibration.

We would be very happy to receive constructive comments and suggestions from readers.

You-Lin Xu and Jia He
The Hong Kong Polytechnic University

Acknowledgements

The writing of this book has been a challenging and time-consuming task that could not have been completed without the help of many individuals. We are grateful to many people who helped in the preparation of this book.

A few PhD students, former and present, at The Hong Kong Polytechnic University or Harbin Institute of Technology (Shenzhen Graduate School) participated in some research works presented in this book: Professor Wen-Shou Zhang, Professor Bo Chen, Dr. Wai-Shan Chan, Dr. Chi-Lun Ng, Dr. Juan Zhang, Dr. Xiao-Hua Zhang, Dr. Zhi-Wei Chen, Dr. Qin Huang, Dr. Wen-Ai Shen, Dr. Yue Zheng, Dr. Feng-Yang Wang, Dr. Chao-Dong Zhang, Dr. Jing-Hua Lin and Miss Yi-Xin Peng.

Several colleagues and research staff, former and present, at The Hong Kong Polytechnic University made contributions to some research works described in this book: Professor Wei-Lian Qu, Professor Jun Teng, Professor Zhi-Chun Yang, Professor Jing Chen, Dr. Bin Li, Dr. Zhe-Feng Yu, Dr. Ting-Ting Liu, Dr. Qi Li, Dr. Yong Xia, Dr. Song-Ye Zhu and Mr. Sheng Zhan.

In particular, the considerable parts of Chapters 3, 5, 7 and 12 in this book are from the book *Structural Health Monitoring of Long-Span Suspension Bridges* written by the first author of this book and Dr. Yong Xia, and the substantial part of Chapter 16 in this book is written by Dr. Song-Ye Zhu and Dr. Wen-Ai Shen, to which the authors of this book are grateful.

We have been influenced by the work of many outstanding scholars and researchers in this field, and most of their names will be found in the reference list attached to each chapter. Particularly, the book *Smart Structures: Analysis and Design* written by Srinivasan and McFarland in 2001 is a main reference book for Chapter 2 of this book, and the book *Smart Structures: Innovative Systems for Seismic Response Control* written by Cheng et al. in 2008 is a main reference book for Chapter 4 of this book. Section 16.3.4 of Chapter 16 briefly introduces the works of Jung et al. (2012), McCullagh et al. (2014) and Elvin et al. (2003). Sections 18.3 through 18.6 of Chapter 18 summarise the works of Łukowski and Adamczewski (2013), Kuang and Ou (2008), Kim et al. (2009) and Adam and Smith (2007). We are grateful for their contributions to this book.

We would also like to acknowledge Professor Ian F.C. Smith of École polytechnique Fédérale de Lausanne and Dr. Imran Rafiq of University of Brighton, who reviewed the proposal of this book, and their comments led to a much better presentation of the materials.

Our work in this research area, some of which has been incorporated into the book, has been largely supported by the Hong Kong Research Grants Council, The Hong Kong Polytechnic University, The Natural Science Foundation of China and the Hong Kong

Highways Department over many years. All the support is gratefully acknowledged. A vote of thanks must go to Mr. Tony Moore, senior editor at CRC Press, for his patience and encouragement from the beginning and during the preparation of this book, and to Ms. Ariel Crockett and Ms. Teresita Munoz for their patience and scrutiny in the editing of this book.

Finally, we are grateful to our families for their help, encouragement and endurance.

Authors

You-Lin Xu is Chair Professor of the Department of Civil and Environmental Engineering, and Dean of the Faculty of Construction and Environment at The Hong Kong Polytechnic University. He is the first author of *Structural Health Monitoring of Long-Span Suspension Bridges*, also published by CRC Press.

Jia He was a Postdoctoral Fellow at the Department of Civil and Environmental Engineering of the Hong Kong Polytechnic University, focusing on smart civil structures through the synthesis of structural health monitoring and vibration control. He is now an Associate Professor of the College of Civil Engineering, Hunan University, China.

Part I

Elements in smart civil structures

Part I of this book includes Chapters 1 through 5. The background materials and the necessity, definition and historical developments of smart civil structures are given in Chapter 1. Chapter 2 introduces various smart materials used in smart civil structures. Chapters 3 through 5, respectively, introduce three important modules in a smart civil structure: sensors and sensory systems, control devices and control systems, processors and processing systems.

Chapter 1

Introduction

1.1 CIVIL STRUCTURES

A structure is an assembly that serves an engineering function. Civil structures refer to civil engineering structures or structures in civil engineering. Typical examples of civil structures are buildings, bridges, towers, stadiums, tunnels, dams, roads, railways and pipelines. Civil engineering is a professional engineering discipline that deals with the design, construction and maintenance of civil structures to fulfil different functions. Civil structures are omnipresent in every society, regardless of culture, religion, geographical location or economic development (Glišić and Inaudi 2007). It is difficult to imagine a society without buildings, roads, railways, bridges, tunnels, dams and power plants.

The most common civil structures are buildings, which serve several needs within society such as shelter from weather, security, privacy, to store belongings and to comfortably live and work. Buildings come in a variety of shapes, sizes and functions. In terms of size, there are low-rise buildings, multi-story buildings and high-rise (tall) buildings. In terms of function, there are residential, commercial and industrial buildings. In terms of materials used, there are timber buildings, brick and stone buildings, steel buildings, concrete buildings and composite buildings. Due to rapid urbanisation and limited land resources in modern cities, tall buildings have been constructed in the past several decades. The structural systems of buildings have also dramatically developed and evolved. The basic components of a building include a vertical load resisting system, a horizontal load resisting system, a floor system, structural joints and energy dissipation systems (Khan et al. 1980). Figure 1.1 shows the top five tallest buildings in the world: Burj Khalifa (829.8 m), Shanghai Tower (632 m), Abraj Al Bait Towers (601 m), One World Trade Center (546.2 m) and CTF Finance Center (530 m).

Bridges are an extremely important civil structure for modern transportation systems. They are designed and constructed to give people or moving vehicles the ability to span physical obstacles such as a body of water, a valley or a road. Bridges can be categorised into several groups according to their different structural systems, for example, arch bridges, beam bridges, truss bridges, cantilever bridges, tied arch bridges, suspension bridges and cable-stayed bridges. Due to their unique force transforming mechanism, suspension bridges and cable-stayed bridges can be built with a long span. Figures 1.2 and 1.3 give the top five longest suspension bridges and cable-stayed bridges in the world, respectively.

Towers are another widely used civil structure and are usually tall and slender. They are specifically distinguishable from 'buildings' in that they are not built to be habitable but to serve other functions, such as television and radio broadcast, power transmission, scenery observation and so forth. Figure 1.4 shows the top five tallest towers for television, sightseeing or both in the world: Tokyo Skytree (634 m), Canton Tower (600 m), CN Tower (553.3 m), Ostankino Tower (540.1 m) and Oriental Pearl Tower (468 m). Power transmission towers are normally not very high but with the demand for high-voltage electric power

(a) (b) (c)

(d) (e)

Figure 1.1 The top five tallest buildings in the world: (a) Burj Khalifa (829.8 m), located in Dubai, United Arab Emirates (2010); (b) Shanghai Tower (632 m), located in Shanghai, China (2013); (c) Abraj Al Bait Towers (601 m), located in Mecca, Saudi Arabia (2012); (d) One World Trade Center (546.2 m), located in New York, USA (2014); and (e) CTF Finance Centre (530 m), located in Guangzhou, China (2014). (From (a) https://en.wikipedia.org/wiki/Burj_Khalifa; (b) https://en.wikipedia.org/wiki/Shanghai_Tower; (c) http://www.shutterstock.com/subscribe?clicksrc=full_thumb; (d) http://vizts.com/one-world-trade-center/one-world-trade-center-the-tallest-building-in-the-us/; (e) https://www.flickr.com/photos/bsterling/23557209723/.)

transmission, the tallest power transmission tower has reached a height of 370 m, as shown in Figure 1.5.

Stadiums are a place or venue usually used for outdoor sports, concerts or other events. Although different sports or events require fields of different sizes and shapes, stadiums basically consist of a flat field or stage either partly or completely surrounded by a tiered structure allowing spectators to stand or sit. The scale or size of a stadium is usually measured by its capacity, which refers to the maximum number of spectators it can normally accommodate. The top five largest stadiums in the world (see Figure 1.6) are the Rungrado May Day Stadium, Michigan Stadium, Beaver Stadium, Estadio Azteca and the AT&T Stadium, with a capacity of 150,000, 109,901, 107,282, 105,064 and 105,000 spectators, respectively.

All civil structures sustain loads while they fulfil their functions. These loads can be static, such as self-weight, or dynamic, such as wind, seismic and vehicle loads, or transitory, such as impact loads caused by an explosion. Civil structures are also exposed to a harsh environment, such as acid rain and humidity. Once built, they deteriorate with time.

Figure 1.2 The top five longest suspension bridges in the world: (a) Akashi Kaikyō Bridge (main span 1991 m), located in Kobe-Awaji island, Japan (1998); (b) Xihoumen Bridge (main span 1650 m), located in Zhejiang, China (2009); (c) Great Belt Bridge (main span 1624 m), located in Korsør-Sprogø, Denmark (1998); (d) Yi Sun-sin Bridge (main span 1545 m), located in Gwangyang-Yeosu, South Korea (2012); and (e) Runyang Bridge (main span 1490 m), located in Jiangsu, China (2005). (From (a) https://en.wikipedia.org/wiki/Akashi_Kaikyō_Bridge; (b) http://www.thinkstockphotos.com/image/stock-photo-xihoumen-bridge-zhoushan-zhejiang-province/166844481; (c) https://en.wikipedia.org/wiki/Great_Belt_Fixed_Link; (d) http://www.iabse.org/IABSE/association/Award_files/Outstanding_Structure_Award/Yi_sun-sin_Bridge.aspx; (e) https://ssl.panoramio.com/photo/85220018.)

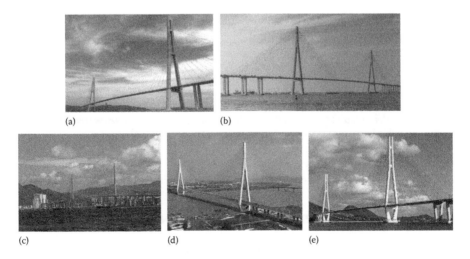

Figure 1.3 The top five longest cable-stayed bridges in the world: (a) Russky Island Bridge (main span 1104 m), located in Vladivostok, Russia (2012); (b) Sutong Bridge (main span 1088 m), located in Jiangsu, China (2008); (c) Stonecutters Bridge (main span 1018 m), located in Hong Kong, China (2009); (d) Edong Bridge (main span 926 m), located in Hubei, China (2010); and (e) Tatara Bridge (main span 890 m), located in Japan (1999). (From (a) http://www.panoramio.com/photo/92738785; (b) http://www.thinkstockphotos.com/image/stock-photo-suspension-bridge/518706498; (c) https://en.wikipedia.org/wiki/Stonecutters_Bridge; (d) http://www.hssjtj.gov.cn/xwllm/zwdt/jttj/201401/t20140124_214587.htm; (e) http://www.thinkstockphotos.com/image/stock-photo-tatara-ohashi-bridge/460474319.)

(a)   (b)   (c)

(d)   (e)

Figure 1.4   The top five tallest towers in the world: (a) Tokyo Skytree (634 m), located in Tokyo, Japan (2012); (b) Canton Tower (600 m), located in Guangzhou, China (2010); (c) CN Tower (553.3 m), located in Toronto, Canada (1976); (d) Ostankino Tower (540.1 m), located in Moscow, Russia (1967); and (e) Oriental Pearl Tower (468 m), located in Shanghai, China (1994). (From (a) http://www.thinkstockphotos.com/image/stock-photo-tokyo-sky-tree/499620453; (b) http://www.iba-bv.com/tvt01.html; (c) http://www.thinkstockphotos.com/image/stock-photo-cn-tower/483465910; (d) https://en.wikipedia.org/wiki/Ostankino_Tower; (e) http://www.thinkstockphotos.com/image/stock-photo-oriental-pearl-tower-in-shanghai/533351850.)

To ensure the functionality, safety and sustainability of civil structures, an understanding of loading conditions and environments around civil structures is essential.

1.2 LOADING CONDITIONS AND ENVIRONMENTS

Owing to their specific functions, different civil structures sustain different types of loads and are located in unique environmental conditions. Only the most common loading conditions and environments are described in this section.

1.2.1 Dead loads

Dead loads are those acting on the structure as a result of the weight of the structure itself and the components that are permanent fixtures on the structure. Dead loads are characterised as having magnitudes and positions. Examples of dead loads are the weight of the

(a)   (b)

Figure 1.5 Power transmission towers: (a) an example of a common power transmission tower and (b) the tallest power transmission tower (370 m, located in Zhejiang, China, 2009). (From (a) http://www. thinkstockphotos.com/image/stock-photo-power-transmission-lines/542198042; (b) http://www. alibaba.com/product-detail/zsst-370-Meter-River-Crossing-High_209385425.html.)

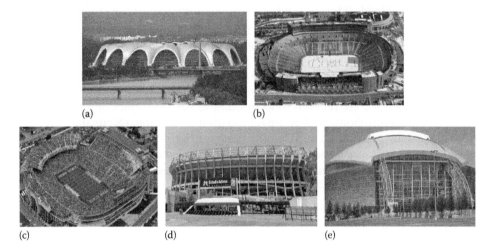

(a)   (b)

(c)   (d)   (e)

Figure 1.6 The top five largest stadiums in the world: (a) Rungrado May Day Stadium (150,000 people), located in Pyongyang, North Korea (1989); (b) Michigan Stadium (109,901 people), located in Ann Arbor, Michigan, USA (1927); (c) Beaver Stadium (107,282 people), located in State College, Pennsylvania, USA (1960); (d) Estadio Azteca (105,064 people), located in Mexico City, Mexico (1966); and (e) AT&T Stadium (105,000 people), located in Texas, USA (2009). (From (a) https://en.wikipedia.org/wiki/Rungrado_1st_of_May_Stadium; (b) http://www.istock-photo.com/photo/university-of-michigan-school-spirit-gm458466515-15631123?st=2d3b5bb; (c) http://greentour.psu.edu/sites/stadium.html; (d) http://www.shutterstock.com/pic-186717815/stock-photo-mexico-city-march-exterior-of-azteca-stadium-in-mexico-city-on-march-azteca.html?src=3Ckp6wkmdS79IAewnD_gbQ-1-0; (e) https://en.wikipedia.org/wiki/AT%26T_Stadium)

structural members themselves in a building, such as beams and columns, the weight of the roof structures, floor slabs, ceilings and permanent partitions, and the weight of the fixed service equipment. The dead loads associated with the structure can be determined if the materials and sizes of the various components are known (West 1993). Standard material unit weights, such as those given in Table 1.1, are used for calculating the dead loads.

Table 1.1 Unit weights of typical building materials

Material	Unit weight (kN/m³)
Aluminium	25.9
Brick	18.9
Concrete	
Reinforced with stone aggregate	23.6
Block, 60% void	13.7
Steel, rolled	77.0
Wood	
Fir	5.0–6.9
Plywood	5.7

Source: Data from West, H.H., *Fundamentals of Structural Analysis*, Wiley, New York, 1993.

1.2.2 Live loads

In a general sense, live loads are considered to include all the loads on the structure that are not classified as dead loads. However, it has become common to narrow the definition of live loads to include only loads that are produced through the construction, use or occupancy of the structure and not to include environmental or dead loads (West 1993). Live loads, where the dynamic nature has significance because of the rapidity with which change in position occurs, are called *moving loads*. Typical moving loads are highway loads and railway loads on bridges, which will be introduced separately. Live loads, in which change occurs over an extended period of time, are referred to as *movable loads*. Examples of movable loads are stored materials in a warehouse and occupancy loads in a building. Occupancy live loads for buildings are usually specified in terms of the minimum values that must be used for design purpose. For more details on these minimum values, the reader can refer to the standard guidelines published by the American Society of Civil Engineers (ASCE 2010).

1.2.3 Wind loads

Wind load is one kind of environmental load acting on a structure. Wind load is particularly important for long-span bridges, high-rise buildings and large-scale spatial structures. In the early 1960s, Davenport established the Alan G. Davenport Wind Loading Chain (see Figure 1.7) and stated clearly that the wind-resistant design of structures should be performed through five links: wind climate, terrain effects, aerodynamic effects, dynamic structural responses and structural design criteria (Davenport 1961). A full description of the wind loading process crosses several scientific disciplines. It relies on an adequate description of wind climate from meteorological observations. It also requires an understanding of atmospheric boundary layers, in which wind speed and direction variation, turbulence, wind and terrain interaction is studied in detail.

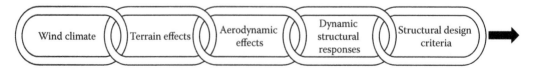

Figure 1.7 The Alan G. Davenport wind loading chain.

Wind pressures and wind loads on building surfaces are influenced by many factors, such as wind speed and direction, turbulence intensity, the shape of the building and the surrounding natural terrains or man-made structures. For the design of low-rise buildings, the traditional approaches to the determination of wind loading assume that wind pressures act in a steady fashion. This is convenient in that it allows the appropriate coefficients of wind pressures to be estimated from wind tunnel tests under a uniform steady velocity (Robertson et al. 1980). Generally, wind pressures act inward on the windward side of a building and outward on most other sides and most roof surfaces. Special concentrations on the outward force, due to aerodynamic lift, occur at building corners and roof edges, particularly at overhangs (Taranath 2009). The overall building structure should be designed for supporting members or components to form a continuous load path to resist the outward and inward wind pressures.

In the case of long-span suspension bridges, as an example, strong wind may induce instability and excessive vibration. Wind effects on bridges are mainly due to static forces induced by mean winds, buffeting excitation, self-excitation and vortex-shedding excitation. These types of instability and vibration may occur alone or in combination (Cai and Montens 2000). The buffeting action on a long-span suspension bridge is caused by fluctuating winds that appear within a wide range of wind speeds. In the wind resistance design of a long-span suspension bridge, the buffeting responses are normally dominant to determine the size of the structural members. In addition to the buffeting action, the self-excited forces induced by wind–structure interaction are also important because the additional energy injected into the oscillating structure by self-excited forces increases the magnitude of vibrations. The buffeting response prediction can be performed in both the frequency domain (Davenport 1962; Scanlan 1978) and the time domain (Bucher and Lin 1988; Chen et al. 2000). Flutter instabilities may occur in several types at very high wind speeds as a result of the dominance of self-excited aerodynamic forces. Flutter instabilities always involve torsional motions, and the most common consideration of the flutter for a long-span suspension bridge is the coupled translational–torsional form of instability. In the design stage, a critical flutter speed of the bridge is determined and will exceed, by a substantial margin, the design wind speed of the bridge site at the deck height (Holmes 2007). Vortex-shedding excitation can induce significant, but limited, amplitude of vibration of a long-span suspension bridge in low wind speed and low turbulence conditions. Scanlan's model can be used for calculating the vortex-shedding force (Simiu and Scanlan 1996). Notably, for large-scale cable-supported bridges (including cable-stayed bridges), appropriate wind tunnel tests are usually required to simulate the wind environment, determine the wind characteristics and examine the responses of the bridge to various winds. Further details on wind loads can be found in Xu (2013).

1.2.4 Earthquake loads

An earthquake is defined as the ground vibration induced by a sudden release of strain energy accumulated in the crust and upper mantle (Robertson et al. 1980). Due to various types of mechanisms of energy release and the complexity of the ground constitution, observed earthquakes on and near the ground surface show apparently random and complex motions. In a region prone to earthquakes, the earthquake-induced ground motion (e.g. El-Centro earthquake as shown in Figure 1.8) would generate very large inertia forces, known as *seismic forces*, on a structure, which probably results in the deficiency, damage or even collapse of the structure. During the construction of the Akashi Kaikyo Bridge, the Kobe earthquake, which occurred in 1995, caused a 1 m movement of the bridge towers.

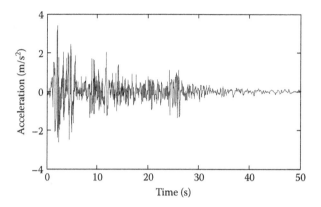

Figure 1.8 Time history of El-Centro earthquake.

It is known that the seismic force is proportional to the mass of the structure and is distributed in proportion to the structural mass at various levels. According to the AASHTO Load and Resistance Factor Design specification (AASHTO 2005) published by the American Association of State Highway and Transportation Officials (AASHTO), earthquake loads are specified as the horizontal force effects and are given by the product of the elastic seismic response coefficient and the equivalent weight of the superstructure, and divided by a response modification factor. The elastic seismic response coefficient is a function of the acceleration coefficient determined from the contour map of the region or nation, period of vibration and site coefficient. It has become more common to use the theories of structural dynamics to analyse a structure subjected to time-dependent earthquake motions.

1.2.5 Highway loads

When a bridge carries a highway, the bridge is subject to a variety of non-stationary loads due to motorcycles, cars, buses, trucks and heavy goods vehicles. Highway loadings are rather complicated. The effect of highway loadings on a bridge is a function of several parameters, such as the axle loads, axle configuration, gross vehicle weight, number of vehicles, speed of vehicles and bridge configuration. The bridge responses under highway loadings can be analysed using the moving load model (Timoshenko et al. 1974), the moving mass model (Blejwas et al. 1979) and the advanced vehicle–bridge interaction models with consideration of the road roughness (Yang and Lin 1995; Cheung et al. 1999). In the AASHTO (2005) specification, Highway Load '93' or HL93 is adopted as the vehicular live loading of highway bridges, which is a combination of a design truck or a design tandem and a design lane load. The AASHTO design truck is shown in Figure 1.9. The axle spacing between the two 145 kN loads can be varied between 4.3 and 9.0 m to create a critical condition for the design of each location in the structure. The AASHTO design tandem consists of two 110 kN axles spaced at 1.2 m apart. The transverse spacing of wheels shall be taken as 1.8 m. The design lane load is equal to a load of 9.3 kN/m uniformly distributed over a 3 m width. In Eurocode 1 (2003), the normal highway load model comprises a tandem axle system acting in conjunction with a uniformly distributed load. Traffic running on bridges produces a stress spectrum, which may cause fatigue. Fatigue load models of vertical forces are defined in Eurocode 1 (2003) and the AASHTO (2005) specification.

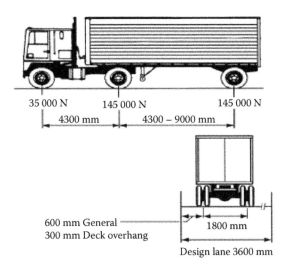

35 000 N   145 000 N   145 000 N

|← 4300 mm →|← 4300 – 9000 mm →|

600 mm General
300 mm Deck overhang

1800 mm

Design lane 3600 mm

Figure 1.9 Design truck in AASHTO load and resistance factor design. (From Xu, Y.L. and Xia, Y., *Structural Health Monitoring of Long-Span Suspension Bridges*, Spon Press, Oxford, UK, 2012.)

1.2.6 Railway loads

When a bridge carries railways, the bridge is subjected to the moving loads of railway vehicles, which include the vertical forces of railway vehicles, the longitudinal forces from acceleration or deceleration of vehicles, the lateral forces caused by irregularities at the wheel-to-rail interface and the centrifugal forces due to track curvature. Figure 1.10 schematically shows a standard eight-car train running on the Tsing Ma Bridge in Hong Kong. Railway vehicles vary greatly with respect to weight, number of axles and axle spacing. This variability requires a representative live load model for design that provides a safe and reliable estimation of the characteristics of railway vehicles within the design life of the bridge (Unsworth 2010). For both bridge safety and vehicle comfort assessment, the interaction between the railway vehicles and the bridge is important (Frýba 1996). In the design of railroad bridges, the Cooper E80 load (AREMA 1997) is the most common design live load. It consists of a series of point loads simulating locomotive wheels, followed by a uniformly distributed load of 14.6 kN/m.

1.2.7 Impact loads

Impact loads that are applied over a very short period of time have a greater effect on the structure than would occur if the same loads were applied statically. The manner in which a load varies with time and the time over which the full load is placed on the structure will determine the factor by which the static response must be increased to obtain the dynamic response. This increase is normally expressed by the use of an impact factor.

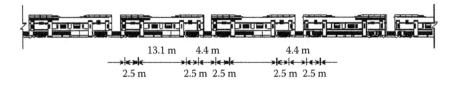

13.1 m   4.4 m   4.4 m

2.5 m   2.5 m   2.5 m   2.5 m   2.5 m

Figure 1.10 Configuration of a train running on the Tsing Ma Bridge.

According to the AASHTO (2005), all bridge components in a navigable waterway crossing, located in design water depths not less than 600 mm, must be designed for vessel impact. The vessel collision loads should be determined on the basis of the bridge importance and the characteristics of the bridge, vessel and waterway. In AASHTO (2005), the head-on ship collision impact force on a pier is taken as a static equivalent force. In Eurocode 1 (2006), a dynamic analysis or an equivalent static analysis is suggested for inland waterways and sea waterways.

1.2.8 Temperature effects

Every structure is subjected to daily and seasonal environmental temperature effects (or thermal effects) induced by solar radiation and ambient air temperature. The variation in temperature of structural components will cause movements and, usually, thermal stress due to indeterminacy and non-uniform distribution of temperature. This is particularly true for large-scale civil structures with various materials involved. The temperature effects on a structure are dependent on the temperature distribution, the structural configuration and boundary conditions and the material mechanical properties of the structure. The local climatic conditions, structural orientation, structural configuration and material thermal properties will affect the structural temperature distribution, which may be divided into three patterns: a uniform temperature, a linearly varying temperature gradient and a nonlinear temperature gradient. Generally, a uniform temperature pattern will result in a change in length for an unrestrained structure. The linearly varying temperature gradient will produce a curvature of the element. The nonlinear temperature gradient will result in self-equilibrated stresses, which have no net load effect on the element. Thermal effects have been studied for a relatively long time. Particularly, in recent years, with the development of structural health monitoring (SHM) technologies, field monitoring of the effect of temperature on bridge performance and structural vibration properties (e.g. frequencies, damping ratios and mode shapes) has been widely carried out (Xia et al. 2006, 2011, 2012; Li et al. 2008; Xu et al. 2010; Zhou et al. 2011).

1.2.9 Corrosion

Corrosion is the deterioration of a metal that results from a reaction with the environment. This reaction is an electrochemical oxidation process that usually produces rust. In bridges, corrosion may occur in structural steels, reinforcing bars and strands in cables. Corrosion of the reinforcing steel is considered to be the primary contributor to the deterioration of highway bridge decks (Mark 1977). For protection against corrosion, structural steels must be self-protecting or have a coating system or cathodic protection. Reinforcing bars in concrete components must be protected by epoxy, galvanised coating, concrete cover or painting. Pre-stressing strands in cable ducts are usually grouted against corrosion.

1.3 MATERIALS USED IN CIVIL STRUCTURES

Over human history, structures have evolved in both their forms and the materials used to construct them. Pre-modern structures were built with natural materials such as stones, bricks, timbers and bamboo. The preference for a particular material to the others was often dictated by factors including local availability, natural environments and cultural considerations. The Great Wall and the Zhaozhou Stone Arch Bridge in China are good examples of these structures. In addition to the use of natural materials, another aspect of these structures is that they were in general not engineered using scientific principles. Most natural materials are brittle materials, which under normal conditions of use are not only much weaker in tension than in compression but also have a tendency to fail abruptly by cracking or splintering.

To date, various types of materials for civil structures have come about and the most widely used ones are steel and concrete. Modern steel and concrete structures are characterised by the use of high-strength, man-made materials and engineering designs based on modern scientific principles. Steel is utilised in a variety of types and forms, such as huge H-columns or small nails and screws, in almost every building in modern society. Some distinguishing properties, such as high tensile strength and ductility, make steel a very desirable structural material. The two major disadvantages of steel for civil structures are its rapid heat gain and the resultant loss of strength when exposed to the intense heat of a fire, and its corrosion when exposed to moisture or corrosive conditions. Several techniques can be employed to handle these shortcomings such as painting or coating. Concrete is a mixture of cement, sand, gravel, crushed rock and water. Water reacts with cement in a chemical reaction known as *hydration*, which sets the cement with other ingredients into a solid mass with high compression strength. It is known that concrete is quite strong in compression but it is very weak in tension. In the practical application of concrete structures, ordinary or pre-stressed steel bars are often placed in the tension zone to compensate for this weakness. There are also many other commonly used materials for civil structures, such as wood, masonry and so forth. Detailed descriptions of these materials can be found in various books (e.g. Chen and Lui 2005; Gupta 2011).

Based on the use of the aforementioned materials, enormous civil structures, including bridges, buildings, tunnels, pipelines, dams and many others, have been built and have provided the most essential living conditions for a modern society. The current approach is to design the structures with sufficient strength to withstand loads and with the ability to deform in a ductile manner. However, such designed structures often have limited capacity owing to a few factors. First, these structures rely on their inherent material damping to dissipate dynamic energy. Nevertheless, the damping of common structural materials, such as steel or reinforced concrete, is small. Second, once the materials for civil structures are selected and used for construction, the capacity of these structures in terms of load resistance and energy dissipation is fixed. Thus, they cannot adapt to sudden or abrupt changes of environmental excitations, such as strong earthquakes or extreme winds, which means damage would probably occur in the component or structural level during the construction or in-service. Third, these civil structures suffer from the problem of deterioration far beyond that initially expected. Therefore, structural engineers should be responsible not only for the design of a structure, but also for its construction and maintenance to ensure the healthy life of the structure and prolong its life span whenever required. Several advanced materials with higher stiffness or more flexible property, for example, high-strength concrete (Aïtcin 2003; Shaha and Ribakovb 2011), fibre-reinforced polymer (FRP) materials (Bakis et al. 2002; Teng et al. 2003; El-Hacha and Soudki 2013), and so forth, have been developed and employed for strengthening or retrofitting of civil structures.

1.4 DESIGN, CONSTRUCTION AND MAINTENANCE

The role of design is to create a structure that fulfils its specific needs over a specified service time with the appropriate level of safety and functionality. The engineering design process encompasses much more than structural design (West 1993). During the conceptual design stage, the specific needs of a structure are identified and the objectives are carefully articulated to meet these needs. During the preliminary design stage, key decisions should be made regarding the positioning of the structure, the structural form to be used and the manner in which the structural components are to be connected. The construction and fabrication aspects must also be considered at this stage because in many cases, the most severe loading conditions occur during construction. The results of the preliminary design stage constitute a starting point for the final design stage. In the final design stage, greater care shall be taken. The loads are determined with greater accuracy than was necessary during the preliminary design, and all plausible loading conditions and combinations are considered. The structural analysis that is required for this stage must be carried out with great precision. The results of the final design stage are encapsulated in a set of complete design drawings accompanied by written specifications. In most countries, the *limit state design* is adopted in the national standards of structural engineering, which requires structures to satisfy a set of performance criteria when the structure is subjected to specific loads. The ultimate limit state and serviceability limit state are two principal criteria. To satisfy the ultimate limit state, the structure must not collapse when subjected to the peak design load. To satisfy the serviceability limit state criteria, a structure must remain functional for its intended use subject to routine loading.

The goal of the construction stage is to bring into existence the structure which was described in the final design stage. Figure 1.11 shows the Tsing Ma Bridge in its two different construction stages, that is, the main towers under construction and the bridge construction completed. The procedure of construction depends on many factors, such as the complexity of the structural system, on-site condition and so forth. Taking a long-span suspension bridge as an example, the construction of such a bridge basically involves sequential construction of the following components: the towers and cable anchorages, main cables and deck structure. The role of the structural engineer is vital here, as is evidenced by the large number of structural failures that occur during construction.

After construction is completed, structures are constantly subjected to aggressive environmental conditions and begin to deteriorate. The maintenance of structures is important to ensure the functionality and safety of the structures during their specified lifetime. The first step in maintenance management is to analyse and identify the deteriorating mechanisms and to indicate the areas susceptible to damage. Next, the prediction of the time-dependent structural performance is given with the consideration of the identified sources of deterioration. The predicted structural performance profile is subsequently used, along with supplementary information such as the cost of interventions, performance thresholds and the effect of maintenance on the structural performance, to find the optimum types and time intervals of inspection, monitoring and maintenance. Optimum intervention scheduling can be performed with various objectives, some of which may conflict. This usually requires the formulation of computationally intensive multi-criteria optimisation problems

(a) (b)

Figure 1.11 Two stages of the Tsing Ma Bridge in the construction procedure: (a) main towers under construction and (b) bridge construction completed.

and a robust optimisation algorithm. After obtaining the optimum inspection, monitoring and/or maintenance schedule that best suits the management goals, the final step is to apply the proposed plan to the investigated structure.

Taking long-span suspension bridges as an example, the frequency, scope and depth of the inspection of bridges can be determined by the bridge owners according to age, traffic, known deficiencies, bridge performance and so forth. A suspension bridge requires a field inspection team led by a qualified professional engineer or certified inspector. The regular inspection interval should not exceed two years. For concrete members, common inspection defects include cracking, scaling, delamination, corrosion and scour. For steel and metal members, common defects include corrosion and cracks, especially fatigue cracking. Both visual and physical examination are commonly employed for inspection. Destructive and non-destructive testing techniques, such as acoustic emissions, computed tomography, ultrasonic and infrared, are also available.

Inspection reports are required to establish and maintain a bridge history file. The findings and results of a bridge inspection must be recorded, preferably on standard inspection forms. In the Pontis bridge management system (Thompson et al. 1998), approximately 120 basic elements are required for inspection. Each element is characterised by discrete condition states, describing the type and severity of element deterioration. Pontis uses a Markovian deterioration model to predict the probability of transitions among the condition states each year. For each state, available actions and associated costs are defined. Network optimisation is performed to minimise expected life-cycle costs while keeping the element beyond risk of failure.

Maintenance measures include preventative procedures and corrective procedures (NYDOT 1997). Planned preventative procedures at appropriate regular intervals can significantly reduce the rate of deterioration of critical bridge elements. Cyclical preventative maintenance procedures include cleaning, sealing cracks, sealing the concrete deck and concrete substructures, replacing the asphalt wearing surface, lubricating bearings and painting steel. Corrective procedures are performed to remedy existing problems. These mainly include repairing the asphalt wearing surface, repairing the concrete deck, repairing or replacing joints, steel members, bearings and concrete substructures, as well as repairing erosion or scour.

A life-cycle cost consists of not only the initial design and construction costs, but also those due to operation, inspection, maintenance, repair and damage consequences during a specified lifetime (Frangopol and Messervey 2007). However, it is noted that as the ageing and deterioration of existing bridges dramatically increase, the current and future needs for bridge maintenance face a major and difficult challenge due to insufficient available funds (Frangopol and Messervey 2007; Frangopol and Liu 2007).

1.5 NECESSITY OF SMART CIVIL STRUCTURES

Civil structures are fundamental to society, and the failure of these structures could be catastrophic not only in terms of the loss of life and economy, but also the subsequent social and psychological impacts. The safety and serviceability of these civil structures are therefore basic elements of a civilised society and a productive economy, and also the ultimate goals of engineering, academic and management communities. However, many civil structures in service are under continuous deterioration and are, in fact, defective owing to many factors such as man-made and natural hazards. The American Society of Civil Engineers estimated that between one-third and one-half of infrastructures in the United States are structurally deficient, and the investment needs about US$1.6 trillion over a 5 year period (ASCE 2005).

Due to the economic boom, a huge number of large-scale and complex civil structures such as long-span bridges, high-rise buildings and large-space structures, have been constructed in China during the past 20 years. However, according to other countries' experiences, enormous cost and effort will be required for the maintenance of these structures and for safeguarding them from damage over the next 20 years (Chang et al. 2009).

1.5.1 Structural health monitoring technology

SHM technology provides a better solution for the problems concerned (Glišić and Inaudi 2007; Çatbaş et al. 2013; Ettouney and Alampalli 2012; Xu and Xia 2012; Farrar and Worden 2013). SHM technology is based on a comprehensive sensory system and a sophisticated data processing system implemented with advanced information technology and structural analysis algorithms (see Figure 1.12). The main objectives of SHM are to monitor the loading conditions of a structure, to assess its performance under various service loads, to verify or update the rules used in its design stage, to detect its damage or deterioration and to guide its inspection and maintenance (Aktan et al. 2000; Ko and Ni 2005; Jiang and Adeli 2005; Brownjohn and Pan 2008; Xu 2008). For example, a long-span suspension bridge is often located in a unique environment. Design loads for a suspension bridge mainly include dead load, traffic load, temperature load, wind load and seismic load. Except the dead load, which can be determined from the design with a higher accuracy, other loads are usually based on design standards or measured from scaled models in laboratories, which may not fully represent the actual loads on each unique bridge. In addition, idealistic laboratory conditions cannot offer the realistic loading environment that a bridge is located in. Moreover, the loading may vary from time to time (e.g. the traffic conditions). With a well-designed sensory system, various types of loadings on-site can be directly or indirectly measured and monitored. Although numerical analysis and laboratory experimental techniques have been extensively developed and employed, their accuracy in predicting the structural responses is limited due to idealistic assumptions used in the mathematical models and the size effect of the scaled physical models. For example, scaled models in wind tunnels for a long-span suspension bridge have a much smaller Reynolds number than the practical one so laboratory results must be used with great caution. By contrast, an SHM system can provide prototype bridge responses which can be employed to accurately evaluate the bridge performance and physical condition. Structural design usually adopts some assumptions and parameters, particularly for long-span bridges and skyscrapers, the geometry of which is beyond design standards. These assumptions need to be verified through on-site structural

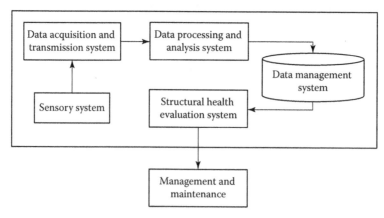

Figure 1.12 The composition of SHM system.

monitoring. The results from SHM can also provide more accurate evidence for designing other structures in the future and updating earlier design specifications.

Civil structures are exposed to natural and man-made hazards, such as typhoons, strong earthquakes, flood, fire and collisions. Failure of civil structures can be catastrophic. Through the online SHM system, structural responses are monitored continuously and a possible abnormality can be detected in the early stages, preventing the potential collapse of the structure and loss of lives. Traditional structural maintenance and management strategies (such as maintenance methods and inspection frequencies) are mainly based on experience. An SHM system can provide the holistic, realistic and latest condition of the structure so the inclusion of an SHM system can significantly improve the efficiency of life-cycle management (LCM) (Okasha et al. 2011). Moreover, the knowledge obtained from an SHM system can give a clear indication of the current condition of the structure under investigation and provide useful information that can be used to update the performance prediction models (Zhu and Frangopol 2013; Soliman and Frangopol 2014). This updated performance prediction will result in updated intervention schedules. Therefore, LCM with the integration of SHM information should be able to help decision makers to make rational and accurate decisions regarding the structures under consideration.

1.5.2 Structural vibration control technology

Although civil structures are designed conservatively according to modern codes and specifications, they still fail on a large scale during strong winds, earthquakes and tsunamis. Typical examples include the Northridge earthquake in 1994, the Kobe earthquake in 1995 (see Figure 1.13a), the Chi-chi earthquake in Taiwan in 1999, Hurricane Katrina in 2005 (see Figure 1.13b), the Wenchuan earthquake in Sichuan in 2008 and the Tohoku earthquake and tsunami in 2011. Each event caused the loss of many lives and billions of US dollars of damage. The poor performance of civil structures under severe hazards can be attributed to the unpredictability of these hazards because traditionally designed structures have deterministic capacities of load resistance and energy dissipation. These structures are actually passive and cannot adapt to ever-changing dynamic loadings and environments.

To withstand stronger loadings or harsher environments, structural vibration control (SVC) technology has been developed. SVC implements energy dissipation devices or control systems in structures to reduce excessive structural vibration, increase human comfort and prevent catastrophic structural failure due to strong winds, severe earthquakes and other disturbances. SVC can also be used to retrofit historic buildings and structures and to

(a)   (b)

Figure 1.13 The failure of civil structure under natural disasters: (a) Kobe earthquake and (b) Hurricane Katrina. (From (a) https://en.wikipedia.org/wiki/File:Hanshin-Awaji_earthquake_1995_343.jpg; (b) https://en.wikipedia.org/wiki/Hurricane_Katrina.)

enhance structural functionality and safety against natural and man-made hazards. By and large SVC can be classified into four types: (1) passive control, (2) active control, (3) hybrid control and (4) semi-active control (Housner et al. 1997).

A passive control system does not require an external power source. Passive control devices impart forces that are developed in response to the motion of the structure.

An active control system is one in which an external source powers control actuators that apply forces to the structure in a prescribed manner (Soong 1990). These forces can be used to add and dissipate energy in the structure. In an active feedback control system, the signals sent to the control actuators are a function of the response of the system measured with physical sensors. The active control system has the advantages of enhanced control effectiveness, adaptability to ground motion, selectivity of control objectives and applicability to various excitation mechanisms. An essential feature of active control systems is that external power is required for the control action. This makes such systems vulnerable to power failure, which is always a possibility during a strong earthquake. The basic configuration of an active control system includes three types of elements: a sensor, an actuator (also known as a *damper*) and a predetermined control algorithm (see Figure 1.14).

A hybrid control system is typically defined as one that is achieved by a combination of passive and active control techniques. With two control techniques in operation, hybrid control systems can alleviate some of the restrictions and limitations that exist when each system is acting alone. As a result, they surpass passive and active systems, and higher levels of performance may be achievable. Additionally, the hybrid control system can be more reliable, more economical and smaller external power is required as compared with a fully active system, although it is also often somewhat more complicated. Significant attention has been paid to the research on hybrid control systems since the 1990s, and much progress has been made (e.g. Cheng et al. 1994; Yang et al. 1995; Branicky et al. 1998; Soong and Spencer 2000; Yang and Agrawal 2002).

A semi-active control system generally originates from a passive control system, which has been subsequently modified to allow for the adjustment of mechanical properties. According to presently accepted definitions, a semi-active control device is one that cannot inject mechanical energy into the controlled structural system (i.e. including the structure and the control device), but has properties that can be controlled to optimally reduce the responses of the system (Housner et al. 1997). Consequently, in contrast to active control devices, semi-active control devices do not have the potential to destabilise the structural system. A semi-active damper system consists of sensors, a control computer, a control actuator and a passive damping device. The excitation and/or the corresponding structural responses are measured by sensors, and then processed by a control computer based on the predetermined

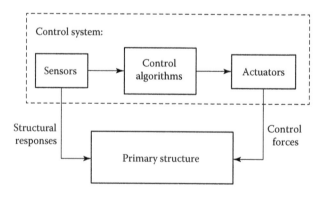

Figure 1.14 The basic configuration of the active control system.

control algorithm. A control signal will subsequently be sent to the actuator to adjust the behaviour of the passive device. Notably, the actuator is used to control the behaviour of the passive damper instead of directly providing force onto the structure. Therefore, the external energy required for a semi-active damper system is usually orders of magnitude smaller than for active systems, and, for example, many can operate on battery power. This is one of the most attractive features of semi-active control systems because during seismic events the main power source to the structure may fail but such systems could still work in this situation.

1.5.3 Definition of smart civil structures

Smart civil structures are defined as civil structures that can mimic biological systems with self-sensing, self-adaptive, self-diagnostic, self-repair and self-powered capacities to perform any intended functions and to preserve the safety and integrity of the structures during strong winds, severe earthquakes and other extreme events. Therefore, a smart civil structure will have the ability to sense, measure, process and diagnose at critical locations any change in selected variables and to command appropriate action using its own power. In this regard, a smart structure shall contain sensors, actuators, signal-conditioning electronics, computers, diagonal strategies, control strategies and a power supply. Some of these elements may be made of smart materials, which can adapt themselves to changes in both the structure and its environment.

Civil structures equipped with SHM systems or SVC systems are called *partially smart structures* because the structure with an SHM system has self-sensing and self-diagnostic abilities whereas the structure with an SVC system has self-sensing and self-adaptive abilities. Although SHM and SVC both require sensors, signal transmission and data acquisition, processing and analysis for their practical implementation, the two have generally been treated separately. Clearly, it is not cost-effective to separate them if a civil structure requires both vibration control and health monitoring. Such separation also hampers the development of smart structures. The authors believe that smart civil structures will involve four systems: (1) SHM system, (2) SVC system, (3) structural self-repairing (SSR) system and (4) structural energy harvesting (SEH) system. The SHM system is capable of automatically sensing and collecting information on a structure, evaluating structural performance, detecting structural deficiencies or damages, predicting remaining lifetime and providing references to decision makers/engineers for maintenance and management. The SVC system has the ability to mitigate the impact of potentially dangerous events such as collisions, earthquakes or typhoons. The excessive structural responses caused by natural and man-made hazards can be significantly reduced with the aid of an SVC system.

With the onset of structural defects and damages, such as the initiation of a crack in concrete or corrosion of steel, the SSR system could be activated by the SHM system to prevent the development of structural damages and recover structural capacity. Smart civil structures have the potential to be independent from external power sources for their required operations. This self-powering capacity can be implemented by equipping the structure with an SEH system, in which some kinds of energy such as kinetic energy, solar energy or thermal energy from the environment and surrounding system are extracted and converted to usable electric energy.

It is worth noting that in these smart civil structures, the aforementioned four element systems should be working closely rather than separately (see Figure 1.15). For example, the signals obtained from the SHM system can be used not only for damage identification and performance assessment but also as feedback for the SVC system for vibration attenuation. By employing the SEH system, the thermal energy created in the SVC system can be

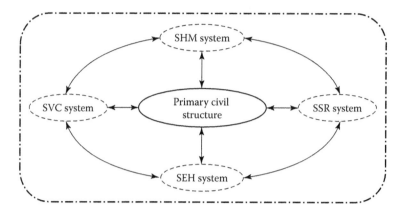

Figure 1.15 Schematic diagram of smart civil structures.

scavenged and converted to electric energy, which can then power the sensors in the SHM system. It can be anticipated that the four element systems, working synergistically and reacting together, will create truly smart civil structures.

1.6 HISTORICAL DEVELOPMENTS OF SMART CIVIL STRUCTURES

The research on the aforementioned four element systems for civil structures has been actively carried on for many years, and much progress has been made. Also, great efforts have been made regarding the application of these systems in real civil structures. Nevertheless, the research and application of these element systems are often conducted separately without integration. Only very recently, the synthesis of two or more element systems on a small structure has been put forward.

Various forms of SHM systems have been implemented in civil structures for a few decades, but it has only rapidly developed in the past two decades with the aid of high-speed computer-based systems for providing timely data processing and information management. The majority of monitoring exercises on buildings and towers have been aimed at improving the understanding of loadings and responses, including those induced by earthquakes and strong winds. For example, three tall buildings in Chicago have been equipped with monitoring systems to compare their wind-induced responses measured by a global positioning system (GPS), and accelerometers with predictions made using wind tunnels and finite element models and discrepancies have been identified (Kijewski-Correa et al. 2006). Brownjohn and Pan (2008) compared data from a decade of monitoring a 280 m office tower in Singapore with design code requirements for both wind and seismic effects, showing that the code provisions for both types of loadings are very conservative. Earlier studies have reported that more than 150 buildings in California, more than 100 buildings in Japan and more than 40 buildings in Taiwan have strong motion monitoring systems for seismic excitation/response measurements and post-earthquake damage assessment (Huang and Shakal 2001; Lin et al. 2005; Huang 2006).

Bridge monitoring can be dated back to the construction of the Golden Gate and Bay Bridges in San Francisco in the 1930s in the United States, in which the dynamic behaviour of the bridges was studied. The collapse of the Tacoma Narrows Bridge in Washington State in the United States in 1940 led to the inspection and modification of other suspension bridges, including strengthening the Golden Gate Bridge. The widespread introduction of systematic bridge inspection programmes was directly attributed to the catastrophic bridge collapse at Point Pleasant, West Virginia, in 1967 (Doebling et al. 1996). Since the 1990s, long-term monitoring systems have been implemented on major bridges in China, Japan, America and Europe. In Hong Kong and mainland China alone, more than 40 long-span bridges had been equipped with long-term monitoring systems by 2005 (Sun et al. 2007). Many attempts have been made to detect structural damage using vibration data in both time domain and frequency domain (Doebling et al. 1996; Sohn et al. 2003). The vibration-based methods have achieved some success in mechanical and aerospace engineering, whereas their successful applications to large-scale civil structures are very limited (Brownjohn 2007) due to the uniqueness of civil structures, significant uncertainties of structures, complicated environmental factors and so forth. A comprehensive knowledge of SHM can be found in many books (e.g. Balageas et al. 2006; Boller et al. 2009; Xu and Xia 2012; Ettouney and Alampalli 2012). Moreover, many literature reviews on damage identification can be found as well (Doebling et al. 1996; Wu et al. 2003; Brownjohn 2007; Farrar and Worden 2007; Ou and Li 2010; Fan and Qiao 2011).

SVC systems offer an attractive means to protect civil structures against natural hazards. Passive control strategies, including base isolation systems, viscoelastic dampers, tuned mass dampers and fluid dampers are well understood and are widely accepted by the engineering community as a means for mitigating the effects of dynamic loading on structures (Housner et al. 1997; Soong and Dargush 1997; Spencer and Nagarajaiah 2003). However, these passive control devices are unable to adapt to structural changes and to varying use patterns and loading conditions. The possibility of using active, hybrid and semi-active control methods has been investigated for more than two decades to improve upon passive approaches to reduce structural responses (Soong 1990; Spencer and Sain 1997; Housner et al. 1997; Kobori 2003). A comprehensive review on this topic was given by Spencer and Nagarajaiah (2003). The first full-scale application of active control to a building was accomplished by the Kajima Corporation in 1989 (Kobori et al. 1991a,b). The building was an 11-story building in Tokyo, having a total floor area of 423 m^2. A control system consisting of two active mass dampers (AMDs) was installed: the primary AMD was used for controlling transverse motion and the secondary AMD was employed to reduce torsional motion. The role of the active system is to reduce building vibration under strong winds and moderate earthquake excitations and consequently to increase the comfort of the occupants of the building.

Hybrid-control strategies have also been investigated by many researchers to exploit their potential to increase the overall reliability and efficiency of the controlled structure (Housner et al. 1994; Kareem et al. 1999; Nishitani and Inoue 2001; Yang and Dyke 2003; Casciati 2003; Faravelli and Spencer 2003). Over 40 buildings and about 10 bridges (during erection) had employed feedback control strategies in full-scale implementations before 2003. The vast majority of these were hybrid control systems. Semi-active control strategies offer the reliability of passive devices yet maintain the versatility and adaptability of fully active systems, able to operate on battery power without requiring the associated large power sources. The Kajima Technical Research Institute in Japan was the first full-scale building structure to be implemented with semi-active hydraulic dampers (Kobori et al. 1993). In the United States, the first full-scale implementation of semi-active control was conducted on the Walnut Creek Bridge on interstate highway I-35 to demonstrate variable-damper technology (Patten et al. 1999). In 2001, the first full-scale implementation of magnetorheological (MR) dampers for civil engineering applications was achieved: the Tokyo National Museum of Emerging Science and Innovation had two 30-tonnes MR fluid dampers installed between the third and fifth floors. The dampers were built by Sanwa Tekki using the Lord Corporation MR fluid. A state-of-the-art review of significant research performed in the area of structural control can be found in the literature (Housner et al. 1997; Symans and Constantinou 1999; Spencer and Nagarajaiah 2003; Fisco and Adeli 2011a,b; Amezquita-Sanchez et al. 2014).

The power supply for operating the SHM systems and SVC systems is externally provided at present. Nevertheless, it is hoped that a smart civil structure will be self-powered. The process of extracting energy from the environment and surrounding system and converting it to usable electric energy is known as *energy harvesting* (Park et al. 2008). The possibility of energy harvesting for providing electric power to small batteries used in wireless sensor networks was actively investigated recently in the health monitoring of civil structures with an attempt to have an autonomous and sustainable operation of the health monitoring system without the periodical replacement of batteries. Solar, wind and structural vibrations are the available energy sources for SHM wireless sensor networks. Miller et al. (2010) developed a solar system to power Imote2 wireless sensor networks and its effectiveness has been validated on a cable-stayed bridge. Spencer et al. (2011) proposed solar or wind energy harvesting as the power supplies for all 113 wireless smart sensors (WSSs) in the wireless monitoring system for the Jindo Bridge in Korea. Energy harvesting systems based on piezoelectric materials are also attractive due to their solid-state nature and the high volumetric density of harvested power. Piezoelectric materials–based transducers are capable of interchanging electrical energy and mechanical motion or force. These materials, therefore, can be used as mechanisms to transfer structural vibration into electrical energy that may be stored and used to power other devices. Some available overviews of the application of piezoelectric transducers as energy harvesters can be found in the references (Sodano et al. 2004; DuToit et al. 2005; Anton and Sodano 2007). However, the piezoelectric harvesters require vibration frequencies in the kilohertz range for an effective power generation. Considering that the vibration frequencies of civil structures, especially long-span bridges, high-rise buildings and bridge stay cables, typically are in the 0.1–100 Hz range, the SEH systems based on an electromagnetic (EM) energy harvesting mechanism is more suitable for application in civil structures. Auge (2003) proposed the concept of energy harvesting via EM dampers, also known as *magnetic induction dampers*, with application to building active control subjected to earthquake ground motions. Scruggs (2004) proposed a regenerative force actuation (RFA) network using a permanent magnet machine in civil structures subjected to seismic excitations. In the proposed RFA network, some devices extract mechanical energy from structural vibration, while others re-inject a portion of that energy back into the structure for suppressing the vibration of other degrees of freedom (DOFs). Jung et al. (2011) proposed the use of EM device to power an MR damper for the vibration mitigation of stay cables, in which the vibration control and energy harvesting functions were fulfilled by the two respective devices. Moreover, the design and experimental characterisation of an electromagnetic transducer for energy harvesting from multi-story buildings and bridges were reported by Cassidy et al. (2011).

In general, self-repairing means a material, a structural component or a structural system that is able to totally or partially recover its undamaged properties with the onset of damage. The concept of self-repairing has been widely investigated and applied in the areas of aerospace, mechanical engineering and life science. Focusing on the field of civil engineering, the majority of researches on the SSR systems have been devoted to the development of self-repairing techniques in concrete structures and bolted joints (Park et al. 2003; Peairs et al. 2004; Faria et al. 2009; Mihashi and Nishiwaki 2012). Concrete structures often suffer from cracking that leads to much earlier deterioration than the designed service life. Generally speaking, cracks in concrete can occur in any stage of the service life of concrete structures due to overloading or volumetric change caused by high temperatures, creep, plastic settlement or shrinkage. The emerging smart materials in recent years and the practical limitations on inspection and maintenance have motivated civil engineers and researchers to investigate and develop the self-repairing techniques for recovering the properties of a concrete structure with the onset of defects such as cracks. By using a shape memory alloy (SMA) as the main reinforcing bars for concrete beams in order that large cracks under loading be mechanically closed after unloading, a crack-closure system was proposed by Sakai et al. (2003). An integrated self-diagnostic and emergency repairing system embedded with SMA bars was proposed by Li et al. (2004). The bolted joint is one of the most common mechanical components in all types of civil structures. Bolted joints are subject to a variety of common modes of failure, such as self-loosening, shaking apart and breaking because of corrosion, stress cracking or fatigue. One of the most frequent modes of failure for bolted joints is self-loosening. To reduce this mode of failure, the concept of a self-sensing and self-repairing bolted joint has been developed (Muntges et al. 2001; Peairs et al. 2001; Park and Inman 2001). This concept combines the impedance-based health monitoring technique with actuators, which are usually included in the joint as SMA washers to restore tension in a loose bolt.

Although great achievements have been made during the past several decades, the majority of attention has been focused on individual element systems such as SHM, SVC, SEH and SSR. The development of smart civil structures with the integration of all SHM, SVC, SEH and SSR is still in its early stage. Nevertheless, the development of smart civil structures by integrating SHM with SVC has been attempted recently (Xu and Chen 2008; Huang et al. 2012; Yang et al. 2013; Lei et al. 2013, 2014; Xu et al. 2014). Recently, SEH systems for powering the wireless or embedded sensors of the SHM system have been developed (Casciati and Rossi 2007; Bogue 2009; Sazonov et al. 2009; Miller et al. 2010). Some attention has also been paid to energy regenerative damping systems for considering both SEH and SVC (Nakano et al. 2003; Lesieutre et al. 2004; Liang and Liao 2009; Shen et al. 2012; Zhu et al. 2012; Shen and Zhu 2015). An SSR system is usually incorporated with an SHM system, by which the SSR system can be activated when the predefined threshold of the 'damage index' is reached according to the information given by the SHM system (Coyle et al. 2004; Li et al. 2004; Mihashi et al. 2008; Kuang and Ou 2008; Kim et al. 2009; Faria et al. 2011).

Clearly, the subject of smart civil structures is interdisciplinary, encompassing a variety of subjects including materials science, electronics, photonics, manufacturing, applied mechanics, random vibration and computing algorithm among others. Some engineering colleges and universities have begun to incorporate smart civil structures into their curricula to nurture future engineers to realise smart civil structures with integrated functions (Zhang and Lu 2008; Hurlebaus et al. 2012). This book aims to serve as an essential book for university students, researchers and professional engineers on the subject of smart civil structures. It covers not only the fundamental knowledge but also state-of-the-art development in the subject of smart civil structures.

1.7 ORGANISATION OF THIS BOOK

Background materials, including loading conditions, materials, design, construction and maintenance of civil structures, have been provided in this chapter. The necessity for, the definition of and the historical developments of smart civil structure have also been given in this chapter. Chapter 2 introduces various smart materials used for smart civil structures, including SMAs, piezoelectric materials, magnetostrictive materials, electrorheological and MR fluids, fibre-opticoptical fibres, bio-inspired materials and nanomaterials. Chapters 3 through 5, respectively, introduce three important modules in a smart civil structure: sensors and sensory systems, control devices and control systems, processors and processing systems. Chapters 1 through 5 form Part I of this book 'Elements in Smart Civil Structures'. Part II of this book is 'Integration for Smart Civil Structures', including Chapters 6 through 11. The multi-scale modelling of civil structures and the associated model updating are described in Chapters 6 and 7, respectively. With the aid of a finite element model, Chapter 8 describes the methods for multi-type sensor placement. Chapter 9 introduces the structural control theory, followed by a description of the optimal placement of control devices in Chapter 10. By combining the optimal sensor placement and optimal control device placement in Chapters 9 and 10, Chapter 11 presents a method for the collective placement of control devices and sensors. Part III of the book is 'Functions of Smart Civil Structures', containing nine chapters from Chapters 12 through 19. Chapters 12 and 13, respectively, introduce structural damage identification and SVC. The synthesis of SHM and SVC in the frequency domain and the time domain is investigated in Chapters 14 and 15, respectively. Chapter 16 introduces the study on the energy harvesting for SHM and vibration control. The integration of SEH, SHM and SVC is given in Chapter 17. The research on the synthesis of SSR and health monitoring is introduced in Chapter 18. Chapter 19 describes the synthesis of structural LCM and health monitoring. Finally, the challenges and prospects of smart civil structures are highlighted in Chapter 20.

REFERENCES

AASHTO. 2005. *AASHTO LRFD Bridge Design Specification*, 3rd edition. American Association of State Highway and Transportation Officials, Washington, DC.

Aïtcin, P.C. 2003. The durability characteristics of high performance concrete: A review. *Cem. Concr. Compos.*, 25(4–5): 409–20.

Aktan, A.E., F.N. Catbas, K.A. Grimmelsman, and C.J. Tsikos. 2000. Issues in infrastructure health monitoring for management. *J. Eng. Mech.*, 126: 711–24.

Amezquita-Sanchez, J.P., A. Dominguez-Gonzalez, R. Sedaghati, R.J. Romero-Troncoso, and R.A. Osornio-Rios. 2014. Vibration control on smart civil structures: A review. *Mech. Adv. Mater. Struct.*, 21: 23–38.

Anton, S.R. and H.A. Sodano. 2007. A review of power harvesting using piezoelectric materials (2003–2006). *Smart Mater. Struct.*, 16(3): R1–21.

AREMA. 1997. *Manual for Railway Engineering*. Landover, MD: American Railway Engineering Maintenance of Way Association.

ASCE. 2005. *2005 Report Card for America's Infrastructure*. Reston, VA: America Society of Civil Engineers.

ASCE. 2010. *Minimum Design Loads for Buildings and Other Structures*. ASCE Standard ASCE/SEI 7-10. Reston, VA: American Society of Civil Engineers.

Auge, L.J. 2003. Structural magnetic induction dampers in buildings. Master thesis, Massachusetts Institute of Technology.

Bakis, C., L. Bank, V. Brown, et al. 2002. Fiber-reinforced polymer composites for construction: State-of-the-art review. *J. Compos. Constr.*, 6(2): 73–87.

Balageas, D., C.P. Fritzen, and A. Güemes. 2006. *Structural Health Monitoring*. London, UK: Wiley, ISTE Ltd.

Blejwas, T.E., C.C. Feng, and R.S. Ayre. 1979. Dynamic interaction of moving vehicles and structures. *J. Sound Vibr.*, 67: 513–22.

Bogue, R. 2009. Energy harvesting and wireless sensors: A review of recent developments. *Sens. Rev.*, 29(3): 194–99.

Boller, C., F.K. Chang, and Y. Fujino. 2009. *Encyclopedia of Structural Health Monitoring*. Chichester, UK: John Wiley & Sons, Ltd.

Branicky, M.S., V.S. Borkar, and S.K. Mitter. 1998. A unified framework for hybrid control: Model and optimal control theory. *IEEE Trans. Autom. Control*, 43(1): 31–45.

Brownjohn, J.M.W. 2007. Structural health monitoring of civil infrastructure. *Phil. Trans. R. Soc. A,* 365(1851): 589–622.

Brownjohn, J.M.W. and T.C. Pan. 2008. Identifying loading and response mechanisms from ten years of performance of monitoring of a tall building. *J. Perform. Constr. Facil.,* 22(1): 24–34.

Bucher, G.C. and Y.K. Lin. 1988. Stochastic stability of bridges considering coupled modes. *J. Eng. Mech.,* 114(12): 2055–71.

Cai, C.S. and S. Montens. 2000. Wind effects on long-span bridges. In *Bridge Engineering Handbook,* eds. W.F. Chen and L. Duan, Chapter 57. Boca Raton, FL: CRC Press.

Casciati, F., ed. 2003. *Proceedings of the 3rd World Conference on Structural Control,* Como, Italy: Wiley.

Casciati, F. and R. Rossi. 2007. A power harvester for wireless sensing applications. *Struct. Control. Health Monit.,* 14(4): 649–59.

Cassidy, I.L., J.T. Scruggs, S. Behrens, and H.P. Gavin. 2011. Design and experimental characterization of an electromagnetic transducer for large-scale vibratory energy harvesting applications. *J. Intell. Mater. Syst. Struct.,* 22(17): 2009–24.

Çatbaş, F.N., T. Kijewski-Correa, and A.E. Aktan. 2013. *Structural Identification of Constructed Systems: Approaches, Methods, and Technologies for Effective Practice of St-Id.* Reston, VA: ASCE Publications. http://ascelibrary.org/doi/book/10.1061/9780784411971.

Chang, S.P., L.Y. Yee, and J. Lee. 2009. Necessity of the bridge health monitoring system to mitigate natural and man-made disasters. *Struct. Infrastruct. Eng.,* 5(3): 173–97.

Chen, W.F. and E.M. Lui. 2005. *Handbook of Structural Engineering,* 2nd edition. Boca Raton, FL: CRC Press.

Chen, X.Z., M. Matsumoto, and A. Kareem. 2000. Time domain flutter and buffeting response analysis of bridges. *J. Eng. Mech.,* 126(1): 7–16.

Cheng, W., W. Qu, and A. Li. 1994. Hybrid vibration control of Nanjing TV tower under wind excitation. In *Proceedings of the First World Conference on Structure Control,* Los Angeles, CA: John Wiley & Sons, Ltd., WP2: 32–41.

Cheung, Y.K., F.T.K. Au, D.Y. Zheng, and Y.S. Cheng. 1999. Vibration of multi-span non-uniform bridges under moving vehicles and trains by using modified beam vibration function. *J. Sound Vibr.,* 228: 611–28.

Coyle, E.A., L.P. Maguire, and T.M. McGinnity. 2004. Self-repair of embedded systems. *Eng. Appl. Artif. Intell.,* 17(1): 1–9.

Davenport, A.G. 1961. The application of statistical concepts to the wind loading of structures. *Proceedings of the Institute of Civil Engineers,* 19(4): 449–72.

Davenport, A.G. 1962. Buffeting of a suspension bridge by storm winds. *J. Struct. Div.,* 88(3): 233–68.

Doebling, S.W., C.R. Farrar, M.B. Prime, and D.W. Shevitz. 1996. Damage identification and health monitoring of structural and mechanical systems from changes in their vibration characteristics: A literature review. Technical report, No: LA-13070-MS, 1-127. Los Alamos, CA: Los Alamos National Laboratory.

DuToit, N.E., B.L. Wardle, and S.G. Kim. 2005. Design considerations for MEMS-scale piezoelectric mechanical vibration energy harvesters. *Integr. Ferroelectr.,* 71(1): 121–60.

El-Hacha, R. and K. Soudki. 2013. Prestressed near-surface mounted fiber reinforced polymer reinforcement for concrete structures: A review. *Can. J. Civ. Eng.,* 40(11): 1127–139.

Ettouney, M.M. and S. Alampalli. 2012. *Infrastructure Health in Civil Engineering: Theory and Components.* Boca Raton, FL: CRC Press.

Eurocode 1. 2003. *Actions on Structures-Part 2: Traffic loads on bridges.* Ref. No. EN 1991-2:2003 E. Brussels, Belgium: European Committee for Standardization.

Eurocode 1. 2006. *Actions on Structures. Part 1–7: General Actions – Accidental Actions.* Ref. No. EN 1991-1-7:2006 E. Brussels, Belgium: European Committee for Standardization.

Fan, W. and P.Z. Qiao. 2011. Vibration-based damage identification methods: A review and comparative study. *Struct. Health Monit.,* 10(1): 83–111.

Faravelli, L. and B.F. Spencer Jr., ed. 2003. *Proceedings of Sensors and Smart Structures Technology.* New York, NY: Wiley.

Faria, C.T., V.L. Junior, and D.J. Inman. 2011. Modeling and experimental aspects of self-healing bolted joint through shape memory alloy actuators. *J. Intell. Mater. Syst. Struct.*, 22(14): 1581–594.

Faria, C.T., V. Lopes Jr., and D.J. Inman. 2009. Overview on self-healing bolted joint using shape memory alloys. In *Proceedings of COBEM, 20th International Congress of Mechanical Engineering*. Gramado, Brazil. Copyrighted by ABCM.

Farrar, C.R. and K. Worden. 2007. An introduction to structural health monitoring. *Phil. Trans. R. Soc. A*, 365(1851): 303–15.

Farrar, C.R. and K. Worden. 2013. *Structural Health Monitoring: A Machine Learning Perspective.* Chichester, UK: John Wiley & Sons, Ltd.

Fisco, N.R. and H. Adeli. 2011a. Smart structures: Part I – Active and semi-active control. *Sci. Iran.*, 18(3): 275–84.

Fisco, N.R. and H. Adeli. 2011b. Smart structures: Part II – Hybrid control systems and control strategies. *Sci. Iran.*, 18(3): 285–95.

Frangopol, D.M. and M. Liu. 2007. Maintenance and management of civil infrastructure based on condition, safety, optimization, and life-cycle cost. *Struct. Infrastruct. Eng.*, 3(1): 29–41.

Frangopol, D.M. and T.B. Messervey. 2007. Integrated life-cycle health monitoring, maintenance, management, and cost of civil infrastructures. In *Proceeding of 2007 International Symposium on Integrated Life-cycle Design and Management of Infrastructure*, eds. L.C. Fan, L.M. Sun, and Z. Sun, 3–20. Shanghai, China: Tongji University Press.

Frýba, L. 1996. *Dynamics of Railway Bridges.* London, UK: Thomas Telford.

Glišić, B. and D. Inaudi. 2007. *Fiber Optical Methods for Structural Health Monitoring.* Chichester, UK: John Wiley & Sons, Ltd.

Gupta, R.S. 2011. *Principles of Structural Design: Wood, Steel, and Concrete.* Boca Raton, FL: CRC Press.

Holmes, J.D. 2007. *Wind Loading of Structures*, 2nd edition. London: Taylor & Francis.

Housner, G.W., L.A. Bergman, T.K. Caughey, et al. 1997. Structural control: Past, present, and future. *J. Eng. Mech.*, 123(9): 897–971.

Housner, G.W., S.F. Masri, and A.G. Chassiakos, eds. 1994. *Proceedings of the First World Conference on Structural Control.* Los Angeles, CA: International Association for Structural Control.

Huang, M.J. 2006. Utilization of strong-motion records for post-earthquake damage assessment of buildings. In *Proceedings of the International Workshop on Structural Health Monitoring and Damage Assessment*, Taichung, Taiwan, IV1-IV29.

Huang, M.J. and A.F. Shakal. 2001. Structure instrumentation in the California strong motion instrumentation program. In *Strong Motion Instrumentation for Civil Engineering Structures*, Volume 373 of the NATO Science Series, eds. M.Ö. Erdik, M. Çelebi, V. Mihailov, N. Apaydin, 17–31. Dordrecht, The Netherlands: Kluwer Academic Publisher.

Huang, Q., Y.L. Xu, J.C. Li, Z.Q. Su, and H.J. Liu. 2012. Structural damage detection of controlled building structures using frequency response functions. *J. Sound Vibr.*, 331(15): 3476–92.

Hurlebaus, S., T. Stocks, and O. Ozbulut. 2012. Smart structures in engineering education. *J. Prof. Issues Eng. Educ. Pract.*, 138(1): 86–94.

Jiang, X. and H. Adeli. 2005. Dynamical wavelet neural network for nonlinear identification of high-rise buildings. *Comput.-Aided Civil Infrastruct. Eng.*, 20(5): 316–30.

Jung, H.J., I.H. Kim, and J.H. Koo. 2011. A multi-functional cable-damper system for vibration mitigation, tension estimation and energy harvesting. *Smart Struct. Syst.*, 7(5): 379–92.

Kareem, A., T. Kijewski, and Y. Tamura. 1999. Mitigation of motions of tall buildings with specific examples of recent applications. *Wind Struct.*, 2(3): 201–51.

Khan, F.R., J. Rankine, W.P. Moore, H.D. Eberhart, and H.J. Cowan. 1980. *Tall Building Systems and Concepts.* New York: American Society of Civil Engineers (ASCE), Council on Tall Buildings and Urban Habitat, volume SC.

Kijewski-Correa, T., J. Kilpatrick, A. Kareem, et al. 2006. Validating wind-induced response of tall buildings: Synopsis of the Chicago full-scale monitoring program. *J. Struct. Eng.*, 132(10): 1509–523.

Kim, J.K., D. Zhou, D.S. Ha, and D. Inman. 2009. A structural health monitoring system for self-repairing. In *Proceedings of SPIE 7295, Health Monitoring of Structural and Biological Systems 2009*, ed. T. Kundu, 729512. San Diego, CA. doi:10.1117/12.816398.

Ko, J.M. and Y.Q. Ni. 2005. Technology developments in structural health monitoring of large-scale bridges. *Eng. Struct.*, 27: 1715–725.

Kobori, T. 2003. Past, present and future in seismic response control of civil engineering structures. In *Proceedings of the 3rd World Conference on Structural Control*, 9–14, Como, Italy: Wiley.

Kobori, T., N. Koshika, K. Yamada, and Y. Ikeda. 1991a. Seismic-response-controlled structure with active mass driver system. Part 1: Design. *Earthq. Eng. Struct. Dyn.*, 20(2): 133–49.

Kobori, T., N. Koshika, K. Yamada, and Y. Ikeda. 1991b. Seismic-response-controlled structure with active mass driver system. Part 2: Verification. *Earthq. Eng. Struct. Dyn.*, 20(2): 151–66.

Kobori, T., M. Takahashi, T. Nasu, N. Niwa, and K. Ogasawara. 1993. Seismic response controlled structure with active variable stiffness system. *Earthq. Eng. Struct. Dyn.*, 22(11): 925–41.

Kuang, Y.C. and J.P. Ou. 2008. Self-repairing performance of concrete beams strengthened using superelastic SMA wires in combination with adhesives released from hollow fibers. *Smart Mater. Struct.*, 17(2): 025020.

Lei, Y., D.T. Wu, and S.Z. Lin. 2013. Integration of decentralized structural control and the identification of unknown inputs for tall shear building models under unknown earthquake excitation. *Eng. Struct.*, 52: 306–16.

Lei, Y., H. Zhou, and L.J. Liu. 2014. An on-line integration technique for structural damage detection and active optimal vibration control. *Int. J. Struct. Stab. Dyn.*, 14(5): 1440003.

Lesieutre, G.A., G.K. Ottman, and H.F. Hofmann. 2004. Damping as a result of piezoelectric energy harvesting. *J. Sound Vibr.*, 269: 991–1001.

Li, D.N., M.A. Maes, and W.H. Dilger. 2008. Evaluation of temperature data of confederation Bridge: Thermal loading and movement at expansion joint. In *Proceedings of the 2008 Structures Congress*, Vancouver, BC, Canada, eds. A. Don, V. Carlos, H. David, and H. Marc, 1–10. Reston, VA: ASCE.

Li, H., Z.Q. Liu, Z.W. Li, and J.P. Ou. 2004. Study on damage emergency repair performance of a simple beam embedded with shape memory alloys. *Adv. Struct. Eng.*, 7(6): 495–502.

Liang, J.R. and W.H. Liao. 2009. Piezoelectric energy harvesting and dissipation on structural damping. *J. Intell. Mater. Syst. Struct.*, 20: 515–27.

Lin, C.C., C.E. Wang, H.W. Wu, and J.F. Wang. 2005. On-line building damage assessment based on earthquake records. *Smart Mater. Struct.*, 14(3): S137–53.

Mark, V.J. 1977. Detection of steel corrosion in bridge corrosion in bridge decks and reinforced concrete pavement. Report HR-156. Ames, IA: Iowa Department of Transportation.

Mihashi, H. and T. Nishiwaki. 2012. Development of engineered self-healing and self-repairing concrete-state-of-the-art report. *J. Adv. Concr. Technol.*, 10: 170–84.

Mihashi, H., T. Nishiwaki, K. Miura, and Y. Okuhara. 2008. Advanced monitoring sensor and self-repairing system for cracks in concrete structures. In *RILEM Symposium on On Site Assessment of Concrete, Masonry and Timber Structures – SACoMaTiS 2008*, eds. L. Binda, M. di Prisco, R. Felicetti, 401–409. Paris, France: RILEM Publications SARL.

Miller, T.I., B.F. Spencer Jr, J. Li, and H. Jo. 2010. Solar energy harvesting and software enhancements for autonomous wireless smart sensor network. NSEL Report No. NSEL-022, Department of Civil and Environmental Engineering, University of Illinois at Urbana-Champaign, IL. https://www.ideals.illinois.edu/handle/2142/16300.

Muntges, D.E., G. Park, and D.J. Inman. 2001. Self-healing bolted joint employing a shape memory actuator. In *Proceedings of SPIE 4327, Smart Structures and Materials 2001: Smart Structures and Integrated Systems*, ed. L.P. Davis, 193–200. Newport Beach, CA. Copyrighted by SPIE Digital Library.

Nakano, K., Y. Suda, and S. Nakadai. 2003. Self-powered active vibration control using a single electric actuator. *J. Sound Vibr.*, 260(2): 213–35.

Nishitani, A., and Y. Inoue. 2001. Overview of the application of active/semiactive control to building structures in Japan. *Earthq. Eng. Struct. Dyn.*, 30(11): 1565–574.

NYDOT. 1997. *Fundamentals of Bridge Maintenance and Inspection*. New York, NY: New York State Department of Transportation (NYDOT).

Okasha, N.M., D.M. Frangopol, D. Saydam, and L.W. Salvino. 2011. Reliability analysis and damage detection in high-speed naval craft based on structural health monitoring data. *Struct. Health Monit.*, 10(4): 361–79.

Ou, J.P. and H. Li. 2010. Structural health monitoring in mainland China: Review and future trends. *Struct. Health Monit.*, 9(3): 219–31.

Park, G. and D.J. Inman. 2001. Smart bolts: An example of self-healing structures. *Smart Mater. Bull.*, 2001(7): 5–8.

Park, G., D.E. Muntges, and D.J. Inman. 2003. Self-repairing joints employing shape-memory alloy actuators. *JOM*, 55(12): 33–37.

Park, G., T. Rosing, M. Todd, C. Farrar, and W. Hodgkiss. 2008. Energy harvesting for structural health monitoring sensor networks. *J. Infrastruct. Syst.*, 14, SPECIAL ISSUE: New Sensors, Instrumentation, and Signal Interpretation, 64–79.

Patten, W.N., J. Sun, G. Li, J. Kuehn, and G. Song. 1999. Field test of an intelligent stiffener for bridges at the I-35 Walnut Creek bridge. *Earthq. Eng. Struct. Dyn.*, 28(2): 109–26.

Peairs, D.M., G. Park, and D.J. Inman. 2001. Self-monitoring and self-healing bolted joint. In *Proceedings of 3rd International Workshop on Structural Health Monitoring*, Stanford, CA: DEStech Publications, Inc. 430–39.

Peairs, D.M., G. Park, and D.J. Inman. 2004. Practical issues of activating self-repairing bolted joints. *Smart Mater. Struct.*, 13(6): 1414–23.

Robertson, L.E., T. Naka, E.H. Gaylord, R.J. Mainstone, and L.W. Lu. 1980. *Tall Building Criteria and Loading*. New York, NY: American Society of Civil Engineers (ASCE), Council on Tall Buildings and Urban Habitat, volume CL.

Sakai, Y., Y. Kitagawa, T. Fukuta, and M. Iiba. 2003. Experimental study on enhancement of self-restoration of concrete beams using SMA wire. In *Proceedings of SPIE 5057, Smart Structures and Materials 2003: Smart Systems and Nondestructive Evaluation for Civil Infrastructures*, ed. C.S. Liu, 178–86, San Diego, CA. Copyrighted by SPIE Digital Library.

Sazonov, E., H. Li, D. Curry, and P. Pillay. 2009. Self-powered sensors for monitoring of highway bridges. *IEEE Sens. J.*, 9(11): 1422–29.

Scanlan, R.H. 1978. The action of flexible bridge under wind, II: Buffeting theory. *J. Sound Vibr.*, 60(2): 201–11.

Scruggs, J.T. 2004. Structural control using regenerative force actuation networks. PhD diss., California Institute of Technology, Pasadena, CA.

Shaha, A.A. and Y. Ribakovb. 2011. Recent trends in steel fibered high-strength concrete. *Mater. Des.*, 32(8–9): 4122–51.

Shen, W.A. and S.Y. Zhu. 2015. Harvesting energy via electromagnetic damper: Application to bridge stay cables. *J. Intell. Mater. Syst. Struct.*, 26(1): 3–19.

Shen, W.A., S.Y. Zhu, and Y.L. Xu. 2012. An experimental study on self-powered vibration control and monitoring system using electromagnetic TMD and wireless sensors. *Sens. Actuators, A*, 180: 166–76.

Simiu, E. and R. Scanlan. 1996. *Wind Effects on Structures*, 3rd edition. New York, NY: Wiley.

Sodano, H.A., D.J. Inman, and G. Park. 2004. A review of power harvesting from vibration using piezoelectric materials. *Shock Vib. Dig.*, 36(3): 197–205.

Sohn, H., C.R. Farrar, F.M. Hemez, et al. 2003. A review of structural health monitoring literature 1996–2001. Technical report, No. LA-13976-MS, 1-301. Los Alamos, CA: Los Alamos National Laboratory.

Soliman, S. and D.M. Frangopol. 2014. Life-cycle management of fatigue-sensitive structures integrating inspection information. *J. Infrastruct. Syst.*, 20(2): 04014001.

Soong, T.T. 1990. *Active Structural Control: Theory and Practice*. New York, NY: Longman Scientific & Technical, UK and Wiley.

Soong, T.T. and G.F. Dargush. 1997. *Passive Energy Dissipation Systems in Structural Engineering*. London, UK: Wiley.

Soong, T.T. and B.F. Spencer Jr. 2000. Active, semi-active and hybrid control of structures. In *WCEE World Conference on Earthquake Engineering No12, Auckland*, 33(3), 387–402. Wellington, New Zealand: New Zealand National Society for Earthquake Engineering.

Spencer Jr., B.F., S. Cho, and S.H. Sim. 2011. Wireless monitoring of civil infrastructure comes of age. *Struct. Mag.*, 2011: 12–15.

Spencer Jr., B.F. and S. Nagarajaiah. 2003. State of the art of structural control. *J. Struct. Eng.*, 129(7): 845–56.

Spencer Jr., B.F. and M.K. Sain. 1997. Controlling buildings: A new frontier in feedback. *Control Syst., IEEE*, 17(6): 19–35.

Sun, L.M., Z. Sun, D.H. Dan, and Q.W. Zhang. 2007. Large-span bridges and their health monitoring systems in China. In *Proceedings of 2007 International Symposium on Integrated Life-Cycle Design and Management of Infrastructure*, ed. L.C. Fan, 79–95, Shanghai, China: Tongji University Press.

Symans, M.D. and M.C. Constantinou. 1999. Semi-active control systems for seismic protection of structures: A state-of-the-art review. *Eng. Struct.*, 21(6): 469–87.

Taranath, B.S. 2009. *Reinforced Concrete Design of Tall Buildings*. Boca Raton, FL: CRC Press.

Teng, J.G., J.F. Chen, S.T. Smith, and L. Lam. 2003. Behaviour and strength of FRP-strengthened RC structures: A state-of-the-art review. *Proceedings of the ICE – Structures and Buildings*, 156(1): 51–62.

Thompson, P.D., E.P. Small, M. Johnson, and A.R. Marshall. 1998. The Pontis bridge management system. *Struct. Eng. Int.*, 4: 303–308.

Timoshenko, S., D.H. Young, and W. Weaver. 1974. *Vibration Problems in Engineering*, 4th edition. New York: Wiley.

Unsworth, J.F. 2010. *Design of Modern Steel Railway Bridges*. Boca Raton, FL: CRC Press.

West, H.H. 1993. *Fundamentals of Structural Analysis*. New York: Wiley.

Wu, Z.S., B. Xu, and T. Harada. 2003. Review on structural health monitoring for infrastructure. *J. Appl. Mech.*, 6: 1043–54.

Xia, Y., B. Chen, S. Weng, Y.Q. Ni, and Y.L. Xu. 2012. Temperature effect on vibration properties of civil structures: A literature review and case studies. *J. Civil Struct. Health Monit.*, 2(1): 29–46.

Xia, Y., H. Hao, G. Zanardo, and A.J. Deeks. 2006. Long term vibration monitoring of a RC slab: Temperature and humidity effect. *Eng. Struct.*, 28(3): 441–52.

Xia, Y., Y.L. Xu, Z.L. Wei, H.P. Zhu, and X.Q. Zhou. 2011. Variation of structural vibration characteristics versus non-uniform temperature distribution. *Eng. Struct.*, 33(1): 146–53.

Xu, Y.L. 2008. Making good use of structural health monitoring systems: Hong Kong's experience. In *Proceedings of the Second International Forum on Advances in Structural Engineering, Structural Disaster Prevention, Monitoring and Control*, Dalian, China, ed. H.N. Li, 159–98. Beijing, China: China Architecture & Building Press.

Xu, Y.L. 2013. *Wind Effects on Cable-Supported Bridges*. Singapore: Wiley.

Xu, Y.L. and B. Chen. 2008. Integrated vibration control and health monitoring of building structures using semi-active friction dampers: Part I – Methodology. *Eng. Struct.*, 30(7): 1789–801.

Xu, Y.L., B. Chen, C.L. Ng, K.Y. Wong, and W.Y. Chan. 2010. Monitoring temperature effect on a long suspension bridge. *Struct. Control. Health Monit.*, 17: 632–53.

Xu, Y.L., Q. Huang, S. Zhan, Z.Q. Su, and H.J. Liu. 2014. FRF-based structural damage detection of controlled buildings with podium structures: Experimental investigation. *J. Sound and Vibr.*, 333(13): 2762–75.

Xu, Y.L. and Y. Xia. 2012. *Structural Health Monitoring of Long-Span Suspension Bridges*. Oxford, UK: Spon Press.

Yang, H.T.Y., J.Z. Shan, C.J. Randall, P.K. Hansma, and W.X. Shi. 2013. Integration of health monitoring and control of building structures during earthquakes. *J. Eng. Mech.*, 140(5): 04014013.

Yang, J.N. and A.K. Agrawal. 2002. Semi-active hybrid control systems for nonlinear buildings against near-field earthquakes. *Eng. Struct.*, 24(3): 271–80.

Yang, J.N. and S.J. Dyke. 2003. Kobori Panel Discussion: Future perspectives on structural control. In *Proceedings of the 3rd World Conference on Structural Control*, 279–86, Como, Italy: Wiley.

Yang, J.N., J.C. Wu, K. Kawashima, and S. Unjoh. 1995. Hybrid control of seismic-excited bridge structures. *Earthq. Eng. Struct. Dyn.*, 24(11): 1437–51.

Yang, Y.B. and H.B. Lin. 1995. Vehicle-bridge interaction analysis by dynamic condensation method. *J. Struct. Eng.*, 121: 1636–43.

Zhang, Y. and L. Lu. 2008. Introducing smart structures technology into civil engineering curriculum: Education development at Lehigh University. *J. Prof. Issues Eng. Educ. Pract.*, 134(1): 41–48.

Zhou, H., Y. Ni, and J. Ko. 2011. Eliminating temperature effect in vibration-based structural damage detection. *J. Eng. Mech.*, 137(12): 785–96.

Zhu, B.J. and D.M. Frangopol. 2013. Incorporation of SHM data on load effects in the reliability and redundancy assessment of ship cross-sections using Bayesian updating. *Struct. Health Monit.*, 12(4): 377–92.

Zhu, S.Y., W.A. Shen, and Y.L. Xu. 2012. Linear electromagnetic devices for vibration damping and energy harvesting: Modeling and testing. *Eng. Struct.*, 34: 198–212.

Chapter 2

Smart materials

2.1 PREVIEW

Materials are the elements, constituents or substances of which something is composed or made. In civil structures, commonly used materials include wood, stone, bronze, steel, aluminium, concrete, composites and plastics. The material itself is not smart, but the materials with unusual, interesting, amazing and useful properties that can be used to design and develop smart structures may be called *smart materials*. Modern materials science is facilitating the creation of diverse smart materials. There are many text books that give detailed summaries of various smart materials (Gandhi and Thompson 1992; Srinivasan and McFarland 2001; Chopra and Sirohi 2014). Nevertheless, only some materials of interest in developing smart civil structures and their basic characteristics are discussed in this chapter. These smart materials include shape memory alloys (SMAs), piezoelectric materials, magnetostrictive materials, electrorheological (ER) and magnetorheological (MR) fluids, optical fibre, bio-inspired materials and nanomaterials.

2.2 SHAPE MEMORY ALLOYS

SMAs possess an interesting property by which the metal 'remembers' its original size or shape and reverts to it at a characteristic transformation temperature. This feature is known as the *shape memory effect*, and can be used to make sensors and actuators for smart civil structures.

2.2.1 Basic characteristics of shape memory alloys

William J. Buehler and his co-researchers at the Naval Ordnance Laboratory (NOL) discovered that nickel-titanium is one of the most useful SMAs and that its metallurgical features give it rather amazing properties (Buehler et al. 1963). This alloy has a low temperature phase and a high temperature phase in which the change occurs between two solid phases that involve the rearrangement of atoms within the crystal lattice (see Figure 2.1). When cooled to below the critical temperature, its crystal structure enters the so-called martensitic phase. In this state, the alloy is plastic and can easily be manipulated with very large strain but little change in the material stress. When the alloy is heated above the critical temperature, it changes to the so-called austenitic phase with a body-centred cubic structure. In this state, the alloy resumes the shape that it formally had within a high temperature range and behaves very much like a normal metal with high strength and modulus. Furthermore, the alloy shrinks by several percentages in the transformation from the low-temperature martensitic phase to the high-temperature austenitic

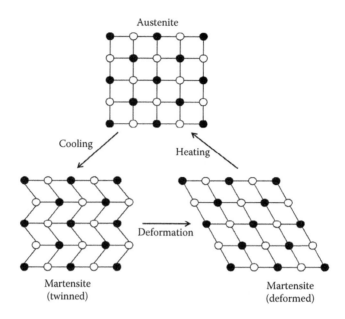

Austenite

Cooling   Heating

Deformation

Martensite
(twinned)

Martensite
(deformed)

Figure 2.1 Phase change of crystal structure in SMAs.

phase (Culshaw 1996). Up to 7% strain can generally be reserved, and in some cases as much as 10%. Typical properties of a nickel-titanium alloy are listed in Table 2.1 as an example. It should be noted that the properties of various SMAs are very complex and depend critically upon their composition and history.

The transformation phase of an SMA wire through a temperature cycle is schematically shown in Figure 2.2 (Srinivasan and McFarland 2001). The critical temperatures where the phase transformations take place are identified as A_s, A_f, M_s and M_f, which, respectively, represent the temperatures at the start of austenite, finish of austenite, start of martensite and finish of martensite transformation. Allow the wire to be started in a fully austenitic phase (i.e. A_f). Now, cool the wire until it reaches the temperature where the material begins to change to the martensite phase (i.e. M_s). Upon further cooling, the martensite plates begin to increase to M_f, at which temperature the wire is in a fully martensitic state. If we now begin to heat the wire, nothing happens to the material phase, but the temperature of the wire increases. The martensite plates begin to rearrange themselves into their original configuration when the temperature reaches A_s, and this arrangement will be completed at A_f. It can be seen from Figure 2.2 that the value of the fraction of the martensite in the material (ξ^{SMA}) is changing from 0 (at A_f and M_s) to 1 (at M_f and A_s). Notably, throughout these transformation, no stress is applied, which means that the cycle is entirely driven by temperature.

The influence of stress on the characteristics of the SMAs is also investigated. Experimental observations indicate that the critical temperatures (i.e. A_s, A_f, M_s and M_f) increase with stress, as schematically shown in Figure 2.3. Generally, it is assumed that $\theta = \gamma$ and the phase transformation occurs without the influence of stress (i.e. $\sigma^{SMA} = 0$) (see Figure 2.3). With $\sigma^{SMA} \neq 0$, higher temperature will be needed to bring about a phase change.

Moreover, the phase transformations that result entirely from stress and lead to particular hysteresis loops, which are different from conventional metals, are another important characteristic of the SMAs, referred to as *superelasticity*. The SMAs behave pseudoelastically at

Table 2.1 Typical properties of NiTi alloys

Physical properties	
Melting point (°C)	1300
Density (g/cm³)	6.45
Electrical resistivity:	
Austenite (μΩ·cm)	≈100
Martensite (μΩ·cm)	≈70
Thermal conductivity:	
Austenite (W/m·°C)	18
Martensite (W/m·°C)	8.5
Corrosion resistance	Similar to 300 series stainless steel or Ti alloys
Mechanical properties	
Young's modulus:	
Austenite (GPa)	≈83
Martensite (GPa)	≈28–41
Yield strength:	
Austenite (MPa)	195–690
Martensite (MPa)	70–140
Ultimate tensile strength (MPa)	895
Transformation properties	
Transformation temperature (°C)	−200–110
Latent heat of transformation (kJ/kg·atom)	167
Shape memory strain	8.5% maximum

Source: Data from Hodgson, D.E. et al., *ASM Handbook, Volume 2: Properties and Selection: Nonferrous Alloys and Special-Purpose Materials*, 897–902, 1990. http://products.asminternational.org/hbk/index.jsp.

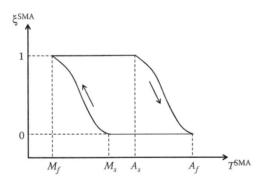

Figure 2.2 Schematic diagram of stress-free phase transformation.

a temperature $T > A_f$. Applying stress induces the transformation of austenite into martensite, resulting in a large amount of inelastic deformation or transformation strain (Brocca et al. 2002). As the stress is reduced, after an initial elastic response, the martensite formed during the loading process transforms back to austenite, the inelastic strain is therefore recovered and the stress–strain diagram exhibits the characteristic hysteretic loop as shown in Figure 2.4.

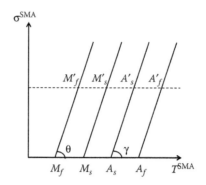

Figure 2.3 Schematic diagram of the influence of stress on critical phase change temperatures.

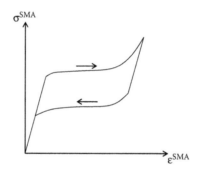

Figure 2.4 Superelasticity for an SMA material.

2.2.2 Constitutive modelling of shape memory effect

The macroscopic mechanical behaviour of SMAs is usually modelled following either a phenomenological or micromechanical approach (Brocca et al. 2002). The phenomenological models for describing the behaviour of SMAs usually contain one-dimensional and three-dimensional models. Since one-dimensional phenomenological models usually utilise measurable quantities as parameters and are relatively simple, they are basically quite accurate in predicting the uniaxial response of SMAs and are particularly suitable for engineering practice. Many researchers have developed various uniaxial phenomenological models (e.g. Tanaka and Nagaki 1982; Liang and Rogers 1990; Brinson 1993; Ivshin and Pence 1994a, b). These models are often composed of a kinetic law for governing the crystallographic transformation and a mechanical law for representing the stress–strain behaviour. Because the transformation strain is very large as compared with the elastic strain, the major factor for distinguishing these one-dimensional models is their particular kinetic law. Since three-dimensional models appear to be capable of capturing the typical features of SMAs, several attempts have been made to extend the one-dimensional models to three dimensions. Based on the use of plasticity models with an internal variable such as the martensite fraction ξ^{SMA}, several three-dimensional models have been developed (Boyd and Lagoudas 1994; Lubliner and Auricchio 1996; Lagoudas et al. 1996; Birman 1997). However, due to the lack of experimental data for a multi-axial response, the performance of these 3-D phenomenological models cannot be directly evaluated, but only compared to the uniaxial experimental data.

By using the laws of thermodynamics to describe the transformation, the micromechanical models attempt to capture the crystallographic phenomena within the material more accurately (e.g. Sun and Hwang 1993; Patoor et al. 1996; Goo and Lexcellent 1997; Lu and Weng 1997; Huang and Brinson 1998; Vivet and Lexcellent 1998; Gao et al. 2000). In these models, the martensitic variants are usually considered as transforming inclusions and the micromechanics is employed for the calculation of the interaction energy caused by the phase transformation in the material. Following this approach, the models are obviously more complicated than the phenomenological models and are usually computationally demanding.

It is known that the stress state in an SMA component is a function of the three primary state variables, which are ξ^{SMA}, the fraction of martensite; T^{SMA}, the temperature at which the component is operating; and ε^{SMA}, the strain at which the component is functioning (Srinivasan and McFarland 2001). With the initial conditions, ε_0^{SMA}, T_0^{SMA} and ξ_0^{SMA}, one can obtain a unified constitutive relation for the general case:

$$\sigma^{SMA} - \sigma_0^{SMA} = E^{SMA}(\varepsilon^{SMA} - \varepsilon_0^{SMA}) + \theta^{SMA}(T^{SMA} - T_0^{SMA}) + \Omega^{SMA}(\xi^{SMA} - \xi_0^{SMA}) \qquad (2.1)$$

where:
- σ^{SMA} is the shear stress
- E^{SMA} is the Young's modulus
- θ^{SMA} is the thermal elastic tensor which is related to the thermal coefficient of expansion for the material
- Ω^{SMA} is the transformation tensor

2.2.3 Applications of shape memory alloys in smart civil structures

Due to their high power density, high damping capacity and solid state actuation, SMAs have been actively investigated and widely applied in various fields, such as aerospace, mechanical engineering and life science. When integrated with civil structures, SMAs can be passive, semi-active or active components to attenuate excessive structural vibration. The passive structural control using SMAs takes advantage of the SMA's damping property to reduce the response and consequent plastic deformation of the structures subjected to severe loadings. SMAs can be effectively used for this purpose via two mechanisms: the ground isolation system and the energy dissipation system (Saadat et al. 2002; Song et al. 2006a). The reported SMA isolation systems include SMA bars for highway bridges (Wilde et al. 2000), SMA wire re-centring devices for buildings (Dolce et al. 2001), SMA spring isolation systems (Khan and Lagoudas 2002) and SMA tendon isolation systems for a multi-degree-of-freedom (MDOF) shear frame structure (Corbi 2003). The SMA energy dissipation devices can be found in the forms of braces for framed structures (Dolce et al. 2000; Saadat et al. 2001; Tamai and Kitagawa 2002; Han et al. 2003), dampers for cable-stayed bridges (Li et al. 2004) or simply supported bridges (Casciati et al. 1998; DesRoches and Delemont 2002), connection elements for columns (Leon et al. 2001; Tamai and Kitagawa 2002) and retrofitting devices for historic buildings (Croci et al. 2000; Indirli et al. 2001).

An actuator is a mechanism by which a control system acts on a structure and thus changes the behaviour of the structure. Due to the shape memory effect, SMAs can function as actuators and several concepts for shape memory actuators in civil engineering have been reported (Amato 1994; Maji and Negret 1998; Li et al. 2006). The use of SMAs as actuators relies upon operating them normally in the low-temperature plastic martensitic phase and constraining them within some structural assembly in this state. When the alloy is heated through to the austenitic phase, the external structure constrains the alloy from returning to its remembered shape. Consequently, significant stresses (probably several hundreds of MPa) are generated within the alloy and these, in turn, can stress the structure within which the alloy is mounted. The basic principle of SMAs for active structural vibration control (SVC) is utilising SMAs to actively tune the resonant frequency of the structure for the mitigation of excessive vibration. Upon heating, SMA actuators that are embedded or installed in structures will increase the stiffness of the host structure so that the natural frequency of the structure changes and the vibration control can be achieved. A comprehensive review of the applications of SMAs in the field of civil engineering for vibration control can be found in the references (Janke et al. 2005; Song et al. 2006a).

Although SMA materials have dramatic properties and much research has been carried out, the application of SMAs in civil structures is still limited in practical situations. The drawbacks of SMAs for practical application mainly lie in high cost, stress-induced martensite in SMAs and thermal actuation, as well as the problem of actuation time (Janke et al. 2005). Moreover, it is known that the greater the magnitude of shape recovery during the phase change, the shorter the fatigue life of the material, which means the disadvantages of SMAs also include their susceptibility to fatigue.

2.3 PIEZOELECTRIC MATERIALS

Piezoelectricity is an electro-deformation phenomenon derived from the Greek word *piezein* for 'pressure'. The first demonstration of the direct piezoelectric effect was in 1880 by the brothers Pierre Curie and Jacques Curie (Curie and Curie 1880). Piezoelectric materials possess an unusual property by which they experience a dimensional change with directionality when an electrical voltage is applied to them (Srinivasan and McFarland 2001). Such materials also possess a converse effect, that is, they generate electricity when pressure is applied. Perhaps the best-known piezoelectric material is lead–zirconate–titanate (PZT); in fact, 'PZT' is commonly used to refer to piezoelectric materials in general, including those of other compositions.

2.3.1 Basic characteristics of piezoelectric materials

When manufactured, a PZT material has electric dipoles arranged in random directions (Srinivasan and McFarland 2001), as shown in Figure 2.5a. The responses of these dipoles to an externally applied electric field would tend to cancel one another, producing no gross change in the dimensions of the PZT specimen. To obtain a useful macroscopic response, a process called *poling* is performed on the material, as shown in Figure 2.5b.

A PZT material actually has a characteristic Curie temperature. When it is heated above this temperature, the dipoles can change their orientation in the material. In poling, the material is heated above its Curie temperature and a strong electric field is then applied. The direction of this field is the polarisation direction, and the dipoles shift into alignment with it (see Figure 2.5b). The material is then cooled below its Curie temperature while the poling field is maintained, with the result that the alignment of the dipoles is permanently fixed. The material is then said to be *poled*.

When the poled material is maintained below its Curie temperature and is subject to a small electric field (compared with that used in poling), the dipoles respond collectively to produce a macroscopic expansion along the poling axis and contraction perpendicular to it (or vice versa, depending on the sign of the applied field). For example, when a PZT that has been poled in the 3-direction is subject to an electric field in this direction, a schematic diagram for the description of its geometry and deformation can be found in Figure 2.6.

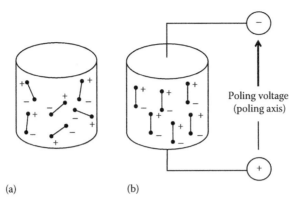

Figure 2.5 The dipole rearrangement of piezoelectric effect: (a) before poling process and (b) after poling process.

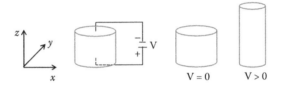

Figure 2.6 Deformation of a poled PZT subjected to an electric field in z-direction.

Notably, the PZT is able to perform well when the working temperature is below its Curie temperature. However, if the material is heated above its Curie temperature when no electric field is applied, the dipoles will revert to random orientations. Even at lower temperatures, the application of too strong a field can cause the dipoles to shift out of the preferred alignment established during poling.

2.3.2 Constitutive modelling of piezoelectric materials

The direct and converse PZT effects basically reflect an interaction between the mechanical and electrical behaviour of PZT materials. The description of the electromechanical properties of PZT materials herein is based on the Institute of Electrical and Electronics Engineers (IEEE) standard for piezoelectricity (IEEE 1988), which is widely accepted as a good representation of PZT material properties. Although PZT materials may show considerable nonlinearity if operated under a high electric field or a high mechanical stress level, the IEEE standard assumes that PZT materials are linear, which has already turned out to be true in the case of a low electric field or a low mechanical stress level.

The constitutive equations describing the PZT properties are based on the assumption that the total strain in the transducer is the sum of the mechanical strain induced by the mechanical stress and the controllable actuation strain caused by the applied electric voltage (Moheimani and Fleming 2006).

The electromechanical equations for a linear PZT material can be expressed as (IEEE 1988; Fuller et al. 1996)

$$\varepsilon_i = S_{ij}^E \sigma_j + d_{mi} E_m \tag{2.2}$$

$$D_m = d_{mi}\sigma_i + \xi^\sigma_{ik}E_k \tag{2.3}$$

where the indexes $i, j = 1, 2, ..., 6$ and $m, k = 1, 2, 3$ refer to different directions within the material coordinate system, as shown in Figure 2.7.

Here, the axes are identified by numerals rather than letters. For the applications that involve sensing, the aforementioned equations are usually rewritten as follows:

$$\varepsilon_i = S^D_{ij}\sigma_j + g_{mi}D_m \tag{2.4}$$

$$E_i = g_{mi}\sigma_i + \beta^\sigma_{ik}D_k \tag{2.5}$$

where:

- σ and ε are the stress vector (unit: N/m²) and the strain vector (unit: m/m), respectively
- E is the vector of applied electric field (unit: V/m)
- ξ and β stand for permittivity (unit: F/m) and impermittivity component (unit: m/F), respectively
- d and g are the matrix of PZT strain constants (unit: m/V) and the matrix of PZT constants (unit: m²/C), respectively
- S is the matrix of compliance coefficients (unit: m²/N)
- D is the vector of electric displacement (unit: C/m²)
- superscripts D, E and σ represent measurements taken at constant electric displacement, constant electric field and constant stress, respectively

Equations 2.2 and 2.3 express the converse PZT effect, which describes the situation when the device is being used as an actuator. This converse effect is usually used to determine the PZT coefficients. However, Equations 2.4 and 2.5 express the direct piezoelectric effect, which deals with the case when the transducer is being used as a sensor.

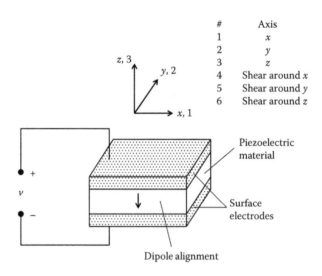

#	Axis
1	x
2	y
3	z
4	Shear around x
5	Shear around y
6	Shear around z

Figure 2.7 Schematic diagram of a piezoelectric transducer and its axis nomenclature.

2.3.3 Applications of piezoelectric materials in smart civil structures

Piezoceramics can be bonded to the surface of a structure in high strain areas with minimal modification of the original structure or they can be easily embedded in composite structures. Since very thin PZT layers or patches can be bonded or embedded in host structures, PZT materials can be utilised for SVC as sensors and actuators with the advantages of being low cost, lightweight and easy to implement. Stack types of piezoceramic actuators can be incorporated into the structure with slight modification or without significantly changing the stiffness of the host structure (Song et al. 2006b). Stack types of piezoceramic actuators are also used to make semi-active friction dampers for controlling the vibration of building structures (Ng and Xu 2007; Xu and Ng 2008). For the active control of small amplitude vibration of very flexible structures such as a cantilever beam, they lead to lightweight, adaptive and high precision control systems (Han et al. 1997; Sugavanam et al. 1998; Blanguernon et al. 1999; Sunar and Rao 1999; Rao and Sana 2001; Irschik 2002). Investigations have also been conducted by various researchers into the vibration control of other types of civil structures, such as truss structures (McClelland et al. 1996; Preumont et al. 2000; Gao et al. 2003), frame structures (Kamada et al. 1997, 1998; Fujita et al. 2001) and cable-stayed structures (Warnitchai et al. 1993; Fujino et al. 1993; Fujino and Susumpow 1995; Achkire and Preumont 1996). It is known that the most piezoceramic materials are very brittle and may fail under operation. Hence, the concept of adding some passive dampers to the structure for the purpose of achieving a more reliable, robust and effective control performance has been actively investigated (Lesieutre and Lee 1996; Benjeddou 2001; Trindade and Benjeddou 2002). Depending on the relative positions of the viscoelastic layer and the PZT actuator, the viscoelastic passive actions and the piezoelectric active actions can operate either separately or simultaneously.

PZT materials can generate a charge in response to mechanical stimulus or, alternatively, provide a mechanical strain when an electric field is applied across them. Because of these unique characteristics, PZT materials can be used for both actuation and sensing (Zou et al. 2000). Therefore, the use of such materials for in situ structural health monitoring (SHM) would lead to not only fewer sensors, but also a reduction in electrical wiring and the associated hardware. Moreover, these materials are low in cost, exhibit excellent mechanical strength and have low acoustic impedance and a broad band dynamic response. From this perspective, PZT materials could be ideal for structural damage detection and on-line health monitoring. The most well-known damage detection method using PZT materials is the electromechanical impedance (EMI) method. This method utilises the electromechanical coupling property of PZT materials and monitors variations in the electric impedance or admittance of the PZT materials bonded to structures in a high-frequency band for detecting the damage. Successful applications of SHM of various types of structures have been reported in the past several decades (Zou et al. 2000; Park et al. 2000a,b, 2003; Giurgiutiu and Cuc 2005; Shin and Oh 2009). The wave-based method could be considered another widely used method for SHM using PZT materials. The basic idea behind this method is that structural damage is associated with the changes or variations in the transmission velocity or energy of elastic waves. Recent studies show that the wave-based method can be successfully utilised in the health monitoring of typical structural members and structures including reinforced concrete beams, columns, steel bridge components and frames under static loading, cyclic reversed loading and seismic loading (Cawley and Alleyne 1996; Kessler et al. 2002; Park et al. 2006; Hu and Yang 2007; Song et al. 2007a,b; Xu et al. 2013).

In addition to the aforementioned applications in SVC and SHM, the use of PZT materials for energy harvesting has also received increased attention due to their ability to directly convert applied strain energy into usable electric energy and the flexibility with which they can be integrated into a system (Anton and Sodano 2007). This energy conversion occurs because the PZT molecular structure is oriented such that the material exhibits a local charge separation, known as an *electric dipole*. When strain energy is applied to the material, it results in the deformation of dipoles and the formation of a charge that can be removed from the material and used to power various devices. PZT materials can be configured in various ways that prove useful in power harvesting applications. The configuration of the power harvesting device can be changed through modification of PZT materials, altering the electrode pattern, changing the poling and stress direction, layering the material to maximise the active volume, adding pre-stress to maximise the coupling and applied strain of the material and tuning the resonant frequency of the device. A large percentage of recent research in power harvesting with PZT materials has focused on improving the efficiency of PZT power harvesting systems. A comprehensive review can be found in the literature (Sodano et al. 2004; Anton and Sodano 2007).

2.4 MAGNETOSTRICTIVE MATERIALS

Generally, magnetostriction is the change in the shape of materials under the influence of an external magnetic field. The basic phenomenon is very similar to electrostriction but magnetostriction materials are capable of exercising high strain levels and large movements. Furthermore, in common with their electrostrictive cousins, magnetostrictive materials have narrow hysteresis loops and therefore are relatively low-loss materials (Culshaw 1996). These characteristics make them suitable for actuators with a relatively large capacity compared with PZT materials.

2.4.1 Basic characteristics of magnetostrictive materials

Magnetostriction is a phenomenon observed in all ferromagnetic materials, which causes them to change their dimensions when they are subject to a magnetic field. Terfenol-D, with Ter for terbium, Fe for iron, NOL for Naval Ordnance Laboratory and D for dysprosium, exhibits the highest magnetostriction among alloys and is the most commonly used engineering magnetostrictive material.

Magnetostrictive materials have a property that can convert magnetic energy into kinetic energy or vice versa. Various physical effects related to the magnetostrictive effect can be found in Figure 2.8. The most understood effect that is related to magnetostriction is the *Joule effect*. This is the expansion, positive magnetostriction or contraction, negative magnetostriction, of a ferromagnetic rod in relation to a longitudinal magnetic field (Olabi and Grunwald 2008). This effect is mainly used in magnetostrictive actuators.

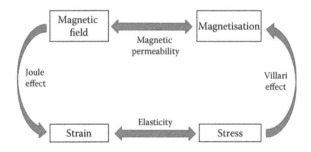

Figure 2.8 Magnetostriction effects.

The *Villari effect* is another widely utilised effect related to magnetostriction. This effect is based on the fact that when a mechanical stress is imposed on a sample, there is a change in the magnetic flux density which flows through the sample as a result of the creation of a magnetic field. The change in flux density can be detected by a pickup coil and is proportional to the level of the applied stress. The Villari effect is reversible and is used in sensor applications (Olabi and Grunwald 2008).

The ΔE-*effect* is the change of Young's modulus as a result of a magnetic field. The $\Delta E/E$ in Terfenol-D is in the range of more than five and can be employed in tunable vibration and broadband sonar systems (Dapino 2002). Due to the change in Young's modulus, there is a change in the velocity of sound inside the magnetostrictive materials, which can be observed.

Another effect related to magnetostriction is the *Wiedemann effect*. The physical background to this effect is similar to that of the Joule effect, but instead of forming a purely tensile or compressive strain as a result of the magnetic field, there is a shear strain, which results in a torsional displacement of the ferromagnetic sample. The inverse Wiedemann effect is called the *Matteuci effect* (Olabi and Grunwald 2008).

2.4.2 Constitutive modelling of magnetostrictive materials

When ferromagnetic materials are subject to a magnetic field, the Joule effect leads to a change of shape in the direction of the magnetic field while maintaining a constant volume. The property can be quantified by the fractional change in length as the magnetisation of the material increases from zero to the saturation value, named magnetostrictive coefficient λ:

$$\lambda = \frac{\Delta L}{L} \tag{2.6}$$

where L and ΔL are the initial length of the material and its variation of the length, respectively. The shape of the materials changes in the direction of the applied magnetic field expands for a positive λ and shrinks for a negative λ, as shown in Figure 2.9.

The material properties change with the conditions during operation, which makes the behaviour of the magnetostrictive materials complex in various applications. In response to an applied magnetic field, the idealised behaviour of length change is shown in Figure 2.10 as butterfly curves. Like a positive magnetic field, a negative magnetic field established in the opposite direction produces the same elongation in the magnetostrictive material.

The two most widely utilised magnetostrictive effects, that is the Joule effect and the Villari effect, can be approximated using the following constitutive equations:

$$B = dT + \mu^T H \qquad \left(H = nI \right) \tag{2.7}$$

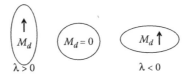

Figure 2.9 Change of shape in joule magnetostriction.

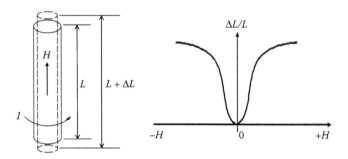

Figure 2.10 Strain against symmetric magnetic field.

$$S = s^H T + dH \qquad (2.8)$$

where:

 B is the magnetic induction
 d is the magneto-mechanical constant
 T is the mechanical stress
 μ^T is the magnetic permeability under constant stress condition
 H is the magnetic field strength
 I is the current
 n is the number of coil turns
 S is the mechanical strain
 s^H is the elastic compliance under constant magnetic field

The observation of the typical S-H and B-H curves reveals that the relationship between the magnetostriction and the applied magnetic field is highly dependent on the magnetic intensity. A polarising field is a field where the magnetic domains are initially aligned. When the intensity of the applied magnetic field is much lower than the polarising intensity, the relationship between the magnetostriction and the applied magnetic field is approximately linear. When the intensity of the applied magnetic field approaches polarising intensity, the nonlinearity begins and the curve gradually flattens out, signifying saturation or completion of all the domain alignments.

2.4.3 Applications of magnetostrictive materials in smart civil structures

The ability to convert an amount of energy between magnetic energy and kinetic energy allows the use of magnetostrictive materials in actuator and sensor applications. Thus, magnetostrictive materials can be employed for vibration control and the non-destructive evaluation of civil structures.

With regard to vibration control, Terfenol-D was first used to reduce and alter unwanted vibration motions in the late 1980s. Actuators employing Terfenol-D were designed and analysed by many researchers and are now used in vibration isolation systems, since the Terfenol-D rod has some distinct advantages over other smart materials, such as high strain, good magneto-mechanical coupling factor, fast response and the magnetostriction property without changing with time (Xu and Li 2006). Theoretical frameworks were established to get a general form of a mathematical model for characterising magnetostrictive-based

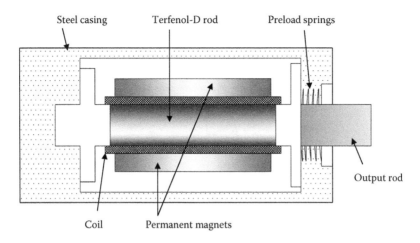

Steel casing   Terfenol-D rod   Preload springs

Output rod

Coil   Permanent magnets

Figure 2.11 Cross section of magnetostrictive actuator.

actuators. Experimental works indicated that magnetostrictive materials were suitable for applications in smart structures (Olabi and Grunwald 2008). A magnetostrictive actuator with Terfenol-D is depicted in Figure 2.11.

Magnetostrictive materials can also be used in sensor applications, such as torque sensors, motion and position sensors, force and stress sensors, material characterisation sensors and magnetic sensors (Hristoforou and Ktena 2007). For a non-destructive evaluation of civil structures, magnetostrictive materials are practically used for emitting and receiving elastic wave as transducers to detect the defect and the depth of concrete structures, and in monitoring the acoustic emission wave in a structure to evaluate the position and propagation of cracks. Sensors based on magnetostrictive materials utilise the magnetostrictive property of a material to convert a mechanical energy into a magnetic field. The basic configuration of a magnetostrictive sensor, including a waveguide, a position magnet, electronics, a strain pulse detection system and a damping module, is shown in Figure 2.12.

Magnetostrictive materials possess many advantages compared with other smart materials applied in smart structures, such as large deformation ability, quick response, high energy density, high courier temperature, high stiffness and wide operation frequency. However, the applications of magnetostrictive materials in civil engineering have been confined due to some drawbacks, for example, severe eddy current losses at high frequency and high cost. These limitations could be overcome, however, by amplifying the force-through lever mechanism and fabricating composites with a polymer matrix.

2.5 ELECTRORHEOLOGICAL AND MAGNETORHEOLOGICAL FLUIDS

In response to an applied electric or magnetic field, ER or MR fluids exhibit the most remarkable change of property in that their ability to support shear stress is tremendously increased. Although application details differ because of the requirements of generating strong electric or magnetic fields, the basic physics describing how ER and MR fluids' material properties change and the design of mechanical devices to capitalise on these changes are similar for the two types of controllable fluid (Srinivasan and McFarland 2001). Therefore, they are simultaneously treated in this section.

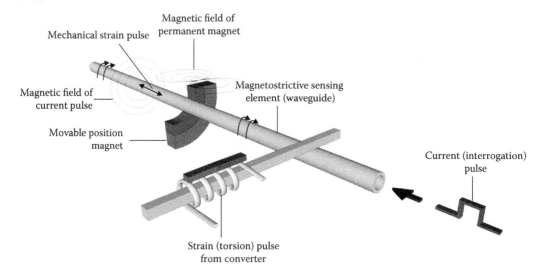

Figure 2.12 Configuration of magnetostrictive sensor. (From http://www.mtssensors.com/fileadmin/media/pdfs/551019.pdf.)

2.5.1 Basic characteristics of ER and MR fluids

Both ER and MR fluids are suspensions of particles, which are typically in size of the order of 1–10 μm in inert carrier liquids, such as mineral oils or silicone oils. Most ER and MR fluids also contain small amounts of additive that affect the polarisation of the particles or stabilise the structure of suspension against settling, but for many engineering purposes these may be neglected in modelling the fluids' mechanical response (Srinivasan and McFarland 2001).

In the absence of an external electric or magnetic field, ER or MR fluids may be characterised as *Newtonian*, that is, the shear stress τ is proportional to the product of the strain rate $\dot{\gamma}$ and viscosity η:

$$\tau = \eta\dot{\gamma} \tag{2.9}$$

Notably, even without the applied field, most ER or MR fluids are not purely Newtonian due to their heavy loading of solid particles and, to some extent, due to the additives they contain. However, in most applications, the field-induced component of the shear stress is much larger than the $\eta\dot{\gamma}$ term, and thus Equation 2.9 can be viewed as an adequate model for the approximation of the rate-dependent part of the total shear stress.

While an electric field is applied to the ER fluid or a magnetic field is applied to the MR fluid, the particles in the fluid are forced to form chains or fibrils in the direction of the field in a few milliseconds. If there is no motion of the fluid or the walls of its container, the fibrils are static structures. When the ER or MR fluid flows, or when there is relative motion between the walls of its container, shear strains occur in the fluid and a shear stress distribution develops across the fluid. This stress distribution can be calculated by using the viscous flow equations of elementary fluid mechanics (Srinivasan and McFarland 2001). When a field oriented normal to the direction of flow or motion is applied, fibrils form across the flow, and because of the motion of the fluid or walls, these fibrils are broken and must reform. The continual breaking and reforming of these particle chains generate a force

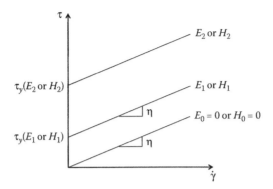

Figure 2.13 Shear stress vs. shear strain rate for the Bingham plastic model.

resisting the motion of the fluid or walls, and give rise to the field-dependent component of the shear stress τ. Generally, with an increase of field strength, the ER or MR shear stress increases accordingly.

2.5.2 Constitutive modelling of ER and MR fluids

One might expect that the formation of fibrils within an ER or MR fluid would increase the fluids' viscosity, but in fact the slope of the shear stress versus shear strain rate curve, and thus the viscosity, changes little if at all. The effect of the fibrils is instead to produce a shear stress commonly referred to as the yield stress (τ_y), which is largely independent of the strain rate. Adding this term to the Newtonian model as shown in Equation 2.9 results in the Bingham plastic model:

$$\tau = \tau_y + \eta \dot{\gamma} \tag{2.10}$$

where τ_y denotes the yield stress, which is a function of the strength of the applied electric or magnetic field. The yield stress in the Bingham plastic model is strongly dependent on the field strength and the response predicted by this model is schematically shown in Figure 2.13. In Figure 2.13, E represents an electric field while H represents a magnetic field. In many practical situations, the dynamic viscosity is determined by a linear regression fit of a line to experimental data, and the intersection of this line with the shear stress is taken as the value of the yield stress τ_y. Till now, the Bingham plastic model and its extensions are the most popular model for use in the design of devices that depend on the post-yield shear resistance of an ER or MR fluid.

According to the Bingham plastic model, stress less than the yield stress τ_y produces no flow of the ER or MR fluid. However, in reality the fluid naturally responds to stress in this range, and for many purposes it may be regarded as a viscoelastic solid (Srinivasan and McFarland 2001). The typical stress–strain characteristics for an ER or MR fluid loaded up to and beyond yield are given in Figure 2.14. Note that yield occurs at approximately the same strain γ_y regardless of the field strength. For clarity, we have shown the yield strain as corresponding to the peak stress on each curve but in practice the correct definition of yield for these materials is not so clear. Weiss et al. (1994) suggest that the onset of nonlinearity in the storage modulus is an accurate indicator of yield and that is correlates well with measured static yield stresses.

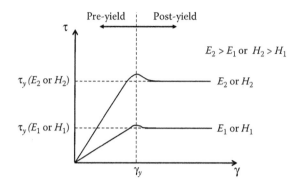

Figure 2.14 Typical stress–strain characteristics for ER or MR fluid.

In many devices where ER or MR fluids are employed, at any time a small portion of the fluid is subject to the applied electric or magnetic field while the remainder is free to flow as a conventional, low-viscosity fluid. Ordinarily, the field is created across a small gap, the surface of which serves as both electrodes (in the case of an ER fluid) or pole pieces (in the case of an MR fluid) and as the walls of a channel confining the fluid. Notably, although these walls are often modelled as parallel, flat plates for the purpose of analysis, the derived equations are frequently applied to annular or other non-flat geometries with acceptable results, such as the analysis of a realistic model for cylindrical dampers (Spencer et al. 1998). As illustrated in Figure 2.15, the shear in the fluid may be generated either by forcing the fluid through the gap under pressure (fixed-plate configuration) or by moving one plate with respect to the other (sliding-plate configuration). It should be noted that the gap is small in the direction of the fluid, often well under 1 mm, and of the order of millimetres in length in the direction of flow or motion.

The analysis of ER or MR fluid flow or plate motion can be carried out by a straightforward extension of elementary fluid mechanics. For the fluid that can be described by the Bingham plastic model, Phillips (1969) derived the following fifth-degree polynomial in 1969 to depict the Poiseuille flow of the Bingham material in a rectangular duct:

$$P^3 - (1 + 3T)P^2 + 4T^3 + P^2 V + \frac{P^2 T V^2}{3A^2} = 0 \tag{2.11}$$

in which

$$P = \frac{bh^3 p'}{12 Q \eta}, T = \frac{bh^2 \tau_y}{12 Q \eta}, V = \frac{bhU}{2Q}, A = P - 2T \tag{2.12}$$

where:

- p' is the pressure gradient
- Q is the volumetric flow rate
- b is the width of the rectangular plate
- h is the gap between two parallel plates
- U is the relative velocity of the two plates

Gavin et al. (1996a,b) found that Equation 2.11, which is based on the simple parallel-plate model, is accurate enough for describing the force-velocity behaviour of cylinder

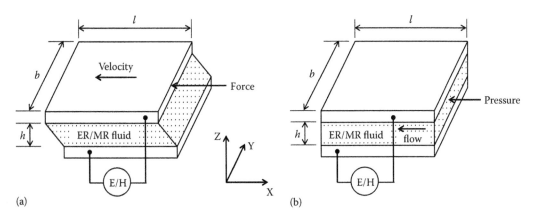

Figure 2.15 ER or MR fluid shear modes: (a) sliding plate and (b) fixed plate.

ER dampers in an axisymmetric flow field. They also found the approximate solutions of Equation 2.11 for either fixed-plate-type smart dampers or sliding-plate-type smart dampers. Upon proper manipulation, the relationship between the force f and velocity v of the smart damper can be expressed as follows:

$$f(t) = C_d v + F_d \operatorname{sgn}(v) \tag{2.13}$$

in which

$$C_d = C_1 \frac{12\eta L A_p}{bh^3} A_p; F_d = C_2 \frac{L\tau_y}{h} A_p + P_y \tag{2.14}$$

For the fixed-plate-type damper:

$$C_1 = 1.0; C_2 = 2.07 + \frac{1.0}{1.0 + 0.4T}; T = \frac{bh^2 \tau_y}{12 A_p \eta v} \tag{2.15}$$

For the sliding-plate-type damper:

$$C_1 = 1.0 - \frac{bh}{2A_p}; C_2 = 2.07 + \frac{1.0}{1.0 + 0.4T} - \frac{1.5V^2}{1.0 + 0.4T^2}; V = \frac{bh}{2A_p} \tag{2.16}$$

where:
   L is the effective axial pole length
   A_p is the cross-sectional area of the piston
   P_y is the mechanical friction force in the damper

Clearly, the damper force f is the function of the yielding shear stress τ_y and it can be controlled through the applied field, but C_d is independent of the applied field. Notably, although the aforementioned equations treat only the case of an ER fluid in principle, most of the results are readily applicable to MR fluids in similar geometries (Spencer et al. 1997, 1998).

2.5.3 Applications of ER and MR fluids in smart civil structures

As mentioned before, ER and MR fluids are made by mixing fine particles into a liquid with low viscosity. The particles will be formed into chain-like fibrous structures in the presence of an electric field or a magnetic field. When the electric field strength or the magnetic field strength reaches a certain value, the suspension will be solidified and has high yield stress. Conversely, the suspension can be liquefied once more by removal of the electric field or the magnetic field (Yao et al. 2002). The process of change is very quick (normally less than a few milliseconds) and is easily controllable with small energy consumption (usually several watts required). After ER and MR fluids were developed in the 1940s (Winslow 1947; Rabinow 1948), they were often utilised in the form of a discrete device (e.g. dampers) in various fields, such as mechanical engineering and civil engineering. Initially, it was ER fluids that received major attention (Petek 1992; Coulter et al. 1993; Carlson 1994; Wu et al. 1994), but they were eventually found to be not as well suited to most applications as the MR fluids, especially for civil structures. In their non-activated or 'off' state, both MR and ER fluids typically have similar viscosity, but MR fluids exhibit a much greater increase in viscosity and therefore yield more strength than their electrical counterparts (Yao et al. 2002). For ER fluid, the maximum yield stress is about 10 kPa but for MR fluid the maximum yield stress can reach about 100 kPa.

A schematic diagram of MR fluid dampers for civil engineering applications is shown in Figure 2.16. The components of a typical MR damper include MR fluid, a piston, a housing, a pair of wires, a magnetic coil and an accumulator, as shown in Figure 2.16. The MR fluid flows through a small orifice within a cylinder. When a current is supplied to the magnetic coil, which is built in the piston or on the housing, the particles are aligned and the fluid changes from the liquid state to the semi-solid state. Depending on the applied current in the damper coil and the piston velocity, the generated damping force can be adjusted.

Only requiring a battery of power, MR dampers are capable of generating a force with a magnitude sufficient for rapid response in large-scale applications. Moreover, MR dampers offer highly reliable operations with relatively lower sensitivity to temperature fluctuations or impurities in the fluid. Consequently, the research on the use of MR dampers for the semi-active control of civil structures under wind and earthquake loadings has been actively conducted for many years. The MR damper was first introduced as an application of civil engineering in 1996 (Dyke et al. 1996). After that, much development and progress has been made on control systems based on MR fluid dampers. To reduce dynamic responses caused by wind and earthquakes, MR fluid dampers were applied to full-scale structures in civil engineering for semi-active control, the first implementations being on the cable-stayed Dongting Lake Bridge in China and the Nihon-Kagaku-Miraikan building in

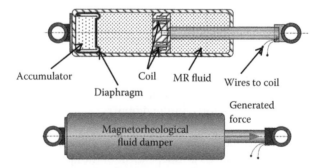

Figure 2.16 MR fluid damper. (From Truong, D.Q. and K.K. Ahn, *Smart Actuation and Sensing Systems: Recent Advances and Future Challenges*, InTech, Rijeka, Croatia, 2012.)

Japan. Examples of the use of state-of-the-art of MR damper-based control systems in civil engineering applications can be found in the literature (Symans and Constantinou 1999; Xu et al. 2000; Jung et al. 2002; Spencer and Nagarajaiah 2003; Muhammad et al. 2006).

2.6 OPTICAL FIBRES

An optical fibre conducts light in much the same way as a copper wire conducts electricity. The most common use of optical fibres is in transmitting data over long distances. In the context of smart civil structures, the role of the optical fibre is different and is in the form of sensing physical parameters such as strain, temperature, pressure and vibration in structural components. By embedding a grid of fibre, a more thorough strain-mapping can be obtained than is possible via discrete strain gauges located on the surface of a structure. Fibre-optic sensors can also be configured to measure the internal chemical states in structures, such as the penetration of corrosion-causing de-icing salts in bridge decks.

2.6.1 Basic characteristics of optical fibres

An optical fibre, as shown in Figure 2.17, is a thread-like material with a circular cross section that makes use of the optical total internal reflection (TIR) to guide light waves (Fang et al. 2012). It is often flexible, transparent and made of a high quality extruded glass (silica). The fabrication of silica fibre is based mainly on modified chemical vapour deposition (MCVD) technology (Akamatsu et al. 1977). The bare drawn fibre is then coated with a plastic jacket and packaged into a cable to enhance its strength. Many kinds of optical cables have been developed, such as the single-fibre cable, multi-fibre cable and so on. Most of them are for telecommunication, but are also necessary and useful for fibre sensor technology.

One of the basic characteristics of an optical fibre is transmission loss. Even in fully pure silica, Rayleigh scattering loss still exists due to the thermal movement of molecules (Born and Wolf 1999). Infrared loss will be dominant in the longer wavelength band, thus a low-loss window is formed in the wave band of 1–2 μm, and the lowest loss appears at 1550 nm for silica fibre (Shibata et al. 1981). It is customary to express the fibre loss in a unit of decibels per kilometre as follows:

$$\alpha_{dB} = \frac{10}{L_f} \lg \frac{P_{in}}{P_{Tr}} \tag{2.17}$$

where:
- L_f is the length of the fibre
- P_{in} is the input power
- P_{Tr} is the transmitted power

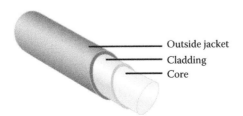

Figure 2.17 Schematic illustration of optical fibres.

Diverse mode is one of the important characteristics of optical fibre. It is important to understand that the condition of the incident angle being larger than the critical angle for guiding light in fibre is just a necessary condition, not a sufficient condition (Fang et al. 2012). The propagating light in the fibre must satisfy phase conditions at the boundary between core and cladding, that is, the phase shift of the light wave between successive reflections keeps an integer multiple of 2π. The requirement results in one of the basic characteristics of a guided wave: only with discrete angles will the light beams propagate in the fibre. Among them, the light with the smallest angle to the axis is termed the *fundamental mode* and others are the *high-order modes*. When the core radius is small enough and/or the index step is low enough, only the fundamental mode can propagate inside. Such fibre is called *single-mode-fibre* and the other type is *multi-mode-fibre*. The mode characteristics depend also on wavelength. For example, fibre can be single-mode-fibre for longer wavelengths but become multi-mode-fibre for shorter wavelengths.

Another important characteristic of fibre is its chromatic dispersion, namely, the dependence of the refractive index on the optical frequency ω. It causes optical signal pulse broadening in fibre communications and also brings about sensed signal impairment in fibre sensors. The dispersion is mainly attributed to two factors: one is the material dispersion and the other is the effect of the waveguide on the propagation constant including the dispersion between modes and the intramodal waveguide dispersion (Fang et al. 2012).

2.6.2 Light propagation in optical fibre

As shown in Figure 2.17, although the basic structure of an optical fibre is almost the same, according to the different ways of light propagation in the fibre, it can be divided into several categories, for example, the step-index fibre and the gradient-index fibre are two kinds of widely used and investigated fibre.

The step-index fibre consists of a core with refractive index n_1 and a cladding layer with index n_2 slightly lower than n_1. The light propagating inside the core will be totally reflected at the interface between the core and cladding layer when the incident angle is larger than the critical angle θ_c, and will be well confined in the core. According to Snell's law, the critical angle is determined by the refractive indexes of the core and cladding:

$$\theta_c = \arcsin\left(n_2 \big/ n_1 \right) \tag{2.18}$$

The gradient-index fibre is another kind of fibre, in which the core index descends with the radial distance r, expressed as

$$n(r) = \begin{cases} n_1\left[1 - \Delta\left(r/a\right)^p\right] & (\text{for} \quad r \leq a) \\ n_1\left[1 - \Delta\right] & (\text{for} \quad r > a) \end{cases} \tag{2.19}$$

where:

p is a positive real number
a is the core radius
Δ $= (n_1 - n_2)/n_1$

In the gradient-index fibre, the optical ray turns towards the axis where the peak index n_1 is located, and takes a waved path.

Equations 2.18 and 2.19 are the basic and simple principles and characteristics for light propagation in fibre through geometric optics theory. Notably, since the fibre size is related to the order of optical wavelength, it is necessary to treat the fibre as a dielectric waveguide and to get insights into wave evolution in optical fibres by using electromagnetic theory. Electromagnetic theories of optical fibres are available in quite a number of publications with different emphases (Snyder and Love 1983; Vassallo 1991; Tsao 1992; Yariv 1997; Agrawal 2004).

2.6.3 Fibre-optic sensors and their applications in smart civil structures

The main parameters of a light wave include amplitude, frequency, phase, polarisation state and intensity. All of them can carry information and thus can be used as sensor parameters. The changes of these parameters, in turn, are related to the measurands through the inter-relationship between the optical phenomena and the specific measurands.

From a practical point of view, fibre-optic sensors can be divided into two broad classes, namely short- and long-gauge length optical fibre sensors, also known as *discrete and distributed* optical fibre sensors. Discrete fibre-optic sensors determine the measurand over a specific segment of optical fibre, and are similar in that sense to those conventional sensors such as resistive foil strain gauges. They typically provide a measure of physical parameters over distances of several millimetres (≤ 20 mm). Distributed sensors make full use of optical fibres, in that each element of the optical fibre could be utilised for the purpose of both measurement and data transmission. Distributed sensors determine the locations and values of measurands along the entire length of the fibre. Due to their multi-point measurement capabilities, these sensors are most appropriate for application in large-scale civil structures. They typically measure physical parameters over distances of a few centimetres to a few metres (40 cm to 5 m).

The optical fibre sensors have unique merit, including but not limited to the following (Fang et al. 2012):

1. Small size and weight
2. Environmental robustness, water- and moist-proof
3. Immunity to electromagnetic interference and radio frequency interference
4. Capability of remote sensing and distributed sensing
5. Safe and convenient, integration with signal transportation
6. Capability of multiplexing and multi-parameter sensing
7. Large bandwidth and higher sensitivity
8. Lower cost and economic effectiveness

With such excellent properties, optical fibre sensors have found many applications in human social activities and daily living, from industrial production to cultural activities, from civil engineering to transportation, from medicine and health care to scientific research. Focusing on the application in civil engineering, the fibre Bragg grating (FBG) sensors could be viewed as the most widely used optical fibre sensors. The basic principle behind the FBG sensor is that its Bragg wavelength can be altered with a change in the grating period or the effective refractive index. The former is the case for strain and the latter for temperature variation. Demodulators or interrogators, roles of which are to extract measurand information from the light signals, are required for FBG sensors. The measurand is typically encoded in the form of a Bragg wavelength change, and hence, the interrogators are typically expected to read the wavelength shift and provide measurand data. FBG sensors have been tested in various fields such as bridges, dams, mines, composite laminates, geotechnical fields and railway systems (Grattan and Sun 2000; Lau et al. 2002; Nellen et al. 2002; Lee 2003; Rao and Huang 2008). More information on the introduction of fibre optical sensors as well as FBG sensors can be found in Chapter 3.

2.7 BIO-INSPIRED MATERIALS

Bio-inspiration is often referred to as *biomimetics* or *biomimicry*, which is a relatively new science that studies nature and systems and then imitates or takes creative inspiration from them to solve human problems (Biggins et al. 2011). Bio-inspired materials mean materials created using the mechanisms and laws operating in biological organisms as inspiration. There are diverse bio-inspired materials. Only some bio-inspired materials of interest in developing smart civil structures, such as self-healing materials and bio-inspired materials for sensing systems, are discussed in this section.

2.7.1 Self-healing materials

Self-healing can be defined as the ability of a material to heal (recover/repair) damages automatically and autonomously, that is, without any external intervention (Ghosh 2009). For example, cracks in buildings that can close on their own lead to mechanical strength being recovered. Many common terms such as self-repairing, autonomic-healing and autonomic-repairing are used to define such a property in materials. Self-healing properties incorporated into man-made materials usually cannot perform the self-healing action without an external trigger. Thus, self-healing can be of the following two types: autonomic (without any intervention) and non-autonomic (needs human intervention or external triggering). Focusing on the self-healing techniques used in civil structures, the materials employed for the repair of concrete and metal are introduced herein.

Concrete is the most widely used man-made building material and it is a well-known quasi-brittle material, that is, strong in compression but relatively weak in tension. For this reason, reinforcement is developed for the improvement of capacity when it is loaded in bending or in tension. Nevertheless, even in reinforced concrete, the formation of cracks is considered an inherent feature and the existence of cracks does not necessarily indicate a safety problem, for example, the crack may have occurred due to a variation in temperature. The presence of cracks, however, may reduce the durability of concrete structures and increase the risk of structural failure, especially in the case where the crack width is beyond the prescribed crack width limit. Therefore, the repair of concrete structures with too many cracks is required. Based on the properties of the materials, self-healing concrete can be categorised as autonomic and autogenic healing. A composite material which exhibits self-healing capabilities due to the release of encapsulated resins or glues as a result of cracking from the onset of damage is categorised as having autonomic-healing properties. If the healing properties of a material are generic to that material, such as cementitious material, the healing process is thus termed *autogenic healing*. To date, the adhesive-based healing of cementitious materials has been widely investigated and has proved to be capable of achieving a successful self-healing mechanism (Schlangen and Joseph 2009).

Metals have played a critical role in the history of human kind, and they continue to play a vital role in today's society. The research on self-healing metals has also received considerable attention recently. According to their micro-structural characteristics, self-healing metals can be broadly classified into two categories: liquid-assisted and solid-state healing (Manuel 2009). These categories describe the mechanism by which the self-healing metal transports matter to the damaged site. Research on the first category is often based on the employment of SMA wires, and the work in the second category focuses on aluminium alloys and steels. Already metal-matrix composites utilising a liquid-assisted healing mechanism have demonstrated the ability to recover tensile strength after large-scale damage, whereas metals using solid-state healing mechanisms have demonstrated high levels of damage tolerance and removal of internal cracks or voids. Both categories of self-healing metals have their own benefits and limitations. It seems that for damage healing at all length scale and under various environmental conditions, a combination of liquid-assisted mechanism and solid-state healing mechanism should perhaps be used for complete self-healing. However, it is important to note that this research area is still in its infancy and although impressive, the healing ability of metals is still far from the characteristic healing behaviour seen in biological materials.

2.7.2 Bio-inspired materials for sensing systems

The studies on bio-inspired materials for sensing systems explore how we can learn from nature to provide new approaches for solving some of the challenges or difficulties we have encountered. The scope for bio-inspired applications to sensors and sensing systems is only limited by our imagination. The development such a system will depend on how it will be deployed and the required level of functionality. For instance, a passive sensor would have no need for power. It could be interrogated from a distance to obtain data. Alternatively, the system might be sedentary and have an engineered suite of sampling, transduction and processing capabilities that have minimal power and energy requirements that could be met by harvesting energy from the environment. Or the system might have a requirement to be highly mobile and carry a programmable range of sensors or be networked to other sensors or sensor systems to provide enhanced area coverage. The power requirements for this system would be considerably greater.

One approach to facilitating the application of biotechnology and bio-inspiration to sensors and sensing systems is to develop a conceptual goal. In the broadest sense, the conceptual goal may never be fully realised. However, the conceptual goal enables a vision to be developed that allows or assists in the identification of enabling technologies that are required or need to be developed for the realisation of the vision (Biggins et al. 2011).

A number of collection and sampling tools that we use in laboratories and on site today might be described as being bio-inspired or biomimetic. Even if the inspiration for these tools was not a natural system, the similarity between them and those found in natural systems is remarkable. For example, echolocation used by many animals for navigation has also been developed by human beings for distance measurement and location determination. Optical fibre sensors could be another simple example of humans understanding the characteristics of light propagation in the materials and utilising it for measuring and transmitting. Actually, many of the tools and concepts we use today have their provenance from the natural world. To fully understand and learn more from nature, interdisciplinary cooperation and research are necessary for the development of new materials and a new artificial system.

2.8 NANOMATERIALS

Recent research on nanomaterials has highlighted the potential use of these materials in various fields because of the special characteristics of materials at the nanoscale (Olar 2011). Nanomaterials can be defined as those physical substances with at least one dimension between 1 and 150 nm (1 nm $= 10^{-9}$ m). The properties of nanomaterials can be very different from those of the same materials at micro (10^{-6} m) or macro scale (10^{-6}–10^{-3} m).

Nanoscience represents the study of phenomena and the manipulation of materials at the nanoscale, and is an extension of common sciences into the nanoscale. Nanotechnology can be defined as the design, characterisation, production and application of structures, devices and systems by controlling shape and size at the nanoscale. Nanotechnology requires advanced imaging techniques for studying and improving the material behaviour and for designing and producing very fine powders, liquids or solids of materials with particle sizes between 1 and 100 nm, known as *nanoparticles* (Gogotsi 2006). Nanotechnologies can generate products with many unique characteristics that can improve the current construction materials: lighter and stronger structural components, low maintenance coatings, better cementitious materials, low thermal transfer rate of fire retardant and insulation, better sound absorption and better reflectivity of glass (Lee et al. 2010).

One of the most important nanomaterials is the carbon tube. Carbon nanotubes are a form of carbon with a cylindrical shape. They can be several millimetres in length and can have one layer or wall or more than one wall (Lee et al. 2010). Nanotubes are members of the fullerene structural family and exhibit extraordinary strength and unique electrical properties, being efficient thermal conductors. For example, they have five times the Young's modulus and eight times the strength of steel, while being one-sixth the density. Another expected benefit of carbon nanotubes is real-time SHM capacity (Mann 2006).

By combining the excellent properties of nanomaterials with smart functional materials, smart nanomaterials can be formed. Smart nanomaterials have caused a profound revolution in sensors and sensing applications. The use of smart nanomaterials in sensing applications enables the altering of texture in conventional sensing models into controlled modes. The unique electronic, magnetic, acoustic and light properties of nanomaterials, coupled with smart materials capable of responsiveness to external stress, electric and magnetic fields, temperature and moisture, make accurate, real-time and modulated analysis possible. Nano- and microelectrical mechanical systems (MEMS) sensors have been developed and used in construction to monitor and/or control the environment condition and the materials/structure performance. Nanosensors range from 10^{-9} to 10^{-5} m. These sensors could be embedded into the structure during the construction process and could monitor internal stresses, cracks and other physical forces in the structures during the structure service life (Lee et al. 2010).

NOTATION

A_p	The cross-sectional area of the piston in ER or MR damper
b	The width of the rectangular plate in ER or MR model
B	The magnetic induction of magnetostrictive material
d	Magneto-mechanical constant
d, g	The matrix of piezoelectric strain constants and matrix of piezoelectric constants
D	The vector of electric displacement of PZT
E	The vector of applied electric field for PZT
E^{SMA}	Young's modulus of SMA
f	Force provided by ER or MR damper
h	The gap between two parallel plates in ER or MR model
H	The magnetic field strength for magnetostrictive material
I	The current applied to magnetostrictive material
L	The effective axial pole length in ER or MR model

L_f	Length of the optical fibre
n	The number of coil turns
n_1, n_2	Refractive index of a core and a cladding layer of optical fibre
p'	Pressure gradient of ER or MR fluid
P_{in}, P_{Tr}	Input power and transmitted power for optical fibre
P_y	The mechanical friction force in ER or MR damper
Q	Volumetric flow rate of ER or MR fluid
r, a	Radial distance and core radius of optical fibre
s^H	The elastic compliance under constant magnetic field
S	The matrix of compliance coefficients of PZT
S, T	The mechanical strain and stress of magnetostrictive material
T^{SMA}, T_0^{SMA}	The operating temperature for SMA and its initial conditions
U	The relative velocity of the two plates in ER or MR model
v	The relative velocity of ER or MR damper
α_{dB}	Transmission loss of optical fibre
γ	Shear strain of ER or MR fluid
$\Delta L, L$	The initial length of magnetostrictive material and its fractional length change
$\varepsilon^{SMA}, \varepsilon_0^{SMA}$	The strain of SMA and its initial conditions
η	Viscosity of ER or MR fluid
θ_c	Critical angle of optical fibre
θ^{SMA}	Thermal elastic tensor for SMA
λ	Magnetostrictive coefficient
μ^T	The magnetic permeability under constant stress condition
ξ, β	The permittivity and impermittivity component of PZT
ξ^{SMA}, ξ_0^{SMA}	The martensite fraction of the SMA and its initial conditions
σ, ε	The stress and strain of PZT
$\sigma^{SMA}, \sigma_0^{SMA}$	The shear stress of SMA and its initial conditions
τ, τ_y	Shear stress and yield stress of ER or MR fluid
Ω^{SMA}	Transformation tensor for SMA

REFERENCES

Achkire, Y., and A. Preumont. 1996. Active tendon control of cable-stayed bridges. *Earthq. Eng. Struct. Dyn.*, 25(6): 585–97.

Agrawal, G.P. 2004. *Nonlinear Fiber Optics*. Singapore: Elsevier Science.

Akamatsu, T., K. Okamura, and Y. Ueda. 1977. Fabrication of long fibers by an improved chemical vapor-deposition method (HCVD method). *Appl. Phys. Lett.*, 31: 174–76.

Amato, I. 1994. The sensual city. *New Sci.*, 144(1947): 33–36.

Anton, S.R. and H.A. Sodano. 2007. A review of power harvesting using piezoelectric materials (2003–2006). *Smart Mater. Struct.*, 16(3): R1–21.

Benjeddou, A. 2001. Advances in hybrid active-passive vibration and noise control via piezoelectric and viscoelastic constrained layer treatments. *J. Vib. Control*, 7(4): 565–602.

Biggins, P.D.E., A. Kusterbeck, and J.A. Hiltz. 2011. *Bio-Inspired Materials and Sensing Systems*. London, UK: Royal Society of Chemistry.

Birman, V. 1997. Review of mechanics of shape memory alloy structures. *Appl. Mech. Rev.*, 50(11): 629–45.

Blanguernon, A., F. Lene, and M. Bernadou. 1999. Active control of a beam using a piezoceramics element. *Smart Mater. Struct.*, 8(1): 116–24.

Born, M. and E. Wolf. 1999. *Principles of Optics*. Seventh Edition. Cambridge, UK: Cambridge University Press.

Boyd, J.G. and D.C. Lagoudas. 1994. Thermomechanical response of shape memory composites. *J. Intell. Mater. Syst. Struct.*, 5(3): 333–46.

Brinson, L.C. 1993. One-dimensional constitutive behavior of shape memory alloys: Thermomechanical derivation with non-constant material functions and redefined martensite internal variable. *J. Intell. Mater. Syst. Struct.*, 4(2): 229–42.

Brocca, M., L.C. Brinson, and Z.P. Bazant. 2002. Three-dimensional constitutive model for shape memory alloys based on microplane model. *J. Mech. Phys. Solids*, 52: 1051–77.

Buehler, W.J., J.V. Gilfrich, and R.C. Wiley. 1963. Effect of low-temperature phase changes on the mechanical properties of alloys near composition TiNi. *J. Appl. Phys.*, 34: 1475–77.

Carlson, J.D. 1994. The promise of controllable fluids. In *Proceedings of the Fourth International Conference on New Actuators, Actuator 94*, eds. H. Borgmann and K. Lenz, 266–70, Axon Technologies Consult GmbH.

Casciati, F., L. Faravelli, and L. Petrini. 1998. Energy dissipation in shape memory alloy devices. *Comput.-Aided Civil Infrastruct. Eng.*, 13: 433–42.

Cawley, P. and D. Alleyne. 1996. The use of Lamb waves for the long range inspection of large structures. *Ultrasonics*, 34: 287–90.

Chopra, I. and J. Sirohi. 2014. *Smart Structures Theory*. New York, NY: Cambridge University Press.

Corbi, O. 2003. Shape memory alloys and their application in structural oscillations attenuation. *Simul. Model. Pract. Theory*, 11: 387–402.

Coulter, J.P., K.D. Weiss, and J.D. Carlson. 1993. Engineering application of electrorheological materials. *J. Intell. Mater. Syst. Struct.*, 4(2): 248–59.

Croci, G., A. Bonci, and A. Viskovic. 2000. The use of shape memory alloys devices in the basilica of St Francis in Assisi. In *Proceedings of the Final Workshop of ISTECH Project*, 117–40, Ispra, Italy, Office for Official Publications of the European Communities.

Culshaw, B. 1996. *Smart Structures and Materials*. Boston, MA: Artech House.

Curie, J. and P. Curie. 1880. Développement par compression de l'électricité polaire dans les cristaux hémièdres à faces inclinées (Development, via compression, of electric polarisation in hemihedral crystals with inclined faces). *Bulletin de la Société minérologique de France*, 3: 90–93.

Dapino, M.J. 2002. On magnetostrictive materials and their use in smart material transducer. *Struct. Eng. Mech. J.*, 1–28.

DesRoches, R. and M. Delemont. 2002. Seismic retrofit of simply supported bridges using shape memory alloys. *Eng. Struct.*, 24(3): 325–32.

Dolce, M., D. Cardone, and R. Marnetto. 2000. Implementation and testing of passive control devices based on shape memory alloys. *Earthq. Eng. Struct. Dyn.*, 29: 945–68.

Dolce, M., D. Cardone, and R. Marnetto. 2001. SMA re-centering devices for seismic isolation of civil structures. In *Proceedings of SPIE 4330, Smart Structures and Materials 2001: Smart Systems for Bridges, Structures, and Highways*, ed. S.C. Liu, 238–49, Newport Beach, CA. Copyrighted by SPIE Digital Library.

Dyke, S.J., B.F. Spencer Jr, M.K. Sain, and J.D. Carlson. 1996. Modeling and control of magnetorheological dampers for seismic response reduction. *Smart Mater. Struct.*, 5(5): 565–75.

Fang, Z.J., K. Chin, R.H. Qu, H.W. Cai, and K. Chang. 2012. *Fundamentals of Optical Fiber Sensors*. Hoboken, NJ: Wiley.

Fujino, Y. and T. Susumpow. 1995. Active control of cables by axial support motion. *Smart Mater. Struct.*, 4(1A): 41–51.

Fujino, Y., P. Warnitchai, and B.M. Pacheco. 1993. Active stiffness control of cable vibration. *J. Appl. Mech.*, 60(4): 948–53.

Fujita, T., M. Enomoto, T. Arikabe, et al. 2001. Active microvibration control of precision manufacturing factories with smart structure using piezoelectric actuators. In *Proceedings of SPIE 4330, Smart Structures and Materials 2001: Smart Systems for Bridges, Structures, and Highways*, ed. S.C. Liu, 449–59, Newport Beach, CA. Copyrighted by SPIE Digital Library.

Fuller, C.R., S.J. Elliot, and P.A. Nelson. 1996. *Active Control of Vibration*. London: Academic Press.

Gandhi, M.V. and B.S. Thompson. 1992. *Smart Materials and Structures*. London: Chapman & Hall.

Gao, W., J.J. Chen, H.B. Ma, and X.S. Ma. 2003. Optimal placement of active bars in active vibration control for piezoelectric intelligent truss structures with random parameters. *Comput. Struct.*, 81(1): 53–60.

Gao, X., M. Huang, and L.C. Brinson. 2000. A multivariant micromechanical model for SMAs Part 1. Crystallographic issues for single crystal model. *Int. J. Plast.*, 16(10–11): 1345–69.

Gavin, G.P., R.D. Hanson, and F.E. Filisko. 1996a. Electrorheological dampers, Part I: Analysis and design. *J. Appl. Mech.*, 63(3): 669–75.

Gavin, G.P., R.D. Hanson, and F.E. Filisko. 1996b. Electrorheological dampers, Part II: Testing and modeling. *J. Appl. Mech.*, 63(3): 676–82.

Ghosh, S.K. 2009. *Self-Healing Materials: Fundamentals, Design Strategies, and Applications*. Weinheim: Wiley-VCH.

Giurgiutiu, V. and A. Cuc. 2005. Embedded non-destructive evaluation for structural health monitoring, damage detection, and failure prediction. *Shock Vib. Dig.*, 37: 83–105.

Gogotsi, Y. 2006. *Nanomaterials Handbook*. Boca Raton, FL: CRC Press.

Goo, B.C. and C. Lexcellent. 1997. Micromechanics based modeling of two-way memory effect of a single crystalline shape memory alloy. *Acta Mater.*, 45(2): 727–37.

Grattan, K.T.V. and T. Sun. 2000. Fiber optic sensor technology: An overview. *Sensor Actuat. A-Phys.*, 82(1–3): 40–61.

Han, J.H., K.H. Rew, and I. Lee. 1997. An experimental study of active vibration control of composite structures with a piezoceramic actuator and a piezo film sensor. *Smart Mater. Struct.*, 6(5): 549–58.

Han, Y.L., Q.S. Li, A.Q. Li, A.Y.T. Leung, and P.H. Lin. 2003. Structural vibration control by shape memory alloy damper. *Earthq. Eng. Struct. Dyn.*, 32(3): 483–94.

Hodgson, D.E., M.H. Wu, and R.J. Biermann. 1990. Shape memory alloys. In *ASM Handbook, Volume 2: Properties and Selection: Nonferrous Alloys and Special-Purpose Materials*, 897–902. http://products.asminternational.org/hbk/index.jsp.

Hristoforou, E. and A. Ktena. 2007. Magnetostriction and magnetostrictive materials for sensing applications. *J. Magn. Magn. Mater.*, 316: 372–78.

Hu, Y. and Y. Yang. 2007. Wave propagation modeling of the PZT sensing region for structural health monitoring. *Smart Mater. Struct.*, 16(3): 706–16.

Huang, M.S. and L.C. Brinson. 1998. A multivariant model for single crystal shape memory alloys. *J. Mech. Phys. Solids*, 46(8): 1379–409.

IEEE. 1988. *IEEE Standard on Piezoelectricity*. New York, NY: Institute of Electrical and Electronics Engineers Inc., ANSI/IEEE Std. 176-1987.

Indirli, M., M.G. Castellano, P. Clemente, and A. Martelli. 2001. Demo application of shape memory alloy devices: The rehabilitation of S. Georgio Church Bell Tower. In *Proceedings of SPIE 4330, Smart Structures and Materials 2001: Smart Systems for Bridges, Structures, and Highways*, ed. S.C. Liu, 262–72, Newport Beach, CA. Copyrighted by SPIE Digital Library.

Irschik, H. 2002. A review on static and dynamic shape control of structures by piezoelectric actuation. *Eng. Struct.*, 24, 5–11.

Ivshin, Y. and T.J. Pence. 1994a. A thermomechanical model for a one variant shape memory material, *J. Intell. Mater. Syst. Struct.*, 5(4): 455–73.

Ivshin, Y. and T.J. Pence. 1994b. A constitutive model for hysteretic phase transition behavior. *Int. J. Eng. Sci.*, 32(4): 681–704.

Janke, L., C. Czaderski, M. Motavalli, and J. Ruth. 2005. Applications of shape memory alloys in civil engineering structures – overview, limits and new ideas. *Mater. Struct.*, 38(5): 578–92.

Jung, H.J., I.W. Lee, and B.F. Spencer Jr. 2002. State-of-the-art of MR damper-based control systems in civil engineering applications. In *Proceedings of US-Korea Workshop on Smart Infra-Structural System*. Busan, Korea: Techno-Press.

Kamada, T., T. Fujita, T. Hatayama, et al. 1997. Active vibration control of frame structures with smart structures using piezoelectric actuators (vibration control by control of bending moments of columns). *Smart Mater. Struct.*, 6(4): 448–56.

Kamada, T., T. Fujita, T. Hatayama, et al. 1998. Active vibration control of flexural-shear type frame structures with smart structures using piezoelectric actuators. *Smart Mater. Struct.*, 7(4): 479–88.

Kessler, S.S., S.M. Spearing, and C. Soutis. 2002. Damage detection in composite materials using Lamb wave methods. *Smart Mater. Struct.*, 11: 269–78.

Khan, M.M. and D. Lagoudas. 2002. Modeling of shape memory alloy pseudoelastic spring elements using Preisach model for passive vibration isolation. In *Proceedings of SPIE 4693, Smart Structures and Materials 2002: Modeling, Signal Processing, and Control*, ed. V.S. Rao, 336–47, San Diego, CA. Copyrighted by SPIE Digital Library.

Lagoudas, D.C., Z. Bo, and M.A. Qidwai. 1996. A unified thermodynamic constitutive model for SMA and finite element analysis of active metal matrix composites. *Mech. Compos. Mater. St.*, 3(2): 153–79.

Lau, K.T., L.M. Zhou, P.C. Tse, and L.B. Yuan. 2002. Applications of composites, optical fiber sensors and smart composites for concrete rehabilitation: An overview. *Appl. Compos. Mater.*, 9(4): 221–47.

Lee, B. 2003. Review of the present status of optical fiber sensors. *Opt. Fiber Technol.*, 9(2): 57–79.

Lee, J., S. Mahendra, and P. Alvarez. 2010. Nanomaterials in construction industry: A review of their applications and environmental health and safety considerations. *ACS Nano*, 4(7): 3580–90.

Leon, R.T., R. DesRoches, J. Ocel, and G. Hess. 2001. Innovative beam column using shape memory alloys. In *Proceedings of SPIE 4330, Smart Structures and Materials 2001: Smart Systems for Bridges, Structures, and Highways*, ed. S.C. Liu, 227–37, Newport Beach, CA. Copyrighted by SPIE Digital Library.

Lesieutre, G.A. and U. Lee. 1996. A finite element for beams having segmented active constrained layers with frequency-dependent viscoelastics. *Smart Mater. Struct.*, 5(5): 615–27.

Li, H., M. Liu, and J.P. Ou. 2004. Vibration mitigation of a stay cable with one shape memory alloy damper. *Struct. Control. Health Monit.*, 11(1): 21–36.

Li, Z.Q., Y.L. Xu, and L.M. Zhou. 2006. Adjustable fluid damper with SMA actuators. *Smart Mater. Struct.*, 15(5): 1483–92.

Liang, C. and C.A. Rogers. 1990. One-dimensional thermomechanical constitutive relations for shape memory materials. *J. Intell. Mater. Syst. Struct.*, 1(2): 207–34.

Lu, Z.K. and G.J. Weng. 1997. Martensitic transformation and stress-strain relations of shape-memory alloys. *J. Mech. Phys. Solids*, 45(11–12): 1905–21, 1923–28.

Lubliner, J. and F. Auricchio. 1996. Generalized plasticity and shape memory alloys. *Int. J. Solids Struct.*, 33(7): 991–1003.

Maji, A.K. and I. Negret. 1998. Smart prestressing with shape memory alloy. *J. Eng. Mech.*, 124(10): 1121–28.

Mann, S. 2006. Nanotechnology and construction. European Nanotechnology Gateway-Nanoforum Report, Institute of Nanotechnology, 2010.

Manuel, M.V. 2009. Principles of self-healing in metals and alloys: An introduction. In *Self-Healing Materials: Fundamentals, Design Strategies, and Applications*, ed. S.K. Ghosh, 251–67. Weinheim, Germany: Wiley-VCH.

McClelland, R., T.W. Lim, A.B. Bosse, and S. Fisher. 1996. Implementation of local feedback of controllers for vibration suppression of a truss using active struts. In *Proceedings of SPIE 2717, Smart Structures and Materials 1996: Smart Structures and Integrated Systems*, ed. I. Chopra, 452–61, San Diego, CA. Copyrighted by SPIE Digital Library.

Moheimani, S.O.R. and A.J. Fleming. 2006. *Piezoelectric Transducers for Vibration Control and Damping*. London, UK: Springer-Verlag.

Muhammad, A., X.L. Yao, and Z.C. Deng. 2006. Review of magnetorheological (MR) fluids and its applications in vibration control. *J. Mar. Sci. Appl.*, 5(3): 17–29.

Nellen, P.M., A. Frank, and A. Kenel. 2002. High strain and high strain gradients measured with fiber Bragg gratings in structural engineering applications. In *Ofs 2002, 2002, 15th Optical Fiber Sensors Conference Technical Digest*. 111–14, Portland, OR: IEEE.

Ng, C.L. and Y.L. Xu. 2007. Semi-active control of a building complex with variable friction dampers. *Eng. Struct.*, 29(6): 1209–25.

Olabi, A.G. and A. Grunwald. 2008. Design and application of magnetostrictive materials. *Mater. Des.*, 29: 469–83.

Olar, R. 2011. Nanomaterials and nanotechnologies for civil engineering. *Buletinul Institutului Politehnic Din Iasi, L VII (LXI)*, 4: 109–16.

Park, G., H.H. Cudney, and D.J. Inman. 2000a. An integrated health monitoring technique using structural impedance sensors. *J. Intell. Mater. Syst. Struct.*, 11(6): 448–55.

Park, G., H.H. Cudney, and D.J. Inman. 2000b. Impedance-based health monitoring of civil structural components. *J. Infrastruct. Syst.*, 6(4): 153–60.

Park, G., H. Sohn, C.R. Farrar, and D.J. Inman. 2003. Overview of piezoelectric impedance-based health monitoring and path forward. *Shock Vib. Dig.*, 35(6): 451–63.

Park, S., C.B. Yun, Y. Roh, and J.J. Lee. 2006. PZT-based active damage detection techniques for steel bridge components. *Smart Mater. Struct.*, 15(4): 957–66.

Patoor, E., A. Eberhardt, and M. Berveiller. 1996. Micromechanical modelling of superelasticity in shape memory alloys. *J. Phys. IV France*, 06: C1-277–C1-292.

Petek, N.K. 1992. An electronically controlled shock absorber using electro-rheological fluid. SAE Technical Paper, Series 920275, doi:10.4271/920275.

Phillips, R.W. 1969. Engineering applications of fluids with a variable yield stress. PhD diss., Berkeley, CA: University of California.

Preumont, A., Y. Achkire, and F. Bossens. 2000. Active tendon control of large trusses. *AIAA J.*, 38(3): 493–98.

Rabinow, J. 1948. The magnetic fluid clutch. *Trans. AIEE.*, 67(2): 1308–15.

Rao, V.S. and S. Sana. 2001. An overview of control design methods for smart structural systems. In *Proceedings of SPIE 4326, Smart Structures and Materials 2001: Modeling, Signal Processing, and Control in Smart Structures*, ed. V.S. Rao, 1–13, Newport Beach, CA.

Rao, Y.J. and S. Huang. 2008. Applications of fiber optic sensors. In *Fiber Optic Sensors, Second Edition*, eds. S. Yin, P.B. Ruffin, and F.T.S. Yu, and S. Yin, 397–435. Boca Raton, FL: CRC Press.

Saadat, S., M. Noori, H. Davoodi, et al. 2001. Using NiTi SMA tendons for vibration control of coastal structures. *Smart Mater. Struct.*, 10: 695–704.

Saadat, S., J. Salichs, M. Noori, et al. 2002. An overview of vibration and seismic application of NiTi shape memory alloy. *Smart Mater. Struct.*, 11: 218–29.

Schlangen, E. and C. Joseph. 2009. Self-healing processes in concrete. In *Self-Healing Materials: Fundamentals, Design Strategies, and Applications*, ed. S.K. Ghosh, 141–83. Weinheim, Germany: Wiley-VCH.

Shibata, S., M. Horiguchi, K. Jinguji, et al. 1981. Prediction of loss minima in infrared optical fibers. *Electron. Lett.*, 17(21): 775–77.

Shin, S.W. and T.K. Oh. 2009. Application of electro-mechanical impedance sensing technique for online monitoring of strength development in concrete using smart PZT patches. *Constr. Build. Mater.*, 23: 1185–88.

Snyder, A.W. and J.D. Love. 1983. *Optical Waveguide Theory*. London, New York: Chapman & Hall.

Sodano, H.A., G. Park, and D.J. Inman. 2004. A review of power harvesting using piezoelectric materials. *Shock Vib. Dig.*, 36: 197–206.

Song, G., H. Gu, Y.L. Mo, T.T.C. Hsu, and H. Dhonde. 2007a. Concrete structural health monitoring using embedded piezoceramic transducers. *Smart Mater. Struct.*, 16(4): 959–68.

Song, G., N. Ma, and H.N. Li. 2006a. Applications of shape memory alloys in civil structures. *Eng. Struct.*, 28(9): 1266–74.

Song, G., C. Olmi, and H. Gu. 2007b. An overheight vehicle–bridge collision monitoring system using piezoelectric transducers. *Smart Mater. Struct.*, 16(2): 462–68.

Song, G., V. Sethi, and H.N. Li. 2006b. Vibration control of civil structures using piezoceramic smart materials: A review. *Eng. Struct.*, 28(11): 1513–24.

Spencer Jr., B.F., S.J. Dyke, M.K. Sain, and J.D. Carlson. 1997. Phenomenological model for magnetorheological dampers. *J. Eng. Mech.*, 123(3): 230–38.

Spencer Jr., B.F. and S. Nagarajaiah. 2003. State of the art of structural control. *J. Struct. Eng.*, 129: 845–56.

Spencer Jr., B.F., G. Yang, J.D. Carlson, and M.K. Sain. 1998. 'Smart' dampers for seismic protection of structures: A full-scale study. In *Proceedings of the Second World Conference on Structural Control*, eds. T. Kobori, Y. Inoue, K. Seto, H. Iemura, and A. Nichitani. Kyoto, Japan: John Wiley.

Srinivasan, A.V. and D.M. McFarland. 2001. *Smart Structures: Analysis and Design*. Cambridge: Cambridge University Press.

Sugavanam, S., V.K. Vardan, and V.V. Vardan. 1998. Modeling and control of a lightly damped T-beam using piezoceramic actuators and sensors. *Smart Mater. Struct.*, 7(6): 899–906.

Sun, Q.P. and K.C. Hwang. 1993. Micromechanics modelling for the constitutive behavior of polycrystalline shape memory alloys – I: Derivation of general relations, *J. Mech. Phys. Solids*, 41(1): 1–17.

Sunar, M. and S.S. Rao. 1999. Recent advances in sensing and control of flexible structures via piezoelectric materials technology. *Appl. Mech. Rev.*, 52(1): 1–16.

Symans, M.D. and M.C. Constantinou. 1999. Semi-active control systems for seismic protection of structures: A state-of-the-art review. *Eng. Struct.*, 21(6): 469–87.

Tamai, H. and Y. Kitagawa. 2002. Pseudoelastic behavior of shape memory alloy wires and its application to seismic resistance member for building. *Comput. Mater. Sci.*, 25: 218–27.

Tanaka, K. and S. Nagaki. 1982. A thermomechanical description of materials with internal variables in the process of phase transitions. *Ingenieur-Archiv*, 51(5): 287–99.

Trindade, M.A. and A. Benjeddou. 2002. Hybrid active-passive damping treatments using viscoelastic and piezoelectric materials: Review and assessment. *J. Vib. Control*, 8: 699–745.

Truong, D.Q. and K.K. Ahn. 2012. MR fluid damper and its application to force sensorless damping control system. In *Smart Actuation and Sensing Systems: Recent Advances and Future Challenges*. eds. G. Berselli, R. Vertechy, and G. Vassura, 383–424. Rijeka, Croatia: InTech.

Tsao, C. 1992. *Optical Fiber Waveguide Analysis*. Oxford, New York, Tokyo: Oxford University Press.

Vassallo, C. 1991. *Optical Waveguide Concepts*. Amsterdam, Oxford, New York, Tokyo: Elsevier.

Vivet, A. and C. Lexcellent. 1998. Micromechanical modeling for tension-compression pseudoelastic behavior of AuCd single crystals. *Eur. Phys. J-Appl. Phys.*, 4(2): 125–32.

Warnitchai, P., Y. Fujino, B.M. Pacheco, and R. Agret. 1993. An experimental study on active tendon control of cable-stayed bridges. *Earthq. Eng. Struct. Dyn.*, 22(2): 93–111.

Weiss, K.D., J.D. Carlson, and D.A. Nixon. 1994. Viscoelastic properties of magneto- and electrorheological fluids. *J. Intell. Mater. Syst. Struct.*, 5: 772–75.

Wilde, K., P. Gardoni, and Y. Fujino. 2000. Base isolation system with shape memory alloy device for elevated highway bridges. *Eng. Struct.*, 22: 222–29.

Winslow, W.M. 1947. Methods and means of translating electrical impulses into mechanical force. US Patent 2417850 A, filed April 14, 1942, and issued March 25, 1947.

Wu, X.M., J.Y. Wong, M. Sturk, and D.L. Russell. 1994. Simulation and experimental study of a semi-active suspension with an electrorheological damper. *Int. J. Mod. Phys. B*, 8(20–21): 2987–3003.

Xu, B., T. Zhang, G.B. Song, and H.C. Gu. 2013. Active interface debonding detection of a concrete-filled steel tube with piezoelectric technologies using wavelet packet analysis. *Mech. Syst. Signal Proc.*, 36(1): 2013, 7–17.

Xu, Y.L. and B. Li. 2006. Hybrid platform for high tech equipment protection against earthquake and microvibration. *Earthq. Eng. Struct. Dyn.*, 35(8): 943–67.

Xu, Y.L. and C.L. Ng. 2008. Seismic protection of a building complex using variable friction damper: Experimental investigation. *J. Eng. Mech.*, 134(8): 637–49.

Xu, Y.L., W.L. Qu, and J.M. Ko. 2000. Seismic response control of frame structures using magneto-rheological/electrorheological dampers. *Earthq. Eng. Struct. Dyn.*, 29: 557–75.

Yao, G.Z., F.F. Yap, G. Chen, W.H. Li, and S.H. Yeo. 2002. MR damper and its application for semi-active control of vehicle suspension system. *Mechatronics*, 12(7): 963–73.

Yariv, A. 1997. *Optical Electronics in Modern Communications*. Fifth Edition. New York, NY: Oxford University Press.

Zou, Y., L. Tong, and G.P. Steven. 2000. Vibration-based model-dependent damage (delamination) identification and health monitoring for composite structures-a review. *J. Sound Vibr.*, 230(2): 357–78.

Chapter 3

Sensors and sensory systems

3.1 PREVIEW

A sensor is a converter that measures a physical quantity and converts it into a signal which can be read by an observer or an instrument. Sensors are one of the most critical components of a smart civil structure since the smart performance depends on the quality of the data collected. There are a variety of physical quantities that must be measured in a smart civil structure, and accordingly, there are many types of sensors required to measure different physical quantities. The physical quantities that need to be measured for civil structures can be categorised as three major types:

1. Loading sources such as gravity force, wind, seismic and traffic loading
2. Structural responses such as strain, displacement, inclination and acceleration and
3. Environmental effects including temperature, humidity, rain and corrosion

This chapter describes the characteristics, functions and installations of the most commonly used sensors, categorised according to the monitoring quantities. Some of these sensors are made of smart materials, introduced in Chapter 2. The general requirements for sensor performance and sensory systems are also discussed. It is noted that many new sensor technologies are being developed, which may not be included in this book.

3.2 WIND SENSORS

3.2.1 Anemometers

The traditional sensors widely used to measure wind speed and direction at the site of a smart civil structure are propeller and ultrasonic anemometers. The propeller anemometer as shown in Figure 3.1 can be directly used to record wind speed and direction. The propeller anemometer is convenient and relatively reliable and sustainable in harsh environments, but it is not sensitive enough to capture the nature of turbulent winds at higher frequencies (Xu and Xia 2012). This is particularly true in situations where the wind speed or direction changes rapidly. The ultrasonic anemometer measures wind velocity through its three orthogonal components (see Figure 3.2). The ultrasonic anemometer is quite sensitive but it is not sustainable in harsh environments. The accuracy of the anemometers in measuring wind velocity must be maintained under heavy rain; that is, there should be no occurrence of spikes during heavy rainstorms. For a long suspension bridge, the anemometers are often installed at a few bridge deck sections on both sides, and along the height of the towers, so that not only wind characteristics at key points can be measured, but also the correlation

Figure 3.1 Propeller anemometer. (From http://www.junyangchina.com/ProductShow.asp?ID=182.)

Figure 3.2 3D ultrasonic anemometer. (From http://www.thiesclima.com/ultrasonic_anemometer_3d_e.html.)

of wind velocity in both horizontal and vertical directions can be determined. For a tall building, the anemometers are often installed at the top of the building. The positions of the anemometers must be selected so as to minimise the effect of the adjacent edges of the bridge deck or the building itself on the airflow towards them. To meet this requirement, anemometer booms or masts are often needed so that the anemometer can be installed a few metres away from the bridge and building edges. The boom or mast may be equipped with a retrievable device to enable retraction of the anemometer in an unrestricted and safe manner for inspection and maintenance.

3.2.2 Pressure transducers

Wind pressure transducers sense differential pressure and convert this pressure difference into a proportional electrical output for either unidirectional or bidirectional pressure ranges. To measure the pressure difference accurately, the location of the reference pressure transducer needs to be selected appropriately to avoid possible disturbances from the surrounding environment. A simple schematic diagram of such a wind pressure measurement system is shown in Figure 3.3, as an example. Pressure transducers are sometimes installed to measure wind pressures and pressure distribution over a particular part of a building. They are also occasionally installed on the surface of the bridge deck to measure pressures and their distribution.

3.2.3 Wind profile measurements

In recent years, Doppler radar, Doppler sodar and global positioning system (GPS) dropsonde have become powerful devices for measuring boundary-layer wind profiles. The basic assumption for the measurement of winds by these profilers is that the turbulent eddies which induce scattering can be carried along by the mean wind. With the aid of sufficient samples, the amplitude of such scattered energy can be clearly identified above the environmental noise, though the energy scattered by these eddies and received by the profiler is much smaller than the energy transmitted. The speed and direction of the mean wind within the volume being sampled can thus be determined accordingly.

Based on the usage of electromagnetic (EM) signals, Doppler radar wind profilers (see Figure 3.4a) can be employed to remotely sense winds aloft. Propagation of radar signals through the atmosphere is strongly dependent on local meteorological conditions, especially in the atmospheric boundary layer (Srinivasulu et al. 2006). Due to their small aperture, ultra-high-frequency (UHF) profilers operating around 900–1300 MHz are most suitable for measuring winds in the boundary layer and lower troposphere regions. Unlike the very

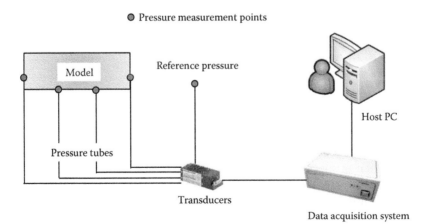

Figure 3.3 Wind pressure measurement system.

(a)   (b)

Figure 3.4 Wind profilers: (a) Doppler radar wind profiler and (b) Doppler sodar wind profiler. (From (a) https://www.eol.ucar.edu/content/pecan-integrated-sounding-arrays-pisas; (b) http://sentrex-wind.com/sodar/.)

high-frequency (VHF) wind profiling radars, UHF radars are very sensitive for hydrometeors due to the small wavelength used. With the utilisation of sound waves, the Doppler sodar wind profiler (see Figure 3.4b) is capable of measuring wind speed at various heights above the ground and the thermodynamic structure of the lower layer of the atmosphere. These sodar profilers can be categorised as mono-static systems, in which the same antenna is used for transmitting and receiving signals, and bi-static systems, in which separate antennas are used. In mono-static systems, the atmospheric scattering is caused by temperature fluctuations, whereas in bi-static systems such scattering is caused by both temperature and wind velocity fluctuations. The GPS dropsonde contains a GPS receiver, along with temperature, humidity and pressure sensors to capture thermodynamic data and atmospheric profiles. The first Omega-based dropsonde system was developed by the Atmospheric Technology Division of the National Center for Atmospheric Research (NCAR) in the early 1970s (Cole et al. 1973). In the following years, a dropsonde based on GPS satellite navigation was further developed. The NCAR GPS dropsonde represents a major advance in both accuracy and resolution for atmospheric measurements over data-sparse oceanic areas of the globe, providing wind accuracies of 0.5–2.0 m/s with a vertical resolution of 5 m (Hock and Franklin 1999).

3.3 SEISMIC SENSORS

Seismometers are instruments that measure motions of the ground, including those of seismic waves generated by earthquakes, nuclear explosions and other sources. As ground motion causes its frame to move, a fixed reference point is required for a seismometer taking measurements. Hence, the installation of a mechanical oscillator can be in the form of a mass-spring system (mobile mass attached to the frame by a spring) or a horizontal pendulum (mobile mass offset from the vertical axis of rotation). Accordingly, there is one way to classify the seismic sensors based on the measurement direction and the aforementioned properties of a mechanical oscillator: either vertical or horizontal. A vertical seismometer, which is used to measure vertical ground motion, utilises an oscillator based on the mass-spring system to compensate for gravity. A horizontal seismometer, which is used to measure horizontal ground motion, is based on the horizontal pendulum principle.

Figure 3.5 Photograph of a seismometer. (From https://en.wikipedia.org/wiki/File:CMG-40T_Triaxial_Broadband_Seismometer.JPG.)

Based on the range of vibration frequencies (or periods) that a seismometer can detect, seismometers can also be categorised as short-period seismometers and long-period, or broadband, seismometers. The short-period seismometer is constructed to have a short natural period and a correspondingly high resonant frequency which is usually higher than most frequencies in a seismic wave. For short-period seismometers, the inertial force produced by a seismic ground motion deflects the mass from its equilibrium position, and the displacement or velocity of the mass is then converted into an electric signal as the output proportional to the seismic ground motion. Long-period or broadband seismometers are built according to the force-balanced principle, in which the inertial force is compensated by an electrically generated force so that the mass moves as little as possible. The feedback force is generated with an electromagnetic force transducer through a servo loop circuit. The feedback force is strictly proportional to the seismic ground acceleration and is converted into an electrical signal as the output. A strong-motion seismometer, which usually measures acceleration, is also built on the force-balanced principle and can be integrated to obtain ground velocities and displacements. Figure 3.5 is a photograph of a seismometer to be buried in a hole.

3.4 LOAD CELLS

3.4.1 Load cells

Load cells are used for the measurement of forces and other structural loads. Most load cells are designed to measure a single component of force and to be insensitive to force components in other directions and to bending moments (Huston 2010). These designs are executed with a variety of specialised, and often proprietary, internal geometries. According to the type of output signal generated (e.g. hydraulic, pneumatic and electric signal), there are various types of load cells such as hydraulic load cells, pneumatic load cells, strain gauge load cells and piezoelectric load cells.

A hydraulic load cell (see Figure 3.6a) operates on a mechanical force-balance principle, which enables it to measure weight based on variations of internal fluid pressure. With the increase of the external force, the pressure of the hydraulic fluid rises accordingly. The output is linear and relatively stable. The accuracy of such load cells can be within 0.25% of full scale or even better, if they have been properly installed and calibrated. Hydraulic load cells

(a)　　　　　　　　(b)

(c)　　　　　　　　(d)

Figure 3.6 Typical load cells: (a) hydraulic load cell, (b) pneumatic load cell, (c) strain gauge–based load cell and (d) piezoelectric load cell. (From (a) http://www.noshok.com/force_5000_series.shtml; (b) http://www.alibaba.com/product-detail/Pneumatic-load-cell-2klb-3klb-5klb_1956925324.html; (c) http://www.dj-sensor.com/Product/Detail.aspx?id=21; (d) http://www.forsentek.com/prodetail_26.html.)

are acceptable for various weighting applications, especially for use in hazardous areas, because they have no electric components.

A pneumatic load cell (see Figure 3.6b) is another kind of force-balance device that operates by measuring changes in air pressure. Air pressure is applied to one end of the diaphragm and it escapes through the nozzle placed at the bottom of the load cell. The deflection of the diaphragm would thus affect the airflow through the nozzle as well as the pressure inside the chamber. To measure the pressure inside the load cell, a pressure gauge is required; it is usually attached to the cell. The major advantages of pneumatic load cells include cleanliness for the environment, safety for utilisation and insensitivity to temperature variations. Thus, they are generally used to measure relatively small weights in industries where cleanliness and safety are of prime concern.

Although there are various types of load cells, strain gauge-based load cells are the most commonly used type (see Figure 3.6c). The basic working principle for strain gauge load cells is that they convert the load acting on them into electrical signals. When the material of the load cells deforms appropriately (usually in the linear elastic range), the strain gauge bonded onto this material deforms accordingly (e.g. stretches or contracts) resulting in the variations of the electrical resistance of the gauges. These variations provide electrical value changes that are relational to the load applied to the load cell.

Another widely used load cell is the piezoelectric load cell (see Figure 3.6d). It works on the same principle of deformation as the strain gauge load cells, but the changes of output voltage are generated through the variations of electrostatic charge, rather than electrical resistance. The electrostatic charge is generated by the quartz crystals of piezoelectric material and is proportional to the applied force. This output is firstly collected on the electrodes sandwiched between the crystals, and then either converted to a low impedance voltage signal within the sensor or routed directly to an external charge amplifier for use. The piezoelectric load cells are most suitable for application in the dynamic loading conditions, where strain gauge load cells may fail due to high dynamic loading cycles.

3.4.2 Weigh-in-motion

Weigh-in-motion (WIM) systems have provided an effective means of data collection for pavement research and facility design, traffic monitoring and weight enforcement. The WIM devices can measure the axle weight of passing vehicles, and thus the sum of the weight of the vehicles, the velocity of the vehicles and the distance between the axles (Xu and Xia 2012). These data can be used to evaluate the traffic load on bridges. Unlike older, static weigh stations, current WIM systems are capable of measuring weight at normal traffic speeds and do not require the vehicle to stop, making them much more efficient. There are a few available WIM devices, such as a bending-plate WIM system with a strain gauge bonded to the underside of the plate scale (McCall and Vodrazka 1997); a piezoelectric WIM system using piezoelectric sensors embedded in the pavement which produce a charge when the tires induce the deformation on the pavement; a load cell WIM system utilising a load cell with two scales to weigh both right and left sides of the axle simultaneously; a capacitive-based WIM system with two or more metal plates; and a fibre-optic sensor-based WIM system.

A dynamic WIM station mainly consists of a metal housing with lightning protection, bending-plate sensors and processing board, induction loop detection and loop processor board, central processing unit and power supply, as schematically shown in Figure 3.7. The metal housing must have sufficient room to house an inner cabinet of suitable size where the control electronics with power supply and maintenance-free backup batteries can be located. Proprietary plug-in circuit boards are designed with individual lightning protection for the bending sensor monitoring, induction loop monitoring, data storage memory and additional serial interfaces. According to the velocity measurement range of passing vehicles, the WIM systems can be categorised as low-speed WIM and high-speed WIM.

As compared with static weight stations, WIM systems can record dynamic axle load information. Moreover, by screening vehicles within a threshold of a maximum permissible weight for the purpose of reduced queuing at weigh stations, WIM systems can provide

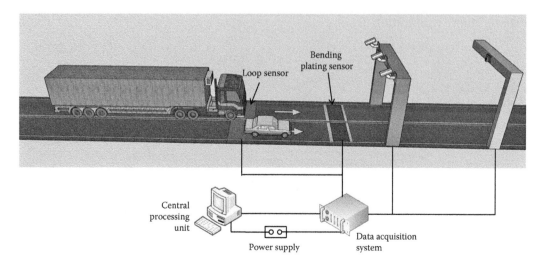

Figure 3.7 Schematic diagram of WIM system.

considerable savings for both truckers and enforcement agencies, thus improving the capacity of weigh station operations. However, WIM systems are less accurate than static scales.

3.5 THERMOMETERS

Temperature, including structural temperature and ambient air temperature, is frequently measured in monitoring systems. It is widely recognised that changes in temperature significantly influence the overall deflection and deformation of bridges and buildings. Restraint of movement can induce stresses within a bridge and a building. Excessive thermal stresses can damage bridges and buildings. The temperature is usually non-uniformly distributed over the entire structure and is different from the ambient temperature, due to heat transfer. The most widely used temperature sensors include thermocouples, thermistors and resistance temperature detectors.

A thermocouple is a temperature-measurement device that takes advantage of a phenomenon known as the Seebeck effect (i.e. whereby an electromagnetic field is generated in a circuit containing junctions between dissimilar metals if these junctions are at different temperatures). It consists of two dissimilar conductors that contact each other at one or more spots. An electrical potential is created when the temperature of one of the spots differs from the reference temperature at other parts of the circuit. This potential is characterised by a coefficient known as the Seebeck coefficient. A simple description of the thermocouple circle is plotted in Figure 3.8. A thermocouple is a widely used type of temperature sensor for measurement and control, and can also convert a temperature gradient into electricity. It has a wide measurement range and is inexpensive and interchangeable, but it is less stable than the other two kinds.

Thermistors and resistance temperature detectors are based on the principle that resistance of a material increases when the temperature goes up. They are generally more accurate and stable than thermocouples. Resistance temperature detectors are usually made of platinum. The most common resistance temperature detectors used in industry have a nominal resistance of 100 ohms at 0°C and are called PT100 sensors. They are often employed in bridge monitoring exercises. Thermistors differ from resistance temperature detectors in that the material used in a thermistor is generally a ceramic or a polymer. Thermistors have a smaller temperature range (−90°C to 130°C) but typically have a higher precision rate than resistance temperature detectors.

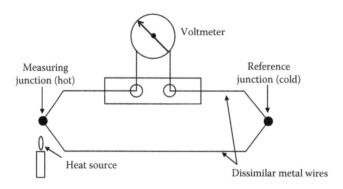

Figure 3.8 Description of the thermocouple circle.

3.6 STRAIN GAUGES

The strain gauge is usually attached to an object by a suitable adhesive or proper installation for measurement of the strain. Foil strain gauges, fibre-optic strain gauges and vibrating wire strain gauges are commonly used sensors measuring strain in civil structures. Fibre-optic strain gauges will be described in Section 3.9.

3.6.1 Foil strain gauge

Foil strain gauges, as shown in Figure 3.9a, are the most common type of strain gauge. They consist of a thin insulating backing which supports a fine metallic foil. The gauge is attached to the object by a suitable adhesive. As the object is deformed, the foil is stretched or shortened causing the change in its electrical resistance in proportion to the amount of strain, which is usually measured using a Wheatstone bridge. Foil strain gauges can be utilised in many situations, and the orientation of the gauges is quite critical in most cases. The preparation of the surface to which the strain gauge is to be attached, using special glue, is also important.

The length of most foil strain gauges is about a few millimetres to a few centimetres. Its full measurement range is about a few millistrain. The foil strain gauges are economical and can measure dynamic strains. However, their long-term performance (e.g. zero-stability) is not as good as that of other alternatives, particularly in a harsh environment. For example, the presence of moisture may result in electrical noise in the measurement and zero-drift. The principle of the self-temperature compensation is that the selected material of the gauge is matched to structural material, and thus the effect of the temperature is compensated by the gauge itself. However, it is usually difficult or impossible to make sure that the properties of the gauge material are similar to the structural material, especially in civil structures, for example, the concrete material. The dummy gauge technique is known as the non-self-temperature compensation method to handle this situation. It should be also noted that in any case, it is better to keep the Wheatstone bridge voltage drive low enough to avoid self-heating of the strain gauge.

3.6.2 Vibrating wire strain gauge

The vibrating wire strain gauge, as shown in Figure 3.9b, consists of a thin steel wire held in tension between two end anchorages. The wire vibrates due to an excitation with a short pulse, and the resonant frequency is measured. Forces within the structural element onto, or in, which the gauge is fixed, cause the length of the gauge to change. When the distance

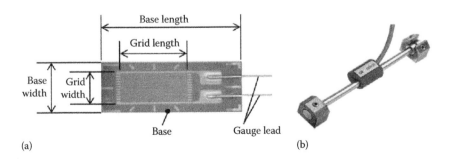

Figure 3.9 Strain gauges: (a) foil strain gauge and (b) vibrating wire strain gauge. (From (b) http://www.geokon.com/4000.)

between the anchorages changes, the tension of the wire changes, and so too does the natural frequency. The change in the vibration frequency of the wire is transferred into the change in strain. The relationship between these terms can be described as follows (Thomson 1993; Neild et al. 2005):

$$f = \frac{1}{2L}\left[\frac{E\varepsilon_w}{\rho}\right]^{1/2} \tag{3.1}$$

where:
f is the fundamental resonance vibration frequency
L is the length of the wire
E is Young's modulus
ρ is the density of the wire
ε_w is the strain of the wire

The captured strain can be transmitted over a relatively long distance (a few hundred metres to a few hundred kilometres) without much degradation. This is one advantage that vibrating wire gauges have over foil gauges.

The cost of the vibrating wire gauges is between that of the foil strain gauges and the fibre-optic strain gauges. Vibrating wire gauges are easy to install on the surface or embed in concrete. A typical vibrating wire gauge is about 100–200 mm long and has a measurement range of 3000 με with a resolution of 1.0 με, which is suitable for the monitoring of civil structures. A significant drawback of vibrating wire gauges is that they can measure the static strain only, as it takes seconds to obtain the frequency of the vibrating wire. Moreover, another disadvantage of vibrating wire gauges is that the gauges tend to be relatively large, requiring bulky cables for power supply and signal transmission.

3.7 DISPLACEMENT SENSORS

Displacement of civil structures serves as an effective indicator of their structural performance condition. Large displacements or deformations may create hazardous conditions for people living in the building, and for traffic moving on the bridge, and excessive displacements may affect the structural integrity. Displacement monitoring is thus needed. Equipment measuring the displacement includes linear variable differential transformers, level-sensing stations, tilt beam sensors, GPS, cameras and so forth.

3.7.1 Linear variable differential transformer

The linear variable differential transformer (LVDT), as shown in Figure 3.10a, is a commonly used electromechanical facility for measuring relative displacements based on the principle of mutual inductance. LVDTs are available in a wide range of linear stroke, ranging from micrometres to 0.5 m. An LVDT consists of a hollow metallic tube containing one primary and two secondary coils and a separate movable ferromagnetic core. The coils produce an electrical signal that is in proportion to the position of the moving core. When the core is at the centre of stoke (named *null position*), the values of these two secondary coils' voltages are equal. Since the two secondary coils are connected in opposite directions, the voltages generated by them have different signs, resulting in the output of the LVDT being zero in this case. When the core is moved away from the centre, the induced voltage

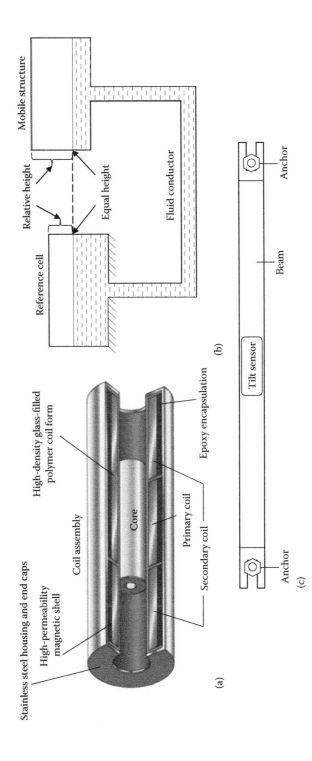

Figure 3.10 Displacement sensors: (a) LVDT (b) level sensing station and (c) tilt beam sensor. (From (a) http://www.te.com/usa-en/industries/sensor-solutions/insights/lvdt-tutorial.html.)

is increased in one of the secondary coils but decreased in the other. This action can thus generate a differential voltage output proportional to the changes of the core position. As the core is moved from one side of null to the other, the phase of output signal alters abruptly by 180°. Notably, to avoid the occurrence of gross nonlinearity of the output signal and overheating of the coils, the core must always be fully within the coil assembly during the operation of LVDT.

The configuration of LVDTs gives it many commendable and outstanding characteristics. The first distinguishing feature is very high resolution, mainly due to its friction-free operation. Normally, an LVDT is a frictionless device because there is no physical contact between the coil assembly and the movable core. This distinct feature is quite useful for many things, such as the fatigue-life testing of materials and structures, and for high-resolution dimensional gaging systems. The second outstanding characteristic is that the null position of an LVDT is stable and repeatable, even over a wide operating temperature range. This makes an LVDT perform well as a null position sensor in closed-loop control systems and high-performance servo balance instruments. Additionally, LVDTs have many other distinguishing features and benefits, such as long mechanical life, environmental robustness and fast dynamic response.

3.7.2 Level sensing station

The measurement of vertical displacement by the level sensors is in principle based on the pressure difference (see Figure 3.10b). The system utilises two or more interconnected fluid-filled cells (the fluid is usually water), and the quasi-static gravity-induced movement of liquids for transduction. A fluid-carrying conductor connects the cells and allows for gravity to equalise the height of the fluids in both cells. Relative vertical movement of the cells causes movement of water, and the variation in the water level is measured. The level-sensing station can be very effective, but the requirement for running long fluid-conducting tubes between the cells is inconvenient. The level-sensing system can detect an elevation difference of about 0.5 mm.

3.7.3 Tilt beams

Tilt-beam sensors, as shown in Figure 3.10c, can be designed to measure differential displacement and angular rotation in civil structures such as bridges, dams, tunnels and buildings. Tilt sensors are also referred to as inclinometers. A tilt beam can be installed in a horizontal direction to monitor settlement and heave, and in a vertical direction to monitor lateral displacement and rotation. It basically consists of a rigid casing (or beam) with an electrolytic tilt sensor mounted at the centre of the beam. The tilt sensor is a precision bubble-level liquid that can conduct electricity and is sensed as a resistance bridge. The bridge circle outputs a voltage that is proportional to the relative inclination of the sensor. The casing, which typically reaches 1–2 m and sometimes 3 m in length, is mounted on a set of bolts that are anchored to the monitored structure. Once the structure is subject to movement, the corresponding displacements change the tilt angle of the casing and the output of the sensor. By subtracting the initial reading from the current reading and multiplying it by the gauge length of the sensor (i.e. the distance between the anchors), the relative displacement can then be determined.

When tilt beams are linked in a series, displacement measurements can be accumulated from anchor to anchor to provide a detailed profile of differential movement or settlement along the components of the structure without using external anchorage points. Tilt beams can be designed to operate within a range as low as ±3 mm/m, but are more commonly designed to operate with a range as high as ±11 mm/m. The operating temperature for most commercially available tilt beams ranges between −20°C and 50°C, with the possibility of reaching 65°C for some products.

3.7.4 Global navigation satellite system

A global navigation satellite system (GNSS) is a satellite navigation system with global coverage. The Global Positioning System (GPS) of the United States was the first global operation GNSS in the world. The Russian GLONASS system and European Union Galileo system are other well-known global operational GNSSs. China is in the process of expanding its regional BeiDou Navigation Satellite System into the global navigation system COMPASS by 2020.

The GPS has revolutionised all disciplines related to navigation, localisation and positioning. GPS, developed by the US Department of Defense in 1973, was originally designed to assist soldiers, military vehicles, planes and ships (Sahin et al. 1999). It consists of three parts: the space segment, the control segment and the user segment. The space segment is composed of 32 satellites in 6 orbital planes. Each satellite operates in circular 20,200 km orbits at an inclination angle of 55°, and each satellite completes an orbit in approximately 11 h and 57.96 min (Hofmann-Wellenhof et al. 2008). The spacing of satellites in orbits is arranged so that at least six satellites are within line of sight from any location on the Earth's surface at all times (Hofmann-Wellenhof et al. 2001). The control segment is composed of a master control station, an alternate master control station and shared ground antennas and monitor stations. The user segment is composed of thousands of military users of the secure precise positioning service, and millions of civil, commercial and scientific users of the standard positioning service.

Basically, a GPS receiver receives the signals sent by the GPS satellites high above the Earth, determines the transit time of each message, computes the distances to each satellite and calculates the position of the receiver. However, even a very small clock error multiplied by the very large speed of light (299,792,458 m/s) results in a large positional error. Therefore, receivers use four or more satellites to improve the accuracy of the positioning. Nevertheless, this accuracy is in the order of a metre and cannot be used for displacement monitoring of bridges, which is about in the order of a centimetre.

In practical application, the real-time kinematic (RTK) technique is used on the basis of carrier-phase measurements of the GPS, where a reference station provides the real-time corrections. A RTK system (see Figure 3.11) usually consists of a base station receiver and a number of mobile units. The base station re-broadcasts the phase of the carrier that it measures, and the mobile units compare their own phase measurements with the ones

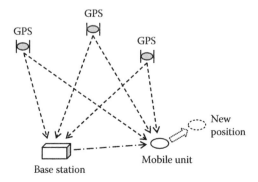

Figure 3.11 A schematic diagram of real-time kinematic (RTK) system.

received from the base station. This system can achieve a nominal accuracy of 1 cm ± 2 parts-per-million (ppm) horizontally and 2 cm ± 2 ppm vertically. In the foreseeable future, this can be improved further.

A few factors affect the accuracy of GPS measurements, in particular atmospheric conditions and multi-path effects. Inconsistencies of atmospheric conditions affect the speed of the GPS signals as they pass through the Earth's atmosphere. The GPS signals are also reflected by surrounding obstacles, causing delay of signals.

3.7.5 Camera

A charge-coupled-device (CCD) camera can also be used to measure displacements of civil structures. CCD cameras use a large number of pixels (dots) to form an image. The CCD camera method then utilises image-processing techniques for pixel identification and subsequent edge detection. The optical sensor used in these cameras is very sensitive to light. Thus, no artificial lighting is required to take an image of a civil structure. However, although a CCD camera has an optical sensor with a large number of pixels, the resolution provided may not be adequate for displacement measurements of a civil structure. The sub-pixel displacement-identification measurement method has been developed to improve accuracy if using digital cameras to levels comparable to those of traditional displacement transducers (Karellas et al. 1992).

3.8 ACCELEROMETERS

Although vibration can be measured in terms of velocity and dynamic displacement as well, acceleration can be measured more accurately. Accelerometers are widely used to measure the acceleration of civil structures induced by force excitation or ambient excitation. The acceleration responses of a structure are closely related to the serviceability and functionality of that structure. In addition, vibration testing can be employed to obtain the natural frequencies, damping ratios and mode shapes of the global structure, which are directly associated with the mass, stiffness and damping characteristics.

In simple terms, an accelerometer is a mass spring-damper system that produces electrical signals in proportion to the acceleration of the base where the sensor is mounted. Selection of accelerometers should consider the following parameters: usable frequency response, sensitivity, base strain sensitivity, dynamic range and thermal transient sensitivity. Installation of accelerometers and cables is also critical for good vibration measurement. There are four main types of accelerometers available: piezoelectric type, piezoresistive type, capacitive type and servo force balance type.

Based on the piezoelectric effect of quartz or ceramic crystals, piezoelectric type accelerometers can generate an electrical output proportional to the applied acceleration when they are compressed, flexed or subject to shear forces. As the body of the accelerometer is subject to vibration, due to inertia, the mass mounted on the crystal would compress or stretch the piezoelectric crystal resulting in the generation of certain charges. According to Newton's law of motion, this inertial force is proportional to the applied acceleration, and thus the generated charges are accordingly related to the applied acceleration. This charge output can then be either converted to a low impedance voltage output by the usage of integral electronics or directly measured as a charge output in a charge output piezoelectric accelerometer. A schematic diagram of a piezoelectric type accelerometer is given in Figure 3.12a. Piezoelectric type accelerometers are very robust and stable in long-term use. However, the major drawback is that they are not capable of measuring a true direct current (dc) (0 Hz)

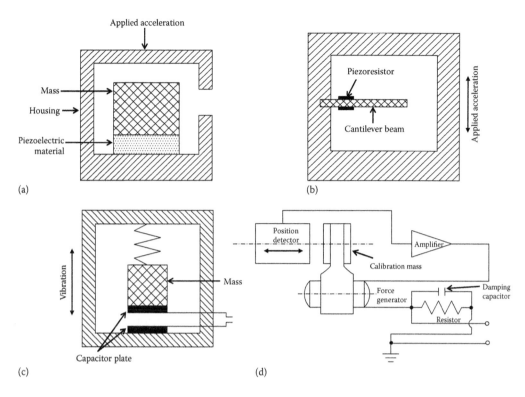

Figure 3.12 Schematic diagrams of accelerometers: (a) piezoelectric type, (b) piezoresistive type, (c) capacitive type and (d) servo force balance type.

response, which makes them unsuitable for some civil structures with very low frequency, for example, around 0.1 Hz. Actually, the lower frequency limit of piezoelectric accelerometers is generally above 1 Hz.

Piezoresistive type accelerometers utilise the piezoresistive effect to realise the signal transduction mechanism. The piezoresistive effect describes the variation of electrical resistivity of a semiconductor or metal when subject to mechanical strain. A schematic diagram of a piezoresistive type accelerometer is shown in Figure 3.12b. According to the topological configurations of the sensing elements, the piezoresistive accelerometers can be basically classified into three categories: single clamped beams, where the seismic mass is suspended with only one beam; double-clamped beams, where the seismic mass is fixed between two (or more) beams; and, finally, axially loaded beams, where the seismic mass is accelerated in the axial direction of the beam (Engesser et al. 2009).

By using capacitive sensing techniques, a capacitive accelerometer is able to measure the acceleration on a surface. It is composed of an oscillator or any stationary component that has the ability to store capacitance. When the accelerometer is subject to vibration, the moving mass (see Figure 3.12c) alters the distance between two capacitor plates, resulting in the changes in their capacitance. This generated capacitance change causes variations of the electrical current or voltage in their connected electrical circuitry. The measured current or voltage is then used for the determination of the intensity and magnitude of the applied acceleration. Both piezoresistive and capacitive accelerometers are adequate for flexible civil structures as they can measure accelerations from dc level. Capacitive type accelerometers are very accurate and appropriate for low frequency and low-level vibration measurement, such as micro-g (gravity acceleration = 9.80 m/s²).

The servo force balance type accelerometer can be divided into two basic groups: the non-pendulous type, having a mass which is displaced linearly, and the pendulous type, having an unbalanced pivoting mass with angular displacement. Generally, the force balance sensor is composed of a position detector, an amplifier and an electromechanical system (see Figure 3.12d). When acceleration is applied to this assembly, a force is exerted on the mass and it will attempt to move from the null position. The varied position of the mass is then monitored by the position detector, and the current is generated accordingly. Since this current is proportional to the restoring force, which is equal to the input force through the calibrated mass, the applied acceleration can then be obtained based on Newton's second law of motion. The force balance sensors have several advantages, including high accuracy, due to their low hysteresis performance, and small internal displacement, high stability and low thermal errors. The force balance sensors are suitable for dc and low-frequency measurement, providing milli-g measurement capability.

3.9 FIBRE-OPTIC SENSORS

Fibre-optic sensor technology uses light to conduct measurement of physical properties in remote sensing applications. Optical fibres can be used as sensors to measure strain, temperature, pressure and other quantities. For such purposes, the sensors modify a fibre so that the quantity to be measured modulates the intensity, phase, polarisation and wavelength of light in the fibre. Accordingly, fibre-optic sensors can be classified into four categories: intensity-modulated sensors, interferometric sensors, polarimetric sensors and spectrometric sensors (Casas and Cruz 2003).

Intensity-modulated optical sensors make use of the variations of the intensity of the transmitted or reflected light for the determination of the physical quantity. In principle, any parameter or variable that can cause intensity losses in the guided light can be measured by this kind of sensor. The intensity-modulated optical sensor can be easily implemented, is low cost, and has the ability of being multiplexed and performed as real distributed sensors. However, such sensors can only measure relative quantity, and a referencing system is usually required for the determination of accurate physical quantity.

Optical interference is a consequence of the wave nature of light. As the phase of a light field can be changed by external perturbations, the fibre-optic sensor can also be built based on the phase variations of a light field. A common case occurs with the splitting of a light beam into two parts, sending both down a separate path, rejoining the beams and then projecting the result onto a detector (Huston 2010). Depending on the relative phase, the rejoined light beams either add or subtract to form fringe patterns of varying intensities. Changes to the effective relative path-lengths alter the phase angles at the detector and cause the fringe patterns to shift. Interferometric sensors measure interference pattern shifts and can provide highly precise and accurate information about changes in physical parameters.

For polarimetric sensors, the modification of polarisation is obtained for the determination of the measurand. A polarisation-based fibre-optic sensor usually consists of a polariser, a polarisation-maintaining (PM) fibre and an analyser (Yin et al. 2008). After passing through the polariser, the light beam becomes a polarised light in the polarimetric sensor. The PM fibre, which is able to generate birefringence effect based on its property of anisotropic refractive index-distribution in the cross section, is then used to make the fibre sensitive to the polarisation direction of the incoming light. When subject to external perturbation (e.g. stress or strain), the output polarisation state would be altered accordingly. Thus, the

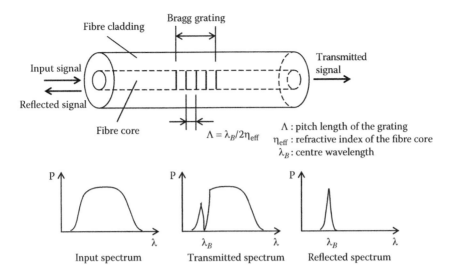

Figure 3.13 Principle of operation of an FBG sensor.

external perturbation can be determined by using an analyser to analyse the variations of the output polarisation state.

The spectrometric sensors monitor changes in the wavelength of the light. These sensors, better known as fibre Bragg grating (FBG) sensors, are the strain transducers of choice for many applications, and their configuration, installation and data processing are extremely straightforward (Casas and Cruz 2003). The FBG sensor uses the Bragg effect for transduction. The operating principle of the FBG is that strain and temperature changes alter the effective Bragg grating spacing and the wavelength of the reflected signal (see Figure 3.13). Therefore, by analysing the varied shift in the central Bragg wavelength, the mechanical or thermal strain variations on the FBG can be determined accordingly. A linear representation is

$$\frac{\Delta \lambda_B}{\lambda_B} = (1 - p_e)\varepsilon + (\alpha_\Lambda + \alpha_n)\Delta T \tag{3.2}$$

where:
λ_B is the centre wavelength
$\Delta \lambda_B$ is the corresponding variant
ε is the mechanical strain
ΔT is the temperature shift
P_e is the strain optic
α_Λ is the fibre thermal expansion
α_n is the thermal-optic coefficient

Notably, the sensed information (shift in wavelength) in FBG sensors is an absolute parameter, and thus an absolute measurement will be obtained, instead of a relative one. Moreover, multiplexing can be easily implemented for FBG sensors due to the wavelength-encoded nature of the output. The fact that the information is wavelength encoded also makes the sensor very stable to ageing, and allows absolute measurements of strain after long periods without recalibration.

A significant advantage of the aforementioned fibre-optic sensors is multiplexing; that is, several fibre-optic sensors can be written at the same optical fibre, and interrogated at the same time via one channel. In addition, fibre-optic sensors are very small in size and immune to electromagnetic interferences. They are also suitable for both static and dynamic measurements with a frequency from hundreds to thousands of hertz. Thus, the fibre-optic technologies are well suited to a very wide range of applications, such as sensing in strong electromagnetic fields, in high-temperature environments, in explosive environments and in very confined spaces. The major drawback of fibre-optic sensors is the high cost of both sensors and the acquisition unit (or readout unit). In addition, the fibres are rather fragile and should be handled very carefully in the field installation.

In civil structure monitoring, FBG sensors can be cost-competitive, especially with the development of affordable readout instrumentation. Thus, they are commonly used for strain measurement in construction materials and bridges (Seim et al. 1999; Ni et al. 2009). The principle applied is that the strain variation causes the shift in the central Bragg wavelength. Consequently, FBG strain sensors monitor changes in the wavelength of the light. Commercially available white light sources have a spectral width of around 40–60 nm. An FBG sensor with the measurement range of 3000 με takes a wavelength of 3 nm. Counting the spectral space between the sensors, one optical fibre can accommodate 6–10 FBG sensors.

Currently, there are many applications of different fibre-optic sensor systems to bridges and other civil structures around the world (Glišić and Inaudi 2007). The fibre-optic sensors are utilised to monitor several of the most important parameters in bridge structures and other civil structures, such as crack (Austin et al. 1999), wear of bearings (Cohen et al. 1999), corrosion (Lo and Shaw 1998), displacement (Vurpillot et al. 1997), temperature (Stewart et al. 2005) and acceleration (Sun et al. 2006).

3.10 NON-CONTACT SENSORS

As the name implies, non-contact sensors mean that there is no physical contact between the sensor and the target material or structure. In civil engineering, the non-contact sensors are usually employed for displacement measurement. Such non-contact displacement sensors mainly work on the following four principles: laser triangulation, eddy current, capacitive and confocal.

Under the laser triangulation principle (see Figure 3.14a), a laser diode projects a visible point of light onto the surface of the target. The back-scattered light reflected from this point is then projected by a high-quality optical lens system. If the position of the target is changed, the movement of the reflected light can then be projected and analysed to determine the corresponding displacement. The eddy current principle (see Figure 3.14b) is an inductive measuring method basing on the changes in the impedance of a coil caused by a moving electrically conducting object. By considering the variations of amplitude and the phase position of the sensor coil, the measurand-induced impedance can be calculated by the processor and the displacement can thus be determined. Under the capacitive principle, sensor and target serve like two plate electrodes to form an ideal parallel plate capacitor (see Figure 3.14c). If an alternating current (ac) with constant frequency is applied to the sensor capacitor, the amplitude of the ac voltage on the sensor would be proportional to the distance between the capacitor electrodes. The technology based on the confocal principle works by focusing polychromatic white light onto the target surface using a multi-lens optical system (see Figure 3.14d). The light reflected from the target surface is collected by a receiver and used to detect the motion-induced spectral changes.

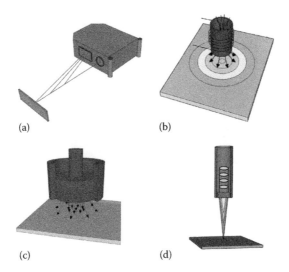

(a)   (b)

(c)   (d)

Figure 3.14 Schematic diagrams of the four principles for non-contact displacement sensors: (a) laser triangulation principle, (b) eddy current principle, (c) capacitive principle and (d) confocal principle.

Other non-contact measurement techniques, such as photogrammetry and videogrammetry techniques, have been developed with the growth in inexpensive and high-performance charge-coupled-device cameras and associated image techniques. Bales (1985) applied a close-range photogrammetric technique to several bridges for the estimation of crack sizes and deflection measurement. Li and Yuan (1988) developed a 3D photogrammetric vision system consisting of video cameras and 3D control points for measuring bridge deformation. Olaszek (1999) incorporated the photogrammetric principle with the computer vision technique to investigate the dynamic characteristics of bridges. Based on the wavelet image edge detection techniques, the presence and location of damage can be identified by using optical measurements (Patsias and Staszewski 2002). Ji and Chang (2008) and Zhou et al. (2010) employed these techniques for cable vibration measurement.

In addition to this, the non-contact sensors can also be used for velocity measurement. A laser Doppler vibrometer (LDV) is an instrument that is used to make non-contact vibration measurements of a surface. The LDV basically uses the Doppler principle to measure velocity at the point to which its laser beam is directed. The reflected laser light is compared with the incident light in an interferometer to give the Doppler-shifted wavelength. This shifted wavelength provides information on surface velocity in the direction of the incident laser beam. An advantage of an LDV over similar measurement devices, such as an accelerometer, is that the LDV can be directed at targets that are difficult to access, or that may be too small or too hot to be attached to a physical transducer. The LDV also makes the vibration measurement without mass-loading the target, which is especially important for very tiny devices such as micro-electromechanical systems. Abe et al. (2001) and Kaito et al. (2001) have applied an LDV to measure vibration of bridge decks and stay cables. When the measurement grid was predetermined, the LDV automatically scanned 45 points at a high frequency such that one LDV could measure the vibration of all points at the same time. With one reference, the modal properties can be extracted. Yan et al. (2008) applied an LDV to measure the vibration of hard disk drives.

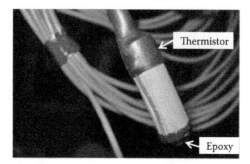

Figure 3.15 Duff gauge sensor with embedded thermistor. (From http://smtresearch.ca/products/embedded-moisture-sensor-ems.)

3.11 WEATHER STATIONS

In some applications, it is desirable to measure the environmental conditions such as ambient temperature, humidity, rainfall, air pressure and solar irradiation. A typical weather station usually integrates a few types of sensors, such as thermometers, barometers and hygrometers, which are used to measure the above-mentioned parameters besides the wind speed and direction. The type of thermometer used in weather stations is discussed in Section 3.5. Solar irradiation intensity, air temperature and wind are important parameters for deriving the temperature distribution of structures. With temperature distribution, the thermal effect on the structural responses can be evaluated quantitatively. The barometer is used to measure air pressure, using whatever material is inside the instrument, mainly water (water-based barometers), mercury (mercury-based barometers), oil (vacuum-pump oil barometers) and aneroid (aneroid barometers). A hygrometer is employed for the measurement of moisture content in the atmosphere. The humidity is usually determined by calibration and calculation of the measurements of some other quantity such as temperature, pressure or mass variation.

Monitoring the moisture levels in structural material such as concrete, masonry and gypsum is important when these materials are subject to freeze–thaw cycles. The presence of moisture within cavities can lead to frost damage and mould growth, which eventually results in the strength degradation of a civil structure. Since it is difficult to monitor moisture in these materials using traditional techniques, the sensors used for these materials indirectly measure levels of moisture. Moisture content–based wood electric resistance sensors, with incorporated thermistors, can be used for this purpose. One type of sensor is an embedded moisture sensor, also known as a Duff gauge sensor (see Figure 3.15). Duff gauge sensors have been used once evidence was produced of their reliable performance in both laboratory and field applications (e.g. Straube and Burnett 2005; Wilkinson et al. 2007; Uddin et al. 2009). This sensor comprises a wood element with known moisture content characteristics. When the sensor is embedded in the material, it gains the same moisture level as the material itself through capillary absorption. The moisture content can then be calculated using the relationship between the moisture content and the electric resistance of the wood. An integrated temperature sensor is used for temperature compensation.

3.12 CHEMICAL AND CORROSION SENSORS

A chemical sensor is a small device that transforms chemical or biochemical information of a quantitative or qualitative type into an analytically useful signal as the result of a chemical

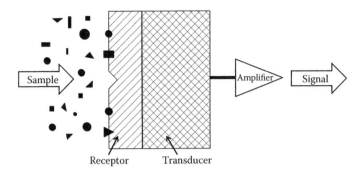

Figure 3.16 Working principle of chemical sensor.

interaction or process between the analyte gas and the sensor device (Stetter et al. 2003). Such chemical information includes composition, presence of a particular element or ion, concentration, chemical activity and partial pressure. Chemical sensors usually contain two basic components connected in a series: a recognition element (receptor) that is sensitive to stimuli produced by various chemical compounds (analyte) and a transduction element (transducer) that generates a signal whose magnitude is functionally related to the concentration of the analyte. The function of the receptor can be fulfilled in many cases by a thin layer which is able to interact with the analyte molecules, catalyse a reaction selectively or participate in a chemical equilibrium together with the analyte. The non-electric signal is then processed and transformed into an electric quantity, voltage, current, impedance/conductance or resistance by the transducer. The basic working principle is illustrated in Figure 3.16.

Chemical sensors also include a special branch referred to as *biosensors* for the recognition of biochemicals and bio-reactions. Although the use of biological elements as receptors – for example cells, tissues, organisms and enzymes – differentiates biosensors from conventional chemical sensors, the biosensor can be still considered a subset of chemical sensors because the transduction methods, sometimes referred to as the sensor platforms, are the same as those for chemical sensors (Stetter et al. 2003). Based on the phases of the analyte, the chemical sensors can be broadly classified into gas, liquid and solid particulate sensors. Piezoelectric sensors can also be considered a kind of chemical sensor. In such sensors, the acoustic measurement is made by finding the resonant frequency of the piezoelectric solid, that is, it looks for the point of maximum admittance between the two electrodes. The resonant frequency is a function of many variables, including mass loading, density, viscosity, pressure and temperature. The optical sensors mentioned before can also be considered as another widely used chemical sensor in civil engineering. If the analyte is placed at the interface of the fibre and a coating, it will have the opportunity to interact with the light. If the conditions are appropriate for either absorption or emission, the intensity and wavelength of the characteristic light provide the opportunity to obtain an analytical signal for quantitative and/or qualitative analysis.

Civil structures, especially steel structures and reinforced concrete structures, may be under threat from a combination of insidious challenges due to environmental conditions, such as accelerated deterioration mechanisms caused by temperature and humidity effect. Up to five environmental measurements are considered essential for monitoring corrosion and corrosive environments (Boller et al. 2009): time-of-wetness (ToW), temperature, relative humidity, chloride level and, possibly, pH. Corrosion sensors can be mechanical, electrical or electrochemical devices. These sensors are an essential element of all corrosion-monitoring systems, and thus the nature of the sensors depends on the various individual techniques used for monitoring. Only several widely used corrosion-monitoring techniques are mentioned herein for examples: corrosion coupons, the electric resistance technique, inductive resistance probes and the corrosion potential monitoring technique.

The simplest and longest-established method of estimating corrosion losses in plant and equipment is weight loss analysis. A sample (coupon) of the metal or alloy of interest is first weighed and introduced into the process, and later removed after a reasonable time interval. The coupon is then cleaned of all corrosion products and is reweighed. The weight loss can thus be easily obtained and converted into a total thickness loss or average corrosion rate using proper conversion equations. The electric resistance technique is a widely used method for measuring material loss occurring in the interior of plant and pipelines. This technique operates by measuring the change in electric resistance of a metallic element immersed in a product media relative to a reference element sealed within the probe body. To minimise the influence of changes caused by the ambient temperature, the procedure of temperature compensation is first required, and the resistance ratio is then measured to determine the metal loss. Inductive resistance probes have many similarities to electric resistance probes but offer significantly improved sensitivity. A coil is located in the sensor element, and the variation of its inductive resistance is used to identify the weight loss caused by the corrosion. For a given sensor element life, these probes will tend to show up a change in corrosion rate much sooner than the equivalent electric resistance version. The measurement of the corrosion potential is a relatively simple concept, with the underlying principle widely used for monitoring reinforcing steel corrosion in concrete and structures. Changes in corrosion potential can also give an indication of active/passive behaviour in stainless steel.

3.13 SENSOR PERFORMANCE AND SENSORY SYSTEMS

As has been demonstrated, different types of sensors are used in measuring different quantities. Equally, various techniques are employed – with different sensors – for measuring the same quantities. The sensor performance depends on many factors, for example the measurement range, sensitivity, resolution and operating condition. Attention is also paid to the reliability of the sensor to continue functioning over an extended period of time. The careful and appropriate selection of these parameters significantly influences the performance of sensors and the result of the measurement (Çatbaş et al. 2013; Farrar and Worden 2013).

A sensory system is a group of specialised sensors with a communication infrastructure intended to monitor and record conditions at diverse locations of a civil structure. It integrates sensors, data processors, data controllers, data links and power supply. Before a sensory system can be operated, it is subject to careful calibration. Calibration is the process of determining the relationship between the quantity to be measured and the signal generated so as to produce meaningful data. The stability of calibration is another critical property of a sensory system.

The capability of a given sensory system can far exceed the sum of the capabilities of individual sensors. There are two basic types of sensory systems usually involved in the application of civil structures: wired sensory systems and wireless sensory systems. Wired systems typically run electrical cables between the sensing nodes for both power supply and communication, whereas wireless systems use wireless data connections and transmissions for connecting network nodes. Both wired sensory systems and wireless sensory systems will be discussed further in Chapter 5. The performance of a sensory system depends not only on the performance of individual sensors but also on other factors, such as the number, type and location of the sensors installed in a civil structure. This important, yet challenging topic will be discussed in Chapters 8 and 11. After a sensory system has been selected, a subsequent step is to define the sampling parameters for the data-acquisition system. The parameters include the sampling rate, the sampling duration and when to sample the data. The selection of these parameters depends on not only on the measured quantities but also on the data storage and processing strategies (Farrar and Worden 2013), which will be further discussed in Chapter 5.

NOTATION

E, ρ	Young's modulus and density of the wire
f	The fundamental resonance vibration frequency of wire
L	The length of the wire for vibrating wire strain gauge
$P_e, \alpha_\Lambda, \alpha_n$	The strain optic, fibre thermal expansion, and thermal-optic coefficients for FBG sensor
ΔT	Temperature shift for FBG sensor
ε	The mechanical strain on FBG
ε_w	The strain of the vibrating wire strain gauge
$\lambda_B, \Delta\lambda_B$	The centre wavelength and its corresponding variant for FBG sensor

REFERENCES

Abe, M., Y. Fujino, and K. Kaito. 2001. Damage detection of civil concrete structures by laser Doppler vibrometry. In *Proceedings of the 19th International Modal Analysis Conference*, eds. A.L. Wicks and R. Singhal, 704–709. Bethel, CT: Society for Experimental Mechanics.

Austin, T.S.P., M.M. Singh, P.J. Gregson, J.P. Dakin, and P.M. Powell. 1999. Damage assessment in hybrid laminates using an array of embedded fibre optic sensors. In *Proceedings of SPIE 3671, Smart Structures and Materials 1999: Smart Systems for Bridges, Structures, and Highways*, Newport Beach, CA, ed. S.C. Liu, 281–88. Bellingham, WA: SPIE.

Bales, F.B. 1985. Close-range photogrammetry for bridge measurement. Transportation Research Report, No. 950, 39–44. Washington, DC: Transportation Research Board.

Boller, C., F.K. Chang, and Y. Fujino. 2009. *Encyclopedia of Structural Health Monitoring*. Chichester, UK: John Wiley & Sons, Ltd.

Casas, J.R. and P.J.S. Cruz. 2003. Fiber optic sensors for bridge monitoring. *J. Bridge Eng.*, 8(6): 362–73.

Çatbaş, F.N., T. Kijewski-Correa, and A.E. Aktan. 2013. *Structural Identification of Constructed Systems: Approaches, Methods, and Technologies for Effective Practice of St-Id*. Reston, VA: ASCE Publications. http://ascelibrary.org/doi/book/10.1061/9780784411971.

Cohen, E.I., S.A. Mastro, C.P. Nemarich, et al. 1999. Recent developments in the use of plastic optical fiber for an embedded wear sensor. In *Proceedings of SPIE 3670, Smart Structures and Materials 1999: Sensory Phenomena and Measurement Instrumentation for Smart Structures and Materials*, eds. R.O. Claus and W.B. Spillman Jr., 256–67. Bellingham, WA: SPIE.

Cole, H.L., S. Rossby, and P.K. Govind. 1973, The NCAR wind finding dropsonde. *Atmos. Tech.*, 2: 19–24.

Engesser, M., A.R. Franke, M. Maute, D.C. Meisel, and J.G. Korvink. 2009. Miniaturization limits of piezoresistive MEMS accelerometers. *Microsyst. Technol.*, 15(2): 1835–44.

Farrar, C.R. and K. Worden. 2013. *Structural Health Monitoring: A Machine Learning Perspective*. Chichester, UK: John Wiley & Sons, Ltd.

Glišić, B. and D. Inaudi. 2007. *Fiber Optical Methods for Structural Health Monitoring*. Chichester, UK: John Wiley & Sons, Ltd.

Hock, T.F. and J.L. Franklin. 1999. The NCAR GPS Dropwindsonde. *Bull. Amer. Meteor. Soc.*, 80(3): 407–20.

Hofmann-Wellenhof, B., H. Lichtenegger, and J. Collins. 2001. *GPS: Theory and Practice*, 5th edition. New York: Springer.

Hofmann-Wellenhof, B., H. Lichtenegger, and E. Wasle. 2008. *GNSS-Global Navigation Satellite Systems: GPS, GLONASS Galileo, and More*. New York: Springer.

Huston, D. 2010. *Structural Sensing, Health Monitoring, and Performance Evaluation*. Boca Raton, FL: CRC Press.

Ji, Y.F. and C.C. Chang. 2008. Non-target image-based technique for bridge cable vibration measurement. *J. Bridge Eng.*, 13(1): 34–42.

Kaito, K., M. Abe, and Y. Fujino. 2001. An experimental modal analysis for RC bridge decks based on non-contact vibration measurement. In *Proceedings of SPIE, the International Society for Optical Engineering*, 4359(2): 1561–67. Society of Photo-Optical Instrumentation Engineers, Bellingham, WA, ETATS-UNIS.

Karellas, A., L. Harris, H. Liu, M. Davis, and J. D'Orsi. 1992. Charge-coupled device detector: Performance considerations and potential for small-field mammographic imaging applications. *Med. Phys.*, 19: 1015–23.

Li, J.C. and B.Z. Yuan. 1988. Using vision technique for bridge deformation detection. In *Proceedings of International Conference on Acoustic, Speech, and Signal Processing*, New York, 912–15. Piscataway, NY: Institute of Electrical and Electronics Engineering.

Lo, Y.L. and F.Y. Shaw. 1998. Development of corrosion sensors using a single-pitch Bragg grating fiber with temperature compensations. In *Proceedings of SPIE 3325, Smart Structures and Materials 1998: Smart Systems for Bridges, Structures, and Highways*, San Diego, CA, ed. S.C. Liu, 64–72. Bellingham, Washington, DC: SPIE.

McCall, B. and W.C. Vodrazka. 1997. *States' Successful Practice Weigh-in-Motion Handbook*. Washington, DC: Department of Transportation, Federal Highway Administration. http://isddc.dot.gov/OLPFiles/FHWA/009585.pdf.

Neild, S.A., M.S. Williams, and P.D. McFadden. 2005. Development of a vibrating wire strain gauge for measuring small strains in concrete beams. *Strain*, 41(1): 3–9.

Ni, Y.Q., Y. Xia, W.Y. Liao, and J.M. Ko. 2009. Technology innovation in developing the structural health monitoring system for Guangzhou new TV tower. *Struct. Control Health Monit.*, 16(1): 73–98.

Olaszek, P. 1999. Investigation of dynamic characteristic of bridge structures using a computer vision method. *Measurement*, 25: 227–36.

Patsias, S. and W.J. Staszewski. 2002. Damage detection using optical measurements and wavelets. *Struct. Health Monit.*, 1(1): 7–22.

Sahin, M., E. Tari, and C.D. Ince. 1999. Continuous earthquakes monitoring with global positioning system (GPS). In *Proceedings of ITU-IAHS International Conference on Kocaeli Earthquake, A Scientific Assessment and Recommendations for Re-Building*, Istanbul, Turkey, eds. M. Karaca and D.N. Ural, 231–38. Istanbul: Istanbul Technical University.

Seim, J.M., E. Udd, W.L. Schulz, and H.M. Laylor. 1999. Health monitoring of an Oregon historical bridge with fiber grating strain sensors. In *Proceedings of SPIE 3671, Smart Structures and Materials 1999: Smart Systems for Bridges, Structures, and Highways*, Newport Beach, CA, ed. S.C. Liu, 128–34. Bellingham, WA: SPIE.

Srinivasulu, P., M.R. Padhy, P. Yasodha, and T.N. Rao. 2006. Development of UHF wind profiling radar for lower atmospheric research applications: Preliminary design report. National Atmospheric Research Laboratory, Department of Space, Government of India, Gadanki – 517112. https://www.narl.gov.in/.

Stetter, J.R., W.R. Penrose, and S. Yao. 2003. Sensors, chemical sensors, electrochemical sensors, and ECS. *J. Electrochem. Soc.*, 150(2): 11–16.

Stewart, A., G. Carman, and L. Richards. 2005. Health monitoring technique for composite materials utilizing embedded thermal fiber optic sensors. *J. Compos Mater.*, 39(3): 199–213.

Straube, J.F. and E.F. Burnett. 2005. *Building Science for Building Enclosure Design*. Westford, MA: Building Science Press.

Sun, R.J., Z. Sun, and L.M. Sun. 2006. Design and performance tests of a FBG-based accelerometer for structural vibration monitoring. In *Proceedings of the 4th China-Japan-US Symposium on Structural Control and Monitoring*, ed. Y.Q. Xiang, 450–51. Hangzhou: Zhejiang University Press.

Thomson, W.T. 1993. *Theory of Vibration with Applications*, 4th edition. London: Chapman and Hall.

Uddin, M., A. Mufti, D. Polyzois, et al. 2009. Monitoring moisture levels in stone masonry using Duff gauge sensors. In *4th International Conference on Structural Health Monitoring on Intelligent Infrastructure (SHMII-4)*, Zurich, Switzerland, Paper No: 473, 1–8.

Vurpillot, S., N. Casanova, D. Inaudi, and P. Kronenburg. 1997. Bridge spatial displacement monitoring with 100 fiber optic sensors deformations: Sensors network and preliminary results. In *Proceedings of SPIE 3043, Smart Structures and Materials 1997: Smart Systems for Bridges, Structures, and Highways*, San Diego, CA, ed. N. Stubbs, 51–57. Bellingham, WA: SPIE.

Wilkinson, J., K. Ueno, D. DeRose, J. Straube, and D. Fugler. 2007. Understanding vapour permeance and condensation in wall assemblies. In *Proceedings of the 11th Canadian Building Science & Technology Conference*. Banff, AL: ABEC.

Xu, Y.L. and Y. Xia. 2012. *Structural Health Monitoring of Long-Span Suspension Bridges*. Oxford: Spon Press.

Yan, T.H., X.D. Chen, and R.M. Lin. 2008. Feedback control of disk vibration and flutter by distributed self-sensing piezoceramic actuators. *Mech. Based Des. Struct. Mech.*, 36(3): 283–305.

Yin, S., P.B. Ruffin, and F.T.S. Yu. 2008. *Fiber Optic Sensors, Second Edition*. Boca Raton, FL: CRC Press.

Zhou, X.Q., Y. Xia, Z.K. Deng, and H.P. Zhu. 2010. Experimental study on videogrammetric technique for vibration displacement measurement. In *Proceedings of the 11th International Symposium on Structural Engineering*, eds. J. Cui, F. Xing, J.P. Ru, and J.G. Teng, 1059–63. Beijing: Science Press.

Chapter 4

Control devices and control systems

4.1 PREVIEW

Control devices are responsible for moving or controlling a civil structure. Control devices discussed in this chapter include not only the actuators operated by a source of energy but also the dampers operated without external energy. Control devices play a critical role in smart civil structures and allow the alternation of structural characteristics as well as the reduction of structural responses in a passive, semi-active and/or active manner. Many types of control devices, using or not using smart materials, have been developed for various applications. They can be classified as base isolation devices, passive energy dissipation devices, active control devices, semi-active control devices and hybrid control devices according to the controlled manner of a civil structure. Although base isolation devices and passive energy dissipation devices have limited intelligence as they are unable to adapt to external excitations and structural responses, they lay a foundation for developing most active, semi-active and hybrid control devices.

This chapter thus provides a brief introduction to the control devices used in civil structures with reference to the excellent book written by Cheng et al. (2008). The basic principle and characteristics of each type of device are presented and discussed. The configuration of a complete control system and its control performance are concisely laid out and detailed information is provided in the subsequent chapters.

4.2 BASE ISOLATION DEVICES

Seismic base isolation is a technique that mitigates the effects of an earthquake using a flexible base composed of a specific material with low lateral stiffness that filters out the energy of high frequencies from potentially dangerous ground motion. Since low- and medium-rise building structures are of relatively high frequencies, base isolation devices are designed to provide seismic protection for these types of buildings in an effective manner. The basic goal of such devices is to prevent the superstructure from absorbing earthquake energy so that the structural responses, including accelerations and inter-story drifts, will be significantly reduced. Earthquake protection of structures using the base isolation technique is generally suitable if the following conditions are fulfilled (Deb 2004):

1. The subsoil does not produce a predominance of long period ground motion.
2. The structure is fairly squat with sufficiently high column load.
3. The site permits horizontal displacements at the base of the order of 200 mm or more.
4. Lateral loads due to wind are less than approximately 10% of the weight of the structure.

The concept of modern seismic base isolation emerged in the early 1970s, but the design and construction of base-isolated buildings increased exponentially until the late 1980s. Early reviews on the seismic isolation technology can be found in many sources (e.g. Kelly 1986; Buckle and Mayes 1990; Taylor et al. 1992; Skinner et al. 1993; Naeim and Kelly 1999). Later, Symans et al. (2002) provided a literature review on the application of base isolation and supplemental damping devices for mitigating the effect of strong earthquakes. Kunde and Jangid (2003) presented a technical review of isolated bridge systems against earthquake excitation with emphasis on their theoretical and parametric studies. The American Society of Civil Engineers (ASCE) prepared a recent primer, which describes fundamentals of seismic isolation, analysis and design of isolated structures, practical issues for implementation and hardware requirements (Taylor and Igusa 2004). A task group (TG44) of the International Council for Research and Innovation in Building and Construction (CIB) presented a state-of-the-art report that introduces various devices for seismic isolation, compares design codes worldwide and describes the current state of seismic isolation practice among countries (Higashino and Okamoto 2006).

Detailed information on the application of base isolation and supplemental damping devices for mitigating the effect of strong earthquakes can also be found in Cheng et al. (2008). More recently, Warn and Ryan (2012) described widely used seismic isolation hardware, the recent development of a full-scale shaking table test and the past efforts to achieve three-dimensional seismic isolation.

Seismic isolation bearings are generally classified into two categories: elastomeric-type bearings and sliding-type bearings (Warn and Ryan 2012). Although some other types of bearings with the integrated characteristics of elastomeric and sliding bearings exist, the aforementioned two basic types are predominantly used in civil structures. Therefore, the following subsections focus on the introduction of these two types of bearings.

4.2.1 Elastomeric bearings

Natural rubber was initially used for the fabrication of elastomeric bearings. Later, synthetic rubber such as neoprene was developed as an alternative to natural rubber for bearing fabrication. For the purpose of reducing the bearing's vertical deformation and keeping the rubber layers from laterally bulging, intermediate steel shim plates are added into bearings, as shown in Figure 4.1a. Moreover, a rubber cover is employed for protecting the steel shim plates and internal rubber layers from stiffness degradation caused by environmental effects, such as corrosion and ozone attack.

Elastomeric bearings can be broadly categorised as

1. Low-damping natural or synthetic rubber of which the equivalent damping ratio ranges between 2% and 3% at 100% shear strain
2. High-damping rubber of which the equivalent damping ratio can reach 20% or even 30% at 100% shear strain

External supplemental damping devices, for example, viscous fluid dampers or yielding steel bars are often utilised in parallel with low-damping rubber bearings to limit or reduce displacements across the isolation interface. By adding some other materials such as carbon black to the raw rubber during the mixing process, a higher level of damping can be achieved and the high-damping rubber bearings (HDRBs) can be produced accordingly. A more detailed description of the HDRBs will be given later.

The shape factors 'S' from 15 to 25 (HITEC 1998a,b) and as high as 30 (Kelly 1991) are traditionally used for the design of elastomeric seismic isolation bearings. The shape factor

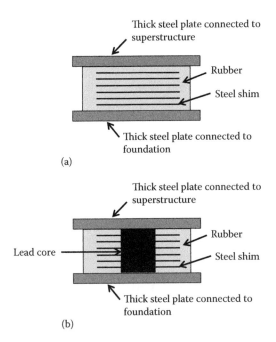

Figure 4.1 Schematic diagrams of elastomeric and lead-plug bearings: (a) elastomeric bearing and (b) lead-plug bearing.

for a single rubber layer is defined as the loaded area over the area free to bulge. By proper transformation, the relationship of the vertical and horizontal stiffness of the elastomeric bearing can be revealed with the aid of the shape factor. For example, for circular bearings, $K_v/K_h = 6S^2$, where K_v and K_h are the vertical and horizontal stiffness of the elastomeric bearing, respectively. It can be seen that with the shape factors in the range of 15–30, the vertical stiffness is thousands of times greater than the horizontal stiffness. Consequently, such bearing isolation systems typically have vertical isolation periods ranging from 0.03 to 0.15 s, and only horizontal isolation is provided by them.

4.2.2 Lead-plug bearings

Lead-plug bearings are elastomeric-type bearings. With an additional lead plug that is tightly fitted into a central hole in the bearing (see Figure 4.1b), lead-plug bearings were developed for enhancing the energy dissipation capabilities of elastomeric bearings. Due to plastic deformation of the lead core, the energy dissipation mechanism of this type of bearing is mainly hysteretic and can be analytically presented by a Bouc–Wen or rate-independent plasticity model (Nagarajaiah et al. 1991). The imposed horizontal force would significantly influence the performance of the lead-plug bearing on seismic isolation. When the horizontal force applied to the lead-plug bearing is small, the movement of the steel shims is restrained by the lead core and thus a higher level of horizontal stiffness as well as an elastic restoring force will be provided as compared with elastomeric bearings. As the horizontal force becomes larger, the lead core is forced to deform or yield by the steel shims, and hysteretic damping is provided by the lead core for energy absorption.

Similar to elastomeric bearings, the vertical stiffness of a lead-plug bearing is typically thousands of times larger than the horizontal stiffness, and thus only horizontal isolation is provided by this type of bearing. Moreover, for both bearings (i.e. elastomeric bearings and lead-plug bearings), uplift or tension is considered undesirable and great efforts have been made in the design process to avoid uplift and control the vertical movements within an acceptable range (Robinson 1982; Griffith et al. 1990; Liu et al. 2009).

4.2.3 High-damping rubber bearings

It is known that the mechanical characteristics of bearings are also significantly dependent on their compounds. Therefore, besides adding a lead plug to form the aforementioned lead-plug bearings, another effective method to improve the performance of the elastomeric bearings is to modify the rubber compounds regardless of whether the rubber is natural or synthetic. HDRBs are also elastomeric-type bearings. In the manufacture of HDRBs, the rubber is vulcanised together with other materials such as plasticizer, oil and carbon black, resulting in specific characteristics, for example, maximum strain dependency of stress evolution, energy-absorbing properties and hardening properties during rubber processing. HDRBs are capable of supporting vertical loads with limited or negligible deflection but horizontal loads with large deflections.

Various models have been developed and employed to represent the relationship between the horizontal force and the deformation of HDRBs. Because of their simplicity, equivalent linear models are commonly used for the design purposes of isolated structures equipped with HDRBs, such as those specified by the Public Works Research Institute of the Japanese Ministry of Construction (Kawashima 1992) and the American Association of State Highway and Transportation Officials (AASHTO 2014). Moreover, to capture the nonlinear mechanical behaviour of the HDRBs more accurately, bilinear model, trilinear models and varieties of nonlinear strain rate–dependent models have also been developed. Notably, some other factors, such as ageing and temperature, somehow affect the mechanical properties of HDRBs.

4.2.4 Friction pendulum bearings

The friction bearing was initially considered using a flat sliding surface. In this type of bearing, the friction force generated on the sliding surface is used to resist the imposed lateral force. However, since the imposed lateral force is basically less than the resistance generated from friction after an earthquake, the building structure cannot return to its original position, which can be considered a major disadvantage of a friction bearing. The accumulation of this kind of shift will finally exceed the bearing's range and result in the failure of the bearings. To overcome this drawback, a friction bearing with a spherical or concave sliding surface, called a *friction pendulum bearing* (FPB), was developed (see Figure 4.2a).

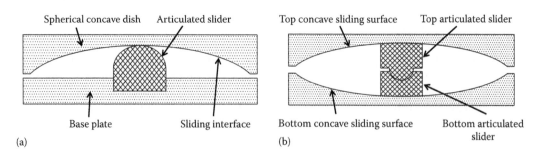

Figure 4.2 Schematic diagrams of friction pendulum bearings: (a) single concave surface and (b) double concave surface.

The resistance of an FPB to horizontal forces is limited due to a low friction coefficient designed on the sliding interface (normally approximately 3%). The stainless steel and polytetrafluorethylene (PTFE)-type materials are usually used as bearing materials at the sliding interface for most friction bearings. As shown in Figure 4.2a, a single FPB is basically composed of a spherical concave dish, an articulated slider and a base plate. The frictional resistance and energy dissipation are provided by the friction of the sliding surface, and the restoring force can be obtained from the radius of curvature of the spherical concave dish if the FPB is not in its equilibrium position after the earthquake attack. From a construction perspective, a multiple FPB differs from a single FPB only because more than one concave surface is being used in the bearing, such as double concave surfaces (Fenz and Constantinou 2006), as schematically shown in Figure 4.2b, or four concave surfaces (Fenz and Constantinou 2008; Becker and Mahin 2012). The benefit of the multiple FPBs over a single FPB is that they allow an adaptive force-deformation behaviour, whereby the stiffness and damping properties of the bearing can change at predetermined displacement amplitudes (Fenz and Constantinou 2008). Moreover, since the horizontal movement of the bearing is contributed by two or more concave surfaces, the multiple FPBs can achieve the same horizontal movement with reduced bearing size as compared with the single FPB.

By using coated Teflon on the stainless steel to protect the sliding surface from corrosion, the FPBs can often operate with low maintenance, which is considered one of the advantages of such bearings. Moreover, the torsional effects caused by the asymmetric building and mass irregularities are naturally balanced by the corresponding spatial variation in the restoring force, such that the associated torsional response is minimal. Therefore, the FPBs are able to effectively reduce the large levels of the superstructure's acceleration under a variety of severe earthquake loading.

4.2.5 Other types of base isolation devices

The combination of elastomeric-type and sliding-type bearings creates some other types of base isolation devices, for example, a pot-type bearing as shown in Figure 4.3. The elastomer in the pot bearing is confined by a pot-like piston coated by Teflon on its surface. The top steel plate behaves like the piston and the elastomer functions like a VF inside a hydraulic jack. Owing to the confinement of the piston, the elastomer is prevented from bulging under high pressure. Besides being able to hold substantially high pressure, the pot-type bearings enable slight rotations under homogeneous compression stress as well. Some other advantages of pot bearings include large movement, high durability, higher allowable compression stress resulting in reduced bearing size and the ability to be designed for vertical tension.

Figure 4.3 Photograph of a pot-type bearing. (From http://www.qiaoliangzhizuo.org/product/psxjzhizuo. html.)

The combined device tested by Earthquake Engineering Research Center (EERC) at Berkeley, California, and the hybrid Taisei shake suspension system (Naeim and Kelly 1999) can be considered as another application of the simultaneous use of elastomeric- and sliding-type bearings by installing both bearings in different locations of the building foundations. The sliding bearings are used to support the vertical loads, whereas the elastomeric bearings are designed to resist lateral force and provide restoring force. The configuration of such base isolation devices is significantly important because the bearings may experience high tensile force with inappropriate placement.

Besides the idea of a combination of elastomeric- and sliding-type bearings, there are other types of base isolation systems based on the concept of passively dissipated energy, such as spring-type systems, resilient-friction base isolation systems and sleeved-pile isolation systems (Cheng et al. 2008). However, some systems are only limited to certain types of building structures and some are not yet commercially available.

4.3 PASSIVE ENERGY DISSIPATION DEVICES

All vibrating structures dissipate energy due to internal stressing, rubbing, cracking, plastic deformations and so on. The larger the energy dissipation capacity, the smaller the amplitude of vibration. Methods of increasing the energy dissipation capacity are thus effective in reducing vibration amplitudes. Passive energy dissipation devices, which encompass a range of materials for enhancing damping and stiffness, can be used for reducing structural vibration. Serious efforts have been undertaken to develop such passive devices for several decades and there is a comprehensive body of literature reviewing these devices (Housner et al. 1997; Soong and Dargush 1997; Soong and Spencer 2002; Cheng et al. 2008). Although no external power or structural response measurements are required for such passive devices, they are able to generate damping forces when the vibration gets higher. However, owing to the properties of the damper itself, passive devices only have a limited control capacity. This section introduces the following typical passive devices: metallic dampers, friction dampers, viscoelastic (VE) dampers, viscous fluid (VF) dampers, tuned mass dampers (TMDs) and tuned liquid dampers (TLDs).

4.3.1 Metallic dampers

One of the effective mechanisms available for the dissipation of structural energy under an earthquake is through inelastic deformation of metals. The traditional seismic-resistant design of civil structures relies upon the post-yield ductility of structural members to provide the required energy dissipation capacity. This concept led to the idea of using metallic hysteretic dampers to absorb a large portion of the seismic energy in the early 1970s (Kelly et al. 1972; Skinner et al. 1974). Since then, many new designs have been proposed, including X-shaped and triangular (or hourglass-shaped) steel plate dampers (Whittaker et al. 1991). A schematic diagram of an X-shaped plate damper or an added damping and stiffness (ADAS) device is given in Figure 4.4 as an example. Other configurations of metallic yield dampers, used mostly in Japan, include a shear panel type as well as a bending type of honeycomb and slit dampers (Soong and Spencer 2002). Alternative materials, such as shape-memory alloys and lead, have been investigated as well (Aiken and Kelly 1992). The distinguishing features of metallic yield devices include their stable hysteretic behaviours, long-term reliability, low-cycle fatigue property and relative insensitivity to environmental temperature. They can be utilised to upgrade the seismic-resistant capacity of existing structures or to improve the seismic-resistant design of new structures.

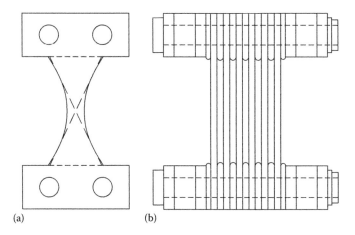

Figure 4.4 Schematic diagram of ADAS device: (a) side view and (b) front view.

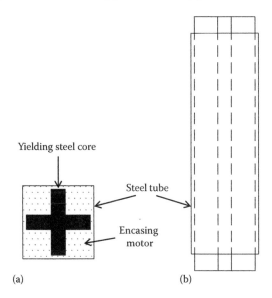

Figure 4.5 Schematic diagram of the un-bonded device: (a) top view and (b) side view.

Different from the aforementioned devices, a tension/compression yielding brace, which is also called an un-bonded brace (Wada et al. 1999; Clark et al. 1999), has a variant type but operates on the same metallic yielding principle. As shown in Figure 4.5, the un-bonded brace is a bracing member consisting of a core steel plate encased in a concrete-filled steel tube (Soong and Spencer 2002). A special coating is provided between the core plate and concrete in order to reduce friction. The core steel plate provides stable energy dissipation by yielding under reversed axial loading, while the surrounding concrete-filled steel tube resists compression buckling.

4.3.2 Friction dampers

Friction provides an effective, reliable and economical mechanism for dissipating kinetic energy by converting it to heat. In the design of friction dampers, two solid bodies that are able to slide relative to each other are required for generating the desired friction force. Since the composition of the interface is of great importance for ensuring the longevity and durability of these devices, compatible materials must be employed to maintain a consistent friction coefficient over the service life of the devices. Moreover, for normal operation it is also important to minimise stick-slip phenomena to avoid introducing high-frequency excitation to the overall structural systems.

Several types of friction dampers have been developed for the purpose of structural vibration suppression. The X-braced friction damper, developed by Pall and Marsh (1982), is an example of such a device (see Figure 4.6). Some other friction dampers, for example, Sumitomo friction damper (Aiken et al. 1993), energy dissipating restraint device (Nims et al. 1993), slotted-bolted connection energy dissipator (FitzGerald et al. 1989; Grigorian et al. 1993) and Tekton friction damper (Li and Reinhorn 1995), have also been investigated. Typically, these devices exhibit good control performance and their behaviour is not significantly affected by loading frequency, loading amplitude or even the number of loading cycles.

In the past several decades, there have been a number of applications of friction dampers to enhance the seismic protection of new and retrofitted structures, including the use of the X-braced friction damper in Canada and the United States (Pall and Pall 1993, 1996), the slotted-bolted connection in the United States (Grigorian et al. 1993), and the Sumitomo friction dampers in Japan (Aiken and Kelly 1990). Ng and Xu (2006a,b) also carried out both analytical and experimental investigations to explore the possibility of using friction dampers to connect a podium structure to a main building in order to reduce the seismic responses of both structures. One challenge for the implementation of a friction damper in practical situations is that it is often difficult to maintain the mechanical properties of these devices over a prolonged time interval. In particular, the friction-generating capacity of the damper would be significantly influenced by corrosion, relaxation of the sliding metal interface and the deformation caused by temperature variations. Moreover, the difficulty in analysing and designing the nonlinearity performance of the friction dampers could be considered another challenge, which limits the application of friction dampers.

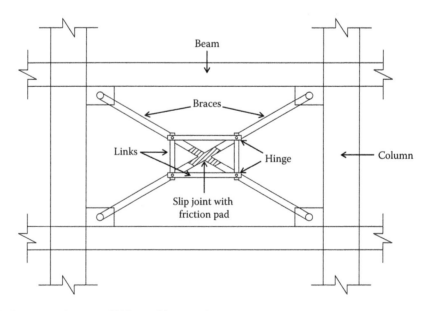

Figure 4.6 Schematic diagram of X-braced friction damper.

4.3.3 Viscoelastic dampers

VE dampers utilise specific high damping VE materials, for example, rubber, copolymers or glassy substances, to dissipate kinetic energy through shear deformation. The application of VE dampers to civil structures began in 1969 when approximately 10,000 viscoelastic dampers were installed in each of the twin towers of the World Trade Center in New York to reduce wind-induced vibration (Mahmoodi 1969).

A typical VE damper (see Figure 4.7) encompasses several VE layers bonded with steel plates. When the relative motion between the outer steel flanges and the centre plates is induced by structural vibration, shear deformation and hence energy dissipation takes place in this type of damper. The behaviour of VE materials under dynamic loading depends on vibrational frequency, strain and ambient temperature. This can present a problem in the design process because the properties of VE materials can only be expressed by shear storage modulus (measure of the energy stored and recovered per cycle) and shear loss modulus (measure of the energy dissipated per cycle). Zhang and Xu (1999) adopted the complex mode superposition method to estimate dynamic characteristics (model frequency, modal shape and modal damping ratio) of adjacent buildings linked by viscoelastic dampers. Moreover, it needs to be pointed out that the VE material is linear over a wide range of strain provided the temperature is constant. However, at large strains, there is considerable self-heating due to the large amount of energy dissipated. The heat generated changes the mechanical properties of the material and, thus, nonlinearity should be considered for the design and analysis of VR dampers.

4.3.4 Viscous fluid dampers

Different from metallic, friction and VE dampers that use the action of solids to dissipate vibration energy, VF dampers use the action of fluids to dissipate energy. Figure 4.8a shows one straightforward design of VF dampers (Makris and Constantinou 1990; Constantinou and Symans 1992, 1993; Soong and Spencer 2002). These dampers generally consist of a piston, which contains a number of small orifices, in the damper housing, filled with a compound of silicone or oil. The piston acts not simply to deform the fluid locally, but rather to force the fluid to pass through the small orifices. As a result, it dissipates energy

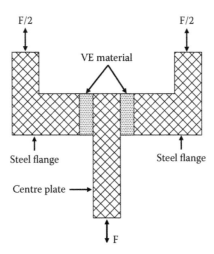

Figure 4.7 Schematic plan view of VE damper.

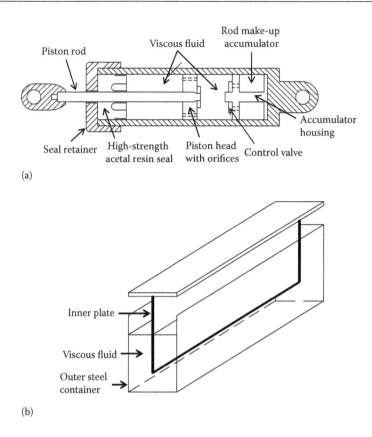

Figure 4.8 Schematic diagrams of two typical VF dampers: (a) VF damper and (b) VF damping wall.

through movement of the piston in the highly VF. If the fluid is purely viscous, then the output force of the damper is directly proportional to the velocity of the piston. Over a large frequency range, the damper exhibits viscoelastic fluid behaviour. The research carried out by Constantinou and Symans (1993) on VF dampers showed that the Maxwell model could be used to describe the viscoelastic fluid behaviour of the fluid damper applicable to civil structures. Zhang and Xu (2000) adopted the complex mode superposition method to estimate the dynamic characteristics (model frequency, modal shape and modal damping ratio) and find the solution for the seismic responses of adjacent buildings linked by the Maxwell model–defined fluid dampers. Xu et al. (1999) and Yang et al. (2003) also carried out experimental studies of adjacent buildings linked by fluid dampers.

An alternative design, developed by the Sumitomo Construction Company in Japan, involves the implementation of the viscous damping wall (see Figure 4.8b). The piston of a VF damping wall is a steel plate constrained to move within a narrow rectangular steel container, filled with highly VF. For typical installation, the piston and the container are, respectively, fixed to the upper floor and lower floor. Inter-story shift shears the fluid and, thus, the friction between the inner plate and the VF provides energy dissipation. Significant efforts have been made towards the application of VF dampers, which include the use of VF dampers in combination with seismic isolation systems (Skinner et al. 1993; Hirai et al. 1994; Kelly 1996; Soong and Dargush 1997) as well as seismic retrofit of reinforced concrete structures (Reinhorn et al. 1995) and steel frame structures (Martinez-Rodrigo and Romero 2003).

4.3.5 Tuned mass dampers

The simplest form of a TMD consists of an auxiliary mass–spring–dashpot device anchored or attached to some floor (usually the top floor) of the main structure, as shown in Figure 4.9. The frequency of the damper is tuned to a specified structural frequency (often the first natural frequency) so that when that frequency is excited, the damper will resonate out of phase with the structural motion. The energy dissipation can thus be achieved by the damper forces acting on the structure. A TMD's effect can also be viewed as equivalent to increase the damping ratio of the structure itself to a larger value (Cheng et al. 2008). For responses of lightly damped structures with a dominant mode, TMD can effectively reduce the peak response or resonant component. Given this characteristic, TMDs are increasingly used for wind-sensitive structures to curb excessive building motion and to ensure occupant comfort.

Successful design and application of a TMD to a building structure is a complicated procedure because many factors for selecting and tuning TMD must be simultaneously considered. For example, the energy-absorbing capacity of the TMD is related to the mass ratio, the stiffness ratio, the frequency (tuning) ratio and the damping ratio of the TMD to the main structure. In the case of a sliding mass arrangement, a low friction bearing surface is required so that the added mass can respond to the building movement at low levels of excitation. This would be quite critical. More details can be found in the literature (Luft 1979; Warburton 1982; Xu et al. 1992; Fujino and Abe 1993; Xu and Kwok 1994).

However, the practical application of a TMD in the real world is still limited by three factors. First, TMDs are effective only for one mode, which means they cannot be employed for seismic vibration control in an efficient way. Second, they are sensitive to mistuning, which means significant effort should be paid for careful tuning. Third, the mass of the TMDs and the place for installing TMDs are both relative large, exacerbating the difficulty of installation and construction. Many researchers are devoted to handling these limitations, and the much progress has been made. For example, multiple tuned mass dampers (MTMDs) and associated optimisation procedures were proposed and investigated (Clark 1988; Jangid 1995; Jangid and Datta 1997; Wu et al. 1999; Park and Reed 2001; Gu et al. 2001; Li 2000; Kwon and Park 2004; Zuo and Nayfeh 2005; Guo and Chen 2007).

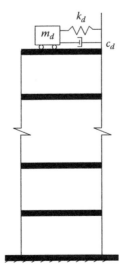

Figure 4.9 Tuned mass damper fixed on the top floor.

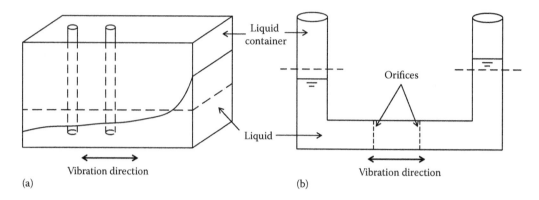

Figure 4.10 Schematic diagrams of tuned liquid dampers and tuned liquid column dampers: (a) tuned liquid damper with rods and (b) tuned liquid column damper.

4.3.6 Tuned liquid dampers and tuned liquid column dampers

Instead of using a solid concrete or metal block as the secondary mass in TMDs for vibration control, TLDs and tuned liquid column dampers (TLCDs) utilise water or some other liquid serving the same purpose, as shown in Figure 4.10. The liquid inside the container is forced to move when the structural vibration occurs, and the restoring force is then generated by gravity. The turbulence of the liquid flow and the friction between the liquid flow and the container convert the dynamic energy of the fluid flow to heat, thus absorbing structural vibration energy. For TLDs, meshes or rods are placed in the liquid to provide damping capacity, and the natural frequency is adjusted by the size of the container or depth of the liquid. The TLCD consists of a partially filled tube containing several internal orifices. It generates high-flow turbulence through the orifice to provide damping capacity, and its natural frequency is dependent on column shape, column length and air pressure (Xu et al. 1992, 2000).

Similar to TMDs, TLDs and TLCDs have been also used to suppress wind-excited vibrations of tall structures, such as airport towers and tall buildings (Xu et al. 1992; Balendra et al. 1995; Tamura 1995). There are two advantages of TLDs: a single TLD can be effective in any direction of lateral vibrations, and water used for TLDs can also be used as part of building protection supply in case of fire. However, compared with TMDs, more space is required for TLDs because liquids have less mass density than the materials for TMDs (e.g. concrete or steel). Moreover, unlike TMDs, which often respond in a purely linear manner, the behaviour of TLDs is generally highly nonlinear due to liquid motion. This inherent nonlinearity complicates the analysis and design process for TLD devices; thus, significant research efforts have focused on their optimum parameters and nonlinearity. On the other hand, TLCDs are similar to TMDs, which are effective only for one mode. Therefore, multiple tuned liquid column dampers (MTLCDs) and the associated optimisation procedures were proposed and investigated as well (Shum and Xu 2002; Xu and Shum 2003).

4.4 ACTIVE CONTROL DEVICES

Active control devices can be used to reduce structural responses under internal or external excitation, such as machinery or traffic noise, wind or earthquakes, where the safety or comfort level of the occupants is of concern. The major advantages of an active control device include enhancement of control effectiveness, adaptability to ground motion, selectivity of control objectives and applicability to various excitation mechanisms. An essential feature of active control devices is that external power is required for the control action. This makes such devices vulnerable to power failure, which will probably occur during a strong earthquake. An active control system is basically composed of three types of elements: sensor, actuator and a controller with a predetermined control algorithm. Excellent reviews on active vibration control are available and can be found in the references (Soong 1988; Housner et al. 1997; Datta 2003). This section provides a brief introduction to some actuators commonly used in active control systems.

4.4.1 Servovalve-controlled hydraulic actuators

A servovalve-controlled hydraulic actuator, as shown in Figure 4.11, is a mechanical device that uses hydraulics to convert energy into useful mechanical work. Servovalve-controlled hydraulic actuators have the advantages of high force capability, good mechanical stiffness, high power per unit weight and volume and high dynamic response. According to the mechanical motion provided by hydraulic actuators, they can be basically divided into three groups: cylinder actuators, limited-rotation actuators and rotary motor actuators.

A cylinder actuator is able to provide a fixed length of straight-line motion. It usually contains a tight-fitting piston moving in a closed cylinder. Xu and Zhang (2002) presented an analytical study that used the cylinder actuators to mitigate seismic responses of adjacent buildings most effectively with an optimum control law. A limited-rotation actuator can be used for opening, closing, lifting, lowering, indexing and transferring movements by producing limited reciprocating rotary force and motion. A rotary motor actuator is an actuator that produces a rotary motion or torque. It should be noted that hydraulic power is only one source of power for rotary motor actuator. The rotary motor actuator can also be powered by electrical or pneumatic power.

4.4.2 Pneumatic actuators

Different from the hydraulic actuator, a pneumatic actuator converts energy (typically in the form of compressed air rather than liquid) into a mechanical motion. It mainly consists of a piston, a cylinder and valves or ports. Similarly, according to the type of movement, pneumatic actuators mainly include tie rod cylinders, rotary actuators, grippers and rodless actuators.

4.4.3 Shape memory alloy actuators

The distinguishing characteristics of shape memory alloys (SMAs) have been described in Chapter 2. This section gives a brief introduction on the use of SMAs as actuators. In

Figure 4.11 A kind of servovalve-controlled hydraulic actuators. (From https://mts.com/ucm/groups/public/documents/library/dev_002093.pdf.)

Figure 4.12 Prototype of SMA actuator within a piston of fluid damper. (From Li, Z.Q. et al., *Smart Mater. Struct.*, 15(5), 1483–92, 2006.)

general, there are three common ways of using SMAs as actuators: extension wires, benders and helicoidal springs (Pons 2005).

By using SMAs in a straight fashion to form an actuator, extension wires are most appropriate when high forces at low displacements are required. Although the maximum recoverable strain can be up to 8% for extension wires, the strain for normal operation is often restricted to 3%–4% to improve their fatigue behaviours. Depending on the cross-sectional area of the actuator, the maximum applicable stress producing recoverable strain is often of the order of 150–250 MPa. Li et al. (2006) used the SMA actuator to make adjustable fluid dampers, as shown in Figure 4.12. Shape memory benders are able to provide moderate displacements at very low forces. When compared with extension wires, the stroke for SMA bending actuators is much larger. This is achieved at the expense of exhibiting virtually zero force. Therefore, these actuators are suitable for applications where a net displacement rather than a net force delivery is required. In practice, the displacement in a bending actuator is usually limited by the maximum strain at the outermost edge of the cross section. If the strain limit is 3%–4%, the ratio of bending radius to wire radius must be limited to approximately 30–25. Helicoidal springs are selected wherever large deformations at relatively low forces are required. They provide the largest stroke among all types of SMA actuators. Moreover, if the spring dimension is properly designed, almost any stroke requirement can be met by such actuators. Actuators of this type are particularly appropriate for developing the concept of thermal actuators; that is, they are best designed to provide some control action in response to environmental temperature changes rather than being included in motion control systems with resistive heating. More detailed information on the SMA actuators, including their basic characteristics, design concepts and their applications, can be found in Waram (1993) and Pons (2005).

4.4.4 Magnetostrictive actuators

The basic characteristics of magnetostrictive (MS) materials were introduced in Chapter 2. This section provides a brief introduction to MS actuators made mainly based on the Joule effect. Terfenol-D is the most useful MS material for making actuators. Such material has a number of notable properties, including

1. A high magneto-mechanical coupling, which enables the efficient conversion of magnetic to mechanical energy for generating great forces
2. A high load-bearing capability

3. High compressional strength
4. Durability under static and dynamic loading
5. Low voltage operation
6. High reliability and unlimited life cycle (Moon et al. 2007)

Significant effort has been made on the design of MS actuators, and various actuator prototypes have been developed (Monaco et al. 2000; Stillesjo et al. 2000; Bartlett et al. 2001; Zhang et al. 2004). Moreover, to effectively capture the hysteresis behaviour of MS actuators, a number of hysteresis models, including physics-based models that, on the one hand, are built on the first principle of physics and phenomenological models that, on the other hand, are used to produce behaviours without necessarily providing physical insight into the problems, have been developed as well. Major attention has been given to the modelling of MS rods, accounting for eddy currents and hysteresis (Venkataraman et al. 1999; Davino et al. 2004; Bottauscio et al. 2008a; Stuebner et al. 2009), but several approaches are also devoted to the modelling of the entire MS devices (Besbes et al. 2001; Bottauscio et al. 2008b).

MS actuators are complex structures requiring careful design with the consideration of several important aspects, such as force, stroke, electric and magnetic circuits. MS actuators can be employed in civil structures by taking the place of passive structural elements (braces or cables). Focusing on the implementation of MS actuators in the context of vibration suppression, they can work in two major modes: stiffeners and dampers. It is known that the Young's modulus for MS material is not constant but is rather a function of the magnetisation state. Therefore, the stiffness provided by MS actuators is controllable if the applied magnetic field is properly adjusted. With this controllable stiffness, the resonance frequency of the overall structure, in which the MS actuator is integrated, can be modified to prevent resonance amplification when subject to external excitations. The MS actuators can also be utilised as dampers in active damping control, in which the objective is to increase the damping of specific vibration modes of the concerned structure.

Xu and Li (2006) explored the possibility of using a hybrid platform to ensure the functionality of high-tech equipment against microvibration under normal working conditions and to protect high-tech equipment from damage when an earthquake occurs. The hybrid platform, on which high-tech equipment is installed, is designed to work as either a single-layer or double-layer passive isolation platform to abate mainly the acceleration response of high-tech equipment during an earthquake and to function as an MS actuator–controlled platform to reduce mainly the velocity response of high-tech equipment under normal working conditions. Figure 4.13 shows a schematic diagram of a Terfenol-D MS actuator.

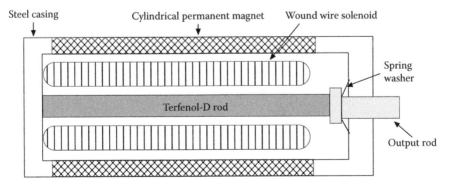

Figure 4.13 Schematic diagram of a Terfenol-D magnetostrictive actuator. (From Xu, Y.L. and Li, B., *Earthq. Eng. Struct. Dyn.*, 35(8), 943–67, 2006.)

4.4.5 Piezoelectric actuators

A piezoelectric (PZT) actuator is able to convert an electrical signal into a controllable physical movement. If this movement is prevented, a usable force will then be generated. PZT actuators have the advantage of precisely controlled movement, compactness, the ability of high force generation, quick response time and no electromagnetic interference. Disadvantages, however, include the fact that PZT actuators have small strains (often only 0.1%–0.2%) and exhibit large hysteresis and creep. The basic types of PZT actuators include stack actuator and stripe (or bending) actuator.

Stack actuator is perhaps the easiest way to produce a linear motion on the basis of the PZT effect (see Figure 4.14a). It is a multilayered structure and each stack is composed of several PZT layers. The required dimensions of the stack can be determined from the requirements of the application in question. The height is determined in respect of the desired movement and the cross-sectional area in respect of the desired force. Typically, a 100 mm long stack with a cross-sectional area of 1 cm^2 provides a free stroke of 100 μm and a blocked force of about 3 kN (Song et al. 2006). It should be noted that in addition to the desired longitudinal movement, some lateral movement usually occurs as well.

A stripe PZT actuator, also called a *bending actuator*, is designed to produce a relatively large mechanical deflection in response to an electrical signal (see Figure 4.14b). This deflection offers a large stroke and a very limited blocking force when compared with a stack actuator. Some benders have only one PZT layer on top of a metal layer (un-bimorph), but generally there are two PZT layers and no metal (bimorph). The basic principle of the bending actuator with two PZT layers is that with the applied electric field, one layer is forced to expand whereas the other one contracts. In this way, the curvature is doubled in comparison to a single layer version. If the number of PZT layers exceeds two, the bender is referred as a multilayer. With thinner piezo layers, a smaller voltage is required to produce the same electric field strength, and therefore, the benefit of the multilayer benders is their lower operating voltage. Bimorph and multilayer benders can be built into either a serial bender or a parallel bender. The major difference between these two types of connections is that there are two electrodes for a serial bender but three electrodes for a parallel bender. The additional electrode in a parallel bender is located between the two parallel-polarised PZT layers and used to receive the control signal.

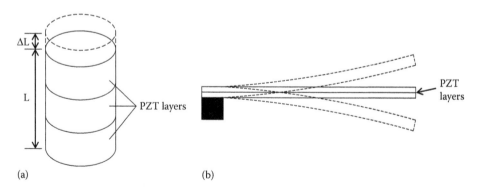

Figure 4.14 Schematic diagrams of piezoelectric actuators: (a) stack actuator and (b) stripe (or bending) actuator.

4.4.6 Active mass damper devices

As mentioned before, TMDs are only effective with a specific tuned frequency (usually the first natural frequency), and therefore is mainly utilised for mitigating wind-induced structural vibration. The development of active mass dampers (AMDs) focuses on seeking seismic vibration suppression with a wide frequency band. The major difference between the aforementioned TMDs and AMDs is that an additional actuator is placed between the main structure and the auxiliary device (i.e. TMD) to augment the control performance (see Figure 4.15). The motion of the auxiliary device would be controlled by such an actuator. Extensive research focuses on investigating the optimum control law to find the appropriate feedback gain of the AMD for suppressing the primary structural vibration in the most effective way (Chang and Soong 1980; Nishimura et al. 1992). Shaking table tests and full-scale implementation of the AMD device can also be found in the literature (Chung et al. 1988; Soong 1990; Soong and Spencer 2002).

The actuator in AMDs is used to drive the auxiliary mass for vibration control, whereas the actuators in other active systems are usually providing controlled force directly to the main structure. From this point of view, AMDs have an economic advantage in full-scale structures because far less control force and a much smaller actuator are required than for other active systems (Cheng et al. 2008). However, the control effectiveness of the AMD device is felt mainly around the fundamental frequency and less so at higher frequencies (Xu 1996).

4.4.7 Active tendon devices

An active tendon control device often includes two coupling parts: a set of pre-stressed tendons and actuators. A basic configuration of active tendon control device is schematically shown in Figure 4.16. Such a control device is usually installed on the inter-story of a building structure. The actuator is located on the lower floor, whereas the active tendons are installed between two stories. The two ends of the tendon are, respectively, connected to

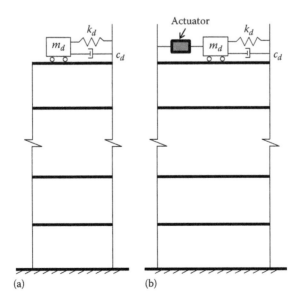

Figure 4.15 Schematic comparison of TMD and AMD: (a) TMD and (b) AMD.

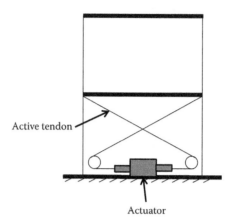

Figure 4.16 Schematic diagram of active tendon control device.

the upper floor and actuator piston. When the structure is subject to earthquake excitation, the inter-story shift induces a relative movement in the tendon device and alters the tension of the pre-stressed tendon accordingly, thus providing dynamic control force for structural vibration mitigation.

Since active tendons can operate in the pulsed and continuous-time modes, both the pulsed and the continuous-time control algorithms can be employed for active tendon control. The advantage of an active tendon device is that such a device can make use of existing structural members, which means few modifications or additions are made.

4.4.8 Active brace devices

The two primary components of an active brace control device are a bracing system and a control device (i.e. actuator). A bracing system mainly includes three types: diagonal, K-braces and X-braces. For the control device, a servovalve-controlled hydraulic actuator is capable of providing a large control force and is often utilised in such an active system. A schematic diagram of an active brace control system with a hydraulic actuator mounted on a K-brace is given in Figure 4.17. The cylinder and piston of the actuator are, respectively, connected to the structural floor and the brace. When the building structure is under earthquake excitation, structural responses are measured by the corresponding sensors and utilised as feedbacks in the predetermined control algorithm. According to the control signal obtained from the algorithm, the control force is then generated by the hydraulic actuator

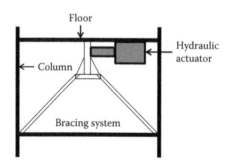

Figure 4.17 Schematic diagram of active bracing system with hydraulic actuator.

for vibration attenuation. Similar to active tendon devices, the active brace device can be installed without too many modifications of the structure.

4.4.9 Pulse generation devices

In pulse generation devices, a pulse generator, which uses a pneumatic mechanism to generate an active control force, is employed instead of a conventional hydraulic actuator. Different from the utilisation of high-pressure fluid in a hydraulic actuator, the actuation force in a pulse generator is produced by compressed air. Since it is a pulse-type force, a trigger is generally required to determine the time for providing control force. For example, when a large relative velocity is detected exceeding the predetermined threshold, the pneumatic actuator is triggered and a control force opposite to the velocity is generated. A drawback of such a control device is that the energy obtained from the compressed gas may not be powerful to drive a full-scale building structure. Moreover, the high nonlinearity derived from compressed gas in the pneumatic actuator makes it difficult to analyse and evaluate.

4.5 SEMI-ACTIVE CONTROL DEVICES

A semi-active control system generally originates from a passive control system, which has been subsequently modified to allow for the adjustment of mechanical properties. According to currently accepted definitions, a semi-active control device is one that cannot inject mechanical energy into the controlled structural system but has properties that can be controlled to optimally reduce the responses of the system (Housner et al. 1997). Consequently, in contrast to active control devices, semi-active control devices do not have the potential to destabilise the structural system. A semi-active damper system consists of sensors, a control computer, a control actuator and a passive damping device. The excitation and/or the corresponding structural responses are measured by sensors and then processed by a control computer based on a predetermined control algorithm. A control signal will subsequently be sent to the actuator for adjusting the behaviour of the passive device. Notably, the actuator is used to control the behaviour of the passive damper instead of directly providing force onto the structure. Therefore, the external energy required for a semi-active damper device is usually orders of magnitude smaller than for active devices, for example, many can operate on battery power. This is one of the most attractive features of semi-active control systems, because during seismic events the main power source to the structure may fail but such a system could still work in this situation. Moreover, preliminary studies show that appropriately implemented semi-active systems are capable of performing better than passive devices, and have the potential to achieve the performance of fully active systems (Dyke et al. 1996a). Comprehensive literature reviews on semi-active control systems are available (Housner et al. 1997; Symans and Constantinou 1999; Spencer and Nagarajaiah 2003; Jung et al. 2004; Fisco and Adeli 2011). This section gives only a brief description of commonly used semi-active control devices.

4.5.1 Semi-active friction dampers

Semi-active friction dampers utilise forces generated by surface friction to dissipate vibratory energy in a structural system. By using electromechanical actuator, Akbay and Aktan (1991) developed a semi-active friction damper that consists of a preloaded friction shaft rigidly connected to the structural bracing. The basic principle of such a system is the friction force used for energy dissipation is proportional to the normal force, which means the damping capacity of the damper could be controlled by adjusting the normal force to the friction interface. The actuator piston is driven by an electric motor to provide normal force. By regulating the normal compression force through an optimal control algorithm, a better control performance could be achieved. A similar device was also investigated at the University of British Columbia (Dowdell and Cherry 1994a,b) but two other control schemes were used.

By using PZT actuators to provide the controllable normal force, Chen and Chen (2004) designed a semi-active friction damper and installed it on the first story of a quarter-scale, three-story building model mounted on a shake table. Such a PZT friction damper mainly consists of PZT stack actuators, preloading units, friction component and a steel box housing other components. When the structure is subject to earthquake excitation, the friction force is generated through the relative movement of the bottom plates and the isolation plate and thus the energy is dissipated. According to the predetermined control algorithm, a controllable friction force can be generated to improve the control performance by adjusting the electric field on the PZT actuators. Xu et al. (2001) explored the possibility of incorporating semi-active friction dampers into wind-excited large truss towers to abate excessive vibration. PZT materials were used to control the clamping force of a friction damper so as to regulate the slip force of the damper through simple feedback of damper motion. A rational analytical method was also developed for determining the wind-induced dynamic response of large truss towers equipped with semi-active friction dampers and performing extensive parametric studies to evaluate their effectiveness. Ng and Xu (2007) and Xu and Ng (2008) also investigated numerically and experimentally seismic protection of a building complex using semi-active friction dampers (see Figure 4.18).

4.5.2 Semi-active hydraulic dampers

The basic principle of a semi-active hydraulic damper is to use a controllable, variable-orifice valve to realise the modification of resistance of a passive hydraulic damper. Semi-active hydraulic dampers are also called semi-active vibration absorbers. Such a damper

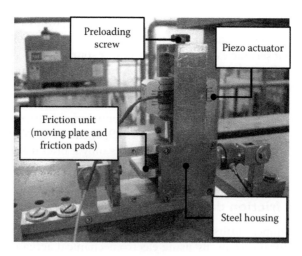

Figure 4.18 Laboratory-scale semi-active friction damper. (From Xu, Y.L. and Ng, C.L., *J. Eng. Mech.*, 134(8), 637–49, 2008.)

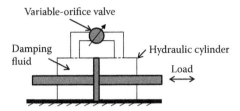

Variable-orifice valve

Damping fluid

Hydraulic cylinder

Load

Figure 4.19 Schematic diagram of semi-active fluid dampers.

typically contains a piston within a cylinder (see Figure 4.19). When the piston is moved, the fluid in the damper is forced to pass through small orifices at high speed. A semi-active hydraulic damper is capable of adjusting both damping and stiffness. It is able to work as a stiffness spring when the valve is closed. However, when the valve is open, the damping fluid can easily flow through the orifices and thus little stiffness would be provided. The damping capacity of the damper can be controlled by adjusting the opening extent of the valve. The control valve may take the form of a solenoid valve for on-off control or a servovalve for variable control. The effectiveness of such a damper in controlling seismically excited buildings and bridges has been demonstrated through both simulation and experimental studies by many researchers (Kawashima et al. 1992; Sack et al. 1994; Symans et al. 1994; Patten et al. 1994, 1996; Kurino et al. 1996; Symans and Constantinou 1997).

4.5.3 Semi-active tuned liquid dampers

The semi-active TLD and TLCD can provide a controllable damping force by simply regulating the behaviour of liquid in operation. Lou et al. (1994) added a set of rotatable baffles in the liquid tank of a sloshing TLD and the control performance could then be improved by adjusting the orientation of these baffles. The liquid tank maintains its original length with the baffles in the horizontal position, whereas the liquid tank is divided into several shorter tanks with the baffles in the vertical position. Thus, by adjusting the rotatable baffles to a desired inclined position, the length of the hydraulic tank is modified accordingly, resulting in variations of natural frequencies of the contained sloshing fluid.

A semi-active TLCD was also developed by using a variable orifice in a TLCD to enhance the control performance. It is known that damping in the TLCD is introduced as a result of the head loss experienced by the liquid column moving through an orifice. In the semi-active TLCD systems, based on the predefined control algorithm, an electro-pneumatic actuator can be utilised to drive a ball valve to change the valve opening angle, leading to variations of the valve orifice area. Therefore, the damping characteristics of the control system can be altered accordingly to achieve better control performance. Research on the investigation of the control performance of the semi-active TLCD systems under seismic and wind loads has been conducted for many years (Kareem 1994; Yalla et al. 1998). Yalla and Kareem (2000) discussed an optimal damping of a controllable TLCD in which the head loss coefficient is changed adaptively in response to excitation variations, and validated the effectiveness of such a system by a shaking table test (Yalla and Kareem 2003). Moreover, based on the clipped-optimal and fuzzy control strategies, the effectiveness of different control algorithms for semi-active TLCD was also investigated by Yalla et al. (2001). Shum et al. (2006) investigated semi-active tuned liquid column damper (SATLCD) with adaptive frequency tuning capacity for buffeting the response control of a long-span cable-stayed bridge during construction. The frequency of the SATLCD is adjusted by the active control of air pressures inside the air chamber at the two ends of the container, as shown in Figure 4.20.

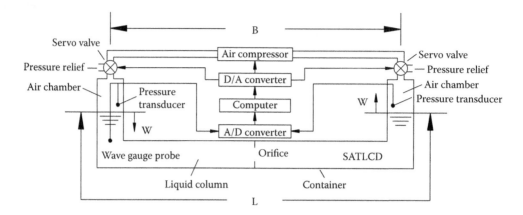

Figure 4.20 Schematic diagram of semi-active tuned liquid column damper. (From Shum, K.M. et al., *Wind Struct.*, 9(4), 271–96, 2006.)

4.5.4 Semi-active stiffness control devices

Semi-active stiffness control devices are primarily used to adjust the stiffness and thus the dynamic characteristics of the main structure to minimise the resonant-type structural response during earthquakes. A full-scale semi-active stiffness control device for seismic response suppression has been developed and described by Kobori et al. (1993). This device is composed of a balanced (double-acting piston rod) hydraulic cylinder with a normally closed solenoid control valve inserted in the tube connecting the two cylinder chambers. Based on the measured acceleration at the base of the structure, a feedforward control system is established for the determination of the solenoid valve in either an open or close state. When the valve is closed, the fluid cannot flow and effectively locks the beam to the braces, thus increasing structural stiffness. In contrast, when the valve is open the fluid flows freely and disengages the beam-brace connection, resulting in a decrease of structural stiffness. This control device has already been installed within the chevron bracing system of a full-scale three-story steel structure in Tokyo, and the control performance under two separate earthquakes has also been reported (Kobori et al. 1993). However, with the use of the on-off mode, the aforementioned device cannot vary stiffness continuously. To overcome this drawback, Nagarajaiah (1997) has developed a semi-active stiffness and damping control device, which is able to modify stiffness continuously and smoothly.

4.5.5 Electrorheological dampers

Electrorheological (ER) dampers typically consist of a hydraulic cylinder containing micro-sized dielectric particles suspended within a fluid (usually oil). In the presence of an electric field, the particles polarise and become aligned, thus offering resistance to flow (Symans and Constantinou 1999). By varying the applied electric field, the behaviour of the ER fluids can reversibly change from that of a VF to that of a yielding solid within milliseconds. Adjustment of the electric field can therefore easily regulate the behaviour of the ER fluids.

ER dampers make use of the smart properties of ER fluid to generate a controllable damping force. According to the predetermined control algorithm, the generated damping force can be adjusted by varying the strength of the electric field. Figure 4.21 gives a schematic diagram of an ER fluid damper, which consists of a cylinder containing a balanced piston rod and a piston head. The piston head pushes the ER fluid through an external bypass

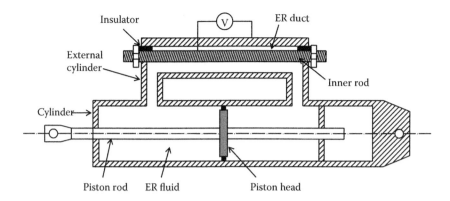

Figure 4.21 Schematic diagram of ER damper.

containing an annular duct. Adjustment of the voltage gradient between the external cylinder and the inner rod can generate a changeable electric field, thus providing a controllable ER damper force for energy dissipation (Makris et al. 1996). Such damping force is developed as a result of both ER effect and viscous effect. Experimental results indicate that as the flow rate increases, the forces due to viscous stresses can dominate over those due to yield stresses. There are three major factors hampering the application of the ER damper for seismic control of large-scale civil structures. First, the yield stress of the ER damper is limited (usually with the maximum value of 5–10 kPa). Second, the capacity of ER fluids is significantly reduced even for common impurities during manufacture. Third, high-voltage power supplies are needed to control ER fluid, and this requirement may result in safety, availability and cost issues. The effectiveness of ER dampers in controlling seismically excited frame structures has been investigated by Xu et al. (2000).

4.5.6 Magnetorheological dampers

Magnetorheological (MR) dampers are essentially magnetic analogues of ER dampers, and typically consist of a hydraulic cylinder containing micron-sized, magnetically polarisable particles suspended within a fluid (usually oil) (Symans and Constantinou 1999). Qualitatively, the behaviour of an MR damper is very similar to the aforementioned ER damper except that the control effect of an MR damper is governed by the application of a magnetic field, whereas an ER damper is governed by an electric field. In the presence of a magnetic field, the fluid behaves as a semisolid, while in the absence of a magnetic field, the MR fluid can flow freely. Under the magnetic field of 150–250 kA/m, MR fluids made from iron particles are able to exhibit a yield strength of 50–100 kPa. By varying the strength of the magnetic field, the characteristics of the MR fluid can be adjusted accordingly, leading to a controllable MR damper force.

A prototype MR damper, having a stroke of ±2.5 cm and a force capacity of 3 kN, was investigated by Spencer et al. (1996, 1997a) and Dyke et al. (1996a). As schematically shown in Figure 4.22, an electromagnet located within the piston head is utilised to generate the magnetic field. The magnetic field is applied perpendicularly to the direction of fluid flow and an accumulator compensates the fluid volume change. Such a damper was installed on the first story of a reduced-scale, three-story steel structure and shaking table tests were performed by Dyke et al. (1996a,b, 1998) for further investigation of the control performance of an MR damper under seismic ground motion. Shaking table tests were conducted by Xu et al. (2005) on a more complex structural system consisting of a 12-story main building

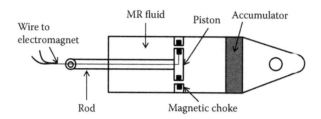

Figure 4.22 Schematic diagram of MR damper.

and a 3-story podium structure to explore the possibility of using MR dampers to connect the podium structure to the main building to reduce the seismic responses of both structures. From a practical application perspective, a large-scale MR damper with a rated force output of 200 kN, a maximum power requirement of about 22 W, a stroke of ±8 cm and a length of about 1 m has also been developed (Carlson and Spencer 1996; Spencer et al. 1997b).

4.6 HYBRID CONTROL DEVICES

A 'hybrid' control device is typically defined as one that is achieved by a combination of passive, active or semi-active control techniques. With two or more control techniques in operation together, hybrid control systems can alleviate some of the restrictions and limitations that exist when each system is acting alone. As a result, they surpass passive, active and semi-active systems, and higher levels of performance may be achievable. Additionally, the hybrid control system can be more reliable, more economic and less external power required as compared with a fully active system, although it is also often somewhat more complicated. Significant attention has been paid to the research on hybrid control systems since the 1990s, and much progress has been made (Yang et al. 1991, 1995; Spencer and Sain 1997; Cheng and Jiang 1998; Yoshioka et al. 2002). In this section, three typical hybrid control systems, that is, the hybrid mass damper (HMD), the hybrid base isolation system and the hybrid bracing control system, will be introduced.

4.6.1 Hybrid mass dampers

The HMD is a widely used control device for full-scale civil structures. It is a combination of a passive TMD and an active control actuator (e.g. using AMD), as schematically shown in Figure 4.23. Since the AMD is placed onto a TMD instead of onto the main structure, the

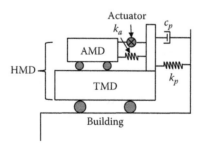

Figure 4.23 Concept of the DUOX HMD system.

mass of the AMD could be small and is often 10%–15% that of the TMD. The structural responses are mainly reduced by the natural motion of the TMD, and the AMD is employed to increase the efficiency and robustness of HMDs. In other words, the TMD is tuned to the structural fundamental mode, whereas the AMD is designed to improve control effectiveness for higher modes of the structure. Therefore, the energy and forces required to operate a typical HMD are far less than those associated with a fully active mass driver system of comparable performance. However, a limitation still exists, in that a relatively large space is required for installing HMD since the TMD is employed. A number of innovative HMDs have been developed and installed on large civil structures, particularly high-rise buildings, for vibration suppression. They include arch-shaped HMD (Tanida et al. 1991; Tanida 1995), V-shaped HMD (Koike et al. 1994), two multi-step pendulum HMDs (Yamazaki et al. 1992) and DUOX HMD (Kobori 1994, 1996; Ohrui et al. 1994; Iemura and Izuno 1994).

4.6.2 Hybrid base isolation devices

A hybrid base isolation device basically consists of a passive base isolation device and an active or semi-active control device, which is used to supplement the effects of the base isolation device. Technically speaking, if the aforementioned various types of active or semi-active control device can be successfully integrated into a passive base isolation device, a new type of hybrid base isolation device would be produced.

Three types of hybrid base isolation device are briefly mentioned herein. The first one consists of a base isolation device connected to an AMD as shown in Figure 4.24a. The performance of this hybrid control device for tall buildings was numerically investigated and evaluated (Yang et al. 1991).

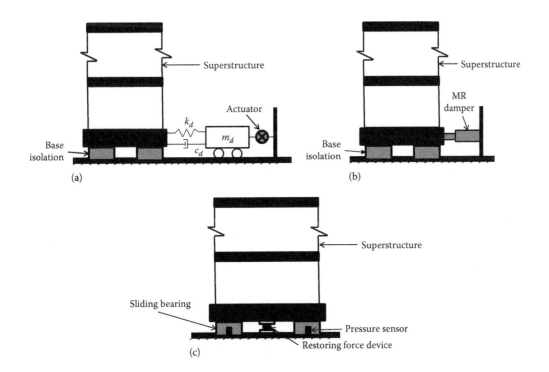

Figure 4.24 Hybrid base isolation devices: (a) base isolation device with AMD, (b) base isolation device with MR damper and (c) hybrid sliding isolation system.

The second one (see Figure 4.24b) uses MR fluid dampers to conduct the hybrid control system (Ramallo et al. 2002). The effectiveness of such a base isolation device was experimentally demonstrated through shaking table tests for both far-field and near-field earthquake excitations (Yoshioka et al. 2002). Yang and Agrawal (2002) presented a similar type of hybrid base isolation system by using two semi-active dampers, that is, resetting semi-active stiffness damper and semi-active electromagnetic friction damper, for seismic protection of nonlinear building structures under near-field earthquakes.

The third type of hybrid base isolation system introduced here can be referred to as a 'hybrid sliding isolation system' as schematically shown in Figure 4.24c. This hybrid system uses a variable friction force that is controlled by a computer which changes the pressure in the fluid chamber of the bearing to enhance the control effect. With the modulation of fluid pressure, the friction coefficient at the sliding bearing interface can be modified, resulting in the generation of a controllable friction force (Feng 1993; Feng et al. 1993). Such semi-active, friction-controllable fluid bearings were further employed by Yang et al. (1995) for reducing the response of seismic-excited bridge structures. Notably, since many base isolation devices are either nonlinear or inelastic or both, nonlinear control strategies developed by many researchers could be employed for consideration of such a hybrid base isolation device, for example, neural network-based control (Venini and Wen 1994), fuzzy control (Nagarajaiah 1994; Symans and Kelly 1999) and adaptive nonlinear control (Rodellar et al. 1994).

Xu et al. (2008) carried out an experimental study to explore the possibility of using a hybrid platform to ensure the functionality of high-tech equipment against microvibration and to protect high-tech equipment from damage when an earthquake occurs. A three-story building model and a hybrid platform model were designed and manufactured. The two-layer hybrid platform, on which high-tech equipment is placed, was installed on the first floor of the building to work as a passive platform, aiming to abate the acceleration response of the equipment during an earthquake, and to function as an actively controlled platform, which tends to reduce velocity response of the equipment under a normal working condition. In the case of the hybrid platform working as a passive platform, it was designed in such a way that its stiffness and damping ratio could be changed, whereas for the hybrid platform functioning as an active platform, a PZT actuator with a sub-optimal velocity feedback control algorithm was used.

A series of shaking table tests, traffic-induced vibration tests and impact tests were performed on the building with and without the platform to examine the performance of the hybrid platform. The experimental results demonstrated that the hybrid platform is feasible and effective for high-tech equipment protection against earthquake and microvibration.

4.6.3 Hybrid bracing control devices

A hybrid bracing control device, as shown in Figure 4.25, was developed by Cheng and Jiang (1998) in the 1990s for the seismic control of building structures. In the hybrid device, an actuator and a damper were mounted on a K-bracing on one floor of a building structure for the consideration of both control objectives and cost saving. Viscous fluid dampers were considered as passive devices for the system, and due to the powerful force-generating capacity hydraulic actuators were employed as the active device. A shaking table test of a three-story structure with this hybrid bracing control device was carried out. For the installation of hybrid bracing control devices, the existing structural braces can be utilised, and thus the active control force can be directly applied to the structure through these braces. The damper and the actuator can either be combined or separated in this device which means it is more flexible for installation.

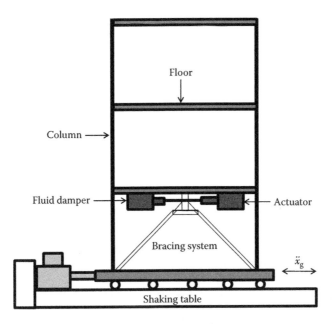

Figure 4.25 Schematic diagram of hybrid bracing control device.

4.7 CONFIGURATION OF CONTROL SYSTEMS AND CONTROL PERFORMANCE

In the previous sections, various dampers/actuators used for passive, active, semi-active and hybrid control of structural vibration were introduced. By using passive control systems, the structural responses are not employed as feedbacks for adjusting the generated damping force, and the properties of the dampers are properly designed according to desired control objectives before the structural construction. For the rest of the control devices, the structural responses are required as feedbacks for the predetermined control algorithms to generate the controllable forces. Therefore, in addition to a group of the aforementioned control devices, various sensors for measuring structural responses, data acquisition and transmission systems, and computers are required together with predefined control algorithms to form a complete structural vibration control system for structural vibration control of a civil structure. The configuration of a complete structural vibration control system will be discussed in Chapter 5.

The control performance of a structural vibration control system for a civil structure is often assessed by the extent of vibration reduction of the structure and the cost of power energy. A series of indices have been proposed for evaluating control performance (Spencer et al. 1998; Ohtori et al. 2004). To achieve good control performance, the number, type and location of the actuators installed in a civil structure shall be selected optimally and the control algorithms shall be designed appropriately. These important yet challenging topics will be discussed in Chapters 9 through 11.

REFERENCES

AASHTO. 2014. *Guide Specifications for Seismic Isolation Design*, 4th edition. American Association of State Highway and Transportation Officials, Washington, DC, https://bookstore.transporta-tion.org/collection_detail.aspx?ID=139

Aiken, I.D. and J.M. Kelly. 1990. Earthquake simulator testing and analytical studies of two energy-absorbing systems for multistory structures. Report No. UCB/EERC-90-03. University of California, Berkeley, CA. https://nisee.berkeley.edu/elibrary/Text/11398

Aiken, I.D. and J.M. Kelly. 1992. Comparative study of four passive energy dissipation systems. *Bull. New Zealand Nat. Soc. Earthq. Eng.*, 25(3): 175–92.

Aiken, I.D., D.K. Nims, A.S. Whittaker, and J.M. Kelly. 1993. Testing of passive energy dissipation systems. *Earthq. Spectra*, 9(3): 335–70.

Akbay, Z. and H.M. Aktan. 1991. Actively regulated friction slip devices. In *Proceedings of 6th Canadian Conference on Earthquake Engineering*, Toronto, Canada, 367–74.

Balendra, T., C.M. Wang, and H.F. Cheong. 1995. Effectiveness of tuned liquid column dampers for vibration control of towers. *Eng. Struct.*, 17(9): 668–75.

Bartlett, P.A., S.J. Eaton, J. Gore, W.J. Metheringham, and A.G. Jenner. 2001. High-power, low frequency magnetostrictive actuation for anti-vibration applications. *Sensors Actuat. A- Phys.*, 91(1–2): 133–36.

Becker, T.C. and S.A. Mahin. 2012. Experimental and analytical study of the bi-directional behaviour of the triple friction pendulum isolator. *Earthq. Eng. Struct. Dyn.*, 41: 355–73.

Besbes, M., Z. Ren, and A. Razek. 2001. A generalized finite element mode of magnetostriction phenomena. *IEEE Trans. Magn.*, 37(5): 3324–28.

Bottauscio, O., M. Chiampi, A. Lovisolo, P.E. Roccato, and M. Zucca. 2008a. Dynamic modelling and experimental analysis of Terfenol-D rods for magnetostrictive actuators. *J. Appl. Phys.*, 103(7): 07F121/1–07F121/3.

Bottauscio, O., A. Lovisolo, P.E. Roccato, M. Zucca, C. Sasso, and R. Bonin. 2008b. Modelling and experimental analysis of magnetostrictive devices: From the material characterisation to their dynamic behaviour. *IEEE Trans. Magn.*, 44(11): 3009–12.

Buckle, I.B. and R.M. Mayes. 1990. Seismic isolation: History, application, and performance – A world view. *Earthq. Spectra*, 6: 161–201.

Carlson, J.D. and B.F. Spencer Jr. 1996. Magneto-rheological fluid dampers for semi-active seismic control. In *Proceedings of Third International Conference on Motion and Vibration Control*, Vol. III, Chiba, Japan, 35–40.

Chang, C.H. and T.T. Soong. 1980. Structural control using active tuned mass dampers. *J. Eng. Mech. Div.*, 106(6): 1091–98.

Chen, C.Q. and G.D. Chen. 2004. Shake table tests of a quarter-scale three-story building model with piezoelectric friction dampers. *Struct. Control Health Monit.*, 11(4): 239–57.

Cheng, F.Y. and H.P. Jiang. 1998. Optimum control of a hybrid system for seismic excitations with state observer technique. *Smart Mater. Struct.*, 7(5): 654–63.

Cheng, F.Y., H.P. Jiang, and K.Y. Lou. 2008. *Smart Structures: Innovative Systems for Seismic Response Control*. CRC Press, Boca Raton, FL.

Chung, L., A. Reinhorn, and T.T. Soong. 1988. Experiments on active control of seismic structures. *J. Eng. Mech.*, 114(2): 241–56.

Clark, A.J. 1988. Multiple passive tuned mass damper for reducing earthquake induced building motion. In *Proceedings of the 9th World Conference on Earthquake Engineering*, 5: 779–84.

Clark, P.W., I.D. Aiken, F. Tajirian, K. Kasai, E. Ko, and I. Kimura. 1999. Design procedures for buildings incorporating hysteretic damping devices. In *International Post-SmiRT Conference Seminar on Seismic Isolation, Passive Energy Dissipation and Active Control of Vibrations of Structures*, Cheju, South Korea. 355–72.

Constantinou, M.C. and M.D. Symans. 1992. Experimental and analytical investigation of seismic response of structures with supplemental fluid viscous dampers. NCEER Technical Report No. 92-0032, State University of New York at Buffalo, Buffalo, NY. http://taylordevices.com/Tech-Paper-archives/literature-pdf/29-ExperimentalAnalytical.pdf

Constantinou, M.C. and M.D. Symans. 1993. Experimental study of seismic response of buildings with supplemental fluid dampers. *Struct. Des. Tall Build.*, 2(2): 93–132.

Datta, T.K. 2003. A state-of-the-art review on active control of structures. *ISET J. Earthq. Technol.*, Paper No. 430, 40(1): 1–17.

Davino, D., C. Natale, S. Pirozzi, and C. Visone. 2004. Rate-dependent losses modeling for magnetostrictive actuators. *J. Magn. Magn. Mater.*, 272–276: E1781–82.

Deb, S.K. 2004. Seismic base isolation: An overview. *Curr. Sci.*, 87(10): 1426–30.

Dowdell, D.J. and S. Cherry. 1994a. Semi-active friction dampers for seismic response control of structures. In *Proceedings of Fifth U.S. National Conference on Earthquake Engineering*, Vol. II, Chicago, IL, 819–28.

Dowdell, D.J. and S. Cherry. 1994b. Structural control using semi-active friction dampers. In *Proceedings of First World Conference on Structural Control*, Los Angeles, CA, FA1-59–FA1-68.

Dyke, S.J., B.F. Spencer Jr., M.K. Sain, and J.D. Carlson. 1996a. Modeling and control of magnetorheological dampers for seismic response reduction. *Smart Mater. Struct.*, 5(5): 565–75.

Dyke, S.J., B.F. Spencer Jr., M.K. Sain, and J.D. Carlson. 1996b. Experimental verification of semi-active structural control strategies using acceleration feedback. In *Proceedings of Third International Conference on Motion and Vibration Control*, Vol. III, Chiba, Japan, 291–96.

Dyke, S.J., B.F. Spencer Jr., M.K. Sain, and J.D. Carlson. 1998. An experimental study of MR dampers for seismic protection. *Smart Mater. Struct.*, 7(5): 693–703.

Feng, M.Q. 1993. Application of hybrid sliding isolation system to buildings. *J. Eng. Mech.*, 119(10): 2090–108.

Feng, M.Q., M. Shinozuka, and S. Fujii. 1993. Friction controllable sliding isolation system. *J. Eng. Mech.*, 119(9): 1845–64.

Fenz, D. and M.C. Constantinou. 2006. Behaviour of the double concave friction pendulum bearing. *Earthq. Eng. Struct. Dyn.*, 35(11): 1403–24.

Fenz, D. and M.C. Constantinou. 2008. Spherical sliding isolation bearings with adaptive behaviour: Theory. *Earthq. Eng. Struct. Dyn.*, 37: 163–83.

Fisco, N.R. and H. Adeli. 2011. Smart structures: Part I – Active and semi-active control. *Sci. Iran.*, 18(3): 275–84.

FitzGerald, T.F., T. Anagnos, M. Goodson, and T. Zsutty. 1989. Slotted bolted connections in aseismic design for concentrically braced connections. *Earthq. Spectra*, 5(2): 383–91.

Fujino, Y. and M. Abe. 1993. Design formulas for tuned mass dampers based on a perturbation technique. *Earthq. Eng. Struct. Dyn.*, 22(10): 833–54.

Griffith, M., I. Aiken, and J. Kelly. 1990. Displacement control and uplift restraint for base-isolated structures. *J. Struct. Eng.*, 116(4): 1135–48.

Grigorian, C.E., T.S. Yang, and E.P. Popov. 1993. Slotted bolted connection energy dissipators. *Earthq. Spectra*, 9(3): 491–504.

Gu, M., S.R. Chen, and C.C. Chang. 2001. Parametric study on multiple tuned mass dampers for buffeting control of Yangpu bridge. *J. Wind Eng. Ind. Aerodyn.*, 89: 987–1000.

Guo, Y.Q. and W.Q. Chen. 2007. Dynamic analysis of space structures with multiple tuned mass dampers. *Eng. Struct.*, 29(12): 3390–403.

Higashino, M. and S. Okamoto. 2006. *Response Control and Seismic Isolation of Buildings*. Taylor & Francis, London, UK.

Hirai, J., H. Harada, H. Abiru, and S. Fukuoka. 1994. Study on the vibration control system for the earthquake applied to the boiler frame. In *Proceedings of the First World Conference on Structural Control*, Los Angeles, CA, 1: WP4-87WP4-96.

HITEC (Highway Innovative Technology Evaluation Center). 1998a. Evaluation findings for Skellerup base isolation elastomeric bearings. Technical Evaluation Report, Civil Engineering Research Foundation, Reston, VA: American Society of Civil Engineers. http://cedb.asce.org/cgi/WWWdisplay.cgi?115608

HITEC (Highway Innovative Technology Evaluation Center). 1998b. Evaluation findings for dynamic isolation systems, Inc. Elastomeric Bearings, Technical Evaluation Report, Civil Engineering Research Foundation, Reston, VA: American Society of Civil Engineers. http://cedb.asce.org/cgi/WWWdisplay.cgi?115604

Housner, G.W., L.A. Bergman, T.K. Caughey, et al. 1997. Structural control: Past, present, and future. *J. Eng. Mech.*, 123(9): 897–971.

Iemura, H. and K. Izuno. 1994. Development of the self-oscillating TMD and shaking table tests. In *Proceedings of the First World Conference on Structural Control*, Los Angeles, CA, WP2: 42–51.

Jangid, R.S. 1995. Dynamic characteristics of structures with multiple tuned mass dampers. *Struct. Eng. Mech.*, 3: 497–509.

Jangid, R.S. and T.K. Datta. 1997. Performance of multiple tuned mass dampers for torsionally coupled system. *Earthq. Eng. Struct. Dyn.*, 26: 307–17.

Jung, H.J., B.F. Spencer Jr., Y.Q. Ni, and I.W. Lee. 2004. State-of-the-art of semi-active control systems using MR fluid dampers in civil engineering applications. *Struct. Eng. Mech.*, 17(3–4): 493–526.

Kareem, A. 1994. The next generation of tuned liquid dampers. In *Proceedings of the First World Conference on Structural Control*, Los Angeles, CA, Vol 1: IASC, 19–28.

Kawashima, K. 1992. Manual for Menshin design of highway bridges. In *2nd US-Japan Workshop on Earthquake Protective Systems for Bridges*, Tsukuba, Japan, 117–39.

Kawashima, K., S. Unjoh, H. Iida, and H. Mukai. 1992. Effectiveness of the variable damper for reducing seismic response of highway bridges. In *Proceedings of Second US–Japan Workshop on Earthquake Protective Systems for Bridges*, PWRI, Tsukuba Science City, Japan, 479–93.

Kelly, J.M. 1986. Aseismic base isolation: Review and bibliography. *Soil Dyn. Earthq. Eng.*, 5(4): 202–16.

Kelly, J.M. 1991. Dynamic failure characteristics of Bridgestone isolation bearings; Report No. UCB/EERC–91/04; Earthquake Engineering Research Center, University of California: Berkeley, CA. http://nisee.berkeley.edu/elibrary/Text/12430

Kelly, J.M. 1996. *Earthquake-Resistant Design with Rubber*, 2nd edition. Springer-Verlag, New York, NY.

Kelly, J.M., R.I. Skinner, and A.J. Heine. 1972. Mechanisms of energy absorption in special devices for use in earthquake resistant structures. *Bull. New Zealand Nat. Soc. for Earthq. Eng.*, 5(3): 63–88.

Kobori, T. 1994. Future direction on research and development of seismic-response-controlled structure. In *Proceedings of the First World Conference on Structural Control*, Los Angeles, CA, Vol. FA2: 19–31.

Kobori, T. 1996. Future direction on research and development of seismic-response-controlled structure. *Comput.-Aided Civil Infrastruct. Eng.*, 11(5): 297–304.

Kobori, T., M. Takahashi, T. Nasu, N. Niwa, and K. Ogasawara. 1993. Seismic response controlled structure with active variable stiffness system. *Earthq. Eng. Struct. Dyn.*, 22(11): 925–41.

Koike, Y., T. Murata, K. Tanida, T. Kobori, K. Ishii, and Y. Takenaka. 1994. Development of V-shaped hybrid mass damper and its application to high-rise buildings. In *Proceedings of the First World Conference on Structural Control*, Los Angeles, CA, FA2: 3–12.

Kunde, M.C. and R.S. Jangid. 2003. Seismic behaviour of isolated bridges: A-state-of-the-art review. *Electron. J. Struct. Eng.*, 3: 140–70.

Kurino, N., T. Kobori, M. Takahashi, et al. 1996. Development and modeling of variable damping unit for active variable damping system. In *Proceeding of Eleventh World Conference on Earthquake Engineering*, Acapulco, Mexico. Paper No. 1521.

Kwon, S.D. and K.S. Park. 2004. Suppression of bridge flutter using tuned mass dampers based on robust performance design. *J. Wind Eng. Ind. Aerodyn.*, 92: 919–34.

Li, C. and A.M. Reinhorn. 1995. Experimental and analytical investigation of seismic retrofit of structures with supplemental damping: Part 2-friction devices. NCEER Technical Report 95-0009, State Univ. of New York at Buffalo, Buffalo, NY.

Li, C.X. 2000. Performance of multiple tuned mass dampers for attenuating undesirable oscillations of structures under the ground acceleration. *Earthq. Eng. Struct. Dyn.*, 29: 1405–21.

Li, Z.Q., Y.L. Xu, and L.M. Zhou. 2006. Adjustable fluid damper with SMA actuators. *Smart Mater. Struct.*, 15(5): 1483–92.

Liu, W., W. He, D. Feng, and Q. Yang. 2009. Vertical stiffness and deformation analysis models of rubber isolators in compression and compression-shear states. *J. Eng. Mech.*, 135(9): 945–52.

Lou, J.Y.K., L.D. Lutes, and J.J. Li. 1994. Active tuned liquid damper for structural control. In *Proceedings of First World Conference on Structural Control*, Los Angeles, CA, TP1-70–TP1-79.

Luft, R.W. 1979. Optimal tuned mass dampers for buildings. *J. Struct. Div.*, 105(12): 2766–72.

Mahmoodi, P. 1969. Structural dampers. *J. Struct. Div.*, 95(8): 1661–72.

Makris, N., S.A. Burton, D. Hill, and M. Jordan. 1996. Analysis and design of ER damper for seismic protection of structures. *J. Eng. Mech.*, 122(10): 1003–11.

Makris, N. and M.C. Constantinou. 1990. Viscous dampers: Testing, modeling and application in vibration and seismic isolation, NCEER Technical Report No. 90-0028, State University of New York at Buffalo, Buffalo, NY. https://ubir.buffalo.edu/xmlui/bitstream/handle/10477/646/90-0028.pdf?sequence=3

Martinez-Rodrigo, M. and M.L. Romero. 2003. An optimum retrofit strategy for moment resisting frames with nonlinear viscous dampers for seismic applications. *Eng. Struct.*, 25(7): 913–25.

Monaco, E., F. Franco, and L. Lecce. 2000. Designing a magnetostrictive actuator by using simple predictive tools and experimental data. In *Proceedings of the Seventh International Conference on New Actuators*, Bremen, 312–17.

Moon, S.J., C.W. Lim, B.H. Kim, and Y. Park. 2007. Structural vibration control using linear magnetostrictive actuators. *J. Sound Vibr.*, 302(4–5): 875–91.

Naeim, F. and J.M. Kelly. 1999. *Design of Seismic Isolated Structures: From Theory to Practice*. Wiley, New York, NY.

Nagarajaiah, S. 1994. Fuzzy controller for structures with hybrid isolation system. In *Proceedings of the First World Conference on Structural Control*, Los Angeles, CA, TA2: 67–76.

Nagarajaiah, S. 1997. Semi-active control of structures. In *Proceedings of Structures Congress XV, ASCE*, Portland, OR, 1574–78.

Nagarajaiah, S., A.M. Reinhorn, and M.C. Constantinou. 1991. Nonlinear dynamic analysis of 3-D-base-isolated structures. *J. Struct. Eng.*, 117(7): 2035–54.

Ng, C.L. and Y.L. Xu. 2006a. Seismic response control of a building complex utilizing passive friction damper: Analytical study. *Struct. Eng. Mech.*, 22(1): 85–106.

Ng, C.L. and Y.L. Xu. 2006b. Seismic response control of a building complex utilizing passive friction damper: Experimental investigation. *Earthq. Eng. Struct. Dyn.*, 35(6): 657–77.

Ng, C.L. and Y.L. Xu. 2007. Semi-active control of a building complex with variable friction dampers. *Eng. Struct.*, 29(6): 1209–25.

Nims, D.K., P.J. Richter, and R.E. Bachman. 1993. The use of the energy dissipating restraint for seismic hazard mitigation. *Earthq. Spectra*, 9(3): 467–89.

Nishimura, I., T. Kobori, M. Sakamoto, N. Koshika, K. Sasaki, and S. Ohrui. 1992. Active tuned mass dampers. *Smart Mater. and Struct.*, 1(4): 306–11.

Ohrui, S., T. Kobori, M. Sakamoto, et al. 1994. Development of active-passive composite tuned mass damper and an application to the high rise building. In *Proceedings of the First World Conference on Structural Control*, Los Angeles, CA, TP1: 100–9.

Ohtori, Y., R.E. Christenson, B.F. Spencer Jr., and S.J. Dyke. 2004. Benchmark control problems for seismically excited nonlinear buildings. *J. Eng. Mech.*, 130(4): 366–85.

Pall, A.S. and C. Marsh. 1982. Response of friction damped braced frames. *J. Struct. Div.*, 108(ST6): 1313–23.

Pall, A.S. and R. Pall. 1993. Friction-dampers used for seismic control of new and existing building in Canada. In *Proceedings of ATC 17-1 Seminar on Isolation, Energy Dissipation and Active Control*, San Francisco, CA, 2: 675–86.

Pall, A.S. and R. Pall. 1996. Friction dampers for seismic control of buildings: A Canadian experience. In *Proceedings of 11th World Conference on Earthquake Engineering*, Oxford, UK, CD-ROM, Paper No. 497.

Park, J. and D. Reed. 2001. Analysis of uniformly and linearly distributed mass dampers under harmonic and earthquake excitation. *Eng. Struct.*, 23: 802–14.

Patten, W.N., C.C. Kuo, Q. He, L. Liu, and R.L. Sack. 1994. Seismic structural control via hydraulic semi-active vibration dampers (SAVD). In *Proceedings of the First World Conference on Structural Control*, Los Angeles, CA, FA2-83–FA2-89.

Patten, W.N., R.L. Sack, and Q.W. He. 1996. Controlled semi-active hydraulic vibration absorber for bridges. *J. Struct. Eng.*, 122(2): 187–92.

Pons, J.L. 2005. *Emerging Actuator Technologies: A Micromechatronic Approach*. Chichester, UK: John Wiley & Sons Ltd.

Ramallo, J.C., E.A. Johnson, and B.F. Spencer Jr. 2002. 'Smart' base isolation systems. *J. Eng. Mech.*, 128(10): 1088–99.

Reinhorn, A.M., C. Li, and M.C. Constantinou. 1995. Experimental and analytical investigation of seismic retrofit of structures with supplemental damping: Part I-fluid viscous damping devices. NCEER Report No. 95-0028, State University of New York at Buffalo, Buffalo, NY.

Robinson, W.H. 1982. Lead-rubber hysteretic bearings suitable for protecting structures during earthquakes. *Earthq. Eng. Struct. Dyn.*, 10(4): 593–604.

Rodellar, J., A.H. Barbat, and N. Molinares. 1994. Response analysis of buildings with a new nonlinear base isolation system. In *Proceedings of the First World Conference on Structural Control*, Los Angeles, CA, TP1: 31–40.

Sack, R.L., C.C. Kuo, H.C. Wu, L. Liu, and W.N. Patten. 1994. Seismic motion control via Semi-active hydraulic actuators. In *Proceedings of Fifth US National Conference on Earthquake Engineering*, Vol. II, Chicago, IL, 311–20.

Shum, K.M. and Y.L. Xu. 2002. Multiple tuned liquid column dampers for torsional vibration control of structures: Experimental investigation. *Earthq. Eng. Struct. Dyn.*, 31(4): 977–91.

Shum, K.M., Y.L. Xu, and W.H. Guo. 2006. Buffeting response control of a long span cable-stayed bridge during construction using semi-active tuned liquid column dampers. *Wind Struct.*, 9(4): 271–96.

Skinner, R.I., J.M. Kelly, and A.J. Heine. 1974. Hysteresis dampers for earthquake-resistant structures. *Earthq. Eng. Struct. Dyn.*, 3(3): 287–96.

Skinner, R.I., W.H. Robinson, and G.H. McVerry. 1993. *An Introduction to Seismic Isolation*. Wiley, New York, NY.

Song, G., V. Sethi, and H.N. Li. 2006. Vibration control of civil structures using piezoceramic smart materials: A review. *Eng. Struct.*, 28(11): 1513–24.

Soong, T.T. 1988. State-of-the-art review: Active structural control in civil engineering. *Eng. Struct.*, 10(2): 74–84.

Soong, T.T. 1990. *Active Structural Control: Theory and Practice*. Longman Scientific & Technical, UK and Wiley, New York, NY.

Soong, T.T. and G.F. Dargush. 1997. *Passive Energy Dissipation Systems in Structural Engineering*. Wiley, Chichester, UK.

Soong, T.T. and B.F. Spencer Jr. 2002. Supplemental energy dissipation: State-of-the-art and state-of-the-practice. *Eng. Struct.*, 24(3): 243–59.

Spencer Jr., B.F., J.D. Carlson, M.K. Sain, and G. Yang. 1997b. On the current status of magnetorheological dampers: Seismic protection of full-scale structures. In *Proceedings of 1997 American Control Conference*, Albuquerque, NM, 458–62.

Spencer Jr., B.F., S.J. Dyke, and H.S. Deoskar. 1998. Benchmark problems in structural control: Part I – Active mass driver system. *Earthq. Eng. Struct. Dyn.*, 27(11): 1127–39.

Spencer Jr., B.F., S.J. Dyke, M.K. Sain, and J.D. Carlson. 1996. Dynamical model of a magnetorheological damper. In *Proceedings of Structures Congress XIV, ASCE*, Chicago, IL, 361–70.

Spencer Jr., B.F., S.J. Dyke, M.K. Sain, and J.D. Carlson. 1997a. Phenomenological model for magnetorheological dampers. *J. Eng. Mech.*, 123(3): 230–38.

Spencer Jr., B.F. and S. Nagarajaiah. 2003. State of the art of structural control. *J. Struct. Eng.*, 129(7): 845–56.

Spencer Jr., B.F. and M.K. Sain. 1997. Controlling buildings: A new frontier in feedback. *IEEE Control Syst.*, 17(6): 19–35.

Stillesjo, F., G. Engdahl, C. May, and H. Janocha. 2000. Design, manufacturing and experimental evaluation of a magnetostrictive actuator for active vibration control and damage analysis. In *Proceedings of the Seventh International Conference on New Actuators*, Bremen, Germany, B 5.1.

Stuebner, M., J. Atulasimha, and R.C. Smith. 2009. Quantification of hysteresis and nonlinear effects on the frequency response of ferroelectric and ferromagnetic materials. *Smart Mater. Struct.*, 18(10): 104019.

Symans, M.D., W.E. Cofer, and K.J. Fridley. 2002. Base isolation and supplemental damping systems for seismic protection of wood structure: Literature review. *Earthq. Spectra*, 18(3): 549–72.

Symans, M.D. and M.C. Constantinou. 1997. Seismic testing of a building structure with a semi-active fluid damper control system. *Earthq. Eng. Struct. Dyn.*, 26(7): 759–77.

Symans, M.D. and M.C. Constantinou. 1999. Semi-active control systems for seismic protection of structures: A state-of-the-art review. *Eng. Struct.*, 21: 469–87.

Symans, M.D., M.C. Constantinou, D.P. Taylor, and K.D. Garnjost. 1994. Semi-active fluid viscous dampers for seismic response control. In *Proceedings of the First World Conference on Structural Control*, Los Angeles, CA, FA4-3–FA4-12.

Symans, M.D. and S.W. Kelly. 1999. Fuzzy logic control of bridge structures using intelligent semi-active seismic isolation systems. *Earthq. Eng. Struct. Dyn.*, 28(1): 37–60.

Tamura, Y. 1995. Effectiveness of tuned liquid dampers under wind excitation. *Eng. Struct.*, 17(9): 609–21.

Tanida, K. 1995. Active control of bridge towers during erection. In *Proceedings of the 3rd Colloquium on Vibration Control of Structures, JSCE*, part A, 173–84. (In Japanese.)

Tanida, K., Y. Koike, K. Mutaguchi, and N. Uno. 1991. Development of hybrid active-passive damper. *Active and Passive Damping, ASME*, PVP-vol. 211: 21–26.

Taylor, A.W. and T. Igusa. 2004. *Primer on Seismic Isolation*. Task Committee on Seismic Isolation, Seismic Effects Committee, Dynamic Effects TA Committee, Structural Engineering Institute (SEI) of ASCE. American Society of Civil Engineers, Reston, VA.

Taylor, A.W., A.N. Lin, and J.W. Martin. 1992. Performance of elastomers in isolation bearings: A literature review. *Earthq. Spectra*, 8: 279–303.

Venini, P. and Y.K. Wen. 1994. Hybrid vibration control of MDOF hysteretic structures with neural networks. In *Proceedings of the First World Conference on Structural Control*, Los Angeles, CA, TA3, 53–62.

Venkataraman, R. 1999. Modeling and adaptive control of magnetostrictive actuators. PhD diss., University of Maryland.

Wada, A., Y.H. Huang, and M. Iwata. 1999. Passive damping technology for buildings in Japan. *Prog. Struct. Eng. Mater.*, 2(3): 335–50.

Waram, T.C. 1993. *Actuator Design Using Shape Memory Alloys*, 2nd edition. Hamilton, Ontario.

Warburton, G.B. 1982. Optimal absorber parameters for various combinations of response and excitation parameters. *Earthq. Eng. Struct. Dyn.*, 10(3): 381–401.

Warn, G.P. and K.L. Ryan. 2012. A review of seismic isolation for buildings: Historical development and research needs. *Buildings*, 2(3): 300–25.

Whittaker, A.S., V.V. Bertero, C.L. Thompson, and L.J. Alonso. 1991. Seismic testing of steel plate energy dissipation devices. *Earthq. Spectra*, 7(4): 563–604.

Wu, J.N., G.D. Chen, and M.L. Lou. 1999. Seismic effectiveness of tuned mass dampers considering soil-structure interaction. *Earthq. Eng. Struct. Dyn.*, 28: 1219–33.

Xu, Y.L. 1996. Parametric study of active mass dampers for wind-excited tall buildings. *Eng. Struct.*, 18(1): 64–76.

Xu, Y.L., J. Chen, C.L. Ng, and W.L. Qu. 2005. Semi-active seismic response control of buildings with podium structure. *J. Struct. Eng.*, 131(6): 890–99.

Xu, Y.L. and K.C.S. Kwok. 1994. Semi-analytical method for parametric study of tuned mass dampers. *J. Struct. Eng.*, 12(3): 747–64.

Xu, Y.L. and B. Li. 2006. Hybrid platform for high tech equipment protection against earthquake and microvibration. *Earthq. Eng. Struct. Dyn.*, 35(8): 943–67.

Xu, Y.L. and C.L. Ng. 2008. Seismic protection of a building complex using variable friction damper: Experimental investigation. *J. Eng. Mech.*, 134(8): 637–49.

Xu, Y.L., W.L. Qu, and Z.H. Chen. 2001. Control of wind-excited truss tower using Semi-active friction damper. *J. Struct. Eng.*, 127(8): 861–68.

Xu, Y.L., W.L. Qu, and J.M. Ko. 2000. Seismic response control of frame structures using magneto-rheological/electrorheological dampers. *Earthq. Eng. Struct. Dyn.*, 29: 557–75.

Xu, Y.L., B. Samali, and K.C.S. Kwok. 1992. Control of along-wind response of structures by mass and liquid dampers. *J. Eng. Mech.*, 118(1): 20–39.

Xu, Y.L. and K.M. Shum. 2003. Multiple tuned liquid column dampers for torsional vibration control of structures: Theoretical investigation. *Earthq. Eng. Struct. Dyn.*, 32(2): 309–28.

Xu, Y.L., Z.F. Yu, and S. Zhan. 2008. Experimental study of hybrid platform for high-tech equipment protection against earthquake and microvibration. *Earthq. Eng. Struct. Dyn.*, 37: 747–67.

Xu, Y.L., S. Zhan, J.M. Ko, and W.S. Zhang. 1999. Experimental investigation of adjacent buildings connected by fluid dampers. *Earthq. Eng. Struct. Dyn.*, 28: 609–31.

Xu, Y.L. and W.S. Zhang. 2002. Closed form solution for seismic response of adjacent buildings with LQG controllers. *Earthq. Eng. Struct. Dyn.*, 31(2): 235–59.

Xue, S.D, J.M. Ko, and Y.L. Xu. 2000. Tuned liquid column damper for suppressing pitching motion of structures. *Eng. Struct.*, 23(11): 1538–51.

Yalla, S.K. and A. Kareem. 2000. Optimal absorber parameters for tuned liquid column dampers. *J. Struct. Eng.*, 126(8): 906–15.

Yalla, S.K. and A. Kareem. 2003. Semi-active tuned liquid column dampers: Experimental study. *J. Struct. Eng.*, 129(7): 960–71.

Yalla, S.K., A. Kareem, and J.C. Kantor. 1998. Semi-active control strategies for tuned liquid column dampers to reduce wind and seismic response of structures. In *Proceedings of Second World Conference on Structural Control*, Kyoto, Japan, 559–68. Wiley, Chichester, UK.

Yalla, S.K., A. Kareem, and J.C. Kantor. 2001. Semi-active tuned liquid column dampers for vibration control of structures. *Eng. Struct.*, 23: 1469–79.

Yamazaki, S., N. Nagata, and H. Abiru. 1992. Tuned active dampers installed in the Minato Mirai (MM) 21 landmark tower in Yokohama. *J. Wind Eng. Ind. Aerodyn.*, 43(1–3): 1937–48.

Yang, J.N. and A.K. Agrawal. 2002. Semi-active hybrid control systems for nonlinear buildings against near-field earthquakes. *Eng. Struct.*, 24(3): 271–80.

Yang, J.N., A. Danielians, and S.C. Liu. 1991. A seismic hybrid control systems for building structures. *J. Eng. Mech.*, 117(4): 836–53.

Yang, J.N., J.C. Wu, K. Kawashima, and S. Unjoh. 1995. Hybrid control of seismic-excited bridge structures. *Earthq. Eng. Struct. Dyn.*, 24(11): 1437–51.

Yang, Z., Y.L. Xu, and X.L. Lu. 2003. Experimental study of seismic adjacent buildings with fluid dampers. *J. Struct. Eng.*, 129(2): 197–205.

Yoshioka, H., J.C. Ramallo, and B.F. Spencer Jr. 2002. 'Smart' base isolation strategies employing magnetorheological dampers. *J. Eng. Mech.*, 128(5): 540–51.

Zhang, T., C. Jiang, H. Zhang, and H. Xu. 2004. Giant magnetostrictive actuators for active vibration control. *Smart Mater. Struct.*, 13(3): 473–77.

Zhang, W.S. and Y.L. Xu. 1999. Dynamic characteristics and seismic response of adjacent buildings linked by discrete dampers. *Earthq. Eng. Struct. Dyn.*, 28: 1163–85.

Zhang, W.S. and Y.L. Xu. 2000. Vibration analysis of two buildings linked by Maxwell model-defined fluid dampers. *J. Sound Vibr.*, 233(5): 775–96.

Zuo, L. and S.A. Nayfeh. 2005. Optimization of the individual stiffness and damping parameters in multiple-tuned-mass-damper systems. *J. Vibr. Acoust.*, 127(1): 77–83.

Chapter 5

Processors and processing systems

5.1 PREVIEW

Signal processors deal with operations on, or analyses of, analogue as well as digital signals, which represent time-varying or spatially varying physical quantities of a smart civil structure, from the sensors and sensory systems (SS) installed in the structure. Signal processors are also responsible for the realisation of control algorithms for structural vibration control. The configuration of a signal processing system depends on the functions expected from a smart civil structure. This chapter first introduces the configurations of a structural health monitoring system, a structural vibration control system and an integrated system for both structural health monitoring and vibration control. The major components involved in the three systems are then presented. The common components for the three systems include SS, data acquisition and transmission systems (DATS), data management systems (DMS) and structural evaluation systems (SES). The specific component is the data processing system (DPS) for the structural health monitoring system, the controller system (CS) for the structural control system, and the combined data processing and controller system (DPCS) for the integrated system for both structural health monitoring and vibration control. Wireless sensors and systems, such as integrated sensing, transmission and processing systems, are described in detail in this chapter. Finally, power supply and energy harvesting as elements of a smart civil structure are commented upon briefly. The high-fidelity design of any one of the foregoing systems requires an integrated analytical/numerical and experimental approach, and therefore this chapter shall be read together with the subsequent chapters in Part II of this book.

5.2 CONFIGURATION OF A STRUCTURAL HEALTH MONITORING SYSTEM

The basic components of a complete structural health monitoring system are illustrated in Figure 5.1, and include SS, DATS, DPS, DMS and SES. The following description of the basic components of a structural health monitoring system is based on the work presented in the book written by Xu and Xia (2012). The SS is composed of various types of sensors, as described in Chapter 3, which are distributed over the structure to capture different signals of interest, such as various types of loadings and structural responses. The DATS is responsible for collecting signals from the sensors and transmitting the signals to the central database server. The DPS is devised to control the data acquisition and transmission, and process, store and display the data. The DMS comprises the database system for temporal and spatial data management. In accordance with monitoring objectives, the SES may have different applications. It may include an on-line structural condition evaluation system

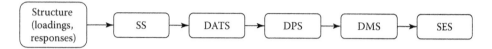

Figure 5.1 Configuration of a structural health monitoring system.

and/or an off-line structural health and safety assessment system. The former (on-line) is mainly to compare the measurement data with the design values, analytical results and predetermined thresholds and patterns to provide a prompt evaluation of the structural condition. The latter (off-line) incorporates varieties of model-based and data-driven algorithms, for example, loading identification, system identification, model updating and damage diagnosis and prognosis.

It can be seen from Figure 5.1 that each component or module is a stand-alone system. This implies that failure of an individual module will have no detrimental effect on the operation of the remaining parts of the entire system (Wong 2004). This concept must be considered during the design of structural health monitoring systems. For example, if the data server in the control office does not work, this should not affect the data acquisition on the structure. This requires the data acquisition system to have capacity for on-board data storage for a specific duration, and the DMS to have a data retrieval function after the server recovers.

The design of a structural health monitoring system requires the designers to understand the needs of monitoring, the characteristics of the structure, environment conditions, hardware performance and economic considerations. First, the designers should have a clear understanding of the structure itself and the objectives of the monitoring. Different structures have different characteristics and unique demands. The designers of the structural health monitoring system should work together with the designers of the structure and know their main concerns. Next, the monitoring items and the corresponding information should be identified, which may include (Aktan et al. 2002)

- The parameters to be monitored, such as temperature, wind, displacement and corrosion
- The nominal value and expected ranges of the parameters
- The spatial and temporal properties of the parameters, for example, the location and variation speed of the measurands
- The accuracy requirement
- The environmental condition of the monitoring
- The duration of the monitoring

After the monitoring parameters are identified, the number and type of sensors should be determined according to the size and complexity of the structure and the monitoring objectives. The types of sensors have been discussed in detail in Chapter 3. They can be selected so that their performance meets the requirements of the monitoring. Important sensor performance characteristics include measurement range, sampling rate, sensitivity, resolution, linearity, stability, accuracy, repeatability, frequency response, durability and so forth. In addition, sensors must be compatible with the monitoring environment, such as temperature range, humidity range, size, packaging, isolation and thermal effect. The optimal sensor placement and number will be discussed in detail in Chapter 8 with help from high-fidelity numerical simulations of a civil structure. Therefore, the design of a structural health monitoring system is actually an integrated analytical/numerical and experimental approach other than an ad hoc procedure (Farrar and Worden 2013), so too are the structural control system and the integrated system for both structural health monitoring and vibration control.

The data acquisition units (DAUs) in the DATS should be compatible with the sensors. The location and number of DAUs should be determined to trade-off the distance from the sensors to the DAUs and the number of channels of the DAUs. Sampling rate, resolution, accuracy and working environment should be taken into account in the selection of hardware.

Besides this, the designers should consider the budget of the project, availability of hardware, wiring, and installation and protection of the hardware. In practical monitoring projects, wires or cables are more easily damaged artificially or naturally than the sensors themselves. Special protection of sensors and wires is worth the effort. Maintenance is also a factor to be considered during the design stage. Important sensors and DAUs should be accessible for inspection and repair after installation.

A Wind and Structural Health Monitoring System (WASHMS) for the Tsing Ma Bridge in Hong Kong has been designed, installed and operated by the Highways Department of the Government of Hong Kong Special Administration Region since 1997 (Wong 2004). The system architecture of the WASHMS is composed of the following integrated modules: the SS, DATS, DPS, DMS, structural health evaluation system (SHES), portable data acquisition system (PDAS) and portable inspection and maintenance system (PIMS). There are 276 sensors installed on the bridge. The layout of the SS for the bridge is illustrated in Figure 5.2. Further details can be found in the Xu and Xia (2012).

5.3 CONFIGURATION OF A STRUCTURAL VIBRATION CONTROL SYSTEM

The basic components of a structural vibration control system are illustrated in Figure 5.3. Unlike a structural health monitoring system, a structural vibration control system has a CS instead of a DPS. For a passive vibration control system, the CS contains mainly passive control devices. These passive control devices dissipate structural vibration energy by virtue of structural motion only, and they do not need to adapt themselves to structural changes and to varying usage patterns and loading conditions. Therefore, other components of the structural vibration control system may not be necessary. Nevertheless, if other components of the structural vibration control system are retained, the structural performance can be monitored, the passive control performance can be evaluated, and the optimal locations and numbers of passive control devices determined numerically can be verified, among others. For an active, semi-active or hybrid control system, the CS has to process the measured information from the SS and DATS to compute necessary control forces or other quantities based on a given control algorithm and has to command the control devices to make corresponding reactions so that the structure can adapt itself to structural changes and to varying usage patterns and loading conditions.

The measured information can be structural responses and/or external excitations. According to the quantities to be measured and used for vibration reduction, controller systems can be divided into three main groups, including feedback control, feedforward control and feedback-feedforward control. In feedback control, only the structural responses are measured and used to make continual corrections to the applied control forces, whereas in feedforward control, the control forces are regulated only by the measured excitation. In the case where the information on both the response quantities and excitation are utilised for control design, the term *feedback-feedforward control* is used (Suhardjo et al. 1990).

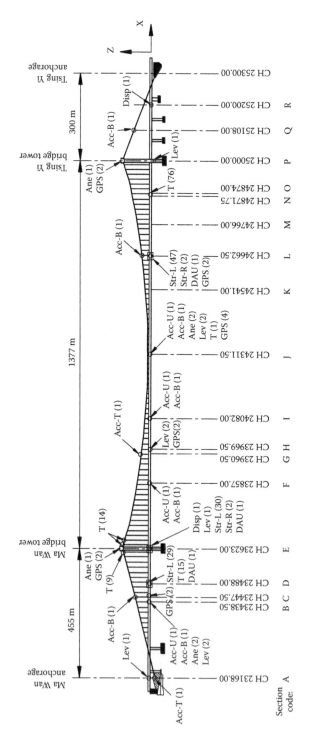

Figure 5.2 Layout of sensors and DAUs of the Tsing Ma Bridge. (1) Number in parentheses are the number of sensors, (2) Lev: Level sensing (9); Ane: Anemometer (6); Acc–U: Uniaxial Accelerometer (4); Acc–B: Biaxial Accelerometer (7); Acc–T: Triaxial Accelerometer (2); Str–L: Linear strain gauge (106); Str–R: Rosette strain gauge (4); T: Temperature sensor (115); Disp: Displacement transducer (2); DAU: Data acquisition unit (3) and (3) Weigh-in-motion sensors are not located on the bridge and not shown in the figure. (From Xu, Y.L. and Xia, Y., *Structural Health Monitoring of Long-Span Suspension Bridges*. Spon Press, 2012.)

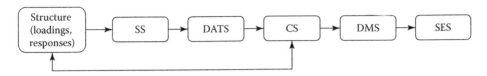

Figure 5.3 Configuration of a structural vibration control system.

Clearly, unlike the structural health monitoring system, in which the measured signals can be analysed and processed off-line, for an active, semi-active and hybrid vibration control system, the controller must analyse and utilise the measured signals immediately during the events. The control force is immediately generated by the actuator. Consequently, the data acquisition and transmission is significantly important for the vibration control system because with inaccurate measurements the control system could not work, or could even work in a detrimental way.

One of the key points for the design of a structural vibration control system is the levels of control objectives: local control or global control. In local control, the objectives can be to augment damping, absorb energy and minimise residual displacements for local problems such as metal fatigue. The objectives of global control can be to stabilise structural response (instability problem), reduce structural response (functionality problem) and avoid collapse (integrity problem). Local control is relatively simple, often with single input and single output. For global control, there are several cases of distributed control. The first one is a centralised controller in which the outputs from all sensors are processed by a centralised processor that provides control outputs to the distributed actuators. The second one is a decentralised controller in which the local control is carried out in an independent manner. However, the decentralised controller is inefficient and uneconomic. Conversely, the computer has to process signals at rates corresponding to the highest mode of interest in the centralised controller. To avoid these issues, a compromise controller is an option, and can feature a centralised controller for overall performance and a distributed processor for local control. This is referred to as a hierarchical or multilevel controller (Chopra and Sirohi 2014).

Another key point for the design of a structural control system is the selection of actuators. Various types of actuators are introduced in Chapter 4 for active, semi-active and hybrid control systems. The actuators can basically be classified into three categories depending on their function: changing stiffness, increasing damping or isolating excitation. The change in structural stiffness can shift the natural frequencies of the structure away from the dominant frequencies of the excitation to avoid resonance. The increase in damping can dissipate more vibration energy to reduce vibration. The isolation of excitation can prevent the propagation of disturbances to sensitive parts of the structure. It is noted from Chapter 4 that some actuators can provide more than one function. Furthermore, a large civil structure may need hundreds of distributed actuators to reduce its vibration. The placement and number of distributed actuators are very important, and will be discussed in detail in Chapter 10. Some applications of active, hybrid and semi-active control technology to civil structures will be presented in Chapter 13.

5.4 CONFIGURATION OF AN INTEGRATED HEALTH MONITORING AND VIBRATION CONTROL SYSTEM

The configuration of an integrated system for both structural health monitoring and vibration control is shown in Figure 5.4. It can be seen that the structural responses and/or dynamic loadings are measured by only one set of SS and one set of DATS for both structural health

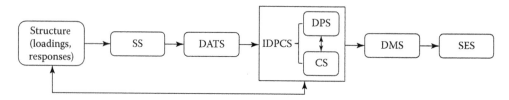

Figure 5.4 Configuration of an integrated system for structural health monitoring and vibration control.

monitoring and vibration control. DMS and SES are also one set only in the integrated system. Nevertheless, an integrated data processing and controller system (IDPCS) is used to override DPS in the structural health monitoring system, and CS in the structural vibration control system.

IDPCS is not a simple combination of DPS and CS. IDPCS couples DPS and CS for more efficient and powerful structural health monitoring and vibration control. For instance, if the dynamic characteristics of a structure are not known accurately or the structural finite element model (FEM) is not well built or updated, the control forces computed according to the control algorithm will not be optimal and the control performance may lose completely. In the integrated system, DPS can be used for system identification to identify the dynamic characteristics of a structure, and if these are used for determining optimal control forces, vibration control performance can be guaranteed or enhanced. This topic will be discussed in detail in Chapter 14. Moreover, Chapter 15 will present a real-time integrated procedure to demonstrate how structural health monitoring and vibration control can be integrated in real time to accurately identify time-varying structural parameters and unknown excitations on the one hand, and to optimally mitigate excessive vibration of the building structure on the other hand.

In the field of civil engineering, the investigation of the optimal placement of sensors often refers to structural damage detection, whereas the study of the optimal placement of actuators always makes reference to structural vibration control. Since the sensory system and actuator system are two tightly interacted parts in the integrated system, it is necessary to develop combinatorial optimisation methods for identifying the locations of both actuators and sensors for integrated structural health monitoring and vibration control of civil structures. This topic will be discussed in detail in Chapter 11.

It is worth pointing out that an integrated structural health monitoring and vibration control system can be extended to include the energy harvesting system, as discussed in Chapters 16 and 17, and the self-repairing system, as investigated in Chapter 18.

5.5 DATA ACQUISITION AND TRANSMISSION SYSTEM

Chapter 3 discussed sensors and SS in detail, and this chapter began with a discussion on DATS. In principle, DATS, as described in Section 5.5.1, can be used for structural health monitoring, structural vibration control and integrated structural health monitoring and vibration control.

5.5.1 Configuration of DATS

Sensors generate analogue signals that represent the physical parameters of a civil structure and its loadings and surrounding environment. Sensor output signals are called

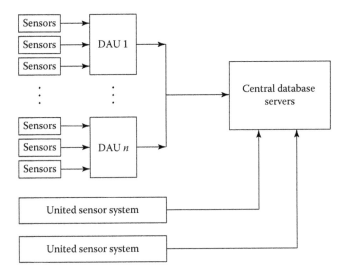

Figure 5.5 Configuration of a data acquisition and transmission system.

measurements. Because the measurements include noises and may be weak, the sensor output signals need to be amplified and/or filtered. The analogue signals also need to be converted into digital signals for further processing by computers. All the aforementioned tasks are implemented by DATS. A DATS is actually an intermediate device between the sensors and computers. It collects the signals generated by the sensors, conditions them, converts them and transmits them to the computers for further processing. For a small laboratory-based experiment, these functions can be achieved using a card-based DAU in a computer. However, the configuration of a DATS for a real smart civil structure is generally complicated. It usually consists of a local cabling network, stand-alone DAUs or substations, and a global cabling network, as illustrated in Figure 5.5. The local cabling network refers to the cables connecting the distributed sensors to the individual DAUs, and the global cabling network refers to the cables connecting the DAUs to central database servers.

Appropriate deployment of DAUs plays a significant role in ensuring the quality and fidelity of the acquired data in a smart civil structure. Long-distance wires cause noise and significant loss in analogue signals (especially for voltage signals) because the distributed sensors may be far from the central control office. It is also inefficient to wire all sensors to one central server. Therefore, DAUs are assigned at a few locations of the structure to collect the signals from surrounding sensors, condition the signals and transmit the digital data into the central database server.

It is also noted that some proprietary sensors, such as global positioning system (GPS), video cameras, fibre-optic sensors and corrosion sensors have their own specific data acquisition systems. These united sensor systems capture corresponding information, transform the information into digital data and directly connect to the central data server for further processing.

5.5.2 Hardware of DAUs

A DAU generally comprises a number of electronic components including signal conditioners, a memory and data storage unit, a microcontroller, a communication device, an uninterruptible power supply, a fan/air conditioner, a lightning conductor and a GPS time synchroniser. All these components are integrated in a waterproof, rugged enclosure or cabinet for the purpose of long-term health monitoring and vibration control.

It is common to have different types of sensors with different output signals and different sampling rates included in one DAU. Consequently, a DAU can facilitate this flexibility and may have more than one signal conditioner. A signal conditioner manipulates an analogue signal such that it meets the requirements of further processing. Signal conditioning usually includes amplification, filtering, analogue-to-digital (A/D) conversion and isolation. Amplification serves to amplify the analogue signal before A/D conversion to utilise the full range of the A/D converter, and thus increases the signal-to-noise ratio and resolution of the input signal. For example, acceleration in an ambient vibration may be very low and the output may be at millivolt level. An amplifier can multiply the voltage signal up to that required by the A/D converter (0–10 V, for instance). The amplifier is preferably placed near to the sensor, but sometimes this placement is difficult to achieve and it is integrated inside the DAU. Filtering is used to remove the unwanted frequency components, for example, alternating current (ac). Most signal conditioners employ low-pass filters. Isolation is used to isolate the possible ground loop and protect the hardware from damage. As the sampling rate in the measurement data of a civil structure is usually low, a single A/D converter can perform A/D conversion by switching between several channels (i.e. multiplexing). This is much less expensive than having a separate A/D converter for each channel, and thus is adopted in most practical structural health monitoring and vibration control systems.

The internal memory serves as a temporary buffer of data for transmission. It is usually integrated with the microcontroller. The data storage unit can save the measurement data relatively longer in case the global cabling network does not work appropriately. The data can be retrieved manually to the external storage devices or automatically to the database server when the global cabling network recovers.

The microcontroller consists of internal electronic circuitry to execute commands sent by the users and to control other hardware components. For example, the sampling rates and acquisition duration of the sensors can be changed by the users.

The communication device is responsible for communication between the DAU and the computer. Usually Ethernet interface is employed.

The power supply provides power to the data acquisition system and to some sensors that require an external power supply. An uninterruptible power supply provides instantaneous or near-instantaneous protection from unexpected power interruption or unstable input voltage.

A fan or air conditioner is used to cool the temperature of the DAU. A lightning conductor can provide protection to the DAU from lightning damage.

Previously, the DAUs were synchronised through a synchronisation signal regularly sent from the central station. Nowadays, GPS time synchronisers have become more popular as they provide an easy way to keep the DAUs and united sensor systems accurately synchronised.

In the DATS, a uniform network communication is crucial to ensure the data can be transmitted over the entire system. Various communication network technologies such as Ethernet, RS-232, RS-485, IEEE-1394 can be employed for the common network.

5.5.3 Network and communication

In a structural health monitoring or vibration control system for a large-scale civil structure, the distance between the DAUs to the central control office may be as far as a few kilometres, in which cases fibre-optic cabling is desirable. In fibre-optic communication, there are basically two types of fibre-optic cables: single mode and multi-mode fibres. Multi-mode fibres generally have a larger core diameter, and are used for shorter distance communication. Single mode fibres are used for communication links longer than 600 m, and are thus preferable for a large civil structure.

Wireless communication and networking have been rapidly developed and employed for data transmission, but its transmission speed and accuracy are still not comparable with that of the cable-based network at the moment. Nevertheless, the wireless network looks promising in the near future. It has advantages in some situations, particularly for construction monitoring when the cable network is not ready. Wireless sensors and SS will be discussed in Section 5.10.

5.5.4 Operation of data acquisition and transmission

After the hardware has been installed, the DATS should be tested or verified through field tests, for example, controlled load tests. This is because the actual performance of the hardware is uncertain under long-term exposure to harsh conditions. Moreover, it is difficult in practice to identify, repair and change the damaged facilities after the structure is put into service, which is usually operated by a different sector from the construction stage. Field verification can help identify problems in hardware, installation, cabling and software, such that the problems can be fixed before normal operation.

During normal operation, data acquisition and transmission are carried out in a systematic and organised manner. Depending on the nature of the monitored parameters, some sensors may work continuously while others may work in the trigger mode, in which the sensor signals are collected only when the parameters are above certain threshold values, usually after some events have occurred (e.g. earthquakes or typhoons). These two modes can be operated simultaneously in one data acquisition system.

The sampling rate (or sampling frequency) is an important factor affecting the data acquisition speed. It relies on the variation speed of the monitored parameters and can be programmed by the users. For example, ambient temperature usually changes slowly with time and can be treated as a static measurand with a low sampling rate. Acceleration, on the other hand, varies more rapidly with time and is usually regarded as a dynamic parameter with a higher sampling rate. If a parameter is not sampled fast enough, which is known as *under-sampling*, the resulting digitised signal will not represent the actual signal accurately. This error is known as *aliasing error*. To avoid this error, the Nyquist criterion requires that the sampling rate should be more than twice the highest frequency component of the original signal of interest. For the low sampling rate, multiplexed sampling rather than simultaneous sampling can be employed. Multiplexed sampling allows different channels to share one A/D converter and to be sampled sequentially. This can reduce cost compared with the simultaneous sampling, in which each signal channel has an individual A/D converter.

After operating for a period in an adverse environment, DAUs are inevitably subject to error or malfunction. It is preferable to carry out periodical calibrations. As mentioned previously, the DAUs should be accessible for maintenance.

5.6 DATA PROCESSING SYSTEMS FOR STRUCTURAL HEALTH MONITORING

The functions of the DPS for structural health monitoring include: (1) control and display of the operation of the data acquisition system; (2) pre-processing of the raw signals received from the data acquisition system; (3) data archive into a database or storage media; (4) post-processing of the data; and (5) viewing the data. Data acquisition control, data pre-processing and data post-processing will be discussed in the following sections. Data archive and display are discussed in Section 5.9 of this chapter.

5.6.1 Data acquisition control

A large-scale structural health monitoring system comprises various types of data acquisition hardware. Therefore, centralised data acquisition control is preferable. As described in Section 5.5.4, the signals can be collected in a long-term or short-term mode. Therefore, the data acquisition control unit should be flexible in handling both continuous monitoring mode and scheduled trigger mode. In practical structural health monitoring systems, the centralised control unit is located in the central control office and is operated by the users carrying out the communication with the local acquisition facilities via a graphical user interface.

The graphical user interface program is an interface between the data acquisition hardware and the hardware driver software. It controls the operation of the DATS, such as how and when the DATS collect data, and where to transmit it. It provides the users with an easy interface as the details of the hardware are very complicated for most users, for example, civil engineers.

5.6.2 Signal pre-processing

Signal quality is determined by the signal's ability to satisfy the requirements of the intended use. There are many factors comprising signal quality, including accuracy, completeness, consistency, timeliness, reliability and interpretability. Thus, signal pre-processing is required for dealing with collected raw signals prior to permanent storage for the purpose of enlarging the storage of useful data in an effective way and being beneficial for further analysis of data in the post-processing process. The data pre-processing has two primary functions: (1) transforming the digital signals into the monitored physical data and (2) removing abnormal or undesirable data. Signal transforming is simply done by multiplying the corresponding calibration factor or sensitivities. There are several data-elimination criteria for removal of typical abnormalities associated with various types of statistical data. The source of abnormal data is possibly derived from malfunction of the measurement instrument. Errors in data transmission can also occur. There may be technology limitations such as limited buffer size for coordinating synchronised data transfer and consumption.

There are a few criteria defining abnormal data. The first criterion is that extremely large or small data are regarded as abnormal. For example, the ambient temperature must be within a certain range for a given place, and an extremely high or low temperature outside this range recorded by the temperature sensor does not have any physical meaning. The second criterion to eliminate the abnormal data is set in terms of the difference between the maximum and minimum values in a specific period. For example, the change in ambient temperature of one day should be within a certain range, and an extremely significant increase or drop in temperature is suspicious. It is also observed that some data from vibration measurements are associated with zero standard deviation. A zero standard deviation physically indicates a steady measurement without any fluctuation within the statistical period. As a result, the statistical values of mean, maximum and minimum have the same magnitude. Having a perfectly flat signal might be considered as an abnormal measurement. Correspondingly, the third criterion to eliminate the abnormal data is set in terms of the zero standard deviation of measured records.

5.6.3 Signal post-processing and analysis

The pre-processed signal will be saved in a database system for future management or onto storage media such as hard disks and tapes after proper packaging and tagging. The stored data are processed for various uses. Here, we only provide examples of a few basic data processing techniques. Other techniques and more advanced analysis can be found in later chapters or other books (Farrar and Worden 2013).

5.6.3.1 Data mining

In a long-term structural health monitoring system, a huge amount of data are recorded from the sensor system. How to extract important features or information is critical to the effective use of structural health monitoring systems for structural condition evaluation. Data mining is a bridge between the data and specific patterns (or features) for decision. Data mining technologies have attracted a great deal of attention in the artificial intelligence community, in which a wider term is called *knowledge discovery* (Fayyad et al. 1996a). Actually, data mining itself can be regarded as a knowledge discovery process. The data mining is defined as 'the nontrivial process of identifying valid, novel, potentially useful, and ultimately understandable patterns in data' (Fayyad et al. 1996b). Data mining mainly has the following several functions (Fayyad et al. 1996a; Duan and Zhang 2006):

- *Regression*: Identify the relationships between a set of variables.
- *Classification*: Classify a data item into one of several predefined classes.
- *Clustering*: Identify a finite set of categories or clusters to describe the data without predefined class labels.
- *Summarisation*: Find a compact description for a subset of data.
- *Outlier detection*: Detect data which do not comply with the general behaviour or model of the data in a database.

A wide variety of data mining methods exist from conventional statistical methods, such as regression analysis, clustering analysis and principal component analysis, to more advanced machine learning methods, such as support vector machine, genetic algorithm, Bayes belief theory, artificial neural networks (ANNs) and colony algorithms. Sohn et al. (2003) reviewed some applications of these techniques to structural damage detection. Here, only the regression analysis is introduced, as it is widely used for preliminary data processing in the context. Interested readers may refer to many textbooks in the area, for example, Kottegoda and Rosso (1997). The regression analysis investigates the relationship between one variable and one or more other variables. The simplest relation is the linear regression as

$$y = \beta_0 + \beta_x x + \varepsilon_y \tag{5.1}$$

where:
x represents the explanatory variable
y is the response variable
β_0 (intercept) and β_x (slope) are regression coefficients
ε_y is the error

With least-squares fitting, the regression coefficients and confidence bounds can be obtained. To examine goodness of fit of the linear relation between x and y, the correlation coefficient, ρ, is defined by

$$\rho = \frac{Cov(x,y)}{\sigma_x \sigma_y} \tag{5.2}$$

where:

σ is the standard deviation

Cov is the covariance

A higher correlation coefficient implies a good linear relation between the two variables. The linear regression can be easily extended to the multiple linear regression where the equation contains more than one explanatory variable.

5.6.3.2 Frequency domain analysis

Frequency domain analysis allows one to examine the data in the frequency domain, rather than in the time domain. It presents the frequency components of a signal and the contributions from each frequency to the signal. The most important concept in the frequency domain analysis is the transformation. Usually, the signal can be converted between the time and frequency domains with a pair of transforms, for example, a Fourier transform and an inverse Fourier transform.

Quite often in structural health monitoring and vibration control, one is interested in the frequency spectrum of loading signals and response signals of a structure to view their frequency components. For example, acceleration responses of a structure can reveal the natural frequencies of the structure. Further, its important vibration characteristics (frequencies, damping and mode shapes) can be obtained via modal analysis tools. In practice, signals are captured at a discrete set of times, say, $1/f_s$, $2/f_s$,..., n/f_s where n is the total number of data and f_s is the sampling rate. Accordingly, the discrete Fourier transform is often used in signal processing, which transforms a series of signal $x(0)$, $x(1)$,..., $x(n-1)$ into n complex numbers as

$$F(j) = \frac{1}{n} \sum_{k=0}^{n-1} x(k) \exp\left(\frac{-i2\pi jk}{n}\right) \tag{5.3}$$

where:

i is the imaginary unit

j $= 0, 1, 2,..., n-1$

and the inverse discrete Fourier transform takes the form

$$x(k) = \sum_{j=0}^{n-1} F(j) \exp\left(\frac{i2\pi jk}{n}\right) \tag{5.4}$$

Equations 5.3 and 5.4 indicate that complex numbers $F(j)$ represent the amplitude and phase of the different sinusoidal components of signal $x(k)$, while signal $x(k)$ is a sum of sinusoidal components. The amplitude or phase of $F(j)$ represents the spectrum of the time series $x(k)$. Due to the symmetric property, usually only the first half spectrum is of interest. It is notable that the jth item is associated with the physical frequency (in hertz) of jf_s/n (or the circular frequency of $2\pi jf_s/n$). The squared amplitude of the Fourier transform, or power, can be obtained as

$$P(j) = |F(j)|^2 \tag{5.5}$$

The resulting plot is referred to as a power spectrum, indicating the averaged power over the entire frequency range. More common in signal processing is power spectrum density, that is, the power component of a signal in an infinitesimal frequency band. According to the Wiener–Khinchin theorem, the power spectrum density is the Fourier transform of the auto-correlation function of the signal (theoretically a random signal does not obey the Dirichlet condition and its Fourier transform does not exist, whereas its autocorrelation function obeys the Dirichlet condition and the Fourier transform is valid). That is

$$S_{xx}(j) = \frac{1}{n} \sum_{k=0}^{n-1} R_{xx}(k) \exp\left(\frac{-i2\pi jk}{n} \right) \tag{5.6}$$

where:
S_{xx} is the auto-power spectrum density
R_{xx} is the autocorrelation function taking the form of

$$R_{xx}(k) = E\left[x(j)x(j-k) \right] \tag{5.7}$$

Similarly, for two discrete signals x and f, their cross-power spectrum can be obtained as the Fourier transform of the cross-correlation function of the two signals. When f and x refer to the input force and output response, respectively, the commonly used frequency response functions (FRFs) can be obtained as

$$H(j) = \frac{S_{fx}(j)}{S_{ff}(j)} \tag{5.8}$$

where S_{ff} and S_{fx} refer to the auto-power spectrum density of the input force and cross-power spectrum density between the input force and response, respectively.

It is recommended that a window is used to minimise the leakage problem during the transform unless the signal is transient and dies away within the record length. The Hanning window function is commonly used, while the exponential window function is suggested for impact testing (Avitabile 2001). The above-mentioned frequency analyses, including the Fourier transform, power spectrum and FRFs, are standard techniques and are available in spectral analysers.

5.6.3.3 Time–frequency domain analysis

An important assumption of the Fourier transform is that the signal is stationary. Some signals in the real world are non-stationary, that is, the signal statistical characteristics vary with time. Examples of this include thunderstorm wind speed signals and earthquake ground motions. Recent advances in the field of signal processing have allowed characterisation of the time–frequency properties of non-stationary signals. There are a few well-known time-frequency distributions and analysis tools, such as short-time Fourier transform (STFT), wavelet transform and Hilbert–Huang transform (HHT).

The STFT computes the time-dependent Fourier transform of a signal using a sliding window (Oppenheim and Schafer 1989). If the time window is sufficiently narrow, each frame extracted can be viewed as stationary so that Fourier transform can be used. The

STFT method thus splits the original signal into overlapping segments and applies the discrete-time Fourier transform to each segment to produce an estimate of the short-time frequency content of the signal over the given time period. With the window moving along the time axis, the relation between the variance of frequency and time can be identified. The two major factors influencing the analysis results of STFT are the size of the selected window and the shape of the selected window. The selection of window size should be carefully conducted with the simultaneous consideration of both time resolution and frequency resolution of STFT, because these two aspects basically act in a conflicting manner. In a general sense, for a short window, the time resolution is high, but the frequency resolution is low. Moreover, it is also important to choose the proper size of overlapping sections for the windows. For instance, a non-zero overlap size usually helps to detect changes between adjacent data frames.

The wavelet transform is a relatively new tool in mathematics and has broad applications (Daubechies 1992). Signal decomposition based on wavelets is very similar to Fourier transforms, which use dilations of sinusoids as the bases. Wavelet functions are composed of a family of basic functions that are capable of describing a signal in a localised time (or space) domain and frequency (or scale) domain. Therefore, using wavelets can perform local analysis of a signal, that is, zooming on any interval of time or space. The wavelet transform can be classified into two broad categories: continuous and discrete. In general, continuous wavelets are better for time–frequency analysis and discrete wavelets are more suitable for decomposition, compression and feature selection. However, it is not always clear which wavelet should be chosen (Staszewski and Robertson 2007).

The HHT was proposed by Huang et al. (1998). It is an adaptive data analysis method designed specifically for analysing data from nonlinear and non-stationary processes. HHT decomposes a signal into a series of intrinsic mode functions (IMFs) with the empirical mode decomposition (EMD), and then uses the Hilbert spectral analysis to obtain instantaneous frequency data. The basic assumption for EMD is that any data consist of different simple oscillatory modes of which each is represented by an IMF. Having obtained the IMF component, the HHT is applied to each IMF component for computing the instantaneous frequency.

5.6.3.4 Data fusion

A structural health monitoring system usually includes various types of sensors located in different spatial positions. Different types of sensors located in the same position may capture different signals. Spatially distributed sensors may also demonstrate different features of the structure. In addition, different methods and different users may reach different results. Therefore, integration of data from different sensors and integration of results made by different algorithms are important to a robust monitoring exercise (Chan et al. 2006). Data fusion is an important data processing tool to achieve this.

Data fusion is a process that integrates data and information from multiple sources to acquire improved information that could not be achieved by using a single source of information alone (Hall 1992). Fusion processes are often categorised as low, intermediate and high level fusion depending on the processing stage at which fusion takes place (Hall and Llinas 1997). The low-level fusion combines raw data from multiple sensors to produce new data that are expected to be more informative and synthetic than the inputs. In the intermediate-level fusion or feature-level fusion, features are extracted from multiple sensors' raw data, and various features are combined into a concatenated feature vector that may be used by further processing. The high-level fusion, also called *decision fusion*, combines decisions coming from several experts to reach a consistent conclusion. A wide range of techniques are involved in feature/decision-level data fusion, including artificial intelligence, pattern recognition, statistical estimation and information theory. Detailed descriptions of these techniques, such as Neural Networks, Bayesian inference and Dempster–Shafer's methods, can be found in corresponding references (Hall and Llinas 1997, 2001; Luo et al. 2002; Esteban et al. 2005; Khaleghi et al. 2013).

5.7 CONTROLLER SYSTEMS FOR STRUCTURAL VIBRATION CONTROL

CSs for structural vibration control are also called *control force generation systems*. In general, a CS contains three major parts: a control device power supplier, a digital control processor (or analogue control console) and a control device, as shown in Figure 5.6 (Chu et al. 2005). The control device power supplier is the source of generating the required control forces. A digital control processor or analogue control console is actually a control command calculator. The primary difference between a digital controller and an analogue controller is the use of a digital control processor as the control command calculator in a digital controller instead of the fixed electronic circuits used in an analogue controller. Owing to the decreasing cost and increasing functionality of computers in both hardware and software, digital controllers have often superseded the analogue controllers. The control command calculator actually first generates digital control signals based on the control algorithm designed. It then converts digital control signals to analogue signals by means of a digital-to-analogue (D/A) converter. The analogue control signals are finally fed to the actuators to generate the control forces with power supply. The calculated command signals are usually different from the outputs actually achieved at the beginning. By employing a feedback control loop, the output measurements of the structural responses from the sensory system and the DATS can be utilised to modify the command signals so as to correct any output errors.

The time lag of the achieved outputs from the command signals can be modelled as a time delay which is unavoidable and should be considered in the theoretical development. In an active control case, to protect the whole system from damage, a fail-safe function, which is able to shut down the control power, is required to monitor the structural response. This control device, as discussed in Chapter 4, mechanically drives a controlled structure. Factors such as power, motion resolution, repeatability and operation bandwidth requirement for a control device can differ significantly, depending on the particular control system. Moreover, the actuator–structure interaction is present in any mechanical control device. In order to include this interaction effect, the response of an actuator can be explicitly modelled and then included in the model of the structural system (Dyke 1996; Riley 1996).

A passive controller system can be defined as a system which utilises the motions of the structure to develop control forces and does not require an external power source for operation. Control forces are developed as a function of the structural response at the location

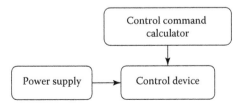

Figure 5.6 Configuration of a controller system.

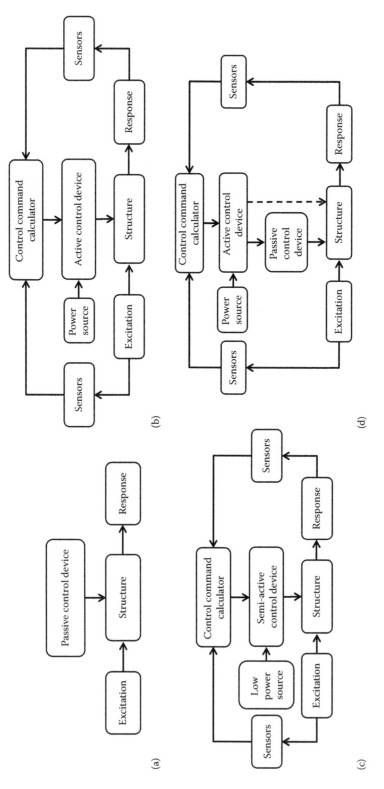

Figure 5.7 Block diagrams of controller systems: (a) passive control system, (b) active control system, (c) semi-active control system and (d) hybrid control system.

of passive control devices without control commands (see Figure 5.7a). Currently available schemes of passive controller systems are very diverse, as discussed in Chapter 4. Passive control systems have proved to be one of the more effective methods in reducing structural vibration. They generally feature four main advantages: (1) they are usually relatively inexpensive; (2) they are inherently stable; (3) they consume no external energy (energy may not be available during a major earthquake); and (4) they could potentially work during any magnitude of earthquake and wind. Although the SS, DATS, DMS and SES for the passive vibration control system are not required in principle, it is preferable to retain these components so that the structural performance can be monitored, the passive control performance can be evaluated and the optimal locations and numbers of passive control devices determined numerically can be verified, among others.

An active controller system can be defined as a system which typically requires a large power source for the operation of electrohydraulic or electromechanical actuators which impart control forces to the structure against external excitations (see Figure 5.7b). The control forces are developed based on signals from the SS that measure the responses of the structure and/or the excitations acting on the structure. The feedback from the structural responses may be measured at locations away from the location of the active controller system, but the DATS can transmit the feedback to the active controller system. The prominent features are the versatility and adaptability of active control systems by means of the adjustable control forces from the actuators based upon feedback information from a measured response of the structure and/or feedforward information from the external excitation.

To actively control the structure, a key task is determining the appropriate control force which counts on control algorithms. The control algorithm is implemented by means of software in the computer for generating control force based on the measurements obtained from the sensors. The control algorithm should achieve the control objective, such as the maximisation of the reduction of structural response with the minimum control energy or control force. However, more reduction of structural response requires more control energy. Hence, a performance index is used in this situation to find a compromise between the need to reduce structural response and the need to minimise control energy. Different quantification of performance indices produces different types of algorithms. Diverse control algorithms are available for active vibration control such as optimal control, modal control, sliding mode control, H_2 and $H\infty$ control, adaptive control and artificial intelligent control. A detailed discussion on these control algorithms can be found in Chapter 9.

The first full-scale implementation of an active control system was the Kyobashi Seiwa Building in Tokyo (Kobori 1994). It is an 11-story building, as shown in Figure 5.8. The active control system consists of two active mass dampers (AMDs) where the primary AMD is used for transverse motion reduction, and the secondary AMD is employed to reduce the torsional motion. Further detailed information can be found in Kobori (1994).

A semi-active controller system can be defined as a system which typically requires a much smaller external power source than active controller systems for adjusting the properties of semi-active control devices. Structural responses and/or external excitations are the feedback and/or feedforward signals for adjustments of control forces (see Figure 5.7c). The acceptance and application of semi-active control systems is growing in recent years primarily because it features advantages of both active and passive control systems. It is promising in offering the reliability of passive control systems while maintaining the versatility and adaptability of active control systems with much lower power requirements, which is a critical design consideration for earthquakes. Examples of semi-active control devices that have been considered to realise semi-active control applications include semi-active hydraulic dampers, semi-active friction dampers, semi-active liquid dampers, semi-active stiffness

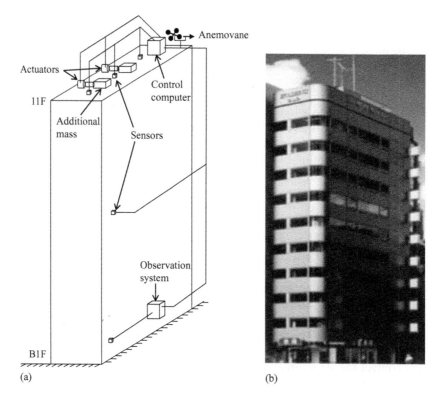

Figure 5.8 Kyobashi Seiwa Building with active mass dampers: (a) a schematic diagram of the configuration of AMD system and (b) a photo of the building. (From Spencer Jr., B.F., *Fourteenth World Conference on Earthquake Engineering,* Beijing, China, 2008.)

control devices and controllable fluid dampers, as discussed in Chapter 4. The semi-active control algorithm is one of the key points for control performance, and will be discussed in Chapter 9.

Semi-active control systems were installed in the Kajima Shizuoka Building in Shizuoka, Japan. As seen in Figure 5.9, semi-active hydraulic dampers were installed inside the walls on both sides of the building to enable it to be used as a disaster relief base in post-earthquake situations. Each damper consists of a flow control valve, a check valve and an accumulator, and each can develop a maximum damping force of 1000 kN. Further detailed information can be found in Kurata et al. (1999) and Kobori (2000).

A hybrid controller system can be defined as a system that employs a combination of passive and active controller systems (see Figure 5.7d). Hybrid control strategies have been investigated to increase the overall reliability and efficiency of the controlled structure. Three major systems in the classification of hybrid control systems are hybrid mass damper systems, hybrid base isolation systems and hybrid bracing control devices, as discussed in Chapter 4. The control algorithms used in the hybrid controller system are similar to those used in the active controller system.

An example of hybrid vibration control is the hybrid mass damper system installed in the Sendagaya INTES building in Tokyo in 1991. As shown in Figure 5.10, the hybrid mass damper was installed on top of the eleventh floor and consists of two masses to control transverse and torsional motions of the structure, while hydraulic actuators provide the active control capabilities. The masses are supported by multi-stage rubber

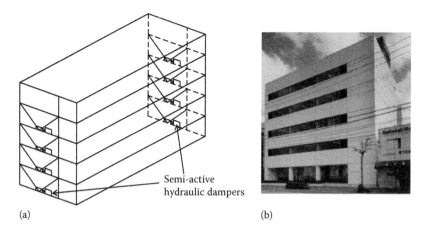

(a) (b)

Figure 5.9 Kajima Shizuoka Building with semi-active hydraulic dampers: (a) the location of semi-active hydraulic dampers and (b) a photo of the building. (From (b) Kobori, T., *Twelfth World Conference on Earthquake Engineering*, New Zealand, 2000.)

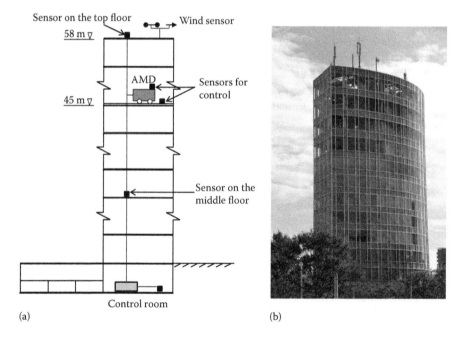

(a) (b)

Figure 5.10 Sendagaya INTES Building with hybrid mass dampers: (a) a schematic diagram of the configuration of hybrid mass damper system and (b) a photo of the building. (From (b) http://www.panoramio.com/photo/62207892.)

bearings intended to reduce the control energy consumed in the hybrid mass dampers, and for ensuring smooth mass movements. Further details can be found in Higashino and Aizawa (1993).

One important consideration in real-time control implementation is time delay and time lag (Chu et al. 2005). In ideal situations, all operations in the control loop can be performed instantaneously. However, in reality, time is required to process measured information, to perform on-line computation and to convert signals between analogue and discrete values. These steps contribute to the time delays in the control feedback loop. Additionally, time lags due to actuator dynamics are often significant and cause unsynchronised application of the control forces if not appropriately accommodated. Improper consideration of these time lags may potentially cause instability in the active control system. Time lags can be accommodated by explicitly incorporating actuator dynamics into the whole system model. Alternatively, time lags may also be modelled as a time delay and, as such, the designer may incorporate them into the discrete-time derivation and compensate for them in the optimisation process.

Another important consideration in real-time control implementation is the discrete-time nature in the application of a control algorithm (Chu et al. 2005). Strictly speaking, continuous-time control algorithms can only be executed in discrete-time since a digital computer is usually used for on-line computation and control execution. As a consequence, response measurements are digitised as feedback signals, and control forces are applied in the form of piecewise step functions. Hence, they are not continuous functions. With this in mind, discrete-time formulation of structural control is an essential requirement. In this formulation, the time delay and sampling period can be considered at the very beginning of the control algorithm. Output feedback can also be accommodated with resulting savings in the number of required sensors.

Optimal control algorithms with full-state feedback are generally used in modern control systems, which will be discussed in Chapter 9. This is because the full-state feedback technique can improve damping at each degree of freedom of the structure and thus effectively reduce structural vibration. However, these algorithms require measurements of all state variables because the control force is generated by feedback of all state variables. In reality, measuring all state variables for effective vibration control is a big challenge for the practical design and construction of large smart civil structures because of the complexity and reliability of the sensory system. One of the ways of solving this problem will be introduced in Chapter 11.

5.8 INTEGRATED DATA PROCESSING AND CONTROLLER SYSTEMS

In an integrated structural health monitoring and vibration control system, an IDPCS is used to override the DPS in the structural health monitoring system and the CS in the structural vibration control system. IDPCS is not a simple combination of DPS and CS. IDPCS couples DPS and CS for more efficient and powerful structural health monitoring and vibration control.

For instance, in vibration control applications, a two-stage model-reduction procedure is generally carried out whereby a civil structure, a distributed-parameter system, is first reduced to a finite-degree-of-freedom system discretised in space as a full-order system. It is then further reduced to a discrete-parameter system with as small a number of degrees of freedom as the reduced-order system (Chu et al. 2005). In practical control design, if analytical and simulated control results are based on the reduced-order system, necessitated by practical limitations as well as computational considerations, modelling errors will result, which may increase the risk of instability. Furthermore, most control developments have been restricted to the consideration of linear structures. In reality, structures will inevitably become nonlinear at some point due to excessive excitation levels. If the nonlinear behaviour of the structure is known, the linear control algorithms can be advanced to deal with nonlinearity. The control laws and the resulting controlled system performance are, in general, functions of structural parameters such as mass, stiffness and damping ratio. In reality, the structural parameters cannot be estimated precisely and their values used in control design may deviate significantly from the actual ones. Thus, parameter uncertainties are another practical concern. With the IDPCS, the structural parameters can be identified, the modelling errors due to the reduced-order system can be quantified and the nonlinear behaviour of the structure can be detected through the structural health monitoring system and the DPS in particular, as demonstrated in Chapters 6 and 7.

From the viewpoint of practicality and economy, the number of sensors and controllers are severely limited for real structural control applications. However, on the other hand, the development of control algorithms has been based on an arbitrary number of controllers and has included the case of an arbitrary number of sensors as long as the structural system is completely controllable and observable. The integration of the structural health monitoring system and structural vibration control system will accommodate the sensor number problem if the response reconstruction technique is used in the sensor placement, as discussed in Chapter 8, and in the collective placement of sensors and controllers, as discussed in Chapter 11.

There are many potential ways to couple DPS and CS for more efficient and powerful structural health monitoring and vibration control, which deserve exploration.

5.9 DATA MANAGEMENT SYSTEMS

The collected data and processed data or results should be stored and managed properly for display, query and further analysis. In addition, relevant information on the structural health monitoring system, structural control system, computational models and design files needs to be documented as well. These tasks are completed by the DMS via a standard database management system, such as MySQL or ORACLE.

5.9.1 Components and functions of data management systems

A standard DMS allows users to store and retrieve data in a structured way so that later assessment is more efficient and reliable. Long-term signal processing systems for large civil structures contain large amounts of different types of data. Therefore, a large signal processing system usually consists of several databases, including a device database, measurement data database, structural analysis data database, structural control performance evaluation data database, health evaluation data database and user data database. The DMS manages each database to fulfil its corresponding function.

The device database records the information on all sensors, actuators and substations. For sensors and actuators, their identification (ID), label, substation, location, specifications, manufacturer, installation time, initial values, sampling rate, thresholds, working condition and maintenance record should be recorded. For substations, their ID, label, sensors, actuators, location, specifications, manufacturer, installation time, working condition, and maintenance record are recorded. These data are necessary to examine the collected measurements in the long term. Various types of sensors and actuators should be labelled properly so that they can be easily identified by users. The DMS has the function of inserting and deleting sensors, actuators and substations, and monitoring their conditions.

The measurement data database records all the data collected from the SS and the CS, including the loads, control forces, structural responses and environmental parameters. For efficiency, the measurement data database usually stores data for a limited period only, for example, one year. The historical data beyond this period are archived in storage media. For data safety, all the measurement data should have a spare backup. A DMS has the function of automatic retrieval and output of the measurement data, and data query from authorised users.

The structural analysis data database records the FEM data, the input parameters of the models and major output data, the design drawings and basic design parameters. For large civil structures, there might be more than one FEM for cross-checking and for different applications. The models can be input into the corresponding analysis software and the output data employed for comparison and evaluation.

The control performance evaluation data database records the excitation, control duration, control force, structural responses, criteria, results and reporting. Once an extreme event occurs, for example, a strong typhoon, a specified evaluation should be carried out immediately. For such an evaluation, the predetermined criteria should be provided.

The structural health evaluation data database records the evaluation time, parameters, objects, criteria, results and reporting. The structural health evaluation may be performed regularly for normal operation. It can also be performed together with the control performance evaluation. Once an extreme event occurs, for example, an earthquake, a specified evaluation should be carried out immediately. For each kind of evaluation, the predetermined criteria should be provided.

The user data database manages the users' information, such as username, ID, user group and personal data and contact information. Different users are assigned different rights by the DMS to log into the structural health monitoring system and/or structural vibration control system.

A DMS should provide security management, which may include network security, data protection, database backup and user operation audit. Finally, a DMS should provide an alarm function. The alarm module can automatically generate warning messages when some predefined criteria are satisfied. Important alarms should be sent to relevant staff through email and short messages until countermeasures are taken.

5.9.2 Maintenance of data management systems

A large signal processing system is operated by authorised staff members who have received basic training in computer technology and civil engineering. When the databases start working, their functions and performance need to be tested. After a period of operation, increases in the size of the databases may cause physical storage malfunctions and reduce their efficiency. Therefore, maintenance of the DMS is necessary. The duties of a DMS administrator include backup and restoration, monitoring and improvement, and reconstruction and reconfiguration of the databases.

5.10 STRUCTURAL HEALTH EVALUATION SYSTEMS

Structural health evaluation system (SHES) can be used regularly during the normal operation of a civil structure based on information from the structural analysis data database and the structural health evaluation data database. The structural health evaluation can also be performed together with the structural control performance evaluation system (SCPES) after an extreme event. The SHES is generally composed of an on-line structural condition evaluation system and an off-line structural health and safety assessment system. The on-line structural condition evaluation system is mainly used to compare the static and dynamic measurement data with the design values, FEM analysis results and predetermined thresholds and patterns to provide a prompt evaluation of the structural condition. The off-line structural health and safety assessment system incorporates varieties of model-based and data-driven damage diagnostic and prognostic algorithms, which mostly require both historical and current monitoring data. The methods and algorithms of system identification, model updating and damage detection, which are discussed in subsequent chapters, are required for health evaluation. Although many SHESs focus on health evaluation after the completion of structural construction, the integration of in-construction monitoring and in-service monitoring seems to be more attractive because of being able to (1) detect anomalies during construction; (2) facilitate the deployment of sensors devised for in-service monitoring; (3) track complete data histories from the onset of construction; (4) enable life-cycle monitoring and assessment of the structure from its 'birth' (Ni et al. 2009). Once an extreme event occurs, for example, an earthquake, a specified evaluation should be carried out immediately. For each kind of evaluation, the predetermined criteria should be provided. It should be noted that the local structural damage in large civil structures is usually difficult to identify in practice using modal properties alone. More reliable structural health evaluation methods are thus required and should combine global modal properties and local information (e.g. strain) through multilevel data fusion techniques and multi-scale health monitoring techniques, all of which will be discussed in Chapters 6, 7, 12 and 19.

One of the most important tasks for a SHES is damage prognosis (DP). DP attempts to forecast system performance by assessing the current damage state of the system (i.e. structural health monitoring), estimating the future loading environments for that system and predicting the remaining useful life of the system through simulation and past experience (Farrar and Lieven 2007; Farrar and Worden 2013). For civil structures, there is a need to perform timely and quantified structural condition assessments and then confidently predict how these structures will respond to future loading such as the inevitable aftershocks that occur following a major seismic event. The DP process begins by collecting as much initial system information as possible, including testing and analyses that were performed during the system design, as well as maintenance and repair information that might be available. Then, based on the data from the operational and environmental sensors, an FEM for structural health monitoring and a future loading model for predicting the future system loading are developed. Finally, the output of the future loading model and FEM will be input into a reliability-based predictive tool that estimates the remaining system life. Note that the definition of 'remaining life' can take on a variety of meanings depending on the specific application, for example, if the system fails or is no longer usable, or the occupants feel uncomfortable. The successful development of a DP capability will require the further development and integration of many technologies including measurement/processing/telemetry hardware and a variety of deterministic and probabilistic predictive modelling capabilities, as well as the ability to quantify the uncertainty in these predictions (Farrar and Lieven 2007). Further discussion on this topic will be presented in Chapter 19.

It is worth noting that a reliable FEM is significantly important for both SHES and SCPES. The FEM to be built in the intact state (e.g. the period just after the completion of construction) could be viewed as a baseline for future dynamic analysis and health evaluation. Although some available data-based structural health monitoring techniques may be able to indicate a change and detect damage through the changes of the measured signals, they are basically performing poorly when trying to classify the nature of the damage without an FEM. Moreover, though uncertainties must be considered in the available predictive techniques, a reliable FEM is still beneficial. Indeed, it is indispensable for DP because the future loading model and FEM are both viewed as input in these techniques. Since the optimal control force is determined based on structural parameters and types, the FEM plays a more important role in SCPES. If the FEM is poorly built , the optimal control force will not be provided and a worse control performance will result. More details on the establishment of a reliable FEM can be found in Chapters 6 and 7.

5.11 WIRELESS SENSORS AND SENSORY SYSTEMS

Advances in micro-electromechanical systems (MEMS) technology, wireless communications and digital electronics have enabled the rapid development of wireless sensor technology since the late twentieth century. This will be discussed here rather than in Chapter 3 because the wireless sensor is actually neither one kind of pure sensing technology nor a new transmission method. Instead it is a new system that can carry out many tasks, including structural health monitoring. A wireless sensory system can comprise all of the components in a wire-based signal processing system described previously, such as SS, DATS, DPS, DMS and SES. However, it has its own unique characteristics when compared with the wire-based signal processing systems.

5.11.1 Overview of wireless sensors

Wireless sensors were developed because a robust system may require a dense network of sensors throughout the system. Traditional SS usually attempts to develop fewer, but increasingly accurate, sensors at optimised locations. A bio-system, however, usually comprises a huge number of distributed sensors, each with limited functions. This philosophy inspires researchers to develop a network of many low-cost small sensors. In addition, the traditional SS is usually wire-based, which has high installation cost. Maintenance of such a monitoring system at a reliable operating level under adverse environment conditions for a long period of time is very difficult. Experiences in monitoring civil structures show that communication wires are more vulnerable to the environment than the sensors themselves. Wireless transmission provides a more flexible communication manner, and sensors can be deployed and scaled easily.

With the support of the US Defense Advanced Research Projects Agency, researchers at the University of California at Berkeley have developed the open platform, well known as 'Berkeley Mote' or 'Smart Dust', whose ultimate goal is to create a fully autonomous system within a cubic millimeter volume (Kahn et al. 1999). Such a system may comprise hundreds or thousands of sensor nodes, each costing as little as about one US dollar.

Berkeley Mote was the first open hardware/software research platform to allow users to customise hardware/software for a particular application. Its first generation was COTS Dust (Hollar 2000), followed by Rene, developed in 2000. The third generation, the Mica, was released in 2001. Subsequent improvements to the Mica platform resulted in Mica2, Mica2Dot and MicaZ. Another commonly used wireless sensor unit is the Intel Mote platform Imote (Kling 2003) and Imote 2 (Kling 2005).

Berkeley Mote, Intel Mote and quite a few others have been used for general purposes in the military, and in the environment, health, home and other commercial areas (Akyildiz et al. 2002; Yick et al. 2008). These systems have been customised for structural health monitoring applications (Kurata et al. 2003; Ruiz-Sandoval et al. 2006; Rice and Spencer 2008). A wireless monitoring system, in which 110 wireless nodes were used, was implemented in the Jindo Bridge, Korea, in June 2009. In the structural discipline, researchers from Stanford University have developed their own wireless sensor unit for structural health monitoring (Lynch et al. 2001, 2002).

5.11.2 Basic architectures and features of wireless sensors

A wireless sensor node usually consists of four basic components, as shown in Figure 5.11: a sensing unit, a processing unit, a transceiver unit and a power unit (Akyildiz et al. 2002). The components are carefully selected to meet specified functions and keep total costs low.

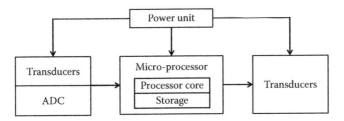

Figure 5.11 Structure of a wireless sensor node.

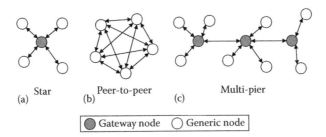

Figure 5.12 Wireless network topology.

The processing unit is a microprocessor (or microcontroller), which controls the sensing, data processing, computation and communication with other sensor nodes or the central station. Unlike traditional sensors, the on-board processor makes the wireless sensor node intelligent. The micro-processor's small storage stores internal programs and processed data.

The sensing unit is usually composed of a few sensors and A/D converters. The analogue signals collected by the sensors are converted to digital signals by the A/D converter, and then sent to the processing unit. It is notable that the A/D converter in most general wireless sensor nodes is only 8 or 10 bits. This is insufficient for vibration monitoring. In the customised Imote 2 (Rice and Spencer 2008), a 16-bits A/D converter is embedded. The wireless sensor unit developed in Stanford University (Lynch et al. 2001) has a 16-bits A/D converter as well.

A transceiver unit connects the node to the network. The transmission distance of most wireless nodes is about 50–500 m in outdoor environments. Consequently, for large civil structures, this transmission range requires that the sensor nodes communicate with peers and send the data to the base station over the network. The wireless network has three kinds of transmission range (Swartz and Lynch 2009): star, peer-to-peer and multi-tier, as shown in Figure 5.12. In the figure, the sensor nodes include generic nodes and gateway nodes. A gateway node, like the substation in the wired systems, gathers data from the adjacent generic nodes and transmits them to the base station. Most of the smart sensors to date adopt radio frequency for wireless communication.

The power unit is an important component in wireless sensor nodes. Currently, most of the available smart sensors rely on battery power supply, which has a finite capacity and a finite life. Several attempts have been made to harvest energy at sensor nodes locally, for example, solar cell, wind turbine, mechanical vibration, fuel cells and mobile supplier. Solar cell is the current mature technique and was used in the wireless monitoring of the Jindo Bridge.

In wireless sensor nodes, communication consumes much more power than other operations, including sensing and processing. Therefore, collected raw data are processed within the sensing unit to reduce the amount of raw data transmitted wirelessly over the network. This also takes advantage of the computational characteristics of the processor board.

Accordingly, this distributed computation and monitoring makes the wireless monitoring different from the tethered monitoring using the traditional wired system.

To facilitate this distributed monitoring, the micro-processor has two types of software: the operating system and engineering algorithms. The operating system controls the nodes and provides device drivers. One popular operating system is TinyOS (http://www.tinyos.net), an open-source operating system designed by the University of California at Berkeley. Both Berkeley Mote and Intel Mote run TinyOS.

Currently, algorithms for distributed monitoring are relatively scarce and simple, mainly in modal analysis. The complicated monitoring algorithms used in centralised monitoring usually need a large amount of memory, heavy computation and data from multiple sensors. Consequently, transplanting the available monitoring algorithms from the wired monitoring system directly is not feasible. It is imperative to develop appropriate algorithms for this distributed monitoring.

5.11.3 Challenges in wireless networks

Although wireless sensors and networks have been developing rapidly, wireless monitoring is not yet mature for continuous health monitoring of large civil structures. Traditional wire-based systems still dominate practical structural health monitoring and vibration control projects, and wireless sensor nodes are mainly used for research purposes or are supplementary to the wired systems. Nevertheless, wireless sensors and networks might be a future direction for structural health monitoring and vibration control. At present, the main challenges in using wireless sensors are power supply and communication bandwidth and range. As mentioned in the previous subsection, lack of mature distributed monitoring algorithms is another big issue.

For many civil structures, ac power outlets are not available adjacent to the sensor nodes. Even if ac power outlets are available, practitioners prefer to adopt the wired data acquisition system adjacent to the power outlets and transmit the collected data to the base station via wireless communication. In any case, wireless communication is just one communication method alternative to wireless sensor networks. For a battery-powered monitoring system, power consumption is critical in maintaining the operation of the sensor nodes in the long term. Currently used battery-powered sensor nodes can only operate for hours in full working state and weeks in stand-by state (or sleep mode). The power restraint requires the sensor components to be energy efficient. However, lower power consumption often comes with reduced functions such as lower resolution, shortened communication range and reduced speed (Swartz and Lynch 2009).

The majority of wireless sensor nodes operate with the unlicensed industrial, scientific and medical radio band, in which the output power is limited, to 1 W, for example, in the United States. The limited radio band restricts the amount of data that can be reliably transmitted within the network during a given time period. In addition, the limitation of the output power restricts the effective communication range of the sensor nodes. For a large civil structure, a few sensor nodes may be insufficient and the sensor network should be designed carefully.

5.12 POWER SUPPLY AND ENERGY HARVESTING

The operation of either a structural health monitoring system or a structural vibration control system needs a power supply. This is also true for the wireless sensor networks. Therefore, it is ideal that a smart civil structure also possesses self-powered capacity.

Extensive research has been done in an attempt to develop a self-powered system capable of generating electrical energy from the operational environment. Recently, there has been a surge of research in the area of energy harvesting, which is known as the process of extracting energy from the environment or from a surrounding system and converting it to usable electrical energy. The sources of typical energies are mechanical vibration, human motion, wind, sunlight, thermal gradient and ambient radio-frequency (RF) energy. Several articles reviewed the possible energy sources for energy harvesting (Glynne-Jones et al. 2004; Roundy 2005; Paradiso and Starner 2005; Shen 2014). The major developments in energy harvesting for low-power embedded SSs and self-powered microsystems can also be found in the literature (Beeby et al. 2006; Park et al. 2008).

Recently, one of the widely investigated methods for the realisation and implementation of an energy harvesting system is the use of mechanical vibration as a source of energy, which is then converted into useful electrical energy through an energy harvesting device. Kinetic energy harvesting requires a transduction mechanism to generate electrical energy from motion, and a corresponding mechanical system in the generator is also required for coupling environmental movements to the transduction mechanism. The design of this mechanical system should maximise the coupling between the kinetic energy source and the transduction mechanism, and will depend entirely upon the characteristics of the environmental motion. There are three main transduction mechanisms: piezoelectric, electromagnetic and electrostatic (Beeby et al. 2006).

Piezoelectric generators utilise active materials, for example, piezoelectric ceramics, to generate a charge when mechanically stressed. Basing on different kinetic energy sources to be employed, various piezoelectric generators have been developed, such as impact coupled devices, human-powered piezoelectric generation, cantilever-based piezoelectric generators, and so on. By using piezoelectric materials, structural vibrations can be directly converted into a voltage output without the requirement of complex geometries and numerous additional components. Since the output impedance of piezoelectric generators is typically very high (>100 kΩ), piezoelectric generators generally produce relatively high output voltages but only at low electrical currents. One drawback for piezoelectric generators is that since the piezoelectric materials must be strained directly, the mechanical properties of materials employed will limit overall performance, lifetime and transduction efficiency.

Based on the relative motion between a magnetic flux gradient and a conductor, electromagnetic generators are capable of generating electricity by electromagnetic transduction mechanism. The conductor typically takes the form of a coil, and the amount of generated electricity relies to a great extent on the number of turns of the coil, the velocity of the relative motion and the strength of the magnetic field. The configuration of permanent magnets, a coil and a resonating cantilever beam could be considered as one of the most effective methods to produce electromagnetic induction in electromagnetic generators for the purpose of energy harvesting. In principle, either the magnets or the coil can be chosen to be mounted on the beam while the other remains fixed. However, since the magnets can also act as the inertial mass, it is generally preferable to have the magnets attached to the beam. In electromagnetic generators, comparatively high output current levels can be achieved at the expense of low voltages (typically <1 V).

Based on the relative movement between electrically isolated charged capacitor plates, the harvested energy can be provided by electrostatic generators through the work done against the electrostatic force between the plates. Electrostatic generators can be classified into three types: in-plane overlap varying, in-plane gap closing and out-of-plane gap closing type (Roundy et al. 2002). By decreasing the capacitor spacing to facilitate miniaturisation, the energy density of the generator could be increased. However, it should also be noted that the energy density would be decreased with the reduced capacitor surface area.

Therefore, simultaneous consideration of both aspects is usually required for the design of electrostatic generators. By incorporating small capacitor gaps and high voltages, high transduction damping of electrostatic generators, at low frequencies, can be achieved. A major drawback of these generators is that their output impedance is often very high, resulting in a limited current-supplying capability and making them less suitable as a power supply.

The three main techniques of harvesting energy from ambient vibrations have been shown to be able to generate output power levels in the range of microwatts to milliwatts. However, only a few commercial solutions are available, because vibration-based energy harvesters are still under development. Most research efforts are still at proof-of-concept stage in laboratory settings.

Besides the aforementioned vibration-based energy harvesting techniques, another more mature technique is obtaining energy from ambient sources through optoelectronic generators that directly utilise sunlight, thermoelectric generators that capitalise on thermal gradients, or wind turbines that use wind energy. In this context, the most well known is arguably solar cells, which are capable of providing excellent power density in direct sunlight. However, they are clearly unsuitable for use in embedded applications where no light may be present, or where the cells may be obscured by contamination. For thermoelectric generators, they use the Seebeck effect, which describes the effect of the current generated when the junction of two dissimilar metals experiences a temperature difference. Using this principle, numerous p-type and n-type junctions are arranged electrically in series and thermally in parallel to construct the thermoelectric generators. One of the drawbacks of thermoelectric generators is low efficiency ($<5\%$) if there is a small temperature gradient present. Furthermore, the fabrication cost is relatively high, and the volume and weight are still too large, especially for micro-scale sensing systems (Park et al. 2008). Wind turbines are able to provide excellent power density on windy days but are not suitable for areas lacking sufficient wind. Synthesis of structural health monitoring, structural control and energy harvesting will be discussed in Chapter 17 by taking wind turbines as an example.

The establishment of a successful energy harvesting system should be a multidisciplinary engineering project. Many factors, including the selection and configuration of energy harvesting materials, the characterisation of the available ambient energy, electronics optimisation, energy storage mechanisms, as well as the power-optimisation and power-awareness designs, are required for simultaneous consideration for maximising the amount of energy harvested. Although the energy harvesting techniques are still in a developmental stage, several conceptual designs for the applications into structural health monitoring and structural vibration control have been proposed and much more attention would be paid to this area (Ha and Chang 2005; Park et al. 2008). Further discussions on this topic can be found in Chapter 16.

NOTATION

Cov	Covariance
f_s	Sampling rate
$F(j)$	Amplitude and phase of the different sinusoidal components of signal
P	Power
R_{xx}	Autocorrelation function
S_{xx}	Auto-power spectrum density
S_{ff}	Auto-power spectrum density of the input force

S_{fx}	Cross-power spectrum density between the input force and response
x	The explanatory variable of the linear regression
y	The response variable of the linear regression
β_0, β_x	Regression coefficients
ε_y	The error
σ	Standard deviation

REFERENCES

Aktan, A.E., F.N. Catbas, K.A. Grimmelsman, and M. Pervizpour. 2002. Development of model health monitoring guide for major bridges. Federal Report, No. DTFH61-01-P-00347. Philadelphia, PA: Drexel Intelligent Infrastructure and Transportation Safety Institute. http://www.di3.drexel.edu/DI3/Events/PaperPresentation/FHWAGuideFull-web.pdf.

Akyildiz, I.F., W. Su, Y. Sankarasubramaniam, and E. Cayirci. 2002. Wireless sensor networks: A survey. *Comput. Netw.*, 38(4): 393–422.

Avitabile, P. 2001. Modal space: Back to basics. *Exp. Tech.*, 25(3): 15–16.

Beeby, S.P., M.J. Tudor, and N.M. White. 2006. Energy harvesting vibration sources for microsystems applications. *Meas. Sci. Technol.*, 17(12): R175–95.

Chan, W.S., Y.L. Xu, X.L. Ding, and W.J. Dai. 2006. Integrated GPS-accelerometer data processing techniques for structural deformation monitoring. *J. Geodesy*, 80(12): 705–19.

Chopra, I. and J. Sirohi. 2014. *Smart Structures Theory*. New York: Cambridge University Press.

Chu, S.Y., T.T. Soong, and A.M. Reinhorn. 2005. *Active, Hybrid and Semi-Active Structural Control: A Design and Implementation Handbook*. Hoboken, NJ: Wiley.

Daubechies, I. 1992. *Ten Lectures on Wavelets*. Philadelphia, PA: Society for Industrial and Applied Mathematics.

Duan, Z.D. and K. Zhang. 2006. Data mining technology for structural health monitoring. *Pac. Sci. Rev.*, 8: 27–36.

Dyke, S.J. 1996. Acceleration feedback control strategies for active and semi-active control systems: Modeling, algorithm development, and experimental verification. PhD diss., Department of Civil Engineering and Geological Science, University of Notre Dame, Notre Dame, Indiana.

Esteban, J., A. Starr, R. Willetts, P. Hannah, and P. Bryanston-Cross. 2005. A review of data fusion models and architectures: Towards engineering guidelines. *Neural Comput. Appl.*, 14(4): 273–81.

Farrar, C.R. and N.A.J. Lieven. 2007. Damage prognosis: The future of structural health monitoring. *Phil. Trans. R. Soc. A*, 365(1851): 623–32.

Farrar, C.R. and K. Worden. 2013. *Structural Health Monitoring: A Machine Learning Perspective*. Chichester, UK: Wiley.

Fayyad, U., G. Piatetsky-Shapiro, and P. Smyth. 1996a. From data mining to knowledge discovery in databases. *AI Mag.*, 17(3): 37–54.

Fayyad, U., G. Piatetsky-Shapiro, P. Smyth, and R. Uthurusamy. 1996b. *Advances in Knowledge Discovery and Data Mining*. Menlo Park, CA: American Association for Artificial Intelligence Press.

Glynne-Jones, P., M.J. Tudor, S.P. Beeby, and N.M. White. 2004. An electromagnetic, vibration-powered generator for intelligent sensor systems. *Sensors Actuat. A-Phys.*, 110(1–3): 344–49.

Ha, S. and F.K. Chang. 2005. Review of energy harvesting methodologies for potential SHM applications. In *Proceedings of the 5th International Workshop on Structural Health Monitoring*, Stanford, CA, ed. F.K. Chang, 1451–60, DEStech Publications, Inc.

Hall, D.L. 1992. *Mathematic Techniques in Multisensor Data Fusion*. Norwood, MA: Artech House.

Hall, D.L. and J. Llinas. 1997. An introduction to multisensor data fusion. *Proc. IEEE*, 85(1): 6–23.

Hall, D.L. and J. Llinas. 2001. *Handbook of Multisensor Data Fusion*. Boca Raton, FL: CRC Press.

Higashino, M. and S. Aizawa. 1993. Application of active mass damper system in actual buildings. In *Proceedings of International Workshop on Structural Control*, Honolulu, HI, eds. G.W. Housner and S.F. Masri, 194–205, USC Publication Number CE-9311.

Hollar, S. 2000. COST dust. Master thesis, Berkeley, CA: University of California at Berkeley.

Huang, N.E., Z. Shen, S.R. Long, et al. 1998. The empirical mode decomposition and the Hilbert spectrum for nonlinear and non-stationary time series analysis. *Proc. Royal Soc. Lond., A*, 454: 903–95.

Kahn, J.M., R.H. Katz, and K.S.J. Pister. 1999. Next century challenges: Mobile networking for "Smart Dust". In *Proceedings of the 5th Annual ACM/IEEE International Conference on Mobile Computing and Networking*, Seattle, WA: IEEE Communications Society, 271–78.

Khaleghi, B., A. Khamis, F.O. Karray, and S.N. Razavi. 2013. Multisensor data fusion: A review of the state-of-the-art. *Inf. Fusion*, 14(1): 28–44.

Kling, R.M. 2003. Intel Mote: An enhanced sensor network node. In *Proceedings of International Workshop on Advanced Sensors, Structural Health Monitoring, and Smart Structures*, Tokyo, Japan, (CD-ROM).

Kling, R.M. 2005. Intel Motes: Advanced sensor network platforms and applications. In *Proceedings of IEEE MTT-S International Microwave Symposium Digest*, Long Beach, CA, USA, 365–68. Piscataway, NY: Institute of Electrical and Electronics Engineers.

Kobori, T. 1994. Future direction on research and development of seismic-response-controlled structure. In *Proceedings of the First World Conference on Structural Control*, Los Angeles, CA, Vol. FA2: 3–18, John Wiley & Sons.

Kobori, T. 2000. Future perspective of structural control in earthquake engineering. In *Twelfth World Conference on Earthquake Engineering*, New Zealand. 2841: 1–4.

Kottegoda, N.T. and R. Rosso. 1997. *Statistics, Probability, and Reliability for Civil and Environmental Engineers*. New York: McGraw-Hill.

Kurata, N., T. Kobori, M. Takahashi, N. Niwa, and H. Midorikawa. 1999. Actual seismic response controlled building with semi-active damper system. *Earthq. Eng. Struct. Dyn.*, 28(11): 1427–47.

Kurata, N., B.F. Spencer Jr., M. Ruiz-Sandoval, Y. Miyamoto, and Y. Sako. 2003. A study on building risk monitoring using wireless sensor network MICA-Mote. In *Proceedings of First International Conference on Structural Health Monitoring and Intelligent Infrastructure*, Tokyo, Japan, (CD-ROM).

Luo, R.C., C.C. Yih, and K.L. Su. 2002. Multisensor fusion and integration: Approaches, applications, and future research directions. *IEEE Sens. J.*, 2(2): 107–19.

Lynch, J.P., K.H. Law, A.S. Kiremidjian, T.W. Kenny, E. Carryer, and A. Partridge. 2001. The design of wireless sensing unit for structural health monitoring. In *Proceedings of the 3rd International Workshop on Structural Health Monitoring*, Stanford, CA, ed. R.A. Livingston, DEStech Publications, Inc.

Lynch, J.P., A. Sundararajan, K.H. Law, A.S. Kiremidjian, T.W. Kenny, and E. Carryer. 2002. Computational core design of a wireless structural health monitoring system. In *Proceedings of Advances in Structural Engineering and Mechanics Conference*, Pusan, Korea, eds. C.K. Choi and W.C. Schnobrich. Daejeon, Korea: Techno Press, http://eil.stanford.edu/publications/jerry_lynch/ASEM02Paper.pdf.

Ni, Y.Q., Y. Xia, W.Y. Liao, and J.M. Ko. 2009. Technology innovation in developing the structural health monitoring system for Guangzhou new TV tower. *Struct. Control Health Monit.*, 16(1): 73–98.

Oppenheim, A.V. and R.W. Schafer. 1989. *Discrete-Time Signal Processing*. Englewood Cliffs, NJ: Prentice-Hall.

Paradiso, J.A. and T. Starner. 2005. Energy scavenging for mobile and wireless electronics. *Pervasive Comput., IEEE*, 4(1): 18–27.

Park, G., T. Rosing, M.D. Todd, C.R. Farrar, and W. Hodgkiss. 2008. Energy harvesting for structural health monitoring sensor networks. *J. Infrastruct. Syst.*, 14(1): 64–79.

Rice, J.A. and B.F. Spencer Jr. 2008. Structural health monitoring sensor development for the Imote2 platform. In *Proceedings of SPIE 6932, Sensors and Smart Structures Technologies for Civil, Mechanical, and Aerospace Systems*, San Diego, CA, ed. M. Tomizuka, 693234, Copyrighted by SPIE Digital Library.

Riley, M.A. 1996. Experimental implementation and design of a hybrid control system with actuator-structure interaction. PhD diss., Department of Civil Engineering, State University of New York at Buffalo, Buffalo, New York.

Roundy, S. 2005. On the effectiveness of vibration-based energy harvesting. *J. Intell. Mater. Syst. Struct.*, 16(10): 809–23.

Roundy, S., P. Wright, and K. Pister. 2002. Micro-electrostatic vibration-to-electricity converters. In *ASME 2002 International Mechanical Engineering Congress and Exposition*, New Orleans, Louisiana, USA, Paper No. IMECE2002-39309, pp. 487–96.

Ruiz-Sandoval, M., T. Nagayama, and B.F. Spencer Jr. 2006. Sensor development using Berkeley Mote platform. *J. Earthq. Eng.*, 10(2): 289–309.

Shen, W.A. 2014. Electromagnetic damping and energy harvesting devices in civil structures. PhD diss., Department of Civil and Environmental Engineering, The Hong Kong Polytechnic University, Hong Kong.

Sohn, H., C.R. Farrar, F.M. Hemez, et al. 2003. A review of structural health monitoring literature: 1996–2001. Report LA-13976-MS, Los Alamos: Los Alamos National Laboratory. https://institute.lanl.gov/ei/shm/pubs/LA_13976_MSa.pdf.

Spencer Jr., B.F. 2008. Structural control in honor of Takuji Kobori. In *The Fourteenth World Conference on Earthquake Engineering*, Beijing, China. http://www.iitk.ac.in/nicee/wcee/article/14_S24-014.pdf.

Staszewski, W.J. and A.N. Robertson. 2007. Time-frequency and time–scale analyses for structural health monitoring. *Phil. Trans. R. Soc. A*, 365(1851): 449–77.

Suhardjo, J., B.F. Spencer Jr., and M.K. Sain. 1990. Feedback-feedforward control of structures under seismic excitation. *Struct. Saf.*, 8(1–4): 69–89.

Swartz, R.A. and J.P. Lynch. 2009. Wireless sensors and networks for structural health monitoring of civil infrastructure systems. In *Structural Health Monitoring of Civil Infrastructure Systems*, eds. V.M. Karbhari and F. Ansari, 72–112. Cambridge: Woodhead Publishing.

Wong, K.Y. 2004. Instrumentation and health monitoring of cable-supported bridges. *Struct. Control Health Monit.*, 11(2): 91–124.

Xu, Y.L. and Y. Xia. 2012. *Structural Health Monitoring of Long-Span Suspension Bridges*. Oxford, UK: Spon Press.

Yick, J., B. Mukherjee, and D. Ghosal. 2008. Wireless sensor network survey. *Comput. Netw.*, 52(12): 2292–330.

Part II

Integration for smart civil structures

Part II of this book includes Chapters 6 through 11. The multi-scale modelling of civil structures and system identification and model updating are described in Chapters 6 and 7, respectively. With the aid of a finite element model, Chapter 8 describes the methods for multi-type sensor placement. Chapter 9 introduces the structural control theory, followed by a description of the optimal placement of control devices in Chapter 10. Combining the optimal control device placement and optimal sensor placement in Chapters 8 and 10, Chapter 11 presents a method for the collective placement of control devices and sensors.

Chapter 6

Multi-scale modelling of civil structures

6.1 PREVIEW

In structural health monitoring (SHM) and structural vibration control (SVC), an accurate finite element (FE) model of a civil structure is often an essential tool to facilitate sensor and control device placement, control performance assessment and damage detection, among many others. A large civil structure, such as a tall building or a long-span bridge, is a complex structure comprising tens of thousands of structural components of different sizes connected to one another in different ways. Local damage often does not significantly affect the global responses of a structure, making global response-based damage analysis inaccurate and sometimes impossible. Moreover, the number of sensors in an SHM system for a large civil structure is limited and the sensors may not directly monitor the locations of structural defects. Therefore, the successful application of damage detection methods to large civil structures is very limited. Multi-scale FE modelling and analysis of a large civil structure have become necessary to provide both global and local structural information to enable a comprehensive assessment of structural safety, including damage detection and damage prognosis. Multi-scale FE modelling can also benefit control vibration performance if both global and local sensors can be used as feedback of state variables, and it may also facilitate the understanding of the interaction between the control devices and the structural components in SVC.

This chapter first gives a brief introduction to the principle of FE modelling and reviews the current status of multi-scale modelling and the mixed-dimensional FE coupling method for large civil structures. A new mixed-dimensional FE coupling method, which can achieve both displacement compatibility and stress equilibrium at the interface between the different element types, is then presented and verified for both linear and nonlinear problems. The new mixed-dimensional FE coupling method is finally used to establish the multi-scale FE model of a transmission tower. A comparison of the numerical results with the experimental results shows that the new mixed-dimensional FE coupling method and the multi-scale modelling method are feasible and necessary.

6.2 INTRODUCTION TO FINITE ELEMENT MODELLING

The FE method has become a widely accepted analysis tool in many disciplines. In SHM and SVC, an accurate FE model of a structure is often essential to facilitate sensor and actuator placement, control performance assessment and damage detection, among many others. This is particularly true for the prognosis of structural performance because the performance predication of the structure under a diverse loading and structural conditions could only be accomplished by an FE model–based computational approach. To obtain an accurate FE model of the structure, appropriate modelling is essential. The theory of the FE method can be found in many references (Zienkiewicz and Taylor 1989; Bathe 1996) and only a brief introduction to the principle of FE modelling of a civil structure is given here.

The vibration of a linear structural system with n degrees of freedom (DOFs) is governed by the equation of motion as

$$\mathbf{M}_E \ddot{\mathbf{x}}(t) + \mathbf{C}_E \dot{\mathbf{x}}(t) + \mathbf{K}_E \mathbf{x}(t) = \mathbf{f}_E(t) \tag{6.1}$$

where:

$\mathbf{M}_E$, $\mathbf{C}_E$ and $\mathbf{K}_E$	are the mass matrix, viscous damping matrix and stiffness matrix of the structural system, respectively
$\mathbf{x}(t)$	is a vector containing nodal displacement
$\mathbf{f}_E(t)$	is the force vector for the system

The matrices and vectors have an order of n in size, corresponding to the number of DOFs used in describing the displacements of the structure. It is noted that linear viscous damping is commonly used because of its convenient form. Hysteretic damping, having a complex form of equation of motion, can lead to a more realistic result in terms of energy loss.

In the FE approach, the structure is assumed to be divided into a system of discrete elements, which are interconnected only at a finite number of nodal points. The properties of the complete structure (the system mass matrix $\mathbf{M}_E$ and the system stiffness matrix $\mathbf{K}_E$) are then assembled by evaluating the properties of the individual FE and superposing them together appropriately (Clough and Penzien 2003). Within each element, each node holds specific DOFs depending on the problem described. Under external loads, the deflected shape of an element follows a specific displacement function, which satisfies nodal and internal continuity requirements. Based on the load–displacement relation, the element stiffness matrix can be established. For mass property, the simplest method is to assume that the mass is concentrated at the nodes where the translational displacements are defined and rotational inertia is null. This is referred to as *lumped mass* and the matrix has a diagonal form. However, following the same method and the same displacement function in deriving the element stiffness matrix, the consistent mass matrix can be calculated.

Although the FE concept could also be used to define the damping coefficients of the system, damping is generally expressed in terms of damping ratios, which can be identified from experiments. If an explicit expression of the damping matrix is needed, it is generally calculated from the mass and stiffness matrices, which is called *Rayleigh damping*:

$$\mathbf{C}_E = \alpha \mathbf{M}_E + \beta \mathbf{K}_E \tag{6.2}$$

Two Rayleigh damping factors, α and β, can be experimentally obtained from the damping ratios of two modes (Clough and Penzien 2003).

The concentrated load acting on the nodes can be directly applied. Other loads such as distributed forces can be similarly evaluated as deriving from the consistent mass matrix with the same displacement function. In engineering practice, the mass and stiffness matrices and load vectors are automatically calculated by structural analysis computer software. However, it is of the utmost importance that the users understand the theories, assumptions and limitations of numerical modelling using the FE method, as well as the limitations of the computation algorithms.

There have been numerous studies on the FE modelling of long-span cable-supported bridges to facilitate static and dynamic analyses (Xu et al. 1997). Most of the studies are based on a simplified spine beam model of equivalent sectional properties to the actual structural components. Such a simplified model is effective in capturing the dynamic characteristics and global structural behaviour of the structure without heavy computational effort. However, local structural behaviour, such as stress and strain concentrations at joints that are prone to cause local damage in static and/or dynamic loading conditions, cannot be estimated directly. Apparently, the spine beam model is not the best option from the perspective of SHM because, in general, the modelling of local geometric features is insufficient. In this regard, an FE model with finer details in highlighting local behaviours of the structural components is needed. However, the rapid development of information technology and the improvement in the speed and memory capacity of personal computers has made it possible to establish a more detailed FE model for a large civil structure. Apparently, a finer FE model will cost more computational resources including computational time and storage memory. To trade off the computational efficiency and capability, one can establish a hybrid model or multi-scale model to cater for the objectives.

6.3 REVIEW OF MULTI-SCALE MODELLING OF LARGE CIVIL STRUCTURES

In practice, a large civil structure is often modelled via the FE method, using a combination of beam, shell, solid and other elements of similar scale for the static and dynamic analyses at a global level. Stress concentration, crack initiation and propagation, fatigue and fracture are local phenomena that are often not represented in the global structural model. However, many types of defects are locally generated at the material points and sectional levels and may evolve into global structural damage and possibly cause structural failure. Thus, a multi-scale FE modelling of large civil structures, the aim of which is to simultaneously simulate and evaluate the structural performance at both the global and local levels, has recently attracted increasing attention in the field of SHM. Furthermore, the multi-scale monitoring system can be best integrated with decentralised wireless communication technologies (see Chapter 5), leading to a powerful SHM system to overcome the limitations and problems encountered by the current SHM system.

Li et al. (2001) used the information-passing multi-scale method for a fatigue analysis of long-span bridges. They used both the simplified global model and the detailed local models. A global structural analysis using the global model was conducted to determine the critical components and extract the results of the critical region at the outer boundary conditions, which were then applied to the local models for fatigue analysis. Obviously, there are inherent difficulties in the accuracies of complicated boundary conditions, which may lead to significant errors if the extraction is not accurate enough. Chan et al. (2003) further pointed out that the information-passing multi-scale method may be successful for linear-static problems, but gives questionable results for dynamic and nonlinear problems. Li et al. (2009) investigated the concurrent multi-scale FE modelling of civil structures, in which the global structural behaviour and nonlinear features of local details in a large civil structure could be concurrently analysed to meet the needs of structural-state evaluation as well as structural deteriorating, where 'large-scale' modelling is adopted for the global structure and 'small-scale' modelling is available for nonlinear analyses of the local welded details. Almansour et al. (2010) also conducted an investigation on long-span cable-stayed bridges made of composite materials using a multi-scale modelling technique.

The FE method has proved to be a successful and powerful tool for various kinds of structural analysis. Several general-purpose commercial software packages such as ANSYS, ABAQUS, AD-INA and SAP have been developed based on the FE method.

Almost all the analysing techniques adopted in the FE method can be easily implemented in the software. However, there is currently no delicate technique provided in the existing software for the concurrent simulation of global and local structural behaviour. The implementation procedure of multi-scale modelling needs to be further studied with reference to the existing FE modelling techniques, such as substructuring and multi-point constraint equations.

6.4 REVIEW OF MIXED-DIMENSIONAL FINITE ELEMENT COUPLING METHODS

Take the frame structure shown in Figure 6.1 as an example to illustrate the concept of multi-scale modelling. The local joints are simulated with shell elements of a small scale while the other components in the frame are simulated with beam elements of a large scale. Such a multi-scale simulation can capture not only the global structural behaviour in terms of displacement and acceleration, but also the local joint behaviour in terms of stress without a huge computation cost.

Since different types of elements (beam, plate, shell and solid) have a different number of DOFs, the multi-scale FE simulation needs a rational FE coupling method to combine mixed-dimensional FEs at their interfaces into a single structural model. The challenging issue in multi-scale FE modelling is therefore how to guarantee the rationality of the coupling method so that it can achieve both displacement continuity and stress equilibrium in the region around the interface between the different types of elements. Broadly speaking, two major coupling methods are currently available: volume coupling and surface coupling (Guidault and Belytschko 2007). Volume coupling refers to a region in which different models co-exist and it is usually realised using the Arlequin method (Dhia and Rateau 2005). The Arlequin method is best suitable for coupling different physical models such as continuum particles (Bauman et al. 2008; Wellmann and Wriggers 2012) among others. In surfacing coupling, there is no overlapping of different models which can be coupled using one of the following methods: (a) the transition element method and (b) the multi-point constraint (MPC) method. Transition elements employing either reduced or full integration can be used for shell–solid transition (Surana 1980, 1982; Cofer and Will 1991; Gmür and Schorderet 1993), beam–solid transition (Gmür and Kauten 1993; Garusi and Tralli 2002) and beam–shell transition (Wagner and Gruttmann 2002; Chavan and Wriggers 2004). Unfortunately, the transition elements have not been widely adopted because of their limitations. Transition elements can only be used with a one-to-one coupling of elements and

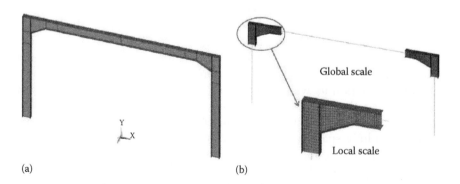

(a)   (b)

Figure 6.1 Multi-scale modelling of a frame structure: (a) frame structure and (b) multi-scale model.

different element transitions require different formulations, which make it difficult and impractical for a commercial FE code.

The MPC method is an attractive coupling method for coupling mixed-dimensional elements by using constraint equations for nodal displacements at the interface. The MPC method can be used for static and dynamic analysis of linear or nonlinear structures (Wang et al. 1996; McCune et al. 2000; Shim et al. 2002; Ho et al. 2010; ANSYS User's Manual 2010; ABAQUS Analysis User's Manual 2010). This method is easy to access in a few commercial FE codes. For instance, the RBE3 MPC method provided in ANSYS code (ANSYS User's Manual 2010) can automatically establish constraint equations for coupling the different types of elements. There are mainly two types of MPC method: the rigid interface method and the deformable interface method. The rigid interface method uses rigid beams to connect nodes of different types of elements, such as CERIG in ANSYS and MPC-BEAM in ABAQUS (ANSYS User's Manual 2010; ABAQUS Analysis User's Manual 2010). The rigid interface method, however, yields stress disturbance at the interface because the interface is defined as a rigid interface. The deformable interface method uses a concept of force distribution at the interface, such as RBE3 in ANSYS, and the distributing coupling method in ABAQUS. The deformable interface method allows the interface deformation with stress distributions at the interface. RBE3 in ANSYS allows the motion of the master node equal to the average of the slave nodes in which only translational DOFs of the slave nodes involve the constraint equations. The force and moment are distributed to the slave nodes by weighting factors and the distance from the centre of the slave nodes times weighting factors, respectively. The distributing coupling method in ABAQUS constrains the motions of the coupling nodes to the motion of a reference node in an average sense. Forces and moments at the reference node are distributed either as a coupling node force only or as a coupling node force and moment. Both the RBE3 and the distributing coupling methods have the sense of force and moment distribution by means of weighting factors, but the accuracy of the stress distribution at the interface resulting from force and moment distribution by means of weighting factors is questionable. However, another deformable MPC method was proposed based on the direct assumption of stress distribution at the interface and the equal work done by the stresses and forces at the interface (McCune et al. 2000). Although the deformable interface method seems more rational, inaccurate constraint equations due to inappropriate stress distribution assumptions may result in stress disturbance at the interface.

The subsequent sections of this chapter present a new deformable MPC coupling method that can achieve both displacement compatibility and stress equilibrium at the interface between the different element types (Wang et al. 2014). The principle of virtual work is first used to derive both linear force and displacement constraint equations for the interface. A numerical method compatible with commercial FE codes is developed to figure out the linear constraint equations, which satisfy both displacement compatibility and stress equilibrium conditions at the interface. The proposed coupling method is then extended to nonlinear mixed-dimensional FE coupling problems. To verify the proposed coupling method, a number of FE test cases are examined. These cases include linear beam-to-plate for linear mixed-dimensional FE coupling, a linear frame structure for multi-scale simulation and a beam-to-shell buckling problem for nonlinear mixed-dimensional FE coupling. The results are also compared with those from the existing methods to demonstrate the accuracy and robustness of the new algorithm. The new mixed-dimensional FE coupling method is finally used to establish the multi-scale FE model of a transmission tower. The numerical results are compared with the experimental results to assess the feasibility and accuracy of the new coupling method and the multi-scale modelling method.

6.5 LINEAR CONSTRAINT EQUATIONS

For static linear structural analysis, the interface coupling for elements of different DOFs can be established using linear constraint equations. An example of interface coupling for a two-dimensional beam and plate connection, as shown in Figure 6.2, is used to illustrate how the constraint equations can achieve both displacement compatibility and stress equilibrium at the interface.

The mixed-dimensional FE model of the beam and plate connection is shown in Figure 6.3, in which plane beam elements are used to model the beam and plane stress elements of 8-nodes are used to model the plate. According to the deformable MPC coupling method proposed in McCune et al. (2000), the displacement constraint equations for coupling the beam and the plate at the interface can be expressed as

$$c\left(u_{\mathrm{B}}, u_{\mathrm{P}}\right) = u_{\mathrm{B}} - Cu_{\mathrm{P}} = 0 \tag{6.3}$$

where C is the coefficient matrix of the displacement constraint equation. For the FE model of the beam and plate connection shown in Figure 6.3, the beam node has three DOFs $\{u_{\mathrm{B}}, v_{\mathrm{B}}, \theta_{\mathrm{B}}\}$ while each node of the plane stress element has two DOFs $\{u_{\mathrm{Pi}}, v_{\mathrm{Pi}}\}$. The term Cu_{p} in Equation 6.3 can be seen as the generalised displacements of the plate, matching with the beam displacements u_{B} at the interface and obtained by weighting the coefficient matrix C over the plate displacement vector u_{P} at the interface. The displacement constraint equation indicates that the displacements of the beam node shall be consistent with the generalised displacements of the plate at the interface. The variation of the displacement constraint equation can be expressed as

$$\delta c\left(u_{\mathrm{B}}, u_{P}\right) = u_{\mathrm{B}} - Cu_{P} = 0 \tag{6.4}$$

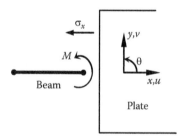

Figure 6.2 Stresses and forces at the interface of beam and plate connection.

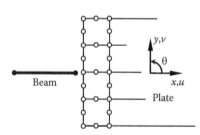

Figure 6.3 Mixed-dimensional FE modelling of beam and plate connection.

in which δu_B and δu_P are the virtual displacement vectors at the interface of the beam and plate, respectively. Equation 6.4 is the constraint equation of virtual displacement at the interface of the beam and plate. For any arbitrary interface coupling, the sum of virtual work done by the corresponding forces at the interface of the two types of elements shall be zero.

$$\delta W_P + \delta W_B = 0 \tag{6.5}$$

where δW_P and δW_B are the virtual works done by the nodal forces at the interface of the plate and beam, respectively. The virtual works done by the node forces at the interface of plate and beam are, respectively,

$$\begin{aligned} \delta W_P &= u_P^T F_P \\ \delta W_B &= u_B^T F_B \end{aligned} \tag{6.6}$$

where F_P and F_B are the nodal forces at the interface of plate and beam, respectively. By substituting Equation 6.6 into Equation 6.5, the following equation can be obtained:

$$\delta u_P^T F_P + \delta u_B^T F_B = 0 \tag{6.7}$$

From Equation 6.4, we have $\delta u_B = \delta u_P$. Equation 6.7 then becomes

$$\delta u_P^T F_P + \delta u_B^T C^T F_B = 0 \tag{6.8}$$

Equation 6.8 holds for any arbitrary virtual displacements and thus the following force constraint equation is derived:

$$F_P = -C^T F_B \tag{6.9}$$

It is interesting to see that the coefficient matrix C^T of the force constraint equation is the transpose of the coefficient matrix C of the displacement constraint equation. The coefficient matrix C^T of the force constraint equation can be regarded as a distribution matrix to distribute the force or moment at the node of the beam to the nodes of the plate at the interface. The satisfaction of both Equations 6.3 and 6.9 can achieve both displacement compatibility and stress equilibrium at the interface between the different element types. Furthermore, Equation 6.9 indicates that the distribution coefficients in the force distribution matrix are the corresponding nodal forces of the plate at the interface under unit force or moment. Hence, the coefficient matrix of the force constraint equation actually refers to the nodal forces along the cross section of the plate under unit force or moment. Based on this principle, a new numerical method compatible with commercial FE codes is developed to figure out the coefficient matrix of the force constraint equation. Once it is obtained, the coefficient matrix of the displacement constraint equation can be easily found, and both the displacement compatibility and the stress equilibrium conditions at the interface are satisfied.

6.6 NUMERICAL METHODS FOR GENERATING CONSTRAINT EQUATIONS

6.6.1 Substructure and nodal force model

The nodal forces, which are used for the coefficient matrix of the force constraint equation, must be compatible with the nodes of the plate at the interface under the unit force or moment.

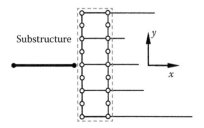

Figure 6.4 Substructure of the plate at the interface.

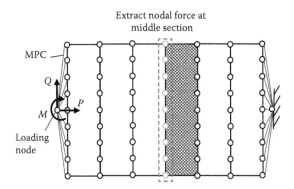

Figure 6.5 Nodal force model.

To this end, a substructure is extracted directly from the FE model of the plate at the interface, as shown in Figure 6.4, so that the substructure has the same mesh and element type with the plate at the interface. The extraction of the substructure from the FE model and the corresponding stiffness matrix can be easily implemented in commercial FE codes. To calculate the nodal forces of the plate at the interface with high accuracy, the substructure is used to assemble a nodal force model. The nodal force model is formed by repeating the substructure with the same mesh and element type, as shown in Figure 6.4. The nodal force model shall be long enough to avoid end effects but over length is not necessary. The nodal forces can then be extracted from the nodes at the middle section of the nodal force model (i.e. the interface of the beam and plate in the multi-scale model) to avoid the influence of boundary conditions.

6.6.2 Application of unit force or moment

To apply the unit force or moment to the two ends of the nodal force model, a loading node is established at the centre of the two end sections, as shown in Figure 6.5. Initially, the rigid connections are used to connect the loading node and all the nodes at the corresponding end section. The rigid connection is a type of MPC which defines a rigid area at the section of the plate. It is deduced from the displacement relationship of an arbitrary two points on a rigid body. To calculate the displacement response of the nodal force model, the unit force or moment is assigned at one of the two loading nodes, and the other loading node on the other end is clamped.

6.6.3 Construction of coefficient matrix

After the nodal displacements of the nodal force model are computed, the nodal displacement vector at the middle section can be extracted, and the nodal force vector of the same substructure at the interface can be expressed as

$$f^{SE} = K^{SE}u^{SE} \tag{6.10}$$

where:

f^{SE} is the nodal force vector of the substructure at the middle section

K^{SE} and u^{SE} are the stiffness matrix and nodal displacement vector of the substructure, respectively

After the nodal force vector is found, the corresponding coefficients in the coefficient matrix of the force constraint equation can be found by using Equation 6.10. By applying the axial force, shear force or bending moment to the loading node one after another, the coefficient matrix of the force constraint equation can be found for the interface of the beam and plate concerned. To enhance the accuracy of the coefficient matrix for the force constraint equation, iteration is applied. The obtained coefficient matrix is used to update the connection between the loading node and the nodes on the end section. The new nodal force vector and then the new coefficient matrix are calculated. Iteration continues until the coefficients meet the convergent requirement.

It should be pointed out that although the proposed method for establishing constraint equations is illustrated using the beam-to-plate connection, this method is also applicable in principle to beam-to-shell, beam-to-solid and other connections. Moreover, conveniently, it can be applied to more than two members of an arbitrary cross section at a structural joint using more than two interfaces, in which the constraint equation can be established for each interface in its local coordinate system (CS) using the aforementioned method. The constraint equations established in the local CS can then be transformed to the global CS for the solution of the entire structure.

6.7 NONLINEAR CONSTRAINT EQUATIONS

The method described in Section 6.6 for the construction of constraint equations is for linear mixed-dimensional FE coupling. In this section, the proposed method is extended to a nonlinear interface coupling, including both geometrical and material nonlinearity. An example of the beam-to-shell coupling of an L-shape member with large deformation, as shown in Figure 6.6, is used to illustrate how to construct nonlinear constraint equations for a nonlinear interface coupling. Considering that the interface moves during the nonlinear solution process, the nonlinear displacement and force constraint equations are first

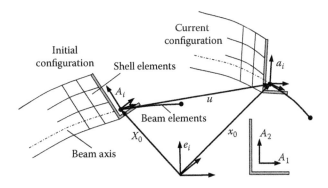

Figure 6.6 Initial and current configurations of an interface.

developed in a local interface CS at the interface. The constraint equations are then transformed to a global CS for a solution. As a result, the constraint equations in the global CS always change during the nonlinear solution process. Furthermore, the matrix coefficients of the constraint equations in the local CS may also change due to material and geometrical nonlinearity during the nonlinear solution process. Therefore, the nonlinear constraint equations for the interface are caused by material and geometrical nonlinearity and the transformation of coordinates from a local CS to a global CS.

6.7.1 Transformation of coordinate systems at interface

The transformation of the CSs is illustrated in Figure 6.6 for the interface coupling of the L-shape member, which is modelled as beam elements and shell elements. A_i and a_i are the local CSs of the interface at the initial and current configurations, respectively, and e_i is the global CS. The three CSs have the following relationship:

$$A_i = R_0 e_i \tag{6.11}$$

$$a_i = R_t e_i = R_d R_0 e_i \tag{6.12}$$

where:

R_0 and R_t are the transformation matrices from the global CS to the initial and current local CS, respectively

R_d is the transformation matrix from the initial local CS to the current local CS

The local CS A_i and a_i can be calculated from the initial and current positions of three nodes at the interface. The transformation matrices R_0 and R_t can be computed using A_i and a_i, respectively. The transformation matrix R_d is then determined by

$$R_d = R_t R_0^T \tag{6.13}$$

For the beam-to-shell coupling, it is also convenient to calculate the transformation matrix R_d by the rotation angle of the interface by means of the Rodrigues' rotation formula (Belytschko et al. 2001). The rotation angle of the interface can be defined by the average rotation of shell nodes at the interface.

6.7.2 Nonlinear constraint equations in a local coordinate system

The relationship between the coefficient matrices of the nonlinear force and displacement constraint equations for the interface is derived in this section. The nonlinearity of the coefficient matrix of the constraint equation is due to the change of stiffness at the interface caused by material and geometrical nonlinearity. The coefficient matrix of the nonlinear displacement constraint equation changes according to the displacement at the interface of the shell. The nonlinear displacement constraint equation for the interface coupling of beam-to-shell can be written as

$$c_{NL}(u_B, u_S) = u_B - C_{NL}(u_S)u_S = 0 \tag{6.14}$$

where:

u_B and u_S are the nodal displacement at the interface of beam and shell, respectively

$C_{NL}(u_S)$ is the coefficient (weighting) matrix varying with u_S

The variation of the nonlinear displacement constraint equation in the current configuration is

$$\delta c_{NL}\left(u_B, u_S\right) = u_B - D\left(u_S\right) u_S = 0 \tag{6.15}$$

where $D\left(u_S\right) = \partial c_{NL}\left(u_B, u_S\right)/\partial u_S$ is the coefficient matrix of the linearised form of the displacement constraint equation expressed by Equation 6.14 used at each increment. In consideration of the principle of virtual work applied to the interface of beam-to-shell coupling, one can obtain $\delta W_S + \delta W_B = 0$. By substituting $\delta u_B = D\left(u_S\right)\delta u_S$, one can achieve the following force constraint equation:

$$F_S = -D^T\left(u_S\right) F_B \tag{6.16}$$

where F_S and F_B are the nodal forces at the interface of shell and beam, respectively. It is interesting to see that the force distribution (coefficient) matrix $D^T(u_S)$ is the transpose of the displacement coefficient matrix $D(u_S)$ of the displacement constraint equation. Therefore, once the force distribution matrix can be determined using the force nodal model in consideration of nonlinearity, the displacement coefficient matrix can be constructed from the force distribution matrix.

6.7.3 Nonlinear constraint equations in a global coordinate system

The force distribution matrix in the local CS is calculated using the proposed method in Section 6.6 in which the stiffness matrix at the interface of the shell is extracted during the nonlinear iteration process. The force distribution matrix in the global CS can be determined through coordinate transformation. The coefficient matrix of the displacement constraint equation in its initial and current configuration in the global CS can then be given by

$$D_0 = T_{i1} C_0 T_{i2}^T \tag{6.17}$$
$$D_t = T_{t1} C_t T_{t2}^T$$

where:

D_0 and D_t are the coefficient matrices of the displacement constraint equation in its initial and current configuration, respectively, in the global CS

C_0 and C_t are the coefficient matrix in its initial and current local CSs, respectively

T_{i1}, T_{i2}, T_{t1} and T_{t2} are the coordinate transform matrices computing from R_0 and R_t, respectively

For instance, if the matrix C_0 is of $6 \times n$ dimensions, T_{i1} and T_{i2} are then formed with the dimensions of 6×6 and $n \times n$, respectively, in terms of R_0.

$$T_{i1} = \begin{bmatrix} R_0 & 0 \\ 0 & R_0 \end{bmatrix}$$

$$T_{i2} = \begin{bmatrix} R_0 & 0 & 0 \\ 0 & \ddots & 0 \\ 0 & 0 & R_0 \end{bmatrix} \tag{6.18}$$

After D_t is determined for the interface at a given step of the nonlinear solution process, the displacement constraint equation for the interface in the global CS can then be written as

$$\Delta U_B - D_t \Delta U_S = 0 \tag{6.19}$$

where ΔU_B and ΔU_S are the displacement increments at the interface of the beam and shell, respectively. The computation of the coefficient matrix C_t in the local CS may need huge computation time. However, if the stiffness of the shell at the interface in the local CS changes very little during the nonlinear solution process, the coefficients of the displacement constraint equation in the current configuration can be obtained directly by the coordinate transformation from its initial configuration.

$$D_t = T_{d1} D_0 R_{d2}^T \tag{6.20}$$

where T_{d1} and T_{d2} are the transformation matrix generated by R_d of a dimension of 6×6 and n × n, respectively, if the matrix D_0 is of $6 \times n$ dimensions.

By combining the nonlinear constraint equations with the nonlinear FE model of the structure, one can use the Newton iteration process to find the solution of the structure with nonlinear constraint equations. However, while the linear constraint equation can be easily implemented using the command CE in the ANSYS code (ANSYS User's Manual 2010) and the command equation in the ABAQUS commercial code, only the ABAQUS provides a user subroutine MPC (Belytschko et al. 2001) for the nonlinear constraint equation. In this chapter, the multi-scale simulation of a linear structure with linear mixed-dimensional coupling is implemented in ANSYS, and that of a nonlinear structure with nonlinear constraint equations is implemented in ABAQUS. For the nonlinear analysis, the derived coefficients of the constraint equation are updated in the user subroutine MPC.

6.8 VERIFICATION OF NEW MIXED-DIMENSIONAL FINITE ELEMENT COUPLING METHODS

This section will examine a number of FE cases using the proposed coupling method and the results will be compared with those obtained from the existing methods to verify the proposed coupling method. These cases include a linear beam-to-plate coupling problem, a linear frame structure for a linear structural multi-scale simulation and a beam-to-shell buckling problem for a nonlinear mixed-dimensional FE coupling.

6.8.1 Linear beam-to-plate coupling

A cantilever of a solid rectangular cross section, as shown in Figure 6.7, is first investigated. It is clamped on the left end and subjected to a concentrated force on the right. The length of the cantilever is 320 mm and its cross section is 40 mm high and 8 mm wide. The multi-scale FE model of the cantilever is established using the ANSYS code, in which the middle part of 100 mm long is modelled using elements of plane 82 (see Figure 6.8) and other parts of the cantilever are modelled using elements of beam3. The proposed method is used to calculate the coefficient matrix of the displacement constraint equation, and the results are listed in Table 6.1 for the left interface. In Table 6.1, node 2 is the beam node at the left interface, and the locations of nodes 42, 62, 63, 64, 65, 66, 67, 68 and 13 can be found in Figure 6.8. CE1, CE2 and CE3 denote the coefficients corresponding to the displacements of the beam node at the interface in the x-direction and y-direction as well as the rotation around the

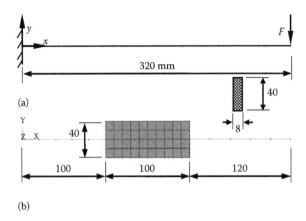

Figure 6.7 Cantilever with a solid rectangular cross section: (a) cantilever and (b) FE model.

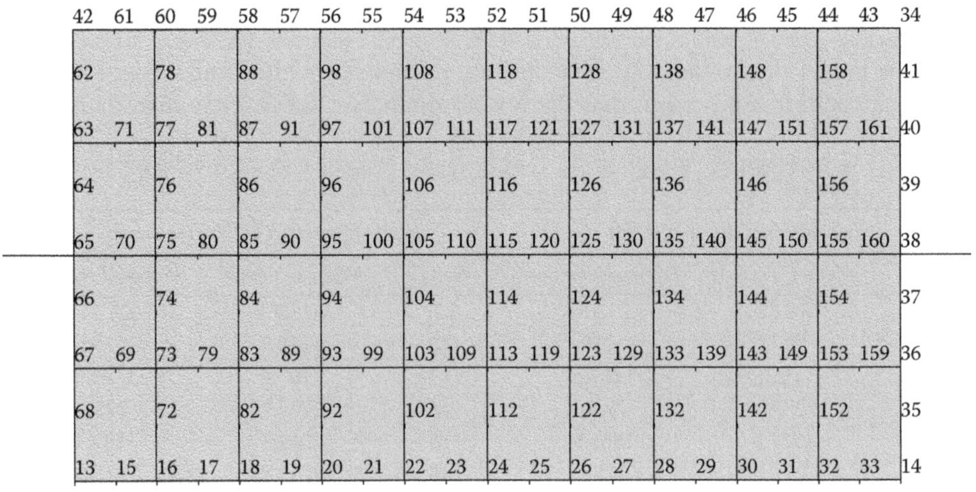

42	61	60	59	58	57	56	55	54	53	52	51	50	49	48	47	46	45	44	43	34
62		78		88		98		108		118		128		138		148		158		41
63	71	77	81	87	91	97	101	107	111	117	121	127	131	137	141	147	151	157	161	40
64		76		86		96		106		116		126		136		146		156		39
65	70	75	80	85	90	95	100	105	110	115	120	125	130	135	140	145	150	155	160	38
66		74		84		94		104		114		124		134		144		154		37
67	69	73	79	83	89	93	99	103	109	113	119	123	129	133	139	143	149	153	159	36
68		72		82		92		102		112		122		132		142		152		35
13	15	16	17	18	19	20	21	22	23	24	25	26	27	28	29	30	31	32	33	14

Figure 6.8 Node number of plate in the cantilever.

Table 6.1 Comparison of displacement CE coefficients between the McCune method and the proposed method

Node number	CE1 McCune method	CE1 New method	CE2 McCune method	CE2 New method	CE3 McCune method	CE3 New method
2	1.00000	1.00000	1.00000	1.00000	1.00000	1.00000
42	0.04167	0.04167	0.00156	0.00260	0.00625	0.00625
62	0.16667	0.16667	0.10625	0.10417	0.01875	0.01875
63	0.08333	0.08333	0.09688	0.09896	0.00625	0.00625
64	0.16667	0.16667	0.23125	0.22917	0.00625	0.00625
65	0.08333	0.08333	0.12813	0.13021	0.00000	0.00000
66	0.16667	0.16667	0.23125	0.22917	−0.00625	−0.00625
67	0.08333	0.08333	0.09688	0.09896	−0.00625	−0.00625
68	0.16667	0.16667	0.10625	0.10417	−0.01875	−0.01875
13	0.04167	0.04167	0.00156	0.00260	−0.00625	−0.00625

z-axis, respectively. For comparison, the coefficients of the displacement constraint equation provided in McCune et al. (2000) for the concerned beam-to-plate coupling problem are also listed in Table 6.1, denoted as the McCune method.

The results listed in Table 6.1 show that the coefficients of the displacement constraint equation obtained from the two methods are very close for CE1 and CE3. However, there are small differences in the coefficients of CE2, which refer to transverse displacements. It is noted that the assumed stress distribution in the plate in the McCune method is based on the beam theory, while the nodal force distribution used in the proposed method is determined numerically from the FE node force model which is consistent with the plate. The proposed method can thus give more accurate results in shear stress distribution in the plate at the interface than the McCune method. Let us consider the concentrated force F of 12,000 N acting at the right end of the cantilever. In principle, the shear stress distribution at the interface of the plate shall be very close to that at any internal cross section of the plate. Internal nodal shear forces calculated by the two methods are listed in Table 6.2 for the left interface and an internal section of the plate model. For the proposed method, the nodal shear forces at the interface are consistent with the nodal shear forces at the internal section. For the McCune method, there is a small difference between the nodal shear forces at the interface and the nodal shear forces at the internal section. The shear stress contours of the plate model are shown in Figure 6.9a for the proposed method and Figure 6.9b for the McCune method. It can be seen that there is no significant shear stress disturbance at the

Table 6.2 Comparison of nodal shear forces between the McCune method and the proposed method

| | Nodal force results at interface | | | Nodal force results at internal section | |
Node	McCune method	Proposed method	Node	McCune method	Proposed method
42	18.75	31.25	52	31.273	31.25
62	1275	1250	118	1249.9	1250
63	1162.5	1187.5	117	1187.6	1187.5
64	2775	2750	116	2749.9	2750
65	1537.5	1562.5	115	1562.6	1562.5
66	2775	2750	114	2749.9	2750
67	1162.5	1187.5	113	1187.6	1187.5
68	1275	1250	112	1249.9	1250
13	18.75	31.25	24	31.273	31.25

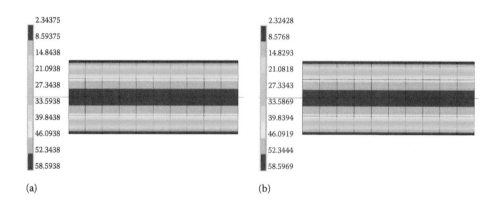

(a)   (b)

Figure 6.9 Shear stress contours: (a) using the proposed method and (b) using the McCune method.

interfaces no matter which method is used. The maximum shear stresses from the proposed method and the McCune method are 58.59 and 58.60 N/mm², respectively. The maximum shear stresses from the two methods are very close to each other for this cantilever with a solid rectangular section.

A cantilever of a box cross section is also used as a numerical example for the detailed calculation and discussion of the linear beam-to-shell coupling problem and the linear beam-to-solid coupling problem (Wang et al. 2014). The results from the proposed method, the McCune method and the RBE3 method are compared with those from the pure shell model and the pure solid model. The comparative results show that the stress results from the proposed method have the best accuracy for beam-to-shell and beam-to-solid couplings among the three methods.

6.8.2 Frame structure

Figure 6.10 shows a frame structure constructed by one beam and two columns of I cross section. The beam is strengthened with a linearly increasing height towards the connection between the beam and the column. There is a uniform load $P = 10$ kN/m acting perpendicularly on the beam. The frame structure is modelled with the pure shell model and the multi-scale model as shown in Figure 6.1, respectively. For the interface connection, the proposed method and the RBE3 method are used to achieve the beam-to-shell coupling. The frame structure is also modelled with the pure beam element for an additional comparison.

The mid-span deflection of the beam is calculated using different methods as listed in Table 6.3. The results from the pure shell model are used as the exact solution for a reference to calculate relative errors. The relative errors of the deflection at the mid span of the

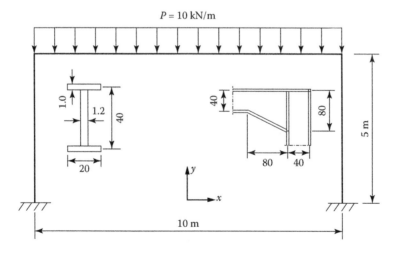

Figure 6.10 Frame structure and member cross section.

Table 6.3 Comparison of the mid-span deflection and the maximum von Mises stress of frame structure among different methods/models

Method	Displacement (mm)	Error (%)	Maximum von Mises stress (MPa)	Error (%)
Proposed method	8.6469	0.81	85.099	0.36
RBE3	8.9370	4.19	153.930	81.53
Pure beam model	8.8098	2.71	70.000	17.45
Pure shell model	8.5775	–	84.795	–

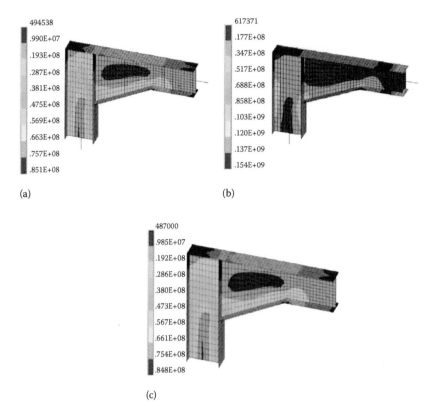

Figure 6.11 von Mises stress contours of frame structure: (a) using the proposed method, (b) using the RBE3 method and (c) using the pure shell model.

beam, calculated by the proposed method and the RBE3 method, are 0.81% and 4.19%, respectively, which indicates that the proposed method has a good accuracy for deflection results.

The von Mises stress contours of the connection part of the frame structure obtained from different methods are shown in Figure 6.11a–c. It can be seen that the von Mises stress contour obtained by the proposed method is almost the same as that obtained from the pure shell model. The stress results from the RBE3 method, however, show serious stress distur-bance at the interface between the beam and the column. The maximum von Mises stresses from the proposed method, the RBE3 method and the pure shell model are 85.1, 154 and 84.8 MPa, respectively. The maximum von Mises stresses at the interface of the horizontal beam from the proposed method, the RBE3 method and the pure shell model are 14.4, 123 and 14.3 MPa, respectively. The proposed method has a better accuracy in both the deflec-tion and stress results and can be used for multi-scale simulation of structures.

6.8.3 Nonlinear analysis of beam–shell coupling

Figure 6.12 shows a column of a box cross section. It is modelled by beam element B31 in ABAQUS for the upper part of the column and by shell element S4R for the bottom part. Nonlinear buckling analysis is implemented by means of a multi-scale model with beam–shell coupling at the interface, the pure beam model and the pure shell model, respectively. The axial loading point is 5 mm from the centre of the section in the x-direction. The axial

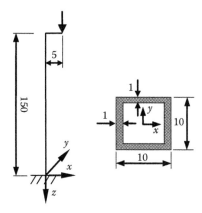

Figure 6.12 Column of a box cross section and its axial load.

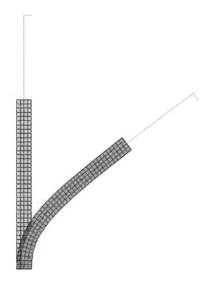

Figure 6.13 Initial and deformed shapes of multi-scale model of the column.

load is increased until the column is deformed to a certain level at which the loading point moves downward for 50 mm in the z-direction. The initial and deformed shapes of the column are shown in Figure 6.13 for reference.

From the load–displacement curves at the loading point in the z-direction, the peak loads are found as 2138.1, 2137.6, 2139.5 and 2139.3 (N), obtained by the proposed method, the distributing coupling method, the pure shell model and the pure beam model, respectively. It can be seen that all the predicted peak loads are similar. At a displacement of 0.05 m, the loads on the column are 350.8, 350.7, 350.6 and 362.1 (N), obtained by the proposed method, the distributing coupling method, the pure shell model and the pure beam model, respectively. The loading results from the multi-scale model using the two coupling methods are very close to the results from the pure shell model. However, the loading result from the beam model is slightly greater than that from the pure shell model. This is because the shell element model can reflect the stiffness change of the column. By taking a segment of 0.006 m long in the z-direction at the interface to compare stress results from different methods, its von Mises stress contours are shown in Figure 6.14a–c. The maximum von Mises stress and the shear

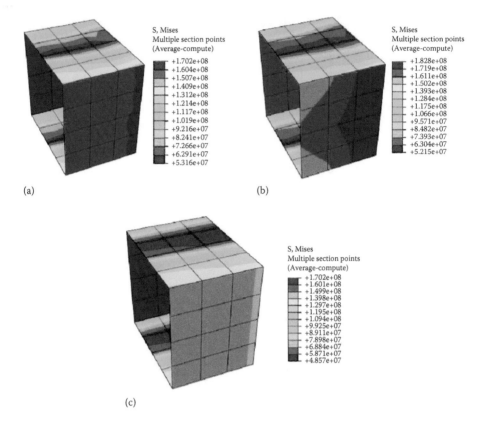

Figure 6.14 von Mises stress contours due to buckling: (a) using the proposed method, (b) using the distributing coupling method and (c) using the pure shell model.

Table 6.4 Comparison for maximum von Mises stress and shear stress for nonlinear analysis

Methods	Maximum von Mises stress (MPa)	Error (%)	Maximum shear stress (MPa)	Error (%)
Proposed method	170.2	0.00	19.5	3.23
Distributing coupling	182.8	7.40	25.9	37.11
Pure shell model	170.2	–	18.89	–

stress from each method are listed in Table 6.4. The maximum von Mises stresses are 170.2, 182.8 and 170.2 (MPa) from the proposed method, the distributing coupling method and the pure shell model, respectively. It can be seen that the buckling loads from the proposed method and the distributing coupling method have a slight difference. However, the stress results at the interface from the proposed method are the same as that from the pure shell model.

6.9 CONCURRENT MULTI-SCALE MODELLING OF A TRANSMISSION TOWER

6.9.1 Background

Transmission towers are vital components of transmission lines. Most transmission tower structures are constructed by thin-walled angle members that are eccentrically connected to each other by bolts directly or through gusset plates. In the global analysis of a transmission tower, its angle members are often modelled using either pin-ended truss elements or fixended beam elements to form a global FE model for the tower (Roy et al. 1984; Albermani and Kitipornchai 2003; Chan and Cho 2008; Rao et al. 2010). Nevertheless, this kind of global model ignores the effects of joint flexibility, local geometric and material nonlinearity, bolt slippage and deformation on the global behaviour of the tower, which make the structural analysis and design of the tower inadequate. To overcome this problem, the joints in the global FE model are sometime modified as semi-rigid linear or nonlinear joints to consider joint effects on the global behaviour of the tower (Knight and Santhakumar 1993; Kitipornchai et al. 1994; Rao and Kalyanaraman 2001; Xu and Zhang 2001; Ungkurapinan et al. 2003; Jiang et al. 2011). In this regard, Rao and Kalyanaraman (2001) presented a nonlinear analysis method for lattice towers in consideration of member eccentricity, material nonlinearity and rotational rigidity of joints. Ungkurapinan et al. (2003) developed some formulas to describe joint slips based on the relevant test data. However, the use of semi-rigid joints cannot guarantee the accuracy of the analysis because there are a variety of joints, in terms of the number of bolts and the shape of gusset plates, making it difficult to determine the structural parameters for semi-rigid joints. For the analysis and design of the local joints of the tower, the local joints are then modelled using solid or shell elements (Cheng and Yam 1994; Salih et al. 2011, 2013; Rosenstrauch et al. 2013). The boundary conditions of a local joint model are often assigned by using the information extracted from the global analysis of the global FE model of the tower. This approach for the analysis and design of the global tower and local joints may be called *the information-passing multi-scale method* (Li et al. 2007). However, it is difficult to determine dynamic boundary conditions for the local joint model, and inaccurate boundary conditions will lead to error in the calculated structural responses. Furthermore, if the solid and shell elements are used to model all the members and joints of the tower, the computational size for the global structure analysis will be too large to be implemented.

This section aims at developing a concurrent multi-scale modelling method for transmission tower structures, in which critical joints of the tower are modelled in great detail using solid elements, while other angle members are modelled with common beam elements. The detailed model for a critical joint includes gusset plates, angle members and bolts. The effects of local geometric and material nonlinearity and the contact problem between the bolts, plates and angles are all taken into consideration. Multi-point constraints for beam-to-solid interface connections developed in Sections 6.5 through 6.7 will be used to ensure the computational accuracy and efficiency at interfaces, so that the critical local joint models can be coupled with the common tower model to form a multi-scale model of the tower. To verify the developed multi-scale modelling method, a physical model of a transmission tower structure at a length scale of 1:10 was constructed and tested. The displacements and strains of the tower model measured from the static tests are compared with the numerical results obtained by the multi-scale modelling method. The dynamic characteristics of the tower identified from the dynamic tests are also compared with the numerical results.

6.9.2 Physical model of a transmission tower structure

To verify the accuracy of multi-scale modelling and analysis of a transmission tower structure, a physical model of a transmission tower structure was built (see Figure 6.15). The prototype of the tower was a cup-type and straight-line tower, with height of 50.50 m and width of 22.02 m. It is used in 500 KV networks of the state grid of China, suitable for areas with heavy icing or high lightning incidences. The tower was assembled from 23 types of angle members, which were connected to each other at joint plates with bolts. Considering

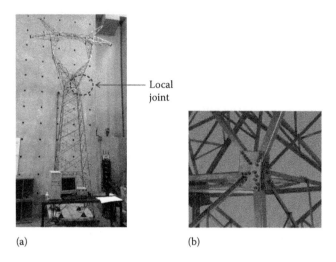

(a)   (b)

Figure 6.15 Physical model of a transmission tower: (a) global tower model and (b) local joint.

the physical modelling of bolts and joint plates as well as the availability of laboratory space, the length ratio of the reduced-scale model to the prototype was selected as 1:10. Stainless steel was chosen to make angle members and joint plates. The scaled model was designed and fabricated following the geometric similarity laws as closely as possible. To guarantee the precision of the local joints of the transmission tower, the components of the local joints, such as bolts and gusset plates, were also fabricated according to the length ratio. The angle members and gusset plates were tailor-made in a factory using stainless steel plates. The completed tower model had 930 angle members, 402 gusset plates and 3649 bolts. The completed tower model assembled in the laboratory is shown in Figure 6.15. Both static and dynamic tests were performed to provide test data for verifying the numerical results from the multi-scale analysis.

6.9.3 Multi-scale modelling of the transmission tower

There are many complex problems to deal with in the process of multi-scale modelling of a transmission tower, such as the interface coupling of mixed-dimensional elements and the contact problem between bolts and plates. Most of the commercial FE software has features to deal with these problems. It is therefore more convenient to build a multi-scale model of the test transmission tower using commercial FE software. The FE software ANSYS is used in this study together with the self-written supplemental programs for multi-scale modelling and analysis of the transmission tower structure.

The main purpose of the multi-scale modelling and analysis of a transmission tower is to obtain its global and local responses at the same time and to facilitate the design of the transmission tower. For the sake of a clear demonstration, only one typical and important tower joint between the crank arm and the tower body is selected to construct a detailed local FE joint model. The selected joint consists of 9 angle members of a shortened length, 3 gusset plates and 40 bolts. To accurately simulate the bolt connection, all the components of the joint are modelled using solid elements. Consequently, the 20-node SOLID95 elements of higher order, which can simulate irregular shapes with no loss in accuracy, are used to model angle members, gusset plate and bolts of the selected joint. Apart from this joint, all other members of the tower are modelled using beam elements and all other joints are

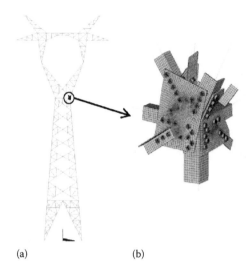

(a)   (b)

Figure 6.16 Multi-scale model of a transmission tower: (a) global tower model and (b) local joint model.

modelled as rigid joints, in which the BEAM188 elements are used to model all other angle members in the global tower. The completed multi-scale model of the transmission tower is shown in Figure 6.16 together with the local joint model. It is noted that there is an interface for each of the nine angle members used in the joint between the global tower model and the local joint model.

One of the most complex problems in the local modelling of the joint is the interaction between different components for bolt connection, such as the contact between the bolt and the angle member, the contact between the bolt and the gusset plate, and the contact between the angle member and the gusset plate. These interactions are achieved by using the contact elements TARGE170 and CONTA174 of a surface-to-surface type, which avoid one element to penetrate into another. Furthermore, these contact elements can simulate friction forces between the two surfaces according to the Mohr–Coulomb law. The friction forces depend on the smoothness of the surfaces and the pretension forces of the bolts. The value of the coefficient of friction is taken as 0.3 in this study. The pretension force elements PRETS179 are used to simulate the pre-tightening of the bolts. In terms of stiffness, the interaction between the angle member and gusset plate is defined as the *flexible-to-flexible* contact problem because both of them are of equal stiffness, while the interaction between the bolt and the gusset plate or angle member is defined as the *rigid-to-flexible* contact problem because the bolt is considered stiffer than the plate or the angle member. In the contact problem considered, the first part refers to the target and the second part to the contact surface. For a rigid-to-flexible contact, the target surface is always more stiff and the contact surface is always less stiff. The contact elements are applied to the joint components by the Augmented Lagrange formulation, and the contact stiffness is updated during equilibrium iteration. Figure 6.17 shows the contact interaction between the angle member, gusset plate and bolt of the local joint. Finally, the interface of the angle member between the solid and beam elements is coupled by using the constraint method.

6.9.4 Validation of multi-scale modelling

A static test was first carried out on the physical model of the transmission tower fixed on the ground of the laboratory. A horizontal concentrated load was applied at the middle of

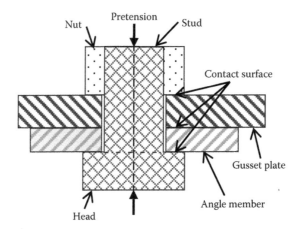

Figure 6.17 Contact between angle member, gusset plate and bolt of the local joint.

the cross arm of the transmission tower in the longitudinal direction (line direction), as shown in Figure 6.18. The concentrated load was increased step-by-step up to 60 N at an incremental load of 10 N. A total of 3 displacement transducers (D1–D3) and 26 strain gauges (S1–S26) were used to measure the static responses of the tower due to the concentrated load. Figure 6.18 shows the locations of the loading point and measurement points. After the static tests were completed, hammer tests were then carried out on the physical model of the transmission tower to identify the natural frequencies and modal shapes of the tower. For the hammer test in the longitudinal direction, a total of 15 accelerometers were arranged and their locations are shown in Figure 6.19.

The beam model and the multi-scale model of the transmission tower are used to calculate the strain and displacement responses of the tower under the concentrated load, and the results from the multi-scale model and beam model are then compared with those from the static test. The comparison results of the displacement responses are listed in Table 6.5. It can be seen that the results obtained from the multi-scale model are very close to the test results. The maximum errors from the beam and multi-scale models are 6.8% and 3.4%, respectively, compared with the test results. The comparison results of the strain responses to the horizontal concentrated load of 60 N are listed in Table 6.6 for the main member of the tower leg. The strain results obtained from the multi-scale model are very close to the test results. Figure 6.20 shows the locations of strain gauges on the gusset plate and the main member of the local joint, which are modelled in great detail using the solid elements in the multi-scale model of the tower. The comparison results of strain responses are listed in Table 6.7. The equivalent strain used for the comparison of strain responses on the gusset plate is computed by

$$\varepsilon_e = \frac{1}{1+v}\left[\frac{(\varepsilon_1 - \varepsilon_2)^2 + (\varepsilon_2 - \varepsilon_3)^2 + (\varepsilon_3 - \varepsilon_1)^2}{2}\right]^{1/2} \tag{6.21}$$

where:

$\varepsilon_1, \varepsilon_2$ and ε_3 are the principal strains
v is the effective Poisson's ratio

In the static test, strain rosettes are used to measure the strain state of the gusset plate, and two principal strains are calculated from the measured strain state. Equation 6.21 is

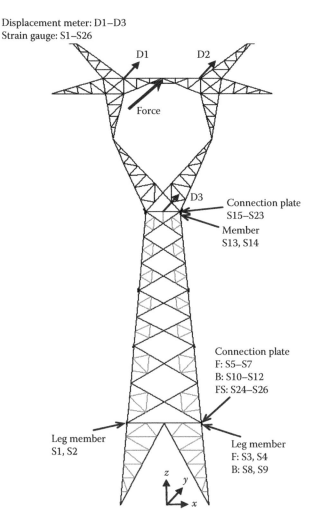

Displacement meter: D1–D3
Strain gauge: S1–S26

D1   D2

Force

D3   Connection plate
S15–S23

Member
S13, S14

Connection plate
F: S5–S7
B: S10–S12
FS: S24–S26

Leg member
S1, S2

Leg member
F: S3, S4
B: S8, S9

Figure 6.18 Locations of loading point and measurement points in static test.

then used to calculate the equivalent strain by combining the two principal strains obtained from the measured results of the strain rosette and a zero principal strain. From Table 6.7, it can be seen that the strain responses of the main member of the local joint calculated from the multi-scale model are more accurate than those from the beam model. The multi-scale model can obtain the strain responses of the gusset plate with a good accuracy of a maximum error less than 7.90%. Comparing the static responses of the displacement and strain, it can be concluded that the multi-scale model can obtain more accurate strain and displacement responses at the region near the local joint modelled by solid elements. It should be pointed out that the multi-scale model of the tower was updated and detailed information on the updating of the multi-scale model of a civil structure can be found in Chapter 7.

A modal analysis is carried out using the beam model and the multi-scale model of the transmission tower. The natural frequencies are obtained and compared with the test results listed in Table 6.8. It can be seen that the maximum error using the multi-scale model occurs at the sixth natural frequency with a relative error of 5.41%. The first, second and third

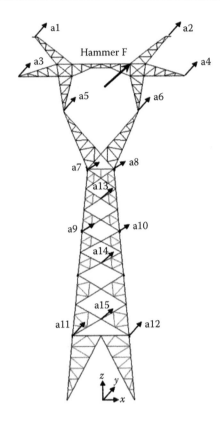

Figure 6.19 Locations of accelerometers in a hammer test.

Table 6.5 Comparison of static displacement responses (mm)

Measure point	Load (N)	Measured displacement	Beam model	Error (%)	Multi-scale model	Error (%)
D1	10.000	0.224	0.228	1.8	0.228	1.8
	20.000	0.468	0.457	−2.4	0.457	−2.4
	30.000	0.701	0.685	−2.3	0.686	−2.1
	40.000	0.915	0.913	−0.2	0.915	0.0
	50.000	1.132	1.142	0.9	1.144	1.1
	60.000	1.378	1.370	−0.6	1.373	−0.4
D2	10.000	0.223	0.228	2.2	0.228	2.2
	20.000	0.462	0.456	−1.3	0.458	−0.9
	30.000	0.689	0.684	−0.7	0.688	−0.1
	40.000	0.907	0.911	0.4	0.917	1.1
	50.000	1.130	1.139	0.8	1.147	1.5
	60.000	1.377	1.367	−0.7	1.377	0.0
D3	10.000	0.059	0.063	6.8	0.061	3.4
	20.000	0.120	0.126	5.0	0.123	2.5
	30.000	0.181	0.189	4.4	0.184	1.7
	40.000	0.247	0.252	2.0	0.245	−0.8
	50.000	0.305	0.315	3.3	0.307	0.7
	60.000	0.367	0.378	3.0	0.368	0.3

Table 6.6 Comparison of strain on main member of tower leg (10^{-6})

Gauge number	Measured strain	Beam model	Error (%)	Multi-scale model	Error (%)
S1	21.59	22.33	3.43	22.34	3.47
S2	26.55	25.73	3.09	25.73	3.09
S3	26.93	25.69	4.60	25.68	4.64
S4	22.52	22.33	0.84	22.33	0.84
S8	−21.09	−22.30	5.74	−22.31	5.76
S9	−25.41	−25.71	1.18	−25.71	1.18

Figure 6.20 Locations of strain gauges on local joint.

mode shapes of the tower obtained from the multi-scale model are shown in Figure 6.21. It can be seen that the first mode of vibration is the translational mode mainly in the longitudinal direction. The second mode of vibration is also the translational mode but mainly in the lateral direction. The third mode of vibration is the torsional mode. The modal assurance criteria (MAC) is calculated as

$$\mathrm{MAC}\left(\phi_i^{\text{test}}, \phi_i^{\text{FE}}\right) = \frac{\left(\phi_i^{\text{test}} \phi_i^{\text{FE}}\right)^2}{\left(\phi_i^{\text{test}} \phi_i^{\text{test}}\right)\left(\phi_i^{\text{FE}} \phi_i^{\text{FE}}\right)} \tag{6.22}$$

where ϕ_i^{test} and ϕ_i^{FE} are the ith modal shape from the test and FE analysis, respectively. The MAC values for the beam model and the multi-scale model are listed in Table 6.9. It can be seen that the multi-scale model and the beam model have the same accuracy for modal shapes. The similar natural frequencies and mode shapes from the beam model and the multi-scale model are understandable because the difference between the beam model and the multi-scale model is one local joint model in the multi-scale model, which does not affect the global dynamic characteristics of the tower.

Table 6.7 Comparison of strain on local joint (10^{-6})

Gauge number	Measured strain	Beam model	Error (%)	Multi-scale model	Error (%)
S13	34.10	28.83	15.45	34.62	1.52
S14	12.27	24.48	99.51	11.70	4.65
S15, S16, S17	21.90	–	–	23.63	7.90
S18, S19, S20	8.20	–	–	8.82	7.56

Table 6.8 Comparison of natural frequencies (Hz)

Mode number	Measured frequency	Beam model	Error (%)	Multi-scale model	Error (%)
1	16.56	16.39	1.03	16.38	1.09
2	16.81	16.46	2.08	16.44	2.20
3	22.50	23.63	5.02	23.49	4.40
4	44.36	44.40	0.09	44.40	0.09
5	44.50	44.62	0.27	44.63	0.29
6	50.44	53.06	5.19	53.17	5.41
7	52.30	53.56	2.41	53.62	2.52
8	59.83	57.54	3.83	57.58	3.76
9	60.56	57.72	4.69	57.75	4.64

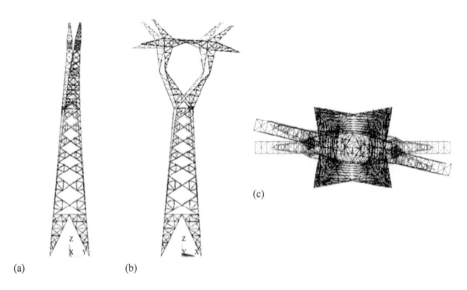

(a)   (b)   (c)

Figure 6.21 First three mode shapes of the tower obtained from the multi-scale model: (a) the first mode (view in x-direction), (b) the second mode (view in y-direction) and (c) the third mode (view in z-direction).

Table 6.9 MAC of modal shapes

Mode number	1	2	3	4	5	6	7	8	9
Beam model	0.9939	0.9958	0.9936	0.9407	0.9653	0.9889	0.9741	0.9089	0.9107
Multi-scale model	0.9940	0.9958	0.9935	0.9377	0.9664	0.9897	0.9719	0.9104	0.9075

NOTATION

A_i, a_i	The local coordinate systems (CSs) of the interface at the initial and current configurations, respectively
C	The coefficient matrix of the displacement constraint equation in the linear domain
$C_{NL}(u_S)$	The coefficient (weighting) matrix varying with u_S in the nonlinear domain
C_0, C_t	The coefficient matrix in its initial and current local CSs, respectively
$D(u_S)$	The coefficient matrix of the linearised form of the displacement constraint equation used at each increment
D_0, D_t	The coefficient matrices of the displacement constraint equation in its initial and current configuration, respectively, in the global CS
e_i	The global CS
$f_E(t)$	The force vector for a structural system
F_B, F_P, F_S	The nodal forces at the interface of beam, plate and shell, respectively
f^{SE}	The nodal force vector of the substructure at the middle section
K^{SE}, u^{SE}	The stiffness matrix and nodal displacement vector of the substructure, respectively
M_E, C_E, K_E	The mass matrix, viscous damping matrix and stiffness matrix of a structural system, respectively
N	The effective Poisson's ratio
R_0, R_t	The transformation matrices from the global CS to the initial and current local CS, respectively
R_d	The transformation matrix from the initial local CS to the current local CS
$T_{i1}, T_{i2}, T_{t1}, T_{t2}$	The coordinate transform matrices computing from R_0 and R_n, respectively
u_B, u_P, u_S	The beam displacements, plate displacement and shell displacement at the interface, respectively
$x(t)$	A vector containing nodal displacement
α, β	Rayleigh damping factors
$\delta u_B, \delta u_P, \delta u_S$	The virtual displacement vectors at the interface of beam, plate and shell, respectively
$\delta W_B, \delta W_P, \delta W_S$	The virtual works done by the nodal forces at the interface of beam, plate and shell, respectively
$\Delta U_B, \Delta U_S$	The displacement increments at the interface of beam and shell, respectively
ε_e	The equivalent strain
$\varepsilon_1, \varepsilon_2, \varepsilon_3$	The principal strains in different directions
$\phi_i^{test}, \phi_i^{FE}$	The ith modal shape from test and finite element analysis, respectively

REFERENCES

ABAQUS Analysis user's manual. 2010. Dassault Systèmes Simulia Corp., Providence, RI. http://www.maths.cam.ac.uk/computing/software/abaqus_docs/docs/v6.12/books/usb/default.htm.

Albermani, F.G.A. and S. Kitipornchai. 2003. Numerical simulation of structural behaviour of transmission towers. *Thin-Walled Struct.*, 41(2–3): 167–77.

Almansour, H.H., M.S. Cheung, and B.Y.B. Chan. 2010. Analysis and design of hybrid long span cable-stayed bridges using multi-scale modelling techniques. In *Proceedings of the International Conference on Computing in Civil and Building Engineering*, ed. W. Tizani. Nottingham University Press, Nottingham, England, Paper No. 267, 533–5.

ANSYS User's Manual. 2010. ANSYS. Inc. PA. http://148.204.81.206/Ansys/readme.html

Bathe, KJ. 1996. *Finite Element Procedures*. Prentice-Hall, Englewood Cliffs, NJ.

Bauman, P.T., H.B., Dhia, N. Elkhodja, J.T. Oden, and S. Prudhomme. 2008. On the application of the Arlequin method to the coupling of particle and continuum models. *Comput. Mech.*, 42(4): 511–30.

Belytschko, T., W.K. Liu, and B. Moran. 2001. *Nonlinear Finite Elements for Continua and Structures*. John Wiley & Sons, New York, NY.

Chan, S.L. and S.H. Cho. 2008. Second-order analysis and design of angle trusses, Part I: Elastic analysis and design. *Eng. Struct.*, 30(3): 616–25.

Chan, T.H.T., L. Guo, and Z.X. Li. 2003. Finite element modelling for fatigue stress analysis of large suspension bridges. *J. Sound and Vibr.*, 261(3): 443–64.

Chavan, K.S. and P. Wriggers. 2004. Consistent coupling of beam and shell models for thermo-elastic analysis. *Int. J. Numer. Methods Eng.*, 59(14): 1861–78.

Cheng, J.J.R. and M.C.H. Yam. 1994. Elastic buckling strength of gusset plate connections. *J. Struct. Eng.*, 120(2): 538–59.

Clough, R.W. and J. Penzien. 2003. *Dynamics of Structures, 3rd edition*. Berkeley, CA: Computers & Structures, Inc.

Cofer, W.F. and K.M. Will. 1991. A three-dimensional shell–solid transition element for general nonlinear analysis. *Comput. Struct.*, 38(4): 449–62.

Dhia, H.B. and G. Rateau. 2005. The Arlequin method as a flexible engineering design tool. *Int. J. Numer. Methods Eng.*, 62(11): 1442–62.

Garusi, E. and A. Tralli. 2002. A hybrid stress-assumed transition element for solid-to-beam and plate-to-beam connections. *Comput. Struct.*, 80(2): 105–15.

Gmür, T.C. and R.H. Kauten. 1993. Three-dimensional solid-to-beam transition elements for structural dynamics analysis. *Int. J. Numer. Methods Eng.*, 36(9): 1429–44.

Gmür, T.C. and A.M. Schorderet. 1993. A set of three-dimensional solid to shell transition elements for structural dynamics. *Comput. Struct.*, 46(4): 583–91.

Guidault, P.A. and T. Belytschko. 2007. On the L2 and the H1 couplings for an overlapping domain decomposition method using Lagrange multipliers. *Int. J. Numer. Methods Eng.*, 70(3): 322–50.

Ho, R.J.S., A. Meguid, Z.H. Zhu, and R.G. Sauve. 2010. Consistent element coupling in nonlinear static and dynamic analyses using explicit solvers. *Int. J. Mech. Mater. Des.*, 6(4): 319–30.

Jiang, W.Q., Z.Q. Wang, G. McClure, G.L. Wang, and J.D. Geng. 2011. Accurate modeling of joint effects in lattice transmission towers. *Eng. Struct.*, 33(5): 1817–27.

Kitipornchai, S., F. Albermani, and A.H. Peyrot. 1994. Effect of bolt slippage on the ultimate strength of latticed structures. *J. Struct. Eng.*, 120(8): 2281–87.

Knight, G.M.S. and A.R. Santhakumar. 1993. Joint effects on behavior of transmission towers. *J. Struct. Eng.*, 119(3): 698–712.

Li, Z.X., T.H.T. Chan, and J.M. Ko. 2001. Fatigue analysis and life prediction of bridges with structural health monitoring data—Part I: Methodology and strategy. *Int. J. Fatigue*, 23(1): 45–53.

Li, Z.X., T.H.T. Chan, Y. Yu, and Z.H. Sun. 2009. Concurrent multi-scale modeling of civil infrastructures for analyses on structural deterioration-Part I: Modeling methodology and strategy. *Finite Elem. Anal. Des.*, 45(11): 782–94.

Li, Z.X., T.Q. Zhou, T.H.T. Chan, and Y. Yu. 2007. Multi-scale numerical analysis on dynamic response and local damage in long-span bridges. *Eng. Struct.*, 29(7): 1507–24.

McCune, R.W., C.G. Armstrong, and D.J. Robinson. 2000. Mixed-dimensional coupling in finite element models. *Int. J. Numer. Methods Eng.*, 49(6): 725–50.

Rao, N.P. and V. Kalyanaraman. 2001. Non-linear behaviour of lattice panel of angle towers, *J. Constr. Steel. Res.*, 57(12): 1337–57.

Rao, N.P., G.M.S. Knight, N. Lakshmanan, and N.R. Lyer. 2010. Investigation of transmission line tower failures. *Eng. Fail. Anal.*, 17(5): 1127–41.

Rosenstrauch, P. L., M. Sanayei, and B.R. Brenner. 2013. Capacity analysis of gusset plate connections using the Whitmore, block shear, global section shear, and finite element methods. *Eng. Struct.*, 48: 543–57.

Roy, S., S. Fang, and E.C. Rossow. 1984. Secondary stresses on transmission tower structures. *J. Energy Eng.*, 10(2): 157–73.

Salih, E.L., L. Gardner, and D.A. Nethercot. 2011. Bearing failure in stainless steel bolted connections. *Eng. Struct.*, 33(2): 549–62.

Salih, E.L., L. Gardner, and D.A. Nethercot. 2013. Numerical study of stainless steel gusset plate connections. *Eng. Struct.*, 49: 448–64.

Shim, K.W., D.J. Monaghan, and C.G. Armstrong. 2002. Mixed dimensional coupling in finite element stress analysis. *Eng. Comput.*, 18(3): 241–52.

Surana, K.S. 1980. Transition finite elements for three-dimensional stress analysis. *Int. J. Numer. Methods Eng.*, 15(7): 991–1020.

Surana, K.S. 1982. Geometrically non-linear formulation for the three dimensional solid-shell transition finite elements. *Comput. Struct.*, 15(5): 549–66.

Ungkurapinan, N., S.R.De S. Chandrakeerthy, R.K.N.D. Rajapakse, and S.B. Yue. 2003. Joint slip in steel electric transmission towers. *Eng. Struct.*, 25(6): 779–88.

Wagner, W. and F. Gruttmann. 2002. Modeling of shell-beam transitions in the presence of finite rotations. *Comput. Assisted Mech. Eng. Sci.*, 9: 405–18.

Wang, F.Y., Y.L. Xu, and W.L. Qu. 2014. Mixed-dimensional finite element coupling for structural multi-scale simulation. *Finite Elem. Anal. Des.*, 92: 12–25.

Wang, J., Z. Lou, X. Min, and J. Zou. 1996. A DOF expanding method for connecting solid and shell element. *Commun. Numer. Methods Eng.*, 12(6): 321–30.

Wellmann, C. and P. Wriggers. 2012. A two-scale model of granular materials. *Comput. Meth. Appl. Mech. Eng.*, 205–208:46–58.

Xu, Y.L., J.M. Ko, and W.S. Zhang. 1997. Vibration studies of Tsing Ma long suspension bridge. *J. Bridge Eng.*, 2(4): 149–56.

Xu, Y.L. and W.S. Zhang. 2001. Modal analysis and seismic response of steel frames with connection dampers. *Eng. Struct.*, 23(4): 385–96.

Zienkiewicz, O.C. and R.L. Taylor. 1989. *The Finite Element Method*, 4th edition volume 1. McGraw-Hill, New York, NY.

Chapter 7

System identification and model updating

7.1 PREVIEW

A finite element (FE) model of a large civil structure, based on design drawings as described in Chapter 6, inevitably has modelling errors compared with its real structure, due to construction variations, uncertainties in boundary conditions, variations in material properties, inaccuracy in FE model discretisation, uncertainties in external excitations and so on. The initial FE model therefore needs to be finely tuned to match references to reduce modelling errors and produce a better FE model to represent the real one.

These references are usually structural responses measured from the structure on site and/or structural dynamic characteristics extracted from measured structural responses. This procedure is known as *model updating* and the identification of structural dynamic characteristics from the measured structural responses is called *modal identification*. Modal identification is part of a system of identification which includes parameter identification as well.

In this regard, this chapter first presents the mathematical model of a dynamic civil structure in three different coordinate systems. Structural dynamic characteristics are introduced through a modal analysis in the modal coordinate system. This chapter then illustrates modal identification methods in the frequency domain, time domain and frequency-time domain, in which how to extract dynamic characteristics of a structure from measurement data is demonstrated. In consideration of the subsequent chapters, this chapter also gives a brief introduction to force identification. The model updating methods, mainly the sensitivity-and-modal-based updating method, are then fully discussed. Finally, the model updating of a multi-scale FE model of a transmission tower is given in detail as a case study.

7.2 MATHEMATICAL DESCRIPTION OF A DYNAMIC CIVIL STRUCTURE

The equilibrium equations of motion of a linear civil structure subjected to dynamic loadings can be expressed in a structural coordinate system, a state-space coordinate system or a modal coordinate system.

Consider a civil structure idealised by an N degrees of freedom (DOFs) system subjected to dynamic loadings. The equations of motion of the structure can be expressed in a structural coordinate system as

$$\mathbf{M}\ddot{\mathbf{x}}(t) + \mathbf{C}\dot{\mathbf{x}}(t) + \mathbf{K}\mathbf{x}(t) = \mathbf{E}\mathbf{f}(t) \tag{7.1}$$

where:

M, C and K   are the $N \times N$ system mass, damping and stiffness matrices of the structure, respectively

$\mathbf{x}, \dot{\mathbf{x}}$ and $\ddot{\mathbf{x}}$ are the displacement, velocity and acceleration response vectors, respectively

f(t)   is an m-dimensional external excitation vector

E   is an $N \times m$ matrix defining the location of m excitations

The equations of motion of the structure can also be reformulated in a state-space coordinate as

$$\dot{\mathbf{z}}(t) = \mathbf{A}\mathbf{z}(t) + \mathbf{B}\mathbf{f}(t) \tag{7.2}$$

where

$$\mathbf{A} = \begin{bmatrix} \mathbf{0} & \mathbf{I} \\ -\mathbf{M}^{-1}\mathbf{K} & -\mathbf{M}^{-1}\mathbf{C} \end{bmatrix}, \ \mathbf{B} = \begin{bmatrix} \mathbf{0} \\ \mathbf{M}^{-1}\mathbf{E} \end{bmatrix} \text{ and } \mathbf{z}(t) = \begin{bmatrix} \mathbf{x}(t) \\ \dot{\mathbf{x}}(t) \end{bmatrix} \tag{7.3}$$

in which 0 and I denote the null matrix and the identity matrix of appropriate dimensions, respectively.

Furthermore, the second-order dynamic equations of motion of the structure can also be expressed in a modal coordinate system by

$$\ddot{\mathbf{q}} + 2\xi\boldsymbol{\omega}_o\dot{\mathbf{q}} + \boldsymbol{\omega}_o^2\mathbf{q} = \boldsymbol{\Phi}^T\mathbf{E}\mathbf{f} \tag{7.4}$$

where:

q   is the vector of modal coordinates

Φ is the mass normalised displacement mode shapes

ξ is the modal damping coefficient matrix

ω_o is the modal frequency matrix

The major advantage of using modal coordinates is that the DOFs can be substantially reduced by truncating high modes that often have minimal and negligible contributions to structural responses under in-service conditions.

7.3 MODAL ANALYSIS AND FREQUENCY RESPONSE FUNCTION

As a theoretical background of modal identification methods, modal analysis and frequency response function (FRF) are introduced here. Let us start from the un-damped free vibration equation:

$$\mathbf{M}\ddot{\mathbf{x}}(t) + \mathbf{K}\mathbf{x}(t) = 0 \tag{7.5}$$

It is known that Equation 7.5 has the solution of

$$\mathbf{x}(t) = \mathbf{X}e^{i\omega t} \tag{7.6}$$

where:

    $\mathbf{X}$ is an $N \times 1$ vector of time-independent amplitudes

    i is the imaginary unit

    ω is the vibration circular frequency

Substituting the solution into Equation 7.5, we have

$$\left(\mathbf{K} - \omega^2 \mathbf{M} \right) \mathbf{X} e^{i\omega t} = 0 \tag{7.7}$$

or

$$\left(\mathbf{K} - \omega^2 \mathbf{M} \right) \mathbf{X} = 0 \tag{7.8}$$

Equation 7.8 is a standard eigenvalue problem. As the trivial solution is of no interest, the determinant of $(\mathbf{K} - \omega^2 \mathbf{M})$ must be zero, resulting in N possible positive real solutions, ω_2^2, ..., ω_N^2 known as the eigenvalues. ω_1, ω_2, ..., ω_N are the un-damped natural frequencies of the system. For each natural frequency ω_r, Equation 7.8 has a set of nonzero vector solution of $\mathbf{X}$, denoted as $\{\phi_r\}$ and known as the rth mode shape (or eigenvector) of the system. Each mode shape vector contains N elements which are relative values only. All mode shape vectors form a mode shape matrix Φ. Because of the orthogonality property of the mode shape matrix, Φ is usually normalised with respect to the mass matrix as

$$\Phi^T \mathbf{M} \Phi = \mathbf{I} \tag{7.9}$$

Equation 7.8 then yields

$$\Phi^T \mathbf{K} \Phi = \Lambda \tag{7.10}$$

where:

    $\mathbf{I}$ is a unit matrix

    Λ is a diagonal matrix consisting of the eigenvalues, ω_1^2, ω_2^2, ..., ω_N^2

By considering a harmonic excitation force vector $\{\mathbf{E}f(t)\} = \mathbf{F} e^{i\omega t}$ in Equation 7.1, the steady-state responses are also harmonic with the same frequency, that is, $\mathbf{x}(t) = \mathbf{X} e^{i\omega t}$, yielding

$$\left(-\omega^2 \mathbf{M} + i\omega \mathbf{C} + \mathbf{K} \right) \mathbf{X} e^{i\omega t} = \mathbf{F} e^{i\omega t} \tag{7.11}$$

and

$$\mathbf{X} = \left(-\omega^2 \mathbf{M} + i\omega \mathbf{C} + \mathbf{K} \right)^{-1} \mathbf{F} = \mathbf{H}(\omega) \mathbf{F} \tag{7.12}$$

where $\mathbf{H}(\omega)$ is the $N \times N$ FRF matrix. In the case of Rayleigh damping, it has

$$\Phi^T \mathbf{H}^{-1} \Phi = \Phi^T \left(-\omega^2 \mathbf{M} + i\omega \mathbf{C} + \mathbf{K} \right) \Phi = -\omega^2 \mathbf{I} + i\omega(a_1 \mathbf{I} + a_2 \Lambda) + \Lambda \tag{7.13}$$

where a_1 and a_2 are the proportional coefficients for constructing the damping matrix. The right-hand side of Equation 7.13 is a diagonal matrix, in which the rth item can be expressed as $-\omega^2 + i\omega(a_1 + a_2\omega_r^2) + \omega_r^2 = -\omega^2 + 2i\omega\omega_r\xi_r + \omega_r^2$. The rth modal damping ratio ξ_r is defined as

$$\xi_r = \frac{a_1}{2\omega_r} + \frac{a_2\omega_r}{2} \tag{7.14}$$

Therefore, from Equation 7.13, the FRF matrix can be associated with the modal frequency, mode shapes and damping ratio as

$$\mathbf{H} = \mathbf{\Phi} \begin{bmatrix} \ddots & & \\ & \dfrac{1}{-\omega^2 + 2i\omega\omega_r\xi_r + \omega_r^2} & \\ & & \ddots \end{bmatrix} \mathbf{\Phi}^T \tag{7.15}$$

The receptance h_{jk}, defined as the response displacement at coordinate j due to a harmonic excitation force at coordinate k with a frequency of ω and all other forces being zero, is then given by

$$h_{jk}(\omega) = \frac{\mathbf{X}_j}{\mathbf{F}_k} = \sum_{r=1}^{N} \frac{\phi_{jr}\phi_{kr}}{-\omega^2 + 2i\omega\omega_r\xi_r + \omega_r^2} \tag{7.16}$$

Equation 7.16 is often expressed as

$$h_{jk}(\omega) = \sum_{r=1}^{N} \left(\frac{A_{jkr}}{i\omega - s_r} + \frac{A_{jkr}^*}{i\omega - s_r^*} \right) \tag{7.17}$$

where:

$s_r = -\omega_r\xi_r + i\omega_r\sqrt{1 - \xi_r^2}$ is the pole
A_{jkr} is the residue for mode r
Superscript *   denotes the complex conjugate

The pole is directly related to the frequency and damping ratio, and the residue is related to the mode shape. Therefore, the FRF (or poles and residues) can be employed to extract the modal parameters. Equations 7.15 and 7.16 show that

1. The FRF matrix is a bridge connecting the modal parameters to the system matrices.
2. It is clear that the FRF matrix is symmetric as $h_{jk} = h_{kj}$, which is called *reciprocity*.
3. The FRF matrix is the summation of the contributions of different vibration modes.
4. One column or one row of the FRF matrix includes all information of modal parameters, and therefore it is sufficient to extract the full modal parameters using one column or one row of the FRF vector only.

7.4 MODAL IDENTIFICATION IN THE FREQUENCY DOMAIN OR THE TIME DOMAIN

In the 1970s and 1980s, a vast range of modal identification methods was developed. According to the domain in which the data is treated, there are frequency-domain and time-domain methods. Maia et al. (1997) and Ewins (2000) described these methods extensively. As far as the excitation method is concerned, field tests can be generally divided into forced vibration tests and ambient vibration tests, although free vibration tests are employed occasionally. Correspondingly, the modal identification methods are called *input-output methods* and *output only methods*, respectively.

In forced vibration tests, the structure is excited by artificial means and both input and output need to be measured. The tests often require very heavy equipment and involve expensive resources in order to provide controlled excitation at a sufficient level (Okauchi et al. 1997). Impulse hammers, drop weights and electro-dynamic shakers are the excitation equipment usually used for relatively small structures.

Quite a few standard techniques of modal parameter estimation have been developed. In brief, these modal parameter identification methods are based on the FRFs in the frequency domain or the impulse response functions (IRFs) in the time domain. An FRF is the ratio of the output (or response) to the input (or force) in the frequency domain. In practice, it can also be calculated as the ratio of the cross-power spectrum between the output and input to the auto-power spectrum of the input, although there are a few variants (Maia et al. 1997).

Modal identification methods that can only handle a single FRF at a time are called *single-input-single-output* (SISO) methods. More commonly, a structure is tested with a set of sensors simultaneously collecting the responses to one excitation (position is fixed). Each pair of input-output forms an FRF and the total pairs are actually one column of the FRF matrix. The methods allowing for several FRFs to be analysed simultaneously are called global or single-input-multiple-output (SIMO) methods. These methods are based on the fact that the natural frequencies and damping ratios of a structure are global properties of the structure and do not vary from one FRF to another (Maia et al. 1997). Finally, the methods that can simultaneously process a few columns of FRFs from the responses to various excitations are called *multiple-input-multiple-output* (MIMO) methods. These type of methods can increase the spatial resolution (the physical measurement locations) of the test as one input may not excite some particular modes sufficiently.

Among many, two very common and useful methods, the peak amplitude method (Bishop and Gladwell 1963) and the rational fraction polynomial method (Richardson and Formenti 1982), are introduced here. The former is a simple, rapid and useful modal identification method during the tests. The natural frequencies are simply taken from the observation of the peaks (or resonances) on the amplitude graphs of the FRFs. The mode shapes are estimated from the ratios of the magnitudes of the imaginary part (or amplitude) of the FRFs at the resonance points (Avitabile 1999) without heavy computational efforts. In the rational fraction polynomial method, the FRF is formulated in a rational fraction form in terms of orthogonal Forsythe polynomials. When the error function between the formulated FRF and the measurement is minimised, the modal parameters can be estimated. The rational fraction polynomial method was expanded to the case of multiple FRFs, named as the *global rational fraction polynomial* method (Richardson and Formenti 1985), such that the frequencies and damping ratios can be obtained from all FRFs consistently and mode shapes from each FRF. This method is now one of the most popular SIMO frequency domain methods and has been adopted by many commercial packages of modal analysis software.

In contrast, ambient vibrations, induced by traffic, winds, water waves and pedestrians, are the natural or environmental excitations of buildings or bridges. An ambient vibration test has a few significant advantages over the forced vibration tests. First, it does not interrupt the operating condition of the structure. Second, it is inexpensive as it does not require excitation equipment. Due to these merits, ambient vibration tests have been widely used in vibration tests of tall buildings and long-span bridges such as the Republic Plaza in Singapore (Brownjohn 2003), the New CCTV Tower in China (Xu et al. 2013), the Golden Gate Bridge in the United States (Abdel-Ghaffer and Scanlan 1985), the Tsing Ma Suspension Bridge in Hong Kong (Xu et al. 1997), the Kap Shui Mun Cable-Stayed Bridge in Hong Kong (Chang et al. 2001) among many others.

In the case of ambient vibration tests, only output data (responses) are measured and actual loading conditions are unknown. Modal parameter identification methods using output-only measurements present a challenge, requiring the use of special identification techniques, which can deal with small magnitude ambient vibration contaminated by noise without knowledge of the input forces. Over the past decades, several output-only modal parameter identification techniques have been developed such as the autoregressive moving-average (ARMA) method (Andersen et al. 1996), the natural excitation technique (NExT) (James et al. 1995) and the stochastic subspace identification (SSI) method (Van Overschee and De Moor 1996; Peeters 2000).

It is noted that these methods are based on a fundamental assumption that the unknown excitation is taken as a white noise process. For example, it can be proved that the cross-correlation function between the responses from a multiple DOFs system excited by multiple white noise random inputs, has the same form as the IRF of the system (Caicedo et al. 2004). Consequently the time-domain modal parameter estimation methods, such as the eigensystem realisation algorithm (ERA) (Juang and Pappa 1985), can be employed to extract the modal parameters.

In practice, each set of measurements in ambient vibration tests is taken for a relatively long period, usually from minutes to hours. One reason is to allow wider frequency components included in the excitation, which conforms to the stationary assumption of the input. In addition, more measurement data results in a better resolution in the frequency domain. Finally, the effect of measurement noise can be reduced by averaging data.

Reference points also need to be considered in practical ambient vibration tests as the number of available sensors is usually insufficient to measure all measurement points at one time. The points can be divided into a few groups and the measurement is taken in several sets, while the reference points should be kept unchanged throughout the entire test.

Using output-only modal identification methods, one cannot obtain an absolute scaling of the identified mode shapes (e.g. mass normalisation) as the input is unknown. This is a big difference, sometimes also a disadvantage, from that in the forced vibration tests.

7.5 MODAL IDENTIFICATION IN THE FREQUENCY-TIME DOMAIN

In the ambient vibration test–based modal identification methods, the fundamental assumption is that the unknown excitation can be represented by a stationary white noise process and that the structural parameters are invariant. In reality, the unknown excitations may be non-stationary and the structural parameters may be varying, particularly when the structure is subject to some damage. In this regard, modal identification methods in the frequency-time domain, capable of yielding a time–frequency representation, have been developed. These methods include short-time Fourier transform (STFT), Wigner–Ville distribution (WVD), wavelet transform (WT) and Hilbert–Huang transform (HHT). These methods can also be used for modal identification of linear structures under stationary excitation.

The STFT is the simplest approach, where the original record is divided into a series of shorter records with the time duration chosen to ensure that the sub records can be considered stationary and the structural parameters can be regarded as linear. Each sub record can then be analysed using the standard fast Fourier transform (FFT) approach and a time-varying estimate of the spectral quantities can be obtained. However, the high resolution for STFT cannot be obtained in both time and frequency domains simultaneously. In contrast, the WVD represents a truly two-dimensional transformation and there is no restriction on the simultaneous resolution in time and frequency domains. However, this approach is much more computationally intensive. Although some approaches based on STFT and WVD have been developed by several researchers for system identification (e.g. Staszewski et al. 1997; Avargel and Cohen 2007), the WT and HHT are the more adaptable candidate techniques as compared with STFT and WVD, especially for civil structures (Gurley and Kareem 1999; Kijewski and Kareem 2003; Yan and Miyamoto 2006). This section focuses on the introduction of WT and HHT for modal identification.

The time–frequency character of WT allows the adaptation of both traditional time- and frequency-domain system identification approaches to examine non-stationary data and nonlinear structural parameters. In the analysis of MDOF systems, WT has the ability to decouple the measured multi-component signals to mono-component signals via some specific mother wavelets, for example, Morlet wavelet, Daubechies wavelet and Symlets wavelet, and represent them with forms of complex-valued signals. According to the instantaneous amplitudes and phase angles of the complex-valued signals, the corresponding modal parameters such as damping ratio and natural frequency can then be identified. In general, the linear least squares fitting technique or the Hilbert transform are applied to the wavelet phases and amplitude curves for effective determination of such modal parameters. Wavelet-based system identification methods through the analysis of free vibration or IRFs from SDOF or MDOF structures have been developed (e.g. Robertson et al. 1998a,b; Hans et al. 2000; Lamarque et al. 2000; Ashory et al. 2013). For many civil structures, the IRF or the free vibration curve is usually difficult to obtain, so there is greater interest in developing WT-based approaches to permit the extraction of modal parameters from ambient vibration data by the combination of some specific techniques, such as the well-known random decrement technique (RDT) (Ruzzene et al. 1997; Lardies and Gouttebroze 2002; Yan et al. 2006; Chen et al. 2009; Nagarajaiah and Basu 2009). Notably, because of the nature of WT, the resolution properties have inevitable influences on the identification accuracies of the modal parameters. Research conducted by Kijewski and Kareem (2003) investigated the impacts of modal separation and end-effect on the time–frequency resolution and suggested guidelines for parameter selection.

The HHT method proposed by Huang et al. (1998, 1999), performs a time adaptive decomposition operation, termed *empirical mode decomposition* (EMD), to decouple the signal into a finite number of intrinsic mode functions (IMF), then the Hilbert transform is applied to each IMF to obtain the time–frequency representation of the signal designated as the Hilbert–Huang spectrum (HHS) in terms of instantaneous amplitude and phase angle. It has been found that the HHT can be implemented to decouple multi-component signals to mono-component signals and represent them with the analogous forms of complex-valued signals (Yan and Miyamoto 2006). According to the instantaneous amplitudes and phase angles of the complex-valued signals, the corresponding modal parameters such as damping ratio, natural frequency and mode shapes can be identified. Many researchers devoted their effort to develop the HHT-based modal identification approaches for the estimation of modal parameters of MDOF linear structures (Yang et al. 2003a,b, 2004; Xu et al. 2003;

Liu et al. 2011). Since nonlinearity widely exists in civil structures, researches on the investigation of HHT-based identification methods for nonlinear structures were also actively carried out (Feldman 1994a,b, 1997, 2007; Pai et al. 2008; Yan et al. 2012). Moreover, the implementation of HHT for the identification of time-varying parameters, for example, the smoothly varying, periodically varying or abruptly varying stiffness, has also developed recently (Shi and Law 2007; Shi et al. 2009; Wang and Chen 2012; Wang et al. 2012). It was emphasised that, especially at the high-frequency region, the EMD may generate some undesired IMFs, each of which may contain more than one frequency component. In such cases, the intermittence check as well as the band-pass filter is employed to separate the close modes (Chen and Xu 2002).

7.6 FORCE IDENTIFICATION

Knowledge on external excitations acting on a structure is very important for the design, control, damage detection and life management of the structure, but it is difficult to obtain in the case of civil structures. Force identification methods have therefore been developed in the past few decades. Some methods rely on direct instrumentation and measurement. For example, wind pressure taps can be arranged on the surface of a building to measure wind forces. Highway vehicle loadings on a bridge can be measured from weigh-in-motion (WIM) systems or an instrumented vehicle (Heywood 1996; Jacob et al. 2002). Other methods require only the measured responses of the structure, for instance, moving force identification methods developed to identify highway or railway vehicle loadings on a bridge. Techniques that identify the forces in motion without disturbing the normal operation have attracted interest from many engineers and researchers. A detailed review has been presented by Zhu and Law (2013) on two main moving force identification methods, namely, those based on the modal superposition method and the FE method.

Moving force identification is not only an inverse problem but also an ill-posed problem because the structural responses are typically continuous vector functions of spatial coordinates and are defined at only a few points. Solutions are frequently unstable in the sense that small perturbations in the responses would result in large changes in the calculated force magnitudes. The solution algorithms of the ill-posed problem have been investigated by many researchers with particular emphasis on different analytical techniques with different types of measurements (Zhu and Law 2013). The modal condensation techniques were found most useful for solving the problem, with only limited measured information required compared with the total number of DOFs of the structure.

7.7 MODEL UPDATING METHODS

Accurate FE models are frequently required in a large number of applications, such as optimisation design, damage identification, structural control and structural performance evaluation. Due to the uncertainties in geometry, material properties and boundary conditions, the dynamic responses of a structure predicted from a highly idealised numerical model usually differ from those measured from the as-built structure. For example, Brownjohn and Xia (2000) reported that the difference between the experimental and numerical modal frequencies of a curved cable-stayed bridge exceeded 10% for most modes of vibration and could reach 40% for some particular modes. In another study (Brownjohn et al. 2003), 18% difference was found between the analytical and measured frequencies. Jaishi and Ren

(2005) discovered that the natural frequencies of a steel arch bridge differed from the FE model predictions by up to 20%, and the modal assurance criterion (MAC) values could be 62%. Similarly, Zivanovic et al. (2007) found that the differences in the natural frequencies for footbridges predicted by a very reasonable FE model before updating could be as large as 29.8%, as compared with the experimental counterparts. Therefore, an effective model updating is necessary to obtain a more accurate FE model to best represent the real structure. Although the measurement noise is also unavoidable, the measurement data obtained from a well-designed test with well-designed data processing are usually regarded as accurate and used as the reference for updating.

Theoretically, many types of measurement data can be employed for model updating, while most exercises in past decades have been based on modal data or their variants, such as frequencies, mode shapes, FRFs, modal flexibility, modal curvatures and modal strain energy (Doebling et al. 1996). Other types of data used include response time histories (Ge and Soong 1998) and static strain data (Sanayei and Saletnik 1996). In this regard, this section focuses on model updating using measured natural frequencies and mode shapes. The next section will introduce the mixed modal and static data–based updating method for the multi-scale FE model of a transmission tower. The FRF-based model updating will be introduced in Chapter 14.

Model updating is intrinsically linked with damage detection. Both aim to determine the difference between two models using measurement data. For model updating, the difference is the modelling errors, while for damage detection, it is structural damage. The model updating methods, which are relevant to damage detection, can be categorised mainly as optimal matrix updating methods, sensitivity-based updating methods and eigenstructure assignment methods (Doebling et al. 1996). The optimal matrix updating method aims to find an updated matrix that is closest to the original matrix and that produces the measured natural frequencies and mode shapes. The drawback is that such a matrix is not able to preserve the physical connectivity of a structure. The eigenstructure assignment method was used in control engineering to dictate the force response of a structure. It is difficult to apply this method for civil structures, particularly large civil structures. The mathematical base of sensitivity-based model updating is derivatives, which means that this method is suitable for dealing with small physical variable changes. The sensitivity-based method often uses the iterative approach to modify the physical parameters of the FE model repeatedly to minimise the discrepancy between the measurement data and the analytical counterparts. Other than changing matrices directly, this approach adjusts the physical parameters in elemental or substructural level. Then, the system stiffness matrix and mass matrix are assembled from all elements in the discrete FE model. Thereby,

1. The matrix properties of symmetry, sparseness and positive-definiteness are guaranteed.
2. The structural connectivity is preserved.
3. The changes in the updated global matrices are represented by the changes in the updated parameters.

Due to these merits, iterative methods have become more popular.

This section therefore focuses on the sensitivity-based updating method with iterative approach in terms of updating algorithms and on the modal-based updating method in terms of measurement data. The sensitivity-and-modal-based FE model updating mainly includes three aspects: the objective function with constraints, updating parameters and optimisation scheme or algorithms. They are introduced separately below in detail in the following sections.

7.7.1 Objective functions and constraints

The objective functions usually comprise the difference between the measured quantities and the modal predictions. The selection of the measurement data type must take into account not only the sensitivity of the data to the structural parameters, but also the inherent uncertainties in the measurement. In general, as found by many researchers, frequencies can be measured with high accuracy, but are not spatially specific and not sensitive to damages. Mode shapes have the advantage of being spatially specific. However, it is difficult to measure high-order mode shapes, which are sensitive to local damages. The advantage of using modal flexibility is that the flexibility matrix converges quickly with the first few modes only. However, measuring the modal flexibility with a sufficient spatial resolution is often experimentally impractical, and so is the modal strain energy method.

The measured quantities concerned in this section are natural frequencies and mode shapes. The eigenvalue equation of an N-DOFs un-damped FE model is given as

$$\left(-\lambda_i \mathbf{M} + \mathbf{K}\right)\mathbf{\Phi}_i = 0 \tag{7.18}$$

where:

λ_i is the ith eigenvalue ($\lambda_i = \omega_i^2$)
$\mathbf{\Phi}_i$ is the ith mass normalised displacement mode shape

When there is change or inaccuracy in the mass and/or stiffness, the natural frequencies and mode shapes will change accordingly. Model updating is to find the change or inaccuracy in the structural properties based on the measured frequencies and mode shapes.

In the objective functions, different types of data can be weighted differently accounting for their importance and measurement accuracy. For example, it is widely accepted that the natural frequencies of a structure can be measured more accurately than mode shapes. Thus, higher weights can be given to the frequency data. An objective function considering all the above factors can be (Hao and Xia 2002)

$$J = \sum_{i=1}^{nm} W_{fi}^2 \left[f_i(\mathbf{\theta})^A - f_i^E \right]^2 + \sum_{i=1}^{nm} W_{\phi i}^2 \sum_{j=1}^{np} \left[\phi_{ji}(\mathbf{\theta})^A - \phi_{ji}^E \right]^2 \tag{7.19}$$

where:

f_i and ϕ_{ji}	are the ith natural frequency and associated mode shape at the jth point, respectively
W_f and W_ϕ	are the weight coefficients of the frequency and mode shape, respectively
nm and np	denote the number of modes and the number of measurement points, respectively
Subscripts A and E	denote the items from the analytical model and the measurement, respectively
θ	includes a vector of the structural parameters to be updated

In Equation 7.19, the relative difference of frequencies can be used instead of the absolute difference (Friswell et al. 1998). From the computational point of view, using eigenvalues as in Equation 7.20 instead of frequencies is preferred, as calculation of the eigenvalue sensitivity is more convenient than that of the frequency sensitivity (Hao and Xia 2002).

$$J = \sum_{i=1}^{nm} W_{\lambda i}^2 \left[\frac{\lambda_i(\boldsymbol{\theta})^A - \lambda_i^E}{\lambda_i^E} \right]^2 + \sum_{i=1}^{nm} W_{\phi i}^2 \sum_{j=1}^{np} \left[\phi_{ji}(\boldsymbol{\theta})^A - \phi_{ji}^E \right]^2 \tag{7.20}$$

One advantage of using Equation 7.20 is that the measured modal data can be compared with the analytical data at the corresponding points and modes directly as

1. The number of identified modes is limited by the frequency range due to the limitation of the excitation level and identification techniques.
2. The number of measurement locations is limited and is always less than that of the analytical model.
3. Some DOFs, such as rotational and internal ones, cannot be measured with present technologies.

Therefore, model reduction or data expansion can be avoided. Special attention, however, should be paid to the following issues:

- The analyst should roughly know the quality of the measurement data. Inclusion of all data does not necessarily lead to more meaningful results because some modes may have higher uncertainty than others.
- Only corresponding modes can be compared in Equation 7.20. Therefore, mode matching is necessary and MAC is usually employed for this purpose. During the updating process, the sequence of some modes may change and thus mode tracking is necessary in each iteration. The MAC value of two mode shapes $\boldsymbol{\Phi}_i^E$ and $\boldsymbol{\Phi}_j^A$ is defined by (Ewins 2000)

$$\mathrm{MAC}\left(\boldsymbol{\Phi}_i^E, \boldsymbol{\Phi}_j^A\right) = \frac{\left| \left\{ \boldsymbol{\Phi}_i^E \right\}^T \boldsymbol{\Phi}_j^A \right|^2}{\left(\left\{ \boldsymbol{\Phi}_i^E \right\}^T \boldsymbol{\Phi}_i^E \right) \left(\left\{ \boldsymbol{\Phi}_j^A \right\}^T \boldsymbol{\Phi}_j^A \right)} \tag{7.21}$$

- The measured mode shapes may have different scales from the analytical ones, which are usually mass normalised. In the case of ambient vibration tests or where the response of the driving point is not measured in force vibration tests, mass-normalised mode shapes are not available. The modal scale factor (MSF) can be employed so that the two pairs of mode shapes have the same phase. This quantity is defined as (Ewins 2000)

$$\mathrm{MSF}\left(\boldsymbol{\Phi}_i^E, \boldsymbol{\Phi}_i^A,\right) = \frac{\left\{ \boldsymbol{\Phi}_i^E \right\}^T \boldsymbol{\Phi}_i^A}{\left\{ \boldsymbol{\Phi}_i^A \right\}^T \boldsymbol{\Phi}_i^A} \tag{7.22}$$

Then, the analytical mode shape is adjusted by multiplying the MSF.

In model updating, the range of the updating parameters usually needs to be constrained with bounds, such that the updated parameters are in an acceptable and reasonable range. For example, many structural parameters should be larger than zero. In the case of damage detection, the damaged structure is usually weaker than the intact one. Optimisation with constraints is generally more difficult to solve than the unconstrained one.

7.7.2 Parameters for updating

The selection of parameters to be updated is very critical to a successful model updating and requires engineering judgements. If incorrect model parameters are selected, the updated model either cannot reproduce the required dynamic properties accurately or does not represent the real structure. In the latter case, although the objective function may reduce to a value below the specified threshold, the updated model has no real meaning and the obtained parameters are, in effect, the compensation for the real model parameters. Therefore, the analyst should first know the type and location of the inaccuracy of the model that needs updating, based on the features of the measurement data and his or her knowledge of the model. Apparently this is closely associated with the modelling of the structure of interest.

The second issue is the number of updating parameters, which depends on the computational resource. For a small structure in terms of the size of system matrices, the updating parameters can be selected at the elemental level and each element may have single or multiple updating parameters. For a large civil structure such as a suspension bridge, the element number of the model is usually more than thousands. In that case, the calculation of eigensolutions and the associated eigen sensitivities is very time-consuming. Updating can then be carried out at the component level or substructure level only with the present computational ability. The substructuring-based model updating (Weng et al. 2011) can be a promising method and merits further development.

A set of updating parameters may include various types of structural parameters whose values may be in a broad range. However, different parameters have different sensitivities. Normalising the parameters can make the changes in these updating parameters at a similar level. Consequently a poor condition during the optimisation is avoided. For example, in the widely used element-by-element model updating, the element bending rigidity is usually employed as the elemental stiffness parameter. Then, the element stiffness matrix is proportional to the parameter as

$$\mathbf{K}_i = \alpha_i \mathbf{K}^e \tag{7.23}$$

where:

$\mathbf{K}_i$ is the ith element stiffness matrix
α_i is the element bending rigidity
$\mathbf{K}^e$ has the identical form and size for the same element type, which is the function of geometry of the element

The global stiffness matrix is the assembly of all ne elements as

$$\mathbf{K} = \sum_{i=1}^{ne} \mathbf{K}_i = \sum_{i=1}^{ne} \alpha_i \mathbf{K}^e \tag{7.24}$$

We assume the geometrical information of the element is accurate and only the bending rigidity contributes to the model inaccuracy and needs to be updated. Consequently, the global stiffness matrix of the updated structure can be similarly assembled as

$$\tilde{\mathbf{K}} = \sum_{i=1}^{ne} \tilde{\mathbf{K}}_i = \sum_{i=1}^{ne} \tilde{\alpha}_i \mathbf{K}^e \tag{7.25}$$

where the symbol '~' stands for the corresponding quantities after updating.

In this manner, the matrix properties and the structural connectivity are preserved after updating, as described previously. One can define a stiffness reduction factor as the actual updating parameter:

$$\theta_i = \frac{\tilde{\alpha}_i - \alpha_i}{\alpha_i} \tag{7.26}$$

Then, all the updating parameters have a value larger than –1 because the updated bending rigidity should be larger than zero.

7.7.3 Optimisation algorithm

The optimisation algorithm is the technique used to minimise the objective functions. According to the nature of the objective functions and constraints, optimisation methods can be classified as linear and nonlinear. Another classification is unconstrained and constrained optimisation problems. Constraints can be classified as equal and unequal.

As there are quite a few different methods in solving general optimisation problems, we do not plan to introduce all of these but to focus on the model updating problem, which can be solved in a more efficient way by utilising its characteristics. The problem of interest in Equation 7.20 is actually a nonlinear least-squares problem, which is the sum of squares of error functions. To exemplify the algorithm, only eigenvalues are considered in Equation 7.20 with unit weights for all items. The equation is rewritten in a vector form as

$$J = \sum_{i=1}^{nm} \left[\lambda_i(\boldsymbol{\theta}) - \lambda_i^E \right]^2 = \left\{ \boldsymbol{\lambda} - \boldsymbol{\lambda}^E \right\}^T \left\{ \boldsymbol{\lambda} - \boldsymbol{\lambda}^E \right\} \tag{7.27}$$

The minimum of J is reached when the gradient is zero. By differentiating J with respect to parameter θ, one has

$$\frac{\partial J}{\partial \boldsymbol{\theta}} = 2 S^T \left(\boldsymbol{\lambda} - \boldsymbol{\lambda}^E \right) = 0 \tag{7.28}$$

where S is the sensitivity matrix of eigenvalues with respect to the parameters and is usually calculated using Nelson's method (Nelson 1976). The eigenvalues can be approximated to the first-order Taylor series expansion with respect to the current point as

$$\boldsymbol{\lambda} = \boldsymbol{\lambda}^k + S^k \left\{ \Delta \boldsymbol{\theta} \right\}^k \tag{7.29}$$

where superscript k denotes the quantity in the kth iteration. For brevity, superscript k on S is omitted hereinafter. Substituting Equation 7.29 into Equation 7.28 leads to

$$S^T S \{ \Delta \boldsymbol{\theta} \}^k = S^T \left(\boldsymbol{\lambda}^E - \boldsymbol{\lambda}^k \right) = S^T \{ \Delta \boldsymbol{\lambda} \}^k \tag{7.30}$$

where $\Delta \lambda$ is the eigenvalue residual.

At each iterate k, the eigenvalue residual can be calculated and then the parameter change can be solved. The aforementioned approach is called the *Gauss–Newton method*. In practice, the updated function J may not be lower than that in the previous iteration. A fraction of the increment τ is employed such that $\{\theta\}^{k+1} = \{\theta\}^k + \tau\{\Delta\theta\}^k$. The search of such an optimal fraction is referred to as a linear search, which is very commonly used in optimisation. If the optimal fraction is close to zero, implying that the increment direction is not effective, the so-called Levenberg–Marquardt algorithm can be employed and the corresponding equation is revised as

$$\left(S^T S + \mu\, I\right)\{\Delta\theta\}^k = S^T\{\Delta\lambda\}^k \tag{7.31}$$

where:

    μ is known as the Marquardt parameter
    I is usually an identity matrix

It is noted that the increment direction is identical to that of the Gauss–Newton method when $\mu = 0$ and the steepest descent direction when μ tends to infinity (Nocedal and Wright 1999). An additional benefit of the Levenberg–Marquardt algorithm is that including a matrix in the left-hand side of Equation 7.31 improves the matrix condition as the sensitivity matrix S might be ill-conditioned.

It is noted that the original nonlinear least-squares problem (see Equation 7.27) has been transferred into a series of linear least-squares problems at each iteration. In practice, one would not solve Equations 7.30 and 7.31 using the inversion of the matrix directly, but via the Cholesky decomposition, QR decomposition, or singular value decomposition.

As mentioned previously, the updated parameters have physical meaning and are bounded, which makes the problem (see Equation 7.27) a constrained optimisation. The upper and lower bounds can then be set beforehand as

$$l^b \leq \theta \leq u^b \tag{7.32}$$

where:

    l^b is the lower bounds on parameter θ
    u^b is the upper bounds on parameter θ

If the parameter is normalised and θ is defined as Equation 7.26 in damage detection, for example, the bounds can be set as $-0.99 \leq \theta \leq 0$ in practice. The constrained optimisation can be solved with active set methods (Bjorck 1996). The active set methods divide the constraints into active and inactive sets. An active constraint refers to the parameter on the lower bound or upper bound, that is, the equality is satisfied, and an inactive constraint implies the parameter falls inside the bound, that is, the inequality is satisfied. Accordingly, the parameters are divided into fixed variables and free variables. The former will not change during the iteration and can be removed from the problem. Consequently, the problem is reduced with a set of free variables without constraints, which can be solved with the Gauss–Newton method or the Levenberg–Marquardt method. The main drawback of the active set methods is that the active and inactive sets need to be updated at each iteration by adding or dropping one constraint. The gradient projection method that allows the active set to change rapidly is more efficient. Readers may refer to the work by More and Toraldo (1989) on the topic.

Among many optimisation algorithms, the evolutionary algorithm, in particular, the genetic algorithm (GA), is worth mentioning as it has been used in a number of model updating and system identification exercises in structural engineering communities (Larson and Zimmerman 1993; Mares and Surace 1996; Friswell et al. 1998; Perry et al. 2006; Hao and Xia 2002). The GA is based on the principle of the evolutionary theory, such as natural selection and evolution (Holland 1975). Compared with the traditional optimisation and search algorithms, the GA differs in several fundamental ways:

- It works with the coding of the decision variables, not the decision variables themselves.
- It searches from a population of points in the region of the whole solution space rather than a single point, and therefore it is a global optimisation method.
- It has the advantage of easy implementation because only an objective function is required and derivatives or other auxiliary information are not necessary.
- It is based on probabilistic transition rules, not deterministic rules.

7.8 MULTI-SCALE MODEL UPDATING

In Chapter 6, the concept of multi-scale modelling of a civil structure was introduced and a multi-scale model of a transmission tower was established to capture both global performance and local behaviour of the tower. The multi-scale model of the transmission tower was updated and validated through the comparison of both dynamic characteristics and static strains between the measurement and numerical computation. This section introduces the model updating method used for updating the multi-scale model of the transmission tower.

7.8.1 Objective functions and updating parameters for multi-scale FE model

Compared with the conventional FE model updating described in Section 7.7, the model updating of a multi-scale FE model requires the updating of both global and local models simultaneously. Accordingly, updating parameters shall be selected from both global and local models. For the multi-scale model of the transmission tower as described in Chapter 6, the global parameters of the structure required for updating refer to the elastic moduli of angle members, the densities of angle members and the lumped masses at the rigid joints. The local parameters refer to the elastic moduli of gusset plates and bolts and the densities of gusset plates and bolts of the selected local joint. Moreover, the selected updating parameters should be sensitive to the residuals. Accordingly, the structural responses used to calculate the residuals are classified as either global responses or local responses. For the concerned multi-scale model of the transmission tower, the dynamic characteristics (natural frequencies and mode shapes) and the static responses of displacement and strain from the global model are regarded as global responses, while the strain responses of the selected local joint are taken as local responses. The residuals between the experimental (test) and numerical quantities can be expressed as follows:

$$R_i^{\text{displacement}} = \frac{u_i^{\text{FE}} - u_i^{\text{test}}}{u_i^{\text{test}}} \times 100$$

$$R_i^{\text{strain}} = \frac{\varepsilon_i^{\text{FE}} - \varepsilon_i^{\text{test}}}{\varepsilon_i^{\text{test}}} \times 100$$

$$R_i^{\text{Frequency}} = \frac{f_i^{\text{FE}} - f_i^{\text{test}}}{f_i^{\text{test}}} \times 100 \tag{7.33}$$

$$R_i^{\text{MAC}} = (1 - \text{MAC}_i) \times 100$$

where:

$R_i^{\text{displacement}}$, R_i^{strain}, $R_i^{\text{Frequency}}$ and R_i^{MAC}	are the ith residual of displacement, strain, frequency and MAC, respectively
u_i^{FE} and u_i^{test}	are the ith displacement responses from FE analysis and test, respectively
$\varepsilon_i^{\text{FE}}$ and $\varepsilon_i^{\text{test}}$	are the ith strain responses from FE analysis and test, respectively
f_i^{FE} and f_i^{test}	are the ith frequencies from FE analysis and test, respectively

The ith MAC is calculated by the MAC of the ith modal shape from the test and FE analysis. Since different types of responses have different units, it is important to ensure that its contribution to the objective function does not depend on the unit. The residuals expressed by Equation 7.33 are thus dimensionless and convenient for comparison.

The residuals shall be further classified as either global or local. For the concerned tower, the residuals of the strain responses from the local joint model constructed by solid elements are defined as local residuals, and other residuals are referred as global residuals. The local residuals include the strains of local angle members and gusset plates. The global residuals consist of frequency residuals, MAC residuals and displacement residuals. The strain residuals of angle members in the global model are also referred to as global residuals. This classification is made according to the results from sensitivity analysis, because the global and local strain residuals are sensitive to different updating parameters. The strain residuals of gusset plates of the local joint are sensitive to the elastic modulus of the gusset plate, while the strain residuals of angle members of the local joint are sensitive to the elastic modulus of the angle member only. The global and local objective functions are therefore constructed in terms of the global and local residuals, respectively, with the following expressions.

$$J_{\text{global}} = \left\{ J^{\text{Frequency}} = \sum_i \left| R_i^{\text{Frequency}} \right|, J^{\text{MAC}} = \sum_i \left| R_i^{\text{MAC}} \right|, J^{\text{displacement}} = \sum_i \left| R_i^{\text{displacement}} \right|, \right.$$

$$\left. J_{\text{global}}^{\text{strain}} = \sum_i \left| R_i^{\text{global strain}} \right| \right\} \tag{7.34}$$

$$J_{\text{local}} = \left\{ J_{\text{local}}^{\text{strain}} = \sum_i \left| R_i^{\text{local strain}} \right| \right\}$$

where J_{global} is the global objective function and J_{local} is the local objective function.

There are four sub-objective functions $\{J^{\text{Frequency}}, J^{\text{MAC}}, J^{\text{displacement}}, J_{\text{global}}^{\text{strain}}\}$ in the global objective function but only one sub-objective function $J_{\text{local}}^{\text{strain}}$ in the local objective function.

7.8.2 Multi-objective optimisation algorithm

The conventional FE model updating employs the optimisation of the single objective function, formed by the weighted summation of all sub-objective functions, which can be expressed as follows:

$$J_{\text{single}} = w_1 J^{\text{Frequency}} + w_2 J^{\text{MAC}} + w_3 J^{\text{displacement}} + w_4 J^{\text{strain}}_{\text{global}} + w_5 J^{\text{strain}}_{\text{local}} \tag{7.35}$$

where w_1, w_2, w_3, w_4 and w_5 are the weighting factors corresponding to the objective functions of frequency, MAC, displacement, global strain and local strain, respectively.

The FE model updating is to find the updated parameters through the minimisation of the objective function. However, the minimum value from the single objective optimisation varies with the selection of weighting factors. It is difficult to select the optimal weighting factors to obtain the preferred FE model. The trial-and-error approach, using varying weighting factors is often used to determine the weighting factors, but it has low efficiency and cannot guarantee to find the preferred FE model. A multi-objective optimisation algorithm was therefore developed to deal with the minimisation of a vector of objective functions subject to a number of constraints or bounds (Konak et al. 2006). A multi-objective optimisation problem (MOP) is formulated as follows:

$$\text{minimise} \quad J(\boldsymbol{\theta}) = \left\{ J_1(\boldsymbol{\theta}), J_2(\boldsymbol{\theta}), ..., J_q(\boldsymbol{\theta}) \right\}$$

$$\text{subject to} \quad \boldsymbol{\theta}_l \leq \boldsymbol{\theta} \leq \boldsymbol{\theta}_u \tag{7.36}$$

where:

    $\boldsymbol{\theta}$ is the vector of updating parameters

    $\boldsymbol{\theta}_l$ is the lower bounds of $\boldsymbol{\theta}$

    $\boldsymbol{\theta}_u$ is the upper bounds of $\boldsymbol{\theta}$

    $J(\boldsymbol{\theta})$ is a vector of objective functions such as $\left\{ J^{\text{Frequency}}, J^{\text{MAC}}, J^{\text{displacement}}, J^{\text{strain}}_{\text{global}}, J^{\text{strain}}_{\text{local}} \right\}$ in the concerned transmission tower

It should be noted that the optimality of MOP is not obvious because a solution $\boldsymbol{\theta}$ that simultaneously minimises all sub-objectives does not generally exist. Instead, the Pareto optimality (also called *non-inferiority*) is used to characterise the objectives (Tomoiaga et al. 2013). The Pareto optimality is a solution in which it is impossible to make any one objective better off without making at least one objective worse off. Compared with single-objective optimisation, the multi-objective approach obtains all the alternative updated models in a single run. The most preferred FE model can be selected from the Pareto optimality with the help of the higher-level information. Here, the preferred FE model is selected by using the maximum residual of each Pareto optimal solution. The maximum of residuals is calculated for each Pareto solution by using the following expression:

$$R^{\text{max}}_j = \max \left\{ R^{\text{displacement}}_i(\boldsymbol{\theta}_j), R^{\text{global strain}}_i(\boldsymbol{\theta}_j), R^{\text{local strain}}_i(\boldsymbol{\theta}_j), R^{\text{Frequency}}_i(\boldsymbol{\theta}_j), R^{\text{MAC}}_i(\boldsymbol{\theta}_j) \right\} \tag{7.37}$$

where $\boldsymbol{\theta}_j$ is the jth Pareto optimal solution. Then, the solution with the minimum R^{max}_j is selected as the preferred FE model with the updated parameters.

In this section, a non-dominated sorting genetic algorithm-II (NSGA-II) is employed for the implementation of multi-objective optimisation to find the Pareto optimal solution. The flowchart of NSGA-II is shown in Figure 7.1. The program of GAMULTIOBJ, which is a prewritten code in MATLAB for the implementation of the NSGA-II method (MATLAB 2010), is employed herein. The GAMULTIOBJ uses a controlled elitist GA. An elitist GA always favours individuals with a better fitness value. A controlled elitist GA also favours

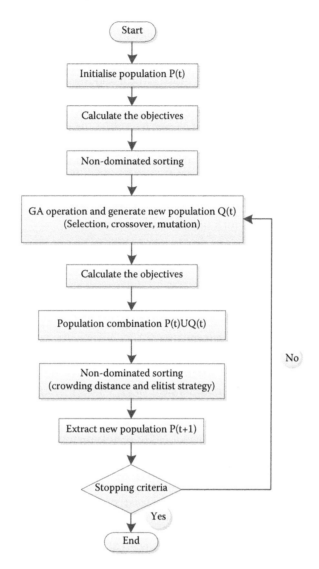

Figure 7.1 Flowchart of NSGA-II.

individuals that can help increase the diversity of the population even if they have a lower fitness value. It is important to maintain the diversity of the population for convergence to the Pareto optimal front. The fast non-dominated sorting is carried out on the combined population to obtain the candidate solutions as the Pareto optimal font. Then, a crowding distance assignment is carried out based on the objective function values to maintain the diversity of new solutions by distributing the solutions evenly on the Pareto optimal front. The process is iteratively carried out every time until a given stopping criteria is met.

7.8.3 Kriging meta-model for multi-objective optimisation

From the flowchart of NSGA-II, it can be seen that the objective functions are calculated for every iteration process. Therefore, a multi-scale analysis of the structure is required to be carried out for each iteration process, which needs huge computation time. To reduce the computation effort of the model updating, the Kriging meta-model was built (Simpson 1998) and used to replace the multi-scale analysis in the iterative optimisation process. In the Kriging meta-modelling, the unknown functions are the residuals of displacements, strains, natural frequencies and MAC. The variables are updating parameters, such as elastic modulus and density.

The selection of sample points in the implementation of the Kriging method is important, and it was decided to extract as much information as possible about the structural responses at hand with as few functional evaluations as possible in the design of experiment (DOE). There are a few schemes proposed in the literature, such as a central composite design and a Latin hypercube design. The central composite design has limited applications, because it needs too many design points as the number of design variables becomes large. The Latin hypercube design scheme is more suitable for the Kriging meta-modelling and is applicable to the DOE with a large number of design variables (Wang et al. 2014). Thus, the Latin hypercube design is employed in the Kriging meta-modelling herein.

Before using the Kriging meta-model in the multi-scale FE model updating, it is required to verify whether the meta-model has enough accuracy or not. The meta-model performance in prediction at untried points is evaluated using the criteria in terms of mean square error (MSE) and R^2:

$$\text{MSE} = \frac{1}{N_C} \sum_{i=1}^{N_C} \left(y_i - \hat{y}_i \right)^2 \tag{7.38}$$

$$R^2 = 1 - \frac{\sum_{i=1}^{N_C} \left(y_i - \hat{y}_i \right)^2}{\sum_{i=1}^{N_C} \left(y_i - \bar{y} \right)^2} = 1 - \frac{\text{MSE}}{\text{variance}} \tag{7.39}$$

where:

$\hat{y}_i$ and y_i are the ith prediction values from the Kriging meta-model and actual function values, respectively

$\bar{y}$ is the mean of all true values

N_C is the total number of the check points

While MSE represents the departure of the meta-model from the actual simulation model, the variance captures how irregular the problem is. The larger the value of R^2 the more accurate the Kriging meta-model.

7.8.4 Implementation procedure of multi-scale model updating method

The multi-scale model updating method discussed in Sections 7.8.1 through 7.8.3 actually consists of two main processes. The first involves the Kriging meta-modelling of the residuals to replace the multi-scale analysis, and the second is an optimisation process, which is carried out using the NSGA-II. The flowchart of the multi-scale model updating based on

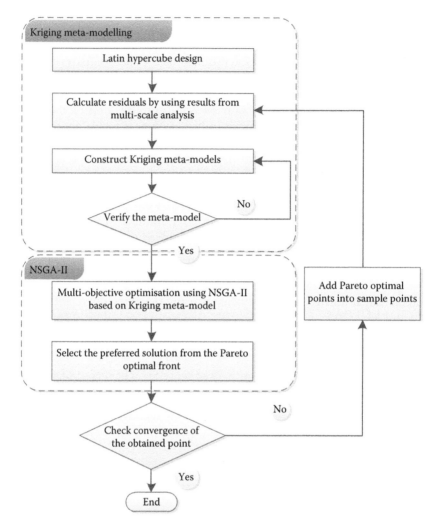

Figure 7.2 Flowchart of multi-scale model updating based on the NSGA-II and the Kriging meta-model.

the NSGA-II and the Kriging meta-model is shown in Figure 7.2. The multi-scale model updating procedures are listed as follows:

1. Latin hypercube design is applied to obtain the sample points.
2. The residuals are calculated by using the results from multi-scale analysis.
3. The Kriging meta-models of the residuals are constructed by determining the regression and correlation models together with their parameters.
4. The performance of the Kriging meta-model is assessed in terms of MSE and R^2 of the predicted values. If the performance of the Kriging meta-model is not good enough, go back to step 3 to renew the Kriging meta-model.
5. NSGA-II is then used to solve the multi-objective optimisation based on the Kriging meta-models, from which the Pareto optimal solutions are obtained.
6. The final solution is chosen from a set of Pareto optimal solutions as the optimal solution that corresponds to the Pareto solution with the minimum defined by Equation 7.37.

7. Convergence shall be checked for the obtained optimal solution. If there is no improvement in minimising the maximum residual after several iteration, stop the iteration and output the solution. Otherwise, the Pareto optimal points are added into the initial sample points and go to step 2 to build a new Kriging meta-model for the next optimisation.

7.9 MULTI-SCALE MODEL UPDATING RESULTS OF A TRANSMISSION TOWER

Detailed information on the multi-scale modelling of the transmission tower can be found in Chapter 6. Both static and dynamic tests were performed to provide test data for multi-scale model updating of the transmission tower. The details of the tests and testing results can be found in Chapter 6 as well. The multi-scale model updating method described in Section 7.8 is now applied to this transmission tower structure. The updating parameters, objective functions and multi-objective optimisation are first discussed. The multi-scale model updating is then performed to find the updated parameters for the transmission tower structure. Finally, the comparison between the Kriging method and the quadratic polynomial response surface (QPRS) method is performed to assess their performance in the multi-objective optimisation.

7.9.1 Multi-objective optimisation model for model updating

Based on the test types and results of the transmission tower, the multi-objective optimisation model was established for multi-scale model updating, which consists of objective functions, constraints and updating parameters. Firstly, the residuals of displacements, strains, natural frequencies and MACs are calculated using Equation 7.33. In this study, there are a total of 31 residuals used in the model updating of the transmission tower. The three objective functions are defined by these residuals as follows:

$$J_1(\boldsymbol{\theta}) = \sum_{i=1}^{9} \left| R_i^{\text{frequency}} \right| + \sum_{i=1}^{9} \left| R_i^{\text{MAC}} \right|$$

$$J_2(\boldsymbol{\theta}) = \sum_{i=1}^{3} \left| R_i^{\text{displacement}} \right| + \sum_{i=1}^{6} \left| R_i^{\text{strain}} \right| \qquad (7.40)$$

$$J_3(\boldsymbol{\theta}) = \sum_{i=7}^{10} \left| R_i^{\text{strain}} \right|$$

where:
- $J_1(x)$ and $J_2(x)$ are the global objective functions
- $J_3(x)$ is the local objective function
- $\boldsymbol{\theta}$ is the vector of updating parameters

The strain residuals are divided into the global residuals $R_1^{\text{strain}} - R_6^{\text{strain}}$ (S1, S2, S3, S4, S8 and S9) and the local residuals $R_7^{\text{strain}} - R_{10}^{\text{strain}}$ (S13–S20). The strain gauges (S13–S20) are all installed on the local joint.

The selection of effective updating parameters is a crucial step in model updating. Since the multi-scale FE model of the transmission tower structure has many elements, it is impractical to associate one updating parameter with each element. Here, one model parameter is used for all the angle members with the same thickness. The sensitivities are then used for the selection of parameters, in which parameters with lower sensitivity are eliminated. Here, a total of 13 updating parameters are selected and listed in Table 7.1. The selected updating parameters refer to seven elastic moduli, four densities and two lumped masses as follows:

$$\text{Elasticity modulus} \quad e_i = \theta_i e_i^0 \quad (i = 1,\ldots,7)$$

$$\text{Density} \quad \rho_i = \theta_i \rho_i^0 \quad (i = 8,\ldots,11) \quad (7.41)$$

$$\text{Lumped mass} \quad M_i = \theta_i M_i^0 \quad (i = 12,\ldots,13)$$

where:

θ_i is the ith updating parameter

e_i and e_i^0 are the ith elastic moduli of updated and initial multi-scale model, respectively

ρ_i and ρ_i^0 are the ith densities of updated and initial multi-scale model, respectively

M_i and M_i^0 are the ith lumped masses of updated and initial multi-scale model, respectively

Moreover, constraints are required for the updating parameters according to physical significance. The constraint of the selected updating parameters is expressed as

$$0.8 \leq \theta_i \leq 1.2 \quad i = 1,\ldots,13 \quad (7.42)$$

Table 7.1 Selected updating parameters

Number	Description of the parameter	Type
1	The equivalent modulus of elasticity of angle members with thickness of 0.4 mm	Global parameter
2	The equivalent modulus of elasticity of angle members with thickness of 0.5 mm	Global parameter
3	The equivalent modulus of elasticity of angle members with thickness of 0.8 mm	Global parameter
4	The equivalent modulus of elasticity of angle members with thickness of 1.0 mm	Global parameter
5	The equivalent modulus of elasticity of angle members with thickness of 1.2 mm	Global parameter
6	The equivalent modulus of elasticity of gusset plates	Local parameter
7	The equivalent modulus of elasticity of angle members at the joint	Local parameter
8	The equivalent mass density of angle members with thickness of 0.4 mm	Global parameter
9	The equivalent mass density of angle members with thickness of 0.5 mm	Global parameter
10	The equivalent mass density of angle members with thickness of 0.8 mm	Global parameter
11	The equivalent mass density of angle members with thickness of 1.0 mm	Global parameter
12	The equivalent lumped mass of the cross arm of the tower	Global parameter
13	The equivalent lumped mass of the tower body	Global parameter

With the defined objective functions, selected updating parameters and their varying ranges, the multi-scale model updating can be performed according to the implementation procedure given in Section 7.8. Firstly, the Kriging meta-model is built for computing the residuals defined by Equation 7.33 using the Latin hypercube sampling. The performance of the Kriging meta-model is verified in terms of MSE and R^2 values defined by Equations 7.38 and 7.39. The optimisation model is then constructed, and the multi-objective optimisation is finally carried out to achieve the updated model of the transmission tower using the NSGA-II method. The parameters used in the NSGA-II method are listed in Table 7.2. The Pareto solution front is shown in Figure 7.3 with three objective functions. The preferred FE model is selected by the Pareto solution with the minimum $R^{\max}(\theta_j)$. The model updating is implemented until there is no improvement in minimising $R^{\max}(\theta_j)$.

7.9.2 Model updating results and discussions

In this section, the updated parameters of the transmission tower structure are first presented. The static responses and dynamic characteristics calculated by the initial and updated FE models are then compared with those from the tests to demonstrate the effectiveness of the proposed multi-scale model updating method.

There are a total of 13 updated parameters. The updated parameters corresponding to the most preferred multi-scaled FE model are listed in Table 7.3. It can be seen that the maximum change in all the updating parameters is the equivalent modulus of elasticity of the gusset plates. This is because the strain responses of the gusset plates predicted by the initial model have the largest error compared with the test results as shown in Table 7.6.

Table 7.2 Parameters used in NSGA-II method

Parameters	Values and operators	Parameters	Values and operators
Population size	1000	Selection	Tournament
Population type	Double	Probability of crossover	0.9
No. of generation	2000	Probability of mutation	0.1
Distance measure	Distance-crowding	Tolerance	10^{-6}

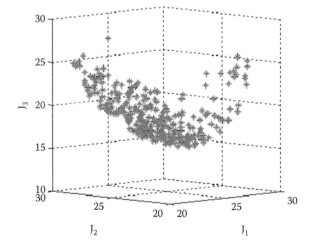

Figure 7.3 Pareto front with three objective functions.

Table 7.3 Updated parameters

Number	Initial values	Updated values
1	1.00	0.96
2	1.00	1.05
3	1.00	1.01
4	1.00	0.95
5	1.00	0.96
6	1.00	1.17
7	1.00	0.97
8	1.00	0.90
9	1.00	1.14
10	1.00	0.91
11	1.00	1.12
12	1.00	1.16
13	1.00	0.85

Furthermore, there is only one updating parameter with a change of more than 10% among the seven updating parameters of elastic modulus, but there are four updating parameters with a change of more than 10% among the six updating parameters of mass (including densities and lumped masses). This is because the masses of the gusset plates and bolts are about 15.6% of the total mass of the transmission tower but it is difficult to accurately estimate the masses of the gusset plates and bolts.

Both the initial and updated multi-scale FE models of the transmission tower are used to calculate the strain and displacement responses of the tower under the concentrated load. The results from the initial and updated multi-scale model are then compared with those from the static tests. The comparison results of the displacement responses are listed in Table 7.4. It can be seen that the results obtained from the updated multi-scale model are closer to the test results compared with those from the initial model. The maximum errors from the initial and updated multi-scale models are 4.39% and 3.03%, respectively, compared with the test results. The comparison results of the strain responses to the horizontal concentrated load of 60N are listed in Tables 7.5 and 7.6 for the main angle member of the tower leg and local joint, respectively. It can be seen that the strain results obtained from the updated multi-scale model are more accurate than those from the initial model, especially for the main angle members of the local joint which is modelled in a great detail using the solid elements in the multi-scale model of the tower. The locations of the strain gauges mentioned in Tables 7.4 through 7.6 can be found in Chapter 6 and will not be detailed herein to avoid redundant description. From the comparison of the static responses of displacement and strain, it can be concluded that the proposed multi-scale model updating method can improve not only the accuracy of the global static response but also the accuracy of the local static response of the multi-scale FE model of the tower.

Table 7.4 Comparison of static displacement responses (mm)

Gauge number	Test result	Initial model	Error (%)	Updated model	Error (%)
D1	1.378	1.384	0.44	1.37	−0.58
D2	1.377	1.38	0.20	1.367	−0.75
D3	0.367	0.383	4.39	0.378	3.03

Table 7.5 Comparison of static strain responses of main angle members of tower leg (10^{-6})

Gauge number	Measured strain	Initial model	Error (%)	Updated model	Error (%)
S1	21.59	22.89	6.02	22.34	3.47
S2	26.55	26.28	1.02	25.73	3.09
S3	26.93	26.23	2.60	25.68	4.64
S4	22.52	22.88	1.60	22.33	0.84
S8	−21.09	−22.86	8.39	−22.31	5.78
S9	−25.41	−26.25	3.31	−25.71	1.18

Table 7.6 Comparison of static strain responses of the local joint (10^{-6})

Gauge number	Measured strain	Initial model	Error (%)	Updated model	Error (%)
S13	34.10	36.47	6.95	34.60	1.47
S14	12.27	11.87	3.26	11.67	4.89
S15, S16, S17	21.90	28.55	30.37	24.04	9.77
S18, S19, S20	8.20	9.82	19.76	8.76	6.83

Table 7.7 Comparison of natural frequencies (Hz)

Mode number	Measured frequency	Initial model	Error (%)	Updated model	Error (%)
1	16.56	16.91	2.11	16.38	1.09
2	16.81	16.97	0.95	16.44	2.20
3	22.50	24.07	6.98	23.49	4.40
4	44.36	45.96	3.61	44.4	0.09
5	44.50	45.98	3.33	44.63	0.29
6	50.44	53.82	6.70	53.17	5.41
7	52.30	54.5	4.21	53.62	2.52
8	59.83	59.41	0.70	57.58	3.76
9	60.56	59.5	1.75	57.75	4.64

Table 7.8 MAC of modal shapes

Mode number	1	2	3	4	5	6	7	8	9
Initial model	0.9939	0.9958	0.9938	0.9224	0.9749	0.9887	0.9734	0.8652	0.8984
Updated model	0.9940	0.9958	0.9935	0.9377	0.9664	0.9897	0.9719	0.9104	0.9075

The modal analysis is carried out using the initial and updated multi-scale models of the transmission tower. The natural frequencies are obtained and compared with the test results listed in Table 7.7. It can be seen that the maximum errors using the updated multi-scale model occurs at the sixth natural frequency with a relative error of 5.41%. The maximum error of the frequencies of the initial model is 6.98% for the third mode of vibration.

According to Equation 7.21, the MAC values can be calculated and the results are listed in Table 7.8 for the comparison of the initial and updated multi-scale models. It can be seen that the accuracy of the eighth and ninth modes of the updated model are better than that of the initial model and that the other modes of the two multi-scale models have almost the same accuracy. The minimum MAC of the initial model is 0.8652 corresponding to the eighth mode of vibration, while the minimum MAC of the updated model is 0.9075 corresponding to the ninth mode of vibration.

7.9.3 Comparison of Kriging method with QPSR method

A comparison between the Kriging method and the QPRS method is carried out in this section to assess their performance in the multi-objective optimisation.

For the sake of clear discussion, only one parameter, θ_5, is selected as a variable while other parameters remain unchanged. Both the Kriging meta-model and the QPRS are constructed. The construction of the two models uses the same four sample (design) points as shown Figure 7.4. The predictions of the four types of residuals from the two models are also shown in Figure 7.4. It can be seen that predictions of the residuals from both models agree well with the true values from the FE analysis for weaker nonlinear residuals, such as the residuals of displacement, strain and frequency. However, the QPRS fails to give a satisfactory prediction to the MAC of modal shape, of which the residual is highly nonlinear, while the Kriging model can still achieve a good prediction of the MAC. The values of R^2 of both the Kriging and QPRS methods are calculated. The ranges of R^2 are [0.9869, 1.0000] and [0.8121, 1.0000] for the Kriging method and QPRS method, respectively. The minimum value of R^2 occurs at the residual of the fourth MAC for both the Kriging method and QPRS method, with 0.9869 from the Kriging method and 0.8121 from the QPRS method.

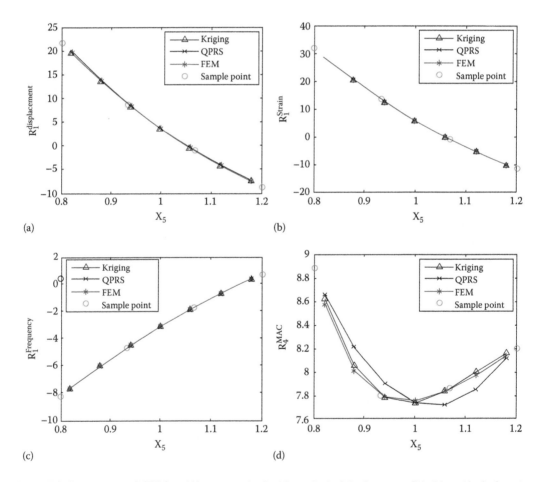

Figure 7.4 Comparison of QPRS and Kriging methods: (a) residual of displacement D1, (b) residual of strain S1, (c) residual of first natural frequency and (d) residual of MAC of fourth-order mode.

The value from the Kriging method is higher than that from the QPRS method, which means that the Kriging method is more accurate than the QPRS method for the present example.

NOTATION

a_1, a_2	The proportional coefficients in Rayleigh damping
e_i, e_i^0	The ith elastic moduli of updated and initial multi-scale model, respectively
$\mathbf{E}$	The matrix defining the location of excitations
$\mathbf{f}(t)$	External excitation vector
f_i, ϕ_{ji}	The ith natural frequency and associated mode shape at the jth point, respectively
f_i^{FE}, f_i^{test}	The ith frequencies from FE analysis and test, respectively
h_{jk}	The response displacement at coordinate j due to a harmonic excitation force at coordinate k with frequency of ω
$\mathbf{H}(\omega)$	Frequency response function (FRF) matrix
$\mathbf{I}$	Identity matrix
J_{global}, J_{local}	The global and local objective function, respectively
l^b, u^b	Lower and upper bounds on parameter θ, respectively
$\mathbf{M, C, K}$	Mass, damping and stiffness matrices of the structure, respectively
MAC	Modal assurance criterion
MSE	Mean square error
MSF	Modal scale factor
N_C	The total number of the check points
nm, np	The number of modes and number of measurement points, respectively
$\mathbf{q}$	The vector of modal coordinates
$R_i^{displacement}, R_i^{strain}, R_i^{Frequency}, R_i^{MAC}$	The ith residual of displacement, strain, frequency and MAC, respectively
$\mathbf{S}$	Sensitivity matrix of eigenvalues with respect to the parameters
u_i^{FE}, u_i^{test}	The ith displacement responses from FE analysis and test, respectively
w_1, w_2, w_3, w_4, w_5	The weighting factors corresponding to the objective functions of frequency, MAC, displacement, global strain and local strain, respectively
W_f, W_ϕ	Weight coefficients of the frequency and mode shape, respectively
$\mathbf{x}, \dot{\mathbf{x}}, \ddot{\mathbf{x}}$	The displacement, velocity and acceleration response vectors, respectively
$\mathbf{X}$	Time-independent amplitudes of structural responses
α_i	Element bending rigidity
$\varepsilon_i^{FE}, \varepsilon_i^{test}$	The ith strain responses from FE analysis and test, respectively
$\mathbf{\Lambda}$	Diagonal matrix consisting of the eigenvalues

λ_i	The ith eigenvalue
μ	The Marquardt parameter
ξ	Modal damping coefficient matrix
ξ_r	The rth modal damping ratio
ρ_i, ρ_i^0	The ith densities of updated and initial multi-scale model, respectively
ω	The vibration circular frequency
ω_0	Modal frequency matrix
Φ	The mass-normalised displacement mode shapes
Θ	Vector of the structural parameters to be updated
$\hat{y}_i, y_i$	The ith prediction values from the Kriging meta-model and actual function values, respectively
$\bar{y}$	The mean of all true values

REFERENCES

Abdel-Ghaffer, A.M. and R.H. Scanlan. 1985. Ambient vibration studies of Golden Gate Bridge: I. Suspended structure. *J. Eng. Mech.*, 111(4): 463–82.

Andersen, P., R. Brincker, and P.H. Kirkegaard. 1996. Theory of covariance equivalent ARMAV models of civil engineering structures. In *Proceedings of the 14th International Modal Analysis Conference*, Dearborn, MI, ed. A.L. Wicks, 518–24. Bethel, CT: Society for Environmental Mechanics.

Ashory, M.R., M.M. Khatibi, M. Jafari, and A. Malekjafarian. 2013. Determination of mode shapes using wavelet transform of free vibration data. *Arch. Appl. Mech.*, 83(6): 907–21.

Avargel, Y. and I. Cohen. 2007. System identification in the short-time Fourier transform domain with crossband filtering. *IEEE Trans. Audio, Speech, Lang. Proc.*, 15(4): 1305–19.

Avitabile, P. 1999. Modal space – In our own little world. *Exp. Tech.*, 5: 17–19.

Bishop, R.E.D. and G.M.L. Gladwell. 1963. An investigation into the theory of resonance testing. *Philos. Trans. R. Soc. Lond. Series A, Math. Phys. Sci.*, 255(1055): 241–80.

Bjorck, A. 1996. *Numerical Methods for Least Squares Problems*. Philadelphia, PA: Society for Industrial and Applied Mathematics.

Brownjohn, J.M.W. 2003. Ambient vibration studies for system identification of tall buildings. *Earthq. Eng. Struct. Dyn.*, 32: 71–95.

Brownjohn, J.M.W., P. Moyo, P. Omenzetter, and Y. Lu. 2003. Assessment of highway bridge upgrading by dynamic testing and finite element model updating. *J. Bridge Eng.*, 8: 162–72.

Brownjohn, J.M.W. and P.Q. Xia. 2000. Dynamic assessment of curved cable-stayed bridge by model updating. *J. Struct. Eng.*, 126(2): 252–60.

Caicedo, J.M., S.J. Dyke, and E.A. Johnson. 2004. Natural excitation technique and eigensystem realization algorithm for phase I of the IASC-ASCE benchmark problem: Simulated data. *J. Eng. Mech.*, 130(1): 49–60.

Chang, C.C., T.Y.P. Chang, and Q.W. Zhang. 2001. Ambient vibration of long-span cable-stayed bridge. *J. Bridge Eng.*, 6(1): 46–53.

Chen, J. and Y.L. Xu. 2002. Identification of modal damping ratios of structures with closely spaced modal frequencies: HHT method. *Struct. Eng. Mech.*, 14(4): 417–34.

Chen, S.L., J.J. Liu, and H.C. Lai. 2009. Wavelet analysis for identification of damping ratios and natural frequencies. *J. Sound Vibr.*, 323(1–2): 130–47.

Doebling, S.W., C.R. Farrar, M.B. Prime, and D.W. Shevitz. 1996. Damage identification and health monitoring of structural and mechanical system from changes in their vibration characteristics: A literature review. Report No. LA-13070-MS, Los Alamos: Los Alamos National Laboratory. https://institute.lanl.gov/ei/shm/pubs/lit_review.pdf

Ewins, D.J. 2000. *Modal Testing: Theory, Practice and Application*. 2nd edition. Baldock, UK: Research Studies Press Ltd.

Feldman, M. 1994a. Non-linear system vibration analysis using Hilbert transform-I: Free vibration analysis method 'Freevib'. *Mech. Syst. Signal Proc.*, 8(2): 119–27.

Feldman, M. 1994b. Non-linear system vibration analysis using Hilbert transform-II: Forced vibration analysis method 'Forcevib'. *Mech. Syst. Signal Proc.*, 8(3): 309–18.

Feldman, M. 1997. Nonlinear free-vibration identification via the Hilbert transform. *J. Sound Vibr.*, 208(3): 475–89.

Feldman, M. 2007. Considering high harmonics for identification of non-linear systems by Hilbert transform. *Mech. Syst. Signal Proc.*, 21(2): 943–58.

Friswell, M.I., J.E.T. Penny, and S.D. Garvey. 1998. A combined genetic and eigensensitivity algorithm for the location of damage in structures. *Comput. Struct.*, 69(5): 547–56.

Ge, L. and T.T. Soong. 1998. Damage identification through regularization method I: Theory. *J. Eng. Mech.*, 124(1): 103–108.

Gurley, K. and A. Kareem. 1999. Applications of wavelet transforms in earthquake, wind and ocean engineering. *Eng. Struct.*, 21(2): 149–67.

Hans, S., E. Ibraim, S. Pernot, C. Boutin, and C.H. Lamarque. 2000. Damping identification in multi-degree of-freedom systems via a wavelet-logarithmic decrement-part 2: Study of a civil engineering building. *J. Sound Vibr.*, 235(3): 375–403.

Hao, H. and Y. Xia. 2002. Vibration-based damage detection of structures by genetic algorithm. *J. Comput. Civil. Eng.*, 16(3): 222–29.

Heywood, R.J. 1996. Influence of truck suspensions on the dynamic response of a short span bridge over Cameron's Creek. *Heavy Vehicle Syst.*, 3(1–4): 222–39.

Holland, J.H. 1975. *Adaption in Natural and Artificial Systems*. Ann Arbor, MI: University of Michigan Press.

Huang, N.E., Z. Shen, and S.R. Long. 1999. A new view of nonlinear water waves: The Hilbert spectrum. *Annu. Rev. Fluid Mech.*, 31: 417–57.

Huang, N.E., Z. Shen, S.R. Long, et al. 1998. The empirical mode decomposition and Hilbert spectrum for nonlinear and nonstationary time series analysis. *Proc. Roy. Soc. Lond. A*, 454: 903–95.

Jacob, B., E.J. Obrien, and S. Jehaes. 2002. Weigh-in-motion of road vehicles, European Specifications of WIM. Final Report of the COST 323 Action, LCPC Publications, Paris. http://www.is-wim.org/doc/wim_eu_specs_cost323.pdf

Jaishi, B. and W.X. Ren. 2005. Structural finite element model updating using ambient vibration test results. *J. Struct. Eng.*, 131(4): 617–28.

James III, G.H., T.G. Carne, and J.P. Lauffer. 1995. The natural excitation technique (NExT) for modal parameter extraction from operating structures. *Int. J. Anal. Exp. Modal Anal.*, 10(4): 260–77.

Juang, J.N. and R.S. Pappa. 1985. An eigensystem realization algorithm for modal parameter identification and model reduction. *J. Guid. Control Dyn.*, 8(5): 620–27.

Kijewski, T. and A. Kareem. 2003. Wavelet transforms for system identification in civil engineering. *Comput.-Aided Civil Infrastruct. Eng.*, 18(5): 339–55.

Konak, A., D.W. Coit, and A.E. Smith. 2006. Multi-objective optimization using genetic algorithms: A tutorial. *Reliab. Eng. Syst. Saf.*, 91: 992–1007.

Lamarque, C.H., S. Pernot, and A. Cuer. 2000. Damping identification in multi-degree-of-freedom systems via a wavelet-logarithmic decrement-part 1: Theory. *J. Sound Vibr.*, 235(3): 361–74.

Lardies, J. and S. Gouttebroze. 2002. Identification of modal parameters using the wavelet transform. *Int. J. Mech. Sci.*, 44(11): 2263–83.

Larson, C.B. and D.C. Zimmerman. 1993. Structural model refinement using a genetic algorithm approach. In *Proceedings of the 11th International Modal Analysis Conference*, Kissimmee, Florida, ed. D.J. Demichele, 1095–101. Bethel, CT: Society for Environmental Mechanics.

Liu, T.Y., W.L. Chiang, C.W. Chen, et al. 2011. Structural system identification for vibration bridges using the Hilbert-Huang transform. *J. Vibr. Control*, 17(4): 589–603.

Maia, N.M.M., J.M.M. Silva, J. He, et al. 1997. *Theoretical and Experimental Modal Analysis*. Baldock, UK: Research Studies Press Ltd.

Mares, C. and C. Surace. 1996. An application of genetic algorithms to identify damage in elastic structures. *J. Sound Vibr.*, 195(2): 195–215.

MATLAB. 2010. Version 7.10.0 (R2010a)[M], Natick, MA: The Math Works Inc. http://www.math-works.com/help/

More, J.J. and G. Toraldo. 1989. Algorithm for bound constrained quadratic programming problems. *Numer. Math.*, 55: 377–400.

Nagarajaiah, S. and B. Basu. 2009. Output only modal identification and structural damage detection using time frequency & wavelet techniques. *Earthq. Eng. Eng. Vib.*, 8(4): 583–605.

Nelson, R.B. 1976. Simplified calculation of eigenvector derivatives. *AIAA J.*, 14(9): 1201–205.

Nocedal, J. and J. Wright. 1999. *Numerical Optimization*. New York: Springer.

Okauchi, I., T. Miyata, M. Tatsumi, and N. Sasaki. 1997. Field vibration test of a long span cable-stayed bridge using large exciters. *J. Struct. Engrg./Earthquake Engrg., Tokyo*, 14(1): 83–93.

Pai, P.F., L. Huang, J.Z. Hu, and R. Langewisch. 2008. Time-frequency method for nonlinear system identification and damage detection. *Struct. Health Monit.*, 7(2): 103–27.

Peeters, B. 2000. System identification and damage detection in civil engineering. PhD diss., Department of Civil Engineering, Catholic University of Leuven, Belgium.

Perry, M.J., C.G. Koh, and Y.S. Choo. 2006. Modified genetic algorithm strategy for structural identification. *Comput. Struct.*, 84: 529–40.

Richardson, M. and D.L. Formenti. 1982. Parameter estimation from frequency response measurements using rational fraction polynomials. In *Proceedings of the 1st International Modal Analysis Conference*, Orlando, FL, ed. D.J. DeMichele, 167–81. Bethel, CT: Society for Experimental Mechanics.

Richardson, M. and D.L. Formenti. 1985. Global curve-fitting of frequency response measurements using the rational fraction polynomial method. In *Proceedings of the 3rd International Modal Analysis Conference*, Orlando, FL, ed. D.J. DeMichele, 390–97. Bethel, CT: Society for Experimental Mechanics.

Robertson, A.N., K.C. Park, and K.F. Alvin. 1998a. Extraction of impulse response data via wavelet transform for structural system identification. *J. Vib. Acoust*, 120(1): 252–60.

Robertson, A.N., K.C. Park, and K.F. Alvin. 1998b. Identification of structural dynamics models using wavelet-generated impulse response data. *J. Vib. Acoust*, 120(1): 261–66.

Ruzzene, M., A. Fasana, L. Garibalidi, and B. Piombo. 1997. Natural frequencies and dampings identification using wavelet transform: Application to real data. *Mech. Syst. Signal Proc.*, 11(2): 207–18.

Sanayei, M. and M.J. Saletnik. 1996. Parameter estimation of structures from static strain measurements, part I: Formulation. *J. Struct. Eng.*, 122(5): 555–62.

Shi, Z.Y. and S.S. Law. 2007. Identification of linear time-varying dynamical systems using Hilbert transform and empirical mode decomposition method. *J. Appl. Mech.*, 74(2): 223–30.

Shi, Z.Y., S.S. Law, and X. Xu. 2009. Identification of linear time-varying mdof dynamic systems from forced excitation using Hilbert transform and EMD method. *J. Sound Vibr.*, 321(3–5): 572–89.

Simpson, T.W. 1998. Comparison of response surface and Kriging models in the multidisciplinary design of an aerospike nozzle. ICASE Report No. 98-16, NASA Langley Research Center, USA. http://www.cs.odu.edu/~mln/ltrs-pdfs/icase-1998-16.pdf

Staszewski, W.J., K. Worden, and G.R. Tomlinson. 1997. Time–frequency analysis in gearbox fault detection using the Wigner–Ville distribution and pattern recognition. *Mech. Syst. Signal Proc.*, 11(5): 673–92.

Tomoiaga, B., M. Chindris, A. Sumper, A. Sudria-Andreu, and R. Villafafila-Robles. 2013. Pareto optimal reconfiguration of power distribution systems using a generic algorithm based on NSGA-II. *Energies*, 6: 1439–55.

Van Overschee, P. and B. De Moor. 1996. *Subspace Identification for Linear Systems: Theory-Implementation-Applications*. Boston, MA: Kluwer Academic Publishers.

Wang, F.Y., Y.L. Xu, and W.L. Qu. 2014. Mixed-dimensional finite element coupling for structural multi-scale simulation. *Finite Elem. Anal. Des.*, 92: 12–25.

Wang, Z.C. and G. Chen. 2012. Recursive Hilbert-Huang transform method for time-varying property identification of linear shear-type buildings under base excitations. *J. Eng. Mech.*, 138(6): 631–39.

Wang, Z.C., W.X. Ren, and G. Chen. 2012. A Hilbert transform method for parameter identification of time-varying structures with observer techniques. *Smart Mater. Struct.*, 21(10): 105007.

Weng, S., Y. Xia, Y.L. Xu, and H.P. Zhu. 2011. Substructuring approach to finite element model updating. *Comput. Struct.*, 89(9–10): 772–82.

Xu, Y.L., S.W. Chen, and R.C. Zhang. 2003. Modal identification of Di Wang Building under Typhoon York using the Hilbert–Huang transform method. *Struct. Design Tall Spec. Build.*, 12(1): 21–47.

Xu, Y.L., J.M. Ko, and W.S. Zhang. 1997. Vibration studies of Tsing Ma suspension bridge. *J. Bridge Eng.*, 2(4): 149–56.

Xu, Y.L., S. Zhan, H. Xia, Y. Xia, and N. Zhang. 2013. Field measurements of the new CCTV tower in Beijing. *Int. J. High-Rise Build.*, 2(3): 171–78.

Yan, B.F. and A. Miyamoto. 2006. A comparative study of modal parameter identification based on wavelet and Hilbert–Huang transforms. *Comput.-Aided Civil Infrastruct. Eng.*, 21(1): 9–23.

Yan, B.F., A. Miyamotoa, and E. Brühwiler. 2006. Wavelet transform-based modal parameter identification considering uncertainty. *J. Sound Vibr.*, 291(1–2): 285–301.

Yan, G.R., A.D. Stefano, and G. Ou. 2012. A general nonlinear system identification method based upon time-varying trend of instantaneous frequencies and amplitudes. *Adv. Struct. Eng.*, 15(5): 781–92.

Yang, J.N., Y. Lei, S. Lin, and N. Huang. 2004. Hilbert-Huang based approach for structural damage detection. *J. Eng. Mech.*, 130(1): 85–95.

Yang, J.N., Y. Lei, S.W. Pan, and N. Huang. 2003a. System identification of linear structures based on Hilbert-Huang spectral analysis, Part 1: Normal modes. *Earthq. Eng. Struct. Dyn.*, 32(9): 1443–67.

Yang, J.N., Y. Lei, S.W. Pan, and N. Huang. 2003b. System identification of linear structures based on Hilbert-Huang spectral analysis, Part 2: Complex modes. *Earthq. Eng. Struct. Dyn.*, 32(10): 1533–54.

Zhu, X.Q. and S.S. Law. 2013. Recent advances on moving force identification in structural dynamics. In *Proceedings of the 6th International Conference on Structural Health Monitoring of Intelligent Infrastructure*, eds. Y.L. Xu, S. Zhu, Y. Xia, Y.Q. Ni, S.S. Law, J.H. Yin, and Z.Q. Su, 347. Hong Kong: The Hong Kong Polytechnic University.

Zivanovic, S., A. Pavic, and P. Reynolds. 2007. Finite element modeling and updating of a lively footbridge: The complete process. *J. Sound Vibr.*, 301: 126–45.

Chapter 8

Multi-type sensor placement

8.1 PREVIEW

As mentioned in Chapter 3, a variety of physical quantities are required to be measured in a smart civil structure, and accordingly many types of sensors are required to be installed in the structure. However, owing to the economic cost associated with sensing systems and the considerable size of a civil structure, sensors are often installed only at a few locations that are much less than the total degrees of freedom (DOFs) of the structure. Therefore, the optimal sensor placement (OSP) becomes a practical and interesting topic, and many methods have been proposed using a variety of techniques and criteria to find the OSP. On the other hand, the lack of information on the responses of a structure at its key locations, including the locations without sensors and the desired locations but not accessible for measurements during its operation, may subsequently hamper the accuracy of system identification and model updating, the reliability of damage detection and the effectiveness of structural control. Accurate response reconstruction at all key structural locations using limited measured responses becomes essential.

This chapter first reviews the traditional OSP methods. A dual-type sensor placement method, in which both strain gauges and displacement transducers are used, is then presented with the aim of the best reconstruction of multi-scale structural responses without knowing external excitations. Experimental work is also carried out to validate the proposed method. The dual-type sensor placement method is further extended by means of the Kalman filter algorithm to multi-type sensors, including accelerometers, displacement transducers and strain gauges, leading to a multi-type sensor placement method with the aim of the best reconstruction of multi-scale structural responses. The multi-type sensor placement and multi-scale response reconstruction method is finally applied to the test bed of a long-span suspension bridge to examine its feasibility and effectiveness.

8.2 REVIEW OF SENSOR PLACEMENT METHODS

As described in Chapter 3, sensors are primarily used to monitor three types of parameters: loading sources such as wind, seismic, highway and railway loading; structural responses such as strain, displacement, inclination and acceleration; and environmental effects including temperature, humidity and corrosion. The corresponding sensing systems thus become increasingly complex in order to provide comprehensive and accurate information for fulfilling smart functions such as structural health monitoring (SHM) and structural vibration control. An optimal configuration of the sensing system can minimise the number of sensors, increase the accuracy of information and provide robustness to the system. Consequently, a great deal of research has been conducted over the last few decades on the

OSP for either system identification and model updating or structural damage detection or structural vibration control.

To provide the maximum information on structural characteristics, a class of information-based approaches has been proposed by researchers (e.g. Shah and Udwadia 1978; Kammer 1991, 1992, 1994, 1996; Unwadia 1994; Kirkegaard and Brincker 1994; Penny et al. 1994; Imamovic 1998; Yuen et al. 2001; Cherng 2003; Kammer and Tinker 2004; Meo and Zumpano 2005; Castro-Triguero et al. 2013; Wang et al. 2013; Friswell and Castro-Triguero 2015). In these methods, optimal sensor locations are so selected that they maximise some norm (determinant or trace) of the Fisher information matrix (FIM) or its variants for the purpose of the best identification of structural characteristics, such as natural frequencies, mode shapes and damping ratios. For example, based on the contribution of each sensor location to the linear independence of the identified modes of vibration, Kammer (1991) proposed a method, referred to as the effective independence (EfI) method, to optimise the locations of sensors for the purpose of on-orbit modal identification and model updating, as discussed in Chapter 7. With the consideration of noise effect, this method was then extended by Kammer (1992) so that the optimal configuration of sensors could be determined to maintain a desired level of signal-to-noise ratio over all the target modes of vibration. Kammer and Tinker (2004) further proposed an EfI-based approach for the optimal placement of triaxial accelerometers. Since uncertainties inevitably exist, Beck and Katafygiotis (1998) proposed a Bayesian statistical framework for OSP to handle the uncertainties in structural model updating. Moreover, Papadimitriou et al. (2000) developed a method to evaluate the uncertainties of model updating by minimising the information entropy (IE) over the set of possible sensor configurations. Later, Papadimitriou (2004, 2005) extended the IE-based OSP method for structural parameter identification by using the forward sequential sensor placement (FSSP) and backward sequential sensor placement (BSSP) algorithm.

Besides the aforementioned information-based approaches for the OSP, other kinds of methods based on modal kinetic energy have been developed and used for the determination of the OSP for system identification and model updating (Salama et al. 1987; Chung and Moore 1993; Heo et al. 1997; Imamovic 1998; Meo and Zumpano 2005; Liu et al. 2008; Debnath et al. 2012; Castro-Triguero et al. 2013). The inherent mathematical connection between the EfI and the kinetic energy methods was revealed by Li et al. (2007). Moreover, with the recent development of intelligent optimisation algorithms, intelligent OSP methods have emerged such as genetic algorithms (GAs) (Yao et al. 1993; Tongpadungrod et al. 2003; Liu et al. 2008) and particle swarm optimisation (PSO) algorithms (Rao and Anandakumar 2007; He et al. 2014; Zhang et al. 2014a). Many other sensor placement approaches for system identification and model updating can be found in the literature, and an excellent review was provided by Barthorpe and Worden (2009).

In the context of structural damage detection, the basic principle of the OSP is that the sensors should be placed appropriately so as to effectively capture the variations in structural characteristics caused by structural damages. By using different criteria or objective functions, a number of approaches have been developed for the OSP and most methods utilise the information obtained from the frequency domain. For example, the mode shapes or natural frequencies of a structure are one of the most widely used variables in the objective functions (Cobb and Liebst 1997; Shi et al. 2000; Kwon et al. 2003; Guo et al. 2004; D'Souza and Epureanu 2008; Shan et al. 2011). Other commonly used variables include modal assurance criterion (Worden and Burrows 2001; Yi et al. 2011), transfer function (Beal et al. 2008; Brehm et al. 2013) and kinetic energy matrix (Hemez and Farhat 1994; He et al. 2013). Several intelligent algorithms, such as generic algorithms (Said and Staszewski 2000; Ma et al. 2008), particular swarm optimisation algorithms (Abdalla and Al-Khawaldeh 2011) and monkey algorithms (Yi et al. 2012a,b, 2014; Jia et al. 2015), have also been employed for the OSP using the aforementioned modal properties. Although the investigation of the OSP using modal parameters has been actively conducted, research on the OSP by directly using the time histories of structural responses is very limited. In the following sections, a response reconstruction-based approach for sensor placement will be presented aiming at the best reconstruction of a time series of multi-scale structural responses. Based on the determined locations of the sensors and the reconstructed responses obtained from the limited sensors, a multi-scale damage detection method has been developed and will be described in detail in Chapter 12.

It is known that if a structural system is fully observable, available measurements may be sufficient to estimate all state variables by means of an observer technique (Cheng et al. 2008). Consequently, for optimal control algorithms requiring full-state feedback, enough sensors must be employed to meet the observability requirement, and this could be considered a basic criterion for the determination of the OSP for a control system. Cheng et al. (2008) presented a trial-and-error procedure for the OSP in an n-story shear building with r active tendons to meet the observability requirement. A detailed description of observability including its definition and properties can be found in Chapter 9. The investigation of the OSP for a control system has been actively conducted in the field of mechanical and aerospace engineering, primarily considering the meticulous requirement of control effectiveness and the wide application of smart materials such as piezoelectric lead-zirconate-titanate (PZT) that is able to act as actuator and sensor simultaneously (Gupta et al. 2010). However, in civil engineering, on the one hand, research on the OSP for the improvement of control effectiveness is limited, and on the other hand, the investigation of the OSP for a control system is basically conducted together with the optimal placement of control devices. An introduction to the collective placements of control devices and sensors, and a response reconstruction-based method for the optimal joint placement of both control devices and sensors will be given in Chapter 11.

Although great efforts have been devoted to the OSP, less research has focused on the design of a sensing system with multiple types of sensors. With the fast advance of sensing technology in the past few decades, the sensing system has become more complex in terms of the number and type of sensors. Although multi-types of sensors can provide more comprehensive information, their distinct properties and limitations considerably complicate the design procedure of such sensing systems. In the following sections, a dual-type sensor placement method and a multi-type sensor placement method will be presented with the aim of minimising the errors of the time histories of reconstructed multi-scale structural responses.

8.3 DUAL-TYPE SENSOR PLACEMENT METHOD

This section presents an integrated method for determining the placement of a sensing system composed of both strain gauges and displacement transducers and at the same time for reconstructing structural responses at all key locations of the structure using the limited measured strain and displacement responses (Zhang et al. 2011). The locations of strain gauges and displacement transducers are so selected that the structural responses at all key locations can be best reconstructed from the limited measured strain and displacement responses with minimum estimation errors.

8.3.1 Strain–displacement relationship

It is noteworthy that the normal strain in the ith element can be expressed as

$$\varepsilon_i = \mathbf{E}_i \, \mathbf{T}_i \, \mathbf{S}_i \, \mathbf{d}_c \tag{8.1}$$

where:
 subscript $\mathbf{c}$ indicates that the vector corresponds to the complete set of DOFs
 $\mathbf{d}_c$ is the displacement vector that includes all translational and rotational DOFs of a structure
 $\mathbf{S}_i$ is a selection matrix that selects the displacements related to the ith element and the number of selected DOFs is dependent on the element type
 $\mathbf{T}_i$ is a transformation matrix that transforms the element nodal displacements in the global coordinate to those in the local coordinate
 $\mathbf{E}_i$ is a matrix representing the relationship between the node displacements of the element and the strains in the element
 ε_i is the normal strain at a location in the ith element

In considering the strains at all the locations of interest in a structure, a strain vector can be obtained as

$$\boldsymbol{\varepsilon}_c = (\varepsilon_1, \ldots, \varepsilon_i, \ldots, \varepsilon_{n^\varepsilon})^T = \mathbf{F} \mathbf{d}_c \tag{8.2}$$

where:
 $\mathbf{F}$ is the transformation matrix between the displacement vector and the strain vector, that is, $\mathbf{F} = ((\mathbf{E}_1 \mathbf{T}_1 \mathbf{S}_1)^T, (\mathbf{E}_2 \mathbf{T}_2 \mathbf{S}_2)^T, \ldots, (\mathbf{E}_{n^\varepsilon} \mathbf{T}_{n^\varepsilon} \mathbf{S}_{n^\varepsilon})^T)^T$
 subscript n^ε denotes the number of strains of interest

In practice, reduced-order mode superposition analysis is widely adopted to calculate the dynamic response of a linear system with a large number of DOFs. A modal expansion of the displacement vector yields

$$\mathbf{d}_c \approx \sum_{r=1}^{k} \varphi_r q_r = \boldsymbol{\Phi}_c \mathbf{q} \tag{8.3}$$

where:
 $\boldsymbol{\Phi}_c$ is the displacement modal matrix consisting of the mass-normalised displacement mode shapes φ_r that can be obtained through the modal analysis of the finite element (FE) model of the structure
 $\mathbf{q}$ is the modal coordinates

The number of mode shapes k considered is usually much less than the total number of DOFs for a large-scale structure. In general, $\boldsymbol{\Phi}_c$ should include all important mode shapes excited by external loads, and the value of k should be determined in consideration of the modal contribution factors, the frequency bandwidth of excitations and the sampling rate of data acquisition. Substituting Equation 8.3 into Equation 8.2, the strain vector can be expressed with the modal coordinates:

$$\boldsymbol{\varepsilon}_c = \sum_{r=1}^{k} \psi_r q_r = \boldsymbol{\Psi}_c \mathbf{q} \tag{8.4}$$

$$\psi_r = \mathbf{F}\varphi_r, \mathbf{\Psi}_c = \mathbf{F}\mathbf{\Phi}_c \tag{8.5}$$

where:

vector ψ_r is the strain mode shape corresponding to the displacement mode shape φ_r

$\mathbf{\Psi}_c$ is the strain modal matrix

The transformation matrices $\mathbf{E}$, $\mathbf{T}$ and $\mathbf{S}$ are all dependent on the element type, as different types of elements have different numbers of nodes, local coordinate systems and shapes functions. An example is presented for the two-dimensional, prismatic and symmetric beam element which will be employed later in the case study. The planar beam is a 6-DOF element, and its normal strain can be divided into deformations due to the axial force, ε_x, and bending moment, ε_b, respectively. The displacement–strain relationship for the planar beam element can be obtained as follows (Ottosen and Petersson 1992):

$$\begin{Bmatrix} \varepsilon_x \\ \varepsilon_b \end{Bmatrix} = \begin{Bmatrix} \dfrac{du}{dx} \\ -y\dfrac{d^2v}{dx^2} \end{Bmatrix} = \begin{Bmatrix} H'_u(x) \\ -yH''_v(x) \end{Bmatrix} \mathbf{G}d_e \tag{8.6}$$

where:

d_e is the nodal displacement vector corresponding to 6 DOFs of the beam element

x and y denote the location in the element coordinate where the strain is calculated

u and v are the shape functions which can be expressed by the functions $H_u(x)$ and $H_v(x)$ as shown in Equations 8.7 and 8.8, respectively

Matrix $\mathbf{G}$ is given in Equation 8.9

$$H_u(x) = \begin{bmatrix} 1 & 0 & 0 & x & 0 & 0 \end{bmatrix} \tag{8.7}$$

$$H_v(x) = \begin{bmatrix} 0 & 1 & x & 0 & x^2 & x^3 \end{bmatrix} \tag{8.8}$$

$$\mathbf{G} = \begin{bmatrix} 1 & 0 & 0 & 0 & 0 & 0 \\ 0 & 1 & 0 & 0 & 0 & 0 \\ 0 & 0 & 1 & 0 & 0 & 0 \\ -1/l & 0 & 0 & 1/l & 0 & 0 \\ 0 & -3/l^2 & -2/l & 0 & 3/l^2 & -1/l \\ 0 & 2/l^3 & 1/l^2 & 0 & -2/l^3 & 1/l^2 \end{bmatrix} \tag{8.9}$$

where l is the length of the beam element. As mentioned before, the actual normal strain is the superposition of two components, that is, $\varepsilon = \varepsilon_x + \varepsilon_b$, and thus the strain–displacement relationship matrix for the planar beam element reads

$$\mathbf{E} = \begin{bmatrix} H'_u(x) - yH''_v(x) \end{bmatrix} \mathbf{G} \tag{8.10}$$

The strain–displacement relationships for other element types can be derived in a similar way, for example three-dimensional beam element or plate element, although they may be more complex.

8.3.2 Theoretical formulations

Assume that the response vector y includes the displacements and strains at the locations of interest of a structure, that is, $y = [\varepsilon \quad d]^T$. For a linear structural system, the response can be expressed as a linear combination of a small subset of mode shapes (Ewins 2000):

$$y = \begin{Bmatrix} \varepsilon \\ d \end{Bmatrix} = \begin{Bmatrix} \Psi \\ \Phi \end{Bmatrix} q = \Gamma q \tag{8.11}$$

where:

    q   is the vector of modal coordinates
    Γ is the general modal matrix which includes both strain mode shapes Ψ and displacement mode shapes Φ

As mentioned before, it is not necessary to include all mode shapes for predicting the structural responses of a large-scale structure and only a subset of k mode shapes corresponding to low frequencies is needed. The dimension of the response vector y, denoted as n, represents the total number of candidate positions, whereas the dimensions of vectors ε and d, denoted as n^ε and n^d, represent the number of candidate locations for strain gauges and displacement transducers, respectively. Therefore, the sizes of the matrices Γ, Ψ and Φ are $n \times k$, $n^\varepsilon \times k$ and $n^d \times k$, respectively. Note that n is much less than the total number of DOFs as only displacements and strains of interest are considered here. Thus, Ψ and Φ are only sub-matrices of the aforementioned modal matrices Ψ_c and Φ_c.

Assume that the total n_m locations are chosen for the placement of strain gauges and displacement transducers. $n_m = n_m^\varepsilon + n_m^d$, where subscript m denotes the measurement and n_m^ε and n_m^d denote the number of strain gauges and displacement transducers, respectively. From Equation 8.11, the measured responses can be expressed as

$$y_m = \begin{Bmatrix} \Psi_m \\ \Phi_m \end{Bmatrix} q + w = \Gamma_m q + w \tag{8.12}$$

where:

    Γ_m, Ψ_m and Φ_m are the partitioned model matrices corresponding to the positions with sensors
    w   represents the noise vector in the measurement

Since Γ_m contains the strain mode shapes, it is no longer orthogonal to stiffness and the mass matrix. Therefore, the modal coordinates could not be solved via modal orthogonality properties. A similar approach to the system equivalent reduction–expansion process method (O'Callahan et al. 1989) is adopted here to estimate the response vector y:

$$y_e = \Gamma \Gamma_m^+ y_m \tag{8.13}$$

where:

subscript e denotes the estimation

Γ_m^+ denotes the pseudo-inverse of the matrix Γ_m and it can be calculated by $\Gamma_m^+ = (\Gamma_m^T \Gamma_m)^{-1} \Gamma_m^T$ provided that Γ_m is of full column rank

It is worth noting that Equation 8.13 provides a promising way for estimating the structural responses on the basis of limited measurements even without knowledge of the external excitations. However, directly using Equation 8.13 is not practical especially for large-scale structures, because the strain responses and the displacement responses have different orders of magnitude resulting in the matrix Γ_m being ill-conditioned.

The approach proposed by Kammer (1991, 1992) is extended herein for the sensor placement of both displacement transducers and strain gauges. According to Equations 8.11 through 8.13, the error between the estimated response and the real response can be obtained as follows:

$$\boldsymbol{\delta} = \mathbf{y}_e - \mathbf{y} = \Gamma\mathbf{q} + \Gamma\Gamma_m^+\mathbf{w} - \Gamma\mathbf{q} = \Gamma\Gamma_m^+\mathbf{w} \tag{8.14}$$

Subsequently, the covariance matrix of the estimation error can be calculated as

$$\boldsymbol{\Delta} = E(\boldsymbol{\delta}\boldsymbol{\delta}^T) = E[\Gamma\Gamma_m^+\mathbf{w}\mathbf{w}^T(\Gamma_m^+)^T\Gamma^T] = \Gamma\Gamma_m^+E(\mathbf{w}\mathbf{w}^T)(\Gamma_m^+)^T\Gamma^T \tag{8.15}$$

where:
$E(\cdot)$ is the expected value operator
$E(\mathbf{w}\mathbf{w}^T)$ is the covariance matrix of the measurement noise

The measurement noise is often assumed as a zero-mean stationary Gaussian noise. It is assumed that the sensor noise is uncorrelated with each other, and each type of sensors is of equal variance, and this yields (Unwadia and Garba 1985):

$$E(\mathbf{w}\mathbf{w}^T) = \begin{bmatrix} \sigma_\varepsilon^2\mathbf{I} & \\ & \sigma_d^2\mathbf{I} \end{bmatrix} = \boldsymbol{\Sigma}_m^2 \tag{8.16}$$

where:
σ_ε^2 and σ_d^2 are the strain gauges noise variance and the displacement transducers noise variance, respectively
$\mathbf{I}$ denotes the identity matrix
subscript m indicates that the dimension of the matrix $\boldsymbol{\Sigma}_m$ is equal to the total number of measurement locations
matrix $\boldsymbol{\Sigma}$ is a diagonal matrix that reads

$$\boldsymbol{\Sigma} = \begin{bmatrix} \sigma_\varepsilon\mathbf{I} & \\ & \sigma_d\mathbf{I} \end{bmatrix} \tag{8.17}$$

Therefore, Equation 8.15 can be rewritten as

$$\boldsymbol{\Delta} = E(\boldsymbol{\delta}\boldsymbol{\delta}^T) = \Gamma\Gamma_m^+\boldsymbol{\Sigma}_m^2(\Gamma_m^+)^T\Gamma^T = (\Gamma\Gamma_m^+\boldsymbol{\Sigma}_m)(\Gamma\Gamma_m^+\boldsymbol{\Sigma}_m)^T \tag{8.18}$$

where $\boldsymbol{\Delta}$ is the covariance matrix of the estimation error, and each diagonal element represents the variance of the estimation error of the corresponding response (strain or displacement). Therefore, the maximum diagonal element denotes the maximum estimation error, while the trace of the matrix $\boldsymbol{\Delta}$ represents the sum of the estimation errors at all locations of interest. Consequently, the OSP can be done by minimising the maximum estimation error, the total estimation error or both.

As mentioned before, however, the magnitudes of strain and displacement responses are of different orders, and as such their absolute estimation errors. The optimisation procedure may considerably bias one type of sensors. In view of this, the relative estimation error, the ratio of the estimation error to the measurement noise, is used here:

$$\tilde{\delta} = \left\{ \begin{matrix} \dfrac{\varepsilon_e - \varepsilon}{\sigma_\varepsilon} \\ \dfrac{d_e - d}{\sigma_d} \end{matrix} \right\} = \Sigma^{-1}\delta \tag{8.19}$$

Similarly, the noise-normalised mode shape matrices are defined as $\tilde{\Psi} = \dfrac{1}{\sigma_\varepsilon} \cdot \Psi$, $\tilde{\Phi} = \dfrac{1}{\sigma_d} \cdot \Phi$, $\tilde{\Psi}_m = \dfrac{1}{\sigma_\varepsilon} \cdot \Psi_m$, $\tilde{\Phi}_m = \dfrac{1}{\sigma_d} \cdot \Phi_m$. Thus

$$\tilde{\Gamma} = \begin{bmatrix} \tilde{\Psi} & \tilde{\Phi} \end{bmatrix}^T = \Sigma^{-1}\Gamma \tag{8.20}$$

$$\tilde{\Gamma}_m = \begin{bmatrix} \tilde{\Psi}_m & \tilde{\Phi}_m \end{bmatrix}^T = \Sigma_m^{-1}\Gamma_m \tag{8.21}$$

and the noise-normalised response vectors:

$$\tilde{y} = \Sigma^{-1} \cdot y = \tilde{\Gamma}q \tag{8.22}$$

$$\tilde{y}_m = \Sigma_m^{-1} \cdot y_m = \tilde{\Gamma}_m q + \Sigma_m^{-1}w \tag{8.23}$$

where '~' denotes the noise-normalised vectors or matrices. Note that the matrices Σ_m and Σ have different dimensions despite their similar format. The estimation of the structural response can be computed by

$$\tilde{y}_e = \tilde{\Gamma}\tilde{\Gamma}_m^+ \tilde{y}_m \tag{8.24}$$

$$y_e = \Sigma \tilde{\Gamma} \tilde{\Gamma}_m^+ \tilde{y}_m = \Gamma \tilde{\Gamma}_m^+ \tilde{y}_m \tag{8.25}$$

Note that the noise-normalised modal matrix $\tilde{\Gamma}_m$ has a much lower condition number and considerably improves the accuracy of the calculation especially for large-scale structures. The covariance matrix of the normalised estimation error vector can then be computed as

$$\tilde{\Delta} = E(\delta\tilde{\delta}\tilde{\delta}^T) = \tilde{\Gamma}\tilde{\Gamma}_m^+ \Sigma_m^{-1} E(ww^T)(\Sigma_m^{-1})^T (\tilde{\Gamma}_m^+)^T \tilde{\Gamma}^T = \tilde{\Gamma}\tilde{\Gamma}_m^+ (\tilde{\Gamma}_m^+)^T \tilde{\Gamma}^T$$

$$= \tilde{\Gamma}(\tilde{\Gamma}_m^T\tilde{\Gamma}_m)^{-1}(\tilde{\Gamma}_m^T\tilde{\Gamma}_m)(\tilde{\Gamma}_m^T\tilde{\Gamma}_m)^{-1}\tilde{\Gamma}^T = \tilde{\Gamma}(\tilde{\Gamma}_m^T\tilde{\Gamma}_m)^{-1}\tilde{\Gamma}^T \tag{8.26}$$

The diagonal elements of the matrix $\tilde{\Delta}$ represent the variances of the normalised estimation error $\text{diag}(\tilde{\Delta}) = \begin{bmatrix} \tilde{\sigma}_1^2 & \tilde{\sigma}_2^2 & \cdots & \tilde{\sigma}_n^2 \end{bmatrix}$ and are of the same order of magnitude for displacement transducers and strain gauges. Therefore, such normalisation enables the simultaneous selection of the optimal locations for displacement transducers and strain gauges. A weight matrix W can be applied to account for the importance of different locations, or the different requirement on the normalised estimation error for strain gauges and displacement transducers.

$$\tilde{\Delta}_w = W\tilde{\Delta}, \text{diag}(\tilde{\Delta}_w) = \begin{bmatrix} W_1\tilde{\sigma}_1^2 & W_2\tilde{\sigma}_2^2 & \cdots & W_n\tilde{\sigma}_n^2 \end{bmatrix} \tag{8.27}$$

The maximum and average estimation errors at all locations can be computed by

$$\tilde{\sigma}^2_{max} = max(diag(\tilde{\mathbf{\Delta}}_w))$$

$$\tilde{\sigma}^2_{avg} = \frac{tr(\tilde{\mathbf{\Delta}}_w)}{n} = \frac{tr\left[\mathbf{W}\tilde{\mathbf{\Gamma}}(\tilde{\mathbf{\Gamma}}^T_m\tilde{\mathbf{\Gamma}}_m)^{-1}\tilde{\mathbf{\Gamma}}^T\right]}{n} = \frac{tr\left[\mathbf{W}(\tilde{\mathbf{\Gamma}}^T\tilde{\mathbf{\Gamma}})(\tilde{\mathbf{\Gamma}}^T_m\tilde{\mathbf{\Gamma}}_m)^{-1}\right]}{n} \tag{8.28}$$

where $tr(\cdot)$ denotes the trace of the enclosed matrix. Here, the weight matrix is taken as an identity matrix, and the optimisation objective can be expressed as

$$min\left(\tilde{\sigma}^2_{avg}\right) \quad \Rightarrow \quad min\left(tr\left[\left(\tilde{\mathbf{\Gamma}}^T\tilde{\mathbf{\Gamma}}\right)\left(\tilde{\mathbf{\Gamma}}^T_m\tilde{\mathbf{\Gamma}}_m\right)^{-1}\right]\right) \tag{8.29}$$

subject to

$$\tilde{\sigma}^2_{max} \le \left[\tilde{\sigma}^2_{max}\right], \tilde{\sigma}^2_{avg} \le \left[\tilde{\sigma}^2_{avg}\right] \tag{8.30}$$

in which $[\tilde{\sigma}^2_{max}]$ and $[\tilde{\sigma}^2_{avg}]$ are the target normalised maximum and average estimation errors, respectively.

The maximum and average estimation errors will increase with the reduction in the number of sensors. The total number of strain gauges and displacement transducers can thus be determined to achieve the prescribed criterion for estimation errors, as shown in constraint functions. A simple iterative procedure is carried out, in which the candidate sensor positions are deleted one by one until the target error level is reached. In each step, one sensor location is removed which results in a minimal trace of the matrix $(\tilde{\mathbf{\Gamma}}^T\tilde{\mathbf{\Gamma}})(\tilde{\mathbf{\Gamma}}^T_m\tilde{\mathbf{\Gamma}}_m)^{-1}$. Once the sensor locations are identified, the response at other locations can be estimated using Equation 8.25.

8.3.3 Numerical example

A cantilever beam with a length of about 2.0 m and a cross section of 50.8×50.8 mm is employed as a numerical example for optimal dual-type sensor placement (see Figure 8.1a). A random excitation is applied vertically at the end of the cantilever beam, which induces flexural vibration of the beam. The beam is divided into 20 elements. Twenty strains in the middle of each element and 20 vertical nodal displacements are taken as the candidate locations for strain gauges and displacements sensors. Here, the noise variances are assumed to be constant for each type of sensors, and they are independent of the magnitude of responses. The following four cases are studied and compared in the numerical example: Case 1: $\sigma_\varepsilon = 25$ µε and $\sigma_d = 0.7$ mm; Case 2: $\sigma_\varepsilon = 25$ µε and $\sigma_d = 0.1$ mm; Case 3: only installing strain gauges and $\sigma_\varepsilon = 25$ µε; Case 4: only installing displacement transducers and $\sigma_d = 0.7$ mm.

The first five mode shapes are considered in the numerical example, and the contribution of higher modes is assumed negligible. In the first two cases, the proposed optimal dual-type sensor placement method is adopted based on the constraints of $\tilde{\sigma}^2_{max} \le 1.0$ and $\tilde{\sigma}^2_{avg} \le 0.5$. In the last two cases, only one type of sensors is installed on the beam, and their performance is compared with the counterpart-Case 1. The single-type sensor systems are placed using the conventional Kammer method.

Figure 8.1 shows the optimal sensor locations for all four cases. Figure 8.2 depicts the variation of the average and maximum normalised estimation error variance in the optimisation

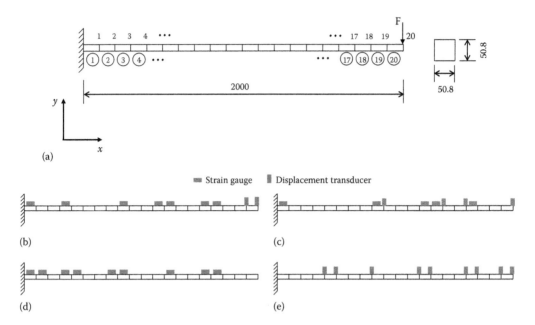

Figure 8.1 Sensor locations determined by optimal dual-type sensor placement method: (a) The finite-element model of the cantilever beam (unit: mm), (b) Case 1, (c) Case 2, (d) Case 3, (e) Case 4.

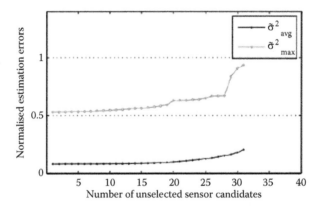

Figure 8.2 Variation of theoretical estimation errors with number of sensors.

procedure, that is, $\tilde{\sigma}^2_{\mathrm{avg}}$ and $\tilde{\sigma}^2_{\mathrm{max}}$. The sensor location which contributes most to minimise the trace of the error variance matrix $\tilde{\mathbf{\Lambda}}$ is removed from the candidate locations in each step. Both $\tilde{\sigma}^2_{\mathrm{max}}$ and $\tilde{\sigma}^2_{\mathrm{avg}}$ increase with the decrease of the sensor number. Therefore, the final sensor number is determined when the aforementioned criteria are reached.

The responses at the remaining locations are then estimated using the sensor measurements contaminated by noise. Taking node 18 and element 3 as an example, Figure 8.3 illustrates the comparison of the time histories of their estimated responses and the real responses. As illustrated in Figure 8.3, both the estimated displacement and strain can match the real response fairly well. Figure 8.4 presents a comparison between the theoretical and actual estimation errors for strains in 20 elements and vertical displacements at 20 nodes, all normalised by sensor noise variances. Here, the theoretical estimation error variances refer to

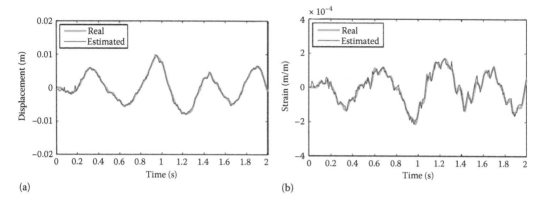

Figure 8.3 Time histories of real response and estimated response (Case 1): (a) displacement time history of node 18 and (b) strain time history of element 3.

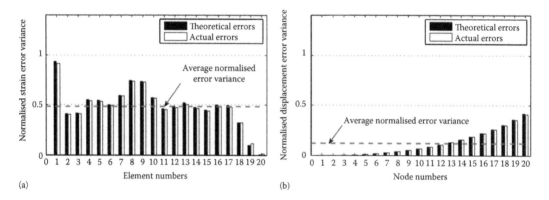

Figure 8.4 Comparison of theoretical and actual estimation errors (Case 1): (a) normalised estimation errors for strains and (b) normalised estimation errors for displacements.

the diagonal elements of error variance matrix $\widetilde{\Lambda}$, while the actual estimation errors are computed based on the difference between the normalised estimated response and the normalised real response. It can be seen that the theoretical values derived using the presented approach can well predict the actual errors in this numerical example. The slight discrepancy that can be observed is caused by truncated higher mode shapes in the theoretical formulation.

Case 2 is presented to demonstrate the effect of measurement noise upon the placement of a dual-type sensor system. The selected sensor locations are plotted in Figure 8.1c. Essentially, the noise covariance matrix serves as a weighting matrix of mode shapes during normalisation, and hence the decrease in the displacement sensor noise actually amplifies its weighting factor relative to that for strain gauges. As a result, more displacement transducers are selected in Case 2 than in Case 1.

Cases 3 and 4 present examples of a single-type sensor system with the noise levels consistent with those specified in Case 1. Figure 8.1d and e demonstrates the sensor locations for these two cases. Figures 8.5 and 8.6 show the distribution of normalised estimation errors for Case 3 and Case 4, respectively. In both cases, the theoretical error levels still accurately predict the actual estimation error levels. Compared with Case 1, Case 3 has similar error levels for strain estimations, whereas the estimation error levels for the displacement

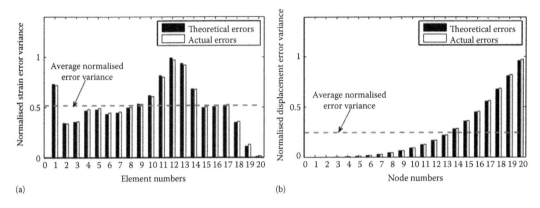

Figure 8.5 Comparison of theoretical and actual estimation errors (Case 3): (a) normalised estimation errors for strains and (b) normalised estimation errors for displacements.

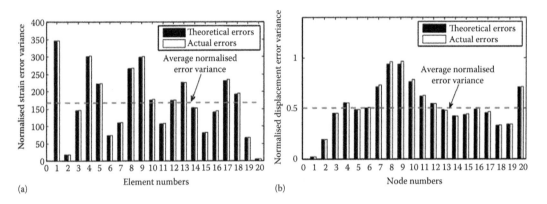

Figure 8.6 Comparison of theoretical and actual estimation errors (Case 4): (a) normalised estimation errors for strains and (b) normalised estimation errors for displacements.

response increase. For Case 4, it can be seen from Figure 8.6 that the estimation errors in the strain response become considerably large, although the estimation of displacements is still acceptable. This is because the displacement responses generally contain less high-frequency components than the strain responses. Through a comparison of Cases 1, 3 and 4, it is seen that the optimal dual-type sensor system contains more information on the entire structural response than the single-type sensor system, provided that the data from dual-type sensors can be appropriately fused.

8.4 EXPERIMENTAL VALIDATION OF THE DUAL-TYPE SENSOR PLACEMENT METHOD

8.4.1 Overhanging beam and FE model

To assess the feasibility and accuracy of the proposed dual-type sensor placement method, dynamic tests on a simply supported overhanging beam were conducted, as shown in Figure 8.7 (Zhang et al. 2014). The beam had a total length of 4 m and a cross section of 50×15.65 mm (width × thickness). The modulus of elasticity of the beam was 2.05×10^{11} N/m^2

Figure 8.7 Experimental arrangement of overhanging beam.

(a)   (b)

Figure 8.8 The details of the roller and hinge bearings: (a) roller bearing and (b) hinge bearing.

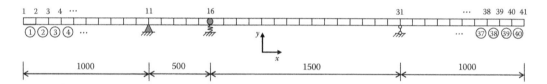

Figure 8.9 Schematic diagram of FE model of overhanging beam.

and the material density of the beam was 7780 kg/m³. The beam was designed and mounted on one roller bearing and one hinge bearing, and the details of these two bearings are shown in Figure 8.8. An FE model was then built using a total of 40 beam elements to represent the overhanging beam, as shown in Figure 8.9. An additional concentrated mass of 1.044 kg and an additional spring of stiffness 17.6 kN/m were added to the FE model to represent the connection between the beam and the exciter at node 16. As described in Chapter 7, model updating is often necessary to minimise modelling errors and ensure the consistency of the FE model and the real structure. Therefore, modal tests were conducted, and the natural frequencies and mode shapes were extracted from the measurement data and used to update the FE model. Detailed results of the FE model of the beam and its model updating can be found in the references (Zhang et al. 2014).

8.4.2 Dual-type sensor placement

Since the overhanging beam is a simple structure, the vertical nodal displacements except for those at the two supports, all the rotational nodal displacements and all element strains are assumed to be key responses. Accordingly, 40 element strains, 39 vertical and 41 rotational nodal displacements are to be reconstructed. Considering the difficulty in measuring rotational nodal displacements, the rotational DOFs are not included as candidates for sensor locations. As a result, 40 possible locations in the middle of each element are available for the strain gauges to measure normal strain responses, whereas 39 possible locations are available for the displacement transducers to measure vertical displacements.

The numbers and locations of both strain gauges and displacement transducers can then be determined by applying the proposed dual-type sensor placement method. The standard deviations of measurement noises from strain gauges and displacement transducers are estimated to be 0.378 μ_ε and 0.03 mm, respectively, based on the measurement results. The thresholds of the normalised maximum and average estimation errors are selected as 1.0 and 0.5, respectively. Based on the updated FE model of the beam (see Figure 8.9) and the dual-type sensor placement method as described in Section 8.3, a total of eight sensor locations are selected, including six strain gauges and two displacement transducers, as shown in Figure 8.10a. It is interesting to see that the locations of the six strain gauges and the two displacement transducers are symmetric with respect to the middle point of the beam as expected. In addition, three more strain gauges, two more laser displacement transducers and two more accelerometers are added to the beam to assess the accuracy of the reconstructed structural responses obtained by the dual-type sensor placement method (see Figure 8.10b).

8.4.3 Validation of reconstructed responses using dual-type sensing system

All the measurement data from the sensors are filtered using a low-pass filter with a cut-off frequency of 80 Hz. The measured responses from the six strain gauges and the two displacement transducers shown in Figure 8.10a are then combined to reconstruct the multiscale structural responses using Equation 8.25. The accuracy of the reconstructed structural responses can be validated by comparing with those measurements obtained from the additional sensors in Figure 8.10b. Taking node 18 as an example, Figure 8.11 gives a comparison of the random excitation–induced time histories of the reconstructed and measured vertical displacement responses. The two time histories are so close that they are almost the same.

Moreover, the relative percentage error (RPE) defined in the following is used as a quantitative measure of the difference between the reconstructed response $\mathbf{Y}^e$ and the real response $\mathbf{Y}$:

$$\text{RPE} = \frac{\text{std}(\mathbf{Y}^e - \mathbf{Y})}{\text{std}(\mathbf{Y})} \times 100\% \tag{8.31}$$

where std represents the standard deviation. The RPE values defined in Equation 8.31 are very stringent and plotted in Figure 8.12a for the case of random excitation. It can be seen

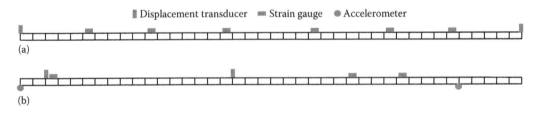

Figure 8.10 Sensor configurations in the overhanging beam experiment: (a) sensor locations based on the dual-type OSP method and (b) additional sensor locations for verification.

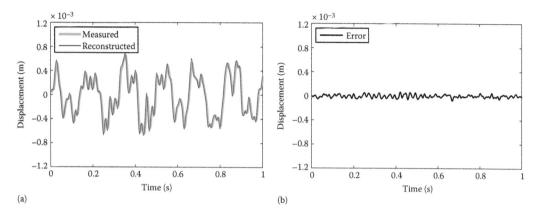

Figure 8.11 Response time histories and reconstruction errors at node 18 (random excitation): (a) response time histories and (b) reconstruction error.

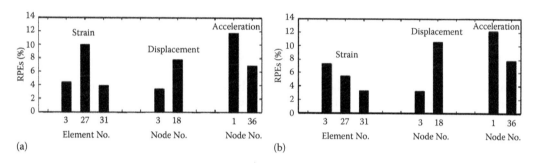

Figure 8.12 Relative percentage errors by using the dual-type sensor placement method: (a) random excitation and (b) hammer excitation.

that the reconstructed responses match well with the measured responses for all three types of structural responses considering that the error definition is very stringent. The maximum RPEs of the reconstructed strain, displacement and acceleration are less than 10%, 8% and 12%, respectively. Another type of excitation, that is, the impact force provided by a hammer, was also applied to the model to validate the accuracy and robustness of the reconstructed responses obtained from the proposed dual-type sensor placement method (Zhang et al. 2014). Similar results were obtained and the RPE values are given in Figure 8.12b.

Although the dual-type sensor placement method is mode-based, the accuracy of the response reconstruction results using the measurements from the selected locations can reach the same level regardless of the type of excitation if the modes excited by this excitation are within the selected frequency range. The good agreement between the reconstructed and measured responses in this section demonstrates that the proposed dual-type sensor placement and multi-scale response reconstruction method can be put into practice.

8.5 MULTI-TYPE SENSOR PLACEMENT METHOD

Based on the Kalman filter algorithm, this section addresses the problem of placing multi-type sensors, which include accelerometers, displacement transducers and strain gauges, in a structure with the objective of the best reconstruction of structural responses (Zhu et al.

2013). By minimising the overall reconstruction error variance at the locations of interest and maintaining reconstruction errors below an allowable level, the initial set of candidate sensor locations can be reduced to a smaller set. The multi-scale structural responses at key locations are then reconstructed from the fusion of measured multi-type sensor data.

8.5.1 State-space equation

The dynamics of a structure can be described by the following time-invariant continuous state space equation (Balageas et al. 2006):

$$\begin{cases} \dot{\mathbf{z}} = \mathbf{A}_c\mathbf{z} + \mathbf{B}_c\mathbf{u} \\ \mathbf{Y} = \mathbf{C}\mathbf{z} + \mathbf{D}\mathbf{u} \end{cases} \tag{8.32}$$

where:

$\mathbf{z}$	is the state vector
$\mathbf{A}_c$ and $\mathbf{B}_c$	are the state matrix and input matrix, respectively
$\mathbf{C}$ and $\mathbf{D}$	represent the output matrix and direct transmission matrix, respectively
$\mathbf{Y}$	is the observation vector
$\mathbf{u}$	denotes the external excitation vectors

In reality, the measurement data are discretely sampled with a time interval of Δt. Moreover, measurement noise and process noise always exist in real applications. Therefore, Equation 8.32 can be converted into the following discrete state-space model:

$$\begin{cases} \mathbf{z}_{k+1} = \mathbf{A}\mathbf{z}_k + \mathbf{B}\mathbf{u}_k + \mathbf{v}_k \\ \mathbf{Y}_k = \mathbf{C}\mathbf{z}_k + \mathbf{D}\mathbf{u}_k + \mathbf{w}_k \end{cases} \tag{8.33}$$

where:

$\mathbf{z}_k$	is the discrete-time state vector
$\mathbf{v}_k$	is the process noise that results from disturbances and modelling inaccuracies
$\mathbf{w}_k$	is the measurement noise of the sensors
$\mathbf{v}_k$ and $\mathbf{w}_k$	are often assumed as zero-mean white noise with variance matrices equal to $\mathbf{Q}$ and $\mathbf{R}$, respectively
$\mathbf{A}$ and $\mathbf{B}$	are the discrete-state matrix and input matrix, respectively

$$\mathbf{z}_k = \mathbf{z}(k\Delta t), \quad k = 1,2,3,\ldots; \quad \mathbf{A} = e^{\mathbf{A}_c\Delta t}; \quad \mathbf{B} = \int_0^{\Delta t} e^{\mathbf{A}_c\tau'}d\tau'\mathbf{B}_c \tag{8.34}$$

8.5.2 Theoretical formulations

The Kalman filter gives an unbiased and recursive algorithm to optimally estimate the unknown state vector of a linear dynamic system from the measurements with Gaussian white noise. The Kalman filter algorithm consists of two sets of equations, that is, time update equations and measurement update equations. More details can be found in the references (Kalman 1960; Welch and Bishop 1995). However, the direct application of the Kalman filter algorithm to a civil structure with high DOFs often presents a computational challenge. Therefore, the modal coordinate is employed and the second-order dynamic equation of a structure can be expressed as

$$\ddot{q} + 2\xi\omega_o\dot{q} + \omega_o^2 q = \Phi_c^T B_u u \qquad (8.35)$$

where:

q is the vector of modal coordinates
Φ_c denotes the mass-normalised displacement mode shapes
ξ is the modal damping coefficient matrix
ω_o is the modal frequency matrix
B_u is the location matrix of excitations

The major advantage of using modal coordinates is that the DOFs can be substantially reduced by truncating high modes that often have minimal and negligible contributions to structural responses under in-service conditions.

Equation 8.35 can be converted into Equation 8.32 in the state-space form, and subsequently the following is obtained:

$$z = \begin{Bmatrix} q \\ \dot{q} \end{Bmatrix}; A_c = \begin{bmatrix} 0 & I \\ -\omega_o^2 & -2\xi\omega_o \end{bmatrix}; B_c = \begin{bmatrix} 0 \\ \Phi_c^T B_u \end{bmatrix} \qquad (8.36)$$

The dynamic responses of common interest include strains, displacements and accelerations in a civil structure. Therefore, the observation equation in Equation 8.32 can be expressed as

$$Y = \begin{bmatrix} \varepsilon & d & a \end{bmatrix}^T; C = \begin{bmatrix} \Psi_c & 0 \\ \Phi_c & 0 \\ -\Phi_c\omega_o^2 & -2\Phi_c\xi\omega_o \end{bmatrix}; D = \begin{bmatrix} 0 \\ 0 \\ \Phi_c\Phi_c^T B_u \end{bmatrix} \qquad (8.37)$$

where:

ε, d and a represent the strain, displacement and acceleration responses, respectively
Ψ_c denotes the strain mode shapes as mentioned in Equation 8.5

Here, the observation equation is used to represent the measurement of multi-type sensors, reconstruct the multi-scale structural responses at key locations based on limited sensor measurements, and represent the real responses at these locations. Therefore, Equation 8.33 is rewritten as

$$\begin{cases} z_{k+1} = A z_k + B u_k + v_k \\ Y_k = C^e z_k + D^e u_k \\ Y_k^m = C^m z_k + D^m u_k + w_k \\ Y_k^e = C^e \hat{z}_k + D^e u_k \end{cases} \qquad (8.38)$$

where Y, Y^m and Y^e represent the real responses of interest, sensor measurements and reconstructed responses at the locations of interest based on the sensor measurements, respectively, and they consist of strains, displacements and accelerations. Matrices C and D should be expressed as Equation 8.37 with the following modifications: in Y^m, the mode shapes Ψ^m and Φ^m include the locations (DOFs) corresponding to measurement locations; in Y^e, the mode shapes Ψ^e and Φ^e depend on the locations where the responses are of interest. Typically, they are incomplete mode shapes if the responses at all DOFs are not interested. For example, the rotational DOFs of structures are often not measurable in practice. Moreover, it is often neither practical nor necessary to consider all modes of vibration in large-scale civil structures. The complete set of modes can be partitioned into a subset of selected modes (typically lower modes with a large contribution to responses) and truncated modes (typically higher modes with a small contribution to responses), that is

$$\boldsymbol{\Psi}_c = \begin{bmatrix} \boldsymbol{\Psi}_s & \boldsymbol{\Psi}_t \end{bmatrix}, \boldsymbol{\Phi}_c = \begin{bmatrix} \boldsymbol{\Phi}_s & \boldsymbol{\Phi}_t \end{bmatrix} \tag{8.39}$$

where subscripts s and t denote the selected and truncated modes, respectively. Considering the negligible effects of truncated modes on structural responses, Equation 8.38 can be simplified as

$$\begin{cases} \mathbf{z}_{k+1} \approx \mathbf{A}_s \mathbf{z}_k + \mathbf{B}_s \mathbf{u}_k + \mathbf{v}_k \\ \mathbf{Y}_k \approx \mathbf{C}_s^e \mathbf{z}_k + \mathbf{D}_s^e \mathbf{u}_k \\ \mathbf{Y}_k^m \approx \mathbf{C}_s^m \mathbf{z}_k + \mathbf{D}_s^m \mathbf{u}_k + \mathbf{w}_k \\ \mathbf{Y}_k^e \approx \mathbf{C}_s^e \hat{\mathbf{z}}_k + \mathbf{D}_s^e \mathbf{u}_k \end{cases} \tag{8.40}$$

Equation 8.40 substantially reduces the DOFs of structural models and it is not only computationally economical but it can also improve the reconstruction accuracy. Therefore, the equation enables the use of the Kalman filter algorithm in civil structures, particularly when real-time SHM is desirable, to reconstruct responses at key locations.

The accuracy of the estimated responses can be measured by the reconstruction error γ_k as follows:

$$\gamma_k = \mathbf{Y}_k^e - \mathbf{Y}_k = \mathbf{C}_s^e (\hat{\mathbf{z}}_k - \mathbf{z}_k) \tag{8.41}$$

The covariance matrix of the estimation error can be calculated as

$$\chi = \mathrm{cov}(\gamma_k) = \mathbf{C}_s^e \, \mathrm{cov}(\mathbf{z}_k - \mathbf{z}_k) \mathbf{C}_s^{eT} = \mathbf{C}_s^e \mathbf{P}_k \mathbf{C}_s^{eT} \tag{8.42}$$

where $\mathbf{P}_k$ is the covariance matrix of the estimation error in the state vector at time step k. Given any initial conditions $\mathbf{P}_0$ and $\hat{\mathbf{z}}_0$, the Kalman filter tends to converge with the iterations. Therefore, the asymptotic covariance matrix of the reconstruction error can be expressed as

$$\chi = \mathbf{C}_s^e \mathbf{P} \mathbf{C}_s^{eT} \tag{8.43}$$

Notably, the output influence matrix $\mathbf{C}_s^e$ or $\mathbf{C}_s^m$ tends to be highly ill-conditioned because the strain, the displacement and the acceleration have significantly different orders of magnitude. Without appropriate pre-treatment of the matrices, the inverse operation for the calculation of the gain matrix in the Kalman filter algorithm may lead to inaccurate results. Therefore, a similar procedure as mentioned in Section 8.3.2 which utilises the standard deviation of the corresponding sensor noise to normalise the mode shape matrix is employed herein to normalise the output matrix $\mathbf{C}_s^m$ or $\mathbf{C}_s^e$. For example, the normalised $\mathbf{C}_s^m$ is given by

$$\tilde{\mathbf{C}}_s^m = \begin{bmatrix} \mathbf{\Psi}_c/\sigma_\varepsilon & 0 \\ \mathbf{\Phi}_c/\sigma_d & 0 \\ -\mathbf{\Phi}_c\omega_o^2/\sigma_a & -2\mathbf{\Phi}_c\xi\omega_o/\sigma_a \end{bmatrix} = (\mathbf{R}^m)^{-1/2}\mathbf{C}_s^m \tag{8.44}$$

where

$$\mathbf{R}^m = E(\mathbf{ww}^T) = \begin{bmatrix} \sigma_\varepsilon^2\mathbf{I} & & \\ & \sigma_d^2\mathbf{I} & \\ & & \sigma_a^2\mathbf{I} \end{bmatrix} \tag{8.45}$$

where:

σ_ε^2, σ_d^2 and σ_a^2 are the strain gauges noise variance, the displacement transducers noise variance and accelerometers noise variance, respectively

$\mathbf{I}$ is the identity matrix

Thus, the optimal Kalman gain can be calculated accordingly:

$$\mathbf{K}_k = \mathbf{P}_{k|k-1}\tilde{\mathbf{C}}_s^{mT}[\tilde{\mathbf{C}}_s^m\mathbf{P}_{k|k-1}\tilde{\mathbf{C}}_s^{mT} + \mathbf{I}]^{-1}(\mathbf{R}^m)^{-1/2} \tag{8.46}$$

Each diagonal element of the χ matrix represents the variance of the reconstruction error for the corresponding response (strain, displacement or acceleration). Therefore, the maximum diagonal element denotes the maximum reconstruction error, whereas the trace of the matrix χ represents the sum of the reconstruction errors at all locations of interest. The OSP can be performed with the objective to minimise the sum of the reconstruction error. Again, the optimisation procedure may be considerably biased to one type of sensors without a proper normalisation because of the different magnitudes of strain, displacement and acceleration responses. Similarly, the measurement noise of displacements, strains and accelerations is utilised here to normalise the reconstruction errors. Thus, Equations 8.41 and 8.43 can be written, respectively, as follows:

$$\tilde{\boldsymbol{\gamma}}_k = (\mathbf{R}^e)^{-1/2}(\mathbf{Y}_k^e - \mathbf{Y}_k) = (\mathbf{R}^e)^{-1/2}\mathbf{C}_s^e(\mathbf{z}_k - \mathbf{z}_k) \tag{8.47}$$

$$\tilde{\boldsymbol{\chi}} = \mathrm{cov}(\tilde{\boldsymbol{\gamma}}_n) = \tilde{\mathbf{C}}_s^e\mathbf{P}\tilde{\mathbf{C}}_s^{eT} \tag{8.48}$$

where $\mathbf{R}^e$ has the same format as $\mathbf{R}^m$ but in a different dimension. The maximum and average estimation errors at all key locations yield

$$\tilde{\sigma}_{\max}^2 = \max\left(\mathrm{diag}\left(\tilde{\boldsymbol{\chi}}\right)\right) \tag{8.49}$$

$$\tilde{\sigma}_{\mathrm{avg}}^2 = \frac{tr(\tilde{\boldsymbol{\chi}})}{n} = \frac{tr\left[\tilde{\mathbf{C}}_s^e\mathbf{P}\tilde{\mathbf{C}}_s^{eT}\right]}{n} \tag{8.50}$$

The objective and constraint functions of the sensor location selection can be expressed as

$$\min\left(tr\left(\tilde{\mathbf{C}}_s^e\mathbf{P}\tilde{\mathbf{C}}_s^{eT}\right)\right) \tag{8.51}$$

subject to

$$\tilde{\sigma}_{\max}^2 \leq \left[\tilde{\sigma}_{\max}^2 \right] \tag{8.52}$$

in which $[\tilde{\sigma}_{\max}^2]$ are the target normalised maximum error.

The deletion of sensor locations increases the maximum and average reconstruction errors. In the iterative procedure for optimisation, the candidate sensor position is deleted in each iteration provided that such deletion results in a minimal trace of the matrix $\tilde{\mathbf{C}}_s^e \mathbf{P} \tilde{\mathbf{C}}_s^{eT}$. The number and locations of the multi-type sensors are determined when the increases in the maximum reconstruction errors reach a prescribed criterion. Subsequently, the multi-scale responses at the key locations can be reconstructed using the last sub-equation in Equation 8.40.

8.6 MULTI-TYPE SENSOR PLACEMENT OF A SUSPENSION BRIDGE MODEL

The reconstructed multi-scale responses can provide comprehensive information for SHM of bridge structures. However, the direct application of the proposed multi-type sensor placement method to a real, long-span suspension bridge is extremely challenging because of its large dimension size and complex structure, harsh operational environment and many uncertainties during the measurement of a long-span suspension bridge. Therefore, prior to real applications, a well-controlled laboratory-based test bed of a long-span suspension bridge and its corresponding updated FE model are utilised to examine the feasibility and accuracy of the proposed method. The experience obtained from this exercise can shed light on the application of the proposed method to real, long-span suspension bridges. It should be pointed out that this laboratory-based test bed was designed not only to verify the proposed method but also to fulfil many other functions of SHM for long-span suspension bridges. For the sake of completion, a brief description of the test bed follows and the details can be found in Xu et al. (2012).

8.6.1 Physical bridge model

The Tsing Ma suspension bridge in Hong Kong is the longest suspension bridge in the world, carrying both highway and railway, as shown in Figure 8.13. The Tsing Ma Bridge has been equipped with a comprehensive SHM system since 1997 (Wong et al. 2001a,b). This bridge was therefore selected as a reference for the design of a physical bridge model as a test bed. All of the major structural components in the bridge were included in the bridge model so that the model could best represent a real, long-span suspension bridge and strain-level

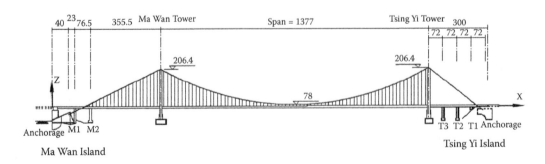

Figure 8.13 Configuration of prototype Tsing Ma Bridge (unit: m).

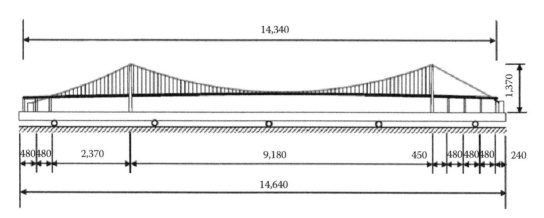

Figure 8.14 Configuration of Tsing Ma Bridge model and shake table (unit: mm).

measurements could be performed on the major structural components. All of the major connections and boundary conditions of the bridge were reproduced in the physical model so that rational damage scenarios could be best simulated. The physical model was installed on a shaking table which could generate proper ground motion as one type of external loading. The size of the physical model was large enough to facilitate the installation of various sensing systems and the measurement of various structural responses. Two kinds of materials were chosen in the bridge model design and fabrication: (1) steel was used for the bridge towers, piers, cables, suspenders and anchorages; and (2) aluminium was used for the bridge deck.

As shown in Figure 8.14, the physical bridge model has a main span 9.18 m with an overall length of 14.34 m. The height of the two bridge tower models is 1.37 m measured from the base level to the tower saddle. The two main cables in the bridge model are 0.24 m apart in the north and south. The width of the bridge deck model in the main span is 0.273 m. The shake table, made of a rectangular hollow-section steel beam, has a length of 14.64 m. The cross section of the steel beam is 400 mm wide, 200 mm high and 6.3 mm thick to facilitate the installation of the bridge model. The beam, weighing 851 kg, sits on five sets of roller bearings with a uniform spacing of 3.16 m. If the two rollers in each set of bearings are arranged along the x-axis of the bridge model, an electromagnetic exciter will be installed in the middle of the table to apply a force normal to the table so that the table will generate the ground motion to the bridge model in the lateral direction (y-axis). Figure 8.15 shows the completed bridge model.

8.6.2 FE model of the Tsing Ma Bridge model

The FE model of the physical bridge model was also established by using the commercial software ANSYS (2009). The three-dimensional tapered spatial beam elements (Beam44) with six DOFs at each end were used to model the tower legs. The portal beams between the two adjacent tower legs were modelled with prismatic spatial beam elements (Beam4). Piers M1, M2, T1, T2 and T3 were all similarly modelled using three-dimensional beam elements (Beam4) with fixed constraints. The bridge deck was a suspended type with two typical structural configurations of 60 mm long segments. Hence, modelling of the entire bridge deck could be conducted by repeatedly assembling the typical suspended deck segments longitudinally along the bridge. Each typical segment was represented by a three-dimensional FE sectional model using beam elements. Besides, elastic shell elements of 0.5 mm thick,

Figure 8.15 Completed Tsing Ma Bridge model.

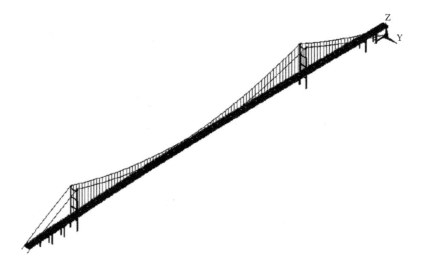

Figure 8.16 Overview of the FE model of the entire bridge model.

with 4 nodes and 24 DOFs, were utilised to model the orthotropic plate lying on the bottom chords of the cross frames. Two-node link elements with six DOFs, having the unique feature of uniaxial tension-only, were applied to model the main cables and the suspenders.

The vertical bearings between the deck and the Man Wan (MW) tower as well as the deck and pier M2 were represented by swing rigid links so that relative motion along the bridge axis was permissible. The vertical bearings between the deck and the Tsing Yi (TY) tower as well as the deck and piers T1, T2, T3 and M1 were simulated as one-dimensional longitudinal spring-damper elements to allow free longitudinal motion of the bridge. The lateral bearings of the deck to the two towers were modelled using one-dimensional longitudinal spring-damper elements. Only the rotation about the lateral axis is permissible in the MW abutment whereas only longitudinal movement and lateral rotation are allowable at the TY abutment. The top surfaces of the cable saddles were simply modelled with the joints coupled with the nodes at the tower leg tops. Similarly, the connection between pier M2 and the main cable was simulated with coupling in both the vertical and lateral directions. The ends of the two main cables were fixed in all DOFs. An overview of the FE model of the entire bridge is plotted in Figure 8.16.

8.6.3 Modal tests and model updating

Geometric measurements and modal tests were conducted on the physical bridge model to identify its geometric configuration and dynamic characteristics. The identified results indicate that the dynamic properties of the physical model can fulfil elementary design requirements. The first natural frequency of the bridge model corresponding to the lateral sway is approximately 3.9 Hz, which ensures that the physical model is not excessively flexible and sensitive to environmental interruptions. The shapes and sequence of the first few vibration modes of the bridge model match those of the prototype bridge. The model updating of the FE model was also performed using the modal test results. The natural frequencies and mode shapes of the updated model are in good agreement with the measured results. The test bed comprising the delicate physical model and the updated FE model of a long-span suspension bridge can serve as a benchmark problem on the SHM of long-span suspension bridges. More detailed information on the test bed can be found in the literature (Xu et al. 2012).

8.6.4 Framework of multi-type sensor placement and multi-scale response reconstruction

The application of the multi-type sensor placement and response reconstruction method to the test bed is explored in the following sections. Only ground motion–induced load is considered because a shaking table is designed for the test bed, as described in Section 8.6.1. Nevertheless, the implementation procedure for the multi-type sensor placement and the multi-scale response reconstruction of the test bed subject to ground motion is also applicable for other loading conditions. Given that the test bed is close to a real structure, practical considerations must be taken in the implementation procedure of the proposed method. The framework of the multi-type sensor placement and multi-scale response reconstruction method for the test bed is first presented, which includes the equation of motion of the test bed subject to ground motion, the objective function for sensor location optimisation, mode selection and multi-scale response reconstruction. A numerical study using the updated test bed FE model is then performed to select the sensor location following the proposed framework. Subsequently, with the identified sensor locations from the numerical study, fibre-Bragg grating (FBG) sensors, laser displacement transducers and accelerometers are mounted on the physical bridge model. Finally, experimental investigations are conducted to examine the accuracy of the selected multi-type sensor locations by comparing some of the reconstructed responses with the measured responses at locations where additional sensors are installed. More details can also be found in Xu et al. (2016).

8.6.4.1 Equation of motion

The dynamic equilibrium equation of motion of a linear structure subject to uniform ground acceleration in terms of the modal coordinate q can be written as

$$\ddot{q} + 2\xi\omega_o\dot{q} + \omega_o^2 q = -\Phi_u{}^T MS\ddot{u}_g \tag{8.53}$$

where:

M is the mass matrix of the bridge model
S is the selection matrix describing the DOFs of the supports with ground motion
$\ddot{u}_g$ denotes the ground acceleration vectors at the bridge supports

Equation 8.53 is the same as Equation 8.35 if the location matrix of excitation $\mathbf{B}_u$ in Equation 8.35 is replaced by the matrix **MS** in Equation 8.53. Therefore, the objective function for multi-type sensor location optimisation expressed by Equation 8.51 and the constraint function described by Equation 8.52 can be applied directly to select the locations of the three types of sensors for the test bed. Before sensor location selection and response reconstruction, the type and number of vibration modes of the test bed must be determined.

8.6.4.2 Mode selection

The proposed multi-type sensor placement method aims to select a set of sensor locations from which the measurement data can be used to best reconstruct the multi-scale responses. The sensor location selection and response reconstruction method is mode-based. Therefore, the selection of mode shapes is an important step. The contribution of individual modes to the overall dynamic behaviour of a structure is different. Not all modes have equally significant functions in the response reconstruction of the bridge, especially when only particular parts of the bridge response shall be reconstructed. Therefore, in the response reconstruction, more emphasis will be placed on the critical mode shapes that make significant contributions to the responses. The definition of RPE can be found in Equation 8.31. Let RPE_i be the relative percentage error of the reconstruction result at the ith DOF, which can be expressed by the following equation:

$$\text{RPE}_i = \frac{\text{std}(\mathbf{Y}_i^r - \mathbf{C}_i \mathbf{z} - \mathbf{D}_i \ddot{u}_g)}{\text{std}(\mathbf{Y}_i^r)}$$

(8.54)

where:

$\mathbf{Y}_i^r = \begin{bmatrix} \mathbf{Y}_i^r(t_1), & \mathbf{Y}_i^r(t_2), & \dots & \mathbf{Y}_i^r(t_k) \end{bmatrix}^T$ is the measured digitised time response series at the ith DOF

$\mathbf{C}_i$ and $\mathbf{D}_i$ denote the ith raw vectors of the output matrix $\mathbf{C}$ and the direct transmission matrix $\mathbf{D}$, respectively

$\hat{z}$ is the estimated state vector

Furthermore, the number of modes used to estimate the state vector $\hat{z}$ is assumed to be NN. By deleting only the jth mode, the reconstructed error at the ith DOF becomes

$$\text{RPE}_i^j = \frac{\text{std}(\mathbf{Y}_i^r - \mathbf{C}_i^j \mathbf{z}^j - \mathbf{D}_i^j \ddot{u}_g)}{\text{std}(\mathbf{Y}_i^r)}$$

(8.55)

where:

$\mathbf{C}_i^j$ denotes the ith raw vector of the output matrix $\mathbf{C}$ where the jth and $(NN + j - 1)$th columns are deleted

$\hat{z}^j$ represents the state vector by deleting the jth and $(NN + j - 1)$th rows

$\mathbf{D}_i^j$ is the ith raw vector of the direct transmission matrix $\mathbf{D}$ by deleting the jth mode

The reconstructed error at the ith DOF caused by the deletion of the jth mode during the response reconstruction is then equal to the difference between RPE_i^j and RPE_i:

$$\Lambda_i^j = \text{RPE}_i^j - \text{RPE}_i$$

(8.56)

Based on the discussion in Section 8.5 and Equations 8.54 and 8.55, it can be stated that the larger the value of Λ_i^j, the more important the jth mode for the response reconstruction at the ith DOF. By contrast, a very small value of Λ_i^j indicates that the jth mode is insignificant and can be neglected in the response reconstruction at the ith DOF.

Let us calculate the reconstructed errors at all DOFs of the key locations in the bridge caused by the deletion of the jth mode.

$$\Lambda^j = \sum_{i=1}^{n} \Lambda_i^j \qquad (8.57)$$

The sum of Λ^j for all the modes of interest in the response reconstruction can then be calculated by

$$\Lambda_{\text{total}} = \sum_{j=1}^{NN} (\Lambda^j) \qquad (8.58)$$

The contribution coefficient of the jth mode to the response reconstruction is defined as

$$c_j = \frac{\Lambda^j}{\Lambda_{\text{total}}} \times 100\% \qquad (8.59)$$

which yields $\sum_{j=1}^{NN} c_j = 1.0$.

One may say that a large value of c_j indicates that the jth mode contributes significantly to the overall response reconstruction of the structure, whereas a small value indicates insignificant contribution. Thus, mode selection can be performed as follows: (1) ranking the contribution coefficient c_j in descending order and (2) selecting the modes with the higher values of c_j so that the sum of c_j reaches a threshold that corresponds to the best reconstruction results.

8.6.4.3 Multi-scale response reconstruction

The selected modes of vibration are used to optimise the locations of the multi-type sensors by using the method proposed in Section 8.5. Once the numbers and locations of the multi-type sensors are determined, the strain, displacement and acceleration responses of the structure $\mathbf{y}^e$ at all key locations, which are considered as important locations for structural monitoring and assessment, can be reconstructed using the responses measured from the sensors $\mathbf{y}^m$ by the following equation:

$$\mathbf{Y}^e \approx \mathbf{C}_s^e \hat{\mathbf{z}} + \mathbf{D}_s^e \ddot{u}_g \qquad (8.60)$$

where:

$\mathbf{C}_s^e$ and $\mathbf{D}_s^e$ are the respective output and direct transmission matrices involving the DOFs of the key locations only for response reconstruction

The state vector $\hat{\mathbf{z}}$ is estimated by using the Kalman filter algorithm with limited measurements from the sensor locations.

8.6.4.4 Implementation procedure

The steps for the multi-type sensor location selection of the bridge model subjected to ground motion can be summarised as follows: (1) calculate the dynamic responses of the structure subjected to ground motion using the Newmark method; (2) select important locations for strain, displacement and acceleration response reconstruction from the viewpoint of structural monitoring and assessment, which could be based on the condition that the standard deviations of the responses in these locations are higher than certain values that significantly exceed the corresponding sensor noise levels; (3) determine all possible locations of the strain and displacement measurement sensors and accelerometers from the potential locations selected in Step 2 based on a number of practical issues such as the accessibility of measurement and the measurability of response; (4) determine the modes by using the mode selection method; (5) delete the key locations determined in Step 3 one by one using Equations 8.51 and 8.52 until the normalised maximum reconstruction errors approach their thresholds; (6) conduct a slight adjustment of the sensor locations by considering a number of practical issues such as convenience in sensor installation; (7) reconstruct the responses at the key locations determined in Step 3 through the polluted responses at the measured locations using Equation 8.60; and (8) compare some of the reconstructed responses with the real or measured ones.

8.6.5 Numerical analysis and results

The FE model of the test bed established using the commercial software ANSYS has a total of 23,700 DOFs and 8,400 elements. Figure 8.17 shows the four longitudinal truss girders of the bridge deck, called girders 1–4. Each girder consists of top and bottom longitudinal beams, diagonal bracings and vertical beams. Each top or bottom longitudinal beam has 242 nodes and 241 elements. Here, only the lateral DOFs of the deck are considered for possible installation of the displacement transducers and accelerometers for response reconstruction, and only the elements of the eight longitudinal beams of the bridge deck are considered for possible placement of strain measurement sensors for strain response reconstruction. The potential total number of locations for response construction is found to be 5776.

A random ground-acceleration time history is transversely applied to the FE model to calculate the dynamic responses of the bridge. The computed strain, displacement and acceleration responses are low-pass filtered with a cut-off frequency of 15 Hz, which includes the first 30 vibration modes of the bridge. The filtered strain, displacement and acceleration responses are then used to select the important locations for response reconstruction as

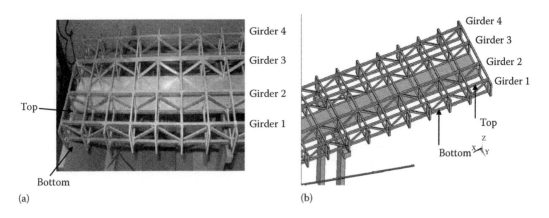

Figure 8.17 Longitudinal girders in the bridge deck: (a) part of the physical model and (b) part of the FE model.

well as the mode shapes for optimising the multi-type sensor locations and final response reconstruction.

The measured responses are inevitably polluted by environmental noise. Structural responses at some locations might be so small that they are overwhelmed by noise. These responses are insignificant in the evaluation of the structure's safety and functionality and therefore shall be ignored. Here, the ratio of the standard deviation of the structural response to that of the corresponding measurement noise is utilised to remove the locations where the ratio is lower than a threshold value. The threshold values are 24 for the strain responses, 7 for the displacement responses and 11 for the acceleration responses of the bridge model investigated. In such a way, 3035 out of 5776 locations are selected for response reconstruction, which include 1195 locations for strains, 1124 locations for displacements and 716 for accelerations.

The candidate locations for OSP are selected from the locations determined in the response reconstruction. First, considering the high correlation of responses at two adjacent nodes or elements and to reduce the computation time, one of every two nodes or elements is selected as a candidate. Second, because the displacement and acceleration responses of the inner girders 2 and 3 are similar to those of the outer girders 1 and 4, and measuring the responses of the inner girders is inconvenient, only the locations in girder 1 and girder 4 are considered as candidates for displacement transducers and accelerometers. However, the elements in girders 1–4 are all considered as a candidate set for strain sensors. As a result, 1053 locations are selected for the sensor candidate locations, including 593 candidate locations for strain sensors, 280 for displacement transducers and 180 for accelerometers.

The first 30 vibration modes from the FE model are considered as candidates for estimating the state vector. The target modes for sensor location optimisation and response reconstruction are selected from these modes based on the method discussed before. Figure 8.18 shows the contribution coefficient c_j of each mode. The 16th mode, which is a torsional mode, contributes most to the reduction of the sum of reconstruction errors. Although the vertical modes of the bridge have a minimal contribution to the lateral response reconstruction, these modes significantly affect the strain response reconstruction. By setting the threshold value of the contribution coefficient $c_j \leq 1.5 \times 10^{-3}$, 22 vibration modes are selected as target modes. Interestingly, all the unselected vibration modes are related only to the main cables.

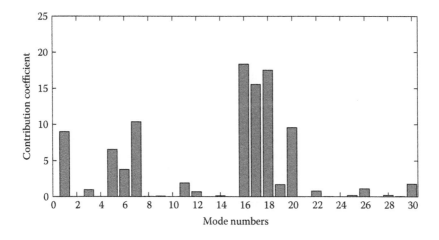

Figure 8.18 Modal contribution coefficients.

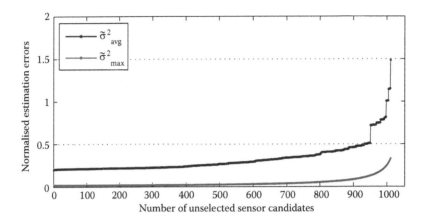

Figure 8.19 Variation of normalised theoretical reconstruction errors with number of sensors.

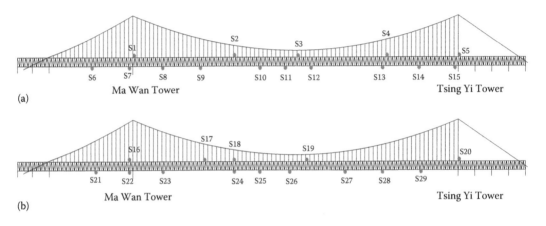

Figure 8.20 Optimal placements of 29 strain sensors (after adjustment): (a) girder 1 and (b) girder 4.

The proposed multi-type sensor placement method is now applied in the selection of sensor locations from all candidate locations. The threshold of the normalised maximum reconstruction error defined by Equation 8.52 is set at 1.5 in terms of the variance in the maximum reconstruction error. The variation in the normalised theoretical reconstruction errors with the number of deleted sensor location candidates is shown in Figure 8.19, which reveals that when 1011 sensor location candidates are deleted, the normalised maximum reconstruction error reaches its threshold value of 1.5. This result indicates that the use of only 42 sensors can satisfy the requirement. Figure 8.19 also shows that if less sensor locations are deleted, for instance, only 950 sensor locations are deleted, the accuracy of the response reconstruction would be higher at the cost of more sensors. The placements of the three types of sensors are obtained from the optimisation. The numbers of strain, displacement and acceleration sensors are 30, 2 and 10, respectively, for a total of 42 sensors.

A close look at the sensor placements decided in the previous paragraph exhibits that the two strain measurement sensors located near the MW tower at the top of girder 1 are very close to each other, and thus one of them can be deleted. Furthermore, one strain measurement sensor location selected in girder 2 is very close to the MW tower, and for the convenience of sensor installation, this location is transferred to a location near the

MW tower but at the bottom of girder 1. The two accelerometers at the bottom of girder 4 are also close to each other. One accelerometer is thus deleted, and the next optimal location is selected from the location of the accelerometer. The locations of strain measurement sensors and accelerometers after the adjustments are shown in Figures 8.20 and 8.21, respectively. The locations of the displacement measurement sensors remain unchanged and are shown in Figure 8.22. In the figures, 'S1' indicates the first strain measurement sensor placed on the deck, whereas 'A1' denotes the first accelerometer mounted on the deck. The total number of sensors is 41, including 29 strain measurement sensors, 2 displacement measurement sensors and 10 accelerometers. The sensor placements after adjustment are the final sensor locations that will be implemented in the subsequent experimental investigation.

By taking the simulated results from the selected sensors as measured ones added with sensor noise, the structural responses can be reconstructed and then compared with the computed results from the FE model. In the response reconstruction, the standard deviation of the sensor noise for the strain and displacement measurement sensors as well as the accelerometers are assumed to be 0.005 με, 0.0006 mm and 0.021 m/s², respectively. Figure 8.23 shows the RPEs between the reconstructed and actual responses of strains, displacements and accelerations after adjustment of the sensor locations. All RPEs are below 10% indicating that the reconstructed responses match the real responses quite well.

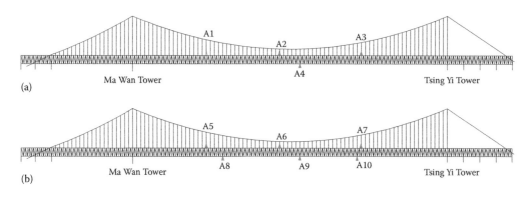

Figure 8.21 Optimal placements of 10 accelerometers (after adjustment): (a) girder 1 and (b) girder 4.

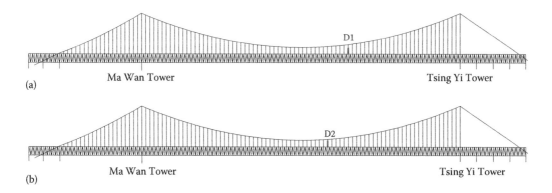

Figure 8.22 Optimal placements of two displacement transducers: (a) girder 1 and (b) girder 4.

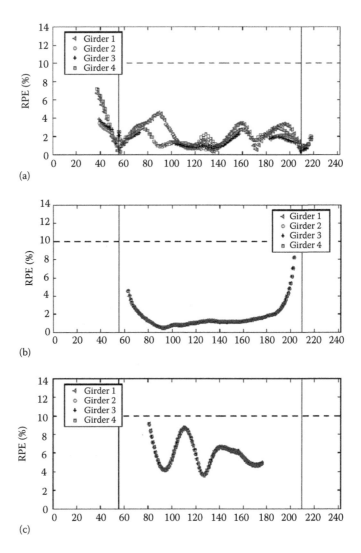

Figure 8.23 RPEs between the reconstructed and actual responses at top of the girder: (a) strain, (b) displacement and (c) acceleration.

8.6.6 Experimental validation

In this section, an experimental investigation was conducted to assess the practicability and accuracy of the proposed multi-type sensor placement and response reconstruction method using the test bed. Strain, displacement and acceleration sensors were installed at the locations of the physical bridge model, as determined before, to measure the structural responses. An additional six strain measurement sensors, two displacement measurement sensors and two accelerometers were installed on the bridge model to validate the reconstruction results.

Considering that a single string of optical fibre can accommodate tens of FBG sensors and that the size of an FBG sensor is small, FBG sensors were utilised in the experiment instead of strain gauges to measure the strain responses. The FBG sensors were attached on the side face of the beam at the middle of the elements to measure the combined axial and flexural deformation of the beam. LK-503 laser displacement transducers and KD1008 accelerometers were mounted laterally at the nodes of the deck. In addition, three strain-free FBG sensors were used to measure the temperature at different sensing points to provide temperature compensation for the strain measurements. These three FBG sensors were, respectively, installed at the MW side, middle span and TY side. To free these sensing points from stress, FBG sensors for temperature compensation were installed on the very short cantilever beams.

To achieve relative displacement and acceleration responses on the deck, ground acceleration and displacement were measured. Four KD1300 accelerometers and two LK-503 laser displacement transducers were installed laterally on the foundation beam. The average ground acceleration attained from the four accelerometers and the average ground displacement attained from the two laser displacement transducers were taken as the ground acceleration and displacement of the physical bridge model. It should be noted that the data acquisition system for recording the displacement and acceleration responses was different from that used for recording strain responses. However, synchronisation of these three types of responses was required for the reconstruction of the corresponding multi-scale responses. Therefore, a synchronous strain hammer was designed to realise the synchronisation of the strain, displacement and acceleration responses. More details on the synchronous strain hammer can be found in Zhang (2013).

A random force with a limited frequency band generated by an LDS V451 exciter was applied to the midpoint of the foundation beam to create ground motion in the bridge model. An overview of the experimental setup is shown in Figure 8.24. To generate a uniform ground motion in the bridge model, a series of tests was implemented, which revealed that the frequency range of the external force should be lower than 6 Hz to avoid resonance with the foundation beam because the first frequency of the foundation steel beam itself is 12.28 Hz.

Two tests at the same frequency range of 1.56–4.69 Hz but at different voltage amplitudes were implemented. In test 1, the excitation amplitude was 140 mV, whereas that of test 2 was 100 mV. Under the input ground motion, all the bridge responses collected from the sensors were combined to reconstruct the responses at the locations where additional sensors were available for validation.

The RPEs of strains, displacements and accelerations between the reconstructed and measured responses are shown in Figure 8.25 for test 1. 'SV', 'DV' and 'AV' in the figure denote the strain sensor, displacement sensor and accelerometer for validation, respectively. It can be seen that the reconstructed responses agree with the measured responses in consideration of the stringent error definition. Similar results can be obtained from test 2. The rotational

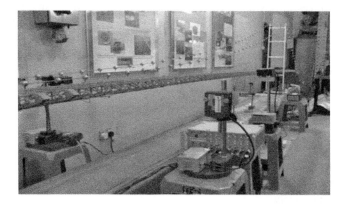

Figure 8.24 Overview of the experimental setup.

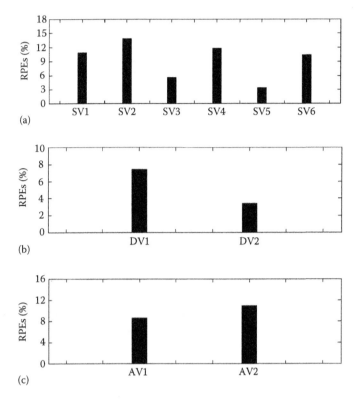

Figure 8.25 RPEs between the reconstructed and measured responses for test I girder: (a) strain, (b) displacement and (c) acceleration.

displacement response can also be reconstructed and more detailed results can be found in Zhang (2013). The two sets of test results show that the reconstructed responses match the measured results well, although the RPEs are over 10% in some cases. An RPE of over 10% is attributed to the fact that the definition of RPE is based on the accumulated error over the entire duration of the time history.

The good agreement of the reconstructed responses between the numerical studies and the experimental investigations validate the effectiveness and practicability of the proposed multi-type sensor placement and multi-scale response reconstruction method. The reconstructed multi-scale responses at key locations provide comprehensive information to the test bed for multi-scale monitoring, which deepens our understanding of both the global and local behaviour of the bridge. The successful application of the proposed method to the test bed also lays a solid foundation for extending the proposed method to a real long-span suspension bridge, which is a more complex structural system characterised by larger size and more uncertainties. This direction is challenging but it is a necessary task in the future.

NOTATION

A, B	State matrix and input matrix in the discrete-state space equation, respectively
$\mathbf{A}_c, \mathbf{B}_c$	State matrix and input matrix in the continuous state space equation, respectively
$\mathbf{B}_u$	Location matrix of excitations

c_j	Contribution coefficient of the jth mode to the response reconstruction
$\mathbf{C}_s^e$, $\mathbf{C}_s^m$	Output matrix for the estimated responses and measured responses on the basis of selected modes, respectively
$\tilde{\mathbf{C}}_s^e$, $\tilde{\mathbf{C}}_s^m$	Noise-normalised output matrix corresponding to $\mathbf{C}_s^e$ and $\mathbf{C}_s^m$, respectively
$\mathbf{C}$, $\mathbf{D}$	Output matrix and direct transmission matrix, respectively
$\mathbf{d}_c$	Displacement vector that includes translations or rotations in all DOFs of a structure
d_e	Nodal displacement vector corresponding to 6 DOFs of the beam element
$\mathbf{D}_s^e$, $\mathbf{D}_s^m$	Direct transmission matrix for the estimated responses and measured responses on the basis of selected modes, respectively
$\mathbf{E}_i$	The matrix representing the relationship between the node displacements of an element and the strains in this element
$\mathbf{F}$	Transformation matrix between the displacement vector and the strain vector
$\mathbf{K}$	Optimal Kalman gain
L	Length of the beam element
$\mathbf{P}_k$	Covariance matrix of the estimation error in state vector at time step k
$\mathbf{q}$	Modal coordinates
RPE	Relative percentage error between the reconstructed response and the real response
$\mathbf{S}_i$	Selection matrix that selects the displacements related to element i
$\mathbf{T}_i$	Transformation matrix that transforms the element nodal displacements in a global coordinate to those in a local coordinate
$\mathbf{u}$	External excitation vectors
$\ddot{u}_g$	Ground acceleration vectors
$\mathbf{W}$	Noise vector in the measurement
$\mathbf{W}$	Weight matrix accounting for the importance of different locations
x, y	Location in the element coordinate where the strain is calculated
$\mathbf{y}, \mathbf{y}_e, \mathbf{y}_m$	The actual response vector, the estimated response vector and the measured response vector in the dual-type sensor placement method, respectively
$\tilde{\mathbf{y}}, \mathbf{y}_e, \tilde{\mathbf{y}}_m$	Noise-normalised actual response vectors, noise-normalised estimated response vector and noise-normalised response measurement vector in the dual-type sensor placement method, respectively
$\mathbf{Y}, \mathbf{Y}^m$ and $\mathbf{Y}^e$	Real responses of interest, sensor measurements and reconstructed responses at the locations of interest in the multi-type sensor placement method, respectively
$\mathbf{z}$	State vector
$\boldsymbol{\gamma}$	Reconstruction error in multi-type OSP method
$\tilde{\boldsymbol{\gamma}}$	Noise-normalised reconstruction error in multi-type OSP method
$\boldsymbol{\Gamma}$	General modal matrix which includes both strain mode shapes and displacement mode shapes
$\boldsymbol{\Gamma}_m, \boldsymbol{\Psi}_m, \boldsymbol{\Phi}_m$	The partitioned model matrices corresponding to the positions with sensors
$\boldsymbol{\Gamma}_m^+$	Pseudo-inverse of the matrix $\boldsymbol{\Gamma}_m$
$\boldsymbol{\delta}$	Error between the estimated response and the real response in dual-type OSP method
$\tilde{\boldsymbol{\delta}}$	Noise-normalised estimation error in dual-type OSP method
$\boldsymbol{\Delta}$	Covariance matrix of the estimation error $\boldsymbol{\delta}$
$\tilde{\boldsymbol{\Delta}}$	Covariance matrix of $\tilde{\boldsymbol{\delta}}$
ε_i	Strain at a location in element i

$\varepsilon_x, \varepsilon_b$	Strain caused by axial force and bending moment in the beam element
$\mathbf{I}$	Identity matrix
Λ_i^j	The reconstructed error at the ith DOF caused by the deletion of the jth mode during the response reconstruction
$\boldsymbol{\xi}$	Modal damping coefficient matrix
$\sigma_\varepsilon^2, \sigma_d^2, \sigma_a^2$	Strain gauges noise variance, displacement transducers noise variance and accelerometers noise variance, respectively
$\tilde{\sigma}_{max}^2, \tilde{\sigma}_{avg}^2$	The maximum and average normalised estimation errors, respectively
$[\tilde{\sigma}_{max}^2], [\tilde{\sigma}_{avg}^2]$	The target normalised maximum and average estimated errors, respectively
$\boldsymbol{\Sigma}$	A diagonal matrix composed of the strain gauges noise variance and displacement transducers noise variance
$\boldsymbol{\Phi}_c$	The complete displacement modal matrix consisting of the mass-normalised displacement modes shapes φ_r
$\boldsymbol{\Phi}_s, \boldsymbol{\Phi}_t$	The selected modes and truncated modes of the displacement modal matrix for multi-type OSP
χ	Covariance matrix of the estimation error γ in multi-type OSP method
$\tilde{\chi}$	Covariance matrix of $\tilde{\gamma}$
$\boldsymbol{\Psi}, \boldsymbol{\Phi}$	Sub-matrices of the modal matrices $\boldsymbol{\Psi}_c$ and $\boldsymbol{\Phi}_c$, respectively
$\boldsymbol{\Psi}_c$	The complete strain modal matrix consisting of strain model shape ψ_r
$\boldsymbol{\Psi}_s, \boldsymbol{\Psi}_t$	The selected modes and truncated modes of the strain modal matrix for multi-type OSP
$\tilde{\boldsymbol{\Psi}}, \tilde{\boldsymbol{\Psi}}_m, \tilde{\boldsymbol{\Phi}}, \tilde{\boldsymbol{\Phi}}_m$	Noise-normalised mode shape matrices corresponding to $\boldsymbol{\Psi}, \boldsymbol{\Psi}_m, \boldsymbol{\Phi}, \boldsymbol{\Phi}_m$, respectively
$\boldsymbol{\omega}_o$	Modal frequency matrix

REFERENCES

Abdalla, M.O. and E. Al-Khawaldeh. 2011. Optimal damage detection sensor placement using PSO. *Appl. Mech. Mater.*, 110–116:5336–41.

ANSYS. 2009. User's Manual. Inc. Documentation for Release 12.1, Swanson Analysis System. http://orange.engr.ucdavis.edu/Documentation12.1/

Balageas, D., C.P. Fritzen, and A. Güemes. 2006. *Structural Health Monitoring.* London, UK: Wiley, ISTE Ltd.

Barthorpe, R.J., and K. Worden. 2009. Sensor placement optimization. In *Encyclopedia of Structural Health Monitoring*, ed. C. Boller, F.K. Chang, and Y. Fujino, Chapter 70. Chichester, UK: John Wiley & Sons, Ltd.

Beal, J.M., A. Shukla, O.A. Brezhneva, and M.A. Abramson. 2008. Optimal sensor placement for enhancing sensitivity to change in stiffness for structural health monitoring. *Optim. Eng.*, 9(2): 119–42.

Beck, J.L., and L.S. Katafygiotis. 1998. Updating models and their uncertainties. I: Bayesian statistical framework. *J. Eng. Mech.*, 124: 455–61.

Brehm, M., V. Zabel, and C. Bucher. 2013. Optimal reference sensor positions using output-only vibration test data. *Mech. Syst. Signal Proc.*, 41(1–2): 196–225.

Castro-Triguero, R., S. Murugan, R. Gallego, and M.I. Friswell. 2013. Robustness of optimal sensor placement under parametric uncertainty. *Mech. Syst. Signal Proc.*, 41(1–2): 268–87.

Cheng, F.Y., H. Jiang, and K. Lou. 2008. *Smart Structures: Innovative Systems for Seismic Response Control.* CRC Press, Boca Raton, FL.

Cherng, A.P. 2003. Optimal sensor placement for modal parameter identification using signal subspace correlation techniques. *Mech. Syst. Signal Proc.*, 17(2): 361–78.

Chung, Y.T., and J.D. Moore. 1993. On-orbit sensor placement and system identification of space station with limited instrumentations. In *Proceedings of the 11th International Modal Analysis Conference*, Kissimmee, Florida, 41–46.

Cobb, R.G., and B.S. Liebst. 1997. Sensor placement and structural damage identification from minimal sensor information. *AIAA J.*, 35(2): 369–74.

Debnath, N., A. Dutta, and S.K. Deb. 2012. Placement of sensors in operational modal analysis for truss bridges. *Mech. Syst. Signal Proc.*, 31: 196–216.

D'Souza, K., and B.I. Epureanu. 2008. Sensor placement for damage detection in nonlinear systems using system augmentations. *AIAA J.*, 46(10): 2434–42.

Ewins, D.J. 2000. *Modal Testing: Theory, Practice and Application. 2nd edition*. Research Studies Press Ltd, Letchworth, England.

Friswell, M.I., and R. Castro-Triguero. 2015. Clustering of sensor locations using the effective independence method. *AIAA J.*, 53(5): 1388–90.

Guo, H.Y., L. Zhang, L.L. Zhang, and J.X. Zhou. 2004. Optimal placement of sensors for structural health monitoring using improved genetic algorithms. *Smart Mater. Struct.*, 13(3): 528–34.

Gupta, V., M. Sharma, and N. Thakur. 2010. Optimization criteria for optimal placement of piezoelectric sensors and actuators on a smart structure: A technical review. *J. Intell. Mater. Syst. Struct.*, 21(12): 1227–43.

He, C., J. Xing, J. Li, Q. Yang, R. Wang, and X. Zhang. 2013. A combined optimal sensor placement strategy for the structural health monitoring of bridge structures. *Int. J. Distrib. Sens. Netw.*, 2013: 820694.

He, L., J. Lian, B. Ma, and H. Wang. 2014. Optimal multiaxial sensor placement for modal identification of large structures. *Struct. Control Health Monit.*, 21(1): 61–79.

Hemez, F.M., and C. Farhat. 1994. An energy based optimum sensor placement criterion and its application to structure damage detection. In *Proceedings of the 12th International Conference. on Modal Analysis*, Society of Experimental Mechanics, Honolulu, 1568–75.

Heo, G., M.L. Wang, and D. Satpathi. 1997. Optimal transducer placement for health monitoring of long span bridge. *Soil Dyn. Earthq. Eng.*, 16 (7–8): 495–502.

Imamovic, N. 1998. Model validation of large finite element model using test data. PhD diss., Imperial College London.

Jia, J., S. Feng, and W. Liu. 2015. A triaxial accelerometer monkey algorithm for optimal sensor placement in structural health monitoring. *Meas. Sci. Technol.*, 26(6): 065104.

Li, D.S., H.N. Li, and C.P. Fritzen. 2007. The connection between effective independence and modal kinetic energy methods for sensor placement. *J. Sound Vibr.*, 305(4–5): 945–55.

Liu, W., W.C. Gao, Y. Sun, and M.J. Xu. 2008. Optimal sensor placement for spatial lattice structure based on genetic algorithms. *J. Sound Vibr.*, 317(1–2): 175–89.

Ma, G., F.L. Huang, and X.M. Wang. 2008. Optimal placement of sensors in monitoring for bridge based on hybrid genetic algorithm. *J. Vibr. Eng.*, 21(2): 191–6.

Meo, M., and G. Zumpano. 2005. On the optimal sensor placement techniques for a bridge structure. *Eng. Struct.*, 27(10): 1488–97.

Kalman, R.E. 1960. A new approach to linear filtering and prediction problems. *J. Basic Eng.*, 82(1): 35–45.

Kammer, D.C. 1991. Sensor placement for on-orbit modal identification and correlation of large space structures. *J. Guid. Control Dyn.*, 14(2): 251–9.

Kammer, D.C. 1992. Effect of noise on sensor placement for on-orbit modal identification of large space structures. *J. Dyn. Sys., Meas., Control*, 114(3): 436–43.

Kammer, D.C. 1994. Enhancement of on-orbit modal identification of large space structures through sensor placement. *J. Sound Vibr.*, 171(1): 119–39.

Kammer, D.C. 1996. Optimal sensor placement for modal identification using system-realization methods. *J. Guid. Control Dyn.*, 19(3): 729–31.

Kammer, D.C., and M.L. Tinker. 2004. Optimal placement of triaxial accelerometers for modal vibration tests. *Mech. Syst. Signal Proc.*, 18(1): 29–41.

Kirkegaard, P.H., and R. Brincker. 1994. On the optimal locations of sensors for parametric identification of linear structural systems. *Mech. Syst. Signal Proc.*, 8(6): 639–47.

Kwon, S.J., S. Shin, H.S. Lee, and Y.H. Park. 2003. Design of accelerometer layout for structural monitoring and damage detection. *KSCE J. Civil Eng.*, 7(6): 717–24.

O'Callahan, J.C., P. Avitabile, and R. Riemer. 1989. System equivalent reduction expansion process. In *Proceedings of the Seventh International Conference on Modal Analysis*, Las Vegas, Nevada, 29–37.

Ottosen, N.S., and H. Petersson. 1992. *Introduction to the Finite Element Method.* New York: Prentice Hall.

Papadimitriou, C. 2004. Optimal sensor placement for parametric identification of structural systems. *J. Sound Vibr.*, 278(4–5): 923–47.

Papadimitriou, C. 2005. Pareto optimal sensor locations for structural identification. *Comput. Meth. Appl. Mech. Eng.*, 194(12–16): 1655–73.

Papadimitriou, C., J.L. Beck, and S.K. Au. 2000. Entropy-based optimal sensor location for structural model updating. *J. Vibr. Control*, 6(5): 781–800.

Penny, J.E.T., M.I. Friswell, and S.D. Garvey. 1994. Automatic choice of measurement location for dynamic testing. *AIAA J.*, 32(2): 407–14.

Rao, A.R.M., and G. Anandakumar. 2007. Optimal placement of sensors for structural system identification and health monitoring using a hybrid swarm intelligence technique. *Smart Mater. Struct.*, 16(6): 2658–72.

Said, W.M., and W.J. Staszewski. 2000. Optimal sensor location for damage detection using mutual information. In *Proceedings of 11th International Conference on Adaptive Structures and Technologies*, Nagoya, 428–35.

Salama, M., T. Rose, and J. Garba. 1987. Optimal placement of excitations and sensors for verification of large dynamical systems. In *Proceedings of the 28th Structures, Structural Dynamics, and Materials Conference*, Monterey, CA, 1024–31.

Shah, P.C., and F.E. Udwadia. 1978. A methodology for optimal sensor locations for identification of dynamic system. *J. Appl. Mech.*, 45(1): 188–96.

Shan, D., Z. Wan, and L. Qiao. 2011. Optimal sensor placement for long-span railway steel truss cable-stayed bridge. In *2011 Third International Conference on Measuring Technology and Mechatronics Automation*, IEEE, Shanghai, China, 2: 795–798.

Shi, Z.Y., S.S. Law, and L.M. Zhang. 2000. Optimum sensor placement for structure damage detection. *J. Eng. Mech.*, 126(11): 1173–9.

Tongpadungrod, P., T.D.L. Rhys, and P.N. Brett. 2003. An approach to optimise the critical sensor locations in one-dimensional novel distributive tactile surface to maximize performance. *Sensor Actuat. A: Phys.*, 105(1): 47–54.

Unwadia, F.E. 1994. Methodology for optimal sensor locations for parameter identification in dynamic system. *J. Eng. Mech.*, 120 (2): 368–90.

Unwadia, F.E., and J.A. Garba. 1985. Optimal sensor locations for structural identification. In *Proceeding of the Workshop on Identification and Control of Flexible Space Structures*, Jet Propulsion Laboratory, California Institute of Technology, Pasadena, CA, 247–61.

Wang, J., S.S. Law, and Q.S. Yang. 2013. Sensor placement methods for an improved force identification in state space. *Mech. Syst. Signal Proc.*, 41(1–2): 254–67.

Welch, G., and G. Bishop. 1995. An introduction to the Kalman filter. Technical report, University of North Carolina at Chapel Hill Chapel Hill, NC, USA. http://citeseerx.ist.psu.edu/viewdoc/summary?doi=10.1.1.117.6808.

Wong, K.Y., K.L. Man, and W.Y. Chan. 2001a. Application of global positioning system to structural health monitoring of cable-supported bridges. In *Proceedings of SPIE 4337, Health Monitoring and Management of Civil Infrastructure Systems*, ed. S.B. Chase, and A.E. Aktan, Newport Beach, CA, 390–401.

Wong, K.Y., K.L. Man, and W.Y. Chan. 2001b. Monitoring Hong Kong's bridges real-time kinematic spans the gap. *GPS World*, 12(7): 10–18.

Worden, K., and A.P. Burrows. 2001. Optimal sensor placement for fault detection. *Eng. Struct.*, 23(8): 885–901.

Xu, Y.L., X.H. Zhang, S. Zhan, et al. 2012. Testbed for structural health monitoring of long-span suspension bridges. *J. Bridge Eng.*, 17(6): 896–906.

Xu, Y.L., X.H. Zhang, S.Y. Zhu, and S. Zhan. 2016. Multi-type sensor placement and response reconstruction for structural health monitoring of long-span suspension bridges. *Science Bulletin*, 61(4): 313–29.

Yao, L., W.A. Sethares, and D.C. Kammer. 1993. Sensor placement for on-orbit modal identification via a genetic algorithm. *AIAA J.*, 31(10): 1922–8.

Yi, T.H., H.N. Li, and M. Gu. 2011. Optimal sensor placement for structural health monitoring based on multiple optimization strategies. *Struct. Design Tall Spec. Build.*, 20(7): 881–900.

Yi, T.H., H.N. Li, M. Gu, and X.D. Zhang. 2014. Sensor placement optimization in structural health monitoring using niching monkey algorithm. *Int. J. Struct. Stab. Dyn.*, 14(5): 1440012.

Yi, T.H., H.N. Li, and X.D. Zhang. 2012a. Sensor placement on Canton tower for health monitoring using asynchronous-climb monkey algorithm. *Smart Mater. Struct.*, 21(12): 125023.

Yi, T.H., H.N. Li, and X.D. Zhang. 2012b. A modified monkey algorithm for optimal sensor placement in structural health monitoring. *Smart Mater. Struct.*, 21(10): 105033.

Yuen, K.V., L.S. Katafygiotis, C. Papadimitriou, and N.C. Mickleborough. 2001. Optimal sensor placement methodology for identification with unmeasured excitation. *J. Dyn. Sys., Meas., Control*, 123(4): 677–86.

Zhang, X.H. 2013. Multi-sensing and multi-scale monitoring of long-span suspension bridges. PhD diss., Department of Civil and Environmental Engineering, The Hong Kong Polytechnic University.

Zhang, X.H., Y.L. Xu, S. Zhu, and S. Zhan. 2014. Dual-type sensor placement for multi-scale response reconstruction. *Mechatronics*, 24(4): 376–84.

Zhang, X.H., S. Zhu, Y.L. Xu, and X.J. Homg. 2011. Integrated optimal placement of displacement transducers and strain gauges for better estimation of structural response. *Int. J. Str. Stab. Dyn.*, 11(3): 581–602.

Zhu, S., X.H. Zhang, Y.L. Xu, and S. Zhan. 2013. Multi-type sensor placement for multi-scale response reconstruction. *Adv. Struct. Eng.*, 16(10): 1779–97.

Chapter 9

Structural control theory

9.1 PREVIEW

As discussed in Chapter 5, for an active, semi-active or hybrid control system, its controller system will utilise and analyse the measured signals from the sensory system, compute the necessary control forces or other quantities based on a given control algorithm, and command the control devices to make corresponding reactions, so that the structure can adapt itself to structural changes and to varying usage patterns and loading conditions. An active, semi-active or hybrid control system will not lead to the desired control performance if the control algorithm is not appropriately designed. The structural control theory is therefore developed to find appropriate control algorithms for active, semi-active or hybrid control systems. There are a number of excellent books introducing various structural control theories (Soong 1990; Ou 2003; Preumont 2011). This chapter provides a concise introduction to some basic concepts of structural control theory and to some commonly used control algorithms in structural vibration control. The basic concepts to be introduced in this chapter include stability, controllability and observability of a control system. The commonly used control algorithms include pole assignment, linear optimal control, independent modal space control (IMSC), sliding mode control (SMC), H_2 and H_∞ control, adaptive control, artificial intelligent control and semi-active control. The application of the control algorithms for vibration control of civil structures will be presented in Chapter 13. Because the performance of the control system also depends on the placement of the control devices and sensors used in the system, the optimal control device placement and the collective placement of sensors and control devices will be discussed in Chapters 10 and 11, respectively.

9.2 STABILITY

Civil structures are normally stable structures designed to serve people. A civil structure equipped with control devices must also be stable otherwise the structure will be useless or dangerous to people. Therefore, stability is a basic requirement for all civil structures equipped with control devices.

Taking a civil structure as a system, the definition of system stability can be referred to whether the free vibration response of the system is bounded or not. Let us consider the following linear time-invariant (LTI) system:

$$\dot{X} = AX \tag{9.1}$$

where:

    $\mathbf{X}$ is the n-dimensional state vector

    $\mathbf{A}$ is the $n \times n$ system matrix

The system described by Equation 9.1 is said to be (Hespanha 2009)

1. Stable (in the sense of Lyapunov or internally stable) if, for every initial condition $\mathbf{X}(t_0) = \mathbf{X}_0$, the homogeneous state response $\mathbf{X}(t) = \Theta(t,t_0) \cdot \mathbf{X}_0$, where Θ is the state transition matrix at $t \geq 0$, is uniformly bounded.
2. Asymptotically stable (in the sense of Lyapunov) if, in addition, for every initial condition $\mathbf{X}(t_0) = \mathbf{X}_0$, one obtains that $\mathbf{X}(t) \to 0$ as $t \to \infty$.
3. Exponentially stable if, in addition, there exist constants a, $b > 0$, such that, for every initial condition $\mathbf{X}(t_0) = \mathbf{X}_0$, one obtains $\|\mathbf{X}(t)\| \leq a \cdot e^{b(t-t_0)} \|\mathbf{X}(t_0)\|$ at $t \geq 0$.
4. Unstable if it is not marginally stable in the sense of Lyapunov.

Most civil structures are asymptotically stable because of their inherent positive structural damping. If a civil structure has negative damping, it becomes unstable.

Stability can also be understood more directly with the aid of Laplace transform and the characteristic equation. Taking a single-input-single-output system as an example, one may construct the impulse response by Laplace inversion of the transfer function. The general solution for the response is the linear combination of the input and a number of exponential terms which are products of the characteristic exponents and the time t. By setting the denominator of the transfer function to zero, the characteristic exponents can be obtained as the roots of the characteristic equation. It is obvious that, if the roots of the characteristic equation are all negative or have negative real parts, all the exponential terms are bounded leading to the response being bounded as well. Therefore, asymptotic stability is guaranteed if all the roots of the characteristic equation are negative or have negative real parts (Vepa 2010).

In the state space representation, the characteristic equation can be expressed as follows:

$$\det(s\mathbf{I} - \mathbf{A}) = 0 \tag{9.2}$$

where:

    s is the Laplace operator

    $\mathbf{I}$ is the identity matrix

  det   is the determinant

The roots of Equation 9.2 are generally referred to as the eigenvalues of the matrix $\mathbf{A}$ and they correspond to the poles of the transfer function. Thus, an LTI system is stable if and only if all the eigenvalues of matrix $\mathbf{A}$ have negative real parts. This stability condition is identical to saying that the corresponding transfer function has all its poles in the open left half of the complex 's' plane.

9.3 CONTROLLABILITY AND OBSERVABILITY

The concepts of controllability and observability, introduced by Kalman (1960), play a significant role in control theory: they appear as necessary and sometimes as sufficient conditions for the existence of a solution to most control problem. Controllability measures the ability of a particular control device configuration to control all the states of the system; conversely, observability measures the ability of a particular sensor configuration to supply all the information necessary to estimate all the states of the system (Preumont 2011).

In the control theory, the concept of controllability, as a coupling between the control vector and the system states, only involves the system matrix $\mathbf{A}$ and the matrix $\mathbf{B}$ which is associated with the control vector. Thus, one can consider the following linear state equation:

$$\dot{\mathbf{X}} = \mathbf{AX} + \mathbf{Bu} \qquad (9.3)$$

where:

- $\mathbf{u}$ is the p-dimensional control vector
- $\mathbf{B}$ is the $n \times p$ influence matrix

The system or the pair $(\mathbf{A}, \mathbf{B})$ is said to be controllable if for any initial condition $\mathbf{X}(t_0) = \mathbf{X}_0$ and any final state $\mathbf{X}(t_1) = \mathbf{X}_1$, there exists an input that is able to transfer $\mathbf{X}_0$ to $\mathbf{X}_1$ in a finite time. It should be noted that both the transferring trajectory of the state and the magnitude of the control vector are not specified in this definition.

The concept of observability is dual to that of controllability. Observability, as a coupling between the system states and the output, only involves the system matrix $\mathbf{A}$ and the output matrix $\mathbf{C}$. Thus, the following observation equation can be considered as

$$\mathbf{Y} = \mathbf{CX} \qquad (9.4)$$

where:

- $\mathbf{Y}$ is the m-dimensional output or observation vector
- $\mathbf{C}$ is the output matrix with the dimension of $m \times n$

The system or the pair $(\mathbf{A}, \mathbf{C})$ is observable at t_0 if the state $\mathbf{X}(t_0)$ can be uniquely determined by the corresponding output $\mathbf{Y}(t)$ for $t \in [t_0, t_1]$. If this is true for all initial moments t_0 and all initial states $\mathbf{X}(t_0)$, the system is completely observable.

There are many criteria for the determination of system controllability and observability. The controllability and observability matrices can be viewed as one of the most widely used indices for the measures of controllability and observability. The controllability matrix is defined as

$$\mathrm{Con} = \begin{bmatrix} \mathbf{B} & \mathbf{AB} & \mathbf{A}^2\mathbf{B} & \dots & \mathbf{A}^{n-1}\mathbf{B} \end{bmatrix} \qquad (9.5)$$

and the observability matrix is defined as

$$\mathrm{Obs} = \begin{bmatrix} \mathbf{C} \\ \mathbf{CA} \\ \mathbf{CA}^2 \\ \vdots \\ \mathbf{CA}^{n-1} \end{bmatrix} \qquad (9.6)$$

The LTI system is completely controllable if and only if the controllability matrix is full-rank, that is, *rank* (**Con**)=*n*. Similarly, the LTI system is completely observable if and only if the observability matrix is full-rank, that is, *rank* (**Obs**)=*n*. Therefore, the controllability and observability matrices provide a simple and direct way for the investigation of system controllability and observability.

However, only using controllability and observability matrices is often not enough for practical engineering problems, in which more quantitative information is required. For example, if a point control force is applied at the centre of a simply supported uniform beam, the structural mode shapes of even orders are not controllable because they have a nodal point at the centre. Similarly, a displacement sensor will be insensitive to the mode shapes having a nodal point where it is located. According to the rank tests, as long as the control device or the sensor is slightly moved away from the nodal point, the rank deficiency of the controllability or observability matrix disappears, indicating that the corresponding mode of vibration becomes controllable or observable. However, it is known that any attempt to control a mode of vibration with a control device located close to a nodal point would inevitably lead to difficulties, because this mode is only weakly controllable or observable. Besides the aforementioned drawback, the direct usage of controllability and observability matrices may lead to computational difficulty especially for large-scale civil structures. Assume, for example, that the system is of dimension $n = 100$. In order to answer the controllability and observability question, one has to find powers of **A** up to 99 and calculate the rank of large-size controllability and observability matrices which may easily result in numerical overflow. To overcome this problem, an alternative approach using the Gramian matrix is developed.

The controllability Gramian is defined as

$$\mathbf{W}_C(t) = \int_0^t e^{\mathbf{A}\tau} \mathbf{B}\mathbf{B}^T e^{\mathbf{A}^T \tau} d\tau \tag{9.7}$$

It should be noted that there appear to be some variations in the literature in the definition of controllability Gramian, but the one given in Equation 9.7 is the most common. It is known that $\mathbf{W}_C(t)$ satisfies

$$\dot{\mathbf{W}}_C(t) = \mathbf{A}\mathbf{W}_C(t) + \mathbf{W}_C(t)\mathbf{A}^T + \mathbf{B}\mathbf{B}^T \tag{9.8}$$

and when **A** is an asymptotically stable matrix, that is, all the eigenvalues of matrix **A** have negative real parts as mentioned before, $\mathbf{W}_C(t)$ reaches a steady state $\mathbf{W}_c$ which is a solution of Lyapunov equation:

$$\mathbf{A}\mathbf{W}_c + \mathbf{W}_c\mathbf{A}^T + \mathbf{B}\mathbf{B}^T = 0 \tag{9.9}$$

In general, the solution $\mathbf{W}_c$ can be expressed as

$$\mathbf{W}_c = \int_0^\infty e^{\mathbf{A}\tau} \mathbf{B}\mathbf{B}^T e^{\mathbf{A}^T \tau} d\tau \tag{9.10}$$

Similarly, the observability Gramian can be defined as

$$\mathbf{W}_O(t) = \int_0^t e^{\mathbf{A}^T \tau} \mathbf{C}^T \mathbf{C} e^{\mathbf{A}\tau} d\tau \qquad (9.11)$$

and it satisfies

$$\dot{\mathbf{W}}_O(t) = \mathbf{A}^T \mathbf{W}_O(t) + \mathbf{W}_O(t)\mathbf{A} + \mathbf{C}^T \mathbf{C} \qquad (9.12)$$

When the system is asymptotically stable, $\mathbf{W}_O(t)$ reaches its steady state $\mathbf{W}_o$ which can be obtained from the following Lyapunov equation:

$$\mathbf{A}^T \mathbf{W}_o + \mathbf{W}_o \mathbf{A} + \mathbf{C}^T \mathbf{C} = 0 \qquad (9.13)$$

as

$$\mathbf{W}_o = \int_0^\infty e^{\mathbf{A}^T \tau} \mathbf{C}^T \mathbf{C} e^{\mathbf{A}\tau} d\tau \qquad (9.14)$$

A detailed description of the controllability and observability Gramians including their properties and theorem proofs can be found in Gawronski (1998), Hespanha (2009) and Chen (2013).

It can be found from the definition of the controllability and observability Gramians that with a given system, that is, the matrix $\mathbf{A}$ is determined, the Gramians are mainly dependent on the matrix $\mathbf{B}$ or $\mathbf{C}$, which is related to the locations of control devices or sensors. Therefore, one can modify the control device and/or sensor locations to obtain the required (or assigned) values of the controllability and observability Gramians. Since Gramian matrices can provide quantitative measures of the degree of the controllability and observability properties, such as using some norm of the Gramians, many researchers investigate optimal control device and/or sensor placement in terms of the Gramian matrices.

9.4 POLE ASSIGNMENT

As mentioned before, the poles of a system are the eigenvalues of the system matrix $\mathbf{A}$. The eigenvalues of the matrix can be either real or complex. When they are complex, it must be in the form of a conjugate complex number, for example, $\lambda_j^{1,2} = \gamma \pm i\omega = -\zeta_j \omega_j \pm i\omega_j \sqrt{1 - \zeta_j^2}$ where ω_j and ζ_j are, respectively, the jth natural frequency and damping ratio. Thus, the eigenvalues of the matrix $\mathbf{A}$ correspond to points in the complex plane (γ, ω), and they represent the structural damping and natural frequency. From this point of view, the characteristics of a dynamic system, to a great extent, rely on the location of the poles in the complex plane.

The procedure of adjusting the location of the poles by using state feedback or output feedback to obtain desirable system characteristics is referred to as pole assignment. Since external disturbance is not related to the system poles, the external disturbance is not involved in the discussion of the pole assignment. Considering the LTI system described by Equations 9.3 and 9.4, two kinds of feedbacks, that is, state feedback and output feedback, are discussed in the following sections.

9.4.1 Pole assignment by state feedback

The necessary and sufficient condition for pole assignment by state feedback is that the system must be completely controllable. In this case, the control force can be given as

$$\mathbf{u} = -\mathbf{GX} \tag{9.15}$$

where $\mathbf{G}$ is the $p \times n$ state feedback gain matrix. By substituting Equation 9.15 into Equation 9.3, one obtains

$$\dot{\mathbf{X}} = (\mathbf{A} - \mathbf{BG})\mathbf{X} \tag{9.16}$$

It can be seen that the modification of the system matrix through active control alters the modal damping ratios and natural frequencies of the system. This can be reflected by the fact that the eigenvalues of $\mathbf{A} - \mathbf{BG}$ are generally different from those of $\mathbf{A}$. Since these closed-loop eigenvalues define the controlled system behaviour, a feasible control strategy is to choose the control gain $\mathbf{G}$ in such a way that the expected poles can be obtained.

Herein, a relatively simple method is introduced for the determination of the gain matrix $\mathbf{G}$ to produce the desirable system poles (Brogan 1974). The characteristic equation of the closed-loop system in Equation 9.16 can be expressed as

$$\left| \lambda \mathbf{I}_n - (\mathbf{A} - \mathbf{BG}) \right| = \left| (\lambda \mathbf{I}_n - \mathbf{A}) + \mathbf{BG} \right|$$

$$= \left| \lambda \mathbf{I}_n - \mathbf{A} \right| \times \left| \mathbf{I}_n + (\lambda \mathbf{I}_n - \mathbf{A})^{-1} \mathbf{BG} \right|$$

$$= \left| \lambda \mathbf{I}_n - \mathbf{A} \right| \times \left| \mathbf{I}_p + \mathbf{G}(\lambda \mathbf{I}_n - \mathbf{A})^{-1} \mathbf{B} \right| = 0 \tag{9.17}$$

where $\mathbf{I}_n$ and $\mathbf{I}_p$ are the $n \times n$ and $p \times p$ identity matrix, respectively.

Since λ is the eigenvalues of the system matrix $(\mathbf{A} - \mathbf{BG})$, $|\lambda \mathbf{I}_n - \mathbf{A}|$ should not be equal to zero. Thus, Equation 9.17 can be reduced to

$$\left| \mathbf{I}_p + \mathbf{G}(\lambda \mathbf{I}_n - \mathbf{A})^{-1} \mathbf{B} \right| = 0 \tag{9.18}$$

It can be seen that if the matrix determinant is equal to zero, at least either one column or one row should be zero. Assume that the selection of the gain matrix $\mathbf{G}$ yields the following condition:

$$\mathbf{e}_j + \mathbf{G}\boldsymbol{\varphi}_j(\lambda_i) = 0 \text{ or } \mathbf{G}\boldsymbol{\varphi}_j(\lambda_i) = -\mathbf{e}_j \tag{9.19}$$

where:

$\boldsymbol{\varphi}(\lambda_i)$ of $n \times p$ is equal to $(\lambda_i \mathbf{I}_n - \mathbf{A})^{-1}\mathbf{B}$ with the dimension of $n \times p$

$\boldsymbol{\varphi}_j(\lambda_i)$ is the n-dimensional vector obtained from the jth column of $\boldsymbol{\varphi}(\lambda_i)$

$\mathbf{e}_j$ is the jth column of the identity matrix $\mathbf{I}_p$

Hence, the state feedback gain matrix can be designed as

$$\mathbf{G} = -e\mathbf{\Gamma}^{-1} \tag{9.20}$$

where:
$$\mathbf{\Gamma} = \begin{bmatrix} \boldsymbol{\varphi}_{j1}(\lambda_1) & \boldsymbol{\varphi}_{j2}(\lambda_2) & \cdots & \boldsymbol{\varphi}_{jn}(\lambda_n) \end{bmatrix} \text{ with the dimension of } n \times n$$
$$\mathbf{e} = \begin{bmatrix} e_{j1} & e_{j2} & \cdots & e_{jn} \end{bmatrix} \text{ with the dimension of } p \times n$$

9.4.2 Pole assignment by output feedback

The necessary and sufficient condition for pole assignment by output feedback is that the system must be controllable and observable. In this case, the control force can be given as

$$\mathbf{u} = -\mathbf{G'Y} \tag{9.21}$$

in which $\mathbf{G'}$ is the $p \times m$ output feedback gain matrix. By substituting Equation 9.21 into Equation 9.3, one can obtain the following closed-loop system:

$$\dot{\mathbf{X}} = (\mathbf{A} - \mathbf{BG'C})\mathbf{X} \tag{9.22}$$

The characteristic equation of the closed-loop system in Equation 9.22 is

$$\left| \lambda\mathbf{I}_n - (\mathbf{A} - \mathbf{BG'C}) \right| = \left| (\lambda\mathbf{I}_n - \mathbf{A}) \right| \times \left| \mathbf{I}_n + (\lambda\mathbf{I}_n - \mathbf{A})^{-1}\mathbf{BG'C} \right|$$

$$= \left| (\lambda\mathbf{I}_n - \mathbf{A}) \right| \times \left| \mathbf{I}_p + \mathbf{G'C}(\lambda\mathbf{I}_n - \mathbf{A})^{-1}\mathbf{B} \right|$$

$$= \left| (\lambda\mathbf{I}_n - \mathbf{A}) \right| \times \left| \mathbf{I}_p + \mathbf{G'}\boldsymbol{\varphi'}(\lambda) \right| = 0 \tag{9.23}$$

where $\boldsymbol{\varphi'}(\lambda) = \mathbf{C}(\lambda\mathbf{I}_n - \mathbf{A})^{-1}\mathbf{B}$. Assuming that the eigenvalues of the open-loop system are different from those of the closed-loop system, one obtains $|\lambda\mathbf{I}_n - \mathbf{A}| \neq 0$ and Equation 9.23 can be thus given as

$$\left| \mathbf{I}_p + \mathbf{G'}\boldsymbol{\varphi'}(\lambda) \right| = 0 \tag{9.24}$$

Similarly, the output feedback gain matrix can then be determined as

$$\mathbf{G'} = -e'\mathbf{\Gamma'}^{-1} \tag{9.25}$$

where:
$$\mathbf{\Gamma'} = \begin{bmatrix} \boldsymbol{\varphi}'_{j1}(\lambda_1) & \boldsymbol{\varphi}'_{j2}(\lambda_2) & \cdots & \boldsymbol{\varphi}'_{jm}(\lambda_m) \end{bmatrix} \text{ with the dimension of } m \times m$$
$$\mathbf{e'} = \begin{bmatrix} e'_{j1} & e'_{j2} & \cdots & e'_{jm} \end{bmatrix} \text{ with the dimension of } p \times m$$

9.5 LINEAR OPTIMAL CONTROL

The control algorithm, which employs an integral of the quadratic function as performance index for simultaneously considering the effect of the system states and control forces, is referred to as linear quadratic optimal control. In this section, the theoretical foundations of two algorithms, that is, linear quadratic regulator (LQR) control and linear quadratic Gaussian (LQG) control, are introduced. The application of LQG control for the mitigation of structural vibration will be given in Chapter 13.

9.5.1 LQR control

Consider an LTI system described by Equation 9.3 and seek a control force in the form of

$$\mathbf{u} = -\mathbf{GX} \tag{9.26}$$

such that the following quadratic cost function is minimised:

$$J = \int_0^\infty \left(\mathbf{X}^T \mathbf{QX} + \mathbf{u}^T \mathbf{Ru} \right) dt \tag{9.27}$$

where:
  $\mathbf{G}$ is the optimal state feedback gain matrix
  $\mathbf{Q}$ is the positive semidefinite weighting matrix
  $\mathbf{R}$ is the positive definite weighting matrix

For a closed-loop system, minimising Equation 9.27 subject to the constraint of Equation 9.3 yields the gain matrix:

$$\mathbf{G} = \mathbf{R}^{-1}\mathbf{B}^T\mathbf{P} \tag{9.28}$$

where $\mathbf{P}$ is the symmetric positive definite matrix by solving the following algebraic Riccati equation:

$$\mathbf{PA} + \mathbf{A}^T\mathbf{P} + \mathbf{Q} - \mathbf{PBR}^{-1}\mathbf{B}^T\mathbf{P} = 0 \tag{9.29}$$

The existence and uniqueness of the solution of the Riccati equation is guaranteed if the pair $(\mathbf{A}, \mathbf{B})$ is controllable and $(\mathbf{A}, \mathbf{Q}^{1/2})$ is observable (Preumont 2011). Under these conditions, the closed loop:

$$\dot{\mathbf{X}} = (\mathbf{A} - \mathbf{BG})\mathbf{X} \tag{9.30}$$

is asymptotically stable.

It can be seen from Equation 9.30 that the poles of the closed-loop system depend on the weighting matrices $\mathbf{Q}$ and $\mathbf{R}$ if the system matrix $\mathbf{A}$ and the matrix $\mathbf{B}$ for the location of the control force are fixed. Multiplying both $\mathbf{Q}$ and $\mathbf{R}$ by a scalar coefficient leads to the same gain matrix $\mathbf{G}$ and thus the same closed-loop poles. Generally, when the elements of weighting matrix $\mathbf{Q}$ are larger, the system response will be reduced more but at the expense of a larger control force. When the elements of weighting matrix $\mathbf{R}$ are larger, the required control energy will be smaller but the structural response may not be sufficiently reduced. Hence, by varying the relative magnitudes of $\mathbf{Q}$ and $\mathbf{R}$, one can synthesise the controllers to achieve a proper trade-off between control effectiveness and control energy consumption. In structural control applications, if the controlled variables are not clearly identified, it may be sensible to choose $\mathbf{Q}$ in such a way that $\mathbf{X}^T\mathbf{Q}\mathbf{X}$ represents the total (kinetic plus strain) energy in the system. The matrix $\mathbf{R}$ can usually be chosen as $\mathbf{R} = \varsigma\mathbf{R}_1$, where $\mathbf{R}_1$ is a constant positive definite matrix and ς is an adjustable parameter whose value is selected to achieve reasonably fast closed-loop poles without excessive values of the control effort.

In principle, the LQR approach allows the design of multivariable state feedbacks which are asymptotically stable. A major drawback of this type of control law is that the full system states $\mathbf{X}$ are assumed to be available for feedbacks. In many practical situations, it is almost impossible to obtain all the system states and thus some form of filtering technique, for example, the Kalman filter, has been employed to compensate for this limitation. LQG control is such an approach which utilises the Kalman filter to estimate the system states and employs the estimated states for vibration control. The details of LQG control are given in the following subsection.

9.5.2 LQG control

Consider the controllable and observable LTI system:

$$\dot{\mathbf{X}} = \mathbf{A}\mathbf{X} + \mathbf{B}\mathbf{u} + \mathbf{w} \tag{9.31}$$

$$\mathbf{Y} = \mathbf{C}\mathbf{X} + \mathbf{v} \tag{9.32}$$

where:
  $\mathbf{w}$ is the process noise vector
  $\mathbf{v}$ is the measurement noise vector

It is assumed that the pair $(\mathbf{A}, \mathbf{C})$ is observable and that $\mathbf{w}$ and $\mathbf{v}$ are uncorrelated white noise processes with zero-mean and variance matrices equal to $\mathbf{W}$ and $\mathbf{V}$, respectively.

Based on the LQR control algorithm mentioned in Section 9.5.1, the optimal control force can be designed in the same format as Equation 9.26. However, only partial system states are available in this case and the control force cannot be directly determined according to Equation 9.26. The Kalman filter technique is thus employed to estimate all the system states for the determination of the control force.

The Kalman filter can be built as

$$\dot{\hat{\mathbf{X}}} = \mathbf{A}\hat{\mathbf{X}} + \mathbf{B}\mathbf{u} + \mathbf{G}_{KF}\left(\mathbf{Y} - \hat{\mathbf{Y}}\right); \ \hat{\mathbf{X}}(t_0) = \hat{\mathbf{X}}_0 \tag{9.33}$$

$$\hat{\mathbf{Y}} = \mathbf{C}\hat{\mathbf{X}} \tag{9.34}$$

where:
  $\hat{\mathbf{X}}$ is the estimated state
  $\mathbf{G}_{KF}$ is the gain of the Kalman filter and can be determined by

$$\mathbf{G}_{KF} = \mathbf{P}_{KF}\mathbf{C}^T\mathbf{V}^{-1} \tag{9.35}$$

where $\mathbf{P}_{KF}$ is the error covariance matrix of the filter and can be obtained from the following Riccati equation:

$$\mathbf{P}_{KF}\mathbf{A}^T + \mathbf{A}\mathbf{P}_{KF} - \mathbf{P}_{KF}\mathbf{C}^T\mathbf{V}^{-1}\mathbf{C}\mathbf{P}_{KF} + \mathbf{W} = 0 \tag{9.36}$$

The estimation error can be defined as $\mathbf{\Delta} = \mathbf{X} - \hat{\mathbf{X}}$, and it has been proved that

$$\lim_{t\to\infty} E\left[\mathbf{\Delta}(t)\right] = 0 \tag{9.37}$$

$$\lim_{t\to\infty} E\left[\mathbf{\Delta}(t)^T \mathbf{\Delta}(t)\right] = \text{trace}\left(\mathbf{P}_{KF}\right) \tag{9.38}$$

Thus, based on the estimated states, the optimal control force can finally be expressed as

$$\mathbf{u} = -\mathbf{G}\hat{\mathbf{X}} \tag{9.39}$$

By substituting Equation 9.39 into Equation 9.33, one obtains the following state equation of the controlled system:

$$\dot{\hat{\mathbf{X}}} = \left(\mathbf{A} - \mathbf{B}\mathbf{G} - \mathbf{G}_{KF}\mathbf{C}\right)\dot{\mathbf{X}} + \mathbf{P}_{KF}\mathbf{Y}; \ \hat{\mathbf{X}}(t_0) = \hat{\mathbf{X}}_0 \tag{9.40}$$

9.6 INDEPENDENT MODAL SPACE CONTROL

In general, the structural responses of a civil structure are dominated by a few mode components, such as the first few modes of vibration. Therefore, consideration of such few modes of vibration is often sufficient for dynamic analysis with acceptable accuracy. Moreover, in practical situations, civil structures are asymptotically stable during their service life. Thus, the structural vibration can be reduced by controlling these dominant mode components, and such a procedure is named *independent modal space control*. Obviously, a prerequisite for IMSC is that the mode components without control are all asymptotically stable.

The equation of motion of a controlled structure of N-degrees of freedom (DOFs) can be expressed as

$$\mathbf{M}_0\ddot{\mathbf{x}}(t) + \mathbf{C}_0\dot{\mathbf{x}}(t) + \mathbf{K}_0\mathbf{x}(t) = \mathbf{B}_0\mathbf{u}(t) \tag{9.41}$$

where:

$\mathbf{M}_0, \mathbf{C}_0$ and $\mathbf{K}_0$ are the mass, damping and stiffness matrix of the structure, respectively

$\ddot{\mathbf{x}}(t), \dot{\mathbf{x}}(t)$ and $\mathbf{x}(t)$ are the acceleration, velocity and displacement response, respectively

$\mathbf{u}(t)$ is the control force

$\mathbf{B}_0$ is the matrix denoting the location of the control devices

As the name implies, the design of a control system based on IMSC takes place in the modal space. By assuming that the damping matrix $\mathbf{C}_0$ is also orthogonal with respect to the mode shape matrix, Equation 9.41 can be decoupled in the generalised modal coordinate by using $\mathbf{x}(t) = \mathbf{\Phi}\mathbf{q}(t)$, resulting in

$$\mathbf{M}_0^*\ddot{\mathbf{q}}(t) + \mathbf{C}_0^*\dot{\mathbf{q}}(t) + \mathbf{K}_0^*\mathbf{q}(t) = \mathbf{u}^*(t) \tag{9.42}$$

in which

$$\left.\begin{aligned} \mathbf{M}_0^* &= \mathbf{\Phi}^T\mathbf{M}_0\mathbf{\Phi} \\ \mathbf{C}_0^* &= \mathbf{\Phi}^T\mathbf{C}_0\mathbf{\Phi} \\ \mathbf{K}_0^* &= \mathbf{\Phi}^T\mathbf{K}_0\mathbf{\Phi} \end{aligned}\right\} \tag{9.43}$$

$$\mathbf{u}^*(t) = \mathbf{\Phi}^T\mathbf{B}_0\mathbf{u}(t) = \mathbf{L}\mathbf{u}(t) \tag{9.44}$$

where:
  $\mathbf{q}(t)$ is the vector of generalised modal coordinates
  $\mathbf{\Phi}$ is the mode shape matrix
  $\mathbf{L}$ is the $N \times p$ matrix in which p denotes the number of control devices

Assume that only N_s modal shapes and modal coordinates ($N_s \leq N$) are considered and a subscript s is added to the corresponding matrices and vectors for clarity of expression. Then, Equations 9.42 through 9.44 can be rewritten as

$$\mathbf{M}_s^*\ddot{\mathbf{q}}_s(t) + \mathbf{C}_s^*\dot{\mathbf{q}}_s(t) + \mathbf{K}_s^*\mathbf{q}_s(t) = \mathbf{u}_s^*(t) \tag{9.45}$$

$$\mathbf{x}(t) = \mathbf{\Phi}_s\mathbf{q}_s(t) \tag{9.46}$$

$$\mathbf{u}_s^*(t) = \mathbf{\Phi}_s^T\mathbf{B}_0u(t) = \mathbf{L}_s\mathbf{u}(t) \tag{9.47}$$

For some control algorithms, such as the aforementioned pole assignment or linear optimal control algorithms, the control force in Equation 9.45 can be derived and shown in the form of

$$\mathbf{u}_s^*(t) = -\mathbf{G}_s\begin{Bmatrix} \mathbf{q}_s(t) \\ \dot{\mathbf{q}}_s(t) \end{Bmatrix} \tag{9.48}$$

where $\mathbf{G}_s$ is the mode feedback gain matrix, and it can be expressed as two parts to match the dimension of the selected modal coordinates. As a result, Equation 9.48 can be rearranged as

$$\mathbf{u}_s^*(t) = -\begin{bmatrix} \mathbf{G}_{s1} & \mathbf{G}_{s2} \end{bmatrix}\begin{Bmatrix} \mathbf{q}_s(t) \\ \dot{\mathbf{q}}_s(t) \end{Bmatrix} = -\mathbf{G}_{s1}\mathbf{q}_s(t) - \mathbf{G}_{s2}\dot{\mathbf{q}}_s(t) \tag{9.49}$$

Since the vectors of generalised modal coordinates are always unmeasurable, the control force cannot be directly determined and realised according to Equation 9.49. Therefore, it is necessary to transform the format of the control force to $\mathbf{u}(t)$ which can be expressed by system states in terms of the structural displacement $\mathbf{x}(t)$ and velocity $\dot{\mathbf{x}}(t)$. With the combination of Equations 9.46, 9.47 and 9.49 and after some deduction, the control force can be expressed by

$$\mathbf{u}(t) = -\bar{\mathbf{L}}_s \left[\mathbf{G}_{s1} \boldsymbol{\Phi}_s^{-1} \mathbf{x}(t) + \mathbf{G}_{s2} \boldsymbol{\Phi}_s^{-1} \dot{\mathbf{x}}(t) \right] \tag{9.50}$$

where

$$\bar{\mathbf{L}}_s = \begin{cases} \mathbf{L}_s^{-1} & N_s = p \\ \left(\mathbf{L}_s^T \mathbf{L}_s \right)^{-1} \mathbf{L}_s^T & N_s > p \\ \mathbf{L}_s^T \left(\mathbf{L}_s \mathbf{L}_s^T \right)^{-1} & N_s < p \end{cases} \tag{9.51}$$

It can be seen that the procedure of IMSC essentially shifts the problem of control design from a coupled n-order structural system to N second-order systems ($n = 2 \cdot N$), leading to a considerably simpler problem with substantial savings in computational efforts. Moreover, it is particularly attractive when only a few critical modes of vibration of a structure need to be controlled.

9.7 SLIDING MODE CONTROL

In principle, the SMC is suitable for both linear and nonlinear systems and it can be designed especially for the system with time-varying parameters. A number of books are available for a detailed description of the SMC including its theory and applications (e.g. Utkin 1992; Edwards and Spurgeon 1998; Bartolini et al. 2008; Shtessel et al. 2014). This section only gives a brief introduction to the SMC for the LTI system as describe by Equation 9.3. In general, the SMC includes two aspects: the design of a sliding surface and SMC controllers. In other words, the theory of SMC is to design controllers to drive the response trajectory into the sliding surface (or switching surface), whereas the motion on the sliding surface is stable (Yang et al. 1995a).

9.7.1 Design of sliding surface

The sliding surface for the LTI system described by Equation 9.3 can be given as

$$\mathbf{S} = \boldsymbol{\Theta} \mathbf{X} = 0 \tag{9.52}$$

where:
  $\mathbf{S}$ is the p-dimensional sliding surface with p sliding variables defined as $\mathbf{s} = \begin{bmatrix} S_1 & S_2 & \cdots & S_p \end{bmatrix}^T$
  $\mathbf{X}$ is the state vector as mentioned before; p is the number of control devices
  $\boldsymbol{\Theta}$ is the $p \times n$ matrix to be determined for sliding surface design

When the full state feedback is considered, the matrix $\boldsymbol{\Theta}$ can be determined on the basis of the aforementioned pole assignment or LQR control algorithm. However, if only partial system states are available, $\boldsymbol{\Theta}$ is determined by pole assignment rather than LQR control algorithm. In this section, the procedure for the determination of $\boldsymbol{\Theta}$ by LQR control is briefly introduced; the readers can find more information on this method in Yang et al. (1994, 1995a,b) and Ou (2003).

One systematic approach for the determination of the matrix $\boldsymbol{\Theta}$ is to convert Equation 9.3 into the so-called regular form by the following transformation (Utkin 1992). Let

$$\boldsymbol{\eta} = \boldsymbol{\Psi X} \text{ or } \boldsymbol{X} = \boldsymbol{\Psi}^{-1} \boldsymbol{\eta} \tag{9.53}$$

in which $\boldsymbol{\Psi}$ is the transformation matrix as

$$\boldsymbol{\Psi} = \begin{bmatrix} \mathbf{I}_{n-p} & -\mathbf{B}_1 \mathbf{B}_2^{-1} \\ 0 & \mathbf{I}_p \end{bmatrix}; \ \boldsymbol{\Psi}^{-1} = \begin{bmatrix} \mathbf{I}_{n-p} & \mathbf{B}_1 \mathbf{B}_2^{-1} \\ 0 & \mathbf{I}_p \end{bmatrix}; \ \mathbf{B} = \begin{bmatrix} \mathbf{B}_1 \\ \mathbf{B}_2 \end{bmatrix} \tag{9.54}$$

where:
$\mathbf{B}_1$ is the $(n-p) \times p$ matrix denoting the location without control devices
$\mathbf{B}_2$ is the $p \times p$ nonsingular matrix denoting the location with control devices

With the transformation matrix $\boldsymbol{\Psi}$, the state equation (Equation 9.3) and the sliding surface (Equation 9.52) become

$$\dot{\boldsymbol{\eta}} = \boldsymbol{\Psi A \Psi}^{-1} \boldsymbol{\eta} + \boldsymbol{\Psi B u} = \hat{\mathbf{A}} \boldsymbol{\eta} + \hat{\mathbf{B}} \mathbf{u} \tag{9.55}$$

$$\mathbf{S} = \hat{\boldsymbol{\Theta}} \boldsymbol{\eta} = 0 \tag{9.56}$$

in which

$$\hat{\mathbf{A}} = \boldsymbol{\Psi A \Psi}^{-1}; \ \hat{\mathbf{B}} = \begin{bmatrix} 0 & \mathbf{B}_2^T \end{bmatrix}^T; \ \hat{\boldsymbol{\Theta}} = \boldsymbol{\Theta} \boldsymbol{\Psi}^{-1} \tag{9.57}$$

It can be seen from Equation 9.55 that only the last p equations involve the equivalent control force $\mathbf{u}$. Thus, the equations of motion on the sliding surface are defined by p equations in Equation 9.56 and $(n-p)$ equations in the upper part of Equation 9.55. Let $\boldsymbol{\eta}, \hat{\mathbf{A}}$ and $\hat{\boldsymbol{\Theta}}$ be partitioned as follows:

$$\boldsymbol{\eta} = \begin{bmatrix} \boldsymbol{\eta}_1 \\ \boldsymbol{\eta}_2 \end{bmatrix}; \ \hat{\mathbf{A}} = \begin{bmatrix} \hat{\mathbf{A}}_{11} & \hat{\mathbf{A}}_{12} \\ \hat{\mathbf{A}}_{21} & \hat{\mathbf{A}}_{22} \end{bmatrix}; \ \hat{\boldsymbol{\Theta}} = \begin{bmatrix} \hat{\boldsymbol{\Theta}}_1 & \hat{\boldsymbol{\Theta}}_2 \end{bmatrix} \tag{9.58}$$

where $\boldsymbol{\eta}_1$ and $\boldsymbol{\eta}_2$ are the $(n-p)$-dimensional and p-dimensional vector, respectively.

With simple transformations and letting $\hat{\boldsymbol{\Theta}}_2 = \mathbf{I}$, one obtains

$$\boldsymbol{\eta}_2 = -\hat{\boldsymbol{\Theta}}_1 \boldsymbol{\eta}_1 \tag{9.59}$$

$$\dot{\eta}_1 = \left(\hat{A}_{11} - \hat{A}_{12}\hat{\Theta}_1 \right) \eta_1 \tag{9.60}$$

The matrix $\hat{\Theta}_1$ can be obtained from Equation 9.60 such that the motion η on the sliding surface is stable. Then, the unknown matrix Θ can be determined from Equation 9.57.

As mentioned before, the LQR method is employed for the determination of the matrix $\hat{\Theta}$. In consideration of Equation 9.53, the design of the sliding surface $S = \Theta X = 0$ can be obtained by minimising the following integral of the quadratic function of the state vector:

$$J = \int_{t_0}^{\infty} X^T Q X dt = \int_{t_0}^{\infty} \begin{bmatrix} \eta_1^T & \eta_2^T \end{bmatrix} T \begin{bmatrix} \eta_1 \\ \eta_2 \end{bmatrix} dt \tag{9.61}$$

where

$$T = \left(\Psi^{-1} \right)^T Q \Psi^{-1} = \begin{bmatrix} T_{11} & T_{12} \\ T_{21} & T_{22} \end{bmatrix} \tag{9.62}$$

in which the dimensions of the matrices T_{11} and T_{22} are $(n-p) \times (n-p)$ and $p \times p$, respectively.

Minimising the performance index J given by Equation 9.61 subjected to the constraint of the equations of motion, one obtains

$$\eta_2 = -0.5 T_{22}^{-1} (\hat{A}_{12}^T \hat{P} + 2 T_{21}) \eta_1 \tag{9.63}$$

where $\hat{P}$ is the solution of the following Riccati equation:

$$\tilde{A}^T \hat{P} + \hat{P} \tilde{A} - 0.5 \hat{P} \hat{A}_{12} T_{22}^{-1} \hat{A}_{12}^T \hat{P} = -2 \left(T_{11} - T_{12} T_{22}^{-1} T_{12}^T \right) \tag{9.64}$$

in which $\tilde{A} = \hat{A}_{11} - \hat{A}_{12} T_{22}^{-1} T_{21}$.

A comparison between Equation 9.59 and Equation 9.63 indicates that

$$\hat{\Theta}_1 = 0.5 T_{22}^{-1} \left(\hat{A}_{12}^T \hat{P} + 2 T_{21} \right) \tag{9.65}$$

The sliding surface can finally be determined from Equation 9.57 as

$$\Theta = \hat{\Theta} \Psi = \begin{bmatrix} \hat{\Theta}_1 & \vdots & I_p \end{bmatrix} \Psi \tag{9.66}$$

9.7.2 Design of controllers using Lyapunov direct method

The controllers are designed to drive the state trajectory into the sliding surface described by Equation 9.52. To achieve this goal, a Lyapunov function, L_v, is considered:

$$L_v = 0.5 S^T S = 0.5 X^T \Theta^T \Theta X \tag{9.67}$$

The sufficient condition for the sliding mode $S = 0$ to occur as $t \to \infty$ is

$$\dot{L}_v = S^T \dot{S} \le 0 \tag{9.68}$$

By combining the state equation of motion (Equation 9.3), one obtains

$$\dot{L}_v = S^T \Theta \dot{X} = S^T \Theta (AX + Bu) = \chi(u - \Lambda) = \sum_{i=1}^{p} \chi_i (u_i - \Lambda_i) \tag{9.69}$$

where:
   u_i is the control force provided by the ith controller

$$\chi = S^T \Theta B$$

$$\Lambda = -(\Theta B)^{-1} \Theta AX$$

For $\dot{L}_v \le 0$, a continuous controller can be given as

$$u_i = \Lambda_i - \delta_i \chi_i \quad \text{or} \quad \mathbf{u} = \Lambda - \delta\chi^T \tag{9.70}$$

where
   $\delta_i \ge 0$ is referred to as the sliding margin
   δ is the $p \times p$ diagonal matrix with diagonal elements $\delta_1, \delta_2, \ldots, \delta_p$

For the discontinuous controller, the control force can be given as

$$u_i = \begin{cases} \Lambda_i - \delta_i H(|\chi| - \varepsilon_0) & \chi_i > 0 \\ \Lambda_i + \delta_i H(|\chi| - \varepsilon_0) & \chi_i < 0 \\ 0 & \chi_i = 0 \end{cases} \tag{9.71}$$

where $H(|\chi| - \varepsilon_0)$ is the unit step function, that is, $H(|\chi| - \varepsilon_0) = 0$ for $|\chi| < \varepsilon_0$ and $H(|\chi| - \varepsilon_0) = 1$ for $|\chi| \ge \varepsilon_0$. In the foregoing expression, $|\chi|$ is any norm of the χ vector and ε_0 is the thickness of the boundary layer of the sliding surface $S = 0$.

9.8 H₂ AND H∞ CONTROL

Consider the system Σ_P shown in Figure 9.1 as

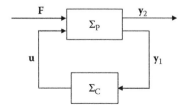

Figure 9.1 Configuration of closed-loop control system.

$$\Sigma_P : \begin{cases} \dot{X} = AX + Bu + DF \\ y_1 = \bar{C}_1 X + \bar{D}_1 F \\ y_2 = \bar{C}_2 X + \bar{B}_2 u \end{cases} \qquad (9.72)$$

where:

   X and u are, respectively, the system state and control input as defined before

     F is the l-dimensional external excitation or disturbance

     D is the $n \times l$ matrix relating to the location of external excitation

     y_1 is the m_1-dimensional measured output

     y_2 is the m_2-dimensional controlled output

Also, consider an arbitrary proper controller Σ_C given by

$$\Sigma_C : \begin{cases} \dot{\theta}_c = A_c \theta_c + B_c y_1 \\ u = C_c \theta_c + D_c y_1 \end{cases} \qquad (9.73)$$

where:

     θ_c is the state of controller

   A_c, B_c, C_c and D_c are the matrices related to the selected controller

The controller Σ_C is said to be admissible if it provides internal stability for the closed-loop system comprising Σ_P and Σ_C. Let $T_{y_2 F}$ denote the closed-loop transfer function from F to y_2 after applying a dynamic controller Σ_C to the system Σ_P. Thus, the H_2 or H_∞ control problem consists of finding a causal controller Σ_C which stabilises the system Σ_P and at the same time minimises the H_2 norm or H_∞ norm of the transfer function $T_{y_2 F}$. This section firstly introduces the definition of the transfer function $T_{y_2 F}$, and then, respectively, describes the H_2 and H_∞ control algorithm.

9.8.1 Transfer function and its norms

With proper transformation on the basis of Equations 9.72 and 9.73, one obtains

$$\left\{ \begin{array}{c} \dot{X} \\ \dot{\theta}_c \end{array} \right\} = \begin{bmatrix} A + BD_c \bar{C}_1 & BC_c \\ B_c \bar{C}_1 & A_c \end{bmatrix} \left\{ \begin{array}{c} X \\ \theta_c \end{array} \right\} + \begin{bmatrix} D + BD_c \bar{D}_1 \\ B_c \bar{D}_1 \end{bmatrix} F$$

$$y_2 = \begin{bmatrix} \bar{C}_2 + \bar{B}_2 D_c \bar{C}_1 & \bar{B}_2 C_c \end{bmatrix} \left\{ \begin{array}{c} X \\ \theta_c \end{array} \right\} + \bar{B}_2 D_c \bar{D}_1 F \qquad (9.74)$$

or

$$\begin{aligned} \dot{\bar{X}} &= \bar{A}_{c1} \bar{X} + \bar{B}_{c1} F \\ y_2 &= \bar{C}_{c1} \bar{X} + \bar{D}_{c1} F \end{aligned} \qquad (9.75)$$

in which $\bar{\mathbf{X}} = \begin{Bmatrix} \mathbf{X} \\ \boldsymbol{\theta}_c \end{Bmatrix}$; $\bar{\mathbf{A}}_{c1} = \begin{bmatrix} \mathbf{A} + \mathbf{BD}_c\bar{\mathbf{C}}_1 & \mathbf{BC}_c \\ \mathbf{B}_c\bar{\mathbf{C}}_1 & \mathbf{A}_c \end{bmatrix}$; $\bar{\mathbf{B}}_{c1} = \begin{bmatrix} \mathbf{D} + \mathbf{BD}_c\bar{\mathbf{D}}_1 \\ \mathbf{B}_c\bar{\mathbf{D}}_1 \end{bmatrix}$;

$\bar{\mathbf{C}}_{c1} = \begin{bmatrix} \bar{\mathbf{C}}_2 + \bar{\mathbf{B}}_2\mathbf{D}_c\bar{\mathbf{C}}_1 & \bar{\mathbf{B}}_2\mathbf{C}_c \end{bmatrix}$; and $\bar{\mathbf{D}}_{c1} = \bar{\mathbf{B}}_2\mathbf{D}_c\bar{\mathbf{D}}_1$.

The transfer function $\mathbf{T}_{y_2F}$ can be obtained as

$$\mathbf{T}_{y_2F}(s) = \bar{\mathbf{C}}_{c1}(s\mathbf{I} - \bar{\mathbf{A}}_{c1})^{-1}\bar{\mathbf{B}}_{c1} + \bar{\mathbf{D}}_{c1} \tag{9.76}$$

The necessary and sufficient condition for the system shown in Figure 9.1 being stable is that all the eigenvalues of matrix $\bar{\mathbf{A}}_{c1}$ have negative real parts.

Let us define the H_2 norm of the transfer function $\mathbf{T}_{y_2F}$ as

$$\left\|\mathbf{T}_{y_2F}\right\|_2 = \frac{1}{2\pi}\int_{-\infty}^{\infty} \text{trace}\left[\mathbf{T}_{y_2F}(j\omega)\,\mathbf{T}_{y_2F}^*(j\omega)\right]d\omega \tag{9.77}$$

where $\mathbf{T}_{y_2F}^*(j\omega)$ is the complex conjugate of $\mathbf{T}_{y_2F}(j\omega)$.

The H_∞ norm of the transfer function $\mathbf{T}_{y_2F}$ is defined as

$$\left\|\mathbf{T}_{y_2F}\right\|_\infty = \sup_{0\leq\omega<\infty} \sigma_{\max}\left[\mathbf{T}_{y_2F}(j\omega)\right] \tag{9.78}$$

where:

$\sigma_{\max}$ denotes the maximum singular value of transfer function $\mathbf{T}_{y_2F}$

The symbol 'sup' in Equation 9.78 stands for supremum.

9.8.2 H_2 control algorithm

For a given system Σ_P, the infimum of the H_2-norm of the closed-loop transfer function $\mathbf{T}_{y_2F}$ over all the stabilising proper controllers Σ_C is denoted by γ_2^*, namely,

$$\gamma_2^* = \inf\left\{\left\|\mathbf{T}_{y_2F}\right\|_2 \;\middle|\; \Sigma_C \text{ internally stabilises } \Sigma_P\right\} \tag{9.79}$$

where symbol 'inf' denotes the infimum. Then, a stabilising proper controller Σ_C is said to be a H_2-optimal controller for Σ_P if $\left\|\mathbf{T}_{y_2F}\right\|_2 = \gamma_2^*$.

For the case of full state feedback, if the control force is

$$\mathbf{u} = -\mathbf{GX} = -\left(\bar{\mathbf{B}}_2^T\bar{\mathbf{B}}_2\right)^{-1}\left(\bar{\mathbf{B}}_2^T\bar{\mathbf{C}}_2 + \mathbf{B}^T\bar{\mathbf{P}}\right)\mathbf{X} \tag{9.80}$$

the transfer function yields the condition of $\left\|\mathbf{T}_{y_2F}\right\|_2 = \gamma_2^*$, in which $\gamma_2^* = \text{trace}\left(\mathbf{D}^T\bar{\mathbf{P}}\mathbf{D}\right)$ and $\bar{\mathbf{P}}$ satisfies the following algebraic Riccati equation:

$$\mathbf{A}^T\bar{\mathbf{P}} + \bar{\mathbf{P}}\mathbf{A} + \bar{\mathbf{C}}_2^T\bar{\mathbf{C}}_2 - \left(\bar{\mathbf{P}}\mathbf{B} + \bar{\mathbf{C}}_2^T\bar{\mathbf{B}}_2\right)\left(\bar{\mathbf{B}}_2^T\bar{\mathbf{B}}_2\right)^{-1}\left(\bar{\mathbf{B}}_2^T\bar{\mathbf{C}}_2 + \mathbf{B}^T\bar{\mathbf{P}}\right) = 0 \tag{9.81}$$

If not all the system states are available, the output feedback would then be adopted. In this case, the following assumption should be followed: the pair $(\mathbf{A}, \mathbf{B})$ is stable; pair $(\mathbf{A}, \bar{\mathbf{C}}_1)$

is observable; matrix $\bar{\mathbf{B}}_2$ is full rank for its column; matrix $\bar{\mathbf{D}}_1$ is full rank for its row; and both $(\mathbf{A}, \mathbf{B}, \bar{\mathbf{C}}_2, \bar{\mathbf{B}}_2)$ and $(\mathbf{A}, \mathbf{D}, \bar{\mathbf{C}}_1, \bar{\mathbf{D}}_1)$ have no zero point in the imaginary axis. If the positive semidefinite matrix $\bar{\mathbf{P}}$ is the unique solution of the Riccati Equation 9.81, and the positive semidefinite matrix $\bar{\mathbf{Q}}$ is the unique solution of the following Riccati equation:

$$\bar{\mathbf{Q}}\mathbf{A}^T + \mathbf{A}\bar{\mathbf{Q}} + \mathbf{D}\mathbf{D}^T - \left(\bar{\mathbf{Q}}\bar{\mathbf{C}}_1^T + \mathbf{D}\bar{\mathbf{D}}_1^T\right)\left(\bar{\mathbf{D}}_1\bar{\mathbf{D}}_1^T\right)^{-1}\left(\bar{\mathbf{D}}_1\mathbf{D}^T + \bar{\mathbf{C}}_1\bar{\mathbf{Q}}\right) = 0 \tag{9.82}$$

the H_2-optimal controller can then be given as

$$\left.\begin{aligned}
\dot{\boldsymbol{\theta}}_c &= \left(\mathbf{A} - \mathbf{B}\mathbf{G} - \bar{\mathbf{K}}_c\bar{\mathbf{C}}_1\right)\boldsymbol{\theta}_c + \bar{\mathbf{K}}_c\mathbf{y}_1 \\
\mathbf{u} &= -\mathbf{G}\boldsymbol{\theta}_c
\end{aligned}\right\} \tag{9.83}$$

in which $\mathbf{G} = \left(\bar{\mathbf{B}}_2^T\bar{\mathbf{B}}_2\right)^{-1}\left(\bar{\mathbf{B}}_2^T\bar{\mathbf{C}}_2 + \mathbf{B}^T\bar{\mathbf{P}}\right)$ and $\bar{\mathbf{K}}_c = \left(\bar{\mathbf{Q}}\bar{\mathbf{C}}_1^T + \mathbf{D}\bar{\mathbf{D}}_1^T\right)\left(\bar{\mathbf{D}}_1\bar{\mathbf{D}}_1^T\right)^{-1}$. This controller satisfies $\left\|\mathbf{T}_{y2F}\right\|_2 = \gamma_2^*$, in which $\gamma_2^* = \left\{\text{trace}\left(\mathbf{D}^T\bar{\mathbf{P}}\mathbf{D}\right) + \text{trace}\left[\left(\mathbf{A}^T\bar{\mathbf{P}} + \bar{\mathbf{P}}\mathbf{A} + \bar{\mathbf{C}}_2^T\bar{\mathbf{C}}_2\right)\bar{\mathbf{Q}}\right]\right\}$.

9.8.3 H∞ control algorithm

The definition of optimal H_∞ control can be given as finding all the admissible controllers Σ_C such that $\left\|\mathbf{T}_{y2F}\right\|_\infty$ is minimised (Zhou and Doyle 1998). It should be noted that the optimal H_∞ controllers as defined are generally not unique for multi-input-multi-output systems. Furthermore, finding an optimal H_∞ controller is often both numerically and theoretically complicated. This is certainly in contrast with the standard H_2 theory, in which the optimal controller is unique and can be obtained by solving the aforementioned Riccati equations without iterations. Therefore, it is usually much easier to obtain controllers that are very close in the norm sense to the optimal ones, which are called *suboptimal controllers*. A suboptimal controller may also have other nice properties (e.g. lower bandwidth) over the optimal ones.

Similarly, for a given system Σ_P, the infimum of the H_∞-norm of the closed-loop transfer function $\mathbf{T}_{y2F}$ over all the stabilising proper controllers Σ_C is denoted by γ_∞^*:

$$\gamma_\infty^* = \inf\left\{\left\|\mathbf{T}_{y2F}\right\|_\infty \mid \Sigma_C \text{ internally stabilises } \Sigma_P\right\} \tag{9.84}$$

If a controller Σ_C stabilises the system Σ_P, and satisfies

$$\left\|\mathbf{T}_{y2F}\right\|_\infty < \gamma \qquad \left(\gamma > \gamma_\infty^*\right) \tag{9.85}$$

such controller can then be called the γ-order suboptimal H_∞ controller.

For the case of full state feedback, if the control force is given as

$$\mathbf{u} = -\mathbf{G}\mathbf{X} = -\left(\bar{\mathbf{B}}_2^T\bar{\mathbf{B}}_2\right)^{-1}\left(\bar{\mathbf{B}}_2^T\bar{\mathbf{C}}_2 + \mathbf{B}^T\bar{\mathbf{P}}\right)\mathbf{X} \tag{9.86}$$

the transfer function will then yield the condition $\left\|\mathbf{T}_{y2F}\right\|_\infty < \gamma$ where the matrix $\bar{\mathbf{P}}$ is the unique solution of the following algebraic Riccati equation:

$$A^T\bar{P} + \bar{P}A + \bar{C}_2^T\bar{C}_2 + \bar{P}DD^T\bar{P}/\gamma^2 - \left(\bar{P}B + \bar{C}_2^T\bar{B}_2\right)\left(\bar{B}_2^T\bar{B}_2\right)^{-1}\left(\bar{B}_2^T\bar{C}_2 + B^T\bar{P}\right) = 0 \qquad (9.87)$$

If the system states are partially available, for a given $\gamma > \gamma_\infty^*$, the output feedback similar to the H$_2$ control algorithm can be used. If the positive semidefinite matrix is the unique solution of the Riccati Equation 9.87, and the positive semidefinite matrix $\bar{Q}$ is the unique solution of the following Riccati equation:

$$\bar{Q}A^T + A\bar{Q} + DD^T + \bar{Q}\bar{C}_2^T\bar{C}_2\bar{Q}/\gamma^2 - \left(\bar{Q}\bar{C}_1^T + D\bar{D}_1^T\right)\left(\bar{D}_1\bar{D}_1^T\right)^{-1}\left(\bar{D}_1D^T + \bar{C}_1\bar{Q}\right) = 0 \qquad (9.88)$$

the γ-order suboptimal H$_\infty$ controller can then be given as

$$\left.\begin{aligned}
\dot{\boldsymbol{\theta}}_c &= \tilde{\mathbf{A}}_c\boldsymbol{\theta}_c + \tilde{\mathbf{B}}_c\mathbf{y}_1 \\
\mathbf{u} &= -\mathbf{G}\boldsymbol{\theta}_c
\end{aligned}\right\} \qquad (9.89)$$

in which $\tilde{\mathbf{A}}_c = \mathbf{A} + \gamma^{-2}DD^T\bar{P} - BG - \left(I - \gamma^{-2}\bar{Q}\bar{P}\right)^{-1}\bar{K}_c\left(\bar{C}_1 + \gamma^{-2}\bar{D}_1D^T\bar{P}\right);$ $\tilde{\mathbf{B}}_c = \left(I - \gamma^{-2}\bar{Q}\bar{P}\right)^{-1}\bar{K}_c;$ $\mathbf{G} = \left(\bar{B}_2^T\bar{B}_2\right)^{-1}\left(\bar{B}_2^T\bar{C}_2 + B^T\bar{P}\right);$ and $\bar{K}_c = \left(\bar{Q}\bar{C}_1^T + D\bar{D}_1^T\right)\left(\bar{D}_1\bar{D}_1^T\right)^{-1}.$

9.9 ADAPTIVE CONTROL

It is well known that 'to adapt' means to change a characteristic or behaviour to conform to new circumstances. Intuitively, an adaptive controller is thus a controller that can modify its behaviour in response to changes in the dynamics of the process and the character of the disturbances (Åström and Wittenmark 1995). An adaptive controller typically consists of an LTI compensator together with an identifier (or tuner) that is used to adjust the compensator parameters; a common approach to tuning is to invoke the certainty equivalence principle, whereby it is assumed at each instance of time that the system parameter estimate is correct and the controller gains are updated accordingly (Miller 2003). Since ordinary feedback and feedforward also try to reduce the effects of disturbances and system uncertainties, difference between feedback/feedforward and adaptive control immediately arises. Over the years, there have been many attempts to define adaptive control formally. Although a meaningful definition of adaptive control is still lacking, there appears to be a consensus that a constant-gain feedback or feedforward system is not an adaptive system.

Though adaptive control has been extensively and actively investigated and developed in both theory and application during the past decades, there are still a number of challenges and problems. For example, most adaptive controllers are nonlinear, which makes their behaviour hard to predict and then results in undesirable action (especially in the case of poor initial parameter estimates), such as large transients or a large control signal. Moreover, other problems also need to be considered in all adaptive control algorithms, such as the problem of impractical control objectives, the transient instability, the suddenly unstable closed loop and the changing experimental conditions (Anderson and Dehghani 2008). Although many obstacles exist in the development of adaptive control, great attention has been paid in this area and much progress has been made. A variety of books with different emphases are available as an introduction to adaptive control (e.g. Åström and

Wittenmark 1995; Feng and Lozano 1999; Landau et al. 2011; Hou and Jin 2014). Notably, adaptive control is a large family, and according to different principles there are different ways of categorising the available algorithms, such as feedforward adaptive control and feedback adaptive control, direct adaptive control and indirect adaptive control, open-loop adaptive control and close-loop adaptive control and so forth. In this section, two broad classes of adaptive control, model-based adaptive control and model-free adaptive control, are briefly introduced.

At present, most adaptive control techniques and methodologies are model-based, which means that they all include the notion of a 'model' of the true system and the existence of an identification mechanism to provide an implicit or explicit system model to the adaptive algorithm. Thus, the system model is the starting point and landmark for the controller design and analysis, as well as the evaluation criterion and control destination for the model-based control methods. In the context of model-based adaptive control, a variety of algorithms have been developed, such as multiple model adaptive control (Narendra and Balakrishnan 1997; Anderson et al. 2000, 2001; Hespanha et al. 2001; Rosa et al. 2009; Baldi et al. 2011), model reference adaptive control (Morse 1996; Blažič et al. 2003; Huang et al. 2004; Shyu et al. 2008; Yucelen and Haddad 2012), feedback linearisation–based adaptive control (Sastry and Isidori 1989; Chen and Khalil 1995; Liu and Zhang 2005; Liu 2007; Yang et al. 2009), sliding mode adaptive control (Chen et al. 2001; Chen 2006; Xie 2007; Guan and Pan 2008), the backstepping adaptive control (Zhou et al. 2004; Zhang et al. 2008; Zhou and Wen 2008; Tong et al. 2009) and so forth. It is recognised that modelling errors always exist due to the complexity of the structural system and the operational environment, and the application of the controllers designed on the basis of an inaccurate mathematical model may cause various practical problems. Therefore, significant effort has been devoted to the development of robust adaptive control to preserve the obvious advantages of model-based control design while increasing robustness against model errors (Ioannou and Sun 1996; Sastry and Bodson 2011; Lavretsky and Wise 2013).

As a counterpart of the model-based adaptive control, another class of adaptive control is model-free. The basic idea of the model-free adaptive control is implemented by building a virtual equivalent dynamical linearisation data model of the nonlinear system at each operation point first, then estimating the system's pseudo partial derivative online by using input and/or output data of the controlled system, and designing the controller according to some weighted one-step-ahead cost functions (Hou and Jin 2014). Since the model-free adaptive control is merely based on the real-time input and/or output measurement data of the controlled system, it is capable of working without prior knowledge of any form of the system model and without the modelling process, which implies that a general controller for a class of practical processes could be designed independently. This may be the most attractive merit for the model-free adaptive control. A number of model-free adaptive control methods, such as unfalsified adaptive control (Safonov and Tsao 1997; Kosut 2001; Dehghani et al. 2007; Jin and Safonov 2012), model-free adaptive predictive control (Tan et al. 2001; Hou 2002; Zhang and Zhang 2006), model-free adaptive iterative learning control (Jin 2008) and so forth can be found in the literature.

Notably, most researches and applications of the adaptive control are in the field of electronic engineering, mechanical engineering and aerospace engineering. Due to the complexity and large size of civil infrastructures, there are few applications of adaptive control in civil engineering, and the corresponding researches are limited as well, focusing on some simple structures, such as shear or frame building models with a single DOF or several DOFs (Burdisso et al. 1994; Faravelli and Yao 1996; Basu and Nagarajaiah 2008; Bitaraf et al. 2010, 2012).

9.10 ARTIFICIAL INTELLIGENT CONTROL

Artificial intelligence (AI) systems are widely accepted as a technology that offers an alternative way to tackle complex and ill-defined problems. They are able to learn from examples, handle noisy and incomplete data, deal with nonlinear problems and once trained can perform prediction and generalisation at high speed. In the field of structural vibration control, it has been recognised that AI algorithms can be incorporated into the conventional control theory to realise more flexible control systems, and such systems have come to be known as intelligent control systems. This section briefly introduces three basic AI control algorithms, that is, neural network control (NNC), fuzzy logic control (FLC) and genetic algorithm control (GAC), and the corresponding hybrid AI control which combines two or more basic AI techniques to perform the specific control task.

Neural networks (NNs) provide a distinctive computational paradigm and have proved effective for a range of problems where conventional computation techniques have not succeeded, such as some highly nonlinear control problems. NNC or neuro-control is a subset of both neural network research and conventional control techniques (Linkens and Nyongesa 1996). The fundamental philosophy of NNC is to view a control system as a mapping between system state and the actuating commands, with learning regarded as the modification of this mapping to improve the control system performance objective. To date, NNs are widely used for a large number of diverse applications leading to various groupings of NNC with different emphases. For example, Hunt et al. (1992) characterised NNs in terms of their learning rules and the network structure. Thus, NNC are classified according to whether they use supervised learning or self-learning, and also whether their structure is feedforward or a recurrent network. Moreover, Agarwal (1997) proposed hierarchical classification, starting the basic level from whether the NNs were used as a controller or as an aid to a controller. A number of books are available providing a detailed description of the design and applications of NNC (e.g. Miller et al. 1995; Hunt et al. 1995; Norgaard et al. 2000; Sarangapani 2006).

FLC has proved effective for complex, nonlinear and imprecisely defined processes where the implementation of standard model-based control techniques is impossible or impractical. However, a number of difficulties exist for the design of an effective FLC. For example, the derivation of fuzzy control rules is often time-consuming and difficult, and relies to a great extent on the so-called process experts who may not be able to transcribe their knowledge into the requisite rule form. Moreover, no formal framework exists for the selection of the parameters of fuzzy systems, which means tuning these parameters and learning models has generally become an important subject of fuzzy control. Thus, from these points of view, when conventional control theory yields a satisfactory result and when an adequate and solvable mathematical model already exists or can easily be created, FLC are not recommended to be used (Kalogirou 2003). For the design of a fuzzy logic controller, a general process is defining the structure and the parameters of a fuzzy controller, designing the rule base and the computational unit, and determining the rules for defuzzification, that is, to transform fuzzy output to control action. A basic requirement for implementing fuzzy control is the availability of a control expert who provides the necessary knowledge for the control problem. More details on fuzzy control and practical applications can be found in a variety of books (e.g. Sugeno 1985; Reznik 1997; Espinosa et al. 2005; Dadios 2012).

Genetic algorithms (GAs) are inspired by the way that living organisms adapt to the harsh realities of life in a hostile world, that is, by evolution and inheritance. The algorithm imitates in the process the evolution of a population by selecting only fit individuals for reproduction. Therefore, GAs are exploratory search and optimisation procedures that are devised on the principles of natural selection and survival of the fittest. There has been widespread interest from the control community in applying GAs to problems in control systems engineering. Compared with traditional search and optimisation procedures, GAs are robust, global and generally more straightforward to apply in situations where there is little or no a priori knowledge about the process to be controlled. Basically, we can encode the structure and the parameters of the controllers into a chromosome, and define a fitness measure as a function over the performance demands, thus formulating the design problem as the minimisation of an objective function with respect to the controller parameters. Since GAs only need a fitness function to guide the optimisation process, they can be employed to execute this search. The creative combination of a variety of pre-existing control methodologies and GAs can result in a powerful tool that is able to address real engineering control problems. More details on the implementation of GAs for control problem can be found in other books (e.g. Man et al. 1997; Jamshidi et al. 2002; Popa 2012).

Based on the aforementioned three types of AI control methods, the hybrid AI control using a combination of two or more AI methods is developed. The classical hybrid AI control is the neuro-fuzzy control (NFC), whereas other types combine GAs and FLC or ANs and GAs as part of an integrated problem solution or in order to perform specific separate tasks of the same problem. Here, taking the famous NFC as an example, a fuzzy system possesses great power in representing linguistic and structured knowledge by fuzzy sets and performing fuzzy reasoning and fuzzy logic in a qualitative manner. However, a common bottleneck in FLC systems is their dependence on the specification of good rules by control experts. NNs, on the other hand, are 'constructed' through training procedures and are particularly effective at representing nonlinear mappings in a computational fashion, but it is not always possible to extract and interpret the learned knowledge contained within them. To avoid the individual drawbacks of these two paradigms, the possibility of integrating the two technologies is considered, leading to a new kind of system called NFC where several strengths of both algorithms are utilised and combined appropriately. The research and application of NFC are actively conducted and can be found in many references (e.g. Brown and Harris; 1994; Fullér 2000; Lewis et al. 2002; Siddique 2014).

9.11 SEMI-ACTIVE CONTROL

Control strategies for semi-active devices appear to combine the best features of both passive and active control systems and to offer the greatest likelihood for near-term acceptance of control technology as a viable means of protecting structural systems against earthquake and wind loading. According to presently accepted definitions, a semi-active control device is one which cannot inject mechanical energy into the controlled structural system (i.e. including the structure and the control device), but has properties that can be controlled to optimally reduce the responses of the system (Spencer and Sain 1997; Spencer and Nagarajaiah 2003). Therefore, in contrast to active control devices, semi-active control devices do not have the potential to destabilise (in the bounded input/bounded output sense) the structural system. In general, the semi-active controller mitigates structural vibration mainly through altering the structural damping, stiffness or, in some cases, both. Various kinds of semi-active control devices have been introduced in Chapter 4, and a variety of control strategies have thus been developed for providing appropriate control signals to command the behaviour of these devices. Several commonly used semi-active control strategies for the control of magnetorheological (MR) fluid dampers, such as simple bang-bang control law, optimal bang-bang control law and clipped optimal control law, are introduced in this section. Some applications of the clipped optimal control law for the vibration suppression of a building complex will be given in Chapter 13, and the implementation of this control algorithm in smart structures for the simultaneous consideration of the damage detection and vibration control in the time domain will be discussed in Chapter 15.

9.11.1 Simple bang-bang control law

A simple bang-bang control law for an MR damper can be expressed as

$$u_d = \begin{cases} c_d \dot{z} + f_d^{\max} \operatorname{sgn}(\dot{z}) & (z\dot{z} > 0) \\ c_d \dot{z} + f_d^{\min} \operatorname{sgn}(\dot{z}) & (z\dot{z} \leq 0) \end{cases} \tag{9.90}$$

where:

$z = z_d$ denotes the relative displacement of the MR damper

u_d is the control force provided by the MR damper

c_d is the viscous damping coefficient

$f_d^{\max}$ and $f_d^{\min}$ are the maximum and minimum value of the adjustable Coulomb damping force, respectively

It is worth noting that the relative displacement of an MR damper is related to the structural motion. For example, with rigid connection, the relative displacement of an MR damper can be easily determined by the inter-story shift of a building structure. Thus, from Equation 9.90, it is obvious that when structural motion is away from its equilibrium position, the maximum control force is provided for energy dissipation; otherwise, the minimum damping force will be given. A schematic diagram of the MR damper force determined by the simple bang-bang control law is shown in Figure 9.2. It can be seen from Figure 9.2 that the maximum damping force is provided in the first and third quadrant.

9.11.2 Optimal bang-bang control law

The optimal bang-bang control law can be expressed as

$$u_d = \begin{cases} c_d \dot{z} + f_d^{\max} \operatorname{sgn}(\dot{z}) & (u\dot{z} < 0) \\ c_d \dot{z} + f_d^{\min} \operatorname{sgn}(\dot{z}) & (u\dot{z} \geq 0) \end{cases} \tag{9.91}$$

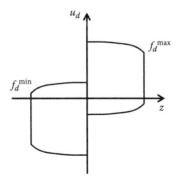

Figure 9.2 Graphical representation of the simple bang-bang control law.

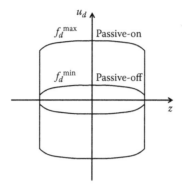

Figure 9.3 Graphical representation of the optimal bang-bang control law.

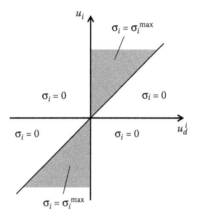

Figure 9.4 Graphical representation of the clipped optimal control algorithm.

In this control law, if the direction of the optimal control force is opposite to the direction of structural motion, the maximum MR damping force will be provided; otherwise, the minimum one will be given. This algorithm also implies that the force provided by an MR damper is merely used for energy dissipation and will not destabilise the structural system.

Figure 9.3 gives two extreme cases of optimal bang-bang control law, that is, the direction of the control force is always identical or opposite to the direction of structural motion. These two extreme cases indicate the MR damper in the so-called passive-off and passive-on state, respectively. Actually, the MR force determined by the optimal bang-bang control law is switched between these two states in many situations.

9.11.3 Clipped optimal control law

It is known that for some semi-active control devices, for example, an MR damper, the voltage or current would be a significant variable for adjusting the damping force. From this point of view, Dyke et al. (1996) proposed a clipped optimal control algorithm for determining the MR damper force. The algorithm for selecting the command signal (i.e. the applied voltage) is graphically represented in Figure 9.4 and can be concisely stated as

$$\sigma_i = \sigma_i^{\max} \cdot \mathrm{H}\left\{\left(u_i - u_d^i\right)u_d^i\right\}$$

(9.92)

where:

    σ_i is the voltage applied to the ith controller

    $\sigma_i^{\max}$ is the maximum voltage to be applied to the ith controller

    u_i denotes the desired optimal force, which can be determined by an appropriate active control algorithm, for example, the aforementioned LQG, H_2 norm control and so forth

    u_d^i is the ith actual control force

    $\mathrm{H}(\cdot)$ is unit step function

When the damper is providing the desired optimal force (i.e. $u_i = u_d^i$), the voltage applied to the damper should remain at the present level. If the magnitude of the force produced by the damper is smaller than the magnitude of the desired optimal force and the two forces have the same sign, the voltage applied to the current driver increases to the maximum level so as to increase the force produced by the damper to match the desired control force. Otherwise, the commanded voltage is set to zero.

NOTATION

$\mathbf{A}$	System matrix
$\mathbf{A}_c, \mathbf{B}_c, \mathbf{C}_c, \mathbf{D}_c$	The matrices related to the selected controller
$\mathbf{B}$	The matrix which is associated with the control vector in the state space equation
$\mathbf{B}_0$	The matrix denoting the location of the control devices
c_d	Viscous damping coefficient of the MR damper
$\mathbf{C}$	Output matrix
$\mathbf{Con}, \mathbf{Obs}$	Controllability matrix, and observability matrix, respectively
$\mathbf{D}$	The matrix relating to the location of external excitation
$f_d^{\max}, f_d^{\min}$	The maximum and minimum value of the adjustable Coulomb damping force, respectively
$\mathbf{F}$	External excitation or disturbance
$\mathbf{G}, \mathbf{G}'$	State feedback gain matrix and output feedback gain matrix, respectively
$\mathbf{G}_s$	Mode feedback gain matrix
$\mathbf{G}_{KF}$	Gain of Kalman filter
$\mathbf{I}$	Identity matrix
$\mathbf{M}_0, \mathbf{C}_0, \mathbf{K}_0$	Mass, damping and stiffness matrix, respectively
$\mathbf{M}_0^*, \mathbf{C}_0^*, \mathbf{K}_0^*$	Modal mass, modal damping and modal stiffness matrix, respectively
$\mathbf{P}$	The solution of the corresponding algebraic Riccati equation
$\mathbf{P}_{KF}$	Error covariance matrix of the Kalman filter
$\mathbf{q}(t)$	Vector of generalised modal coordinates
$\mathbf{Q}$	Positive semidefinite weighting matrix
$\mathbf{R}$	Positive definite weighting matrix
s	Laplace operator
$\mathbf{S}$	Sliding surface
$\mathbf{T}_{y_2 F}$	The closed-loop transfer function from $\mathbf{F}$ to $\mathbf{y}_2$ after applying a dynamic controller to the system

$\mathbf{T}_{y_2F}^*(j\omega)$	The complex conjugate of $\mathbf{T}_{y_2F}(j\omega)$.
$\mathbf{u}$	Control force
u_d	Control force provided by the MR damper
u_d^i	The ith actual control force
$\mathbf{W}_C(t)$, $\mathbf{W}c$	Controllability Gramian, and its steady state
$\mathbf{W}_O(t)$, $\mathbf{W}o$	Observability Gramian, and its steady state
$\ddot{\mathbf{x}}(t)$, $\dot{\mathbf{x}}(t)$, $\mathbf{x}(t)$	Structural acceleration, velocity and displacement, respectively
$\mathbf{X}$	System state vector
$\mathbf{y}_1$	Measured output
$\mathbf{y}_2$	Controlled output
z	The relative displacement of the MR damper
γ_2^*	The infimum of the H_2-norm of the closed-loop transfer function $\mathbf{T}_{y_2F}$ over all the stabilising proper controllers
γ_∞^*	The infimum of the H_∞-norm of the closed-loop transfer function $\mathbf{T}_{y_2F}$ over all the stabilising proper controllers
Δ	The estimation error for Kalman filter
η	A regular form for the transformation of system state
θ_c	State of the controller
Θ	A matrix to be determined for sliding surface design
λ_j	The jth eigenvalue of the system in the complex form
Θ	State transition matrix
σ_i	The voltage applied to the ith controller
σ_i^{max}	The maximum voltage to be applied to the ith controller
Φ	Mode shape matrix
Ψ	Transformation matrix
ω_j, ζ_j	The jth natural frequency and damping ratio, respectively

REFERENCES

Agarwal, M. 1997. A systematic classification of neural-network-based control. *Control Syst., IEEE*, 17(2): 75–93.

Anderson, B.D.O., T. Brinsmead, F. De Bruyne, et al. 2000. Multiple model adaptive control. Part 1 – Finite controller coverings. *Int. J. Adapt. Control Signal Process.*, 10(11): 909–29.

Anderson, B.D.O., T. Brinsmead, D. Liberzon, and A.S. Morse. 2001. Multiple model adaptive control with safe switching. *Int. J. Adapt. Control Signal Process.*, 15(5): 445–70.

Anderson, B.D.O. and A. Dehghani. 2008. Challenges of adaptive control – Past, permanent and future. *Annu. Rev. Control*, 32(2): 123–35.

Åström, K.J. and B. Wittenmark. 1995. *Adaptive Control*, second edition. Reading, MA: Addison-Wesley.

Baldi, S., P. Ioannou, and E. Mosca. 2011. Multiple model adaptive mixing control: The discrete-time case. *IEEE Trans. Autom. Control*, 57(4): 1040–45.

Bartolini, G., L. Fridman, A. Pisano, and E. Usai. 2008. *Modern Sliding Mode Control Theory: New Perspectives and Applications*. Berlin: Springer-Verlag.

Basu, B. and S. Nagarajaiah. 2008. A wavelet-based time-varying adaptive LQR algorithm for structural control. *Eng. Struct.*, 30(9): 2470–77.

Bitaraf, M., L.R. Barroso, and S. Hurlebaus. 2010. Adaptive control to mitigate damage impact on structural response. *J. Intell. Mater. Syst. Struct.*, 21: 607–19.

Bitaraf, M., S. Hurlebaus, and L.R. Barroso. 2012. Active and semi-active adaptive control for undamaged and damaged building structures under seismic load. *Comput.-Aided Civil Infrastruct. Eng.*, 27: 48–64.

Blažič, S., I. Škrjanc, and D. Matko. 2003. Globally stable direct fuzzy model reference adaptive control. *Fuzzy Sets Syst.*, 139(1): 3–33.

Brogan, W.L. 1974. Applications of a determinant identity to pole-placement and observer problems. *IEEE Trans. Autom. Control*, 19(5): 612–14.

Brown, M. and C. Harris. 1994. *Neurofuzzy Adaptive Modelling and Control*. Prentice Hall, Hemel Hempstead.

Burdisso, R.A., L.E. Suarez, and C.R. Fuller. 1994. Feasibility study of adaptive control of structures under seismic excitation. *J. Eng. Mech.*, 120(3): 580–92.

Chen, C.T. 2013. *Linear System Theory and Design*, fourth edition. New York: Oxford University Press.

Chen, F.C. and H.K. Khalil. 1995. Adaptive control of a class of nonlinear discrete-time systems. *IEEE Trans. Autom. Control*, 40(5): 791–801.

Chen, X.K. 2006. Adaptive sliding mode control for discrete-time multi-input multi-output systems. *Automatica*, 42(3): 427–35.

Chen, X.K., T. Fukuda, and K.D. Young. 2001. Adaptive quasi-sliding-mode tracking control for discrete uncertain input output systems. *IEEE Trans. Ind. Electron.*, 48(1): 216–24.

Dadios, E.P. 2012. *Fuzzy Logic – Controls, Concepts, Theories and Applications*. Rijeka, Croatia: InTech.

Dehghani, A., B.D.O. Anderson, and A. Lanzon. 2007. Unfalsified adaptive control: A new controller implementation and some remarks. In: *Proceedings of the European Control Conference*, Kos, Greece, 709–706.

Dyke, S.J., B.F. Spencer Jr., M.K. Sain, and J.D. Carlson. 1996. Modeling and control of magneto-rheological dampers for seismic response reduction. *Smart Mater. Struct.*, 5(5): 565–75.

Edwards, C. and S.K. Spurgeon. 1998. *Sliding Mode Control: Theory and Applications*. CRC Press.

Espinosa, J., J. Vandewalle, and V. Wertz. 2005. *Fuzzy Logic, Identification and Predictive Control*. Springer-Verlag London Limited, London.

Faravelli, L. and T. Yao. 1996. Use of adaptive networks in fuzzy control of civil structures. *Comput.-Aided Civil Infrastruct. Eng.*, 11(1): 67–76.

Feng, G. and R. Lozano. 1999. *Adaptive Control Systems*. Oxford, UK: Reed Educational and Professional Publishing Ltd.

Fullér, R. 2000. *Introduction to Neuro-Fuzzy Systems*. Berlin: Springer-Verlag.

Gawronski, W.K. 1998. *Dynamics and Control of Structures*. Springer-Verlag, New York.

Guan, C. and S. Pan. 2008. Adaptive sliding mode control of electro-hydraulic system with nonlinear unknown parameters. *Control Eng. Practice*, 16(11): 1275–84.

Hespanha, J.P. 2009. *Linear Systems Theory*. Princeton University Press, Princeton, NJ.

Hespanha, J.P., D. Liberzon, A.S. Morse, et al. 2001. Multiple model adaptive control. Part 2 – Switching. *Int. J. Adapt. Control Signal Process.*, 11(5): 479–96.

Hou, Z.S. 2002. The model-free direct adaptive predictive control for a class of discrete-time non-linear system. In: *Proceedings of the 4th Asian Control Conference*, Singapore, eds. Z.T. Ding, J.X. Xu, and K.C. Yow, 519–24, Causal Productions.

Hou, Z.S. and S.T. Jin. 2014. *Model Free Adaptive Control: Theory and Application*. CRC Press, Boca Raton, FL.

Huang, K.Y., H.C. Chin, and Y.C. Huang. 2004. A model reference adaptive control strategy for interruptible load management. *IEEE Trans. Power Syst.*, 19(1): 683–89.

Hunt, K.J., G.R. Irwin, and K. Warwick. 1995. *Neural Network Engineering in Dynamic Control Systems*. Springer-Verlag London Limited, London.

Hunt, K.J., D. Sbarbaro, R. Żbikowski, and P.J. Gawthrop. 1992. Neural networks for control systems: A survey. *Automatica*, 28(6): 1083–112.

Ioannou, P.A. and J. Sun. 1996. *Robust Adaptive Control*. PTR Prentice-Hall, Upper Saddle River, NJ.

Jamshidi, M., R.A. Krohling, L.S. Coelho, and P.J. Fleming. 2002. *Robust Control Systems with Genetic Algorithms*. CRC Press, Boca Raton, FL.

Jin, H. and M.G. Safonov. 2012. Unfalsified adaptive control: Controller switching algorithms for nonmonotone cost functions. *Int. J. Adapt. Control Signal Process*, 26: 692–704.

Jin, S.T. 2008. On model free learning adaptive control and applications. PhD diss., Beijing Jiaotong University.

Kalman, R.E. 1960. On the general theory of control systems. In: *Proceeding of 1st International Congress on Automatic Control*, Moscow, 16, 481–92.

Kalogirou, S.A. 2003. Artificial intelligence for the modeling and control of combustion processes: A review. *Prog. Energy Combust. Sci.*, 29(6): 515–66.

Kosut, R.L. 2001. Uncertainty model unfalsification for robust adaptive control. *Annu. Rev. Control*, 25: 65–76.

Landau, I.D., R. Lozano, M. M'Saad, and A. Karimi. 2011. *Adaptive Control: Algorithms, Analysis and Applications*, second edition. Springer-Verlag London Limited, London.

Lavretsky, E. and K.A. Wise. 2013. *Robust and Adaptive Control: With Aerospace Applications*. Springer-Verlag, London.

Lewis, F.L., J. Campos, and R. Selmic. 2002. *Neuro-Fuzzy Control of Industrial Systems with Actuator Nonlinearities*. Society for Industrial and Applied Mathematics, Philadelphia, PA.

Linkens, D.A. and H.O. Nyongesa. 1996. Learning systems in intelligent control: An appraisal of fuzzy, neural and genetic algorithm control applications. *IEEE Proc.-Control Theory Appl.*, 143(4): 367–86.

Liu, Y.G. 2007. Output-feedback adaptive control for a class of nonlinear systems with unknown control directions. *Acta Autom. Sin.*, 33(12): 1306–12.

Liu, Y.G. and J.F. Zhang. 2005. Output-feedback adaptive stabilization control design for non-holonomic systems with strong non-linear drifts. *Int. J. Control*, 78(7): 474–90.

Man, K.F., K.S. Tang, and S. Kwong. 1997. *Genetic Algorithms for Control and Signal Processing*. Springer-Verlag London Limited, London.

Miller, D.E. 2003. A new approach to adaptive control: No nonlinearities. *Syst. Control Lett.*, 49: 67–79.

Miller, W.T., R.S. Sutton, and P.J. Werbos. 1995. *Neural Networks for Control*. MIT Press, Cambridge, MA.

Morse, A.S. 1996. Overcoming the obstacle of high relative degree. *Eur. J. Control*, 2(1): 29–35.

Narendra, K.S. and J. Balakrishnan. 1997. Adaptive control using multiple models. *IEEE Trans. Autom. Control*, 42(2), 171–87.

Norgaard, M., O. Ravn, N.K. Poulsen, and L.K. Hansen. 2000. *Neural Networks for Modelling and Control of Dynamic Systems*. Springer-Verlag, London.

Ou, J.P. 2003. *Structural Vibration Control: Active, Semiactive, and Intelligent Control*. Science Press, Beijing.

Popa, R. 2012. *Genetic Algorithms in Applications*. InTech.

Preumont, A. 2011. *Vibration Control of Active Structures: An Introduction*, third edition. Berlin: Springer-Verlag.

Reznik, L. 1997. *Fuzzy Controllers*. Oxford, UK: Newnes, a division of Reed Educational and Professional Publishing Ltd.

Rosa, P., C. Silvestre, J.S. Shamma, and M. Athans. 2009. Multiple-model adaptive control with set-valued observers. In: *Proceedings of the 48th IEEE Conference on Decision and Control, 2009 held jointly with the 28th Chinese Control Conference*, Shanghai, China. 2441–47.

Safonov, M.G. and T.C. Tsao. 1997. The unfalsified control concept and learning. *IEEE Trans. Autom. Control*, 42(6): 843–47.

Sarangapani, J. 2006. *Neural Network Control of Nonlinear Discrete-Time Systems*. CRC Press, Boca Raton, FL.

Sastry, S. and M. Bodson. 2011. *Adaptive Control: Stability, Convergence and Robustness*. Dover Publications, Mineola, NY.

Sastry, S. and A. Isidori. 1989. Adaptive control of linearizable systems. *IEEE Trans. Autom. Control*, 34(11): 1123–31.

Shtessel, Y., C. Edwards, L. Fridman, and A. Levant. 2014. *Sliding Mode Control and Observation*. Springer Science+Business Media, New York.

Shyu, K.K., M.J. Yang, and Y.F. Lin. 2008. Model reference adaptive control design for a shunt active-power-filter system. *IEEE Trans. Ind. Electron.*, 55(1): 97–106.

Siddique, N. 2014. *Intelligent Control: A Hybrid Approach based on Fuzzy Logic, Neural Networks and Genetic Algorithms*. Springer International Publishing, Switzerland.

Soong, T.T. 1990. *Active Structural Control: Theory and Practice*. Longman Scientific & Technical, UK and Wiley, New York.

Spencer Jr., B.F. and S. Nagarajaiah. 2003. State of the art of structural control. *J. Struct. Eng.*, 129(7): 845–56.

Spencer Jr., B.F. and M.K. Sain. 1997. Controlling buildings: A new frontier in feedback. *IEEE Control Syst.*, 17(6): 19–35.

Sugeno, M. 1985. *Industrial Applications of Fuzzy Control*. Elsevier, New York.

Tan, K.K., T.H. Lee, S.N. Huang, and F.M. Leu. 2001. Adaptive-predictive control of a class of SISO nonlinear systems. *Dyn. Control*, 11(2): 151–74.

Tong, S., C. Li, and Y. Li. 2009. Fuzzy adaptive observer backstepping control for MIMO nonlinear systems. *Fuzzy Sets Syst.*, 160(19): 2755–75.

Utkin, V. I. 1992. *Sliding Modes in Control Optimization*. Springer-Verlag, New York.

Vepa, R. 2010. *Dynamics of Smart Structures*. Wiley, Hoboken, NJ.

Xie, W.F. 2007. Sliding-mode-observer-based adaptive control for servo actuator with friction. *IEEE Trans. Ind. Electron.*, 54(3): 1517–27.

Yang, C., S.S. Ge, and T.H. Lee. 2009. Output feedback adaptive control of a class of nonlinear discrete-time systems with unknown control directions. *Automatica*, 45(1): 270–76.

Yang, J.N., J.C. Wu, and A.K. Agrawal. 1995a. Sliding mode control for seismically excited linear structures. *J. Eng. Mech.*, 121(12): 1386–90.

Yang, J.N., J.C. Wu, and A.K. Agrawal. 1995b. Sliding mode control for nonlinear and hysteretic structures. *J. Eng. Mech.*, 121(12): 1330–39.

Yang, J.N., J.C. Wu, A.K. Agrawal, and Z. Li. 1994. Sliding mode control for seismic-excited linear and nonlinear civil engineering structures. Technical Report NCEER-94-0017, National Center for Earthquake Engineering Research (NCEER), State University of New York at Buffalo, New York. https://ubir.buffalo.edu/xmlui/handle/10477/705.

Yucelen, T. and W.M. Haddad. 2012. Low-frequency learning and fast adaptation in model reference adaptive control. *IEEE Trans. Autom. Control*, 58(4): 1080–85.

Zhang, B. and W.W. Zhang. 2006. Adaptive predictive functional control of a class of nonlinear systems. *ISA Trans.*, 45(2): 175–83.

Zhang, Y., W.H. Chen, and Y.C. Soh. 2008. Improved robust backstepping adaptive control for nonlinear discrete-time systems without overparameterization. *Automatica*, 44(3): 864–7.

Zhou, J. and C.Y. Wen. 2008. Decentralized backstepping adaptive output tracking of interconnected nonlinear systems. *IEEE Trans. Autom. Control*, 53(10): 2378–84.

Zhou, J., C.Y. Wen, and Y. Zhang. 2004. Adaptive backstepping control of a class of uncertain nonlinear systems with unknown backlash-like hysteresis. *IEEE Trans. Autom. Control*, 49(10): 1751–59.

Zhou, K. and J.C. Doyle. 1998. *Essentials of Robust Control*. Prentice Hall, Upper Saddle River, NJ.

Chapter 10

Control device placement

10.1 PREVIEW

The sensor, control device and processor that are described, respectively, in Chapters 3 through 5 can be viewed as hardware devices for the establishment of an entire control system, whereas the control algorithms introduced in Chapter 9 can be viewed to some extent as software for the realisation of the optimal control of civil structures. To enhance control performance, one can target the improvement of the quality of hardware devices and develop more efficient and reasonable control algorithms. Nevertheless, with the given hardware and software, control effectiveness will strongly depend on the locations of control devices in the structure, because it is impractical and uneconomical to install control devices in all the possible locations of a large-scale civil structure. Therefore, it is highly desirable to investigate control device placement in a civil structure for the purpose of improving control efficiency and cost. A control device is defined in this book as a type of device that is responsible for moving or controlling a civil structure, including not only the actuator operated by a source of energy, but also the damper operated without external energy. This chapter first reviews some existing control device placement methods for passive control systems, active control systems, semi-active control systems and hybrid control systems. Then, by taking the linear quadratic performance index as an objective function, an increment-based algorithm is presented in this chapter for the optimal placement of active/passive control devices in terms of the sequence of the calculated performance index increments and the number of control devices to be used. With the control devices in their optimal places, the seismic response of the building is computed using the suboptimal control gain derived on the basis of the minimum error principle, from which the equivalent optimal parameters of passive devices can also be determined.

10.2 REVIEW OF CONTROL DEVICE PLACEMENT METHODS

One of the important issues in the controlled structures is how to place a minimum number of discrete control devices at their respective optimal locations to achieve a desirable control performance. This issue is particularly important for large civil structures, because of their huge sizes and complex structural systems. Extensive studies have thus been performed, and much progress has been made in optimal control device placement. However, technical reviews on optimal control device placement for civil structures are very limited. The aim of this section is to provide a brief introduction to some existing methods for optimal placement of passive control devices, active control devices, semi-active control devices and hybrid control devices in civil structures.

Previous studies (e.g. Chang et al. 1995; Miyamoto and Singh 2002; Yang et al. 2003) showed that passive control devices can be used to reduce seismic responses of civil structures effectively. Nevertheless, the control performance of passive control devices, such as the energy dissipation devices described in Chapter 4, strongly depends on their locations within a structure. A great amount of research has been conducted to find the optimal locations of passive control devices by defining appropriate objective functions (performance indices) and designing efficient optimisation algorithms in either the frequency domain or the time domain.

For example, an efficient and systematic procedure was proposed to find the optimal damper placement with the aim of minimising the sum of the transfer function amplitudes of the inter-story drifts evaluated at the un-damped fundamental natural frequency of a structural system subject to a constraint on the sum of the damping coefficients of added dampers (Takewaki 1997, 2000; Takewaki and Yoshitomi 1998; Takewaki et al. 1999). Some other objective functions regarding the usage of transfer functions for passive control device placement include minimising the amplitude of a transfer function of the tip deflection evaluated at one of the un-damped natural frequencies (Takewaki 1998), minimising the transfer function amplitude of the base shear force evaluated at the un-damped fundamental natural frequency (Aydin et al. 2007; Aydin and Boduroglu 2008; Sonmez et al. 2013), minimising the transfer function amplitude of the elastic base moment evaluated at the first natural circular frequency of the structure (Aydin 2012), minimising the norm absolute acceleration transfer function evaluated at the un-damped fundamental natural frequency (Cimellaro 2007; Cimellaro and Retamales 2007) and minimising the sum of the square of the absolute values of the transfer matrix elements (Mousavi and Ghorbani-Tanha 2012). For the methods using the objective functions defined in the time domain, most of them find the optimal damper configurations in terms of minimisation of the peak and/or root-mean-square (RMS) responses of displacement and/or acceleration. For example, based on the RMS value of the inter-story drift of a multi-story building frame, the idea of the controllability index was applied to find optimal damper placement (Shukla and Datta 1999). Four optimal location indices, which were related to RMS responses of acceleration, were presented by Chen and Wu (2001) for the optimal placement of multiple tuned mass dampers in building structures subject to ground excitations. On the basis of H∞ and H_2 performances, Yang et al. (2002) presented two optimal design methodologies for passive energy dissipation devices, which were capable of determining not only the optimal passive damper locations but also their corresponding optimal capacities. With the linear combination of story-drift angle, acceleration and story displacement, Ou and Li (2012) presented an objective function for optimal placement of passive energy dissipation devices in seismically excited building structures by using genetic algorithms (GAs). Besides the aforementioned investigations conducted in the shear building structures, the research on optimal damper configuration has also been conducted for some other structural systems, such as three-dimensional frame structures in consideration of translation–torsion coupling effects (Wu et al. 1997), long-span suspension bridges (Wang et al. 2010), adjacent building systems (Bigdelia et al. 2012), high reinforced-concrete frame structures (Martínez et al. 2014) and flexural steel frame structures (Estekanchi and Basim 2011; Amouzegar et al. 2014). Moreover, some comparative studies have also been carried out and demonstrated that different objective functions or performance indices often lead to different optimal damper placements (Agrawal and Yang 1999; Liu et al. 2004, 2005; Whittle et al. 2012).

To realise the defined objective function, an optimisation algorithm must be developed to seek global or local optimal solutions for passive damper placement. There are three types of optimisation algorithms: enumerative, random and calculus-based search algorithms

(Cheng et al. 2008). Theoretically speaking, enumerative search algorithms are able to find global optimal solutions, but there is no guarantee that random and calculus-based search algorithms can do so, especially for problems with a high degree of nonlinearity and discontinuity. Enumerative search algorithms are fairly straightforward. Within the given search space, such algorithms assess the value of the objective function at every point, one at a time. Since all the possible cases will be evaluated and compared with others, global optimal solutions can be guaranteed. However, enumerative algorithms are rather time-consuming and inefficient, especially for large-scale civil structures. For example, the selection of k damper locations out of m possible story units ($m > k$) leads to the total number of cases being $m!/[k!(m-k)!]$, which means that the placement of 20 dampers in a 40-story building will have 1.38×10^{11} possible cases. Random search algorithms use random choices as a tool to guide a highly exploitative search through the given search space. The integer heuristic programming is one kind of random algorithm to choose a set of k discrete (integer) optimal locations out of m possible locations ($m > k$) for the optimal placement of dampers (Agrawal and Yang 1999). Heuristic techniques for optimal damper placement include the worst-out-best-in (WOBI) algorithm and the exhaustive single point substitution (ESPS) algorithm (Haftka and Adelman 1985), the sequential search algorithm (SSA) (Zhang and Soong 1992) and a simplified SSA method (Lopez Garcia 2001; Lopez Garcia and Soong 2002). Other well-known random search algorithms, such as GAs (Singh and Moreschi 2002; Bishop and Striz 2004; Wongprasert and Symans 2004) and simulated annealing algorithms (Chen et al. 1991; Milman and Chu 1994) have also been utilised for the optimal placement of passive control devices. Different from random search algorithms, calculus-based algorithms seek the solution by setting the gradient of the objective function equal to zero. Thus, the direction that reduces or increases the objective function most effectively under the predetermined constraint condition should first be found, and the design or optimisation is then updated in that direction. The steepest direction search algorithm, which is implemented on the basis of sensitivities of the objective function with respect to the designed variables, can be viewed as one of the widely used calculus-based algorithms for optimal placement of passive control devices (Takewaki 1997, 2000; Singh and Moreschi 2001; Bilbao et al. 2009; Fujita et al. 2010).

For an active control system, the placement of discrete active control devices should first consider the stability and controllability of the control system, as mentioned in Chapter 9. It is of particular importance in the control of a civil structure, where the placement of control devices near a nodal point of a vibration mode may result in significant difficulty or even inability to control this mode. Cheng and Pantelides (1988) computed the controllability index associated with each story of a multi-story building, from which the optimal location of active control devices was provided. By maximising the controllability Gramian, a performance criterion was proposed for the optimal placement of piezoelectric patch actuator locations on a flexible plate structure (Peng et al. 2005). As described in Chapter 9, a variety of control algorithms have been developed for adjusting the optimal control forces of actuators to achieve a desirable control performance. Therefore, the optimal locations of active control devices are inevitably related to the control algorithms selected. For example, within the framework of modal control of a large system, Chang and Soong (1980) proposed an approach for optimal placement of a limited number of actuators such that the total energy required for control is minimised. By using output feedback control to enhance the damping of the critical modes of a structure, Lu et al. (1994) chose the controllability and robustness of the tendon control system as the criterion for optimal tendon placement. Abdullah (2000) presented an algorithm for the optimisation of output feedback gains and actuator placement for slender civil structures. On the basis of certain control logic, a practical sequential procedure was developed to suboptimally place active piezoelectric friction dampers to control the seismic responses of a nonlinear 20-story steel building (Chen and Chen 2002). Based on full-state feedback control, a general statistical method was proposed for optimal actuator placement of seismically excited structures (Cheng et al. 2002, 2008). On the basis of the LQG algorithm, Agranovich and Ribakov (2010) developed a simple alternative method for efficient placement of active control devices in seismically excited structures. Moreover, with the rapid development of computing technologies, some intelligent algorithms, such as GAs (Rao et al. 1991; Yan and Yam 2002; Liu et al. 2003; Li et al. 2000, 2004; Tan et al. 2005; Rao and Sivasubramanian 2008) and neural network (Amini and Tavassoli 2005), have also been actively investigated and employed for the optimal placement of active control devices.

As compared with passive and active control systems, the research on optimal control device placement in semi-active control systems is relatively limited. Although a number of semi-active control devices have been developed for vibration control, as described in Chapter 4, most researchers focus on the optimal placement of one kind of well-known semi-active control device; that is, magnetorheological (MR) dampers. For example, by using a multi-objective binary GA, Kwok et al. (2007) presented an efficient algorithm for the optimal placement of MR dampers in a high-rise building model. Based on a stochastic linearisation method and a multi-objective GA, Ok et al. (2008) proposed an optimal design method for the determination of locations and characteristics of nonlinear hysteretic dampers to enhance the seismic performance of two adjacent building structures. Li et al. (2010) proposed a two-phase optimisation process for the placement of MR dampers for a nonlinear benchmark structure with the aid of GAs. An improved version of the ant colony optimisation algorithm was presented by Amini and Ghaderi (2012, 2013) for choosing optimal locations of MR dampers in civil structures. Moreover, the investigation of optimal placement of semi-active friction dampers has also been conducted. A step-by-step heuristic approach was employed by Chen and Chen (2004) for the optimal placement of piezoelectric friction dampers to improve the semi-active control of a 20-story benchmark building.

As defined in Chapter 4, hybrid control devices are those achieved by the combination of passive, active or semi-active control devices. Thus, technically speaking, the optimal placement of hybrid control devices would be much more complicated if the cost, performance and characteristics of different control devices were required to be taken into consideration simultaneously. To date, there have been few studies on the optimal placement of hybrid control devices. However, in general, there is still some consensus for the determination of locations of hybrid control devices. For example, the hybrid mass dampers are usually located on the top or the upper floor of the building structure, whereas the hybrid base isolation devices are basically placed near the bearings or on the bottom floor of the building structure.

In principle, the design of control devices involves two aspects: one is to determine the optimal number and placement of control devices in a structure; the other is to find the optimal control law for active/hybrid/semi-active control devices or the optimal damper parameter for passive control devices. Most previous studies, as reviewed in this section, have focused on the investigation of the optimal number and placement of control devices in a structure, but with either the control law or the damper parameter predetermined. Moreover, a great number of researchers have been dedicated to developing various types of control devices and improving the efficiency of the associated control systems, as described in Chapter 4. A variety of control algorithms for active and semi-active control systems have also been developed, as introduced in Chapter 9. However, very little information is available on the determination of both the optimal placement and the optimal parameters of control devices in a structure at the same time. In this chapter, a relatively simple increment-based algorithm is presented for the optimal placement of active/passive control devices in a structure (Xu and Teng 2002). The linear relation between the increment of the performance index and the change in the position matrix of the control devices is first established, based on the assumption that the control gain remains unchanged. The optimal placement of control devices is then determined in terms of the sequence of the calculated performance index increments and the number of control devices to be used. With the control devices in their optimal places, the seismic response of the building is finally computed using the suboptimal control gain derived using the minimum error principle, from which the equivalent optimal parameters of passive devices can also be determined. The applicability of the aforementioned approach and its limitations are carefully examined through numerical examples. The results from the numerical examples show that the approach is quite accurate and effective in determining the optimal placement and optimal parameters of control devices if the number of control devices removed is limited to a certain range.

10.3 INCREMENT ALGORITHMS FOR CONTROL DEVICE PLACEMENT

Consider a seismically excited tall building modelled by an n-degree-of-freedom (DOF) lumped mass–spring–dashpot system with active/passive control devices. The matrix equation of motion of the controlled building can be written as

$$\mathbf{M}\ddot{\mathbf{x}}(t) + \mathbf{C}\dot{\mathbf{x}}(t) + \mathbf{K}\mathbf{x}(t) = -\mathbf{M}\mathbf{1}\ddot{x}_g + \mathbf{D}\mathbf{U}(t) \tag{10.1}$$

where:

$\mathbf{x}(t)$ is the n-dimensional displacement vector relative to the ground

$\mathbf{M}, \mathbf{K} \text{ and } \mathbf{C}$ are, respectively, the $n \times n$ mass, stiffness and damping matrix of the building structure

$\mathbf{1}$ is an n-dimensional vector, with all elements being unity

$\ddot{x}_g$ is the ground acceleration

$\mathbf{U}(t)$ is the m-dimensional control force vector

$\mathbf{D}$ is the $n \times m$ matrix denoting the location of the control force

Equation 10.1 can also be converted to the following continuous state-space equation:

$$\dot{\mathbf{Z}} = \mathbf{A}\mathbf{Z} + \mathbf{B}\mathbf{U} + \mathbf{W} \tag{10.2}$$

where

$$\mathbf{A} = \begin{bmatrix} 0 & \mathbf{I}_n \\ -\mathbf{M}^{-1}\mathbf{K} & -\mathbf{M}^{-1}\mathbf{C} \end{bmatrix}; \mathbf{B} = \begin{bmatrix} 0 \\ \mathbf{M}^{-1}\mathbf{D} \end{bmatrix}; \mathbf{Z} = \begin{Bmatrix} \mathbf{x} \\ \dot{\mathbf{x}} \end{Bmatrix}; \mathbf{W} = \begin{Bmatrix} 0 \\ -1 \end{Bmatrix}\ddot{x}_g \tag{10.3}$$

The following linear quadratic performance index, as introduced in Chapter 9, is usually chosen for study in structural vibration control under random disturbance (Housner et al. 1997; Soong 1990):

$$J = \frac{1}{2}E\left\{ \int_0^{t_f} \left(\mathbf{Z}^T\mathbf{Q}\mathbf{Z} + \mathbf{U}^T\mathbf{R}\mathbf{U} \right)dt \right\} \tag{10.4}$$

where:

Q is a $2n \times 2n$ positive semi-definite weighting matrix for the structure response

R is an $m \times m$ positive definite weighting matrix for the control force

E is the expectation operator

t_f is the duration defined to be longer than that of the earthquake

For a closed-loop control configuration with the ground motion being a white noise random process, minimising Equation 10.4 subject to the constraint of Equation 10.2 results in the following optimal control force:

$$\mathbf{U} = -\mathbf{GZ} \tag{10.5}$$

where $\mathbf{G}$ is the control gain, given by

$$\mathbf{G} = \mathbf{R}^{-1}\mathbf{B}^T\mathbf{P} \tag{10.6}$$

The matrix $\mathbf{P}$ is the solution of the classical Riccati equation:

$$\mathbf{A}^T\mathbf{P} + \mathbf{P}\mathbf{A} - \mathbf{P}\mathbf{B}\mathbf{R}^{-1}\mathbf{B}^T\mathbf{P} + \mathbf{Q} = 0 \tag{10.7}$$

The substitution of Equation 10.5 into Equations 10.2 and 10.4 leads to

$$\dot{\mathbf{Z}} = (\mathbf{A} - \mathbf{BG})\mathbf{Z} + \mathbf{W} \tag{10.8}$$

$$J = \frac{1}{2}E\left\{\int_0^{t_f} (\mathbf{Z}^T\bar{\mathbf{Q}}\mathbf{Z})dt\right\} \tag{10.9}$$

in which

$$\bar{\mathbf{Q}} = \mathbf{Q} + \mathbf{G}^T\mathbf{RG} \tag{10.10}$$

The solution of Equation 10.8 with an initial condition $\mathbf{Z}(t=0) = \mathbf{Z}_0$ can be expressed as

$$\mathbf{Z}(t) = \psi(t)\mathbf{Z}_0 + \int_0^t \psi(t-\tau)\mathbf{W}(\tau)d\tau \tag{10.11}$$

where

$$\psi(t) = \exp\{(\mathbf{A} - \mathbf{BG})t\} = \mathbf{\Phi}\,\mathrm{diag}(e^{\lambda_1 t}, e^{\lambda_2 t}, ..., e^{\lambda_{2n} t})\mathbf{\Phi}^{-1} \tag{10.12}$$

where λ_1, λ_2, ..., λ_{2n} are the eigenvalues of matrix $(\mathbf{A} - \mathbf{BG})$ and $\mathbf{\Phi}$ consists of all the eigenvectors.

Assume that the initial condition of the structure and the ground motion are completely uncorrelated and note that the mean value of the ground excitation is zero. Then, the substitution of Equation 10.11 into Equation 10.9, with some mathematical manipulation, yields

$$J = \frac{1}{2} \int_0^{t_f} \mathbf{Z}_0^T \psi(t)^T \overline{\mathbf{Q}} \psi(t) \mathbf{Z}_0 dt + \frac{1}{2} E \left\{ \int_0^{t_f} \int_0^t \mathbf{W}(\tau)^T \psi(t-\tau)^T \overline{\mathbf{Q}} \psi(t-\tau) \mathbf{W}(\tau) d\tau dt \right\}$$

$$= \frac{1}{2} \mathbf{Z}_0^T \mathbf{S} \mathbf{Z}_0^T + \frac{1}{2} E \left\{ \int_0^{t_f} \mathbf{W}(\tau)^T \mathbf{S} \mathbf{W}(\tau) d\tau \right\} \tag{10.13}$$

in which

$$\mathbf{S} = \int_0^{t_f} \mathbf{S}_0(t) dt \tag{10.14}$$

$$\mathbf{S}_0(t) = \psi(t)^T \overline{\mathbf{Q}} \psi(t) \tag{10.15}$$

Consider that only a few control devices are moved away from a tall building having a large number of story units equipped with control devices, and note that some uncertainties will be involved in the trial-and-error selection of the new weighting matrices for the controlled building with a few control devices removed. The position matrix of control devices **B** will then have a change $\Delta \mathbf{B}$, but the control gain **G** may be assumed to remain unchanged. As a result,

$$\Delta \mathbf{S}_0(t) = \Delta \psi(t)^T \overline{\mathbf{Q}} \psi(t) + \psi(t)^T \overline{\mathbf{Q}} \Delta \psi(t) = -t \left[\mathbf{G}^T \Delta \mathbf{B}^T \mathbf{S}_0(t) + \mathbf{S}_0(t) \Delta \mathbf{B} \mathbf{G} \right] \tag{10.16}$$

$$\Delta \mathbf{S}(t) = \int_0^{t_f} \Delta \mathbf{S}_0(t) dt = -\mathbf{G}^T \Delta \mathbf{B}^T \mathbf{S}_m - \mathbf{S}_m \Delta \mathbf{B} \mathbf{G} \tag{10.17}$$

in which

$$\mathbf{S}_m = \int_0^{t_f} t[\mathbf{S}_0(t)] dt \tag{10.18}$$

The change of the position matrix of control devices **B** with $\Delta \mathbf{B}$ leads to the increment of the performance index ΔJ. Then, use of the trace theorem of a matrix yields

$$\Delta J = \frac{1}{2} \text{tr} \left\{ \left[\mathbf{Z}_0 \mathbf{Z}_0^T + E \left(\int_0^{t_f} \mathbf{W}(\tau) \mathbf{W}(\tau)^T d\tau \right) \right] \Delta \mathbf{S} \right\} = -\text{tr} \{ (\mathbf{R}_0 + \mathbf{F}) \mathbf{S}_m \Delta \mathbf{B} \mathbf{G} \} \tag{10.19}$$

in which

$$\mathbf{R}_0 = \mathbf{Z}_0 \mathbf{Z}_0^T \tag{10.20}$$

$$\mathbf{F} = E\left\{\int_0^{t_f} \mathbf{W}(\tau)\mathbf{W}(\tau)^T \, d\tau\right\} \tag{10.21}$$

If the ith control device is removed, the increment of the performance index is then

$$\Delta J_i = -\mathrm{tr}\left\{(\mathbf{R}_0 + \mathbf{F})\mathbf{S}_m \Delta \mathbf{B}_i \mathbf{G}\right\} \quad (i = 1, 2, 3, \ldots, m) \tag{10.22}$$

The optimal performance index expressed by Equation 10.13 is the result of minimising the performance index expressed by Equation 10.4 with the constraint of Equation 10.2. It implies that the maximum response reduction of the structure with reasonable control power can be obtained. Then, the increment of the optimal performance index due to the removal of the ith control device, calculated by Equation 10.22, reflects the sensitivity of the ith device to the optimal performance index. A small increment of the index means a low sensitivity and less importance of the ith control device for the total control performance, whereas a high increment implies that the ith control device has great influence over the total control performance. Therefore, based on the calculated increments from the removal of each control device, the sequence of importance of all the control devices can be achieved.

Furthermore, if only a few (say s) control devices are moved away from a tall building having a large number of story units equipped with control devices, and the control gain $\mathbf{G}$ can be assumed to remain constant, one may obtain the following linear relations with respect to a change in the position matrix of control devices and in the increments of the performance index due to the change of position matrix of control devices:

$$\Delta \mathbf{B}_s = \sum_{i=1}^{s} \Delta \mathbf{B}_i \tag{10.23}$$

$$\Delta J_s = \sum_{i=1}^{s} \Delta J_i \tag{10.24}$$

These linear relations provide great convenience for determining the optimal placement of control devices.

10.4 SUBOPTIMAL CONTROL GAIN AND RESPONSE

Once the optimal placement of control devices is determined, one may use Equation 10.8 to compute the controlled building response if only a few (say s) control devices are removed from the building having a large number of story units equipped with control devices (say, m control devices or a fully controlled building). However, the use of the control gain $\mathbf{G}$ determined for the fully controlled building may not be good enough to determine the seismic response of the controlled building with k ($=m-s$) control devices in their optimal placements. In this connection, a suboptimal control gain $\mathbf{G}_d$ of dimension $k \times 2n$ is presented herein to replace $\mathbf{G}$ of dimension $m \times 2n$.

The relation between the position matrix $\mathbf{B}$ of dimension $2n \times m$ for the fully controlled building and the reduced-order position matrix $\mathbf{B}_d$ of dimension $2n \times k$ for the k control devices at their optimal placement can be written as

$$\mathbf{B}_d = \mathbf{B}\mathbf{C}_{sd} \tag{10.25}$$

For a building with k control devices at their optimal placement, the governing equation for determining the controlled building response vector $\mathbf{Z}_d$ is

$$\dot{\mathbf{Z}}_d = (\mathbf{A} - \mathbf{B}_d\mathbf{G}_d)\mathbf{Z}_d + \mathbf{W} \tag{10.26}$$

The matrix $\mathbf{G}_d$ in this equation is selected in such a way that the response vector $\mathbf{Z}_d$ of the building with k control devices approaches the response vector $\mathbf{Z}$ of the fully controlled building as closely as possible. Comparing Equation 10.26 with Equation 10.8, the response error can be written as

$$\dot{\mathbf{\Delta}}(t) = (\mathbf{A} - \mathbf{B}_d\mathbf{G}_d)\mathbf{\Delta}(t) + \mathbf{B}\mathbf{e} \tag{10.27}$$

in which

$$\mathbf{\Delta}(t) = \mathbf{Z}_d(t) - \mathbf{Z}(t) \tag{10.28}$$

$$\mathbf{e} = (\mathbf{G} - \mathbf{C}_{sd}\mathbf{G}_d)\mathbf{Z} \tag{10.29}$$

Clearly, the response error depends on the excitation error $\mathbf{e}$. However, the excitation error $\mathbf{e}$ is not equal to zero in general. Thus, one may find the suboptimal control gain $\mathbf{G}_d$ to minimise the excitation error, that is, to minimise the following objective function with the identity matrix $\mathbf{R}_e$:

$$J_e = \frac{1}{2}\int_0^{t_f} \mathbf{e}^T \mathbf{R}_e \mathbf{e} \, dt = \frac{1}{2}\int_0^{t_f} \mathbf{Z}^T (\mathbf{G} - \mathbf{C}_{sd}\mathbf{G}_d)^T (\mathbf{G} - \mathbf{C}_{sd}\mathbf{G}_d)\mathbf{Z} \, dt \tag{10.30}$$

By using the Lyapunov direct method, Equation 10.30 can be expressed as

$$J_e = \frac{1}{2}\left[\mathbf{Z}_0^T \bar{\mathbf{P}}\mathbf{Z}_0 - \mathbf{Z}_{t_f}^T \bar{\mathbf{P}}\mathbf{Z}_{t_f}\right] \tag{10.31}$$

where the matrix $\bar{\mathbf{P}}$ should satisfy the following equation:

$$W(\bar{\mathbf{P}}, \mathbf{G}_d) = (\mathbf{A} - \mathbf{B}\mathbf{G})^T \bar{\mathbf{P}} + \bar{\mathbf{P}}(\mathbf{A} - \mathbf{B}\mathbf{G}) + (\mathbf{G} - \mathbf{C}_{sd}\mathbf{G}_d)^T (\mathbf{G} - \mathbf{C}_{sd}\mathbf{G}_d) = 0 \tag{10.32}$$

Then, according to the method given by Levine and Athans (1970), the problem becomes one of finding the suboptimal control gain $\mathbf{G}_d$ to minimise $\text{tr}(\bar{\mathbf{P}})$ subject to the constraint of Equation 10.32:

$$\overline{J}_e = \text{tr}[\overline{\mathbf{P}}] + \text{tr}\left\{\mathbf{H}^T W(\overline{\mathbf{P}}, \mathbf{G}_d)\right\} \tag{10.33}$$

where $\mathbf{H}$ is a Lagrange multiplier matrix. As a result, by carrying out the first-order partial differentiation

$$\frac{\partial \overline{J}_e}{\partial \mathbf{G}_d} = 0 \tag{10.34}$$

one may have the suboptimal control gain

$$\mathbf{G}_d = \left(\mathbf{C}_{sd}{}^T \mathbf{C}_{sd}\right)^{-1} \mathbf{C}_{sd}{}^T \mathbf{G} \tag{10.35}$$

Once the suboptimal control gain $\mathbf{G}_d$ is obtained by Equation 10.35, the suboptimal response of the controlled building with k control devices at their optimal placements can be obtained by solving Equation 10.26.

10.5 EQUIVALENT OPTIMAL PARAMETERS OF CONTROL DEVICES

Let us consider an n-story two-dimensional linear elastic shear building subject to ground motion, as shown in Figure 10.1. The mass of the building is concentrated at its floor, and the stiffness is provided by its massless columns. By using the increment algorithm proposed in Section 10.3 and the suboptimal control gain suggested in Section 10.4, the optimal placement of the k control devices and the response of the controlled building with the k control devices at their optimal placement can be determined. Assume that these control devices are passive dampers described by the Voigt model, that is, a combination of a linear and elastic spring and a viscous dashpot connected in parallel. Now let us determine the equivalent optimal parameters for these passive dampers.

For the jth passive damper located at the ith floor of the building, the displacement and velocity of the jth damper can be determined by

$$\begin{Bmatrix} \mathbf{w}_{ji} \\ \dot{\mathbf{w}}_{ji} \end{Bmatrix} = \begin{Bmatrix} (\mathbf{x}_i - \mathbf{x}_{i-1})\cos\alpha_i \\ (\dot{\mathbf{x}}_i - \dot{\mathbf{x}}_{i-1})\cos\alpha_i \end{Bmatrix} \tag{10.36}$$

where α_i is the angle of the damper to the floor. Introduce a position vector $\mathbf{L}_{ni}$ as

$$\mathbf{L}_{ni}{}^T = \{0 \quad \cdots \quad 0 \quad 1 \quad 0 \quad \cdots \quad 0\}^T \tag{10.37}$$

where the subscript n is the dimension of the vector, and the subscript i means that only the ith element is unity while all the other elements are zero. Then, Equation 10.36 can be expressed as

$$\begin{Bmatrix} \mathbf{w}_{ji} \\ \dot{\mathbf{w}}_{ji} \end{Bmatrix} = \mathbf{\Gamma}_{ji} \mathbf{Z}_d \tag{10.38}$$

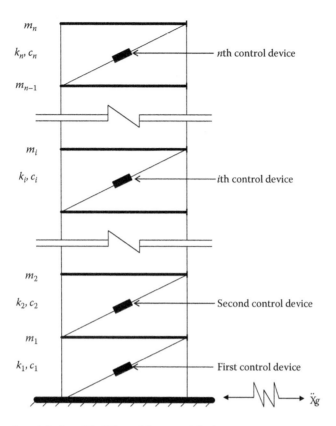

Figure 10.1 Structural model of a tall building with control devices.

in which

$$\mathbf{\Gamma}_{ji} = \begin{bmatrix} \left(\mathbf{L}_{ni}^{\ T} - \mathbf{L}_{n(i-1)}^{\ T} \right) \cos \alpha_i & \mathbf{0}^T \\ \mathbf{0} & \left(\mathbf{L}_{ni}^{\ T} - \mathbf{L}_{n(i-1)}^{\ T} \right) \cos \alpha_i \end{bmatrix}$$ (10.39)

Equation 10.38 can also be written as

$$\mathbf{w}_{ji} = \mathbf{L}_{21}^{\ T} \mathbf{\Gamma}_{ji} \mathbf{Z}_d$$ (10.40)

$$\dot{\mathbf{w}}_{ji} = \mathbf{L}_{22}^{\ T} \mathbf{\Gamma}_{ji} \mathbf{Z}_d$$ (10.41)

The standard deviations of the displacement and the velocity of the jth damper are thus

$$E\{\mathbf{w}_{ji}^2\} = \mathbf{L}_{21}^{\ T} \mathbf{\Gamma}_{ji} \mathbf{V}_z \mathbf{\Gamma}_{ji}^{\ T} \mathbf{L}_{21}$$ (10.42)

$$E\{\dot{\mathbf{w}}_{ji}^2\} = \mathbf{L}_{22}^{\ T} \mathbf{\Gamma}_{ji} \mathbf{V}_z \mathbf{\Gamma}_{ji}^{\ T} \mathbf{L}_{22}$$ (10.43)

in which $\mathbf{V}_z$ is the matrix of the correlation function of the response vector $\mathbf{Z}_d$.

The suboptimal control force vector $\mathbf{U}_d = -\mathbf{G}_d \mathbf{Z}_d$ obtained from Section 10.4 can be resolved as two components: the stiffness force vector and the damping force vector. For the jth damper at the ith floor, the stiffness force and damping force are given by, respectively,

$$\mathbf{u}_{ji} = -\mathbf{L}_{ki}{}^T \mathbf{G}_d \mathbf{E}_d \mathbf{Z}_d \tag{10.44}$$

$$\mathbf{v}_{ji} = -\mathbf{L}_{ki}{}^T \mathbf{G}_d \mathbf{E}_v \mathbf{Z}_d \tag{10.45}$$

in which

$$\mathbf{E}_d = \begin{bmatrix} \mathbf{I}_n & 0 \\ 0 & 0 \end{bmatrix}; \ \mathbf{E}_v = \begin{bmatrix} 0 & 0 \\ 0 & \mathbf{I}_n \end{bmatrix} \tag{10.46}$$

The standard deviations of the stiffness force and damping force of the jth damper at the ith floor are, respectively,

$$E(\mathbf{u}_{ji}^2) = \mathbf{L}_{ki}{}^T \mathbf{G}_d \mathbf{E}_d \mathbf{V}_z \mathbf{E}_d \mathbf{G}_d{}^T \mathbf{L}_{ki} \tag{10.47}$$

$$E(\mathbf{v}_{ji}^2) = \mathbf{L}_{ki}{}^T \mathbf{G}_d \mathbf{E}_v \mathbf{V}_z \mathbf{E}_v \mathbf{G}_d{}^T \mathbf{L}_{ki} \tag{10.48}$$

The equivalent optimal stiffness and damping coefficients of the jth damper at the ith floor are, therefore, given by

$$K_{ji} = \left(\frac{E\{\mathbf{u}_{ji}^2\}}{E\{\mathbf{w}_{ji}^2\}} \right)^{1/2} \tag{10.49}$$

$$C_{ji} = \left(\frac{E\{\mathbf{v}_{ji}^2\}}{E\{\dot{\mathbf{w}}_{ji}^2\}} \right)^{1/2} \tag{10.50}$$

10.6 NUMERICAL EXAMPLE FOR ACTUATOR PLACEMENT

A six-story two-dimensional linear elastic shear building is selected to evaluate the applicability and limitation of the increment algorithm and the suboptimal control gain suggested in this chapter. The structural parameters of the building are listed in Table 10.1. The Kanai–Tajimi filtered white noise spectrum is used as the earthquake excitation spectrum. The parameters in the filtered white noise spectrum are selected as $\omega_g = 15.0$ rad/s, $\xi_g = 0.65$ and $S_0 = 4.0 \times 10^{-3}$ m^2/rad·s^3. There is no building motion when the time is equal to zero.

Table 10.1 Structural parameters of the six-story building

Floor number	Mass (kg)	Stiffness (N/m)	Damping (N·s/m)
1	2793.0	4137.0×10^4	34.40×10^4
2	2793.0	4137.0×10^4	34.40×10^4
3	2793.0	3014.0×10^4	20.50×10^4
4	2793.0	3014.0×10^4	20.50×10^4
5	2793.0	1943.0×10^4	10.40×10^4
6	2793.0	1943.0×10^4	10.40×10^4

For a fully controlled building, every story is equipped with one actuator (active control device). Then, Equation 10.22 is used to calculate the increments of performance index, ΔJ_i ($i = 1, 2, \ldots, 6$), owing to the removal of actuators one by one. The results are shown in Figure 10.2a, from which one may see that the optimal sequence of actuator locations is (3, 1, 5, 2, 4, 6). As mentioned in Section 10.3, Equation 10.22 is derived based on the assumption that the full control gain **G** remains unchanged even though only one actuator is removed. To examine the effect of the assumption, the increment of the performance index due to the removal of the ith actuator is calculated based on Equation 10.13. That is, the removal of the ith actuator first results in the new control position matrix $\mathbf{B}_i$. Then, the use of Equations 10.6 and 10.7 leads to the new control gain matrix $\mathbf{G}_i$, and the use of Equation 10.10 yields the new matrix $\bar{\mathbf{Q}}_i$. Finally, the use of Equations 10.12 through 10.15 gives the performance index J_i. The increment ΔJ_i of the performance index due to the removal of the ith actuator is equal to $(J_i - J)$, where J is the performance index of the building with the actuators fully installed. In this way, the increments of the performance index ΔJ_i ($i = 1, 2, \ldots, 6$) due to the removal of the actuators one by one are calculated and are displayed in Figure 10.2b. Comparing these with the results presented in Figure 10.2a, one may see that, even though the increments of the performance indices may have slight differences, the optimal sequence of actuator locations is the same (3, 1, 5, 2, 4, 6). Therefore, the use of Equation 10.22 to calculate the increment of the performance index due to the removal of one actuator and to obtain the optimal sequence of actuator location is quite satisfactory for the six-story building. Certainly, it would be even better for a tall building.

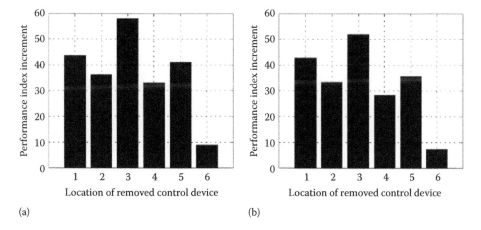

(a)   (b)

Figure 10.2 Effect of control gain on performance index increment: (a) constant control gain, (b) optimal control gain.

In a similar way to the evaluation of Equation 10.22, the linear relations Equations 10.23 and 10.24 are also examined using the six-story shear building. Based on the optimal sequence of actuator location, the sixth actuator is removed first and the increment of the performance index is calculated using the two methods described. The sixth and fourth actuators are then removed and the corresponding increment of the performance index is calculated, also using these two methods. Then, the sixth, fourth, second and fifth actuators are removed, following the removal of the sixth, fourth and second actuators. Finally, all actuators are taken away, after the removal of the sixth, fourth, second, fifth and first actuators. The calculated increments of the performance index from the two methods for each case are depicted in Figure 10.3. It can be seen that the use of linear Equations 10.23 and 10.24 gives satisfactory results for the removal of actuators up to the first three. The further removal of actuators may lead to some significant errors, but this may be solved by implementing the proposed increment algorithm in several steps.

Now, let us evaluate the suboptimal control gain in terms of both the optimal performance index and the maximum seismic displacement response of the building. One method of doing so is to use the suboptimal control gain, and the other method is to follow the conventional optimal control theory for the given actuators at the given locations. The comparison of optimal performance indices is displayed in Figure 10.4a. It can be seen that, with more actuators removed, the difference between the optimal performance indices of the two methods becomes greater. However, with the first three actuators removed, the difference in the optimal performance indices is quite acceptable. The comparison of the maximum displacement response of the building is depicted in Figure 10.4b. It can be seen that the displacement responses of the building using the suboptimal control gain are close to those using the optimal control gain, following the conventional control theory for the building with the three actuators on the first, third and fifth floors. The displacement responses of the building with the three actuators obtained by the two methods described both approach those of the fully controlled building. The control performance of the building with the three actuators at their optimal placements determined by the proposed approach is quite satisfactory if compared with the maximum displacement responses of the building without control.

To demonstrate further the applicability of the described approach, the optimal placement of the actuators and the optimal response of the 18-story two-dimensional linear

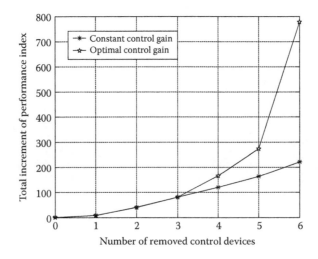

Figure 10.3 Effect of control gain on total increment of performance index.

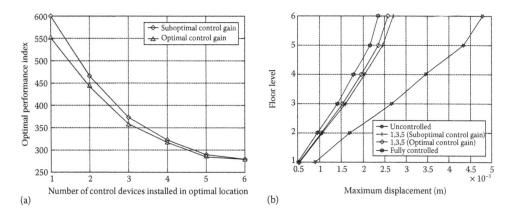

Figure 10.4 Evaluation of suboptimal control gain: (a) optimal performance index, (b) maximum displacement.

elastic shear building are investigated under either the N-S 1940 El-Centro ground excitation or the artificial Shanghai ground excitation, which contains significant energy at low frequencies. The structural parameters of the building are listed in Table 10.2, in which the damping coefficients given imply that the first two modal damping ratios of the building are about 3.4%. The initial motion of the building is assumed to be zero.

The computed increments of the performance index due to the removal of actuators one by one using Equation 10.22 are plotted in Figure 10.5 for a building originally equipped

Table 10.2 Structural parameters of the 18-story building

Floor number	Mass (kg)	Stiffness (N/m)	Damping (N·s/m)
1–4	5278.0	8956.0×10^4	51.3×10^4
5–8	5278.0	5542.0×10^4	42.1×10^4
9–12	5278.0	3837.0×10^4	34.4×10^4
13–16	5278.0	3014.0×10^4	26.5×10^4
17–18	5278.0	2176.0×10^4	10.4×10^4

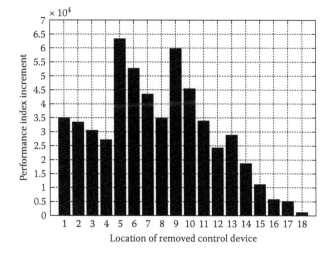

Figure 10.5 Optimal sequence of control devices (El-Centro wave).

with actuators at every floor under the El-Centro wave. The optimal sequence of actuator locations is seen to be (5, 9, 6, 10, 7, 1, 8, 11, 2, 3, 13, 4, 12, 14, 15, 16, 17, 18). By arranging nine actuators according to the optimal sequence (5, 9, 6, 10, 7, 1, 8, 11, 2) and using the suboptimal control gain, the maximum displacement responses of the building are computed and compared with those of the fully controlled building and those of the uncontrolled building (see Figure 10.6). It is clear that the control performance of the building with the nine actuators at their optimal placements determined by the approach described is close to that of the fully controlled building. The maximum displacement responses of the uncontrolled building are reduced tremendously because of the installation of the nine actuators at their optimal placements. This can be also seen from the time histories of displacement response at the top of the building with the nine actuators and without control, as shown in Figure 10.7.

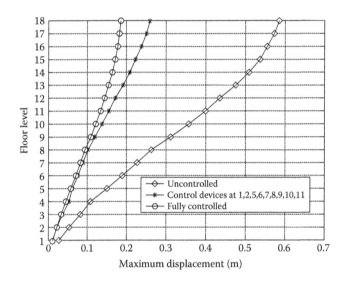

Figure 10.6 Comparison of maximum displacement responses (El-Centro wave).

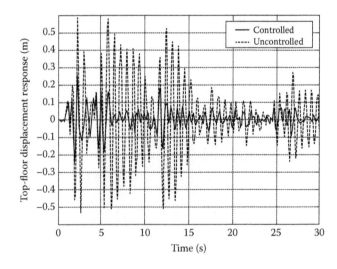

Figure 10.7 Time histories of top-floor displacement response (El-Centro wave).

A similar exercise is applied to the same building but under the Shanghai artificial wave. The computed increments of the performance index are plotted in Figure 10.8. The optimal sequence of actuator locations becomes (9, 5, 6, 10, 7, 11, 8, 13, 1, 2, 3, 12, 4, 14, 15, 16, 17, 18), indicating that ground motion will affect the optimal sequence of actuator locations. However, if nine actuators are selected, one may find that the optimal placements of the actuators are almost the same, except that one actuator is moved from the second floor to the 13th floor. The control performance of the building with the nine actuators at their optimal locations is again close to that of the fully controlled building (see Figure 10.9). The maximum displacement responses of the uncontrolled building are significantly reduced, as shown in Figures 10.9 and 10.10.

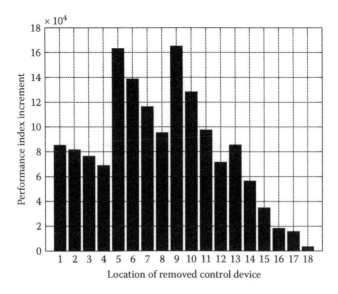

Figure 10.8 Optimal sequence of control devices (Shanghai wave).

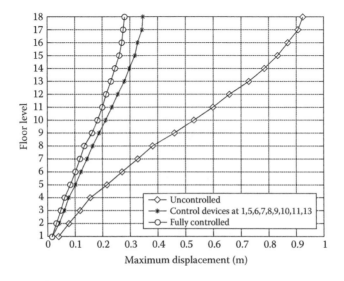

Figure 10.9 Comparison of maximum displacement response (Shanghai wave).

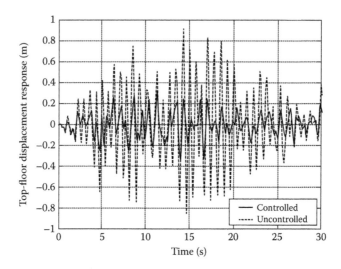

Figure 10.10 Time histories of top-floor displacement response (Shanghai wave).

10.7 NUMERICAL EXAMPLE FOR PASSIVE DAMPER PLACEMENT

The 18-story shear building used as the numerical example for active control is now employed as a numerical example for passive control. In consideration of determining the matrix of correlation function of the building response, V_z, the Kanai–Tajimi filtered white noise spectrum, is again utilised as the excitation spectrum of ground motion. The parameters in the filtered white noise spectrum remain as $\omega_g = 15.0$ rad/s, $\xi_g = 0.65$ and $S_0 = 4.0 \times 10^{-3}$ m^2/rad·s^3. There is no building motion when the time is equal to zero.

The increments of the performance index due to the removal of control devices one by one are computed using Equation 10.22 and depicted in Figure 10.11. The sequence of control

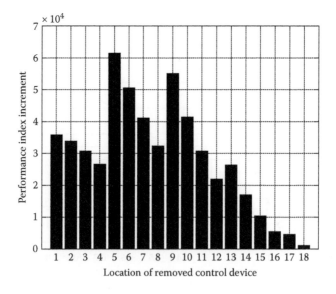

Figure 10.11 Optimal sequence of control devices (filtered white noise wave).

device locations is optimally arranged as (5, 9, 6, 10, 7, 1, 2, 8, 11, 3, 4, 13, 12, 14, 15, 16, 17, 18), which is slightly different from that under the El-Centro wave or the Shanghai artificial wave. However, if only nine control devices are selected, the optimal sequence becomes (5, 9, 6, 10, 7, 1, 2, 8, 11) and the stories involved are the same as the case under the El-Centro wave. The matrix of correlation function of the building response with the nine control devices at their optimal positions is then computed. The use of Equations 10.49 and 10.50 with the computed matrix of correlation functions then provides the equivalent optimal parameters of the nine passive dampers. Table 10.3 lists the computed equivalent optimal stiffness coefficients and damping coefficients for each damper at its optimal position. The maximum displacement responses of the uncontrolled building, the fully controlled building, the building with the nine active control devices (actuators) at their optimal positions and the building with the nine passive dampers at their optimal positions with the equivalent optimal parameters are computed and plotted in Figure 10.12. It is interesting to see that the maximum displacement responses of the building with the nine passive dampers are very close to those of the building with the nine active control devices. The control performances of the nine active control devices and the nine passive dampers approach that of the fully active control. Nevertheless, these results depend on the feasibility of making the passive dampers of the equivalent optimal parameters in real implementation.

Table 10.3 Equivalent optimal parameters of passive dampers

Floor level	Stiffness (N/m)	Damping (N·s/m)
1	3.0515×10^5	9.5564×10^4
2	3.2523×10^5	8.5246×10^4
5	1.6067×10^6	1.9485×10^6
6	1.7865×10^5	7.7434×10^4
7	6.1499×10^4	7.1942×10^4
8	1.4796×10^6	1.7958×10^6
9	1.3846×10^6	1.4906×10^6
10	2.9115×10^5	1.0631×10^5
11	1.3403×10^5	8.6617×10^4

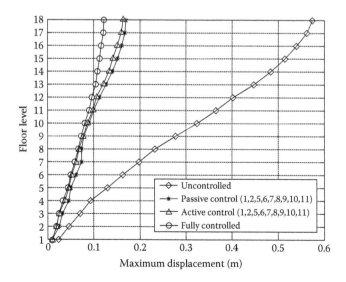

Figure 10.12 Effectiveness of passive dampers with equivalent optimal parameters (filtered white noise wave).

NOTATION

A	The system matrix in the state-space equation
B	The position matrix of control devices in the state-space equation
$\mathbf{B}_d$	The reduced-order position matrix for the control devices at their optimal placement
C_{ji}	The equivalent optimal damping coefficient of the jth passive damper at the ith floor
D	The matrix denoting the location of the control force
E	Excitation error
G	Optimal control gain matrix
$\mathbf{G}_d$	Suboptimal control gain matrix
H	Lagrange multiplier matrix
J	The performance index
K_{ji}	The equivalent optimal stiffness coefficient of the jth passive damper at the ith floor
$\mathbf{L}_{ni}$	The position vector where the subscript n is the dimension of the vector and the subscript i means that only the ith element is unity while all the other elements are zero
M, K, C	Mass, stiffness and damping matrix of the building structure, respectively
Q	The positive semi-definite weighting matrix for the structure response
R	The positive definite weighting matrix for the control force
$\mathbf{u}_{ji}$	The stiffness force provided by the jth damper at the ith floor
U	Control force vector
$\mathbf{v}_{ji}$	The damping force provided by the jth damper at the ith floor
$\mathbf{w}_{ji}$	The displacement vector of the jth passive damper located at the ith floor of the building
$\dot{\mathbf{w}}_{ji}$	The velocity vector of the jth passive damper located at the ith floor of the building
W	The input matrix associated with the ground excitation in the state-space equation
x	Displacement vector relative to the ground
$\ddot{\mathbf{x}}_g$	Ground acceleration vector
Z	The system states composed of displacement and velocity vectors
$\mathbf{Z}_d$	The system states for a building with the reduced number of control devices at their optimal placement
$\Delta\mathbf{B}$	The change of the position matrix of control devices $\mathbf{B}$
ΔJ_i	The increment of the performance index caused by the removal of the ith control device
λ_i	The ith eigenvalue of matrix $(\mathbf{A}-\mathbf{BG})$
Φ	The matrix consisting of all the eigenvectors

REFERENCES

Abdullah, M.M. 2000. Optimal placement of output feedback controllers on slender civil structures at discrete locations. *Eng. Struct.*, 22(8): 1042–7.

Agranovich, G. and Y. Ribakov. 2010. A method for efficient placement of active dampers in seismically excited structures. *Struct. Control Health Monit.*, 17(5): 513–31.

Agrawal, A.K. and J.N. Yang. 1999. Optimal placement of passive dampers on seismic and wind-excited buildings using combinatorial optimization. *J. Intell. Mater. Syst. Struct.*, 10(12): 997–1014.

Amini, F. and P. Ghaderi. 2012. Optimal locations for MR dampers in civil structures using improved ant colony algorithm. *Optim. Control Appl. Methods*, 33(2): 232–48.

Amini, F. and P. Ghaderi. 2013. Hybridization of harmony search and ant colony optimization for optimal locating of structural dampers. *Appl. Soft. Comput.*, 13(5): 2272–80.

Amini, F. and M.R. Tavassoli. 2005. Optimal structural active control force, number and placement of controllers. *Eng. Struct.*, 27(9): 1306–16.

Amouzegar, H., H.T. Riahi, and M. Daei. 2014. Evaluation of endurance time analysis in optimization of viscous damper placement for seismic design. *J. Seismolog. Earthq. Eng.*, 16(1): 25–37.

Aydin, E. 2012. Optimal damper placement based on base moment in steel building frames. *J. Constr. Steel. Res.*, 79: 216–25.

Aydin, E. and M.H. Boduroglu. 2008. Optimal placement of steel diagonal braces for upgrading the seismic capacity of existing structures and its comparison with optimal dampers. *J. Constr. Steel. Res.*, 64(1): 72–86.

Aydin, E., M.H. Boduroglu, and D. Guney. 2007. Optimal damper distribution for seismic rehabilitation of planar building structures. *Eng. Struct.*, 29(2): 176–85.

Bigdelia, K., W. Hareb, and S. Tesfamariam. 2012. Configuration optimization of dampers for adjacent buildings under seismic excitations. *Eng. Optimiz.*, 44(12): 1491–509.

Bilbao, A., R. Avilés, J. Aguirrebeitia, and I.F. Bustos. 2009. Eigensensitivity-based optimal damper location in variable geometry trusses. *AIAA J.*, 47(3): 576–91.

Bishop, J.A. and A.G. Striz. 2004. On using genetic algorithms for optimum damper placement in space trusses. *Struct. Multidiscip. Optim.*, 28(2–3): 136–45.

Chang, K.C., T.T. Soong, S.T. Oh, and M.L. Lai. 1995. Seismic behavior of steel frame with added viscoelastic dampers. *J. Struct. Eng.*, 121(10): 1418–26.

Chang, M.I.J. and T.T. Soong. 1980. Optimal controller placement in modal control of complex systems. *J. Math. Anal. Appl.*, 75(2): 340–58.

Chen, C.Q. and G.D. Chen. 2002. Nonlinear control of a 20-story steel building with active piezoelectric friction dampers. *Struct. Eng. Mech.*, 14(1): 21–38.

Chen, G.D. and C.Q. Chen. 2004. Semiactive control of the 20-story benchmark building with piezoelectric friction dampers. *J. Eng. Mech.*, 130(4): 393–400.

Chen, G.D. and J.N. Wu. 2001. Optimal placement of multiple tune mass dampers for seismic structures. *J. Struct. Eng.*, 127(9): 1054–62.

Chen, G.S., R.J. Rruno, and M. Salama. 1991. Optimal placement of active/passive members in truss structures using simulated annealing. *AIAA J.*, 29(8): 1327–34.

Cheng, F.Y., H.P. Jiang, and K.Y. Lou. 2008. *Smart Structures: Innovative Systems for Seismic Response Control*. Boca Raton: CRC Press.

Cheng, F.Y., H.P. Jiang, and X.Z. Zhang. 2002. Optimal placement of dampers and actuators based on stochastic approach, *Earthq. Eng. Eng. Vib.*, 1(2): 237–49.

Cheng, F.Y. and C.P. Pantelides. 1988. Optimal placement of actuators for structural control. Technical Report NCEER-88-0037, National Center for Earthquake Engineering Research, State University of New York, Buffalo, NY.

Cimellaro, G.P. 2007. Simultaneous stiffness-damping optimization of structures with respect to acceleration, displacement and base shear. *Eng. Struct.*, 29(11): 2853–70.

Cimellaro, G.P. and R. Retamales. 2007. Optimal softening and damping design for buildings. *Struct. Control. Health Monit.*, 14(6): 831–57.

Estekanchi, H.E. and M.C. Basim. 2011. Optimal damper placement in steel frames by the endurance time method. *Struct. Des. Tall Spec. Build.*, 20(5): 612–30.

Fujita, K., A. Moustafa, and I. Takewaki. 2010. Optimal placement of viscoelastic dampers and supporting members under variable critical excitations. *Earthq. Struct.*, 1(1): 43–67.

Haftka, R.T. and H.M. Adelman. 1985. Selection of actuator locations for static shape control of large space structures by heuristic integer programming. *Comput. Struct.*, 20(1–3): 578–82.

Housner, G.W., L.A. Bergman, T.K. Caughey, et al. 1997. Structural control: Past, present and future. *J. Eng. Mech.*, 123(9): 897–971.

Kwok, N.M., Q.P. Ha, and B. Samali. 2007. MR damper optimal placement for semi-active control of building using an efficient multi-objective binary genetic algorithm. In: *24th International*

Symposium on Automation and Robotics in Construction, Construction Automation Group, I.I.T. Madras, 361–7.

Levine, W.S. and M. Athans. 1970. On the determination of the optimal constant output feedback gains for linear multivariable systems. *IEEE Trans. Autom. Control*, 15(1): 44–8.

Li, L., G. Song, and J. Ou. 2010. A genetic algorithm-based two-phase design for optimal placement of semi-active dampers for nonlinear benchmark structure. *J. Vib. Control*, 16(9): 1379–92.

Li, Q.S., D.K. Liu, J.Q. Fang, and C.M. Tam. 2000. Multi-level optimal design of buildings with active control under winds using genetic algorithms. *J. Wind Eng. Ind. Aerodyn.*, 86 (1): 65–86.

Li, Q.S., D.K. Liu, J. Tang, N. Zhang, and C.M. Tam. 2004. Combinatorial optimal design of number and positions of actuators in actively controlled structures using genetic algorithms. *J. Sound Vibr.*, 270(4–5): 611–24.

Liu, D.K., Y.L. Yang, and Q.S. Li. 2003. Optimum positioning of actuators in tall buildings using genetic algorithm. *Comput. Struct.*, 81(32): 2823–7.

Liu, W., M. Tong, and G.C. Lee. 2005. Optimization methodology for damper configuration based on building performance indices. *J. Struct. Eng.*, 131(11): 1746–56.

Liu, W., M. Tong, Y. Wu, and G.C. Lee. 2004. Optimized damping device configuration design of a steel frame structure based on building performance indices. *Earthq. Spectra*, 20(1): 67–89.

Lopez Garcia, D. 2001. A simple method for the design of optimal damper configurations in MDOF structures. *Earthq. Spectra*, 17(3): 387–98.

Lopez Garcia, D. and T.T. Soong. 2002. Efficiency of a simple approach to damper allocation in MDOF structures. *J. Struct. Control*, 9(1): 19–30.

Lu, J., J.S. Thorp, B.H. Aubert, and L.B. Larson. 1994. Optimal tendon configuration of a tendon control system for a flexible structure. *J. Guid. Control Dyn.*, 17(1): 161–9.

Martínez, C.A., O. Curadelli, and M.E. Compagnoni. 2014. Optimal placement of nonlinear hysteretic dampers on planar structures under seismic excitation. *Eng. Struct.*, 65: 89–98.

Milman, M.H. and C.C. Chu. 1994. Optimization methods for passive damper replacement and tuning. *J. Guid. Control Dyn.*, 17(4): 848–56.

Miyamoto, H.K. and J.P. Singh. 2002. Performance of structures with passive energy dissipaters. *Earthq. Spectra*, 18(1): 105–19.

Mousavi, S.A. and A.K. Ghorbani-Tanha. 2012. Optimum placement and characteristics of velocity-dependent dampers under seismic excitation. *Earthq. Eng. Eng. Vib.*, 11(3): 403–14.

Ok, S.Y., J. Song, and K.S. Park. 2008. Optimal design of hysteretic dampers connecting adjacent structures using multi-objective genetic algorithm and stochastic linearization method. *Eng. Struct.*, 30(5): 1240–9.

Ou, J. and H. Li. 2012. Optimal placement of passive energy dissipation devices by genetic algorithms. *Math. Probl. Eng.*, 2012: 474282.

Peng, F.J., A. Ng, and Y.R. Hu. 2005. Actuator placement optimization and adaptive vibration control of plate smart structures. *J. Intell. Mater. Syst. Struct.*, 16(3): 263–71.

Rao, A.R.M. and K. Sivasubramanian. 2008. Optimal placement of actuators for active vibration control of seismic excited tall buildings using a multiple start guided neighbourhood search (MSGNS) algorithm. *J. Sound Vibr.*, 311(1–2): 133–59.

Rao, S.S., T.S. Pan, and V.B. Venkayya. 1991. Optimal placement of actuators in actively controlled structures using genetic algorithms. *AIAA J.*, 29(6): 942–3.

Shukla, A. and T. Datta. 1999. Optimal use of viscoelastic dampers in building frames for seismic force. *J. Struct. Eng.*, 125(4): 401–9.

Singh, M.P. and L.M. Moreschi. 2001. Optimal seismic response control with dampers. *Earthq. Eng. Struct. Dyn.*, 30(4): 553–72.

Singh, M.P. and L.M. Moreschi. 2002. Optimal placement of dampers for passive response control. *Earthq. Eng. Struct. Dyn.*, 31(4): 955–76.

Sonmez, M., E. Aydin, and T. Karabork. 2013. Using an artificial bee colony algorithm for the optimal placement of viscous dampers in planar building frames. *Struct. Multidiscip. Optim.*, 48(2): 395–409.

Soong, T.T. 1990. *Active Structural Control: Theory and Practice*. Longman: London; Wiley: New York.

Takewaki, I. 1997. Optimal damper placement for minimum transfer function. *Earthq. Eng. Struct. Dyn.*, 26(11): 1113–24.

Takewaki, I. 1998. Optimal damper positioning in beams for minimum dynamic compliance. *Comput. Meth. Appl. Mech. Eng.*, 156(1–4): 363–73.

Takewaki, I. 2000. Optimum damper placement for planar building frames using transfer functions. *Struct. Multidiscip. Optim.*, 20(4): 280–7.

Takewaki, I., and S. Yoshitomi. 1998. Effects of support stiffnesses on optimal damper placement for a planar building frame. *Struct. Des. Tall Build.*, 7(4): 323–36.

Takewaki, I., S. Yoshitomi, K. Uetani, and M. Tsuji. 1999. Non-monotonic optimal damper placement via steepest direction search. *Earthq. Eng. Struct. Dyn.*, 28(6): 655–70.

Tan, P., S.J. Dyke, A. Richardson, and M. Abdullah. 2005. Integrated device placement and control design in civil structures using genetic algorithms. *J. Struct. Eng.*, 131(10): 1489–96.

Wang, H., A.Q. Li, C.K. Jiao, and B.F. Spencer Jr. 2010. Damper placement for seismic control of super-long-span suspension bridges based on the first-order optimization method. *Sci. China Technol. Sci.*, 53(7): 2008–14.

Whittle, J.K., M.S. Williams, T.L. Karavasilis, and A. Blakeborough. 2012. A comparison of viscous damper placement methods for improving seismic building design. *J. Earthq. Eng.*, 16(4): 540–60.

Wongprasert, N. and M.D. Symans. 2004. Application of a genetic algorithm for optimal damper distribution within the nonlinear seismic benchmark building. *J. Eng. Mech.*, 130(4): 401–6.

Wu, B., J.P. Ou, and T.T. Soong. 1997. Optimal placement of energy dissipation devices for three-dimensional structures. *Eng. Struct.*, 19(2): 113–25.

Xu, Y.L. and J. Teng. 2002. Optimum design of active/passive control devices for tall buildings under earthquake excitation. *Struct. Des. Tall Build.*, 11(2): 109–27.

Yan, Y.J. and L.H. Yam. 2002. Optimal design of number and locations of actuators in active vibration control of a space truss. *Smart Mater. Struct.*, 11(4): 496–503.

Yang, J.N., S. Lin, J.H. Kim, and A. K. Agrawal. 2002. Optimal design of passive energy dissipation systems based on H_∞ and H_2 performances. *Earthq. Eng. Struct. Dyn.*, 31(4): 921–36.

Yang, Z., Y.L. Xu, and X.L. Lu. 2003. Experimental seismic study of adjacent buildings with fluid dampers. *J. Struct. Eng.*, 129(2): 197–205.

Zhang, R.H. and T.T. Soong. 1992. Seismic design of viscoelastic dampers for structural applications. *J. Struct. Eng.*, 118(5): 1375–92.

Chapter 11

Collective placement of control devices and sensors

11.1 PREVIEW

Sensor placement and control device placement have been discussed in Chapter 8 and Chapter 10, respectively. The optimal sensor locations are often determined for the purpose of structural health monitoring, such as system identification and damage detection without consideration of control performance, whereas the optimal locations of control devices always refer to control performance by assuming that structural responses, which act as feedbacks in control algorithms and indices in performance functions, can be measured by the sensors without consideration of their availability in both numbers and locations. However, it is impractical and uneconomical to install the sensors at all the required locations when full-state feedbacks are used for the control of a large civil structure. It is also not economical to install two sensory systems, one for structural health monitoring and the other one for structural vibration control, when a civil structure needs both structural health monitoring and vibration control. Therefore, for a smart civil structure, it is highly desirable to develop the techniques to locate sensors and control devices collectively and cost-effectively to give the structure self-sensing, self-adaptive and self-diagnostic ability. In this chapter, some existing collective placement methods of both sensors and control devices for civil structures are first reviewed. In terms of the response reconstruction-based sensor placement method as described in Chapter 8 and the increment-based algorithm for control device placement as introduced in Chapter 10, a collective placement method for the determination of the minimal number and optimal location of both control devices and sensors is then presented in this chapter for vibration control of building structures under earthquake excitation. The feasibility and accuracy of the proposed method are finally investigated numerically through a 20-story shear building structure under the El-Centro ground excitation and the Kobe ground excitation. The number and location of sensors and control devices determined by this collective placement method lay the foundation for the synthesis of structural health monitoring and vibration control in the time domain, which will be introduced in Chapter 15.

11.2 REVIEW OF COLLECTIVE PLACEMENT METHODS FOR SENSORS AND CONTROL DEVICES

As discussed in Chapters 8 and 10, the optimal placement of sensors often refers to structural health monitoring, such as system identification and damage detection, whereas the optimal placement of control devices always makes reference to structural vibration control performance. Since the sensors and the control devices are two tightly interacting parts, it will be beneficial and cost-effective to consider the collective placement of both sensors and control devices for simultaneous damage detection and vibration attenuation if the structure needs both structural health monitoring and vibration control. However, there has been very limited study on the collective placement of sensors and control devices with the aim of improving damage detection accuracy and enhancing vibration control efficiency at the same time, although a few studies on integrated structural vibration control and system identification/damage detection can be found. Most of the accessible studies consider the joint placement of control devices and sensors solely for control cost-effectiveness and performance. Reviews of this topic in the fields of mechanical and aerospace engineering can be found in the references Padula and Kincaid (1999), Frecker (2003) and Gupta et al. (2010). A review of this topic, the collective placement of control devices and sensors for control cost-effectiveness and performance, in the field of civil engineering is provided in this chapter.

It is well known that controllability and observability are two basic characteristics of a controlled structural system. Controllability measures the ability of the particular control device configuration to control the system states, whereas observability measures the ability of the particular sensor configuration to supply the observations for estimating the system states. Although directly using the definition of controllability and observability for optimal control device and sensor placement is not applicable, the controllability and observability Gramians, which act as quantitative measures of the degree of controllability and observability, respectively, provide a promising means for the optimal placement of control devices and sensors in accordance with some criteria or objective functions, such as the minimisation of control energy or the maximisation of output energy. For example, in order to establish measures of the system controllability and observability with respect to differing locations of sensors and actuators, a scheme was proposed by Ibidapo-Obe (1985) for the optimal spatial placement of a limited number of sensors and actuators under a minimum energy requirement for the active control of flexible structures. By computing the eigenvalues of the controllability and observability Gramians, Hac and Liu (1993) found the optimal location of actuators and sensors to maximise the values of the preset performance indices. This approach was also further investigated and developed by Bruant and Proslier (2005). With the aid of the controllability and observability Gramians, the locations of sensors and actuators were determined such that the Hankel singular values of a structure from actuator inputs to sensor outputs were as close as possible to the Hankel singular values of the structure from the disturbance inputs to performance outputs (Gawronski 1997).

Since many control algorithms are employed to determine the optimal control forces using the information measured by sensors as feedbacks, the optimal placement of control devices and sensors is inevitably related to the selected control algorithms. In general, two types of methods are developed in this context: the placement of collocated and non-collocated control devices and sensors. For the collocated case, sensors and actuators are placed in the same position, based on the selected control algorithms and the predetermined performance indices. A well-designed sensor/actuator collocation configuration is able to provide excellent control performance and stability, especially when velocity feedback is adopted (Lee 2011). With the aid of two explicit solutions of generalised algebraic Riccati equations, Hiramoto et al. (2000) found the optimal locations of collocated actuators/sensors that minimise the H_2 norm of the transfer function matrix of the closed-loop system for a simply supported beam. This approach was further developed by Güney and Eşkinat (2008) for determining the location of actuator/sensor pairs with consideration of signal weightings and damping. By using genetic algorithms (GAs) in conjunction with gradient-based optimisation techniques, Abdullah et al. (2001) proposed an algorithm for the placement of collocated actuators and sensors on the basis of velocity feedback control law, and validated it numerically using a 40-story structure. The efficiency of this algorithm was then improved by Richardson and Abdullah (2002) based on a real-coded GA. Based on balanced reduced models, an optimisation procedure, which relied on H_2 and H_∞ norms as well as on controllability and observability Gramians, was proposed for optimal placement of collocated piezoelectric patches and numerically illustrated using a cantilever bending beam (Nestorović and Trajkov 2013). Since piezoelectric materials are able to act as actuators and sensors at the same time, most researchers concentrated on investigating the optimal placement of collocated piezoelectric actuator/sensor pairs, especially in mechanical and aerospace engineering (Gupta et al. 2010).

For the non-collocated case, the locations of control devices and sensors are usually different due to different performance indices or objective functions being used. Some commonly used objective functions for control device placement include the minimisation of control energy and the maximisation of control force, whereas the widely used objective function for sensor placement is the maximisation of system output energy. Since it is basically a multi-objective optimisation problem, some algorithms are required to balance each aspect and find the desired solutions. For example, based on an extension of the multi-objective linear quadratic Gaussian (LQG) method, Brown et al. (1999) developed an algorithm to synthesise the Pareto optimal controllers in the form of Pareto optimal trade-off curves to determine the preferred actuator and sensor locations, and they illustrated the proposed method using a lumped-mass shear building model under stochastic wind and earthquake loads. On the basis of the linear quadratic control method, an approach with the predetermined optimisation criteria was proposed and used to find the optimal location of one actuator and one sensor installed on a cantilever beam and a three-beam structure (Bruant et al. 2001). An H_2 norm-based approach was proposed by Ambrosio et al. (2012) for the actuator and sensor placement with the aim of not only maximising the norms of the controlled modes but also reducing spillover problems. Cha et al. (2012, 2013a) proposed a multi-objective GA for optimal placements of actuators and sensors in large civil structures under seismic excitations. Moreover, Cha et al. (2013b) proposed a gene manipulation, multi-objective GA to optimise the placement of active devices and sensors in seismically excited civil structures to reduce active control cost and increase the structural control strategy's effectiveness.

The number of the sensors is often limited compared with the degrees of freedom of a large civil structure in practice, and therefore response reconstruction using the measured responses from the limited sensors is necessary for better structural health monitoring, as discussed in Chapter 8. Furthermore, if the unmeasured responses can be reconstructed with acceptable accuracy and in a very short time from the limited measured responses, these reconstructed responses can be utilised for not only structural health monitoring but also structural vibration control. The idea of how to use the response reconstruction method to link structural health monitoring and vibration control together will be discussed in this chapter and the subsequent chapters. Based on the concept of response reconstruction, this chapter presents a method for the determination of the minimal number and optimal location of both control devices and sensors for vibration control of building structures under earthquake excitation (He et al. 2015). The number and location of control devices are first determined in terms of the sequence of increments of performance index and the predetermined control performance as described in Chapter 10. The response reconstruction method presented in Chapter 8 is then extended to the controlled building structure for the determination of the minimal number and optimal placement of sensors, with the objective that the reconstructed structural responses can be used as feedbacks for the vibration control while the predetermined control performance can be maintained. The use of the reconstructed structural responses for structural damage detection will be introduced in Chapter 12, whereas the use of the control devices and sensors for integrated structural vibration control and damage detection will be discussed in Chapters 14 and 15.

11.3 COLLECTIVE PLACEMENT OF CONTROL DEVICES AND SENSORS

It is well known that the equation of motion of a controlled building structure of multi-degrees of freedom (MDOFs) under earthquake excitation can be given by

$$M\ddot{x} + C\dot{x} + Kx = -M1\ddot{x}_g + H_cU \tag{11.1}$$

where:

x, $\dot{x}$ and $\ddot{x}$ are the displacement, velocity and acceleration response vector, respectively

M, K and C are the mass, stiffness and damping matrix of the building structure, respectively

U is the control force vector

H_c is the matrix denoting the location of the control force

$\ddot{x}_g$ is the ground acceleration

Equation 11.1 can also be converted to the continuous state-space equation

$$\dot{X} = A_cX + B_c\ddot{x}_g + D_cU \tag{11.2}$$

where

$$X = \begin{bmatrix} x & \dot{x} \end{bmatrix}^T \tag{11.3}$$

$$A_c = \begin{bmatrix} 0 & I \\ -M^{-1}K & -M^{-1}C \end{bmatrix}; B_c = \begin{bmatrix} 0 \\ -1 \end{bmatrix}; D_c = \begin{bmatrix} 0 \\ M^{-1}H_c \end{bmatrix} \tag{11.4}$$

Since it is almost impossible to install control devices and sensors at all the possible locations of a building structure, especially for large building structures, it is necessary to find an optimal or suboptimal way to install limited control devices and sensors at their appropriate locations to achieve best/better performance in vibration control and/or health monitoring.

11.3.1 Increment-based approach for optimal placement of control devices

The increment-based approach for optimal placement of control devices for achieving best control performance and cost-effectiveness has been introduced in Chapter 10 in detail. This approach is employed in this chapter for optimal control device placement. The basic formulas involved in the increment-based approach are not repeated in this section, but the contribution percentage (CP) index is rewritten as follows:

$$CP_i = \frac{\Delta J_i}{\sum\limits_{i=1}^{j} \Delta J_i} \tag{11.5}$$

The increment of the control performance index due to the removal of the ith control device, ΔJ_i, calculated by Equation 10.22 reflects the sensitivity of the ith control device to the performance index. Therefore, based on the calculated increment from the removal of each control device, the sequence of importance of all the control devices can be obtained. It can be seen from Equation 11.5 that a larger value of CP_i indicates more important influence of the ith control device on the total control performance. Consequently, the CP index provides great convenience for determining the number and location of control devices according to the predetermined control performance. Unlike the classical integer heuristic programming methods such as the sequential search algorithm (SSA) (Zhang and Soong 1992) or the Worst-Out-Best-In (WOBI) algorithm (Haftka and Adelman 1985), the control device placement method based on the sequence of the calculated CP_i is relatively simple, because the value of CP_i does not need to be re-calculated when one of the control devices is removed. A comparative study of these methods will be conducted in the subsequent numerical example in this chapter.

After the number and location of the control devices have been determined according to the sequence of CP_i, the control algorithm is decided and the number and location of sensors are selected by which the structural responses measured by the sensors can be used as the feedbacks for vibration control. However, it is often difficult in practice to install enough sensors to obtain the required structural responses as feedbacks for vibration control of a large civil structure. This is particularly true if linear quadratic regulator (LQR) or LQG control algorithms are selected with complete structural responses required for feedbacks. In this regard, the determination of the number and location of limited sensors of the controlled structure to fulfil the vibration control task becomes necessary.

11.3.2 Response reconstruction-based approach for optimal placement of sensors

The response reconstruction-based method for the sensor placement of an uncontrolled structure has been discussed in Chapter 8 in detail. Since the unmeasured structural responses can be reconstructed from the limited structural response measurements, these reconstructed responses could also be employed as feedbacks for vibration control so as to improve the applicability of the control system to civil structures. In this regard, the minimal number and optimal placement of sensors will be determined in this section for a controlled structure, with the objective that the reconstructed structural responses can be used as feedbacks for vibration control while the predetermined control performance is maintained. It is noted that although the increment-based approach mentioned in Chapter 10 is capable of optimally locating active and passive control devices, only active control devices are considered in this chapter, because the feedbacks are not required for passive control devices.

By minimising the linear quadratic performance index, the optimal control force vector can be expressed as

$$\mathbf{U} = -\mathbf{GX} \tag{11.6}$$

where $\mathbf{G}$ is the control gain, expressed by Equation 10.6. The substitution of Equation 11.6 into Equation 11.2 yields

$$\dot{\mathbf{X}} = [\mathbf{A}_c - \mathbf{D}_c\mathbf{G}]\mathbf{X} + \mathbf{B}_c\ddot{\mathbf{x}}_g = \mathbf{A}_1\mathbf{X} + \mathbf{B}_c\ddot{\mathbf{x}}_g \tag{11.7}$$

In practice, the responses to be measured as feedbacks for vibration control of a building structure under earthquake excitation are often the absolute acceleration responses, which can be directly measured by accelerometers. The corresponding observation equation of the controlled building structure can then be written as

$$Y = C_c X + F_c U \tag{11.8}$$

where Y denotes the measured absolute acceleration responses of the building structure. C_c and F_c can be found as follows:

$$C_c = \begin{bmatrix} -M^{-1}K & -M^{-1}C \end{bmatrix}; F_c = \begin{bmatrix} M^{-1}H_c \end{bmatrix} \tag{11.9}$$

By using Equation 11.6, Equation 11.8 can be rewritten as

$$Y = \begin{bmatrix} C_c - F_c G \end{bmatrix} X = C_1 X \tag{11.10}$$

In reality, the measured responses are discretely sampled with a time interval of Δt. Moreover, measurement noise and process noise always exist. Consequently, the state-space Equation 11.7 and the observation Equation 11.10 shall be converted to the discrete forms

$$X_{k+1} = A_2 X_k + B_d \ddot{x}_{g,k} + w_k \tag{11.11}$$

$$Y_k = C_1 X_k + v_k \tag{11.12}$$

where:

X_{k+1} is the discrete state vector

w_k and v_k are the process noise and measurement noise, respectively, which are assumed as zero-mean white noise processes with variance matrices equal to Q_1 and R_1, respectively

C_1 is defined in Equation 11.10

A_2 denotes the discrete-state control matrix

B_d denotes the discrete-state input matrix

They can be expressed as

$$X_k = X(k \cdot \Delta t) \, (k = 1, \ 2, \ 3,...) \tag{11.13}$$

$$A_2 = e^{A_1 \Delta t}; B_d = \int_0^{\Delta t} e^{A_1 t} dt \cdot B_c \tag{11.14}$$

where A_1 and B_c are defined in Equations 11.7 and 11.4, respectively.

The Kalman filter provides an unbiased and recursive algorithm to optimally estimate the unknown state vector. It is employed here for the optimal placement of sensors as well as the response reconstruction. For the active control of a building structure, the Kalman filter algorithm involves two sets of equations. The first set of equations is the time update equations

$$\hat{X}_{k+1|k} = A_2 \hat{X}_k + B_d \ddot{x}_{g,k}$$ (11.15)

$$P_{k+1|k} = A_2 P_k A_2^T + Q_1$$ (11.16)

where $\hat{X}_{k+1|k}$ and $P_{k+1|k}$ denote a priori state estimate and a priori error covariance matrix, respectively. The second set of equations is the measurement update equations

$$\hat{X}_{k+1|k+1} = \hat{X}_{k+1|k} + K_{k+1}^{KF} \left\{ Y_{k+1} - C_1 \hat{X}_{k+1|k} \right\}$$ (11.17)

$$P_{k+1|k+1} = \left[I - K_{k+1}^{KF} C_1 \right] \cdot P_{k+1|k}$$ (11.18)

$$K_{k+1}^{KF} = P_{k+1|k} C_1^T \cdot \left[C_1 P_{k+1|k} C_1^T + R_1 \right]^{-1}$$ (11.19)

where:

$\hat{X}_{k+1|k+1}$ is the posteriori state estimate matrix
$P_{k+1|k+1}$ is the posteriori error covariance matrix
K_{k+1}^{KF} is the optimal Kalman gain matrix

However, it will be computationally prohibited to directly apply the aforementioned Kalman filter algorithm to civil structures with a large number of degrees of freedom (DOFs) involved, as described by Equation 11.1 or 11.2. Considering that under earthquake excitation, the structural responses are mainly denominated by the first several modes of vibration of the structure, and the contribution of the modes of vibration with high frequencies can be ignored, the mode superposition method can be employed to lift the computation prohibition. Letting $x = \Phi_s \cdot q_s$, Equation 11.2 in the state space can then be expressed by

$$\dot{Z} = A_c' Z + B_c' \ddot{x}_g + D_c' U$$ (11.20)

where:

$$Z = \begin{bmatrix} q_s & \dot{q}_s \end{bmatrix}^T ; q_s = \begin{bmatrix} q_1 & q_2 & \cdots & q_s \end{bmatrix}^T$$ (11.21)

$$A_c' = \begin{bmatrix} 0 & I \\ -\omega_s^2 & -2\xi_s\omega_s \end{bmatrix} ; B_c' = \begin{bmatrix} 0 \\ -\Phi_s^T M1 \end{bmatrix} ; D_c' = \begin{bmatrix} 0 \\ \Phi_s^T H_c \end{bmatrix}$$ (11.22)

where:

q is the vector of modal coordinates
subscript s denotes the number of the selected modes of vibration
Φ_s is the selected mass-normalised displacement mode shape matrix
ξ_s and ω_s are the modal damping ratio matrix and modal frequency matrix with respect to the selected modes of vibration, respectively

In the modal domain, the control force shown in Equation 11.6 could be rearranged as

$$\mathbf{U} = -\mathbf{GX} = -\mathbf{G} \begin{bmatrix} \mathbf{\Phi}_s & 0 \\ 0 & \mathbf{\Phi}_s \end{bmatrix} \mathbf{Z} \tag{11.23}$$

Consequently, Equation 11.20 can be rewritten as

$$\dot{\mathbf{Z}} = \left[\mathbf{A}'_c - \mathbf{D}'_c\mathbf{G} \begin{bmatrix} \mathbf{\Phi}_s & 0 \\ 0 & \mathbf{\Phi}_s \end{bmatrix} \right] \mathbf{Z} + \mathbf{B}'_c \ddot{x}_g = \mathbf{A}'_1\mathbf{Z} + \mathbf{B}'_c \ddot{x}_g \tag{11.24}$$

As mentioned in Equation 11.8, the measured responses of the building structure are the absolute accelerations, and Equation 11.10 can then be rewritten as

$$\mathbf{Y} = \mathbf{C}_1 \begin{bmatrix} \mathbf{\Phi}_s & 0 \\ 0 & \mathbf{\Phi}_s \end{bmatrix} \mathbf{Z} = \mathbf{C}'_1\mathbf{Z} \tag{11.25}$$

To implement the Kalman filter algorithm in the modal domain, the matrices $\mathbf{A}_1$, $\mathbf{B}_c$ and $\mathbf{C}_1$ in Equations 11.11 through 11.19 should be substituted by $\mathbf{A}'_1$, $\mathbf{B}'_c$ and $\mathbf{C}'_1$, respectively, in Equations 11.24 and 11.25. Furthermore, by comparing Equation 11.7 with Equation 11.24, one finds that the dimension of the state vectors in the modal domain is significantly reduced. Since the higher-frequency modes of vibration, which may be falsely excited by noise, are truncated in this procedure, it is not only computationally economic but also likely to improve the estimation accuracy to some extent. From this point of view, the modal domain provides a promising means for the use of the Kalman filter algorithm in the vibration control of large civil structures.

In this study, the observation equation is used not only to represent the measured responses but also to reconstruct structural responses at unmeasured key locations. Therefore, three types of structural responses in the discrete forms are introduced according to Equation 11.25 as

$$\mathbf{Y}_{e,k} = \mathbf{C}'_e\mathbf{Z}_k; \ \hat{\mathbf{Y}}_{e,k} = \mathbf{C}'_e\hat{\mathbf{Z}}_k; \ \mathbf{Y}_{m,k} = \mathbf{C}'_m\mathbf{Z}_k \tag{11.26}$$

where:

$\mathbf{Y}_e$ represent the real structural responses at the locations of interest
$\hat{\mathbf{Y}}_e$ represents the reconstructed structural responses at the locations of interest
$\mathbf{Y}_m$ represents the measured structural responses from the sensors

The matrix $\mathbf{C}'_e$ depends on the locations where the responses are of interest, and the matrix $\mathbf{C}'_m$ depends on the limited number of sensors for measurements. The accuracy of the reconstructed responses can be measured by the reconstruction error δ_k:

$$\delta_k = \hat{\mathbf{Y}}_{e,k} - \mathbf{Y}_{e,k} = \mathbf{C}'_e\left(\hat{\mathbf{Z}}_k - \mathbf{Z}_k\right) \tag{11.27}$$

Therefore, the asymptotic covariance matrix of the reconstruction error can be expressed as

$$\mathbf{\Delta} = \text{cov}(\mathbf{\delta}) = \mathbf{C}'_e \mathbf{P} \mathbf{C}'^T_e \tag{11.28}$$

Notably, the output influence matrices $\mathbf{C}'_e$ and $\mathbf{C}'_m$ probably tend to be ill conditioned or badly scaled, especially when only a few responses are measured, because the absolute acceleration responses of the building structure at different locations may have different orders of magnitude. Without appropriate pre-treatment of the matrix, the inverse operation for the determination of optimal Kalman gain matrix may lead to inaccurate results. Consequently, the standard deviation of the corresponding sensor noise is employed to normalise the matrices $\mathbf{C}'_e$ and $\mathbf{C}'_m$ as

$$\bar{\mathbf{C}}'_e = \mathbf{R}_e^{-1/2} \mathbf{C}'_e; \bar{\mathbf{C}}'_m = \mathbf{R}_m^{-1/2} \mathbf{C}'_m \tag{11.29}$$

where $\mathbf{R}_e$ and $\mathbf{R}_m$ are the signal noise matrices with different dimensions. For example, if the measurements are absolute acceleration responses, $\mathbf{R}_m$ can be expressed as

$$\mathbf{R}_m = E(\mathbf{v}\mathbf{v}^T) = \sigma_a^2 \mathbf{I} \tag{11.30}$$

where σ_a^2 is the measurement noise variance matrix of acceleration responses. Hence, the reconstruction error and the corresponding covariance matrix in Equations 11.27 and 11.28 should be normalised accordingly in consideration of the un-bias estimation:

$$\bar{\mathbf{\delta}}_k = \bar{\mathbf{C}}'_e \left(\hat{\mathbf{Z}}_k - \mathbf{Z}_k \right) \tag{11.31}$$

$$\bar{\mathbf{\Delta}} = \text{cov}(\bar{\mathbf{\delta}}) = \bar{\mathbf{C}}'_e \mathbf{P} \bar{\mathbf{C}}'^T_e \tag{11.32}$$

Moreover, since the normalised output influence matrix is used, the optimal Kalman gain shown in Equation 11.19 should be updated and given by

$$\mathbf{K}^{KF}_{k+1} = \mathbf{P}_{k+1|k} \bar{\mathbf{C}}'^T_m \cdot \left[\bar{\mathbf{C}}'_m \mathbf{P}_{k+1|k} \bar{\mathbf{C}}'^T_m + \mathbf{I} \right]^{-1} \mathbf{R}_m^{-1/2} \tag{11.33}$$

It can also be seen that each diagonal element of the $\bar{\mathbf{\Delta}}$ matrix in Equation 11.32 represents the normalised variance of the reconstruction error for the corresponding response. Therefore, the maximum diagonal element denotes the maximum reconstruction error, whereas the trace of the matrix $\bar{\mathbf{\Delta}}$ represents the sum of the reconstruction errors at all the locations of interest. From this point of view, the optimal sensor placement can be performed with the objective to minimise the sum of the normalised reconstruction error.

Object function:

$$\min tr(\bar{\mathbf{\Delta}}) \tag{11.34}$$

subject to

$$\bar{\sigma}_{\max} \le \sigma_{\max} \tag{11.35}$$

where $\bar{\sigma}_{max}$ is the maximum estimation error and defined as

$$\bar{\sigma}_{max} = \max\left(\mathrm{diag}(\bar{\Delta})\right) \tag{11.36}$$

σ_{max} is the preset allowable error.

It is understood that the maximum value of reconstruction error as well as the trace of the matrix $\bar{\Delta}$ will be increased when the number of sensors is reduced. A simple iterative procedure can then be conducted, in which the candidate sensors are removed one by one until the target error level is reached. In each step, only one sensor location, the removal of which leads to a minimal trace of the matrix $\bar{\Delta}$, will be deleted. Thus, the sensor with minimal contribution to the response reconstruction will be removed at each step, and this procedure for sensor location is thus suboptimal. Nevertheless, this suboptimal procedure is beneficial and applicable for large and complex civil structures, for the dimension of the state vectors in the modal domain is significantly reduced when only the first several modes of vibration are used.

It can be seen that the controlled system with optimal locations of both actuators and sensors can be established according to the approaches presented in this section, and a schematic diagram is plotted in Figure 11.1 to show the establishment and application of the proposed control system. The number of the control devices and sensors in this control system would be rather small. The feasibility and accuracy of the proposed method are investigated in the following section by using a 20-story shear building under earthquake excitation as a numerical example.

11.4 CASE STUDY

A 20-story shear building, as shown in Figure 11.2, is employed to investigate the feasibility and accuracy of the presented method. The mass and stiffness coefficients of the 20-story shear building are listed in Table 11.1. The Rayleigh damping assumption with

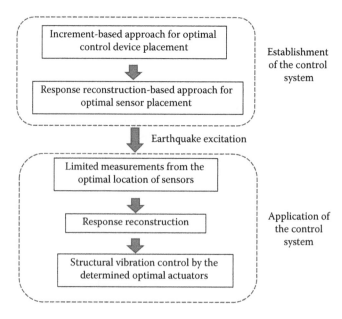

Figure 11.1 Schematic diagram for the establishment and application of the proposed control system.

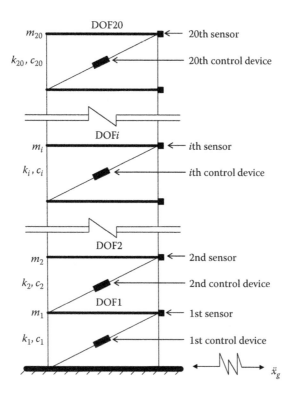

Figure 11.2 Structural model of a tall building with control devices and sensors.

Table 11.1 Structural parameters of the 20-story shear building

Floor number	Mass (kg)	Stiffness (N/m)	Floor number	Mass (kg)	Stiffness (N/m)
1–4	8000	9.2×10^7	13–16	8000	7.5×10^7
5–8	8000	8.8×10^7	17–20	8000	7.0×10^7
9–12	8000	8.0×10^7	–	–	–

a proportional coefficient of 0.8 for the mass matrix and 1×10^{-5} for the stiffness matrix is used to construct the damping matrix. The increment of performance index for vibration control of the building is first computed for the optimal control device placement. The eigenvalue analysis is then performed to extract the mode shapes for the optimal sensor placement and response reconstruction. The El-Centro ground excitation with a peak acceleration of 0.34 g and the Kobe ground excitation with a peak acceleration of 0.81 g are finally selected as the input to assess the control performance of the building equipped with the selected control system. The comparison of the control performance subject to two more earthquakes, the Northridge earthquake and the Hachinohe earthquake, is also conducted. It should be noted that the building model is linear, and the nonlinear effect on the structural dynamics is not considered and analysed in this study.

11.4.1 Determination of the configurations of the control system

For the determination of the optimal locations of control devices and sensors, a ground excitation composed of white noise random signals is applied to the shear building. The control devices are initially installed on each floor together with the braces. As the control devices are removed one by one, the increment of performance index is calculated, and then the contribution percentage of every control device can be obtained from Equation 11.5, as shown in Figure 11.3. It can be easily found from Figure 11.3 that the sequence of the control devices' locations is 1, 2, 4, 3, 5, 7, 16, 13, 19, 9, 10, 11, 12, 15, 18, 14, 8, 6, 17 and 20. Moreover, the relationship between the summation of CP value and the displacement reduction of the top floor is given in Figure 11.4. It can be seen that with the increase of the summation of CP value, the structural responses are reduced accordingly. The dashed line indicates the response reduction when the 20 control devices are all installed. It can also be found from Figure 11.4 that when the summation reaches 70%, the trend for vibration attenuation is becoming slow. Therefore, for the consideration of both effectiveness and economy, the summation of the CP value is assumed to be 70% in this study. According to the calculated sequence and the desired summation of the CP value, the first ten control devices in the locations of 1, 2, 4, 3, 5, 7, 16, 13, 19 and 9 shall be retained and used for the control.

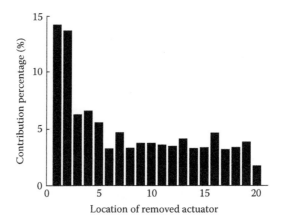

Figure 11.3 Optimal sequence of control devices.

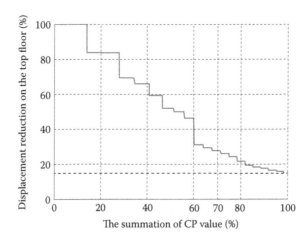

Figure 11.4 Relationship between the summation of CP value and displacement reduction.

It is known that in practical situations, structural parameter uncertainties resulting from modelling errors often exist. Thus, the influence of such uncertainties on the number and location of the control devices is discussed here. Since it is relatively time-consuming to consider the uncertainties existing in all the structural parameters, only two structural parameters with uncertainties (e.g. k_2 and k_6) are considered for demonstration purposes. The values of k_2 and k_6 are approximated as a normal distribution with a mean value of 9.2×10^7 and 8.8×10^7, respectively, and the standard deviation of 5% of the corresponding mean value. Although some variations exist in the calculated CP values, for example, in some cases CP_{16} becomes larger than CP_7, the aforementioned ten control devices are still retained when the two random parameters are involved. It can thus be concluded, to some extent, from these results that the increment-based approach is not very sensitive to the structural parameter variations.

Furthermore, two integer heuristic programming methods, i.e. the sequential search algorithm (SSA) (Zhang and Soong 1992) and the WOBI algorithm (Haftka and Adelman 1985), are employed for determination of the optimal control device location and comparison with the results obtained from the proposed method. In these two algorithms, 10 optimal locations out of 20 possible locations are selected for the placement of the control devices. In this regard, the two performance indices of the building structure are considered. The first one is the maximum peak acceleration at the top floor, described by

$$PI_1 = \max\left\{|\ddot{x}_{\text{top}}|\right\} \tag{11.37}$$

The second one is the maximum peak inter-story drift:

$$PI_2 = \max\left\{|\bar{x}_{pi}|\right\}(i = 1, 2, \ldots, 20) \tag{11.38}$$

where $|\ddot{x}_{\text{top}}|$ and $|\bar{x}_{pi}|$ denote the absolute value of peak acceleration at the top floor and the absolute value of peak inter-story drift of the ith story, respectively. The objective function to find the optimal location of the control devices is to minimise the maximum peak acceleration response at the top floor (PI_1) and the maximum peak inter-story drift (PI_2), respectively.

The control device locations determined by the SSA and WOBI methods are shown in Table 11.2. The results obtained from the increment-based approach are also listed in Table 11.2. For ease of comparison, the locations are listed in ascending order. Although there are several different control device locations, the results obtained by the three algorithms are close to each other. It can also be seen from Table 11.2 that even for the identical algorithm, the determined locations are not the same if different objective functions are used. It

Table 11.2 Comparison of the control device locations determined by three algorithms

Algorithm	Objective function	Control device locations
SSA	Minimise (PI_1)	1, 2, 4, 5, 7, 9, 10, 13, 15, 19
	Minimise (PI_2)	1, 3, 5, 7, 9, 13, 14, 16, 18, 20
WOBI	Minimise (PI_1)	2, 3, 5, 6, 9, 10, 13, 16, 19, 20
	Minimise (PI_2)	1, 3, 5, 7, 10, 13, 15, 17, 19, 20
Increment-based approach	Maximise (CP)	1, 2, 3, 4, 5, 7, 9, 13, 16, 19

should be noted that for r control devices to be placed in n possible locations, there would be $nr-[r(r-1)/2]$ configurations for SSA and n evaluations of the objective function in each iteration for WOBI. This means that 155 combinations are required to be considered for SSA, and 20 location strategies are required to be evaluated in each iteration for WOBI. Moreover, if all of the possible locations are considered, it would be more time-consuming. For example, in our case, i.e. 10 control devices being placed in 20 possible locations, the total number of possible combinations of the control device locations is $n!/[r!(n-r)!]=20!/(10!\times10!)=184,756$ (Agrawal and Yang 1999).

To determine the optimal placement of the sensors, the approach introduced in Section 11.3.2 is employed. The first five modes of vibration with the corresponding natural frequencies of 1.26, 3.65, 6.05, 8.43 and 10.74 Hz are employed. Twenty accelerometers, which are used to measure the acceleration response of each floor, are used as initial candidate locations for sensors. For practical consideration, all the measured structural responses are simulated by the numerically computed structural responses superimposed on a white noise at a 5% noise-to-signal ratio in terms of the root mean square (RMS). The predefined threshold value of the maximum estimation error is applied to determine the number of sensors, and it is defined as the ratio of the standard deviation of reconstruction error variance to that of noise, which is used to quantify the estimation accuracy. A smaller value of the ratio corresponds to higher estimation accuracy and represents a more stringent criterion on the estimation error, and of course, more sensors are required. It should also be noted that if the value of the ratio is too large, the allowable reconstruction error will be too large, which may result in the reconstructed responses being probably incorrect or too contaminated by the noise. In this study, the target maximum estimation error σ_{max} in Equation 11.36 is set to 3.0, which means that the standard deviation of reconstruction error variance is three times that of noise. By following the procedure for the determination of optimal sensor placement, four accelerometers, which are located on the 3rd floor, 12th floor, 16th floor and 20th floor, respectively, are finally selected. It is found that the number of sensors is rather reduced as compared with the initial candidate set. Since the locations of the control devices and sensors are both determined, the optimal control system for this shear building is established, which includes ten control devices and four accelerometers.

As mentioned Section 11.3.2, the standard deviation of the sensor noise is employed to normalise the reconstruction error and the corresponding covariance matrix. It is thus anticipated that the number and location of the selected sensors will be altered if the measurement noise covariance matrix R_1 is changed. However, in many cases, the measurement noise covariance is evaluated prior to the actual operation of the Kalman filter. Thus, for the same sensors under the same conditions, the variation of the measurement noise covariance matrix R_1 is small, and the influence of R_1 matrix can be controlled in a reasonable range. The Q_1 matrix defined in Equation 11.11 is the process noise covariance matrix, which is mainly used for the consideration of modelling errors. The structural parameter uncertainties mentioned before are used for investigating the effect of the modelling errors on the optimal sensor placement. Likewise, the values of k_2 and k_6 are approximated as a normal distribution with a mean value of 9.2×10^7 and 8.8×10^7, respectively, and a standard deviation of 5% of the corresponding mean value. With the consideration of these parameter uncertainties, the determined optimal location of the sensors was found to remain unchanged. It should be noted that one basic premise of the presented response reconstruction-based technique is that the structural model is relatively accurate and updated using the appropriate model updating technique. From this point of view, although the process noise covariance Q_1 matrix is generally difficult to estimate, the influence of the Q_1 matrix on the number and location of sensors should be relatively small. Some statements can also be found in Zhang (2012).

11.4.2 Investigation of the control performance with El-Centro ground excitation

To show the efficiency of the control system with limited control devices and sensors, the El-Centro ground excitation is applied to the shear building with and without the control system. Only the acceleration responses at the 3rd, 12th, 16th and 20th floors of the building are assumed to be measured. The measured responses are then contaminated by 5% white noise. It is noted that although only four acceleration responses are measured, the remaining structural responses can be reconstructed and used for vibration control. Four cases are considered: Case 1: the control devices and sensors are installed on each floor of the shear building for vibration control; Case 2: the control devices are installed in the determined optimal position (the aforementioned ten locations), but the sensors are installed on each floor of the building for feedbacks; Case 3: the control devices and sensors are both installed in the determined positions (the aforementioned ten and four positions, respectively), and the reconstructed responses are used as feedbacks for vibration control; and Case 4: no control devices and sensors are installed in the building (without control).

The time histories of acceleration response of the building on the top floor are computed and plotted in Figure 11.5 for the four cases. For the sake of clarification, only the time segment of the acceleration responses from 8 to 16 s is given in Figure 11.5, although the time duration of the displacement response is from 0 to 30 s. On one hand, it can be seen that the structural responses of the uncontrolled building are significantly reduced with the control system. On the other hand, the control performance of the optimal control system, defined as Case 3, is close to that of the fully controlled building, defined as Case 1. Moreover, it can also be seen that the control performance in Case 3 is in good agreement with that in Case 2, which indicates that the reconstructed responses can be employed for vibration control with acceptable accuracy. Though only the displacement and acceleration responses of the building at the top floor are plotted in Figure 11.5, similar results for the remaining building floors can be obtained as well. For a more comprehensive comparison of the control performance, the maximum displacement and acceleration responses of the building at each floor are depicted in Figure 11.6 for the four cases. It can be seen that the maximum responses are significantly reduced when the control devices are employed. It can also be seen that the results of Case 2 and Case 3 are close to each other, implying that the utilisation of the reconstructed responses for control is reliable. Though the maximum displacement and acceleration responses of the building at several upper floors from Case 3 are relatively larger than those from Case 1, the control performance of the optimal control system with only ten control devices and four sensors is still acceptable in terms of both control effectiveness and cost-effectiveness.

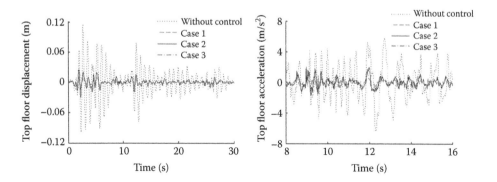

Figure 11.5 Time histories of displacement and acceleration responses at the top floor (El-Centro ground excitation).

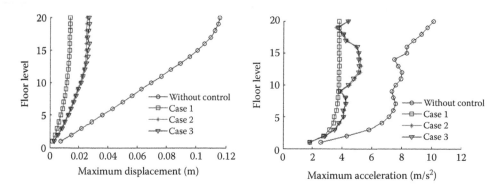

Figure 11.6 Comparison of the maximum displacement and acceleration responses (El-Centro ground excitation).

Moreover, Figure 11.7 shows the time histories of the reconstructed displacement and acceleration responses (dashed lines) and the corresponding actual ones (solid lines). Only the displacement and acceleration responses of the building on the top floor are shown in Figure 11.7 as an example, and only the time segments from 8 to 16 s are demonstrated for clarification. It is clear that the reconstructed responses are in good agreement with the corresponding actual responses, confirming that the reconstructed responses could be reliably employed as feedbacks for vibration control. Moreover, since the reconstructed responses are very close to the actual ones, they could be used for the purpose of damage detection as well. More details of the damage detection algorithms based on the response reconstruction technique will be given in Chapter 12.

The control performances of the control devices determined by SSA and WOBI as shown in Table 11.2 are also computed and compared with those from the increment-based approach. The maximum displacement and acceleration responses of the building at each floor are shown in Figure 11.8. The cases of the SSA with the objective function of minimising PI_1 and PI_2 are denoted as SSA1 and SSA2, respectively, in Figure 11.8. Similar definitions of WOBI1 and WOBI2 can also be found in Figure 11.8. It can be seen that although several control device locations are different according to these algorithms, the control performances are still close to each other. It should also be noted that the complete structural responses, such as the displacement and velocity responses at all floors, are assumed to be known for control in the SSA and WOBI algorithms, whereas only four accelerometers

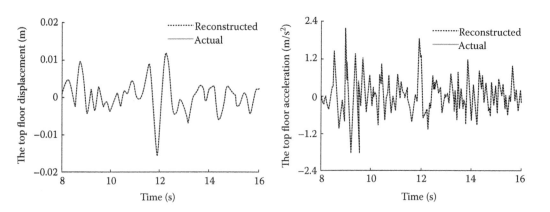

Figure 11.7 Comparison of the reconstructed responses with the actual ones (El-Centro ground excitation).

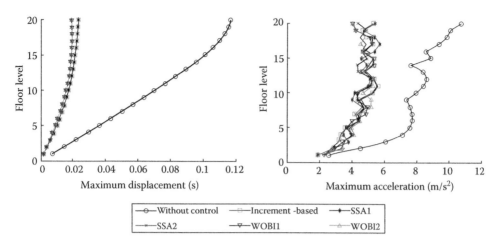

Figure 11.8 Comparison of the control performance with different algorithms (El-Centro ground excitation).

are required in the proposed control system. Moreover, based on the peak value of the control force, one more performance index is considered (Spencer et al. 1998):

$$PI_3 = \left\{ \frac{\max |\{U(t)\}|}{W} \right\} \tag{11.39}$$

where:

$U(t)$ is the control force
W is the total weight of the building

The values of PI_3 for the aforementioned cases of full control (i.e. the control devices being installed on each floor of the building), increment-based approach, SSA1, SSA2, WOBI1 and WOBI2 are 0.018, 0.042, 0.044, 0.033, 0.037 and 0.032, respectively. The total energy consumed in the full control, increment-based approach, SSA1, SSA2, WOBI1 and WOBI2 is 1.313×10^4, 1.288×10^4, 1.291×10^4, 1.297×10^4, 1.296×10^4 and 1.295×10^4 kN·m, respectively. By comparison with the case of full control, it can be found that with the reduction of the number of control devices, the control forces of the retained control devices are increased in order to achieve an acceptable control performance. However, from the viewpoint of energy consumption, with identical input energy (e.g. identical earthquakes applied to the system), the larger the vibration reduction achieved in the system, the more energy is consumed by the control devices. Therefore, although the control force for each control device in the case of full control is the smallest, it can be seen that with 20 control devices, in this case the total energy consumption is the largest.

11.4.3 Investigation of the control performance with Kobe ground excitation

To demonstrate the robustness of the optimal control system, the strong Kobe earthquake, which was measured in the near field (the station KJMA) with the epicentral distance of 18.27 km, is applied to the shear building. Similarly, four cases are taken into consideration: (1) the control devices and sensors are installed on each floor of the shear building for vibration control (Case 1); (2) the control devices are installed in the determined optimal positions (the aforementioned ten locations), but the sensors are installed on each floor of the building for feedback (Case 2); (3) the control devices and sensors are both installed in the determined positions (the aforementioned ten and four positions, respectively), and the reconstructed responses are used as feedbacks for vibration control (Case 3); and (4) no control devices and sensors are installed in the building (without control). For each case, white noise with 5% noise-to-signal ratio in terms of RMS is superimposed on the numerically calculated structural responses.

The time histories of the displacement and acceleration responses of the building on the top floor are shown in Figure 11.9 for the four cases. Only the time segments of the acceleration responses from 8 to 16 s are given in Figure 11.9 for the clarification of comparison, although the time periods of the displacement responses calculated are from 0 to 30 s. Moreover, the comparison of the maximum displacement and acceleration responses of each floor is shown in Figure 11.10 for the four cases. It can be seen from Figures 11.9 and 11.10 that the structural vibration is significantly reduced when the control devices are employed. It can also be found from these two figures that the response time series as well as the maximum structural responses of the building in Case 3 are rather close to those in Case 2, implying that the reconstructed responses are reliable and could be employed as feedbacks

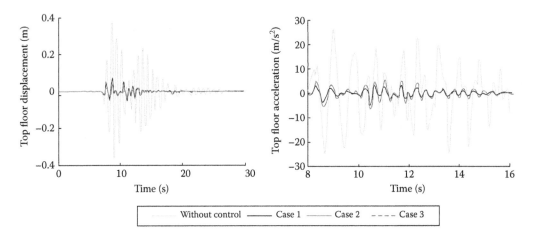

Figure 11.9 Time histories of displacement and acceleration responses on the top floor (Kobe ground excitation).

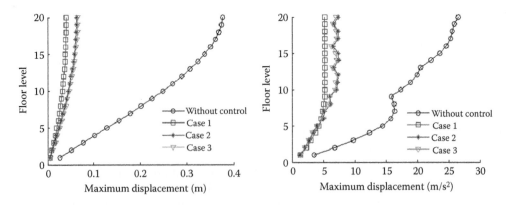

Figure 11.10 Comparison of the maximum displacement and acceleration responses (Kobe ground excitation).

for vibration control. Though the maximum displacement responses of the building on the upper floors and the maximum acceleration responses of the building on a few floors from Case 3 are relatively larger than those from Case 1, the control performance of the optimal control system with only ten control devices and four sensors is still acceptable in terms of both control effectiveness and cost-effectiveness. As a confirmation, the time histories of the reconstructed acceleration and displacement responses of the building on the top floor are compared with the actual structural responses, and the comparative results are shown in Figure 11.11. It is clear that the reconstructed responses are close to the actual ones. Similar results can be obtained for the rest of the building floors.

Moreover, the comparison of control performance among the SSA and WOBI methods and the increment-based approach is also considered, and the results are shown in Figure 11.12. As mentioned in Section 11.4.2, the SSA and WOBI with the objective function of minimising PI_1 and PI_2 are respectively denoted as SSA1, SSA2, WOBI1 and WOBI2. It can be seen that the results obtained from these three algorithms are close to each other. Furthermore, the index PI_3 defined in Equation 11.39 is also investigated. The corresponding values for the full control, increment-based approach, SSA1, SSA2, WOBI1 and WOBI2 are 0.043, 0.084, 0.088, 0.079, 0.082 and 0.075, respectively. It can be derived that, even under the Kobe ground excitation with a peak acceleration of 0.81 g, the maximum control force is around 140 kN, which can probably be realised by several hydraulic actuators.

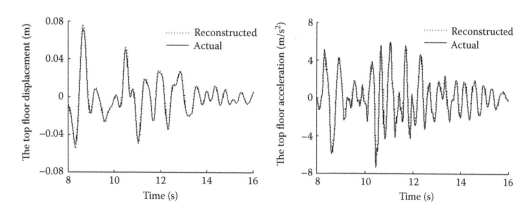

Figure 11.11 Comparison of the reconstructed responses with the actual ones (Kobe ground excitation).

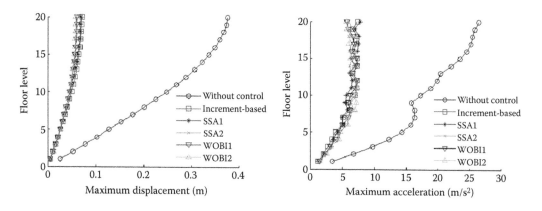

Figure 11.12 Comparison of the control performance with different algorithms (Kobe ground excitation).

Table 11.3 Control performance of the system under other earthquakes

Approaches for control device locations	Northridge earthquake				Hachinohe earthquake			
	PI_1 (m/s^2)	PI_2 (mm)	PI_3	Total energy consumption $(kN \cdot m)$	PI_1 (m/s^2)	PI_2 (mm)	PI_3	Total energy consumption $(kN \cdot m)$
Without control	4.227	2.854	–	–	15.51	14.467	–	–
Full control	1.515	0.711	0.007	6.556×10^2	3.08	3.207	0.026	2.775×10^4
Increment based	1.977	0.942	0.017	6.364×10^2	4.96	5.267	0.053	2.706×10^4
SSA1	1.759	0.926	0.018	6.388×10^2	4.47	4.904	0.054	2.715×10^4
SSA2	1.932	0.878	0.014	6.415×10^2	4.68	4.494	0.047	2.727×10^4
WOBI1	1.719	0.901	0.016	6.412×10^2	4.32	4.769	0.051	2.726×10^4
WOBI2	1.805	0.859	0.013	6.406×10^2	4.55	4.335	0.048	2.725×10^4

Moreover, the total energy consumed in the full control, increment-based approach, SSA1, SSA2, WOBI1 and WOBI2 is 8.931×10^4, 8.783×10^4, 8.801×10^4, 8.832×10^4, 8.827×10^4 and 8.828×10^4 kN·m, respectively.

Two more ground excitations, the Northridge earthquake, with a peak acceleration of 0.15 g, and the Hachinohe earthquake, with a peak acceleration of 0.55 g, are considered for further investigation of the robustness of the control system. The corresponding results are given in Table 11.3. It can be seen that the control performance of the proposed approach is close to that of the SSA and WOBI methods. Although better control performance can be achieved in the case of full control, the corresponding total energy consumption and the number of control devices in this case are the highest. Furthermore, the purchase, installation and maintenance of 20 control devices in the full control would be much more expensive than the case of only 10 control devices involved in the proposed control system. From this point of view, it can be concluded that the proposed approach is relatively cost-effective in terms of both energy consumption and the number of control devices.

NOTATION

CP	Contribution percentage index
G	Optimal control gain matrix
H_c	The matrix denoting the location of the control force
K_{k+1}^{KF}	Optimal Kalman gain matrix
M, K, C	The mass, stiffness and damping matrix of the building structure, respectively
$P_{k+1|k}$	The priori error covariance matrix
$P_{k+1|k+1}$	The posteriori error covariance matrix
Q	The vector of modal coordinates
Q_1, R_1	The variance matrices of process noise and measurement noise, respectively
R_e, R_m	The signal noise matrices with different dimensions
U	Control force
w_k, v_k	The process noise and measurement noise, respectively
$\ddot{x}_g$	Ground acceleration

$\mathbf{x}, \dot{\mathbf{x}}, \ddot{\mathbf{x}}$	The displacement, velocity and acceleration response vector, respectively
$\mathbf{X}$	System state composed of displacement and velocity
$\hat{\mathbf{X}}_{k+1\vert k}$	The priori state estimate
$\hat{\mathbf{X}}_{k+1\vert k+1}$	The posteriori state estimate
$\mathbf{Y}_e$	The real structural responses at the locations of interest
$\hat{\mathbf{Y}}_e$	The reconstructed structural responses at the locations of interest
$\mathbf{Y}_m$	The measured structural responses from the sensors
$\mathbf{Z}$	System state in the modal coordinate
$\delta, \bar{\delta}$	Reconstruction error and the corresponding normalised reconstruction error, respectively
$\mathbf{\Delta}, \bar{\mathbf{\Delta}}$	The asymptotic covariance matrix of the reconstruction error and the corresponding normalised matrix, respectively
ΔJ	The increment of the performance index
Δt	Time interval
$\mathbf{\xi}_s, \mathbf{\omega}_s$	The modal damping ratio matrix and modal frequency matrix with respect to the selected modes of vibration, respectively
σ_a^2	The measurement noise variance matrix of acceleration responses
$\bar{\sigma}_{max}$	The maximum estimation error
σ_{max}	The preset allowable error
$\mathbf{\Phi}_s$	The selected mass-normalised displacement mode shape matrix

REFERENCES

Abdullah, M.M., A. Richardson, and J. Hanif. 2001. Placement of sensors/actuators on civil structures using genetic algorithms. *Earthq. Eng. Struct. Dyn.*, 30(8): 1167–84.

Agrawal, A.K. and J.N. Yang. 1999. Optimal placement of passive dampers on seismic and wind-excited buildings using combinatorial optimization. *J. Intell. Mater. Syst. Struct.*, 10(12): 997–1014.

Ambrosio, P., F. Resta, and F. Ripamonti. 2012. An H2 norm approach for the actuator and sensor placement in vibration control of a smart structure. *Smart Mater. Struct.*, 21: 125016.

Brown, A.S., S. Ankireddi, and H.T.Y. Yang. 1999. Actuator and sensor placement for multiobjective control of structures. *J. Struct. Eng.*, 125(7): 757–65.

Bruant, I., G. Coffignal, F. Lene, and M. Verge. 2001. A methodology for determination of piezoelectric actuator and sensor location on beam structures. *J. Sound Vibr.*, 243(5): 861–82.

Bruant, I. and L. Proslier. 2005. Optimal location of actuators and sensors in active vibration control. *J. Intell. Mater. Syst. Struct.*, 16(3): 197–206.

Cha, Y.J., A.K. Agrawal, Y. Kim, and A.M. Raich. 2012. Multi-objective genetic algorithms for cost-effective distributions of actuators and sensors in large structures. *Expert Syst. Appl.*, 39(9): 7822–33.

Cha, Y.J., Y. Kim, A.M. Raich, and A.K. Agrawal. 2013a. Multi-objective optimization for actuator and sensor layouts of actively controlled 3D buildings. *J. Vibr. Control*, 19(6): 942–60.

Cha, Y.J., A. Raich, L. Barroso, and A. Agrawal. 2013b. Optimal placement of active control devices and sensors in frame structures using multi-objective genetic algorithms. *Struct. Control Health Monit.*, 20(1): 16–44.

Frecker, M.I. 2003. Recent advances in optimization of smart structures and actuators. *J. Intell. Mater. Syst. Struct.*, 14(4–5): 207–16.

Gawronski, W. 1997. Actuator and sensor placement for structural testing and control. *J. Sound Vibr.*, 208(1): 101–9.

Güney, M. and E. Eşkinat. 2008. Optimal actuator and sensor placement in flexible structures using closed-loop criteria. *J. Sound Vibr.*, 312(1–2): 210–33.

Gupta, V., M. Sharma, and N. Thakur. 2010. Optimization criteria for optimal placement of piezoelectric sensors and actuators on a smart structure: A technical review. *J. Intell. Mater. Syst. Struct.*, 21(12): 1227–43.

Hac, A. and L. Liu. 1993. Sensor and actuator location in motion control of flexible structures. *J. Sound Vibr.*, 167(2): 239–61.

Haftka, R.T. and H.M. Adelman. 1985. Selection of actuator locations for static shape control of large space structures by heuristic integer programming. *Comput. Struct.*, 20(1–3): 578–82.

He, J., Y.L. Xu, C.D. Zhang, and X.H. Zhang. 2015. Optimum control system for earthquake-excited building structures with minimal number of actuators and sensors. *Smart Struct. Syst.*, 16(6): 981–1002.

Hiramoto, K., H. Doki, and G. Obinata. 2000. Optimal sensor/actuator placement for active vibration control using explicit solution of algebraic Riccati equation. *J. Sound Vibr.*, 229(5): 1057–75.

Ibidapo-Obe, O. 1985. Optimal actuators placements for the active control of flexible structures. *J. Math. Anal. Appl.*, 105(1): 12–25.

Lee, Y.S. 2011. Comparison of collocation strategies of sensor and actuator for vibration control. *J. Mech. Sci. Technol.*, 25(1): 61–8.

Nestorović, T. and M. Trajkov. 2013. Optimal actuator and sensor placement based on balanced reduced models. *Mech. Syst. Signal Proc.*, 36(2): 271–89.

Padula, S.L. and R.K. Kincaid. 1999. Optimization strategies for sensor and actuator placement. Report NASA/TM-1999-209126, National Aeronautics and Space Administration Langley Research Center, Langley, Virginia 23681, USA.

Richardson, A. and M.M. Abdullah. 2002. Sensor/actuators placement on civil structures using a real-coded genetic algorithm. In: *Proceedings of SPIE 4696, Smart Structures and Materials 2002: Smart Systems for Bridges, Structures, and Highways*, San Diego, CA. ed. S.C. Liu and D.J. Pines, 244. doi:10.1117/12.472560

Spencer Jr, B.F., S.J. Dyke, and H.S. Deoskar. 1998. Benchmark problems in structural control: Part II–active tendon system. *Earthq. Eng. Struct. Dyn.*, 27(11): 1141–7.

Zhang, R.H. and T.T. Soong. 1992. Seismic design of viscoelastic dampers for structural applications. *J. Struct. Eng.*, 118(5): 1375–92.

Zhang, X.H. 2012. Multi-sensing and multi-scale monitoring of long-span suspension bridges. PhD diss., Department of Civil and Environmental Engineering, The Hong Kong Polytechnic University, Hong Kong.

Part III

Functions of smart civil structures

Part III of the book contains nine chapters from Chapters 12 through 20. Chapters 12 and 13, respectively, introduce the two main functions of smart civil structures: structural damage detection and structural vibration control. The synthesis of structural health monitoring and vibration control in the frequency domain and the time domain is investigated in Chapters 14 and 15, respectively. Chapter 16 introduces the study on energy harvesting for structural health monitoring and vibration control. Chapter 17 investigates the synthesis of energy harvesting, structural control and health monitoring. The research on the synthesis of structural self-repairing and health monitoring is introduced in Chapter 18. Chapter 19 describes the synthesis of structural life cycle management and health monitoring. Finally, the challenges and prospects for smart civil structures are highlighted in Chapter 20.

Chapter 12

Structural damage detection

12.1 PREVIEW

Many civil structures in service are in fact deficient due to many factors. They deteriorate due to environmental corrosion and long-term fatigue after many years in service. They may degrade due to strong winds, severe earthquakes, terrorist attacks and other abnormal events. The failure of deficient civil structures could be catastrophic in terms not only of loss of life and the economy, but also of subsequent social and psychological impacts. It is therefore imperative to detect early damage and enable maintenance of the structure prior to its complete failure and consequently, to save lives and assets. Damage detection has been a challenging task in structural health monitoring (SHM) technology, particularly for large civil structures subject to multiple loadings. Although many problems in damage detection of civil structures have not been solved, this chapter introduces the basic concepts and current status of structural damage detection with a relatively complete coverage of the subject.

After presenting the general concepts of structural damage detection, this chapter first introduces non-destructive testing (NDT) methods, which are commonly used damage detection methods for civil structures and are regarded as the local approaches. This chapter then presents dynamic characteristics-based damage detection methods, which are often regarded as the global approaches. In recognition of the shortcomings of dynamic characteristics-based damage detection methods, dynamic response–based damage detection methods have been developed, which directly utilise the fact that the measured structural responses are direct functions of the physical properties of the structure. Following these three methods, this chapter further introduces multi-scale damage detection methods by considering both global and local structural responses and using multiple types of sensors. Last but not least, this chapter presents statistical approaches with consideration of the uncertainties involved in structural damage detection. The fatigue damage prognosis and life-cycle management of civil structures will be addressed in Chapter 19.

12.2 INTRODUCTION TO STRUCTURAL DAMAGE DETECTION

In general, damage can be defined as changes introduced into a system that adversely affect its current or future performance, such as the degradation of stiffness or the looseness of connectivity of a structure. In terms of the length scale of damage, all damage begins at a material level, which can be referred to as a *defect* or *flaw*. Under certain loading scenarios, the defects or flaws grow at various rates to cause structural component damage. At this stage, the damage does not necessarily imply a total loss of structural functionality, but rather, that the structure is no longer operating in its normal manner. As the damage further grows and accumulates, it will reach a point, namely failure, where it affects the structure such that its operation is no longer acceptable to the user. Although recent improvements in design methodologies and advanced construction technologies have increased the reliability and safety of structures, it is still not possible to build structures with no probability of failure. In fact, as soon as they are built, civil structures start to deteriorate due to environmental corrosion and fatigue. In 2005, the American Society of Civil Engineers (ASCE) estimated that between one-third and one-half of infrastructures that were built 30 or 40 years ago in the United States are structurally deficient, and an investment of about US$1.6 trillion over a 5 year period is needed (ASCE 2005). A huge number of large and complex civil structures, such as long-span bridges, high-rise buildings and large-space structures, have been constructed in China during the past 20 years. It can be predicted, according to the ASCE's experience, that enormous cost and effort will be required for the maintenance of deficient structures in China in the next 10 or 20 years (Chang et al. 2009). Therefore, accurate and reliable damage detection methods are required for cost-effective maintenance of civil structures.

The commonly used damage detection methods for civil structures are visual inspection and NDT techniques, such as acoustic or ultrasonic methods, magnetic field methods, radiographs, eddy current methods and thermal field methods. These NDT techniques require that the vicinity of the damage is known a priori and that the portion of the structure being inspected is readily accessible. Subject to these limitations, these NDT methods can only detect damage on or near the surface of the structure. The need for better damage detection methods that can be applied to complex structures by means of the SHM system has led to many other damage detection methods. These include dynamic characteristics-based damage detection methods, which have been very popular for the past three decades. These methods are based on the premise that the measured dynamic characteristics of a structure are functions of the physical properties of the structure, and that changes in the physical properties will cause detectable changes in the dynamic characteristics. The dynamic characteristics-based damage detection methods are usually regarded as the global methods. Although these methods have demonstrated various degrees of success, the damage detection of civil structures still remains a challenging task. The main obstacles to the dynamic characteristics-based methods include the insensitivity to local damage and the high sensitivity to measurement noise. In recognition of the shortcomings involved in the dynamic characteristics-based methods, dynamic response-based damage detection methods have been developed. The fundamental principle behind the dynamic response-based methods is that the measured structural responses are direct functions of the physical properties of the structure and that the damage-induced structural responses can be captured and reflected by the defined damage indices even without the extraction of structural dynamic characteristics. The wavelet transform (WT) and the Hilbert–Huang transform (HHT) are two kinds of widely used techniques in dynamic response-based damage detection methods. In consideration of the fact that dynamic structural responses can be on multiple scales and measured by multiple types of sensors, such as strain gauges and accelerometers, multi-scale damage detection methods have also been developed. However, the procedures for damage detection in civil structures often involve a significant number of uncertainties. Consequently, the damage detection results are, in effect, of uncertainties. If the uncertainties are large, the compensation for the damage detection will distort the actual results and lead to false damage identification. It is thus imperative to analyse the source of the uncertainties, quantify the uncertainties and their effects, and evaluate the reliability of the damage detection results. It is also worth mentioning that the model updating methods introduced in Chapter 7 can also be used for damage detection. This is because model updating is intrinsically linked with damage detection. Both aim to determine the difference between two models using measurement data. For model updating, the difference is the modelling errors, while for damage detection, it is structural damage.

In recognition of the difficulties of damage detection in civil structures, the damage detection of civil structures can be performed at four levels (Rytter 1993):

Level 1: Determination that damage is present in the structure
Level 2: Determination of the geometric location of the damage
Level 3: Quantification of the severity of the damage
Level 4: Prediction of the remaining service life of the structure

Most of the damage detection methods developed to date limit themselves to Levels 1 through 3. Level 4 requires knowledge associated with the disciplines of structural design, fracture mechanics and structural reliability. The relevant applications are very limited and will be discussed in Chapter 19.

With regard to the algorithms used, damage detection methods can be classified into two main categories: the 'inverse-problem' or 'model-based' approach and the 'data-based' approach (Farrar and Worden 2013). The model-based approach is often implemented by building a physical or numerical model of the structure of interest, as discussed in Chapter 6. Once the model is built, it is usually updated on the basis of measured data from the real structure, as discussed in Chapter 7. The updating step actually adjusts the built model in such a way as to make it conform better to the real structure using data from the real structure. When data from a subsequent monitoring phase become available, and if any deviations from the normal condition are observed, a further update of the model will indicate the location and extent of where structure changes have occurred, and this provides a damage diagnosis. Via model updating methods, both damage location (Level 2) and severity (Level 3) can be identified.

The data-based approach, as the name suggests, does not proceed from a numerical model. One establishes training data from all the possible healthy and damage states of interest for the structure and then uses pattern recognition to assign measured data from the monitoring phase to the relevant diagnostic class label. In order to carry out the pattern recognition, one needs to build a statistical model of the training data, for example, to characterise their probability density functions. This approach depends on the use of machine learning algorithms. This book emphasises the model-based approach, and the data-based approach will be briefly introduced. More details on the data-based approach can be found in Farrar and Worden (2013).

12.3 NON-DESTRUCTIVE TESTING METHODS

Many NDT methods have been developed to detect damage/change to the materials or structural components and to evaluate their condition. The process is also frequently called *non-destructive evaluation* (NDE) or *non-destructive inspection* (NDI). There are a few books published in this area, such as Hellier (2001) and Shull (2002). The American Society for Nondestructive Testing has published a series of handbooks on various NDT methods: leak testing, liquid penetrant testing, infrared and thermal testing, radiographic testing, electromagnetic testing, acoustic emission testing, ultrasonic testing, magnetic testing and visual testing. Interested readers may refer to these handbooks for details. The American Society for Testing and Materials (ASTM) has standardised some NDT methods. It is noted that various methods and techniques, due to their particular nature, may lend themselves well to certain applications while being of little value in other applications. Therefore, choosing the appropriate methods and techniques is important for NDT. In addition, some methods and techniques may need to be combined to obtain a more comprehensive condition assessment.

The following sections will summarise some commonly used NDT techniques suitable for civil structures, including the ultrasonic pulse velocity method, the impact-echo method, the acoustic emission method, the radiography method and the eddy current method. The basic principles, equipment, applications, advantages and limitations of each technique will be described.

12.3.1 Ultrasonic pulse velocity method

The ultrasonic pulse velocity method has been used to assess the quality of concrete for many years. The method is based on the fact that the velocity of a pulse of compressive waves through a medium is a function of the elastic properties and density of the medium. Its applications include estimating concrete strength (Anderson and Seals 1981), determining the homogeneity of concrete, monitoring the setting and hardening process of cement, detecting cracking and deterioration (Knab et al. 1983), and determining the dynamic modulus of elasticity.

The instrument consists of a pulse generator for producing a wave pulse into the concrete and a receiver for sensing the pulse arrival and measuring the travel time of the pulse. Transducers with frequencies of 25–100 kHz are usually used for testing concrete. High-frequency transducers (above 100 kHz) may be used for small-size specimens with relatively short path lengths, whereas low-frequency transducers (below 25 kHz) may be used for large specimens with relatively longer path lengths. There are three possible configurations in which the transducers may be arranged: direct transmission, semi-direct transmission and indirect transmission (Naik et al. 2004), as shown in Figure 12.1.

The pulse velocity method is an excellent means for investigating the uniformity of concrete. The test procedure is simple and easy to use. The testing procedures have been standardised by ASTM C597 (ASTM 2009). Besides the concrete properties, other factors, such as transducer contact, temperature of concrete, path length and reinforcing steels, also affect the pulse velocity. Therefore, the pulse velocity method should be used with care, so that the pulse velocity is affected only by the properties of the concrete and other adverse factors can be eliminated.

12.3.2 Impact-echo/impulse-response methods

The impact-echo method uses a mechanical impact to generate a high-energy stress pulse. The stress pulse propagates into the object along spherical wave fronts as P- and S-waves, which are reflected by internal interfaces (for example, flaws) or external boundaries. The reflected waves, or echoes, can be measured by a receiver and used to detect the depth of the flaws. The principle of the impact-echo method is illustrated in Figure 12.2.

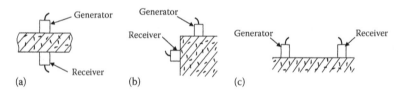

Figure 12.1 Pulse velocity measurement arrangements: (a) direct transmission, (b) semi-direct transmission, (c) indirect transmission.

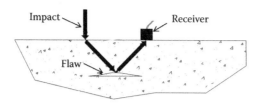

Figure 12.2 Principle of the impact-echo method.

As the energy propagates into the object in all directions, and reflections may arrive from many directions, the impact-echo methods are primarily used for testing piles, whose boundary confines most of the energy within the pile. This is generally referred to as *low strain integrity testing* and standardised by ASTM D5882 (ASTM 2007b). Other than piles, the method has also been applied to detect cracks, voids and delamination in concrete slabs or slab-like structures. The ASTM C1383 Standard (ASTM 2004) specifies the use of the impact-echo method to measure the thickness of slab-like concrete members.

An impact-echo test system is composed of three components: an impact source, a receiving transducer and a data acquisition system with appropriate software for signal analysis and data management. The selection of the impact source is a critical aspect of a successful impact-echo test system. The force-time history of an impact may be approximated as a half-cycle sine curve. The duration of the impact, that is, contact time, determines the frequency content of the stress pulse generated by the impact. A shorter contact time indicates that higher-frequency components are contained. Therefore, smaller defects or shallow defects can be detected. Geophones (velocity transducers) or accelerometers can be used as the receiver.

A variant of the impact method is known as the *impulse-response method*, *transient response method* or *impedance testing method* (Davis and Hertlein 1991). Unlike in the impact-echo method, the time history of the impact force is recorded in the impulse-response method by an instrumented hammer. Given the time history of the impact force and the structural response, the transfer function can be calculated. The transfer function represents the characteristics of a structure, including material properties, geometry, support conditions and the existence of defects. It is noted that velocity is usually measured, and the corresponding transfer function is known as *mobility*. The impulse-response method is primarily applied for testing the integrity of piles. Other successful applications have also been reported: for example, Davis et al. (1997) evaluated the integrity of a concrete tank using the impulse-response method and other NDT methods.

12.3.3 Acoustic emission method

Acoustic emission refers to the phenomenon that stress waves are generated when a material undergoes a rapid release of energy, for example, cracking. These waves can be picked up by sensors attached to the surface of the material and used to evaluate the health condition of the material. The acoustic emission method has been widely used for highway structural assessment, particularly monitoring cracking and crack development, debonding and corrosion (Rens et al. 1997; Yoon et al. 2000). The method is also effective for monitoring the wire fracture in stay cables, main cables and pre-stressed tendons (Elliot 1996; Li and Ou 2008; Vogel et al. 2006). A recent review on acoustic emission monitoring of bridges is presented by Nair and Cai (2010).

The selection of appropriate transducers is based on the purpose and sensitivity required for the investigation. Resonant sensors are preferable, as they are highly sensitive to typical acoustic emission sources. In bridge monitoring, unidirectional sensors and sensors more sensitive to in-plane wave modes may be beneficial (Nair and Cai 2010).

The acoustic emission method is applicable for local, global and continuous monitoring purposes without interrupting traffic over bridges. It can detect and locate flaws but cannot determine the size of flaws. Quantitative analysis of damage is still difficult in practical bridge applications. In addition, extraneous noise may provide uncertain results in real bridges. For example, slight movement of bolted joints can also generate acoustic signals and may cause false damage detection (Chang and Liu 2003).

12.3.4 Radiographic method

Radiography is typically composed of a radiation source and an image collector. Radiation passes directly through a test specimen and exposes the film on the other side of the specimen. Different materials may attenuate the radiation differently. For example, steel attenuates x-rays and gamma rays much more than concrete does. The differences in attenuation can therefore yield a photographic image of the internal structure of the specimen. X-rays and gamma rays are typically used as the radiation source. The radiographic methods are primarily used for examining welded products and castings for defects.

A more powerful technique called *computed tomography* could be used to produce a three-dimensional representation of the internal structure of an object, such as dimensions, shape, internal defects and density. The object is essentially radiographed at various orientations, and then a computer is used to construct the three-dimensional image.

Conventional radiographic techniques can provide rapid and accurate information on the internal characteristics that is not available via other NDT methods. However, the equipment is generally heavy, and the power consumption is large. Power sources of 250 kV–4 MV are often needed to penetrate the thick and dense materials used in civil structures (Chang and Liu 2003). Portability of the equipment and accessibility of the object are two major problems for field implementation.

12.3.5 Eddy current method

Eddy current testing uses electromagnetic induction to detect flaws in conductive materials. In this method, a circular probe coil carrying current is placed in proximity to the test specimen. The alternating current in the coil generates a changing magnetic field, which induces eddy currents in the test specimen. The presence of a flaw will cause a change in the eddy current, which can be sensed by the probe coil. Eddy current testing is primarily used for surface or subsurface crack detection. It is also used for detecting corrosion in thin materials and measuring the thickness of paints and other coatings.

Eddy current instruments have a large variety of configurations depending on the application. A basic eddy current testing instrument consists of an alternating current source, a coil of wire connected to this source and a voltmeter/ammeter to measure the voltage/current change across the coil. An appropriate coil is the most important part to achieve accurate signals from the probe.

Eddy current testing can detect very small cracks in or near the surface of the material, and is capable of inspecting complex shapes and sizes. The limitations of the method include: (1) only conductive materials can be inspected; (2) the surface of the material must be accessible to the probe; (3) the depth of penetration into the material is limited, as the eddy current density decreases with depth; and (4) flaws such as delamination parallel to the probe are undetectable.

12.3.6 Infrared thermographic method

Infrared radiation has a wavelength longer than visible light, or greater than 700 nm. Any object whose temperature is above $0°K$ ($273.15°C$) radiates infrared energy, which is not visible to the human eye but can be detected by an infrared camera. Infrared thermography is a testing technique measuring the temperature or temperature differences of an object. It can be simply used for inspecting electronic components or mechanical systems, in which a defect usually causes an increase in temperature.

In bridge inspection applications, the temperature measurements may be taken by day or night, as long as heat transfer between the bridge and the environment is taking place. Solid concrete is a reasonably good conductor of heat, and the convection within the concrete can be considered negligible. If the concrete has voids, the conduction paths will be disrupted. The disruptions in the flow of thermal energy lead to temperature differences on the surface, which can be detected by the infrared camera. Infrared thermography has been found to be an economical and accurate method for the determination of pavement and bridge deck conditions (Zachar and Naik 1992). The procedures are standardised by ASTM D4788 (ASTM 2007a).

Various parameters affect the surface temperature measurements: solar radiation, cloud, ambient temperature, wind speed and surface moisture (Weil 2004). Therefore, thermographic testing should be carried out on days with no solid cloud cover, with the wind speed below 15 mph and with the surface dry (Kunz and Eales 1985).

A complete thermographic data collection and analysis system includes the infrared sensor head, infrared scanning system, data collection system, and image recording and retrieving devices. The whole system can be installed in a specially equipped van in bridge deck assessment applications (Zachar and Naik 1992).

Infrared thermography is an area-testing rather than a point-testing technique. It is more efficient than other invasive methods when testing large areas. One limitation of the technique is that the depth or thickness of a void cannot be determined.

12.4 DYNAMIC CHARACTERISTICS-BASED DAMAGE DETECTION METHODS

Dynamic characteristics-based damage detection methods are often called *vibration-based damage detection methods*. The change of the name from vibration-based methods to dynamic characteristics-based methods in this book is due to two considerations: (1) the current vibration-based methods are based on dynamic characteristics of civil structures; and (2) there is a distinction between dynamic response-based methods and dynamic characteristics-based methods.

Whereas the NDT methods are regarded as local approaches, dynamic characteristics-based damage detection methods are regarded as global approaches. These methods have been developed on the premise that commonly measured dynamic characteristics, such as natural frequencies, mode shapes and damping ratios, are functions of the physical properties of the structure (mass, stiffness, damping and boundary conditions). Therefore, changes in the physical properties, such as the reduction in stiffness resulting from the onset of cracks or the loosening of a connection, will cause detectable changes in the dynamic characteristics (Doebling et al. 1996). Identifying the damage from the changes in the dynamic characteristics is the main task of dynamic characteristics-based damage detection methods.

The civil engineering community has studied dynamic characteristics-based damage detection since the early 1980s. Doebling et al. (1996) conducted a comprehensive review of dynamic characteristics-based damage detection methods. After that, Sohn et al. (2003) reviewed the literature between 1996 and 2001. More recently, other literature reviews have been published (e.g. Brownjohn 2007; Farrar and Worden 2007; Fassois and Akellariou 2007; Ou and Li 2010; Fan and Qiao 2011). The majority of dynamic characteristics-based methods fall into the frequency domain by using natural frequency, mode shape, damping, frequency response function (FRF), mode shape curvature, modal flexibility and modal strain energy as damage indices, which will be described in the following sections.

12.4.1 Natural frequency changes

Because the natural frequency is the most fundamental dynamic characteristic, the method of directly measuring the shifts in natural frequency (or eigenvalue) has been widely used. Salawu (1997) reviewed the damage detection methods with frequency shifts and the problems faced.

Cawley and Adams (1979) may have been the first researchers to provide a formulation for damage detection from frequency changes before and after damage. For an un-damped system, the eigenvalue equation is

$$\left(-\lambda_i \mathbf{M} + \mathbf{K}\right) \Phi_i = 0 \tag{12.1}$$

where:

$\mathbf{M}$ and $\mathbf{K}$ are the mass and stiffness matrices, respectively
λ_i is the ith eigenvalue
Φ_i is the associated mode shape vector

Assuming that the effect of damage causes a change in the stiffness matrix, $\Delta \mathbf{K}$, while the mass is unchanged, Equation 12.1 then becomes

$$\left[-\left(\lambda_i + \Delta\lambda_i\right)\mathbf{M} + \left(\mathbf{K} + \Delta\mathbf{K}\right)\right]\left(\Phi_i + \Delta\Phi_i\right) = 0 \tag{12.2}$$

where $\Delta\lambda_i$ and $\Delta\Phi_i$ are the changes in eigenvalue and mode shape vector, respectively, due to the damage. Left-multiplying Φ_i^T and substituting Equation 12.1 into Equation 12.2, the following equation can be derived by disregarding the high-order terms:

$$\Delta\lambda_i = \frac{\Phi_i^T \Delta\mathbf{K}\Phi_i}{\Phi_i^T \mathbf{M}\Phi_i} \tag{12.3}$$

where the superscript T represents the transposed vector. Then, the ratio between eigenvalue shifts for modes i and j, $\Delta\lambda_i/\Delta\lambda_j$, can be obtained from the undamaged state of the structure. An error index is defined as the difference between the analytical $\Delta\lambda_i/\Delta\lambda_j$ and the measured one, due to possible damage at position r, that is,

$$e_{rij} = \begin{cases} \dfrac{\left(\Delta\lambda_{ri}/\Delta\lambda_{rj}\right)^A}{\left(\Delta\lambda_{ri}/\Delta\lambda_{rj}\right)^E} - 1, & \text{if} \quad \left(\Delta\lambda_{ri}/\Delta\lambda_{rj}\right)^A \geq \left(\Delta\lambda_{ri}/\Delta\lambda_{rj}\right)^E \\ \\ \dfrac{\left(\Delta\lambda_i/\Delta\lambda_j\right)^E}{\left(\Delta\lambda_{ri}/\Delta\lambda_{rj}\right)^A} - 1, & \text{if} \quad \left(\Delta\lambda_{ri}/\Delta\lambda_{rj}\right)^A < \left(\Delta\lambda_i/\Delta\lambda_j\right)^E \end{cases} \tag{12.4}$$

where the subscripts E and A represent the experimental and analytical data, respectively. The total error, with the assumption of the damage at position r, is the sum of the errors in all the mode pairs. The lowest error indicates the location of the actual damage.

Stubbs et al. (1990) and Stubbs and Osegueda (1990a, 1990b) presented a sensitivity method for damage identification that is based on the work by Cawley and Adams (1979). Hearn and Testa (1991) developed a similar damage detection method at the elemental level. Friswell et al. (1994) developed Cawley and Adams's work. Other works using frequency changes include Rizos et al. (1990), Narkis (1994), Choy et al. (1995) and Xu et al. (2004).

The effects of temperature on structural natural frequencies were discussed by Adams and Coppendale (1976). This factor was considered in Adams et al. (1991) to detect various damages in a space-truss structure using natural frequency changes.

The advantages of damage detection methods based on frequency changes are that the frequency can be measured using very few sensors and has relatively higher precision than other parameters. However, frequencies are not spatially specific and are not very sensitive to damage.

12.4.2 Mode shape changes

In this method, the modal assurance criterion (MAC) and its variations are usually used. The MAC value gives an indication of the closeness between two sets of mode shapes. It is necessary to identify the correlated mode pairs of structural model and experiments, or mode pairs before and after damage. The COMAC (COordinate MAC) is related to the degrees of freedom (DOFs) of the structure rather than to the mode numbers. They are expressed as

$$MAC\left(\Phi_i,\Phi_j\right) = \frac{\left|\Phi_i^T\Phi_j\right|^2}{\left(\Phi_i^T\Phi_i\right)\left(\Phi_j^T\Phi_j\right)} \tag{12.5}$$

$$COMAC\left(\Phi^u,\Phi^d,q\right) = \frac{\left(\sum_{i=1}\left|\left(\Phi_i^u\right)_q\left(\Phi_i^d\right)_q\right|\right)^2}{\left(\sum_{i=1}\left(\Phi_i^u\right)_q^2\right)\left(\sum_{i=1}\left(\Phi_i^d\right)_q^2\right)} \tag{12.6}$$

where q is one DOF, and the superscripts 'u' and 'd' represent the undamaged and damaged states, respectively.

For a good mode correlation or coordinate correlation, the value should be near to 1.

COMAC has been shown to be capable of damage location (Kim et al. 1992). Ko et al. (1994) also pointed out that COMAC could be used to locate the damage, but MAC could not. Salawu and Williams (1995) conducted a full-scale test on a multi-span reinforced concrete highway bridge before and after structural repairs. MAC and COMAC were used to indicate the presence and location of repairs. The authors suggested that a MAC threshold value of 0.8 and at least 5% change in frequency would imply the presence of damage with confidence.

It is simple to use COMAC for detecting damage or locating error in a structural model, but it has less physical basis as compared with other methods. It is advisable to use this method in cases where the structure is tested and modelled in a free-free configuration (Maia et al.

1997). Direct comparison of the pre- and post-damage mode shapes has been found ineffective in identifying the damaged region unless damage is severe (Farrar and Jauregui 1996).

12.4.3 Modal damping changes

As the extraction of structural modal damping ratios is usually not as accurate as frequencies and mode shapes, modal damping ratio is not a commonly used damage indicator. Brownjohn and Steele (1979) may be among the first researchers detecting damage with damping ratios. Salane and Baldwin (1990) tested a composite bridge model and a composite highway bridge. They found that modal damping ratios in the laboratory model decreased after the flange was cut, whereas those in the bridge increased initially and subsequently decreased. Ndambi (2002) investigated the principle of structure modal damping ratio and used it as a damage index for pre-stressed concrete beam structures. Keye (2006) detected delamination damages in a carbon fibre-reinforced polymer composite panel using damping ratios. With the Rayleigh damping assumption, the damping ratio changes between the undamaged and damaged models were calculated, and the maximum correlation with the measured damping ratio changes indicated the possible damage.

12.4.4 FRF changes

Ibrahim (1993) used an error indicator computing the Euclidean norm of the FRF vectors measured at discrete frequencies as

$$e\left(\mathbf{H}_{ij}\right) = \frac{\left\|\mathbf{H}_{ij}^{A} - \mathbf{H}_{ij}^{E}\right\|}{\left\|\mathbf{H}_{ij}^{A}\right\|} \tag{12.7}$$

where $\mathbf{H}$ is the FRF matrix. Pascual et al. (1996) proposed a correlation index to measure the closeness between the measured and analytical FRFs, similarly to the MAC technique, using the following frequency domain assurance criterion (FDAC):

$$FDAC\left(\omega_{A},\omega_{E},j\right) = \frac{\left|\left\{\mathbf{H}_{j}^{A}\left(\omega_{A}\right)\right\}^{T}\left\{\mathbf{H}_{j}^{E}\left(\omega_{E}\right)\right\}\right|^{2}}{\left(\left\{\mathbf{H}_{j}^{A}\left(\omega_{A}\right)\right\}^{T}\left\{\mathbf{H}_{j}^{A}\left(\omega_{A}\right)\right\}\right)\left(\left\{\mathbf{H}_{j}^{E}\left(\omega_{E}\right)\right\}^{T}\left\{\mathbf{H}_{j}^{E}\left(\omega_{E}\right)\right\}\right)} \tag{12.8}$$

where ω denotes the circular frequency.

Samman et al. (1991) investigated the change in FRF signals of a scaled highway bridge model caused by cracks in the girders. A pattern recognition method was introduced by utilising the integer slope and curvature values of FRF wave forms, rather than peak magnitudes. Only one FRF reading per girder was required to detect and locate relatively minor cracks. Wang and Liou (1991) presented a new method to identify joint parameters using the two sets of measured FRFs of a substructure with and without the effect of joints. Some strategies were applied to overcome the measurement noise problem, which might result in false identification. Numerical simulation and experiments verified the accuracy of the proposed technique. Biswas et al. (1994) developed the modified chain-code method for the rapid detection of a small fault in a structure. Slope and curvature-based signatures were derived from the averaged composite FRF signature. Comparison of the intact signatures with the cracked signatures can detect cracks as small as 4 mm in hammer vibration tests of a bridge model consisting of three steel girders supporting a concrete deck. The method was robust even when noisy data were present. Xia et al. (2007) adopted two damage indicators based on FRFs to detect possible damage to shear connectors in a slab-girder bridge model. Their method is based on the fact that damage to the shear connectors leads to the slab separating from the girders to a certain extent, and so the nearby points on the slab respond differently from those on the girders. One damage index is similar to Equation 12.8, in which the vertical FRFs of the girders and those of the slab were employed. The other damage index is the Euclidean norm of difference of the FRFs. Both indicators could be used to identify damage in shear connectors accurately and consistently. One advantage of the method is that it is a reference-free method; that is, the undamaged data are not required. In addition, it is not affected by environmental variation, as the slab and girders were measured simultaneously and under identical environmental conditions.

In the FRF-based methods, however, the input information is required to form the FRFs. This is quite difficult for large civil structures.

12.4.5 Mode shape curvature changes

An alternative to obtaining spatial damage information is using mode shape derivatives, such as curvature, instead of mode shapes. The logic is based on the fact that a decrease in the flexural stiffness causes an increase in curvature. The mode shape curvatures are obtained by using a central difference approximation as

$$\phi''_{q,i} = \frac{\phi_{q+1,i} - 2\phi_{q,i} + \phi_{q-1,i}}{h^2} \tag{12.9}$$

where:

  $\phi_{q,i}$ is the ith modal displacement at measurement point q
  h is the length of elements

Pandey et al. (1991) demonstrated that the absolute change in mode shape curvatures could be a good indicator of damage for beam structures. Chance et al. (1994) detected artificial cracks in a beam and concluded that curvatures were locally sensitive to the fault, but mode shapes were not. It is noted that the accuracy of this method is subject to numerical estimation difficulties resulting from the need for differentiation. Moreover, it is error sensitive, because a small noise in the mode shapes may lead to a very different result.

12.4.6 Modal strain energy changes

The strain energy of a Bernoulli–Euler beam is given by Stubbs et al. (1995):

$$U = \frac{1}{2} \int_0^L EI \left(\frac{\partial^2 w}{\partial x^2} \right)^2 dx \tag{12.10}$$

where:

  EI is the flexural rigidity of the beam
  w denotes the transverse displacement
  L is the total length of the beam

For a particular ith mode shape, $\Phi_i(x)$, the jth member contribution to the mode is

$$U_{ij} = \frac{1}{2} \int_{a_j}^{a_{j+1}} (EI)_j \left(\frac{\partial^2 \Phi_i}{\partial x^2} \right)^2 dx \tag{12.11}$$

where a_j and a_{j+1} are the nodal coordinates of the beam element. The fractional energy of the element can be used as a damage indicator. Cornwell et al. (1997, 1999) extended it to plate-like structures characterised by two-dimensional curvatures.

Shi et al. (1998) defined the modal strain energy (MSE) of the jth element and ith mode as

$$MSE_{ij} = \mathbf{\Phi}_i^T \mathbf{K}_{e,j} \mathbf{\Phi}_i \tag{12.12}$$

where $\mathbf{K}_{e,j}$ denotes the jth elemental stiffness matrix. The change ratio of MSE before and after damage was used as an indicator of the damage location.

Although damage detection methods based on mode shape changes and their derivatives can provide spatial information regarding the location of structural damage, they have several limitations in application. First, a dense array of measurement points is required for an accurate estimate of mode shapes and mode shape curvatures. Second, the mode shape has larger statistical variation than modal frequencies. Third, the mode shape curvature methods are not readily applicable to structures with complex configurations. Finally, it is necessary to select a mode shape, yet it is a priori unknown which mode suffers from a significant change due to a particular damage.

12.4.7 Flexibility changes

With mass-normalised mode shapes, the modal flexibility matrix can be obtained as

$$\mathbf{F} = \mathbf{\Phi} \mathbf{\Lambda}^{-1} \mathbf{\Phi}^T = \sum_{i=1}^{n} \omega_i^{-2} \, \mathbf{\Phi}_i \, \mathbf{\Phi}_i^T \tag{12.13}$$

where:
  $\mathbf{F}$ is the flexibility matrix
  $\mathbf{\Phi}$ is the complete mode shape matrix
  $\mathbf{\Lambda}$ $= \mathrm{diag}(\omega_i^2)$
  ω_i is the ith circular frequency

From Equation 12.13, it can be seen that the mode contribution to the flexibility matrix decreases as the frequency increases. Therefore, a good estimate of the flexibility matrix can be made from a few lower modes rather than the higher-frequency modes, which are difficult to measure in practice.

Raghavendrachar and Aktan (1992) demonstrated that flexibility was more sensitive than frequencies and mode shapes to local damage. Pandey and Biswas (1994) proposed a method based on flexibility changes in damage detection of a free-free steel beam. This method was applied to detect artificial cuts in the I-40 bridge over the Rio Grande in New Mexico, together with several other methods, by Farrar and Jauregui (1996). The results demonstrated that this method was not successful in practice.

12.4.8 Comparison studies

Wang and Zhang (1987) investigated the sensitivity of the dynamic characteristics to faults in structures so as to choose sensitive characteristics as the observing properties. The results showed that modal parameters (damped frequencies, natural frequencies and mode shapes) were not sensitive to structural damage, but FRFs were when the excitation frequency was near the natural frequencies.

Five damage detection algorithms, including the 1-D strain energy method (Stubbs et al. 1995), the mode shape curvature method (Pandey et al. 1991), the flexibility change method (Pandey and Biswas 1994), a method combining mode shape curvature and flexibility change (Zhang and Aktan 1995) and the stiffness change method (Zimmerman and Kaouk 1994), were applied to detect artificial cuts in the I-40 bridge over the Rio Grande in New Mexico (Farrar and Jauregui 1996). In general, all methods could identify the damage location correctly for the severest damage, one cut from the mid-web completely through the bottom flange. The methods were inconsistent and did not clearly identify the damage location when they were applied to the three less severe damage cases. Detection results generally showed that the strain energy method performed best.

Cornwell et al. (1998) applied two non-model-based methods to detect damage. One is based on the changes in flexibility and the other on the changes in the strain energy. They were tested on an aluminium free-free I-beam and a clamped aluminium plate before and after the cut. The results with the two methods showed that changes for serious damage cases could be located successfully, but smaller levels of damage could not. Zhao and DeWolf (1999) reported that modal flexibilities were more sensitive to damage than either natural frequencies or mode shapes.

12.4.9 Challenges in dynamic characteristics-based damage detection methods

Although the dynamic characteristics-based damage detection methods have achieved some degree of success in aerospace and mechanical engineering communities, their application to civil structures is still limited due to several inherent weaknesses.

One issue is the dependence on prior analytical models and/or prior test data. Many dynamic characteristics-based damage detection methods assume that the initial finite element (FE) model of a structure can represent the intact structure. But its accuracy is doubtful, because there are many uncertainties in civil structures, for example, existence of non-structural members, boundary conditions, nonlinear response and damping. The solution to these problems is to update the FE model properly. This needs measurement data in the intact state, which are usually not available in practice.

In addition to FE modelling error, measurement noise is also severe for civil structures. Civil structures with very large size and inertia are typically excited under ambient vibration. The frequency components are generally low (below 100 Hz or even a few hertz), and the bandwidth is relatively narrow compared with aerospace structures. This means that dynamic characteristics-based methods are not sensitive to local damage in civil structures, which has a more significant influence on high modes. Unless the damage becomes very severe, or the sensors are correctly distributed around it, the damage is difficult to detect reliably using limited sensors.

Some kinds of damage in civil structures occur by a long, gradual and slow process. Dynamic characteristics do not have an abrupt change, in general. In addition, civil structures are located under varying environmental conditions, such as temperature and operational loadings. These environmental conditions affect structural dynamic characteristics, on many occasions, more significantly than structural damage does. Although the temperature effect can be eliminated from the relation between the temperature and the dynamic characteristics, there are a few difficulties in practice: (1) the establishment of this relation requires a long period of field measurement; (2) the global dynamic characteristics are associated with the temperature distribution of the entire structure, not only the air temperature and/or surface temperatures (Xia et al. 2011); and (3) the temperature effect is combined with the operational loading effect. Therefore, environmental variation is a critical and complicated factor for reliable damage detection in civil structures. Finally, civil structures are unique in terms of geometry, boundary conditions, configuration and loadings. Therefore, the methods may be applicable to one structure only but not to others.

12.5 DYNAMIC RESPONSE-BASED DAMAGE DETECTION METHODS

Dynamic response-based damage detection methods directly employ structural responses without the extraction of structural dynamic characteristics for damage diagnosis. The fundamental principle behind the dynamic response-based methods is that the measured structural responses are direct functions of the physical properties of the structure and that the damage-induced structural responses can be captured and reflected by the defined damage indices even without the extraction of structural dynamic characteristics. WT and HHT are two kinds of widely used techniques in dynamic response-based damage detection methods, which will be reviewed in this section. The applicability of empirical mode decomposition (EMD) for identifying structural damage caused by a sudden change of structural stiffness (Xu and Chen 2004) will then be presented in detail as an example. Another example to be introduced in this section is a statistical moment-based damage detection method (Zhang et al. 2008; Xu et al. 2009).

12.5.1 Review of response-based damage detection methods using WT and HHT

There are three major damage indices considered in the WT-based methods for damage detection: (1) variation of wavelet coefficients before and after damage; (2) local perturbation of wavelet coefficients; and (3) reflective wave caused by local damage (Kim and Melhem 2004). The main idea behind the use of wavelets for damage detection is based on the fact that the presence of damage, such as cracks, introduces small discontinuities in the measured structural response in the vicinity of the damage location. Often, these discontinuities cannot be observed from direct examination of structural response, but they may be detectable from the distribution and analysis of the wavelet coefficients obtained by the continuous wavelet transform (CWT) or discrete wavelet transform (DWT). Hou et al. (2000) proposed a wavelet-based approach to identify the location and onset of damage in a simple structural model with breakage springs. Sun and Chang (2002) proposed a wavelet-packet-transform-based method for damage assessment, in which dynamic signals measured from a structure were first decomposed into wavelet packet components, whose energies were then calculated and used as inputs into neural network models for damage assessment. Based on a CWT, Li et al. (2006) proposed a damage detection method and applied it to analyse flexural wave in a cracked beam to identify the damage location and extent precisely. Huang et al. (2009) developed a distributed two-dimensional (2-D) CWT algorithm that can use data from discrete sets of nodes and provide spatially continuous variation in the structural response parameters to monitor structural degradation. Some experimental studies were carried out by Wu and Wang (2011) for damage detection of a beam structure with different crack depths by employing spatial WT. Many researchers have also devoted themselves to developing WT-based damage detection methods in bridge engineering with the consideration of a moving load (Zhu and Law 2006; Nguyen and Tran 2010; Hester and González 2012). Literature reviews on this research area can also be found in Kim and Melhem (2004), Ovanesova and Suárez (2004) and Reda Taha et al. (2006).

HHT is an empirically based data-analysis method and consists of two parts: EMD and Hilbert spectral analysis (HSA) (Huang and Shen 2005). Similarly to wavelet analysis,

HHT has been demonstrated to be capable of precisely composing a signal in the frequency-time domain (Huang et al. 1998). Consequently, it is capable of detecting damage-induced discontinuity in the recorded response data and has been used for damage detection by many researchers (Yang et al. 2004; Loutridis 2004; Chen et al. 2007; Hsu et al. 2012). Similarly to WT-based approaches, the basic concept of the HHT-based methods is that sudden breakage of a structural element will cause discontinuity in the response signal measured in the vicinity of the damage location. When the vibration signal is decomposed using EMD, the discontinuity will form a signal feature, termed a *damage spike*, through HSA. The damage time instant can then be identified in terms of the occurrence time of the spike, and the damage location can be determined by the spatial distribution of the observed spikes.

It is noted that dynamic characteristics-based damage detection methods are also based on structural response measurements, because the measured dynamic characteristics are often extracted from the measured structural responses. The cornerstone for distinguishing dynamic characteristics-based methods from response-based methods is to see the damage indices. If the structural dynamic characteristics extracted from the response measurements are used as damage indices, such damage detection methods can be classified as dynamic characteristics-based methods. If the damage indices are obtained directly from the analysis of structural responses, such as the variation of wavelet coefficients or the spike in the response signal, these approaches can be viewed as response-based methods. From this point of view, some WT- or HHT-based methods can also be considered as dynamic characteristics-based methods if dynamic characteristics, such as mode shape or natural frequencies, are extracted by WT or HHT methods and used as damage indices (Yang et al. 2003a,b; Cao and Qiao 2008; Fan and Qiao 2009; Tang et al. 2010).

12.5.2 Experimental investigation of damage detection using EMD

The EMD method, developed by Huang et al. (1998), is a signal processing method, which can decompose any data set into several intrinsic mode functions (IMFs) by a procedure called a *sifting process*. To illustrate the application of EMD for detecting structural damage due to a sudden change in structural stiffness, a comprehensive experimental study has been carried out (Xu and Chen 2004) and will now be described.

A three-story building model was designed and constructed as shown in Figure 12.3. The steel frame of the three-story building consisted of three steel plates of $850 \times 500 \times 25$ mm and four equally sized rectangular columns with a cross section of 9.5×75 mm. The plates and columns were properly welded to form rigid joints. The building model was then welded on a steel base plate of 20 mm thickness. The steel base plate was in turn bolted firmly on the shaking table using a total of eight bolts of high tensile strength. The overall dimensions of the building model were measured as $850 \times 500 \times 1450$ with a clear story-height of 450 mm. All the columns were made of high-strength steel of 435 MPa yield stress and 200 GPa modulus of elasticity. The 9.5×75 mm cross section of the column was arranged in such a way that the first natural frequency of the building was much lower in the x-direction than in the y-direction. Each steel floor could be regarded as a rigid horizontal plate, leading to a shear-type building model in the x-direction. To simulate the inherent energy dissipation capacity of a real structure, dashpots were installed between every two floors. The geometrical scale of the building model was assumed to be 1/5. With additional mass of 135 kg placed on each floor of the building model, the resulting similarity scales for displacement, velocity, acceleration and time were approximately 0.20, 0.64, 2.03 and 0.32, respectively.

To simulate a sudden change of structural stiffness, on each side of the first-floor plate of the building a spring was horizontally installed along the x-direction with one end connected

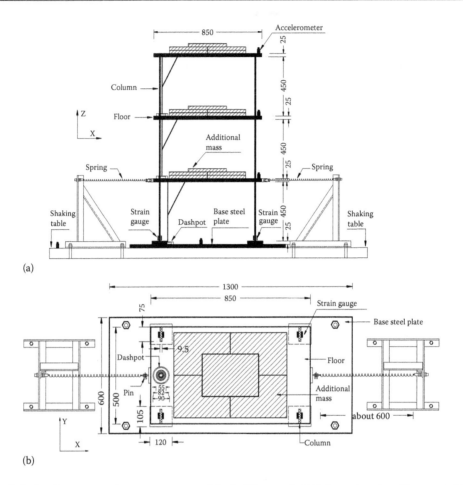

Figure 12.3 Configuration and instrumentation of the building model (all dimensions in millimetres): (a) elevation, (b) plan.

Table 12.1 Stiffness calibration results for each spring group (N/mm)

Spring group1		Spring group2		Spring group3		Spring group4	
Spring k₁	Spring k₂	Spring k₃	Spring k₄	Spring k₅	Spring k₆	Spring k₇	Spring k₈
82	73	54.9	60.6	46.25	44.25	28.81	29.24

to the floor plate in the middle through a pin and the other end connected to a steel frame made by steel angles and fixed on the shaking table (see Figure 12.3). The connecting pin between the floor plate and the spring was so designed that it could be easily and quickly pulled out to release the pretensioned spring during the vibration of the building, thereby simulating a sudden change of building horizontal stiffness on the first floor. The arrangement of springs at the first floor was because damage to a building under seismic excitation most likely occurs in the lowest story.

To simulate different damage severities, four groups of springs were used in the experiment. Each group consisted of two springs of the same length, and the stiffness of the two springs was calibrated and listed in Table 12.1. The first group of springs (spring group 1) had the highest stiffness, and the last group of springs (spring group 4) had the lowest stiffness. The damage severity simulated by releasing each spring group was quantified by the ratio of the total stiffness of each spring group to the story stiffness of the first story. The first-story stiffness was computed as 581.14 N/mm, based on the assumption of a shear building, and as a result, the damage severities simulated by spring groups 1 through 4 were 26.7%, 19.8%, 15.6% and 10.0%, respectively.

To fully utilise the springs for generating different damage patterns, the two springs in each group were released either simultaneously or successively during the vibration of the building. The symbols D1 and D2 are used hereafter to represent the simultaneous damage pattern and the successive damage pattern, respectively. For each test, the two springs were stretched to such an extent as to ensure the springs were always in tension even after one spring was released.

Each building floor was equipped with two B&K 4370 accelerometers in the x-direction. The signals from the accelerometers were transferred to B&K 2635 signal conditioners, by which one of them was converted into a displacement signal using the built-in electronic circuit. Two more B&K 4370 accelerometers were installed on the base plate of the building and on the shaking table, respectively, to measure and cross-check the acceleration in the x-direction as a direct ground motion to the building. At the bottom of each column of the first story of the building, a TML PFC-10-11 strain gauge of 10 mm length was stuck on each side of the column to measure the bending strain of each column. The locations of the accelerometers and strain gauges are illustrated in Figure 12.3.

To accurately record the time instant of releasing the springs (the damage time instant) in the experiment, an electrical circuit was formed for each spring, in which the spring served as an electric resistance and the pin served as a switch. The voltage was supplied to the electrical circuit using a battery, and the signal was conditioned and input into the computer. In such an arrangement, the measured voltage would drastically change at the moment of the pin being pulled out, by which the damage time instant could be accurately recorded. For convenience of description, the two special channels are referred to hereafter as Channels 17 and 18.

The experiment was carried out in two phases with a total of 28 test cases. Phase 1 was a free vibration test in which the building was excited to vibrate by hand shaking on the top floor and then released for free vibration in the x-direction. For each spring group, the two test cases, representing the damage pattern D1 and the damage pattern D2, were conducted and repeated five times with a sampling frequency of 500 Hz. In the repeated tests, the time instant of releasing the springs was different to examine its effects on damage detection. The recording time duration was 40 s for each case. To investigate the effects of sampling frequency, four more test cases were performed on spring group 3 for the damage pattern D1 with the sampling frequency of 100–400 Hz at an interval of 100 Hz. For each sampling frequency, the test case was repeated five times but with different releasing times of the springs.

Phase 2 consisted of random excitation and earthquake simulation tests that were carried out using the unidirectional shake table facility at The Hong Kong Polytechnic University. Two types of input excitation were used in Phase 2: one was white noise random excitation and the other was the El-Centro 1940 earthquake ground acceleration (N-S component). The original El-Centro earthquake time history was first compressed in time by a factor of 3.0 to meet the similitude requirements. The white noise random excitation of frequency range from 0.5 to 30 Hz was automatically generated by the control system of the shaking table. The input peak ground acceleration for both input excitations was scaled to 0.1 g to facilitate the analysis of the measurement results. Eight test cases for the building under the El-Centro earthquake excitation were performed with four spring groups and two damage patterns included. Each test case was repeated four times. For white noise random excitation, six test cases were carried out with four spring groups and two damage patterns considered, and each test case was repeated twice. The sampling frequency for all test cases in

Table 12.2 Summary of test cases

Phase and case	Sampling frequency(Hz)	Spring group	Number of repeated tests
Phase 1: 12 cases Free vibration test	500	Group 1, 2, 3, 4	5 for D1 and 5 for D2 for each spring group
	400	Group 3	5 for D1
	300	Group 3	5 for D1
	200	Group 3	5 for D1
	100	Group 3	5 for D1
Phase 2: 8 cases Random excitation	500	Group 1, 2, 3, 4	2 for D1 and 2 for D2 for each spring group
Phase 2: 8 cases El-Centro earthquake	500	Group 1, 2, 3, 4	3 for D1 and 3 for D2 for each spring group

Phase 2 was set as 500 Hz, with the recording duration being 40 and 60 s for the El-Centro earthquake excitation and the white noise random excitation, respectively. All the test cases are summarised in Table 12.2.

The measured structural response time history from each test case was processed using the EMD approach with intermittency check to obtain the first IMF component with an identical intermittency frequency of 50 Hz. Software (EMD V1.0) issued by Princeton Satellite System Inc. was used to implement the EMD approach. The first IMF components obtained in this way from each test case were used to identify the damage time instant and damage location of the building through observation of the presence and distribution of damage spikes. The results identified from the response time histories were then compared with the damage time instants recorded by Channels 17 and 18, and it was checked whether the damage occurred at the first story.

Let us first consider the damage pattern D1, that is, when the two springs were simultaneously released. Only the test cases with the spring group 3, the damage pattern D1 and the sampling frequency of 500 Hz are discussed in this section based on the structural acceleration response. Figures 12.4 through 12.6 show the measured acceleration response time history at each floor and its first IMF component obtained by EMD for the building under free vibration, white noise random excitation and earthquake excitation, respectively. Figures 12.4 and 12.6 also display the time histories of voltage recorded by Channels 17 and 18 for damage time instants.

It can be seen from Figure 12.4g and h that in this case of free vibration, the two springs were simultaneously released at $t = 19.5$ s. Though it is hard to identify this damage event directly from the acceleration response time histories (see Figures 12.4a–c), it is easy to observe a sharp spike in the first IMF of acceleration response of the first floor (see Figure 12.4f) at time $t = 19.5$ s, which is exactly the moment when the two springs were suddenly released and the damage event occurred. For the test case shown in Figure 12.5 for the building under white noise random excitation, a sharp spike can also be identified in the first IMF of acceleration response of the first floor at $t = 14.7$ s, which is the same as the time instant when the stiffness of the first floor was suddenly changed. The same observation can be made for the building under the El-Centro earthquake excitation. The only difference between the three test cases is the pattern of the damage spike, which is very sharp in the free vibration test but has a short decaying tail in both the white noise random test and the El-Centro earthquake simulation test. This difference is mainly due to the type of excitation and the structural vibration energy suddenly released at the damage time instant. The repeated tests for each test case also demonstrated the same phenomena. Thus, one may

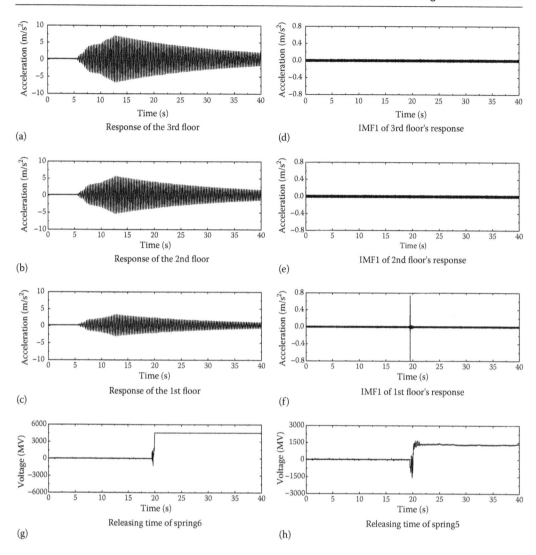

Figure 12.4 Building responses and damage time instants: free vibration (pattern D1).

conclude that the EMD approach can identify the occurrence of a damage event and the damage time instant from the measured acceleration response time history by the identification of a damage spike in its first IMF component of very high frequency. Moreover, for the determination of damage location, it can also be seen from Figures 12.4 through 12.6 that a sharp damage spike appears clearly only in the first IMF of the acceleration response of the first floor. No damage spike emerges in the first IMFs of the acceleration responses of the second and third floors of the building under free vibration, while there are very small spikes appearing in the first IMFs of the acceleration responses of the second and third floors of the building under either random excitation or earthquake excitations. Therefore, by analysis of the distribution of damage spikes along the height of the building, the damage location can be easily identified at the first story of the building. The repeated tests for each test case support this observation.

To assess the potential application of EMD for detecting successive damage events from the measurement data, the test cases of the building with spring group 3 and damage pattern

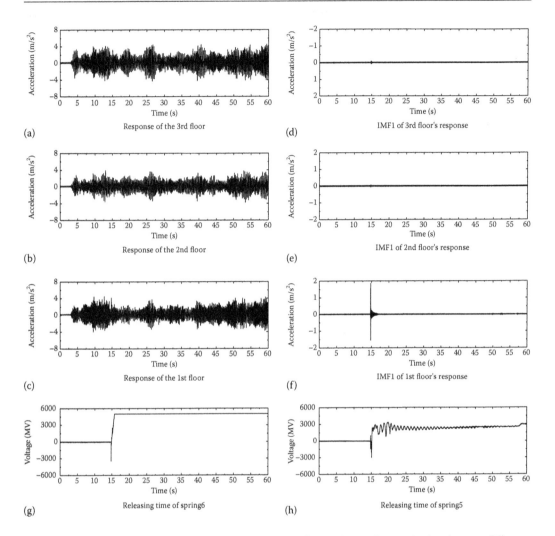

Figure 12.5 Building responses and damage time instants: white noise random excitation (pattern DI).

D2 are investigated in this section as well. Figure 12.7 displays the damage time instants when the two springs were successively released and the acceleration responses of the building under free vibration. It can be seen that the two spikes appearing in the first IMF of the acceleration response of the first floor (see Figure 12.7f) clearly confirm that the two damage events occurred at 16.9 s and 20 s successively. Similar observations and conclusions can be obtained in the cases of white noise random excitation and earthquake excitation. As for the damage localisation, it is straightforward to conclude that the damage occurs at the first story, because the damage spikes appear in the first IMF of the acceleration response of the first floor only, and there are no obvious spikes arising in the first IMFs of the acceleration responses of the second and third floors. The repeated tests for each case also support the statement that the EMD approach can be applied to detect successive damage events and their locations for the building concerned.

It is clear from this discussion that the EMD approach is capable of detecting the damage time instant and damage location using the measured acceleration responses of the building with either a single damage event or successive damage events. To further investigate

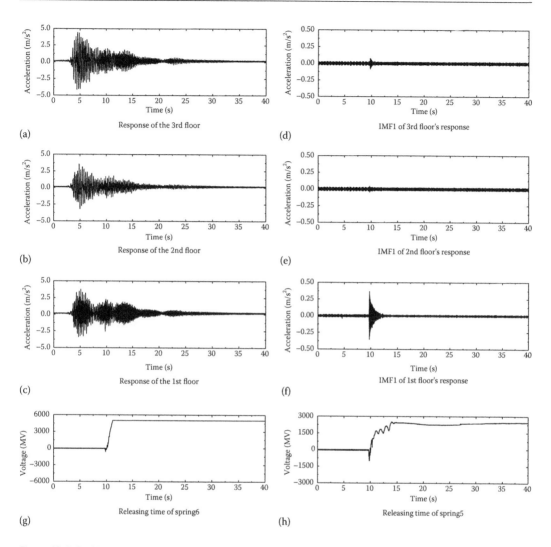

Figure 12.6 Building responses and damage time instants: earthquake excitation (pattern DI).

the performance of EMD for damage detection, several factors, including damage severity, sample frequency and measurement quantity, were investigated. Regarding the effects of damage severity, it seemed that higher damage severity led to higher amplitude of the damage spike. However, since the damage spike was a transient signal feature, its amplitude would easily be affected by many factors. No quantitative relationship can be found at this stage, and further investigation is needed on this aspect. Regarding the effects of sample frequency, it seems that a sampling frequency higher than the frequency of the damage event is necessary for the successful application of the EMD approach for damage detection. Regarding the measurement quantity, it was found that the measurement of an acceleration response other than the displacement and strain responses is necessary for damage detection by EMD. This is because the displacement and strain responses change very little during the abrupt change of structural stiffness compared with the acceleration response. For more details, readers can refer to Xu and Chen (2004).

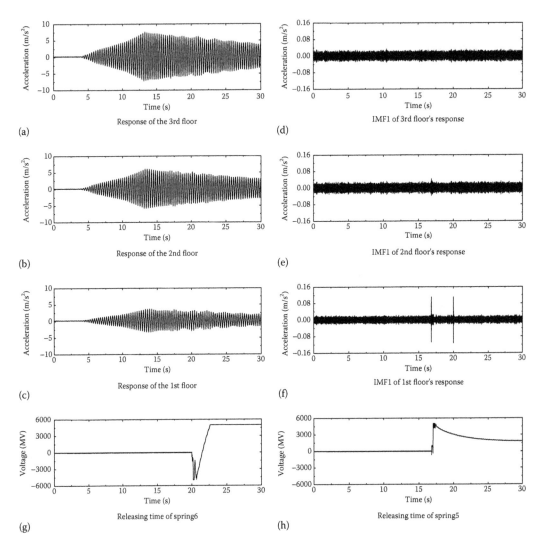

Figure 12.7 Building responses and damage time instants: free vibration (pattern D2).

12.5.3 Statistical moment-based damage detection method

Structural damage, such as stiffness losses in a structure, will cause changes in both the statistical moments and the probability density functions (PDFs) of the structure under random excitation. Therefore, the changes in statistical moments can be used as damage indices. In this regard, the principle of a statistical moment-based damage detection method (Zhang et al. 2008) is put forward in terms of a single-degree-of-freedom (SDOF) system in this section. Further information on the statistical moment-based damage detection method for multi-degrees of freedom (MDOF) systems, as well as the experimental validation of the method, can be found in Xu et al. (2009).

Let us consider a single-story shear building under zero-mean white noise ground acceleration of Gaussian distribution. The equation of motion of the linear-elastic shear building can be expressed as

$$m\ddot{x} + c\dot{x} + kx = -m\ddot{x}_g \tag{12.14}$$

or

$$\ddot{x} + 2\xi\omega_o\dot{x} + \omega_o^2 x = -\ddot{x}_g$$

where:

    m, c and k are the mass, damping and stiffness coefficient of the building, respectively

    x, $\dot{x}$ and $\ddot{x}$ are the relative displacement, velocity and acceleration responses of the building to the ground, respectively

    $\ddot{x}_g$ is the white noise ground acceleration

    ξ is the damping ratio of the building

    ω_o is the circular natural frequency of the building and is equal to $\sqrt{k/m}$

If a structure is a linear system, the power spectrum $S(\omega)$ and the variance σ^2 of the structural response can be obtained by

$$S(\omega) = |H(\omega)|^2 S_f(\omega) \tag{12.15}$$

$$\sigma^2 = \int_{-\infty}^{+\infty} S(\omega)d\omega \tag{12.16}$$

where:

    $S_f(\omega)$ is the power spectrum of ground excitation

    $H(\omega)$ stands for the FRF

If the ground excitation is an ideal white noise, $S_f(\omega)$ is then a constant S_0 over the whole frequency zone from $-\infty$ to $+\infty$. For an SDOF system, the module of the displacement FRF, $H_d(\omega)$, the velocity FRF, $H_v(\omega)$, and the acceleration FRF, $H_a(\omega)$, can be obtained as follows:

$$|H_d(\omega)| = \frac{1}{m\sqrt{(\omega_0^2 - \omega^2)^2 + (2\xi\omega_0\omega)^2}} \tag{12.17}$$

$$|H_v(\omega)| = \frac{\omega}{m\sqrt{(\omega_0^2 - \omega^2)^2 + (2\xi\omega_0\omega)^2}} \tag{12.18}$$

$$|H_a(\omega)| = \frac{\omega^2}{m\sqrt{(\omega_0^2 - \omega^2)^2 + (2\xi\omega_0\omega)^2}} \tag{12.19}$$

Substituting Equation 12.17 into Equation 12.15 and then Equation 12.16 leads to the variance or second-order moment of displacement response σ_d^2:

$$M_2^{dis} = \sigma_d^2 = \int_{-\infty}^{+\infty} |H_d(\omega)|^2 S_f(\omega)d\omega = \frac{\pi S_0}{2\xi\sqrt{mk^3}} \tag{12.20}$$

In a similar way, the variance or second-order moment of velocity, σ_v^2, and acceleration, σ_a^2, can be obtained as follows:

$$M_2^v = \sigma_v^2 = \frac{\pi S_0}{2\xi\sqrt{m^3 k}} \tag{12.21}$$

$$M_2^a = \sigma_a^2 = \frac{\pi S_0 (1 - 4\xi^2)\sqrt{k}}{2\xi\sqrt{m^5}} \tag{12.22}$$

The fourth-order and sixth-order moments of displacement, velocity and acceleration can then be given as

$$M_4^{\text{dis}} = \frac{3\pi^2 S_0^2}{4m\xi^2 k^3} \tag{12.23}$$

$$M_4^v = \frac{3\pi^2 S_0^2}{4m^3\xi^2 k} \tag{12.24}$$

$$M_4^a = \frac{3\pi^2 S_0^2}{4m^5\xi^2}(1 - 4\xi^2)^2 k \tag{12.25}$$

$$M_6^{\text{dis}} = \frac{15\pi^3 S_0^2}{8\xi^3 \sqrt{m^3 k^9}} \tag{12.26}$$

$$M_6^v = \frac{15\pi^3 S_0^3}{8\xi^3 \sqrt{m^9 k^3}} \tag{12.27}$$

$$M_6^a = \frac{15\pi^3 S_0^3}{8\xi^3 \sqrt{m^{15}}}(1 - 4\xi^2)^3 \sqrt{k^3} \tag{12.28}$$

Based on Equations 12.21 through 12.28, the following sensitivity equations can be derived:

$$\frac{dM_2^{\text{dis}}}{M_2^{\text{dis}}} = -\frac{3}{2}\frac{dk}{k}, \quad \frac{dM_2^v}{M_2^v} = -\frac{1}{2}\frac{dk}{k}, \quad \frac{dM_2^a}{M_2^a} = \frac{1}{2}\frac{dk}{k} \tag{12.29}$$

$$\frac{dM_4^{\text{dis}}}{M_4^{\text{dis}}} = -3\frac{dk}{k}, \quad \frac{dM_4^v}{M_4^v} = -\frac{dk}{k}, \quad \frac{dM_4^a}{M_4^a} = \frac{dk}{k} \tag{12.30}$$

$$\frac{dM_6^{\text{dis}}}{M_6^{\text{dis}}} = -\frac{9}{2}\frac{dk}{k}, \quad \frac{dM_6^v}{M_6^v} = -\frac{3}{2}\frac{dk}{k}, \quad \frac{dM_6^a}{M_6^a} = \frac{3}{2}\frac{dk}{k} \tag{12.31}$$

From Equations 12.29 through 12.31, it can be observed that the relative changes of the statistical moments of displacement and velocity are negatively proportional to the relative change of stiffness, while the relative change of the statistical moment of acceleration is positively proportional to the relative change of stiffness. The ratios of the relative change of the displacement moments to the relative change of stiffness are three times those of velocity moments positively and three times those of acceleration moments negatively. This result reflects that the relative change of the statistical moment of displacement is twice as sensitive to the relative change of stiffness as those of velocity and acceleration. Thus, the statistical moment of displacement will be selected as the damage index. Furthermore, it can be observed that the relative change of higher-order moments of displacement is more sensitive to the relative change of stiffness.

Based on Equations 12.20, 12.23 and 12.26, the stiffness of the structure can be obtained from the second-order, fourth-order and sixth-order statistical moments of displacement response, respectively.

$$k = \sqrt[3]{\frac{\pi^2 S_0^2}{4m\xi^2 (M_2^{dis})^2}} \tag{12.32}$$

$$k = \sqrt[3]{\frac{3\pi^2 S_0^2}{4m\xi^2 M_4^{dis}}} \tag{12.33}$$

$$k = \sqrt[9]{\frac{225\pi^6 S_0^6}{64\xi^6 m^3 (M_6^{dis})^2}} \tag{12.34}$$

The statistical moment-based damage detection on an SDOF system can therefore be carried out in the following steps: (1) measure the displacement response of the SDOF system to the ground acceleration, which is a zero-mean Gaussian white noise random process; (2) calculate the even statistical moments; (3) calculate the structural stiffness of the undamaged structure $\hat{k}^u$ using Equations 12.32 through 12.34; (4) use the same procedure to calculate the structural stiffness of the damaged structure $\hat{k}^d$ under another random excitation; and (5) identify the structural damage severity $\hat{\mu}$ by

$$\hat{\mu} = \frac{\hat{k}^d - \hat{k}^u}{\hat{k}^u} \times 100\% \tag{12.35}$$

where the symbol '^' represents 'estimated' in contrast with 'theoretical', and the superscripts 'u' and 'd' stand for 'undamaged' and 'damaged', respectively.

There are many sources causing measurement noise to pollute desirable signals during either laboratory tests or field measurements of civil structures. The type and intensity of measurement noise depend on the type and size of a structure, the measurement system and the environment surrounding the structure. Assume that the measurement noise is white noise, indicating that the measurement noise is caused by many sources of equal importance. The measurement noise intensity η is defined as the ratio of the root mean square (RMS) of measurement noise ε to the RMS of displacement response x:

$$\eta = \frac{\text{RMS}(\varepsilon)}{\text{RMS}(x)} \tag{12.36}$$

The effect of measurement noise on damage detection is measured in terms of the noise effect ratio γ:

$$\gamma = \frac{\hat{k}_n - \hat{k}}{\hat{k}} \times 100\% \qquad (12.37)$$

where:
$\hat{k}_n$ is the identified structural stiffness considering the effect of measurement noise
$\hat{k}$ is the identified structural stiffness without considering the effect of measurement noise

By defining the structural response with measurement noise as y, there is a relationship

$$y = x + \varepsilon \qquad (12.38)$$

where:
x is the actual structural response
ε is the measurement noise independent of x

Then, by taking a noise intensity of 15%, for example, one may have

$$\sigma_y^2 = \sigma_x^2 + \sigma_\varepsilon^2 = 1.0225\sigma_x^2 \qquad (12.39)$$

As a result, the statistical moments of the structural responses with measurement noise can be obtained as

$$M_{2y} = 1.0225 M_{2x} \qquad (12.40)$$

$$M_{4y} = 3(\sigma_y^2)^2 = 3 \times (1.0225)^2 \sigma_x^4 = (1.0225)^2 M_{4x} \qquad (12.41)$$

$$M_{6y} = 15(\sigma_y^2)^3 = 15 \times (1.0225)^3 \sigma_x^6 = (1.0225)^3 M_{6x} \qquad (12.42)$$

where:
M_{2y}, M_{4y} and M_{6y} are the second-order, fourth-order and sixth-order moments, respectively, of y
M_{2x}, M_{4x} and M_{6x} are the second-order, fourth-order and sixth-order moments, respectively, of x

Using Equations 12.32 through 12.34 leads to

$$k'' = \left\{ \begin{array}{c} \sqrt[3]{\dfrac{\pi^2 S_0^2}{4m\xi^2 M_{2y}^2}} = \dfrac{1}{\sqrt[3]{1.0225^2}} \sqrt[3]{\dfrac{\pi^2 S_0^2}{4m\xi^2 M_{2x}^2}} \\[3ex] \sqrt[3]{\dfrac{3\pi^2 S_0^2}{4m\xi^2 M_{4y}}} = \dfrac{1}{\sqrt[3]{1.0225^2}} \sqrt[3]{\dfrac{3\pi^2 S_0^2}{4m\xi^2 M_{4x}}} \\[3ex] \sqrt[9]{\dfrac{225\pi^6 S_0^6}{64\xi^6 m^3 M_{6y}^2}} = \dfrac{1}{\sqrt[3]{1.0225^2}} \sqrt[9]{\dfrac{225\pi^6 S_0^6}{64\xi^6 m^3 M_{6x}^2}} \end{array} \right\} = 0.9853k \qquad (12.43)$$

where:
  k^n is the theoretically identified stiffness considering measurement noise
  k is the corresponding stiffness without considering measurement noise

It can be seen that the theoretical noise effect ratio is only 1.47% even at the noise intensity of 15%. This result indicates that the statistical moment of displacement response is not sensitive to measurement noise.

Consider a numerical example, in which the mass, damping ratio and horizontal stiffness of a single-story shear building model are taken as 230.2 kg, 1% and 5.46×10^5 N/m, respectively. The ground excitation is taken as a zero-mean white noise stationary process. Five damage cases, stiffness reduction of 2%, 5%, 10%, 20% and 30% (Scenario 1 through Scenario 5), are considered. The structural displacement, velocity and acceleration responses are obtained, but the displacement responses are utilised hereinafter to effectively conduct damage detection.

The second-order (M_2), fourth-order (M_4) and sixth-order (M_6) statistical moments, which represent the characteristics of distribution, are computed for the undamaged building and the damaged building with the aforementioned five scenarios and listed in Table 12.3. The corresponding circular frequencies (ω_0) of the undamaged building and the damaged building are also computed and shown in Table 12.3. Superscripts d and u in the table stand for damage and the undamaged state, respectively.

It can be found from Table 12.3 that statistical moments are more sensitive to structural damage than circular natural frequency. Furthermore, it seems that the higher-order statistical moment would be a good index for structural damage detection, because higher-order statistical moments are more sensitive to structural damage than lower-order moments. However, the statistical moments are random variables, and higher statistical moments are less stable in the actual numerical calculation due to the effects of limited time duration. Therefore, as far as a damage index is concerned, the fourth-order moment may be a good choice, which represents a compromise measure between sensitivity and stability. In the study described here, the fourth-order moments are adopted for damage detection.

The procedure of damage detection is then carried out without consideration of noise. The identification results are given in Table 12.4. It can be seen that even for a damage severity of 2%, the proposed statistical moment-based damage detection method produces a satisfactory result if measurement noise is not considered. Random white measurement noises are subsequently introduced into the structural displacement responses to investigate the effect of measurement noise on damage detection. Five noise intensities are considered: 1%, 2%, 5%, 10% and 15%. Table 12.5 displays the noise effect ratio γ obtained for the aforementioned five damage cases and five noise intensities. As shown in Table 12.5, the noise effect ratio has almost nothing to do with damage severity, which means that the measurement

Table 12.3 Change ratios of natural frequency and statistical moments (SDOF system)

Scenario	$\dfrac{\omega_0^d - \omega_0^u}{\omega_0^u}$ (%)	$\dfrac{M_2^d - M_2^u}{M_2^u}$ (%)	$\dfrac{M_4^d - M_4^u}{M_4^u}$ (%)	$\dfrac{M_6^d - M_6^u}{M_6^u}$ (%)
1	−1.00	3.08	6.25	9.52
2	−2.53	8.00	16.64	25.96
3	−5.13	17.12	37.17	60.66
4	−10.55	39.75	95.31	172.96
5	−16.33	70.75	191.55	397.80

Table 12.4 Damage detection results and theoretical values (SDOF system)

Scenario	$\hat{M}_4^d$ $(10^{-13}m^4)$	M_4^d $(10^{-13}m^4)$	$\hat{k}^d$ (N/m)	k^d (N/m)	$\dfrac{\hat{k}^d - \hat{k}^u}{\hat{k}^u}$ (%)	$\dfrac{k^d - k^u}{k^u}$ (%)
1	2.5516	2.5530	535,218	535,080	−2.20	−2
2	2.1644	2.2310	524,683	518,700	−4.12	−5
3	3.2873	3.3646	495,871	491,400	−9.38	−10
4	3.8272	3.8841	438,702	436,800	−19.83	−20
5	7.2024	7.3511	385,054	382,200	−29.64	−30

Table 12.5 Noise effect ratio γ (SDOF system)

Noise level	Scenario 1	Scenario 2	Scenario 3	Scenario 4	Scenario 5
1%	−0.01%	−0.01%	−0.01%	−0.01%	−0.01%
2%	−0.03%	−0.03%	−0.03%	−0.03%	−0.03%
5%	−0.21%	−0.19%	−0.19%	−0.13%	−0.17%
10%	−0.90%	−0.80%	−0.76%	−0.57%	−0.71%
15%	−1.88%	−1.63%	−1.72%	−1.39%	−1.65%

noise has only small effects on the damage detection. It can be concluded that the fourth statistical moment is a sensitive measure, but it is insensitive to measurement noise. By using the fourth moment as a damage index, the proposed statistical moment-based damage detection method can provide not only reliable damage detection results but also explicit estimation of noise effects on damage detection results.

12.6 MULTI-SCALE DAMAGE DETECTION METHOD

This section provides a multi-scale damage detection method by combining the multi-scale response reconstruction technique with the radial-basis-function (RBF) network and the response sensitivity-based FE model updating method (Zhang and Xu 2016). The multi-scale response reconstruction technique, based on the optimal placement of multi-type sensors, has been discussed in Chapter 8 in detail for an intact structure. If damage occurs in the structure, the trained RBF network is then employed to predict the displacement and strain mode shapes of the damaged structure using the modal parameters extracted from the measurement data through experimental modal analysis (EMA). The predicted displacement and strain mode shapes are further used to reconstruct the responses of the damaged structure by virtue of the Kalman filter. The reconstructed responses are finally used to identify the damage in terms of sensitivity-based FE model updating. In each iteration for model updating, sparse regularisation is employed to overcome the ill condition of the problem. A simply supported overhanging steel beam is experimentally investigated to demonstrate the feasibility and superiority of the multi-scale damage detection method.

12.6.1 RBF network for response reconstruction of damaged structure

As mentioned in Section 12.4, the structural responses measured from the limited sensors installed on a civil structure may not contain enough information to identify local damage. An intuitive idea is to reconstruct the responses at the locations that are in the vicinity of the damage. However, it is not straightforward to reconstruct the response of a damaged structure using the Kalman filter, as introduced in Chapter 8. Nevertheless, it is noted from Equations 8.35 through 8.52 for the optimal sensor location and the best response reconstruction that the system matrices A_s, B_s, C_s^m and D_s^m in Equation 8.40 are actually composed of the modal parameters, including frequencies, damping ratios and mode shapes. It is possible to extract these modal parameters from the limited measurement data by the EMA method, although the mode shapes extracted from the limited measurement points may not be good or detailed enough for a structure with local damage. Several well-established techniques are reported in the literature to extract the modal parameters from the measured noise-corrupted responses. The subspace identification method is used in this study because of its computational efficiency and strong capacity for modal identification.

Once the modal parameters are extracted from the noise-corrupted responses measured by the limited sensors at their optimal locations, the artificial neutral network (ANN) is then adopted to estimate the strain and displacement mode shapes at the unmeasured locations in the damaged structure so as to provide the best response reconstruction and accurate damage detection. ANNs are computational tools inspired by biological neural networks, in that they imitate the parallel and distributed processing nature of the biological neurons to approach a specific problem by using certain rules to achieve favourable results. ANNs have found extensive applications, including function approximation, time series prediction, classification and system control. The application of ANNs to mode shape prediction has been investigated in Goh et al. (2013). The basic idea of ANN applications in mode shape prediction in this study is to build a model to establish the relationships between the measured modal parameters and the complete displacement and strain mode shapes of the damaged structure through a training process. The RBF network is a kind of multi-hidden-layer feedforward ANN, with sigmoid and linear functions as activation functions for the neurons in the hidden layers and the output layer, respectively. It can be proved that the RBF network is capable of providing arbitrarily good approximation to any prescribed function using only a finite number of parameters (Park and Sandberg 1991). The output vector of the RBF network is a function of the input vector, and it is given by

$$\tilde{\mathbf{y}}(\tilde{\mathbf{x}}) = \sum_{i=1}^{\tilde{N}} \tilde{w}_i \tilde{\psi}(\|\tilde{\mathbf{x}} - \tilde{c}_i\|) \tag{12.44}$$

where:

$\tilde{\mathbf{x}}$ is the input vector
$\tilde{N}$ is the number of neurons in the hidden layer
$\tilde{c}_i$ is the centre vector for the neuron i
$\tilde{w}_i$ is the weight of the neuron i in the linear output neuron
$\tilde{\psi}$ is the radial basis function, which depends only on the distance from a centre vector and is radially symmetric about that vector

In this study, the input of the network is the frequencies and the displacement and strain mode shapes of the damaged structure at the sensor locations, which can be extracted by EMA from the measured data. These data are then used to predict the strain and displacement mode shapes of the damaged structure at the unmeasured locations (the output vector) through the well-trained RBF network expressed by Equation 12.44.

To avoid the mass-normalisation in EMA, COMAC, defined in Equation 12.6, is utilised. The calculated result of COMAC is a symmetric matrix indicating the spatial comparison of mode shapes before damage and after damage, and only its lower triangular parts are utilised. If the strain mode shapes before damage $\mathbf{\Psi}_u^m$ and after damage $\mathbf{\Psi}_d^m$ are used in Equation 12.6, the strain COMAC can be obtained. The superscript m denotes measurement and the subscripts u and d denote the undamaged and the damaged state, respectively. The same definition will be used hereafter. In this study, the frequencies, displacement COMAC and strain COMAC consist of the input vector to the trained RBF network to predict the mass-normalised displacement mode shapes of the damaged structure. The strain mode shapes of the damaged structure can be worked out as $\mathbf{\Psi}^e = \mathbf{B}'\mathbf{\Phi}^e$, where $\mathbf{B}'$ is the element-dependent matrix characterising the relationship between displacement and strain, and the superscript e denotes the estimated results.

As can be found in Equation 12.44, the RBF network must be trained before it can be used to predict the mass-normalised displacement and strain mode shapes of the damaged structure. In this study, the RBF network is trained through the extensive numerical analysis of a structure for a series of damage scenarios. The experimental example to be presented in Section 12.6.3 will provide the details regarding the training procedure. Once the mass-normalised displacement and strain mode shapes of the damage structure are predicted from the trained RBF network, the response reconstruction method presented in Chapter 8 is finally used to estimate the responses of the damaged structure at the unmeasured locations to make the complete responses available.

12.6.2 Response sensitivity-based FE model updating and damage detection

Damage can be viewed as structural defects that cause discrepancies in the structure between the damage-free and damaged states. FE model updating is commonly used for damage detection to minimise the discrepancies as far as possible (Friswell and Mottershead 1995). The measured response of a damage structure, as a nonlinear function of damage parameter, can be expanded on its damage-free state with respect to damage parameters using the Taylor series as

$$\mathbf{y}_d^m = \mathbf{y}_u^m + \frac{\partial}{\partial \overline{\theta}}\left(\mathbf{y}_u^m\right)\Delta\overline{\theta} + \mathrm{O}\left(\Delta\overline{\theta}^2\right) \tag{12.45}$$

where:

$\mathbf{y}_u^m$ and $\mathbf{y}_d^m$ are the $(n_s \cdot n_m)$-dimensional vector of the measured responses of the structure before and after damage occurrence, respectively

$\overline{\theta}$ is the n_e-dimensional vector of damage parameters

n_s, n_m and n_e are the number of time instants, sensors and elements, respectively

$\partial\mathbf{y}_u^m/\partial\overline{\theta}$ is the first-order derivative of the measured response $\mathbf{y}_u^m$ with respect to damage parameters $\overline{\theta}$

If the higher-order term is neglected, Equation 12.45 can be written as

$$\mathbf{S}^m\Delta\overline{\theta} = \Delta\mathbf{y}^m \tag{12.46}$$

where:

$\mathbf{S}^m = \partial\mathbf{y}_u^m/\partial\overline{\theta}$ is usually termed a *sensitivity matrix* with the dimension of $(n_s \cdot n_m) \times n_e$ herein

$\Delta\mathbf{y}^m = \mathbf{y}_d^m - \mathbf{y}_u^m$ is the response discrepancy between the damaged and intact states

If the dynamic response $\mathbf{y}(\overline{\theta})$ is a highly nonlinear function of damage parameter, an iterative procedure using a gradient-based optimisation such as the Gaussian–Newton iteration shall be employed to obtain the identified damage parameter as

$$\mathbf{S}^{m,k}\Delta\overline{\theta}^{k+1} = \Delta\mathbf{y}^{m,k} \qquad (12.47)$$

where:

$\mathbf{S}^{m,k} = \mathbf{S}^m(\overline{\theta}^k) = \partial\mathbf{y}^m(\overline{\theta}^k)/\partial\overline{\theta}$ is the sensitivity matrix for the kth iteration with $\mathbf{S}_{ij}^{m,k} = \partial\mathbf{y}_i^m(\overline{\theta}^k)/\partial\overline{\theta}_j$ and can be calculated using the numerical central difference method

$\Delta\mathbf{y}^{m,k} = \mathbf{y}_d^m - \mathbf{y}_a^{m,k} = \mathbf{y}_d^m - \mathbf{y}_a^m(\overline{\theta}^k)$ becomes the discrepancy between the measured response and the analysed response from the FE analysis of the damaged structure

$\overline{\theta}^k = \sum_{l=1}^{k}\Delta\overline{\theta}^l$ is the cumulative damage parameter in the kth iteration

When the structure is free of damage, it is assumed that the equations $\overline{\theta}^0 = 0$ and $\mathbf{y}_a^{m,0} = \mathbf{y}_a^m(\overline{\theta}^0) = \mathbf{y}_u^m$ hold. The sensitivity matrix $\mathbf{S}^k$ and the response discrepancy vector $\Delta\mathbf{y}^k$ are assembled by stacking the columns of sensitivity sequences and response discrepancy sequences, respectively:

$$\left(\mathbf{S}^{m,k}\right)^T = \left[\left(\mathbf{S}_1^{m,k}\right)^T \quad \cdots \quad \left(\mathbf{S}_i^{m,k}\right)^T \quad \cdots \quad \left(\mathbf{S}_{n_m}^{m,k}\right)^T\right]$$
$$\left(\Delta\mathbf{y}^{m,k}\right)^T = \left[\left(\Delta\mathbf{y}_1^{m,k}\right)^T \quad \cdots \quad \left(\Delta\mathbf{y}_i^{m,k}\right)^T \quad \cdots \quad \left(\Delta\mathbf{y}_{n_m}^{m,k}\right)^T\right] \qquad (12.48)$$

$$\mathbf{S}_i^{m,k} = \begin{bmatrix} \dfrac{\partial\mathbf{y}_i^{m,k}(t_1)}{\partial\overline{\theta}_1} & \dfrac{\partial\mathbf{y}_i^{m,k}(t_1)}{\partial\overline{\theta}_2} & \cdots & \dfrac{\partial\mathbf{y}_i^{m,k}(t_1)}{\partial\overline{\theta}_{n_e}} \\ \vdots & \vdots & \vdots & \vdots \\ \dfrac{\partial\mathbf{y}_i^{m,k}(t_j)}{\partial\overline{\theta}_1} & \dfrac{\partial\mathbf{y}_i^{m,k}(t_j)}{\partial\overline{\theta}_2} & \cdots & \dfrac{\partial\mathbf{y}_i^{m,k}(t_j)}{\partial\overline{\theta}_{n_e}} \\ \vdots & \vdots & \vdots & \vdots \\ \dfrac{\partial\mathbf{y}_i^{m,k}(t_{n_s})}{\partial\overline{\theta}_1} & \dfrac{\partial\mathbf{y}_i^{m,k}(t_{n_s})}{\partial\overline{\theta}_2} & \cdots & \dfrac{\partial\mathbf{y}_i^{m,k}(t_{n_s})}{\partial\overline{\theta}_{n_e}} \end{bmatrix}, \Delta\mathbf{y}_i^{m,k} = \begin{bmatrix} \mathbf{y}_{d,i}^m(t_1) - \mathbf{y}_{a,i}^{m,k}(t_1) \\ \vdots \\ \mathbf{y}_{d,i}^m(t_i) - \mathbf{y}_{a,i}^{m,k}(t_i) \\ \vdots \\ \mathbf{y}_{d,i}^m(t_{n_s}) - \mathbf{y}_{a,i}^{m,k}(t_{n_s}) \end{bmatrix} \qquad (12.49)$$

In this study, the reconstructed multi-type responses are regarded as the complete measured responses used in the modal updating procedure. The right-hand side of Equation 12.47 is replaced with

$$\Delta\mathbf{y}^k = \mathbf{y}_d^e - \mathbf{y}_a^k \qquad (12.50)$$

where:

$\mathbf{y}_d^e$ is the $n_s\cdot(n_e + 2\,n)$-dimensional vector of the reconstructed responses from the limited measured responses of the damaged structure at the optimal sensor locations

n denotes the number of DOFs

$\mathbf{y}_a^k$ represents the analytical responses of the structure of damage parameters

Since different types of responses (such as displacement, acceleration and strain) have different units and different orders of amplitude, a standardisation procedure is performed on Equation 12.50:

$$\overline{\mathbf{S}}^k \Delta \overline{\boldsymbol{\theta}}^{k+1} = \Delta \overline{\mathbf{y}}^k \tag{12.51}$$

$$\Delta \mathbf{y}^k = \Sigma^{-1/2} \Delta \overline{\mathbf{y}}^k, \quad \overline{\mathbf{S}}^k = \Sigma^{-1/2} \mathbf{S}^k \tag{12.52}$$

where Σ is the $(n_m \times n_m)$ variance matrix of dynamic response of the structure.

Sparsity refers to the fact that only very few entries in a vector are non-zero. As damage usually occurs at very few structural components compared with the total discretised elements in the entire structure, it is sensible that there exists a sparse solution to damage identification. It is thus advantageous to utilise the sparse properties for better identification results (Hernandez 2014; Zhou et al 2015; Zhang and Xu 2016). If the sparsity restriction is imposed, the solution of Equation 12.51 is equivalent to solving the following optimisation problem:

$$\min_{\Delta \overline{\boldsymbol{\theta}}^{k+1}} \left(\left\| \overline{\mathbf{S}}^k \Delta \overline{\boldsymbol{\theta}}^{k+1} - \Delta \overline{\mathbf{y}}^k \right\|_2^2 + \overline{\lambda} \left\| \Delta \overline{\boldsymbol{\theta}}^{k+1} \right\|_p^p \right) \tag{12.53}$$

where $\left\| \Delta \overline{\boldsymbol{\theta}}^{k+1} \right\|_p = \sqrt[p]{\Sigma |\Delta \overline{\boldsymbol{\theta}}^{k+1}|^p}$ is the p-norm of vector $\Delta \overline{\boldsymbol{\theta}}^{k+1}$. Mathematically, the regularisation norms $\left\| \Delta \overline{\boldsymbol{\theta}}^{k+1} \right\|_p^p$ with $0 \leq p \leq 1$ can all enforce sparsity in solution. Due to the non-smoothness of the regularisation norms $\left\| \Delta \overline{\boldsymbol{\theta}}^{k+1} \right\|_p^p (0 \leq p \leq 1)$, methods capable of handling the non-smoothness are necessary for computing minimisers involving these norms. Here, the sparse regularisation is specified as ℓ_1 norm regularisation, that is, $p = 1$ in Equation 12.53. The main reason is that although the ℓ_1 norm is weaker than the $p < 1$ norm in ensuring sparsity, the ℓ_1 norm regularisation is a convex problem and admits an efficient solution via linear programming techniques. As a result, Equation 12.53 can be expressed as

$$\min_{\Delta \overline{\boldsymbol{\theta}}^{k+1}} \left(\left\| \overline{\mathbf{S}}^k \Delta \overline{\boldsymbol{\theta}}^{k+1} - \Delta \overline{\mathbf{y}}^k \right\|_2^2 + \overline{\lambda} \left\| \Delta \overline{\boldsymbol{\theta}}^{k+1} \right\|_1^1 \right) \tag{12.54}$$

in which the term $\left\| \overline{\mathbf{S}}^k \Delta \overline{\boldsymbol{\theta}}^{k+1} - \Delta \overline{\mathbf{y}}^k \right\|_2^2$ forces the residual to be small, whereas the term $\left\| \Delta \overline{\boldsymbol{\theta}}^{k+1} \right\|_1^1$ enforces sparsity of the solution. The parameter $\overline{\lambda} > 0$ controls the trade-off between the sparsity of the solution and the residual norm. Care must be taken to choose the right proper regularisation parameter to come up with an acceptable solution. Sometimes, multiple runs are needed to determine the right penalty parameter. However, this difficulty can be relieved by using a re-weighting strategy, as explained in Overschee and De Moo (1994). There are a number of solvers with theoretical guarantees that are available to solve the problem. The primal-dual interior point algorithm proposed by Boyd and Vandanberghe (2004) is employed in this study. If the following convergence criteria are met, iteration could be finished:

$$\left\| \Delta \overline{\boldsymbol{\theta}}^k \right\| / \left\| \overline{\boldsymbol{\theta}}^k \right\| \leq \text{Tol} \tag{12.55}$$

where $\|\cdot\|$ denotes the Frobenius norm.

The flowchart of the proposed multi-scale damage detection method is shown in Figure 12.8. Figure 12.9 shows the schematic illustration of the response reconstruction

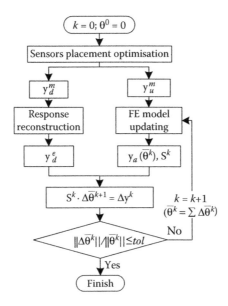

Figure 12.8 Flowchart of multi-scale damage detection method.

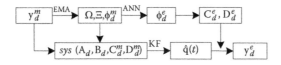

Figure 12.9 Response reconstruction strategy of damaged structure.

process based on the mode shape prediction using the RBF network. The matrices Ω and Ξ in Figure 12.9 are diagonal matrices containing the natural frequencies and modal damping ratios, respectively. For a detailed description of the system matrices, $\mathbf{A}_d$, $\mathbf{B}_d$, $\mathbf{C}_d^m$ and $\mathbf{D}_d^m$, the reader can refer to the response reconstruction method presented in Chapter 8.

12.6.3 Experimental studies

Experimental studies of a simply supported overhanging steel beam were conducted in a structure dynamics laboratory as shown in Figure 12.10. The total length of the beam was 4 m, and its cross-sectional area was 50 (width) $\times$ 15 mm (height). An LDS V451 electromagnetic vibrator, working together with a B&K signal generator and an LDS PA500 power amplifier, was mounted on the beam at 1.8 m from the left end of the beam, generating

Figure 12.10 Laboratory test setup.

a white noise excitation vertically with a bandwidth from 2 to 102 Hz. The vibrator was connected to the beam through a spring with a small stiffness to reduce the side effect of the vibrator on the stiffness of the beam. An FE model of the beam was established as shown in Figure 12.11. It was modelled by 40 Euler beam elements with 41 nodes and a total of 123 DOFs. The beam was constrained by a hinge support on the node 11 and a roller support on the node 31, forming 120 effective DOFs. The effect of the exciter on the beam was modelled by a mass element and a spring element in the FE model. By using the multi-type sensor placement method as introduced in Chapter 8 and considering the first 6 displacement and strain modes for response reconstruction, 11 sensor locations were selected, including 5 locations for strain gauges, 2 for displacement transducers and 2 for accelerometers, as shown in Figure 12.12.

To validate the accuracy of reconstructed responses obtained by the proposed response reconstruction method, three more strain gauges and five more accelerometers were added to the beam. As a result, the multi-type sensors, including eight BX120-55AA strain gauges, two LK-503 laser displacement transducers and nine KD1008 accelerometers, were installed on the beam and used in the test, as shown in Figure 12.13.

All the strain gauges were stuck on the upper surface of the beam in the middle of the elements concerned. The acceleration signal was amplified by a KD5008C charge amplifier. All three types of responses were collected by the Kyowa EDX-100A data recorder and stored in a personal computer. The sampling frequency was 500 Hz, and the sample duration was 40 s with zero initial condition. The original data passed through a band-pass filter of 2–82 Hz before utilisation. It is vital that the FE model of the beam in its intact state is of high

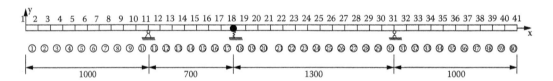

Figure 12.11 FE model of simply supported overhanging beam (all dimensions in millimetres).

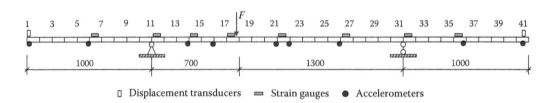

Figure 12.12 Optimal sensor locations of simply supported overhanging beam.

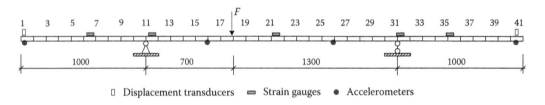

Figure 12.13 Total number and location of multi-type sensors used in laboratory test.

Table 12.6 Two damage scenarios

Damage scenario	Damage description
Single damage	20% at element 23
Double damage	20% at element 23; 10% at element 7

quality before the implementation of the multi-scale damage detection method. Thus, the model updating of the intact beam was performed in the time domain to refine the FE model by minimising the discrepancies between the dynamic responses from the FE model and the measured responses.

Two typical damage scenarios listed in Table 12.6, including single damage and double damage with different damage severities, were designed and accomplished in a laboratory test to validate the effectiveness of the proposed method. The damage was induced by reducing the width of the concerned elements, generating equal reduction of both the element stiffness matrix and the element mass matrix. It should be noticed that there were no sensors installed at the damaged locations. This was arranged to demonstrate the superiority of the proposed damage detection method. After damage occurrence, the modal parameters of the damaged beam were extracted from the measured responses at the optimal sensor locations using EMA. Then, the frequencies as well as the displacement COMAC and the strain COMAC were imported to the trained RBF network to predict the mass-normalised strain and displacement mode shapes of the damage beam.

The RBF network was trained through numerical analysis of the updated FE model of the beam for a series of simulated damage scenarios. The simulated damage scenarios covered the cases of both single damage and double damage, with the damage extent varying from 0.0 to −0.2 at an increment of −0.1. This led to a total of 3200 training cases. For each case, one damage scenario was introduced to the intact FE model of the beam. Then, the numerical dynamic responses at the optimal sensor locations were added by the noises of normally distributed sequences based on the laboratory test conditions. The modal identification was performed to derive the modal parameters. The identified modal parameters, as well as the complete displacement mode shapes of the damaged beam, were used to train the RBF network. The sum-squared error threshold in the RBF network training was set as 5×10^{-8} for displacement mode shape prediction. The strain mode shapes of the damaged beam were predicted as $\mathbf{\Psi}^e = \mathbf{B}' \mathbf{\Phi}^e$.

Once the mass-normalised displacement and strain mode shapes of the damage structure were predicted from the trained RBF network, the response reconstruction method presented in Chapter 8 was used to estimate the responses of the damaged structure at the unmeasured locations to form the reconstructed responses. To check the accuracy of the response reconstruction method, the reconstructed responses at the additional sensor locations were compared with the measured responses as well as the numerically simulated responses, as shown in Figure 12.14. It can be seen that the reconstructed responses match well with the measured responses and the numerically simulated responses.

The sensitivity-based FE model updating method was then used for damage detection. According to the damage scenario created in the test, the element stiffness matrix and the element mass matrix were degraded to the same extent. Thus, the damage could be modelled as

$$\mathbf{K}^d = \mathbf{K}^u + \sum_{i=1}^{n_e} \bar{\theta}_i \mathbf{K}_i \left(-1 \leq \bar{\theta}_i \leq 0\right), \quad \mathbf{M}^d = \mathbf{M}^u + \sum_{i=1}^{n_e} \bar{\theta}_i \mathbf{M}_i \left(-1 \leq \bar{\theta}_i \leq 0\right) \quad (12.56)$$

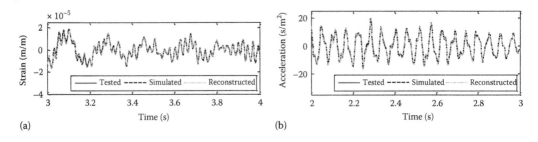

(a)   (b)

Figure 12.14 Comparison of reconstructed responses with measured and simulated ones: (a) bending strain at element 26, (b) vertical displacement at node 22.

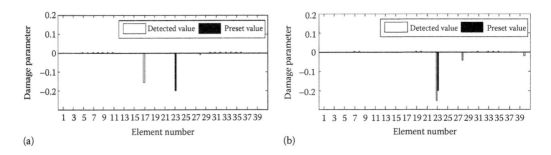

(a)   (b)

Figure 12.15 Test and predicted results of single damage scenario: (a) without response reconstruction, (b) with response reconstruction.

where:

$\mathbf{K}^u$ and $\mathbf{M}^u$ are the global stiffness and mass matrix, respectively, of the structure in the intact state

$\mathbf{K}_i$ and $\mathbf{M}_i$ are the ith element stiffness and mass matrix in the global coordinate

$\bar{\theta}_i$ is the fractional stiffness damage parameter of $\mathbf{K}_i$

The classical Rayleigh damping model is adopted here for construction of the damping matrix, that is, $\mathbf{C} = \alpha_1 \mathbf{M} + \alpha_2 \mathbf{K}$, where α_1 and α_2 are the mass and stiffness proportional Rayleigh damping coefficients, respectively. The damping coefficients α_1 and α_2 were determined by the first two measured mode damping ratios in this study.

The reconstructed responses of the beam for each damage scenario were used for the sensitivity-based model updating with sparsity regulation to find both damage location and extent. The results of identification are depicted in Figure 12.15 for the single damage scenario and Figure 12.16 for the double damage scenario. In the single damage scenario, if only the responses measured by the optimal sensors are used for damage detection without response reconstruction, the damage location identified, which should be element 23 rather than element 17, is not correct. This result indicates that the responses measured by the optimal sensors cannot provide enough information for accurate damage location. Nevertheless, by using the damage detection with response reconstruction, the damage location and severity can be identified correctly to some extent. In the double damage scenario, it can be seen that there is a noticeable false-positive error for element 8, and the damage severities of elements 7 and 23 are not correctly predicted when the responses measured

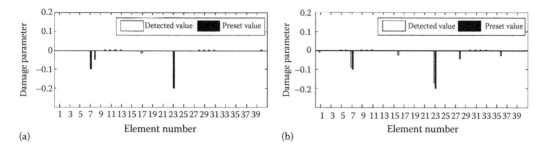

Figure 12.16 Test and detected results of double-damage scenario: (a) without response reconstruction, (b) with response reconstruction.

from the optimal sensors are used directly without response reconstruction. When the proposed damage detection method with response reconstruction is used, there are small false-positive errors, but these small errors are acceptable compared with the detected damage of 10% at element 7 and 20% at element 23.

12.7 DAMAGE DETECTION METHODS WITH CONSIDERATION OF UNCERTAINTIES

12.7.1 Uncertainties in damage detection

Damage detection in civil structures involves a significant amount of uncertainty. Consequently, the damage detection results are, in effect, the compensation for the uncertainties. If the uncertainties are large, the compensation for the damage detection will distort the actual results and lead to two kinds of false damage identification (Farrar et al. 1998): (1) false-positive damage identification (identifying the intact element as damaged) and (2) false-negative damage identification (failure to identify the damaged elements). The second category of false damage detection can have serious life and safety implications. False-positive readings can also erode confidence in the damage detection process. It is imperative to analyse the source of the uncertainties, quantify the uncertainties and their effects, and evaluate the reliability of the damage identification results.

The uncertainties in damage identification are mainly attributed to (1) inaccuracy in the FE model discretisation; (2) uncertainties in geometry and boundary conditions; (3) variations in material properties; (4) environmental variability (such as temperature, wind and traffic); (5) errors associated with measured signals; (6) errors in post-processing techniques and (7) improper methods employed in damage identification.

According to their sources, these uncertainties can be classified into three groups: methodology errors, modelling errors and measurement noise. Methodology errors are generated by the limitation of the method itself in damage identification. Modelling errors are related to the uncertainties in modelling the actual structure, mainly including discretisation errors, configuration errors and mechanical parameter errors. Measurement noise mainly comes from procedures and equipment related to the response measurements. It can be classified into system errors and random errors. Random errors include measurement noise from environmental sources, calibration error in the sensors and calibration error in the actuators. System errors include the imprecise placement and orientation of the sensors and the misalignment of force actuators (Peterson et al. 1996), errors such as the added mass or stiffness due to sensors, and errors that result from the signal processing or identification techniques.

Typical error distributions adopted to simulate the characteristics of random data are the uniform and normal PDFs (Sanayei et al. 1992). The latter case is more common for the uncertainties of the FE modelling and the measurement data. Based on the prior PDFs of the uncertainties, different techniques have been developed to estimate the PDFs of the damage or updated structural parameters.

12.7.2 Perturbation approach

When the uncertainties are considered as normally distributed random variables with zero means and given covariance, the measured quantities (structural parameters and modal properties) are regarded as the true values plus the random noise:

$$\theta_j = \theta_j^0 + \theta_j^0 W_{\theta j} = \theta_j^0 (1 + W_{\theta j}) \quad (j = 1, 2, \ldots, n_e) \tag{12.57}$$

$$\begin{aligned} \lambda_i &= \lambda_i^{0}(1 + W_{\lambda i}) \\ \Phi_i &= \Phi_i^{0}(1 + W_{\Phi i}) \end{aligned} \quad (i = 1, 2, \ldots, n_d) \tag{12.58}$$

where:

$$\begin{array}{rl} \theta & \text{is the structural parameter} \\ \lambda_i & \text{is the } i\text{th eigenvalue} \\ \Phi_i & \text{is the } i\text{th eigenvector} \\ n_e \text{ and } n_d & \text{are the number of elements and modes, respectively} \\ \text{superscript 0} & \text{represents the corresponding true value} \\ W_{\theta j}, W_{\lambda i} \text{ and } W_{\Phi i} & \text{are the corresponding proportional uncertainties with zero means} \end{array}$$

With the perturbation method (Liu 1995; Xia et al. 2002; Xia and Hao 2003), parameters in the governing equation are expanded as a Taylor series in terms of the uncertainties. For example, in sensitivity-based model updating, it can be written as

$$e = S \cdot \Delta\theta \tag{12.59}$$

where:

e   is an error vector containing the differences between the eigenvalues and mode shapes at the measured DOFs of the structure before and after updating

S   is the modal sensitivity matrix

Equation 12.59 is expanded as a second-order Taylor series in terms of W_i (including $W_{\theta j}$, $W_{\lambda i}$ and $W_{\Phi i}$):

$$e = e^0 + \sum_{i=1}^{N} \frac{\partial e^0}{\partial W_i} W_i + \frac{1}{2} \sum_{i=1}^{N} \sum_{j=1}^{N} \frac{\partial^2 e^0}{\partial W_i \partial W_j} W_i W_j \tag{12.60}$$

$$S = S^0 + \sum_{i=1}^{N} \frac{\partial S^0}{\partial W_i} W_i + \frac{1}{2} \sum_{i=1}^{N} \sum_{j=1}^{N} \frac{\partial^2 S^0}{\partial W_i \partial W_j} W_i W_j \tag{12.61}$$

$$\Delta\theta = \Delta\theta^0 + \sum_{i=1}^{N} \frac{\partial \Delta\theta^0}{\partial W_i} W_i + \frac{1}{2} \sum_{i=1}^{N} \sum_{j=1}^{N} \frac{\partial^2 \Delta\theta^0}{\partial W_i \partial W_j} W_i W_j \tag{12.62}$$

where N is the total number of uncertainties. By substituting the above equations into Equation 12.59 and comparing the terms of 1, W and $W_i W_j$, the unknown quantities can be solved one by one as

$$\Delta\theta^0 = \left(S^0\right)^+ e^0 \tag{12.63}$$

$$\frac{\partial \Delta\theta}{\partial W_i} = \left(S^0\right)^+ \left(\frac{\partial e}{\partial W_i} - \frac{\partial S}{\partial W_i} \Delta\theta^0 \right) \tag{12.64}$$

$$\frac{\partial^2 \Delta\theta}{\partial W_i \partial W_j} = \left(S^0\right)^+ \left(\frac{\partial^2 e}{\partial W_i \partial W_j} - \frac{\partial^2 S}{\partial W_i \partial W_j} \Delta\theta^0 - 2 \frac{\partial S}{\partial W_i} \frac{\partial \Delta\theta}{\partial W_j} \right) \tag{12.65}$$

where + refers to the pseudo-inverse. As the modal sensitivity matrix S is often ill conditioned, the direct solutions to Equations 12.63 through 12.65 may yield poor estimates. Hua et al. (2008) employed regularisation techniques to handle the problem.

From Equation 12.62, the mean values of $\Delta\theta$ are, noting that $E(W_i) = 0$,

$$E\left(\Delta\theta\right) = E\left(\Delta\theta^0\right) + \frac{1}{2} \sum_{i=1}^{N} \frac{\partial^2 \Delta\theta}{\partial W_i^2} COV\left(W_i, W_i\right) \tag{12.66}$$

where $E(\cdot)$ and $COV(\cdot)$ refer to the mean value and covariance matrix, respectively. It is noted that when the correlation between the updating parameters and the measurement is disregarded, the second-order derivatives in Equation 12.66 are not necessary. This has been verified by Khodaparast et al. (2008). The covariance matrix of $\Delta\theta$ is

$$COV\left(\Delta\theta, \Delta\theta\right) = \left[\frac{\partial \Delta\theta}{\partial W} \right] COV\left(W, W\right) \left[\frac{\partial \Delta\theta}{\partial W} \right]^T \tag{12.67}$$

During these procedures, the calculation of the derivatives of S to noise vector W requires the calculation of the eigenvalue derivatives and the eigenvector derivatives. These can be obtained using Nelson's method (Nelson 1976).

Subsequently, the statistics of the updated parameters, $\tilde{\theta}_i$, can be obtained as

$$E(\tilde{\theta}_i) = E\left(\theta_i\right) + E\left(\Delta\theta_i\right) = \theta_i^0 + E\left(\Delta\theta_i\right) \tag{12.68}$$

$$COV\left(\tilde{\theta}_i, \tilde{\theta}_j\right) = COV\left(\theta_i + \Delta\theta_i, \theta_j + \Delta\theta_j\right)$$

$$= COV\left(\theta_i, \theta_j\right) + COV\left(\theta_i, \Delta\theta_j\right) + COV\left(\Delta\theta_i, \theta_j\right) + COV\left(\Delta\theta_i, \Delta\theta_j\right) \tag{12.69}$$

From Equations 12.57 and 12.62,

$$COV\left(\theta_i, \theta_j\right) = \theta_i^0 \theta_j^0 COV\left(W_{\theta i}, W_{\theta j}\right) \tag{12.70}$$

$$COV\left(\theta_i, \Delta\theta_j\right) = E\left[\left(\theta_i - \theta_i^0\right)\left(\Delta\theta_j - \Delta\theta_j^0\right)\right] = \theta_i^0 \sum_{k=1}^{N} \frac{\partial \Delta\theta_j}{\partial W_k} COV\left(W_{\theta i}, W_k\right) \tag{12.71}$$

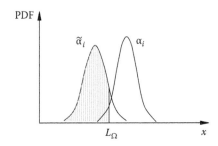

Figure 12.17 Probability density functions of the structural parameters.

and similarly,

$$COV\left(\Delta\theta_i,\theta_j\right) = COV\left(\theta_j,\Delta\theta_i\right) = \theta_j^0 \sum_{k=1}^{N} \frac{\partial\Delta\theta_i}{\partial W_k} COV\left(W_{\theta j}, W_k\right) \qquad (12.72)$$

From Equations 12.68 through 12.72, the standard deviation of $\tilde{\theta}_i$, that is, $\sigma(\tilde{\theta}_i)$, can be derived as the square root of the corresponding diagonal elements of the covariance matrix. With the assumption of normal distribution, the statistical distributions of the stiffness parameters in the updated state can be derived. Without losing generality, the PDFs of a structural parameter in the undamaged and damaged states are illustrated in Figure 12.17. The interval of the healthy stiffness parameter, $\Omega(\theta_i, \mu)$, is defined such that the probability of θ_i contained within the interval is μ:

$$P\left(x_\theta \in \Omega\left(\theta_i,\mu\right)\right) = P(L_\Omega \leq x_\theta < \infty) = \mu \qquad (12.73)$$

where L_Ω is the lower bound of the interval $\Omega(\theta_i, \mu)$. The probability of damage existence (PDE) is then defined as that of $\tilde{\theta}_i$ not being within the healthy interval at the confidence level of μ:

$$PDE = P(-\infty < x_{\tilde{\theta}} \leq L_\Omega) \qquad (12.74)$$

The PDE is a value between 0 and 1, which depends on the PDFs and the confidence interval. It is apparent that if the PDE of an element is close to 1, then most likely the element is damaged; and on the other hand, if the PDE is close to 0, damage of the element is very unlikely. Papadopoulos and Garcia (1998) proposed a few different definitions of the PDE.

12.7.3 Bayesian approach

Bayesian updating has been developed by Beck and his co-workers (Beck and Katafygiotis 1998; Katafygiotis and Beck 1998; Beck and Au 2002; Yuen and Katafygiotis 2005).

Let H_j, represented with a prior probability $P(H_j)$, denote a hypothesis for a damage event that can contain any number of substructures as damaged. Using the Bayes theorem, the posterior probability $P(H_j|\psi)$, given a set of observed modal parameters ψ, can be represented as

$$P\left(H_j|\psi\right) = \frac{P\left(\psi|H_j\right)}{P\left(\psi\right)} P\left(H_j\right) \qquad (12.75)$$

The substructures most likely to be damaged are the ones included in the hypothesis $H_{\max}$ that has the largest posterior probability (Sohn and Law 1997), that is,

$$P\left(H_{\max}|\psi\right) = \max_{\forall H_j}\left(P\left(H_j|\psi\right)\right) \tag{12.76}$$

The distribution of the measurement noise and modelling error are explicitly considered within the Bayesian probabilistic framework. To avoid permuting all possible damage events H_j, a branch-and-bound method was devised to search the results. Sohn and Law (2000) applied this algorithm to detect damage in a bridge column. The proposed probabilistic damage detection method was able to locate the damaged region, whereas two deterministic methods were not.

12.7.4 Statistical pattern recognition

Pattern recognition is the assignment of some sort of output values to a given input value, according to some specific algorithms. There are a number of distinct approaches to pattern recognition, the main ones being the statistical, neural and syntactic approaches (Schalkoff 1992). As civil structures are subject to various degrees of uncertainty, the statistical approach to pattern recognition appears to stand out as a natural approach to damage detection. A general statistical pattern recognition for damage detection can be implemented through the integration of four procedures: (1) operational evaluation; (2) data acquisition; (3) feature selection and (4) statistical modelling for feature discrimination (Farrar and Worden 2013). Statistical pattern recognition is often implemented through machine learning algorithms. The idea of machine learning is to learn the relationship between some features derived from the measured data (Step 3) and the damaged state of the structure. If such a relationship between these two quantities exists, but is unknown, the learning problem is to estimate the function that describes this relationship using training data acquired from the test structure (Step 4). Learning problems naturally fall into two categories (Watanabe 1985): (1) supervised learning, in which the training data come from multiple classes, and the labels for the data are known; and (2) unsupervised learning, in which the training data do not have class labels, and one can only attempt to learn intrinsic relationships within the data. In the context of damage detection, unsupervised learning can be applied to the case without damaged data and consequently is limited to Level 1 or Level 2 damage classification, which identifies the presence of damage only. When data are available from both the undamaged and the damaged structure, the supervised learning approach can be employed in higher-level damage identification such as damage quantification (Sohn et al. 2003).

Sohn et al. (2003) reviewed some applications of statistical pattern recognition to structural damage detection. The supervised learning approaches include response surface analysis, fisher discrimination, neural networks, genetic algorithms and support vector machines. The unsupervised learning approaches include control chart analysis, outlier detection, neural networks, hypothesis testing and principal component analysis.

12.7.5 Monte Carlo simulation

The Monte Carlo simulation method is a commonly used numerical method for uncertainty analysis. The basic concept of the method can be explained as follows: (1) given a probability distribution of the stochastic parameters, a large number of samples are generated; (2) for each set of parameters, a deterministic analysis is performed to obtain the corresponding solution and (3) statistics are eventually estimated from these individual solutions. The technique is usually employed when the analytical solution is unfeasible or impossible to obtain.

As it requires a large number of simulations to obtain accurate and reliable statistics, the Monte Carlo simulation method is computationally intensive. Its accuracy is closely related to the size of the samples. Some variance reduction techniques can be applied to improve the variance of simulation results without increasing the sample size (Kottegoda and Rosso 1997). In this context, Agbabian et al. (1988) employed the Monte Carlo simulation method to identify the statistical properties of stiffness coefficients in a linear system. They computed the time histories of the applied excitation as well as the accelerations, velocities and displacements of the system. The calculated data were then corrupted with a set of Gaussian noise. By separately applying the model updating procedure to different time segments, they identified stiffness coefficients. Statistics such as mean, variance and PDFs were subsequently obtained. This work has since been extended to the statistical identification of a nonlinear system approximated by an equivalent linear one (Smyth et al. 2000). Banan et al. (1994), Sanayei and Saletnik (1996), Yeo et al. (2000) and Zhou et al. (2003) adopted similar approaches for studying the effect of measurement noise on identification results.

Researchers from Los Alamos National Laboratory employed the Monte Carlo simulation method to estimate the statistical confidence intervals on modal parameters identified from measured vibration data (Farrar et al. 1998) and then studied the effect of the measurement noise on damage detection (Doebling et al. 1997). The Monte Carlo simulation method was also used to verify other approximate techniques (Xia et al. 2002; Xia and Hao 2003).

12.7.6 Stochastic damage detection method with parametric uncertainties

The stochastic damage detection method with parameter uncertainties contains two steps. The first step is to determine the PDFs of the structural stiffness parameters before and after damage occurrence by integrating the statistical moment-based damage detection method (Section 12.5.3) with the probability density evolution method. In the second step, new damage indices are proposed for identification of both damage locations and damage severities based on a special probability function calculated from the PDFs obtained (Xu et al. 2011).

Without loss of generality, the equation of motion of an MDOF shear building structure with random parameters in the matrix form can be expressed as

$$\mathbf{M}\ddot{\mathbf{x}}(t) + \mathbf{C}(\mathbf{\Theta})\dot{\mathbf{x}}(t) + \mathbf{K}(\mathbf{\Theta})\mathbf{x}(t) = \mathbf{f}(t) \tag{12.77}$$

with the deterministic initial condition

$$\mathbf{x}(t_0) = \mathbf{x}_0, \quad \dot{\mathbf{x}}(t_0) = \dot{\mathbf{x}}_0 \tag{12.78}$$

where:

$\ddot{x}, \dot{x}$ and $\mathbf{x}$ are the structural acceleration, velocity and displacement vector of n order, respectively

$\mathbf{f}(t)$ is the external excitation and $\mathbf{f}(t) = [f_1(t), f_2(t),\ldots, f_n(t)]$

$\mathbf{M}, \mathbf{C}$ and $\mathbf{K}$ are the $n \times n$ mass matrix, damping matrix and stiffness matrix, respectively

$\mathbf{\Theta}$ is the random parameter vector of n_θ order, which reflects the uncertainties in the structural identification procedure, with the known PDF $p_\Theta(\theta)$

To calculate the PDFs of the structural elemental stiffness parameters before and after damage, the random parameter vector $\mathbf{\Theta}$ is discretised into representative points in the domain Ω_Θ. For every given value θ_q of the random parameter vector Θ, the structural stiffness parameters can be identified by the statistical moment-based damage detection method as described in Section 12.5.3. Let us denote the fourth-order moment vector of the story drifts as $\mathbf{M}_{4\theta_q}(\mathbf{k})$. In addition, the fourth-order moment vector of the measured quantities can be directly calculated from the measured displacement responses, denoted as $\hat{\mathbf{M}}_4$. Hence, the residual vector between $\mathbf{M}_{4\theta_q}(\mathbf{k})$ and $\hat{\mathbf{M}}_4$ can be written as

$$\mathbf{F}_\theta(\mathbf{k}) = \mathbf{M}_{4\theta_q}(\mathbf{k}) - \hat{\mathbf{M}}_4 \tag{12.79}$$

The system identification of the undamaged or damaged building structure can then be converted into a nonlinear least-squares problem. Giving $\mathbf{k}$ an initial value $\mathbf{k}_0$ and minimising $\|\mathbf{F}_\theta(\mathbf{k})\|^2$, the structural elemental stiffness parameters can be identified through optimisation algorithms for the specific value θ_q of the random parameter vector. Evidently, due to the random nature of Θ, the identified structural stiffness parameters are also stochastic and dependent on the random parameter Θ denoted as $\mathbf{k}(\Theta)$. The probability density evolution method (Li and Chen, 2004) that has been used successfully in many stochastic systems is employed here to obtain the PDFs of $\mathbf{k}(\Theta)$ because of its versatility and less computationally intensive nature. Construct a virtual random vector process for every elemental stiffness parameter:

$$Z_l(t) = k_l(\Theta) \cdot t \tag{12.80}$$

where $k_l(\Theta)$ is the lth elemental stiffness parameter or the lth element of $\mathbf{k}(\Theta)$. Clearly, there is

$$\dot{Z}_l = k_l(\Theta) \tag{12.81}$$

For a building structure, the lth element stiffness parameter identified is existent, unique for every given value of Θ and dependent on the random parameters Θ, so is the virtual random vector process $Z_l(t)$. According to the principle of preservation of probability, the joint PDF of $(Z_l(t), \Theta)$, denoted as $p_{Z_l\Theta}(z, \theta, t)$, satisfies the following probability density evolution equation:

$$\frac{\partial p_{Z_l\Theta}(z, \theta, t)}{\partial t} + \dot{Z}_l(\theta, t)\frac{\partial p_{Z_l\Theta}(z, \theta, t)}{\partial z} = 0 \tag{12.82}$$

Substituting Equation 12.81 into Equation 12.82 yields

$$\frac{\partial p_{Z_l\Theta}(z, \theta, t)}{\partial t} + k_l(\theta)\frac{\partial p_{Z_l\Theta}(z, \theta, t)}{\partial z} = 0 \tag{12.83}$$

with the initial condition

$$p_{Z_l\Theta}(z, \theta, t)\,|_{t=0} = \delta(z)p_\Theta(\theta) \tag{12.84}$$

where $\delta(\cdot)$ is the Dirac's function. After solving the initial-value problem of Equations 12.83 and 12.84, the PDF of $Z_l(t)$ can be given by

$$p_{Z_l}(z, t) = \int_{\Omega_\Theta} p_{Z_l\Theta}(z, \theta, t)\mathrm{d}\theta \tag{12.85}$$

where Ω_Θ is the distribution domain of Θ. Note that

$$k_l(\Theta) = Z_l(t)\big|_{t=1} \tag{12.86}$$

Therefore, the PDF of $Z_l(t)$ at time $t = 1$ is just the PDF of $k_l(\Theta)$, which we aimed to obtain. In such a way, PDFs of all structural elemental stiffness parameters can be obtained (Xu et al. 2011).

Although structural damage locations can be determined by comparing the PDFs of structural elemental stiffness parameters before and after damage, they have to be identified manually, and it will be time-consuming if there are a large number of structural elements to be investigated. Besides, when the distributions of random parameters are more complex and are not only normal or log-normal distributions, it will not be easy to determine the damage location by comparing the PDFs of structural elemental stiffness parameters before and after damage. Furthermore, damage severities can only be qualitatively given in the first step. In this regard, a new damage index is proposed in the second step to automatically determine the damage locations and to quantitatively determine damage severities according to the value of $P(K_u - K_d) \geq 0$ once the PDFs of structural elemental stiffness parameters before and after damage are obtained. For simplicity, the lth elemental stiffness parameter is denoted as $\hat{K}$ in the following expression. First, a probability function is defined and calculated as follows:

$$P\left\{(\hat{K}^u - \hat{K}^d) \geq \alpha \times \hat{K}^u\right\} = P\left\{\hat{K}^d \leq (1-\alpha) \times \hat{K}^u\right\} = \int_0^\infty \left[\int_0^{(1-\alpha)\times\hat{K}^u} p(\hat{K}^d)\, d\hat{K}^d\right] p(\hat{K}^u)\, d\hat{K}^u \tag{12.87}$$

The probability function $P\left\{(\hat{K}^u - \hat{K}^d) \geq \alpha \times \hat{K}^u\right\}$ is the function of α, denoted as $G(\alpha)$. Its value decreases with the increase of α. The coefficient α is a variable ranging from 0% to 100%. It is assumed that $\hat{K}^u$ and $\hat{K}^d$ are independent of each other. As introduced in Equation 12.81, the PDFs of structural stiffness parameters are dependent on the PDFs of the random parameters Θ. Therefore, if there is no damage at the associated location, the identified $\hat{K}^u$ and $\hat{K}^d$ should have the same PDF, since the same uncertainties or random parameters are considered before and after damage occurrence. Under this condition, $\hat{K}^u$ and $\hat{K}^d$ are both denoted as $\hat{K}$, and the value of $P\left\{(\hat{K}^u - \hat{K}^d) \geq \alpha \times \hat{K}^u\right\}$ at $\alpha=0$, that is, the value of $P\left\{(\hat{K}^u - \hat{K}^d) \geq 0\right\}$, can be derived as follows:

$$P\left\{(\hat{K}^u - \hat{K}^d) \geq 0\right\} = P\left\{\hat{K}^d \leq \hat{K}^u\right\} = \int_0^\infty \int_0^{\hat{K}^u} p(\hat{K}^d, \hat{K}^u)\, d\hat{K}^d d\hat{K}^u$$

$$= \int_0^\infty \left[\int_0^{\hat{K}} p(\hat{K})\, d\hat{K}\right] p(\hat{K})\, d\hat{K} = \frac{1}{2}\left[F(\hat{K})\right]^2\Big|_0^\infty = \frac{1}{2} \tag{12.88}$$

where $F(\hat{K})$ is the distribution function of $\hat{K}$. According to Equation 12.88, it can be concluded that when there is no damage at the location investigated, the value of $P\left\{(\hat{K}^u - \hat{K}^d) \geq 0\right\}$ should equal 0.5. Otherwise, if there is damage at the location investigated, the PDF of $\hat{K}^d$ should offset towards the negative abscissa compared with the PDF of the stiffness parameter in the undamaged state, $\hat{K}^u$. Hence, the value of $P\left\{(\hat{K}^u - \hat{K}^d) \geq 0\right\}$ should be larger than 0.5. Therefore, whether structural damage occurs or not can be determined according to the values of $P\left\{(\hat{K}^u - \hat{K}^d) \geq 0\right\}$. When the value of $P\left\{(\hat{K}^u - \hat{K}^d) \geq 0\right\}$ is larger than 0.5, there should be damage occurrence at the corresponding location, and the larger the value of $P\left\{(\hat{K}^u - \hat{K}^d) \geq 0\right\}$, the higher the probability of damage occurrence at this place.

In addition, once the probability function $P\{(\hat{K}^u - \hat{K}^d) \geq \alpha \times \hat{K}^u\}$ is obtained, the derivative of $P\{(\hat{K}^u - \hat{K}^d) \geq \alpha \times \hat{K}^u\}$ in terms of α, which is also the function of the variable α, can further be calculated. Rearrange the probability function $G(\alpha)$ as

$$G(\alpha) = P\{(\hat{K}^u - \hat{K}^d) \geq \alpha \times \hat{K}^u\} = P\left\{\left(\frac{\hat{K}^u - \hat{K}^d}{\hat{K}^u}\right) \geq \alpha\right\} = 1 - P\left\{\left(\frac{\hat{K}^u - \hat{K}^d}{\hat{K}^u}\right) \leq \alpha\right\} \quad (12.89)$$

Since $(\hat{K}^u - \hat{K}^d)/\hat{K}^u$ is the definition of structural damage severity, $P\{[(\hat{K}^u - \hat{K}^d)/\hat{K}^u] \leq \alpha\}$ is just the distribution function of structural damage severity, denoted as $F(\alpha)$. Therefore,

$$-G'(\alpha) = g(\alpha) \quad (12.90)$$

The negative derivative of $G(\alpha)$ is just the PDF of structural damage severity, denoted as $g(\alpha)$. Therefore, the value of α corresponding to the maximum of $g(\alpha)$ should be the most likely value of structural damage severity, denoted as β, which is straightforwardly set as the index of damage severity for structures with uncertainties. Therefore, according to these deductions, not only structural damage locations but also their corresponding damage severities can be identified by the proposed stochastic damage detection method. The flowchart of the stochastic damage detection method is presented in Figure 12.18.

To evaluate the effectiveness of the stochastic damage detection method proposed for building structures with uncertainties or of random parameters, a three-story shear building model is investigated here. The mass and horizontal stiffness coefficients of the building are 350,250 kg and 4,728,400 kN/m, respectively, for the first story, 262,690 kg and 315,230 kN/m for the second story, and 175,130 kg and 157,610 kN/m for the third story. The mass of each floor is assumed to be invariant. The structural damping ratio is a very difficult parameter to ascertain because of its complex nature. In the following numerical

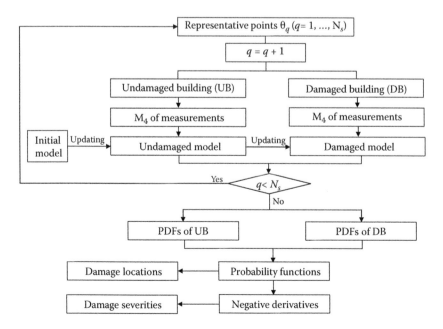

Figure 12.18 Flowchart of the stochastic damage detection method.

investigation, the first modal damping ratio is selected as a random parameter due to its uncertainty, or the random error in the identification of this parameter. Actually, it is not clear what kind of probability distribution can best reflect the uncertainty of the structural damping ratio. Nevertheless, the structural damping ratio shall be non-negative, and therefore, the first damping ratio is approximated as a log-normal distribution with mean value $\bar{\xi} = 1\%$ and standard deviation σ of 10% of the mean value. The second modal damping ratio is fixed as 2.14%. The third modal damping ratio is calculated from the first two modal damping ratios according to the Rayleigh damping assumption. The ground acceleration is simulated as a colour white noise corresponding to the Kanai–Tajimi spectrum and applied to the building.

Three damage scenarios, named Scenarios 1, 2 and 3, are created for the shear building. Both Scenarios 1 and 2 have single damage at the second story, but with different damage severities of 10% and 20%, respectively. Scenario 3 has multi-damage at the first and third stories, with damage severities of 20% and 10%, respectively. The first modal damping ratio is discretised into 21 representative points in the domain $[\bar{\xi} - 3\sigma, \bar{\xi} + 3\sigma]$. For every given representative point, the horizontal stiffness parameters of the undamaged and damaged shear building can be identified using the statistical moment-based damage detection method. Then, the PDF of every story's horizontal stiffness parameter for the undamaged and damaged structure can be obtained.

Take Scenario 3 as an example. Figure 12.19 gives the comparison of the identified PDFs of horizontal stiffness parameters before and after damage. The solid lines stand for the PDFs of the undamaged shear building, while the dotted lines stand for the counterparts of the damaged building. It can be seen that, for a damaged story, the stiffness values corresponding to the peak values of the PDFs after damage are apparently smaller than those before damage occurrence. The larger the damage severity, the more backward offset occurs between the PDFs of the damaged state and those of the undamaged state. Therefore, the damage locations can be qualitatively determined according to the identified PDFs of structural elemental stiffness parameters before and after damage.

Subsequently, the damage locations and their corresponding damage severities can be quantitatively identified based on the proposed probability functions $P\left\{(K^u - K^d) \geq \alpha \times K^u\right\}$ according to Equation 12.90. Also, taking Scenario 3 as an example, the corresponding probability function is presented in Figure 12.20. In this damage case, the values of $P\left\{(K^u - K^d) \geq \alpha \times K^u\right\}$ at $\alpha = 0$, or the values of $P\{(K^u - K^d) \geq 0\}$, for the first, second and third story, are 99.99%, 50.34% and 95.58%, respectively. It is apparent that the values of the first and third stories are much larger than 0.5, which means that the damage exists on these two stories. To provide more information about the structural damage for these damage scenarios, the negative derivatives of the probability functions $P\{K^u - K^d\} \geq \alpha \times K^u$ or the PDFs of structural damage severity for every story are calculated. Here, only the calculated result of Scenario 3 is plotted in Figure 12.21. As seen from Figure 12.21, the identified damage severity values of the first and third stories in this case are 20.5% and 10%, respectively, which are almost the same as the actual values, 20% and 10%. Also, it can be seen that the second story should have no damage, according to the profile of Figure 12.21b. Clearly, the identified results are identical with or very close to the real values. In summary, the example numerical results show that the proposed stochastic damage detection method can accurately detect both damage locations and their corresponding damage severities when uncertainty or random parameters of the building structure are taken into account.

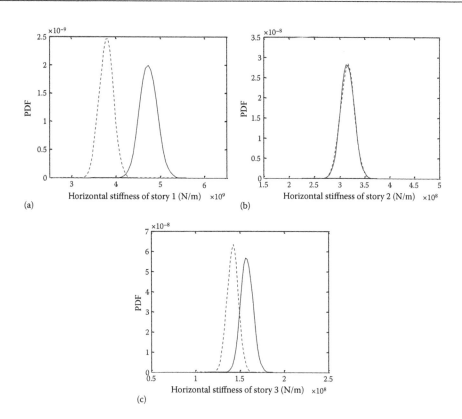

Figure 12.19 Comparison of identified PDFs of horizontal stiffness before and after damages.

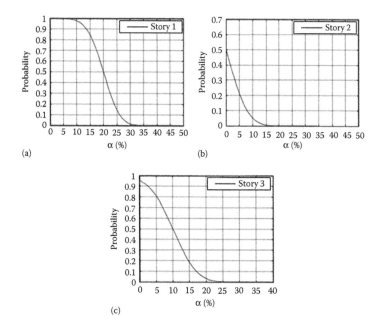

Figure 12.20 Probability function of every story in Scenario 3: (a) story 1, (b) story 2, (c) story 3.

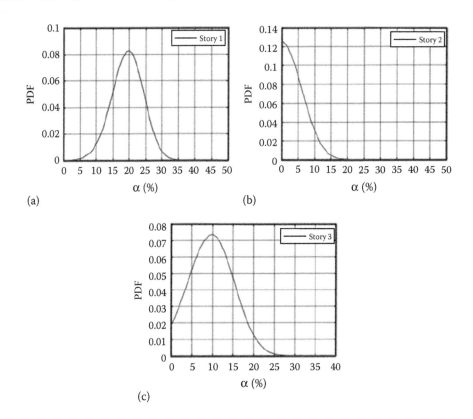

Figure 12.21 Derivatives of probability functions in Scenario 3: (a) story 1, (b) story 2, (c) story 3.

NOTATION

B′	The element-dependent matrix characterising the relationship between displacement and strain
COMAC	CO ordinate MAC
e	An error vector containing the differences between the eigenvalues and mode shapes at the measured DOFs of the structure before and after updating
EI	Flexural rigidity of the beam
$f(t)$	External excitation
F	Modal flexibility matrix
$F(\alpha)$	The distribution function of structural damage severity
$g(\alpha)$	The negative derivative of $G(\alpha)$
$G(\alpha)$	A probability function
H	The FRF matrix
H	The length of elements
$H_d(\omega), H_v(\omega), H_a(\omega)$	The displacement FRF, the velocity FRF and the acceleration FRF, respectively
$K_{e,j}$	The jth elemental stiffness matrix
$k_l(\Theta)$	The lth elemental stiffness parameter
L	The total length of the beam
L_Ω	The lower bound of the interval $\Omega(\theta i, \mu)$

m, c, k	Mass, damping and stiffness coefficient in single DOF structure, respectively
$\mathbf{M}_{4\theta_q}(\mathbf{k})$	The fourth-order moment vector of the story drifts
$\hat{\mathbf{M}}_4$	The fourth-order moment vector of the measured quantities
$M_2^{\mathrm{dis}}, M_2^{v}, M_2^{a}$	The second-order moment of displacement, velocity and acceleration, respectively
$M_4^{\mathrm{dis}}, M_4^{v}, M_4^{a}$	The fourth-order moment of displacement, velocity and acceleration, respectively
$M_6^{\mathrm{dis}}, M_6^{v}, M_6^{a}$	The sixth-order moment of displacement, velocity and acceleration, respectively
MAC	Modal assurance criterion
$\mathbf{M, C, K}$	Mass, damping and stiffness matrices, respectively
MSE	Modal strain energy
n_s, n_m, n_e	The number of time instants, sensors and elements, respectively
$\tilde{N}$	The number of neurons in the hidden layer
$p_{Z|\Theta}(z, \theta, t)$	The joint PDF of $(Z_l(t), \Theta)$
PDE	The probability of damage existence
PDF	Probability density function
$P(H_j|\psi)$	The posterior probability
$P(H_{\max}|\psi)$	The largest posterior probability
$\mathbf{S}$	Sensitivity matrix
$S(\omega)$	The power spectrum of the structural response
$S_f(\omega)$	The power spectrum of ground excitation
w	The transverse displacement
$\tilde{w}_i$	The weight of the neuron i in the linear output neuron
$W_{\theta j}, W_{\lambda}, W_{\Phi i}$	The proportional uncertainties of the structural parameters, eigenvalues and eigenvectors, respectively
$\ddot{\mathbf{x}}, \dot{\mathbf{x}}, \mathbf{x}$	The structural acceleration, velocity and displacement vector, respectively
$\ddot{x}, \dot{x}, x$	The structural acceleration, velocity and displacement response in single DOF structure, respectively
$\tilde{\mathbf{x}}, \tilde{\mathbf{y}}$	The input and output vector of the RBF network
$\ddot{x}_g$	Ground excitation
y	Structural response measurement with noise
$\mathbf{y}_d^e$	The $n_s \cdot (n_e + 2n)$-dimensional vector of the reconstructed responses from the limited measured responses of the damaged structure at the optimal sensor locations
$\mathbf{y}_a^k$	The analytical responses of the structure of damage parameters
$\mathbf{y}_u^m, \mathbf{y}_d^m$	The $(n_s \cdot n_m)$-dimensional vector of the measured responses of the structure before and after damage occurrence, respectively
$Z_l(t)$	A virtual random vector process
α	A variable ranging from 0% to 100%
γ	Noise effect ratio
Δ	The changes of the stiffness matrix caused by damages
Δ^m	The response discrepancy between the damaged and intact states
Σ	The $(n_m \times n_m)$ variance matrix of dynamic response of the structure
η	The measurement noise intensity

θ	The structural parameter
$\boldsymbol{\theta}$	The n_e-dimensional vector of damage parameters
Θ	The random parameter vector
$\bar{\lambda}$	The parameter used for controlling the trade-off between the sparsity of the solution and the residual norm
λ	The ith eigenvalue
μ	The probability of θ_i contained within the interval $\Omega(\theta_i, \mu)$
$\hat{\mu}$	The structural damage severity
Ξ	The damping ratio of a building
σ^2	The variance of the structural response
$\phi_{q,i}$	The ith modal displacement at measurement point q
$\boldsymbol{\Phi}_i$	The ith mode shape vector
$\tilde{\psi}$	The radial basis function that depends only on the distance from a centre vector
$\boldsymbol{\Psi}_u^m, \boldsymbol{\Psi}_d^m$	The strain mode shapes before damage and after damage
$\Omega(\theta_i, \mu)$	The interval of the healthy stiffness parameter
Ω_Θ	The distribution domain of Θ
ω_i	The ith circular frequency
$\tilde{c}_i$	The centre vector for the neuron i
subscript or superscript u	Undamaged state
subscript or superscript d	Damaged state
subscript or superscript m	Measurements

REFERENCES

Adams, R.D., J.M.W. Brownjohn, and P. Cawley. 1991. The detection of defects in GRP lattice structures by vibration measurements. *NDT E Int.*, 24(3): 123–34.

Adams, R.D. and J. Coppendale. 1976. Measurement of the elastic moduli of structural adhesives by a resonant bar technique. *J. Mech. Eng. Sci.*, 18(3): 149–58.

Agbabian, M.S., S.F. Masri, R.K. Miller, and T.K. Caughey. 1988. A system identification approach to the detection of changes in structural parameters. In *Proceedings of Workshop on Structural Safety Evaluation Based on System Identification Approach*, ed. H.G. Natke and J.T.P. Yao, 341–56. Wiesbaden: Friedrick Vieweg & Son.

Anderson, D.A. and R.K. Seals. 1981. Pulse velocity as a predictor of 28 and 90 day strength. *ACI J. Proc.*, 78(9): 116–22.

ASCE. 2005. 2005 report card for America's infrastructures. Reston, VA: American Society of Civil Engineers. http://ascelibrary.org/doi/book/10.1061/9780784478851.

ASTM. 2004. *Standard C1383: Standard Test Method for Measuring the P-Wave Speed and the Thickness of Concrete Plates Using the Impact-Echo Method.* West Conshohocken, PA: ASTM International.

ASTM. 2007a. *Standard D4788: Standard Test Method for Detecting Delaminations in Bridge Decks Using Infrared Thermography.* West Conshohocken, PA: ASTM International.

ASTM. 2007b. *Standard D5882: Test Method for Low Strain Integrity Testing of Piles.* West Conshohocken, PA: ASTM International.

ASTM. 2009. *Standard C597: Standard Test Method for Pulse Velocity through Concrete.* West Conshohocken, PA: ASTM International.

Banan, M.R., M.R. Banan, and K.D. Hjelmstad. 1994. Parameter estimation of structures from static response I: Computational aspects. *J. Struct. Eng.*, 120: 3243–58.

Beck, J.L. and S.K. Au. 2002. Bayesian updating of structural models and reliability using Markov chain Monte Carlo simulation. *J. Eng. Mech.*, 128(4): 380–91.

Beck, J.L. and L.S. Katafygiotis. 1998. Updating models and their uncertainties I: Bayesian statistical framework. *J. Eng. Mech.*, 124: 455–61.

Biswas, M., A.K. Pandey, and S. Bluni. 1994. Modified chain-code computer vision techniques for interrogation of vibration signatures for structural fault detection. *J. Sound Vibr.*, 175(1): 89–104.

Boyd, S. and L. Vandanberghe. 2004. *Convex Optimization*. New York: Cambridge University Press.

Brownjohn, J.M.W. 2007. Structural health monitoring of civil infrastructure. *Phil. Trans. R. Soc. A*, 365(1851): 589–622.

Brownjohn, J.M.W. and G.H. Steele. 1979. Non-destructive testing using measurements of structural dampings. PhD diss., University of Bristol, Bristol, UK.

Cao, M.S. and P.Z. Qiao. 2008. Integrated wavelet transform and its application to vibration mode shapes for the damage detection of beam-type structures. *Smart Mater. Struct.*, 17(5): 055014.

Cawley, P. and R.D. Adams. 1979. The locations of defects in structures from measurements of natural frequencies. *J. Strain Anal. Eng. Des.*, 14(2): 49–57.

Chance, J., G.R. Tomlinson, and K. Worden. 1994. A simplified approach to the numerical and experimental modeling of the dynamics of a cracked beam. In *Proceedings of the 12th International Modal Analysis Conference*, ed. D.J. DeMichele, 778–85. Bethel, CT: Society for Experimental Mechanics.

Chang, P.C. and S.C. Liu. 2003. Recent research in nondestructive evaluation of civil infrastructures. *J. Mater. Civ. Eng.*, 15(3): 298–304.

Chang, S.P., J.Y. Yee, and J. Lee. 2009. Necessity of the bridge health monitoring system to mitigate natural and man-made disasters. *Struct. Infrastruct. Eng.*, 5(3): 173–97.

Chen, H.G., Y.J. Yan, W.H. Chen, et al. 2007. Early damage detection in composite wingbox structures using Hilbert-Huang transform and genetic algorithm. *Struct. Health Monit.*, 6(4): 281–97.

Choy, F.K., R. Liang, and P. Xu. 1995. Fault identification of beams on elastic foundation. *Comput. Geotech.*, 17: 157–76.

Cornwell, P.J., S.W. Doebling, and C.R. Farrar. 1997. Application of the strain energy damage detection method to plate-like structures. In *Proceedings of the 15th International Modal Analysis Conference*, ed. A.L. Wicks, 1312–18. Bethel, CT: Society for Experimental Mechanics.

Cornwell, P.J., S.W. Doebling, and C.R. Farrar. 1999. Application of the strain energy damage detection method to plate-like structures. *J. Sound Vibr.*, 224(2): 359–74.

Cornwell, P.J., M. Kam, B. Carlson, et al. 1998. Comparative study of vibration-based damage ID algorithms. In *Proceedings of the 16th International Modal Analysis Conference*, ed. A.L. Wicks and D.J. DeMichele, 1710–16. Santa Barbara, CA. Bethel, CT: Society for Experimental Mechanics.

Davis, A.G., J.G. Evans, and B.H. Hertlein. 1997. Nondestructive evaluation of concrete radioactive waste tanks. *J. Perform. Constr. Facil.*, 11(4): 161–7.

Davis, A.G. and B.H. Hertlein. 1991. Development of nondestructive small-strain methods for testing deep foundations: A review. *Transp. Res. Record*, 1331: 15–20.

Doebling, S.W., C.R. Farrar, and R.S. Goodman. 1997. Effects of measurement statistics on the detection of damage in the Alamosa Canyon Bridge. In *Proceedings of the 15th International Modal Analysis Conference*, ed. A.L. Wicks, 919–29. Orlando, FL. Bethel, CT: Society for Experimental Mechanics.

Doebling, S.W., C.R. Farrar, M.B. Prime, and D.W. Shevitz. 1996. Damage identification and health monitoring of structural and mechanical systems from changes in their vibration characteristics: A literature review. Los Alamos National Laboratory report LA-13070-MS.

Elliot, J.F. 1996. Monitoring of prestressed structures. *Civil Eng.*, 66(7): 61–3.

Fan, W. and P.Z. Qiao. 2009. A 2-D continuous wavelet transform of mode shape data for damage detection of plate structures. *Int. J. Solids Struct.*, 46(25–26): 4379–95.

Fan, W. and P.Z. Qiao. 2011. Vibration-based damage identification methods: A review and comparative study. *Struct. Health Monit.*, 10(1): 83–111.

Farrar, C.R., S.W. Doebling, and P.J. Cornwell. 1998. A comparison study of modal parameter confidence intervals computed using the Monte Carlo and Bootstrap techniques. In *Proceedings of the 16th International Modal Analysis Conference*, ed. A.L. Wicks and D.J. DeMichele, 936–44, Santa Barbara, CA. Bethel, CT: Society for Experimental Mechanics.

Farrar, C.R. and D.V. Jauregui. 1996. Damage detection algorithms applied to experimental and numerical modal data from the I-40 bridge. Report LA-13074-MS, Los Alamos: Los Alamos National Laboratory.

Farrar, C.R. and K. Worden. 2007. An introduction to structural health monitoring. *Phil. Trans. R. Soc. A*, 365(1851): 303–15.

Farrar, C.R. and K. Worden. 2013. *Structural Health Monitoring: A Machine Learning Perspective*. Hoboken, NJ: Wiley.

Fassois, S.D. and J.S. Akellariou. 2007. Time-series methods for fault detection and identification in vibrating structures. *Phil. Trans. R. Soc. A*, 365(1851): 411–48.

Friswell, M.I. and J.E. Mottershead. 1995. *Finite Element Model Updating in Structural Dynamics*. Dordrecht: Kluwer Academic.

Friswell, M.I., J.E.T. Penny, and D.E.L. Wilson. 1994. Using vibration data and statistical measures to locate damage in structures. *Modal Anal.: Int. J. Anal. Exp. Modal Anal.*, 9(4): 239–54.

Goh, L.D., N. Bakhary, A.A. Rahman, and B.H. Ahmad. 2013. Application of neural network for prediction of unmeasured mode shape in damage detection. *Adv. Struct. Eng.*, 16(1): 99–113.

Hearn, G. and R.G. Testa. 1991. Modal analysis for damage detection in structures. *J. Struct. Eng.*, 117(10): 3042–63.

Hellier, C.J. 2001. *Handbook of Nondestructive Evaluation*. New York: McGraw-Hill.

Hernandez, E.M. 2014. Identification of isolated structural damage from incomplete spectrum information using ℓ_1 norm minimization. *Mech. Syst. Signal Proc.*, 46(1): 59–69.

Hester, D. and A. González. 2012. A wavelet-based damage detection algorithm based on bridge acceleration response to a vehicle. *Mech. Syst. Signal Proc.*, 28: 145–66.

Hou, Z., M. Noori, and R. St. Amand. 2000. Wavelet-based approach for structural damage detection. *J. Eng. Mech.*, 126(7): 677–83.

Hsu, W.K., D.J. Chiou, C.W. Chen, et al. 2012. Sensitivity of initial damage detection for steel structures using the Hilbert-Huang transform method. *J. Vib. Control*, 19: 857–878.

Hua, X.G., Y.Q. Ni, Z.Q. Chen, and J.M. Ko. 2008. An improved perturbation method for stochastic finite element model updating. *Int. J. Numer. Methods Eng.*, 73(13): 1845–64.

Huang, N.E. and S.S. Shen. 2005. *Hilbert-Huang Transform and Its Applications*. Toh Tuck Link, Singapore: World Scientific Publishing Co. Pte. Ltd.

Huang, N.E., Z. Shen, S.R. Long, et al. 1998. The empirical mode decomposition and the Hilbert spectrum for nonlinear and non-stationary time series analysis. *Proc. R. Soc. Lond. A*, 454(1971): 903–95.

Huang, Y., D. Meyer, and C. Nemat-Nasser. 2009. Damage detection with spatially distributed 2D continuous wavelet transform. *Mech. Mater.*, 41(10): 1096–107.

Ibrahim, S.R. 1993. Correlation and updating methods: Finite element dynamic model and vibration test data. In *Proceedings of International Conference on Structural Dynamics Modelling, Test, Analysis and Correlation*, 323–47, Cranfield, UK. Glasgow, Scotland: Bell & Bain.

Katafygiotis, L.S. and J.L. Beck. 1998. Updating models and their uncertainties II: Model identifiability. *J. Eng. Mech.*, 124: 463–7.

Keye, S. 2006. Improving the performance of model-based damage detection methods through the use of an updated analytical model. *Aerosp. Sci. Technol.*, 10(3): 199–206.

Khodaparast, H.H., J.E. Mottershead, and M.I. Friswell. 2008. Perturbation methods for the estimation of parameter variability in stochastic model updating. *Mech. Syst. Signal Proc.*, 22(8): 1751–73.

Kim, H. and H. Melhem. 2004. Damage detection of structures by wavelet analysis. *Eng. Struct.*, 26(3): 347–62.

Kim, J.H., H.S. Jeon, and C.W. Lee. 1992. Application of the modal assurance criteria for detecting and locating structural faults. In *Proceedings of the 10th International Modal Analysis Conference*, ed. D.J. DeMichele, 536–40. Bethel, CT: Society for Experimental Mechanics.

Knab, L.I., G.V. Blessing, and J.R. Clifton. Laboratory evaluation of ultrasonics for crack detection in concrete. *ACI J. Proc.*, 80(1): 17–27.

Ko, J.M., C.W. Wong, and H.F. Lam. 1994. Damage detection in steel framed structures by vibration measurement approach. In *Proceedings of the 12th International Modal Analysis Conference*, ed. D.J. DeMichele, 280–6. Bethel, CT: Society for Experimental Mechanics.

Kottegoda, N.T. and R. Rosso. 1997. *Statistics, Probability, and Reliability for Civil and Environmental Engineers*. New York: McGraw-Hill.

Kunz, J.T. and J.W. Eales. 1985. Evaluation of bridge deck condition by the use of thermal infrared and ground penetrating radar. In *Proceedings of the 2nd Annual International Bridge Conference*, 121–7, Pittsburgh, PA. North Vancouver, British Columbia: Buckland & Taylor.

Li, D.S. and J.P. Ou. 2008. Acoustic emission monitoring and critical failure identification of bridge cable damage. In *Proceedings of SPIE Nondestructive Characterization for Composite Materials, Aerospace Engineering, Civil Infrastructure, and Homeland Security*, ed. P.J. Shull, H.F. Wu, A.A. Diaz, and D.W. Vogel, 1–5. Bellingham, Washington: SPIE 6934.

Li, J. and J.B. Chen. 2004. Probability density evolution method for dynamic response analysis of structures with uncertain parameters. *Comput. Mech.*, 34: 400–9.

Li, Z., S.M. Xia, J. Wang, and X.Y. Su. 2006. Damage detection of cracked beams based on wavelet transform. *Int. J. Impact Eng.*, 32(7): 1190–200.

Liu, P.L. 1995. Identification and damage detection of trusses using modal data. *J. Struct. Eng.*, 121(4): 599–608.

Loutridis, S.J. 2004. Damage detection in gear systems using empirical mode decomposition. *Eng. Struct.*, 26(12): 1833–41.

Maia, N.M.M., J.M.M. Silva, J. He, et al. 1997. *Theoretical and Experimental Modal Analysis*. Hertfordshire, UK: Research Studies Press Ltd.

Naik, T.R., V.M. Malhotra, and J.S. Popovics. 2004. The ultrasonic pulse velocity method. In *Handbook on Nondestructive Testing of Concrete*, ed. V.M. Malhotra and N.J. Carino. Boca Raton, FL: CRC.

Nair, A. and C.S. Cai. 2010. Acoustic emission monitoring of bridges: Review and case studies. *Eng. Struct.*, 32(6): 1704–14.

Narkis, Y. 1994. Identification of crack location in vibrating simply supported beams. *J. Sound Vibr.*, 172(4): 549–58.

Ndambi, M.J.M. 2002. Damage assessment in reinforced concrete beams by damping analysis. Ph.D. Diss., Vrije University, Brussels, Belgium.

Nelson, R.B. 1976. Simplified calculation of eigenvector derivatives. *AIAA J.*, 14: 1201–5.

Nguyen, K.V. and H.T. Tran. 2010. Multi-cracks detection of a beam-like structure based on the on-vehicle vibration signal and wavelet analysis. *J. Sound Vibr.*, 329(21): 4455–65.

Ou, J.P. and H. Li. 2010. Structural health monitoring in mainland China: Review and future trends. *Struct. Health Monit.*, 9(3): 219–31.

Ovanesova, A.V. and L.E. Suárez. 2004. Applications of wavelet transforms to damage detection in frame structures. *Eng. Struct.*, 26(1): 39–49.

Overschee, P. and B. De Moo. 1994. N4SID: Subspace algorithms for the identification of combined deterministic-stochastic systems. *Automatica*, 30(1): 75–93.

Pandey, A.K. and M. Biswas. 1994. Damage detection in structures using changes in flexibility. *J. Sound Vibr.*, 169(1): 3–17.

Pandey, A.K., M. Biswas, and M.M. Samman. 1991. Damage detection from changes in curvature mode shapes. *J. Sound Vibr.*, 145(2): 321–32.

Papadopoulos, L. and E. Garcia. 1998. Structural damage identification: A probabilistic approach. *AIAA J.*, 36(11): 2137–45.

Park, J. and I.W. Sandberg. 1991. Universal approximation using radial-basis-function networks. *Neural Comput.*, 3(2): 246–57.

Pascual, R., J.C. Golinval, and M. Razeto. 1996. Testing of FRF based model updating methods using a general finite element program. In *Proceedings of the 21st International Seminar on Modal Analysis: Noise and Vibration Engineering*, Katholieke Universiteit Leuven, Departement Werktuigkunde, 1933–45. Leuven, Belgium. https://www.isma-isaac.be/past/conf/.

Peterson, L.D., S.J. Bullock, and S.W. Doebling. 1996. The statistical sensitivity of experimental modal frequencies and damping ratios to measurement noise. *Modal Anal.: Int. J. Anal. Exp. Modal Anal.*, 11(1): 63–75.

Raghavendrachar, M. and A.E. Aktan. 1992. Flexibility by multireference impact testing for bridge diagnostics. *J. Struct. Eng.*, 118(8): 2186–203.

Reda Taha, M.M., A. Noureldin, J.L. Lucero, and T.J. Baca. 2006. Wavelet transform for structural health monitoring: A compendium of uses and features. *Struct. Health Monit.*, 5(3): 267–95.

Rens, K.L., T.J. Wipf, and F.W. Klaiber. 1997. Review of nondestructive evaluation techniques of civil infrastructure. *J. Perform. Constr. Facil.*, 11(4): 152–60.

Rizos, P.F., N. Aspragathos, and A.D. Dimarogonas. 1990. Identification of crack location and magnitude in a cantilever beam from the vibration modes. *J. Sound Vibr.*, 138(3): 381–8.

Rytter, A. 1993. Vibration based inspection of civil engineering structures. PhD diss., Aalborg University, Denmark.

Salane, H.J. and J.W. Baldwin. 1990. Identification of modal properties of bridges. *J. Struct. Eng.*, 116(7): 2008–21.

Salawu, O.S. 1997. Detection of structural damage through changes in frequency: A review. *Eng. Struct.*, 19(9): 718–23.

Salawu, O.S. and C. Williams. 1995. Bridge assessment using forced-vibration testing. *J. Struct. Eng.*, 121(2): 161–73.

Samman, M.M., M. Biswas, and A.K. Pandey. 1991. Employing pattern recognition for detecting cracks in a bridge model. *Modal Anal.: Int. J. Anal. Exp. Modal Anal.*, 6(1): 35–44.

Sanayei, M., O. Onipede, and S.R. Babu. 1992. Selection of noisy measurement locations for error reduction in static parameter identification. *AIAA J.*, 30(9): 2299–309.

Sanayei, M. and M.J. Saletnik. 1996. Parameter estimation of structures from static strain measurements II: Error sensitivity analysis. *J. Struct. Eng.*, 122: 563–72.

Schalkoff, R.J. 1992. *Pattern Recognition: Statistical, Structural and Neural Approaches*. New York: Wiley.

Shi, Z.Y., S.S. Law, and L.M. Zhang. 1998. Structural damage localization from modal strain energy change. *J. Sound Vibr.*, 218(5): 825–44.

Shull, P.J. 2002. *Nondestructive Evaluation: Theory, Techniques, and Applications*. New York: Marcel Dekker.

Smyth, A.W., S.F. Masri, T.K. Caughey, and N.F. Hunter. 2000. Surveillance of mechanical systems on the basis of vibration signature analysis. *J. Appl. Mech.*, 67: 540–51.

Sohn, H., C.R. Farrar, F.M. Hemez, et al. 2003. A review of structural health monitoring literature: 1996–2001. Report LA-13976-MS, Los Alamos: Los Alamos National Laboratory.

Sohn, H. and K.H. Law. 1997. A Bayesian probabilistic approach for structure damage detection. *Earthq. Eng. Struct. Dyn.*, 26: 1259–81.

Sohn, H. and K.H. Law. 2000. Bayesian probabilistic damage detection of a reinforced-concrete bridge column. *Earthq. Eng. Struct. Dyn.*, 29: 1131–52.

Stubbs, N., T.H. Broome, and R. Osegueda. 1990. Nondestructive construction error detection in large space structures. *AIAA J.*, 28(1): 146–52.

Stubbs, N., J.T. Kim, and C.R. Farrar. 1995. Field verification of a nondestructive damage localization and sensitivity estimator algorithm. In *Proceedings of the 13th International Modal Analysis Conference*, ed. A.L. Wicks, 210–18. Bethel, CT: Society for Experimental Mechanics.

Stubbs, N. and R. Osegueda. 1990a. Global non-destructive damage evaluation in solids. *Modal Anal.: Int. J. Anal. Exp. Modal Anal.*, 5(2): 67–79.

Stubbs, N. and R. Osegueda. 1990b. Global damage detection in solids: Experimental verification. *Modal Anal.: Int. J. Anal. Exp. Modal Anal.*, 5(2): 81–97.

Sun, Z. and C. Chang. 2002. Structural damage assessment based on wavelet packet transform. *J. Struct. Eng.*, 128(10): 1354–61.

Tang, J.P., D.J. Chiou, C.W. Chen, et al. 2010. A case study of damage detection in benchmark buildings using a Hilbert-Huang Transform-based method. *J. Vib. Control*, 17: 623–636.

Vogel, T., B. Schechinger, and S. Fricker. 2006. Acoustic emission analysis as a monitoring method for prestressed concrete structures. In *Proceedings of the 9th European Conference on NDT*, ed. M. Farley, 281–98, Berlin, Germany. Copyrighted by NDT.net. http://www.ndt.net/article/ecndt2006/doc/We.4.4.3.pdf.

Wang, J.H. and C.M. Liou. 1991. Experimental identification of mechanical joint parameters. *J. Vib. Acoust.*, 113: 28–36.

Wang, W.J. and A.Z. Zhang. 1987. Sensitivity analysis in fault diagnosis of structures. In *Proceedings of the 5th International Modal Analysis Conference*, London, ed. D.J. DeMichele, 496–501, London. Bethel, CT: Society for Experimental Mechanics.

Watanabe, S. 1985. *Pattern Recognition: Human and Mechanical*. New York: Wiley.

Weil, G.J. 2004. Infrared thermographic techniques. In *Handbook on Nondestructive Testing of Concrete*, ed. V.M. Malhotra and N.J. Carino. Boca Raton, FL: CRC.

Wu, N. and Q. Wang. 2011. Experimental studies on damage detection of beam structures with wavelet transform. *Int. J. Eng. Sci.*, 49(3): 253–61.

Xia, Y. and H. Hao. 2003. Statistical damage identification of structures with frequency changes. *J. Sound Vibr.*, 263(4): 853–70.

Xia, Y., H. Hao, J.M.W. Brownjohn, and P.Q. Xia. 2002. Damage identification of structures with uncertain frequency and mode shape data. *Earthq. Eng. Struct. Dyn.*, 31(5): 1053–66.

Xia, Y., H. Hao, and A.J. Deeks. 2007. Dynamic assessment of shear connectors in slab-girder bridges. *Eng. Struct.*, 29(7): 1475–86.

Xia, Y., Y.L. Xu, Z.L. Wei, H.P. Zhu, and X.Q. Zhou. 2011. Variation of structural vibration characteristics versus non-uniform temperature distribution. *Eng. Struct.*, 33(1): 146–53.

Xu, Y.L. and J. Chen. 2004. Structural damage detection using empirical mode decomposition: Experimental investigation. *J. Eng. Mech.*, 130(11): 1279–88.

Xu, Y.L., J. Zhang, J. Li, and X.M. Wang. 2011. Stochastic damage detection method for building structures with parametric uncertainties. *J. Sound Vibr.*, 330(20): 4725–37.

Xu, Y.L., J. Zhang, J.C. Li, and Y. Xia. 2009. Experimental investigation on statistical moment-based structural damage detection method. *Struct. Health Monit.*, 8: 555–75.

Xu, Y.L., H.P. Zhu, and J. Chen. 2004. Damage detection of mono-coupled multistory buildings: Numerical and experimental investigations. *Struct. Eng. Mech.*, 18(6): 709–29.

Yang, J.N., Y. Lei, S. Lin, and N. Huang. 2004. Hilbert-Huang based approach for structural damage detection. *J. Eng. Mech.*, 130(1): 85–95.

Yang, J.N., Y. Lei, S.W. Pan, and N. Huang. 2003a. System identification of linear structures based on Hilbert–Huang spectral analysis. Part 1: normal modes. *Earthq. Eng. Struct. Dyn.*, 32(9): 1443–67.

Yang, J.N., Y. Lei, S.W. Pan, and N. Huang. 2003b. System identification of linear structures based on Hilbert–Huang spectral analysis. Part 2: complex modes. *Earthq. Eng. Struct. Dyn.*, 32(10): 1533–54.

Yeo, I., S. Shin, H.S. Lee, and S.P. Chang. 2000. Statistical damage assessment of framed structures from static responses. *J. Eng. Mech.*, 126: 414–21.

Yoon, D.J., W.J. Weiss, and S.P. Shah. 2000. Assessing damage in corroded reinforced concrete using acoustic emission. *J. Eng. Mech.*, 126(3): 273–83.

Yuen, K.V. and L.S. Katafygiotis. 2005. Model updating using noisy response measurements without knowledge of the input spectrum. *Earthq. Eng. Struct. Dyn.*, 34: 167–87.

Zachar, J. and T.R. Naik. 1992. Principles of infrared thermography and application for assessment of the deterioration of the bridge deck at the zoo interchange. In *Proceedings of the Materials Engineering Congress*, ed. T.D. White, Atlanta, Georgia, 107–115. New York: American Society of Civil Engineers.

Zhang, C.D. and Y.L. Xu. 2016. Comparative studies on damage identification with Tikhonov regularization and sparse regularization. *Struct. Control Health Monit.*, 23(3): 560–579.

Zhang, J., Y.L. Xu, Y. Xia, and J. Li. 2008. A new statistical moment-based structural damage detection method. *Struct. Eng. Mech.*, 30(4): 445–66.

Zhang, Z. and A.E. Aktan. 1995. The damage indices for the constructed facilities. In *Proceedings of the 13th International Modal Analysis Conference*, ed. D.J. DeMichele, 1520–9, Nashville, TN. Bethel, CT: Society for Experimental Mechanics.

Zhao, J. and J.T. DeWolf. 1999. Sensitivity study for vibrational parameters used in damage detection. *J. Struct. Eng.*, 125(4): 410–16.

Zhou, J., X. Feng, and Y.F. Fan. 2003. A probabilistic method for structural damage identification using uncertain data. In *Proceedings of the 1st International Conference on Structural Health Monitoring and Intelligent Infrastructure*, ed. Z.S. Wu and M. Abe, 487–92. Lisse, Netherlands: A.A. Balkema.

Zhou, X.Q., Y. Xia, and S. Weng. 2015. ℓ_1 regularization approach to structural damage detection using frequency data. *Struct. Health Monit.*, 14(6): 571–82.

Zhu, X.Q. and S.S. Law. 2006. Wavelet-based crack identification of bridge beam from operational deflection time history. *Int. J. Solids Struct.*, 43(7–8): 2299–317.

Zimmerman, D.C. and M. Kaouk. 1994. Structural damage detection using a minimum rank update theory. *J. Vib. Acoust.*, 116: 222–31.

Chapter 13

Structural vibration control

13.1 PREVIEW

Structural vibration control is one of the most important functions of a smart civil structure subjected to dynamic loadings such as strong winds and severe earthquakes. Control devices and control systems were described in Chapter 4 and a variety of control algorithms were introduced in Chapter 9. Passive control strategies, including base isolation systems, viscoelastic dampers, tuned mass dampers (TMDs) and fluid dampers, have been developed into workable technologies in the past four decades for mitigating the effects of dynamic loadings on civil structures. However, these passive control systems are unable to adapt to structural changes and to varying use patterns and loading conditions. Serious efforts have been undertaken in the past two decades to develop active, semi-active and hybrid control concepts into workable technologies. This chapter first gives a brief introduction to the full-scale implementations of passive, active, semi-active and hybrid control systems. The theoretical investigation of the active control of adjacent buildings using hydraulic actuators is then presented as a potential full-scale implementation case. The experimental investigations of the semi-active control of a complex structure using semi-active friction dampers and magnetorheological (MR) dampers are, respectively, carried out and introduced in this chapter as two promising real implementation cases. This chapter finally introduces the multi-objective hybrid control of high-tech equipment in a high-tech facility using both passive dampers and smart actuators to ensure the functionality of high-tech equipment against microvibration under normal working conditions and to protect high-tech equipment from damage when an earthquake occurs.

13.2 INTRODUCTION TO FULL-SCALE IMPLEMENTATIONS

As introduced in Chapter 4, structural vibration control systems are often classified as base-isolation systems, passive-energy dissipation systems, active control systems, semi-active control systems and hybrid control systems according to the controlled manner of a civil structure.

The concept of seismic base isolation emerged in the early 1970s, but the design and construction of base-isolated buildings did not begin until the early 1980s. The William Calyton Building in Wellington, New Zealand, started in 1978 and completed in 1981, was the first building in the world to incorporate lead-rubber bearings (Skinner et al. 1993). The first seismically isolated building in Japan was the Yachiyodai Residential Dwelling, which was mounted on six laminated-rubber bearings and completed in 1982 (Skinner et al. 1993). The Foothill Communities Law and Justice Centre was the first new building in the United States, which was mounted on elastomeric bearings, and was completed in 1985 (Skinner et al. 1993). In China, over 5000 buildings were built with seismic isolators in 2014, and some railway and highway bridges have also been built with seismic isolation (Zhou 2015).

The full-scale implementations of passive energy dissipation systems for reducing structural vibration can be found in Soong and Dargush (1997). For example, the application of viscoelastic dampers to civil structures began in 1969 when approximately 10,000 viscoelastic dampers were installed in each of the twin towers of the World Trade Center in New York to reduce wind-induced vibration (Mahmoodi 1969). The 305 m high Sydney Tower in Australia is the first slender structure with the installation of a large-scale TMD to mitigate wind-induced vibrations (Housner et al. 1997). The earliest applications of metallic yield dampers to structural systems occurred in New Zealand for seismic protection (Skinner et al. 1980). The earliest application of friction dampers to structural systems for seismic protections in Canada can be found in Aiken and Kelly (1990) and Pall and Pall (1993).

The full-scale implementation of active control systems has been accomplished in several structures, mainly in Japan (Spencer and Nagarajaiah 2003). The first full-scale implementation of an active control system was in the Kyobashi Seiwa Building in Tokyo (Kobori 1994). It is an 11-story building with two active mass dampers (AMDs) installed on the top floor for the reduction of both transverse motion and torsional motion. The AMD was also designed to reduce wind-induced vibration of the Nanjing communication tower (Cao et al. 1998; Reinhorn et al. 1998). More recently, a pair of AMD systems was installed on the top of the Canton Tower in China (Xu et al. 2014). However, cost-effectiveness and reliability considerations have limited the widespread acceptance of AMD systems.

Due to their mechanical simplicity, low power requirements and large controllable force capacity, semi-active control systems provide an attractive alternative to active control systems for vibration attenuation. The Kajima Technical Research Institute in Japan was the first full-scale building structure implemented with semi-active variable stiffness (SAVA) systems (Kobori et al. 1993). The first full-scale implementation of semi-active control systems in the United States was conducted on the Walnut Creek Bridge (Patten 1998; Patten et al. 1999) using SAVA systems. Semi-active hydraulic damper systems were installed in the Kajima Shizuoka Building in Shizuoka, Japan (Kobori 1998; Kurata et al. 1999). In 2001, the first full-scale implementation of MR dampers for civil structures was achieved; the Tokyo National Museum of Engineering Science and Innovation was equipped with two 30-ton MR dampers (Spencer and Nagarajaiah 2003). More recently, a new semi-active TMD with an MR damper (MR-STMD) was installed on the Volgograd Bridge in 2011 in Russia (Weber et al. 2013). The main feature of the MR-STMD is to replace a passive oil damper in the TMD with a real-time controlled MR damper, which is used to adjust both the natural frequency and the damping of the MR-STMD to the actual frequency of the bridge.

A hybrid control system is defined as one that is achieved by a combination of the aforementioned passive, active or semi-active control techniques. Since multiple control devices are operating together, hybrid control systems are able to alleviate some of the restrictions and limitations that exist when each system is acting alone. The hybrid mass damper (HMD) is the most common hybrid control system employed in full-scale civil engineering applications. A number of buildings and bridges have been equipped with HMDs for vibration reduction. The first full-scale HMD was implemented to the Shimizu Technical Laboratory in Tokyo in 1991. The relevant information can be found in the references (Soong and Spencer 2002; Spencer and Nagarajaiah 2003).

Most of the full-scale implementations previously mentioned were carried out for a single structural system to mitigate its own vibration. However, buildings and structures in modern cities are becoming more and more close to each other due to limited land availability. From the viewpoint of achieving both less control cost and more effective control capacity, it is desirable to consider the use of control devices to couple adjacent buildings to reduce wind- or seismic-induced responses of both buildings simultaneously. The idea of employing control devices for the vibration control of two adjacent buildings can also be applied to a building complex, which is typically composed of a main building and a podium structure in modern urban cities. This chapter therefore will present a theoretical investigation of the active control of adjacent buildings using hydraulic actuators with linear quadratic Gaussian (LQG) controllers as a potential full-scale implementation case. The experimental investigations of the semi-active control of a complex structure using semi-active friction dampers and MR dampers are also carried out and introduced in this chapter as two promising real implementation cases. Finally, multi-objective hybrid control is investigated for high-tech equipment in a high-tech facility using both passive dampers and smart actuators to ensure the functionality of high-tech equipment against microvibration under normal working conditions and to protect high-tech equipment from damage when an earthquake occurs.

13.3 ACTIVE CONTROL OF ADJACENT BUILDINGS USING HYDRAULIC ACTUATORS

This section presents a general yet simple closed-form solution for actively controlled adjacent buildings linked by hydraulic actuators with LQG controllers under earthquake excitation (Xu and Zhang 2002). The derivation of a closed-form solution is naturally fulfilled by combining the complex modal superposition method with the pseudo-excitation method and the residue theorem. The derived closed-form solution is then used to perform parametric studies of adjacent buildings, connected by LQG controllers, and to assess the effectiveness of LQG controllers in reducing the seismic responses of both buildings.

13.3.1 Equations of motion

Consider a two-dimensional system consisting of two shear buildings connected by active hydraulic actuators, as shown in Figure 13.1. Both buildings are assumed to be subjected to the same base acceleration. The total degrees of freedom (DOFs) of the two adjacent buildings are N, whereas the number of DOFs of the left building (building 1) and right building (building 2) is N_1 and N_2, respectively. The first floor is designated as the first DOF in either building. The equations of motion of the building-control device system can be expressed as

$$\mathbf{M}\ddot{\mathbf{x}}(t) + \mathbf{C}\dot{\mathbf{x}}(t) + \mathbf{K}\mathbf{x}(t) = -\mathbf{M}\mathbf{E}\ddot{x}_g(t) + \mathbf{H}\mathbf{u}(t) \qquad (13.1)$$

where:

    $\mathbf{M}, \mathbf{C}$ and $\mathbf{K}$ are the mass, damping and stiffness matrices of the adjacent buildings, respectively

    $\mathbf{x}(t)$ is the vector of relative displacement response with respect to the ground

    $\mathbf{E}$ is the index vector with all its elements equal to one

    $\mathbf{u}(t)$ is an r-dimensional vector consisting of r active control forces

    $\mathbf{H}$ is an $N \times r$ matrix denoting the location of r actuators

    $\ddot{x}_g(t)$ is the ground acceleration

The details of mass, damping and stiffness matrices, $\mathbf{M}, \mathbf{C}$ and $\mathbf{K}$ can be found in Zhang and Xu (1999). Here, the Kanai–Tajimi filtered white noise spectrum is considered as the ground acceleration spectrum:

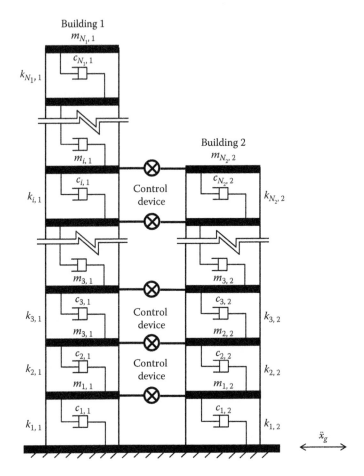

Figure 13.1 Structural model of adjacent buildings.

$$S_{\ddot{x}_g \ddot{x}_g}(\omega) = \frac{1 + 4\xi_g^2 \left(\dfrac{\omega}{\omega_g}\right)^2}{\left[1 - \left(\dfrac{\omega}{\omega_g}\right)^2\right]^2 + 4\xi_g^2 \left(\dfrac{\omega}{\omega_g}\right)^2} S_0 \qquad (13.2)$$

in which ω_g, ξ_g and S_0 may be regarded as the characteristics and the intensity of an earthquake in a particular geological location. This spectrum can be represented in the time domain with the following state equations:

$$\dot{\boldsymbol{\Gamma}}(t) = \mathbf{A}_c \boldsymbol{\Gamma}(t) + \mathbf{D}_c w(t) \qquad (13.3)$$

$$\ddot{x}_g(t) = \mathbf{C}_c \boldsymbol{\Gamma}(t) \qquad (13.4)$$

where:
  $w(t)$ is the zero-mean Gaussian white noise with intensity S_0
  $\boldsymbol{\Gamma}(t)$ is the state of the seismic excitation model

$$\mathbf{A}_c = \begin{bmatrix} 0 & 1 \\ -\omega_g^2 & -2\xi_g\omega_g \end{bmatrix}, \mathbf{C}_c = \begin{bmatrix} \omega_g^2 & 2\xi_g\omega_g \end{bmatrix}, \mathbf{D}_c = \begin{bmatrix} 0 \\ 1 \end{bmatrix} \tag{13.5}$$

Writing Equation 13.1 as a state equation and then combining it with Equations 13.3 and 13.4 yields the equation of motion of the system:

$$\dot{\mathbf{q}}(t) = \mathbf{Aq}(t) + \mathbf{Bu}(t) + \mathbf{G}w(t) \tag{13.6}$$

where

$$\mathbf{A} = \begin{bmatrix} 0 & \mathbf{I} & 0 \\ -\mathbf{M}^{-1}\mathbf{K} & -\mathbf{M}^{-1}\mathbf{C} & -\mathbf{EC}_c \\ 0 & 0 & \mathbf{A}_c \end{bmatrix}, \ \mathbf{B} = \begin{bmatrix} 0 \\ \mathbf{M}^{-1}\mathbf{H} \\ 0 \end{bmatrix}, \ \mathbf{G} = \begin{bmatrix} 0 \\ 0 \\ \mathbf{D}_c \end{bmatrix}, \ \mathbf{q}(t) = \begin{bmatrix} \mathbf{x}(t) \\ \dot{\mathbf{x}}(t) \\ \mathbf{\Gamma}(t) \end{bmatrix} \tag{13.7}$$

13.3.2 LQG controller

In reality, the structural states, that is, displacements and velocities relative to the ground at each DOF, cannot be fully measured, and the measurement is often limited to absolute accelerations. The absolute accelerations $\ddot{\mathbf{x}}_a(t)$ can be related to the relative displacements and relative velocities through the following equation:

$$\ddot{\mathbf{x}}_a(t) = -\mathbf{M}^{-1}\mathbf{C}\dot{\mathbf{x}}(t) - \mathbf{M}^{-1}\mathbf{Kx}(t) + \mathbf{M}^{-1}\mathbf{Hu}(t) \tag{13.8}$$

If output measurements are selected as $\mathbf{m}(t) = \ddot{\mathbf{x}}_a(t) - \mathbf{M}^{-1}\mathbf{Hu}(t)$, then the measured output vector $\mathbf{m}(t)$ can be expressed as

$$\mathbf{m}(t) = \mathbf{C}_m\mathbf{q}(t) + \mathbf{v}_m(t) \tag{13.9}$$

where $\mathbf{v}_m(t)$ are the random signals known as *measurement noises* and

$$\mathbf{C}_m = \begin{bmatrix} -\mathbf{M}^{-1}\mathbf{K} & -\mathbf{M}^{-1}\mathbf{C} & 0 \end{bmatrix} \tag{13.10}$$

In the practical application, sensors may not be placed on every floor and thus only a subset of $\mathbf{m}(t)$, denoted as $\mathbf{y}(t)$, is used.

$$\mathbf{y}(t) = \mathbf{C}_L\mathbf{q}(t) + \mathbf{v}(t) \tag{13.11}$$

where:
 N_m denotes the dimension of $\mathbf{y}(t)$ ($N_m \le N$)
 $\mathbf{C}_L$ is the matrix obtained by eliminating the rows related to the floors without sensors in the matrix $\mathbf{C}_m$
 $\mathbf{v}(t)$ is the measurement noise vector of N_m dimension

If the measurement noise is regarded as the white noise of the same intensity at each measurement point and independent of each other, its covariance matrix can be expressed as

$$E\left[\mathbf{v}(t)\mathbf{v}^T(t+\tau)\right] = \mathbf{I}S_v\delta(\tau) \tag{13.12}$$

where:

E   is the expectation operator
I   is the identity matrix
S_ν is the intensity of noise
$\delta(\tau)$ is the Dirac delta function

Once the state space Equation 13.6 and the observation Equation 13.11 are determined, a reasonable controller can be designed on the basis of LQG control law as introduced in Chapter 9, leading to the following optimal control force:

$$\mathbf{u}(t) = -\mathbf{K}_c \hat{\mathbf{q}}(t) \tag{13.13}$$

where:

$\mathbf{K}_c$ is the optimal control gain matrix
$\hat{\mathbf{q}}(t)$ is the system state estimate generated by the Kalman filter (KF) technique

$$\dot{\hat{\mathbf{q}}}(t) = \mathbf{A}\hat{\mathbf{q}}(t) + \mathbf{B}\mathbf{u}(t) + \mathbf{K}_f\left(\mathbf{y}(t) - \mathbf{C}_L\hat{\mathbf{q}}(t)\right) \tag{13.14}$$

in which $\mathbf{K}_f$ is the estimator gain matrix in the KF algorithm.

13.3.3 Closed-form solution for dynamic characteristics

Equations 13.6, 13.11, 13.13 and 13.14 can be rearranged as

$$\dot{\mathbf{p}}(t) = \hat{\mathbf{A}}\,\mathbf{p}(t) + \mathbf{f}(t) \tag{13.15}$$

where:

$$\hat{\mathbf{A}} = \begin{bmatrix} \mathbf{A} - \mathbf{B}\mathbf{K}_c & \mathbf{B}\mathbf{K}_c \\ 0 & \mathbf{A} - \mathbf{K}_f\mathbf{C}_L \end{bmatrix}; \quad \mathbf{f} = \begin{bmatrix} \mathbf{G} & 0 \\ \mathbf{G} & -\mathbf{K}_f \end{bmatrix}\begin{Bmatrix} w(t) \\ v(t) \end{Bmatrix}; \quad \mathbf{p} = \begin{Bmatrix} \mathbf{q}(t) \\ \mathbf{e}(t) \end{Bmatrix}; \quad \mathbf{e}(t) = \mathbf{q}(t) - \hat{\mathbf{q}}(t)$$

$$\tag{13.16}$$

The solution of the homogeneous form of Equation 13.15 can then be taken as

$$\mathbf{p}(t) = \boldsymbol{\varphi} e^{\lambda t} \tag{13.17}$$

The associated complex eigenvalue problem of Equation 13.15 becomes

$$\hat{\mathbf{A}}\boldsymbol{\varphi} = \lambda\boldsymbol{\varphi} \tag{13.18}$$

where:

λ is the eigenvalue
$\boldsymbol{\varphi}$ is the associated eigenvector

The solution of Equation 13.18 comprises a set of $2N'$ $(N' = 2N + 2)$ eigenvalues and eigenvectors that exist in either complex conjugate pairs (underdamped mode) or real pairs (overdamped mode).

For complex conjugate pairs:

$$\boldsymbol{\varphi}_j = \boldsymbol{\varphi}_{j+N'}^* \text{ and } \lambda_j = \lambda_{j+N'}^* \quad (j = 1, 2, \ldots, N') \tag{13.19}$$

The eigenvalue is usually written under the form:

$$\lambda_j = \lambda_{j+N'}^* = -\omega_j \xi_j + i\omega_{dj} \quad (j = 1, 2, \ldots, N') \tag{13.20}$$

where:
* means the conjugation
'i' is the imaginary unit

ω_j, ω_{dj} and ξ_j denote the modal frequency, the damped modal frequency and the modal damping ratio respectively, which are associated with mode j and can be given as follows:

$$\omega_j = |\lambda_j|, \qquad \xi_j = -\mathrm{Re}(\lambda_j)/|\lambda_j|, \qquad \omega_{dj} = \omega_j\sqrt{1 - \xi_j^2} \tag{13.21}$$

For real pairs, it is convenient to express the real pairs λ_j in the following form analogous to Equation 13.20:

$$\lambda_j = -\omega_j \xi_j + \omega_{dj}, \lambda_{j+N'} = -\omega_j \xi_j - \omega_{dj} \tag{13.22}$$

in which ω_j, ω_{dj} and ξ_j are determined by

$$\omega_j = \sqrt{\lambda_j \lambda_{j+N'}}, \quad \xi_j = -(\lambda_j + \lambda_{j+N'})/(2\omega_j), \quad \omega_{dj} = \omega_j\sqrt{\xi_j^2 - 1} = (\lambda_j - \lambda_{j+N'})/2 \tag{13.23}$$

13.3.4 Closed-form solution for seismic response

To find the closed-form solution for seismic responses of adjacent buildings with LQG controllers, the pseudo-excitation method (Zhang and Xu 1999) is used in conjunction with the complex modal superposition method. The seismic input and the measurement noise vector are assumed to be independent in this study. The spectral density matrix $\mathbf{S}_{wv}$ of both the ground excitation and the measurement noise is thus given by

$$\mathbf{S}_{wv} = \begin{bmatrix} S_0 & 0 \\ 0 & S_v\mathbf{I} \end{bmatrix} = S_0 \begin{bmatrix} 1 & 0 \\ 0 & \eta^2\mathbf{I} \end{bmatrix} \tag{13.24}$$

in which

$$\eta^2 = \frac{S_v}{S_0} \tag{13.25}$$

Note that $\mathbf{S}_{wv} = S_0 \begin{bmatrix} 1 & 0 \\ 0 & \eta\mathbf{I} \end{bmatrix} \begin{bmatrix} 1 & 0 \\ 0 & \eta\mathbf{I} \end{bmatrix}$. Thus, letting $\mathbf{L} = \begin{bmatrix} 1 & 0 \\ 0 & \eta\mathbf{I} \end{bmatrix}$, the pseudo-excitation vectors for the system expressed by Equation 13.15 can be constituted as

$$\begin{Bmatrix} w(t) \\ \mathbf{v}(t) \end{Bmatrix}_k = \mathbf{L}_k\sqrt{S_0}\exp(i\omega t) \quad (k = 1, 2, \ldots, N_m + 1) \tag{13.26}$$

where L_k is the k th column of L. By considering the k th pseudo-excitation vector $\left\{ \begin{smallmatrix} w(t) \\ v(t) \end{smallmatrix} \right\}_k$, Equation 13.15 becomes

$$\dot{p}_k(t) = \hat{A} p_k(t) + F_k \sqrt{S_0} \exp(i\omega t) \tag{13.27}$$

where

$$F_k = \begin{bmatrix} G & 0 \\ G & -K_f \end{bmatrix} L_k \tag{13.28}$$

To decouple Equation 13.27, the following coordinate transformation is adopted:

$$p_k(t) = \Phi z_k(t) \tag{13.29}$$

where z_k is the $2N'$-dimensional generalised coordinate vector and Φ is the $2N' \times 2N'$ right modal matrix.

$$\Phi = [\varphi_1, \varphi_2, ..., \varphi_{2N'}] \tag{13.30}$$

By using this transformation, Equation 13.27 can be reduced to $2N'$ decoupled modal equations with the jth modal equation being

$$\dot{z}_{kj}(t) = \lambda_j z_{kj}(t) + r_{kj} \sqrt{S_0} \exp(i\omega t) \tag{13.31}$$

where

$$r_{kj} = \psi_j^T F_k \tag{13.32}$$

Ψ_j is the jth column of Ψ; and the matrix Ψ is equal to $= \Phi^{-T}$, the left modal matrix. The solution of the first-order jth Equation 13.31 to the kth pseudo excitation vector is

$$z_{kj}(\omega, t) = \frac{r_{kj}}{i\omega - \lambda_j} \sqrt{S_0} \exp(i\omega t) \quad (j = 1, 2, ..., 2N') \tag{13.33}$$

Denoting the mth components of φ_j as φ_{mj} and the mth component of p_k as p_{km}, then the pseudo response p_{km} is given by

$$p_{km}(\omega, t) = \sum_{j=1}^{2N'} \varphi_{mj} z_{kj}(\omega, t) = \sum_{j=1}^{2N'} \varphi_{mj} \frac{r_{kj}}{i\omega - \lambda_j} \sqrt{S_0} \exp(i\omega t) \tag{13.34}$$

Since the eigenvectors are in pairs for either underdamped mode or overdamped mode, Equation 13.34 can be reduced to

$$p_{km} = \sum_{j=1}^{N'} H_j(\omega)(i\omega\alpha_{kmj} + \beta_{kmj}) \sqrt{S_0} \exp(i\omega t) \tag{13.35}$$

in which $H_j(\omega)$ is the frequency response function for the jth mode.

$$H_j(\omega) = \frac{1}{\omega_j^2 - \omega^2 + i\,2\xi_j\omega_j\omega} \tag{13.36}$$

The pseudo response p_{km} can be the pseudo displacement response, pseudo velocity response and pseudo acceleration response. The proper use of the pseudo displacement responses of adjacent buildings can result in the pseudo shear force responses of both buildings. Also, the proper use of the pseudo state estimator responses in conjunction with Equation 13.13 can lead to the pseudo control forces. For instance, the mth ($m \le N$) pseudo displacement, velocity or acceleration response can be obtained from Equation 13.35 if the coefficients α_{kmj} and β_{kmj} are calculated by the following equations:

When the jth mode is an underdamped mode:

$$\alpha_{kmj} = \begin{cases} 2\,\mathrm{Re}\left(\varphi_{mj}r_{kj}\right) & \text{For displacement} \\ 2\,\mathrm{Re}\left(\lambda_j\varphi_{mj}r_{kj}\right) & \text{For velocity} \\ 2\,\mathrm{Re}\left(\lambda_j^2\varphi_{mj}r_{kj}\right) & \text{For acceleration} \end{cases} \tag{13.37}$$

$$\beta_{kmj} = \begin{cases} -2\,\mathrm{Re}\left(\varphi_{mj}r_{kj}\lambda_j^*\right) & \text{For displacement} \\ -2\omega_j^2\,\mathrm{Re}\left(\varphi_{mj}r_{kj}\right) & \text{For velocity} \\ -2\omega_j^2\,\mathrm{Re}\left(\lambda_j\varphi_{mj}r_{kj}\right) & \text{For acceleration} \end{cases} \tag{13.38}$$

When the jth mode is an overdamped mode:

$$\alpha_{kmj} = \begin{cases} \varphi_{mj}r_{kj} + \varphi_{mj+N'}r_{kj+N'} & \text{For displacement} \\ \lambda_j\varphi_{mj}r_{kj} + \lambda_{j+N'}\varphi_{mj+N'}r_{kj+N'} & \text{For velocity} \\ \lambda_j^2\varphi_{mj}r_{kj} + \lambda_{j+N'}^2\varphi_{mj+N'}r_{kj+N'} & \text{For acceleration} \end{cases} \tag{13.39}$$

$$\beta_{kmj} = \begin{cases} -\left(\varphi_{mj}r_{kj}\lambda_{j+N'} + \varphi_{mj+N'}r_{kj+N'}\lambda_j\right) & \text{For displacement} \\ -\omega_j^2\left(\varphi_{mj}r_{kj} + \varphi_{mj+N'}r_{kj+N'}\right) & \text{For velocity} \\ -\omega_j^2\left(\lambda_j\varphi_{mj}r_{kj} + \lambda_{j+N'}\varphi_{mj+N'}r_{kj+N'}\right) & \text{For acceleration} \end{cases} \tag{13.40}$$

According to the principle of the pseudo-excitation method, the response spectral density of p_{km} can then be obtained by

$$S_{p_{km}p_{km}}(\omega) = p_{km}(\omega)p_{km}^*(\omega)$$

$$= \sum_{i=1}^{N'}\sum_{j=1}^{N'} H_i(\omega)H_j^*(\omega)\left(\beta_{kmi}\beta_{kmj} + i\,\omega\left(\alpha_{kmi}\beta_{kmj} - \alpha_{kmj}\beta_{kmi}\right) + \omega^2\alpha_{kmi}\alpha_{kmj}\right)S_0 \tag{13.41}$$

The variance response of p_{ki} under the kth pseudo-excitation can be evaluated as

$$\sigma_{p_{km}}^2 = \int_{-\infty}^{+\infty} S_{p_{km}p_{km}}(\omega)\,d\omega \tag{13.42}$$

The above integration in the complex plane can be accomplished using the residue theorem to have the closed-form solution as

$$\sigma_{p_{km}}^2 = \sum_{i=1}^{N'}\sum_{j=1}^{N'} \rho_{0,ij} u_{kmi} u_{kmj} + \sum_{i=1}^{N'}\sum_{j=1}^{N'} \rho_{01,ij} u_{kmi} v_{kmj} + \sum_{i=1}^{N'}\sum_{j=1}^{N'} \rho_{1,ij} v_{kmi} v_{kmj} \tag{13.43}$$

where

$$u_{kmi} = \sqrt{\frac{\pi}{2}} \frac{\beta_{kmi}\sqrt{S_0}}{\omega_i\sqrt{\xi_i\omega_i}}, u_{kmj} = \sqrt{\frac{\pi}{2}} \frac{\beta_{kmj}\sqrt{S_0}}{\omega_j\sqrt{\xi_j\omega_j}}, v_{kmi} = \sqrt{\frac{\pi}{2}} \frac{\alpha_{kmi}\sqrt{S_0}}{\sqrt{\xi_i\omega_i}}, v_{kmj} = \sqrt{\frac{\pi}{2}} \frac{\alpha_{kmj}\sqrt{S_0}}{\sqrt{\xi_j\omega_j}} \tag{13.44}$$

$$\rho_{0,ij} = \frac{8\sqrt{\xi_i\xi_j}\left(\xi_i + \gamma\xi_j\right)\gamma^{3/2}}{\left(1-\gamma^2\right)^2 + 4\xi_i\xi_j\gamma\left(1+\gamma^2\right) + 4\left(\xi_i^2 + \xi_j^2\right)\gamma^2} \tag{13.45}$$

$$\rho_{01,ij} = \frac{8\sqrt{\xi_i\xi_j}\left(\gamma^2 - 1\right)\gamma^{3/2}}{\left(1-\gamma^2\right)^2 + 4\xi_i\xi_j\gamma\left(1+\gamma^2\right) + 4\left(\xi_i^2 + \xi_j^2\right)\gamma^2} \tag{13.46}$$

$$\rho_{1,ij} = \frac{8\sqrt{\xi_i\xi_j}\left(\xi_j + \gamma\xi_i\right)\gamma^{3/2}}{\left(1-\gamma^2\right)^2 + 4\xi_i\xi_j\gamma\left(1+\gamma^2\right) + 4\left(\xi_i^2 + \xi_j^2\right)\gamma^2} \tag{13.47}$$

$$\gamma = \frac{\omega_j}{\omega_i} \tag{13.48}$$

The final variance response of p_m can be determined by a summation with respect to the $n+1$ pseudo excitation vectors.

$$\sigma_{p_m}^2 = \sum_{k=1}^{n+1} \sigma_{p_{km}}^2 \tag{13.49}$$

The closed-form solution derived above makes it possible to carry out extensive parametric studies and to evaluate the performance of both buildings with LQG controllers.

13.3.5 Application of closed-form solutions

For the application of the closed-form solutions, two 20-story buildings having the same floor elevations with hydraulic actuators connecting two neighbouring floors are used. The mass, shear stiffness and internal and external damping coefficients of the left building are uniform for all stories with the mass of 1.29×10^6 kg, the shear stiffness of 4.0×10^9 N/m, the internal damping coefficient of 3.0×10^6 N·s/m and the external damping coefficient of 8.0×10^4 N·s/m. For the right building, the mass, shear stiffness, internal damping coefficient and external damping coefficient are also uniform for all stories with the same mass, internal and external damping coefficients as the left building but with the shear stiffness of 2.0×10^9 N/m only. Hence, the two buildings have the same height but the right building is more slender than the left building, this is because if the dynamic properties of both buildings are the same or close to each other, the use of control devices linking the two buildings will not function or not function properly (Xu et al. 1999). The damping coefficients selected ensure the first modal damping ratio in both building is around 1%. The parameters for the ground motion are selected as $\omega_g = 15.0$ rad/s, $\xi_g = 0.65$, and $S_0 = 4.65 \times 10^{-4}$ m²/rad·s³, respectively, to reflect firm-soil conditions (Heredia-Zavoni and Vanmarcke 1994). The intensity ratio of the measurement noise to the ground motion, η in Equation 13.25, is selected as 1:7. The active control devices and sensors are arranged first at every floor and then at selected locations for a comparison.

13.3.5.1 Selection of weighting matrices

As mentioned in Chapter 9, the performance of LQG controllers strongly depends on the weighting matrices **Q** and **R**. Here, **Q** is the positive semi-definite matrix associated with system states, whereas **R** is the positive definite matrix associated with the control forces. To achieve the beneficial performance of LQG for the maximum reduction of key structural responses of both buildings with reasonable control forces, several potential weighting matrices are selected. The key structural responses are then computed in terms of the closed-form solution from which the best weighting matrices can be identified. The basic configurations of the weighting matrices in this study are taken as

$$\mathbf{R} = 10^{-8}\mathbf{I}, \mathbf{Q} = \mu \begin{bmatrix} \mathbf{Q}_{dL} & 0 & 0 & 0 & 0 \\ 0 & \mathbf{Q}_{dR} & 0 & 0 & 0 \\ 0 & 0 & \mathbf{Q}_{vL} & 0 & 0 \\ 0 & 0 & 0 & \mathbf{Q}_{vR} & 0 \\ 0 & 0 & 0 & 0 & 0 \end{bmatrix} \tag{13.50}$$

where:

$\mathbf{Q}_{dL}$ and $\mathbf{Q}_{dR}$ are the submatrices assigned to the displacement responses of the left and right buildings, respectively

$\mathbf{Q}_{vL}$ and $\mathbf{Q}_{vR}$ are the submatrices assigned to the velocity responses of the left and right buildings, respectively

$\mathbf{I}$ is an identity matrix

μ is a proportional coefficient

Clearly, by varying the coefficient μ, a proper trade-off between control effectiveness and control energy consumption can be achieved. Herein, the following four cases are selected for finding the balanced submatrices $\mathbf{Q}_{dL}$, $\mathbf{Q}_{dR}$, $\mathbf{Q}_{vL}$ and $\mathbf{Q}_{vR}$:

Case A: $\mathbf{Q}_{dL} = \mathbf{Q}_{dR} = \mathbf{Q}_{vL} = \mathbf{Q}_{vR} = \mathbf{I}$

Case B: $\mathbf{Q}_{dL} = \mathbf{Q}_{dR} = \mathbf{I}$; $\mathbf{Q}_{vL} = \mathbf{Q}_{vR} = 0$

Case C: $\mathbf{Q}_{dL} = \mathbf{Q}_{dR} = 0$; $\mathbf{Q}_{vL} = \mathbf{Q}_{vR} = \mathbf{I}$

Case D: $\mathbf{Q}_{dL} = \mathbf{Q}_{dR} = \mathbf{I}$; $\mathbf{Q}_{vL} = \mathbf{Q}_{vR} = 0.1\mathbf{I}$

For each case, the key responses of the actively controlled adjacent buildings are computed against the parameter μ. Then, by comparing the results among all the cases, the beneficial case and parameter μ can be found for achieving the maximum or beneficial response reduction of both buildings with reasonable control forces.

Figures 13.2 and 13.3 depict the variations of the top floor displacement and the base shear force responses of the adjacent building, respectively, with the parameter μ. Figure 13.4 shows how the control force at the top of the building varies with the parameter μ. It can be seen that for all cases the top floor displacement and the base shear force responses of both buildings are rapidly reduced until μ reaches a value about 2×10^5. After that, the gradients of the response reduction become small in Cases A, B and D, but not in Case C where the further increase of μ makes the displacement and shear force responses become larger with larger control force required. Thus, Case C, where the velocity response reduction is maximised regardless of the displacement response reduction, is disregarded in the subsequent computation. Furthermore, if the weighting matrices in Case A are adopted, one may benefit from the response reduction of the left building but not the right building. The situation is reversed if the weighting matrices in Case B are used. As a result, a compromise is made herein to select the weighting matrices of Case D.

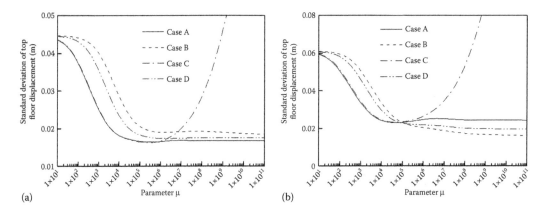

Figure 13.2 Top floor displacement responses of adjacent building vs. parameter: (a) left building (b) right building.

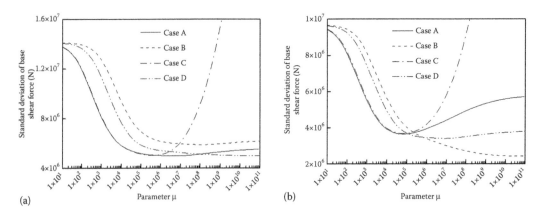

Figure 13.3 Base shear force responses of adjacent building vs. parameter: (a) left building (b) right building.

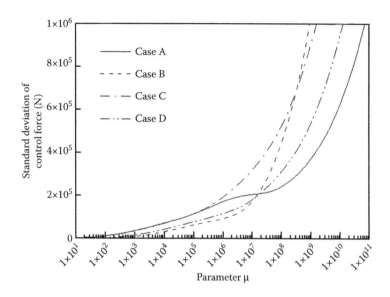

Figure 13.4 Control force of adjacent building vs. parameter.

13.3.5.2 Modal properties

By using the weighting matrix $\mathbf{Q}$ in Case D, the first five natural frequencies and modal damping ratios of the actively controlled adjacent buildings are computed against the parameter μ. The results are depicted in Figures 13.5 and 13.6 for the natural frequencies and modal damping ratios, respectively. It can be seen from Figure 13.5 that the first five natural frequencies do not vary with the parameter μ when μ is in the range from 10 to 2×10^5. Further analysis shows that the first, third and fifth natural frequencies of the actively controlled adjacent buildings are almost the same as the first three natural frequencies of the uncontrolled right building of 3.02, 9.03 and 14.99 rad/s, respectively. The second and fourth natural frequencies of the actively controlled adjacent buildings are almost

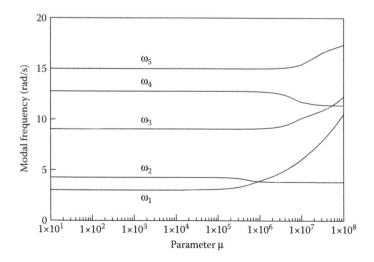

Figure 13.5 Variations of modal frequencies with parameter μ.

the same as the first two natural frequencies of the uncontrolled left building of 4.27 and 12.77 rad/s, respectively.

The first five modal damping ratios of the actively controlled adjacent buildings increase with the increasing value of μ when μ is in the range from 10 to 6×10^5, as shown in Figure 13.6. Since the first, third and fifth modes of vibration of the system are dominated by the right building and the second and fourth modes of vibration are dominated by the left building, the pattern of the curves in Figure 13.6a is different from that in Figure 13.6b. When μ is further increased from 2×10^5, the second modal damping ratio starts to decrease but the fourth modal damping ratio still increases until μ reaches 8×10^6. The first, third and fifth modal damping ratios, which are dominated by the right building, are always increased with increasing value of μ, and eventually the modes of vibration are overdamped. The selection of parameter μ could have been compromised by the information on the key structural response reductions and the modal properties as well as control forces. This study

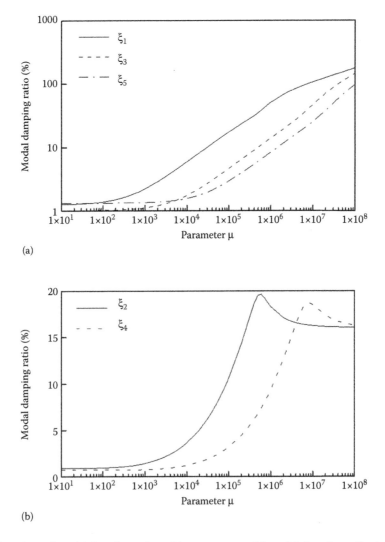

(a)

(b)

Figure 13.6 Variations of modal damping ratios with parameter μ: (a) modal damping ratios in right building and (b) modal damping ratios in left building.

selects μ of 2×10^5 as a beneficial value for the subsequent computation of seismic response using the derived closed from solution.

Further modal analysis of the actively controlled adjacent buildings with the beneficial weighting matrix **Q** and beneficial parameter μ, as selected, shows that the real parts of all eigenvalues of $\hat{\mathbf{A}}$ are negative. Therefore, according to the Lyapunov's criterion about the stability of linear systems, the present actively controlled adjacent buildings form a stable system.

13.3.5.3 Seismic response

To demonstrate the overall performance of LQG controllers, the standard deviations of displacement, shear force and acceleration responses at each floor for each building with and without active control devices are computed. Figures 13.7 and 13.8 show the variations of the standard deviation of displacement and shear force response, respectively, with the height of the buildings. The reduction of responses from the LQG controllers is significant

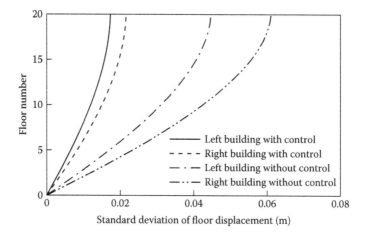

Figure 13.7 Variations of displacement response of adjacent buildings with height.

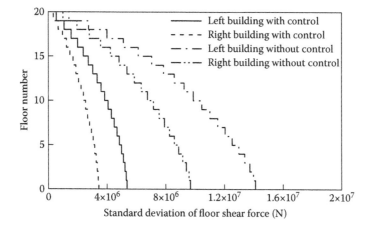

Figure 13.8 Variations of shear force response of adjacent buildings with height.

for all floors in either building. In particular, the top floor displacement standard deviation of the unlinked left building is 44.7 mm but with the active control devices installed, it is reduced to 17.8 mm, leading to a 60% reduction of the response. For the right building, the top floor displacement standard deviation is 61.1 mm for the unlinked building and 22.1 mm for the linked building, resulting in a 64% reduction. For base shear force response, without control the bottom shear force standard deviation is 1.41×10^7 N in the left building and 9.68×10^6 N in the right building. With the LQG controllers, the base shear force standard deviation is reduced to 5.65×10^6 N in the left building and 3.57×10^6 N in the right building, leading to a 60% and a 63% reduction, respectively.

The variations of acceleration response with the building height, as shown in Figure 13.9, are different from displacement and shear force response profiles shown in Figures 13.7 and 13.8. The acceleration response for each unlinked building does not vary monotonically with the building height. This is due to the contributions from higher modes of vibration. The top floor acceleration standard deviation of the unlinked left building is 1.11 m/s² but with the LQG controllers installed it is reduced to 0.519 m/s², leading to a 53% reduction of the response. For the right building, the top floor acceleration standard deviation is 0.807 m/s² for the unlinked building and 0.427 m/s² for the linked building, resulting in a 47% reduction.

To further enhance the understanding of the seismic response of the adjacent buildings with and without control devices, the spectral density functions of the top floor displacement response, base shear force response and top floor acceleration response of both buildings with and without control devices are computed and plotted in Figures 13.10 through 13.12, respectively. It is clearly seen that all the peaks in the response spectra of both buildings are significantly reduced when the LQG controllers are used. The effects of the higher modes of vibration on the acceleration response are much larger than on the top displacement and base shear force responses. From Figure 13.11, it can also be seen that there are only very small changes in the natural frequencies of each building after the installation of the active control devices.

The values of the control forces required for the achievement of significant vibration reduction of the adjacent buildings are important for the design of the hydraulic actuators and adjacent buildings. In terms of the closed-form solution, the control forces can be easily calculated. Figure 13.13 shows the variations of control force with the building height.

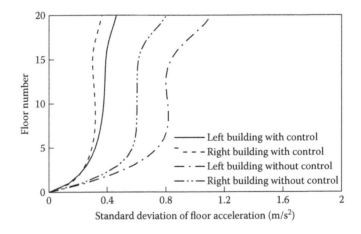

Figure 13.9 Variations of acceleration response of adjacent buildings with height.

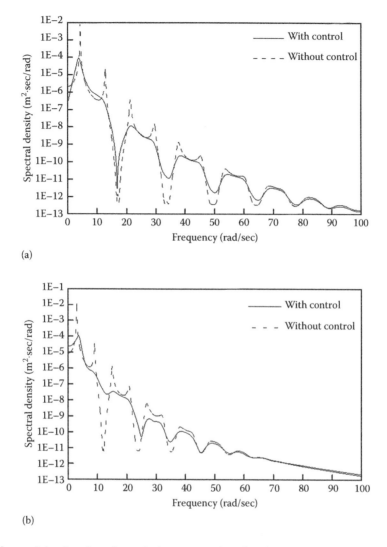

Figure 13.10 Spectral density of top floor displacement with and without active controller: (a) left building, (b) right building.

The maximum control force occurs at the top of the building as expected. The fact that very small control forces occur at the floors near the ground indicates that the corresponding actuators may be removed. To this end and also to demonstrate the capacity of the present formulation and computer program, the actuators and sensors at the bottom 10 floors are removed (Case II) and the seismic responses of both buildings are computed using the same weighting matrices. Some of the major seismic responses of both buildings and the control force at the top floor are listed in Table 13.1 and compared with the case where the actuators and sensors are installed at all the floors (Case I). It can be seen from Table 13.1 that while the major seismic responses of both buildings remain almost the same for the two cases, the control force at the top floor in Case II increases about 31% compared with Case I. Thus, a proper trade-off between control effectiveness and control cost should be considered in the design of controllers.

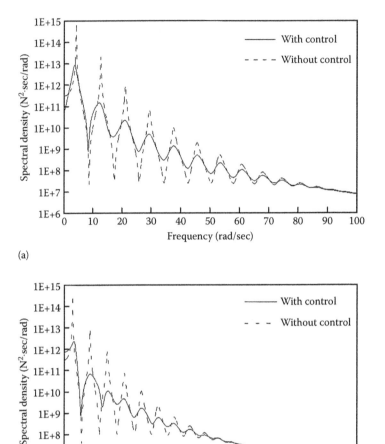

Figure 13.11 Spectral density of base shear force with and without active controller: (a) left building, (b) right building.

13.4 SEMI-ACTIVE CONTROL OF A BUILDING COMPLEX USING FRICTION DAMPERS

A main building and a podium structure that are grouped as a building complex are often seen in modern urban cities because it is a functionally and architecturally appealing scheme. However, such a building complex, which is a kind of setback structure, is prone to greater damage at the level of setback compared to other locations because there are abrupt changes in stiffness and mass (Shahrooz and Moehle 1990). To overcome this problem while keeping a favourable building complex configuration, passive friction dampers (PFDs) were recently proposed to link the podium structure to the main building (Ng and Xu 2007). However, a major drawback of using a PFD is that the control effectiveness as a result of the fixed friction force level would fluctuate under different magnitudes of earthquakes. It has been concluded

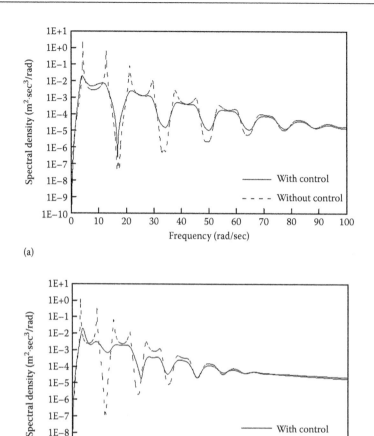

(a)

(b)

Figure 13.12 Spectral density of top floor acceleration with and without active controller: (a) left building, (b) right building.

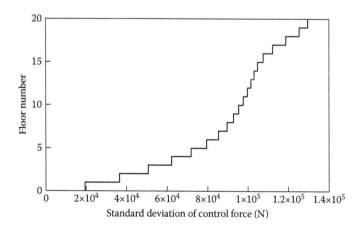

Figure 13.13 Variations of control force of adjacent buildings with height.

Table 13.1 Comparison of major seismic responses of both buildings and control force

Standard deviation	Installed at all floors (Case I)	Installed at top 10 floors only (Case II)
Top floor displacement of left building (mm)	17.8	18.1
Top floor displacement of right building (mm)	22.1	22.2
Bottom shear force of left building (N)	5.65×10^6	5.71×10^6
Bottom shear force of right building (N)	3.57×10^6	3.74×10^6
Top floor acceleration of left building (m/s^2)	0.519	0.554
Top floor acceleration of right building (m/s^2)	0.427	0.453
Control force at top floor (N)	8.79×10^4	1.15×10^5

that a higher level of friction force is preferable under strong earthquakes, whereas moderate earthquakes require a lower level of friction force. In addition, the optimal reduction of inter-story drift is attained at a larger friction force but floor acceleration requires a smaller friction force. To improve the adaptability of the control system to the variation of earthquakes for the steady response reduction of a building complex, the performance of the vibration attenuation on the basis of global- and local-feedback control strategy associated with the application of variable friction dampers is investigated in this section.

13.4.1 Modelling of building complex with variable friction dampers

The analytical model of a building complex can also be depicted by Figure 13.1, in which a main building (building 1) and a podium structure (building 2) have N_1 and N_2 stories $(N_1 > N_2)$, respectively. Only one DOF is assigned to each mass, which is concentrated at each floor of both buildings under unidirectional horizontal earthquake excitation. The equation of motion of such a building complex can also be expressed by Equation 13.1. However, different from Section 13.3 using LQG controllers for active control of adjacent buildings, the variable friction dampers on the basis of global- and local-feedback control strategy are employed in this section for the semi-active control of building complex.

The basic mechanism of a variable friction damper, which employs a piezoelectric actuator as a clamping force regulator to subsequently adjust friction force is shown in Figure 13.14. The operation of a piezo-type variable friction damper is to manipulate the level of input voltage to the piezoelectric actuator so as to control the clamping force magnitude on friction materials. Normally, the clamping force is proportional to the voltage level.

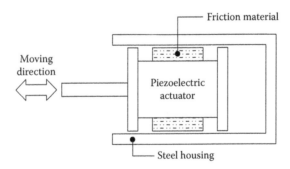

Figure 13.14 Schematic diagram of variable friction dampers.

According to the evaluated properties of a variable friction damper, the damper is modelled with two components: a linear spring and a variable friction slider connected in series. The linear spring simulates the elastic member connected to the friction device, which is modelled by friction slider, for mounting purpose. The behaviour of the friction slider is assumed to follow the Coulomb friction model, which is consistently revealed in the developed devices. According to the Coulomb friction model, the dynamic friction coefficient μ' is considered to be constant and independent of sliding velocity and displacement. Likewise, semi-active friction force f_{dk} of the kth damper is linearly proportional to the clamping force. Because a variable friction damper consists of a linear spring element connected in series with a variable friction slider, the semi-active control forces $\mathbf{u}(t) = \begin{bmatrix} u_1(t) & u_2(t) & \cdots & u_{r'}(t) \end{bmatrix}^T$ (subscript r' denotes the number of variable friction dampers involved) depend on either the sticking or slipping state of each damper, which can be described by

$$u_k(t) = \begin{cases} k_{dk}\left[x_k^{rel}(t) - e_k(t) \right] & \text{if } \left| \bar{f}_k(t) \right| \leq \left| f_{dk}(t) \right| \quad \text{(sticking)} \\ f_{dk}(t) & \text{if } \left| \bar{f}_k(t) \right| > \left| f_{dk}(t) \right| \quad \text{(slipping)} \end{cases} \tag{13.51}$$

$$\bar{f}_k(t) = k_{dk}\left[x_k^{rel}(t) - \bar{e}_k \right] \tag{13.52}$$

where:

$f_{dk}(t)$ is the semi-active friction force provided by the kth variable friction damper in its slipping state

k_{dk} is the spring stiffness of the kth variable friction damper if it is in sticking state

$x_k^{rel}(t)$ denotes the relative displacement between the two buildings at the floor where the kth damper is mounted

The slip deformation of the kth friction damper is represented by e_k and it is given by the expression as follows:

$$e_k = \bar{e}_k + \left| \left| x_k^{rel} - \bar{e}_k \right| - \frac{|u_k|}{k_{dk}} \right| \text{sgn}\left[\dot{x}_k^{rel} \right] \tag{13.53}$$

where:

$\bar{e}_k$ is the previously cumulated slip deformation of the kth damper

$\dot{x}_k^{rel}$ is the relative velocity between the two buildings

Semi-active friction forces can generally be given by

$$f_{dk}(t) = \mu' N_k(t) \text{sgn}\left[\dot{x}_k^{rel}(t) \right] \tag{13.54}$$

in which the clamping force $N_k(t)$ (≥ 0) is time dependent and it is determined by the feedback controller. The semi-active control force $u_k(t)$ is exactly the same as the semi-active friction force $f_{dk}(t)$ if the friction damper slips continuously without sticking. A flowchart regarding the simulation of seismic response of a building complex with variable friction dampers is depicted in Figure 13.15. The friction force is updated at every time step according to the feedback controller. Because the variable friction damper can also stick if the determined feedback control force is larger than the pseudo-friction force, an iterative calculation of slip deformation is therefore required at each time step.

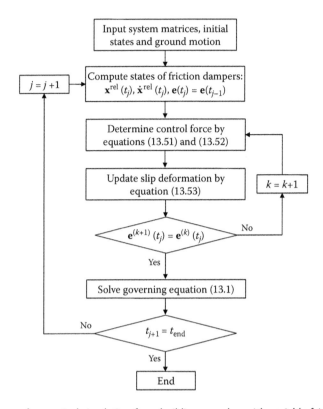

Figure 13.15 Flowchart of numerical simulation for a building complex with variable friction dampers.

13.4.2 Control strategy

13.4.2.1 Global-feedback control strategy

For global-feedback control strategy, the well-known LQG controller is considered here. The design procedure of the LQG controller can be found in Chapter 9. However, the obtained closed-loop feedback control force (active control force) can be achieved by a variable friction damper only when the force direction opposes the velocity direction of the damper. For magnetorheological (MR) dampers, Dyke et al. (1996) proposed a clipped control strategy to let semi-active MR dampers replicate the desired control force as close as possible. However, this clipped control strategy is not applicable to friction-based semi-active dampers because the desired control force can be imitated correctly only when the damper is in its slipping state. Switching the clamping force $N_k(t)$ to its maximum level will very likely make the friction damper stick and then the elastic member force will be the output control force instead of the slipping friction force according to Equation 13.51. Therefore, a modified clipped control strategy is developed for the friction-based semi-active damper (Ng and Xu 2007). In the proposed control strategy, it is assumed that the friction force will follow the Coulomb friction model and the friction coefficient μ' is independent of velocity. Based on this assumption, the desired clamping force $N_k(t)$ can be tuned into compliance with the following decision criteria:

$$N_k(t) = \begin{cases} u_k^{LQG}(t)/\mu' & \text{if } u_k^{LQG}(t) \cdot \ddot{x}_k^{rel}(t) < 0 \\ 0 & \text{if } u_k^{LQG}(t) \cdot \ddot{x}_k^{rel}(t) \geq 0 \end{cases} \tag{13.55}$$

In case the active control force and the motion of the semi-active friction damper are in opposite directions, the clamping force $N_k(t)$ is set to the magnitude of the desired active control force $u_k^{LQG}(t)$ divided by the friction coefficient. Otherwise, a zero clamping force is applied to produce a zero friction force. According to these two switching criteria, the semi-active friction force can be determined by substituting Equation 13.55 to Equation 13.54. The actual control force $u_k(t)$ governed by Equation 13.51 can be subsequently decided by identifying the current states of the friction damper. It is noted that the desired control force cannot be replicated completely unless the slipping of the friction damper is always assured.

13.4.2.2 Local-feedback control strategy

The aforementioned global-feedback control law normally requires many sensors to provide full-state feedbacks of a building complex. It is also rather sensitive to the accuracy of the mathematical model of the building complex and the design appropriateness of weighting matrices for the objective functions targeted. Instead of controlling a building complex using the global-feedback controller, there are some available local-feedback controllers but there have been fewer studies for coupling building control. One merit of a local-feedback controller is that fewer sensors are required, which is particularly important for a building complex. However, it is not certain whether a local-feedback controller can achieve control performance as effectively as a global-feedback controller because available feedbacks in a local-feedback controller are much less than those in a global-feedback controller. Here, three local-feedback control algorithms, which will be used in the experimental investigation are introduced: viscous and Reid friction (VRF) controller, modulated homogeneous friction (MHF) controller, and non-sticking friction (NSF) controller.

The VRF control algorithm, proposed by Chen and Chen (2002, 2004a,b), utilises real-time states of actuated displacement and velocity of a semi-active friction damper in determination of appropriate feedback control force as expressed in the following:

$$N_k(t) = N_k\left(x_k^{\mathrm{rel}}, \dot{x}_k^{\mathrm{rel}}\right) = g_k^1 \left|x_k^{\mathrm{rel}}(t)\right| + g_k^2 \left|\dot{x}_k^{\mathrm{rel}}(t)\right| \tag{13.56}$$

where g_k^1 and g_k^2 are the positive gain coefficients for the slip $x_k^{\mathrm{rel}}(t)$ and the slip rate $\dot{x}_k^{\mathrm{rel}}(t)$, respectively, of the kth variable friction damper. It is obvious that the proposed control logic essentially combines both features of viscous damper and nonlinear Reid damper (Caughey and Vijayaraghavan 1970), which can be seen by taking either control gain $g_k^1 = 0$ or $g_k^2 = 0$, respectively. The advantage of the VRF control logic is that the control force is responsive to both slip and slip rate, which is deemed to be vibration indication of a building. However, this control logic will still cause sticking of the damper if the control gains g_k^1 and g_k^2 are designed too large or the structural vibration is small due to weak ground motion.

The MHF controller was first proposed by Inaudi (1997) and has been modified by He et al. (2003) to guarantee a continuous slipping of semi-active friction damper in the application of base isolation system. Previous actuated peak displacement of the damper is the key signal to determine feedback force as written by

$$N_k(t) = N_k\left(x_k^{\mathrm{rel}}, \dot{x}_k^{\mathrm{rel}}\right) = \beta_k^1 \left|x_k^{\mathrm{rel}}(t-s)\right| \tanh\left[\beta_k^2 \dot{x}_k^{\mathrm{rel}}(t)\right]$$

$$\tag{13.57}$$

where:

β_k^1 is the positive control gain

β_k^2 is the velocity control gain that defines the velocity region that the friction force starts transiting to zero magnitude

s is the time interval between the time instant with the closest previous local maximum (peak or trough) and the current time t, which can be expressed as $s = \left\{ \min \bar{t} \geq 0 : \dot{x}_k^{rel}(t - \bar{t}) = 0 \right\}$

Employing the previous maximum deformation signal for clamping force regulation produces energy dissipation more efficiently than utilising the current deformation signal. This is because the latter one will only yield a triangular hysteresis loop which encloses only half of the area enclosed by a rectangular hysteresis loop if the former type of signal is fed back. Similar to the VRF controller, the MHF controller will also make the damper lock at the moment of force reversal ($\dot{x}_k^{rel} = 0$).

The NSF controller proposed by Ng and Xu (2007) employs actuated velocity in the current state of damper to adjust control force to maintain continuous slipping and energy dissipation by semi-active friction damper. The feedback control force is formulated as

$$N_k(t) = N_k\left(\dot{x}_k^{rel}\right) = N_k^{\max} \left| \tanh\left[\gamma_k^2 \dot{x}_k^{rel}(t) \right] \right|$$ (13.58)

where:

γ_k^2 is the velocity control gain holding the same physical meaning as β_k^2 in the MHF controller

$N_k^{\max}$ is the maximum clamping force

This control algorithm can be interpreted as a controller which tries to drive a friction force to its largest possible magnitude yet maintains continuous slipping of the damper. Therefore, the control device could be used mostly to its full capacity to increase energy dissipation. Another merit of the NSF controller is that the parameter $N_k^{\max}$ sets a clear upper bound of control force of the variable friction damper and therefore the load-bearing capacity of the damper. The design of the damper capacity, based on the VRF and MHF controller, may not be so certain because the magnitude of control force depends on the vibration level of the buildings and the severity of earthquakes for a particular set of control gains. It is possible that the load-bearing capacity of the damper is under designed when a stronger earthquake is experienced in reality. In contrast, the force limit of a damper can be easily considered in the design stage for the NSF controller. Furthermore, the optimal friction force identified in the passive control strategy could be a good indicative value for $N_k^{\max}$ in the NSF controller, which leads to a more intuitive and explicit design.

13.4.3 Experimental investigation of building complex with variable friction damper

For numerical studies on the vibration control of a building complex based on the aforementioned global- and local-feedback control strategies see Ng and Xu (2007) or Section 14.5 of Chapter 14 as a reference. This section aims to experimentally demonstrate the viability of the seismic control of a building complex using variable friction dampers (Xu and Ng 2008).

13.4.3.1 Experimental setup

To examine the seismic mitigation ability of variable friction dampers for a building complex, a 12-story building model and a 3-story building model subject to a series of shaking table tests were accordingly designed to form a building complex model as shown in Figures 13.16 and 13.17. The 3-story building model (Building 2) was considered to imitate a typical podium structure, and a main building (Building 1) with 12 stories was selected based upon a height difference of 4 times. For the effective demonstration of control feasibility, the 12-story building model was designed with a fundamental frequency that could lead to moderate vibration under considered historical earthquake records. The 3-story building was designed stiffer than the main building to maintain a fundamental frequency difference in a ratio of 2.5:1.0 approximately, as would be the case in reality. However, the details of both building models were not similitude-scaled from a prototype building complex. Caution should thus be taken when using the test results directly. Furthermore, only the elastic behaviour of buildings was concerned in this study with the justification that the control scheme employed could effectively keep the buildings within the elastic and linear range.

The overall heights of the two equal-floor-height buildings were 2400 and 600 mm, respectively, for Building 1 and Building 2. The plan dimension and the total weight of the 12-story building were, respectively, 600×400 mm and 382 kg. The corresponding geometries and weight were 1500×710 mm and 547 kg for the 3-story building. Four and eight rectangular columns in a cross section size of 6×50 mm were used in the 12-story and 3-story buildings, respectively, to support the respective rigid plates. The plates and columns were properly welded together to form rigid connections. There was a rectangular opening of 700×500 mm at the centre of each rigid plate of the 3-story building so that the 12-story building could be arranged inside the middle of the 3-story building with a 50 mm gap between the 4 sides of the two buildings. The two buildings were then welded on a steel

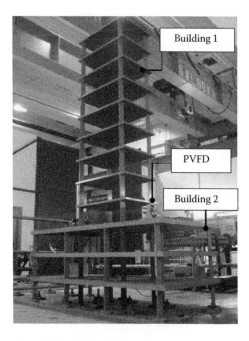

Figure 13.16 Test building complex with variable friction damper.

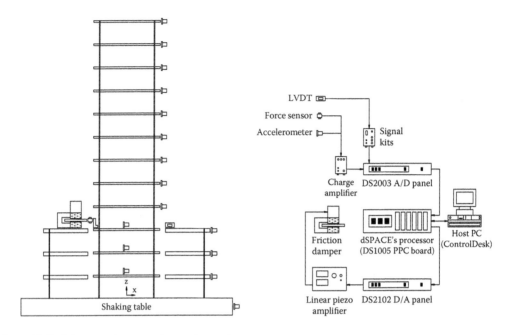

Figure 13.17 Schematic diagram of experimental arrangement.

base plate, which was then bolted firmly on a 3 × 3 m shaking table. The 6 × 50 mm columns were arranged in such a way as to lead to one-directional vibration, and thus the two buildings were effectively reduced to planar frames. In the aforementioned arrangement, the fundamental frequency of the 12- and 3-story buildings is 3.7 and 9.9 Hz, respectively. To simulate a rigid coupling case of the buildings, the bottom three floors of the buildings were linked by high strength steel bars. In the case of semi-active control, a variable friction damper linked the two buildings at their third floor.

Based on an numerical study of the required control force capacity, a laboratory-scale piezo-driven variable friction damper (PVFD) that could provide a force capacity of 250 N and suit for the building complex model was developed and employed as a joint damper, as shown in Figure 13.18, to interconnect the two building models. The damper is simply a mechanical integration of a piezoelectric actuator and a friction generation unit. The

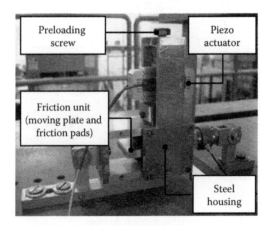

Figure 13.18 Laboratory-scale variable friction damper.

damper consists of a longitudinal part and a vertical part. The longitudinal part is basically a friction generation unit with dimensions of 100 (length) $\times$ 40 (width) $\times$ 54 mm (height). One 8-mm thick steel plate, which allows a stroke range of ± 25 mm, is sandwiched between the two frictions pads to generate friction force as the plate slides. The friction pad is a common car braking pad. Because the developed damper is to laboratory scale and the car braking pad is normally designed to resist larger braking force than required for this study, the reliability and longevity of the device in the present scale of force is apparently not a major issue. The bottom friction pad is fixed to the steel housing whereas the top one is attached to the piezoelectric actuator. The piezoelectric actuator, which acts as the normal clamping force regulator, is allocated in the vertical part, which has dimensions of 30 (length) $\times$ 54 (width) $\times$ 80 mm (height). The steel housing of the PVFD is of high stiffness to prevent loss of clamping force, and it is made with mild steel with a 210 GPa modulus of elasticity and 8 mm thickness.

The PVFD was mounted on the third floor of the two buildings to provide an interactive control force for both buildings. The magnitude of the friction control force was regulated continuously in the semi-active control case and kept constant in the passive control case. The effective range of the friction force was designed roughly between 5 and 340 N at an input voltage from 0 to 150 V, correspondingly. The input voltage was supplied by a linear amplifier, which amplified the command signal from a controller by 20 times. The corresponding range of voltage command from the digital controller was therefore between 0 and 7.5 V.

The instrumentation used in the semi-active control test consisted of sensors for building response measurements and computational hardware for data processing (see Figure 13.17). Regarding sensors, a total of 16 accelerometers were employed to measure the absolute accelerations at all floors of the two test buildings as well as the ground acceleration of the earthquake simulator. Equal numbers of charge amplifiers of model type 2635 produced by Brüel and Kjær were jointly utilised with the accelerometers. The relative displacement between two buildings at the level where the variable friction damper was installed, which was the feedback signal in the semi-active control case, was obtained by a linear variable displacement transducer (LVDT). The LVDT was installed in the same alignment as, but on opposite side from, the PVFD. A load cell, which was attached at one end of the sliding plate of the damper being firmly pin-supported on the third floor of 12-story building, recorded the real-time friction control force actuated from the PVFD. For some control algorithms, relative velocity between two buildings was also needed for feedback. This was determined from the subtraction between the absolute velocities of the two buildings. The absolute velocities were measured by the two accelerometers with their charge amplifiers switched to velocity measurement.

The data processing was performed by a dSPACE real-time data acquisition and feedback control system. This system was operated with a processor board of DS1005, which features a PowerPC FX processor at 800 MHz to provide computing power for real time as well as to function as an interface to the I/O hardware. Quantisation of measured analogy signals (A/D) from various sensors was performed by passing through the I/O board of DS2003. The digital command signal sent out from a complied controller was converted back to an analogy signal (D/A) with the aid of I/O board DS2102 before conveying to the linear piezo-electric amplifier. The programming of real-time controllers and virtual simulation of the closed-loop controlled building system were developed by Simulink, which is an interactive environment integrated in MATLAB. The generation of C code, which is in compliance with the real-time interface (RTI) of dSPACE was executed by a provided function of the Real-Time Workshop in Simulink. The visualisation of measured signals and manipulation of control parameters over the real-time controlled system were solely managed by software of ControlDesk. In this closed-loop control system, sampling and controlling rates were equally set to 500 Hz for all test cases to provide high-quality results for later analysis.

13.4.3.2 Seismic control tests of building complex with PVFD

Performance tests of the piezoelectric actuator and the variable friction damper were first conducted to investigate the characteristics of PVFD, for details refer to Xu and Ng (2008). The application of the PVFD for the seismic control of the building complex was then carried out through a series of shaking table tests.

The earthquakes employed for input ground motions in this study are the El-Centro NS (1940, PGA = 3.417 m/s²), the Hachinohe NS (1968, PGA = 2.250 m/s²), the Kobe NS (1995, PGA = 8.178 m/s²) and the Northridge NS (1994, PGA = 8.267 m/s²). The first two and the last two ground acceleration records are the typical far-field and near-field earthquakes, respectively. The time span of the earthquake records was compressed by three times in the test to produce moderate vibration of the main building as a result of frequency matching between the ground motion and the building model. The level of peak acceleration considered in the test is 0.1, 0.15 and 0.2 g for the four ground motions in an attempt to investigate the adaptability of various control algorithms in response to changes of ground motion intensities.

There were four interconnection configurations of the building complex considered in this experimental study: (1) Case 1, the uncoupling of two buildings; (2) Case 2, the rigid-coupling of two buildings at common floors; (3) Case 3, the coupling control of two buildings with a PFD at the third floor; and (4) Case 4, the coupling control of two buildings with variable friction damper at the third floor.

For Case 1, the two buildings were totally separated and could vibrate individually under the considered four earthquakes. This configuration was one of the uncontrolled cases and provided a comparison basis for control performance. To provide rigid coupling of the building complex in Case 2, three box-section steel members of equal size and strength were used for interconnection. The rigid-couple configuration is a common practice for the construction of a main building with low-rise podium structure, and it was another uncontrolled case considered. In Case 3, the variable friction damper worked as a PFD by maintaining a constant level of voltage supply to the piezoelectric actuator. Three friction forces $f_d = [50, 80, 110]$ N were considered in the passive control. In Case 4, varying voltage was input to the piezoelectric actuator according to the control algorithms employed.

13.4.3.3 Evaluation of control performance

The control effectiveness of various control strategies is evaluated based on the reduction of inter-story drift and absolute acceleration of both buildings. The maximum peak and root mean square (RMS) reductions of the two types of responses are considered, and the corresponding performance indices are given by

$$J_1 = \frac{\max_{t,i}|d_i(t)|}{d_{UCS}^{\max}}, J_2 = \frac{\max_{t,i}|\ddot{x}_i(t)|}{\ddot{x}_{UCS}^{\max}}, J_3 = \frac{\max_i\|d_i(t)\|}{\|d_{UCS}^{\max}\|}, J_4 = \frac{\max_i\|\ddot{x}_i(t)\|}{\|\ddot{x}_{UCS}^{\max}\|} \tag{13.59}$$

where $d_i(t)$ is the ith inter-story drift response time history of either building in different coupling configurations under each earthquake. Similarly, the absolute acceleration response of either building is represented by $\ddot{x}_i(t)$. The maximum peak response of the coupled building complex is normalised by the maximum peak response of the uncoupled building denoted by d_{UCS}^{max} and $\ddot{x}_{UCS}^{max}$. The uncoupled structures (UCS), which are noted in the subscript of uncoupled response, are the reference structural configuration for the comparison of other coupling cases. The evaluation on response reduction in terms of RMS magnitude is given by indices J_3 and J_4. The norms of the structural responses are effective measures of control performance. The norm response of ith floor of either building is determined by $\|z_i(t)\| = \sqrt{1/T \int_0^T z_i^2(t)dt}$. $\|d_{UCS}^{max}\|$ and $\|\ddot{x}_{UCS}^{max}\|$ are the maximum norm inter-story drift and absolute acceleration of the uncoupled building under each of the earthquakes, respectively.

The control performance indices J_1 to J_4 of the 12-story building under the four earthquakes are listed in Tables 13.2 and 13.3. The control gains in the aforementioned various control algorithms are given in Table 13.4. The gains of VRF, MHF and NSF controllers are selected on the basis that a balance performance can be obtained at a similar control force level. In general, the inter-story drift of the main building under different types and intensities of earthquakes is fairly reduced by four semi-active control algorithms. The maximum reduction of peak and RMS inter-story is approximately 4%–34% and 33%–62%,

Table 13.2 Performance indices (J_1 and J_2) of 12-story building

		J_1					J_2				
		PFD	VRF	MHF	NSF	LQG	PFD	VRF	MHF	NSF	LQG
El-Centro	0.1 g	0.63	0.74	0.78	0.70	0.80	0.77	0.66	0.67	0.57	0.62
	0.15 g	0.74	0.74	0.71	0.70	0.80	0.76	0.69	0.65	0.56	0.63
	0.2 g	0.81	0.74	0.73	0.73	0.80	0.79	0.69	0.80	0.62	0.68
Hachinohe	0.1 g	0.85	0.87	0.88	0.78	0.85	0.92	0.88	0.72	0.73	0.73
	0.15 g	0.90	0.89	0.88	0.83	0.86	0.86	0.92	0.78	0.78	0.75
	0.2 g	0.90	0.87	0.91	0.85	0.86	0.90	0.92	0.87	0.75	0.74
Northridge	0.1 g	0.79	0.95	0.88	0.75	0.87	0.99	1.00	0.96	0.88	1.00
	0.15 g	0.81	0.95	0.92	0.83	0.91	0.95	1.03	0.99	0.98	0.98
	0.2 g	0.83	0.96	0.96	0.88	0.92	0.98	1.03	1.04	0.99	1.01
Kobe	0.1 g	0.68	0.80	0.79	0.66	0.80	0.69	0.78	0.74	0.62	0.72
	0.15 g	0.69	0.78	0.77	0.76	0.80	0.74	0.84	0.81	0.75	0.77
	0.2 g	0.75	0.79	0.77	0.76	0.83	0.78	0.84	0.83	0.76	0.76

Note: PFD, Passive friction damper.

Table 13.3 Performance indices (J_3 and J_4) of 12-story building

		J_3					J_4				
		PFD	VRF	MHF	NSF	LQG	PFD	VRF	MHF	NSF	LQG
El-Centro	0.1 g	0.51	0.60	0.67	0.54	0.64	0.60	0.56	0.63	0.51	0.57
	0.15 g	0.56	0.59	0.56	0.53	0.64	0.60	0.53	0.54	0.50	0.56
	0.2 g	0.60	0.57	0.57	0.54	0.60	0.61	0.55	0.61	0.53	0.57
Hachinohe	0.1 g	0.50	0.63	0.68	0.54	0.65	0.61	0.59	0.65	0.52	0.60
	0.15 g	0.59	0.64	0.63	0.57	0.65	0.66	0.60	0.63	0.55	0.60
	0.2 g	0.66	0.62	0.62	0.58	0.64	0.70	0.60	0.62	0.58	0.61
Northridge	0.1 g	0.32	0.51	0.59	0.38	0.50	0.44	0.50	0.57	0.40	0.47
	0.15 g	0.35	0.51	0.53	0.41	0.50	0.46	0.50	0.52	0.42	0.48
	0.2 g	0.42	0.51	0.53	0.43	0.50	0.51	0.55	0.57	0.43	0.49
Kobe	0.1 g	0.39	0.52	0.55	0.42	0.53	0.46	0.51	0.54	0.51	0.48
	0.15 g	0.46	0.51	0.49	0.47	0.53	0.53	0.52	0.57	0.49	0.52
	0.2 g	0.50	0.49	0.47	0.46	0.51	0.59	0.55	0.56	0.52	0.54

Table 13.4 Averaged peak and RMS control forces in different controllers

Control algorithm and parameter		$\frac{1}{N}\sum_{i}^{N}\|\hat{u}_i\|$	$\frac{1}{N}\sum_{i}^{N}\|u_i\|$
PFD	$f_d = 110$	130.5	51.8
VRF	$g^1 = 15.6, g^2 = 10.7$	100.6	20.5
MHF	$\beta^1 = 39.0, \beta^2 = 20$	98.3	19.5
NSF	$F_{max} = 150$	99.8	19.9
LQG	$\mathbf{Q} = \mathbf{I}, \mathbf{R} = 0.01$	77.3	17.0

respectively. The reduction in acceleration response is also noted: there are about 1%–44% and 35%–60% reduction in the maximum peak and RMS acceleration, respectively. The stability of control efficiency in various earthquake intensities is also satisfactorily maintained and it is particularly convinced by RMS reduction. However, a higher reduction is normally found under far-field earthquakes compared to near-field earthquakes. It is also observed that the semi-active control strategy is not always effective for peak acceleration reduction as noted in the Northridge earthquake although the corresponding reduction in terms of RMS magnitude is fairly outstanding. Nevertheless, control performance is relatively stable when employing the NSF control algorithm compared with other semi-active controllers at a similar control force level. This is because the NSF control algorithm could effectively reduce the control force to zero when the slip velocity of the friction damper is zero.

A performance comparison is also made between the passive and semi-active control strategies under the four earthquakes with peak acceleration of 0.2 g, and the results are highlighted in Figure 13.19. The NSF control algorithm is selected to compare with the PFD as well as the two uncontrolled cases, which are a UCS and a rigid-coupled structure (RCS), respectively. The selected control parameters for the two control cases are the passive friction force $f_d = 110$ N and the maximum control force $F_{max} = 150$ N. It is noted in Figure 13.19 that the rigid-coupled building is the worst configuration. Peak inter-story drift and absolution acceleration of Building 1 are amplified in most of the earthquakes compared to the uncoupled case. In contrast, a satisfactory performance in resisting magnification of the peak inter-story drift of Building 1 by the PFD is shown. It is, however, not equally outstanding in reducing the peak acceleration response. By implementing the NSF controller, it is not only effective to reduce the inter-story drift of Building 1 to a comparable degree, but also stable in controlling the acceleration response for most of the earthquakes. Furthermore, as the large friction force in the passive control is required (see Table 13.4) the passive control efficiency is not as high as the NSF controller. A relatively stable control performance is maintained by employing the NSF controller when the intensity of the same type earthquake alters. The NSF controller is thus more effective in minimising both inter-story drift and acceleration response of the building complex compared with the passive control.

To confirm the adequacy of the analytical model of the variable friction damper and the realisation of the various semi-active controllers, the numerical simulation of the seismic control of the building complex model is additionally carried out. Results regarding the local-feedback controllers of VRF, MHF, NSF and the global-feedback controller of LQG in the aforementioned control settings under the 0.15 g El-Centro earthquakes are highlighted in Figures 13.20 and 13.21. It is clearly shown that the simulated absolute acceleration and control force time histories due to the four control algorithms are all in good agreement with

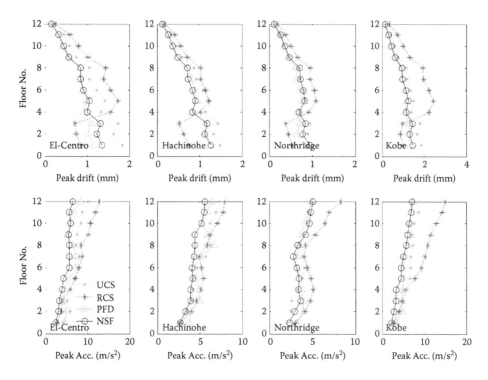

Figure 13.19 Comparison of peak response profiles of uncontrolled and controlled 12-story building.

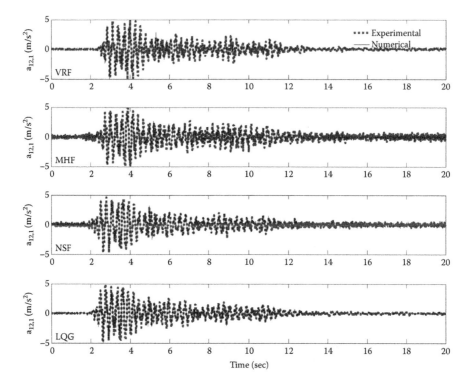

Figure 13.20 Comparison of experimental and numerical acceleration time histories.

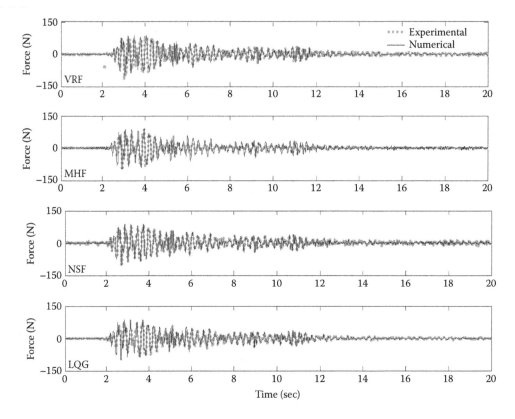

Figure 13.21 Comparison of experimental and numerical control force time histories.

the experimental results. Such an agreement is equally good under the four earthquakes for each controller. The satisfactory comparison of the control force confirms the promptness of the PVFD as well as the appropriateness of the proposed operating scheme and the associated forced feedback control strategy. The reliability of the semi-active control technology, including the examined controllers and the developed control device, is therefore assured.

13.5 SEMI-ACTIVE CONTROL OF BUILDING COMPLEX USING MR DAMPERS

Under earthquake excitation, a tall building with a large podium structure may suffer from a whipping effect due to the sudden change of building lateral stiffness and mass at the top of the podium structure. The experimental investigation of the semi-active control of a building complex using MR dampers as a connection to prevent the whipping effect is introduced in this section (Xu et al. 2005).

13.5.1 Experimental arrangement

The model of the building complex as shown in Figure 13.16 was used again. It should be noted in this experiment that the MR damper was used for seismic control instead of the variable friction damper. To compare the control performance of the MR damper, four connection cases were considered: (1) the two buildings separated, (2) the two buildings rigidly connected, (3) the two building connected by the MR damper in the passive off mode and (4) the two buildings connected by the MR damper in the semi-active control mode.

To create a rigid connection between the 12-story building and the 3-story building, high strength steel bars were used to link the two buildings along the middle line of the two buildings in the x-direction at the first, second and third floor, respectively. Each bar was fixed on the two building floors by a total of four high strength bolts. The assumption of a rigid connection was verified by examining the measured response of each floor of each building to see if there was any relative movement between the two buildings.

One MR damper (RD-1097-01X) manufactured by the Lord Corporation, was used to link the two buildings at their third floors along the middle line of the two buildings in the x-direction. The maximum allowable input current to the damper is 0.5 A and 1.0 A, respectively, for continuous and intermittent application. More than 150 N force can be produced by the damper at 1.0 A, whereas an inherent damper force at 0.0 A is less than 9 N. To achieve the best performance of the MR damper, the Rheonetic Wonder Box device controller kit (RD-3002-03) designed by the Lord Corporation was used together with the MR damper.

To identify the dynamic characteristics of the buildings and to form a semi-active control system for the two buildings linked by the MR damper, a total of 12 accelerometers (B&K 4370) were installed on the two buildings in the x-direction to measure their responses. Each of the first, second, third, fourth, sixth, eighth, tenth and twelfth floors of the 12-story building had one accelerometer placed at its floor centre, whereas each floor of the 3-story building had one accelerometer mounted at the floor centre. To monitor the motion of the simulator in the x-direction, an accelerometer was mounted on the table surface of the simulator to directly measure the ground acceleration at the base of the buildings. To obtain the displacement response of the building, the real-time integration of the acceleration response was performed and then passed through a high-pass filter with the lowest frequency set as 0.64 Hz. A force transducer was placed in series with the MR damper to measure the control force applied to the buildings.

13.5.2 Multi-level logic control algorithm

Considering that the motion of the MR damper in this application was related to the relative motion of the two buildings at the third floor, a semi-active logic control algorithm was implemented in the experiment (Chen et al. 2002).

The semi-active logic control algorithm has the advantage of quick and simple control decisions and avoids the requirement of accurate mathematical models for the control system and the MR damper. Only the relative velocity and displacement between the two buildings at the third floor, where the MR damper was installed, are considered as feedbacks for the determination of the control force by complying with the logic rules of PanBoolean algebra. The core concept of the semi-active logic control is to switch the control damper force to a corresponding pre-specified actuated force region, based on different states of feedback responses by either the increase or decrease of applied current. Obviously, the magnitude of control force should be a function of the level of deviation of the structure from its static equilibrium, which is reflected by the relative velocity and displacement of the two buildings at the third floor. The larger the magnitude of deviation, the larger the damper force that should be applied. Let us define the relative floor displacement and velocity as x_r and $\dot{x}_r$. The pre-specified static equilibrium of the two buildings corresponding to x_r and $\dot{x}_r$ are denoted x_0' and $\dot{x}_0'$, where both x_0' and $\dot{x}_0'$ are a real number with value greater than or equal to zero. There are three possible states concerning the relative floor displacement x_r.

They are x_1^1, x_1^2 and x_1^3, corresponding to $x_r < -x_0'$, $|x_r| \le x_0'$ and $x_r > x_0'$. Similarly, there are three different states of relative floor velocity $\dot{x}_r$, denoted by x_2^1, x_2^2 and x_2^3 corresponding to $\dot{x}_r < -\dot{x}_0'$, $|\dot{x}_r| \le \dot{x}_0'$ and $\dot{x}_r > \dot{x}_0'$, respectively. Therefore, there are nine combinations for the states of x_r and $\dot{x}_r$ and the different combinations of states disclose the level of deviation of the buildings. When both x_r and $\dot{x}_r$ are within the range of x_0' and $\dot{x}_0'$, no additional current (damper force) will be applied. Once either one of the responses, x_r or $\dot{x}_r$, falls outside of the pre-specified equilibrium range while the other still falls inside that range, the control force will be augmented by moderately increasing the applied current. In the situation that both x_r and $\dot{x}_r$ have already deviated from the pre-specified static equilibrium region but they have the opposite sign, the damper force will remain the same magnitude as the previous one. On the contrary, when both x_r and $\dot{x}_r$ depart from the equilibrium region and they have the same sign, the largest control force will be applied by supplying a higher current. As a result, there are in total four states of control force region y_1^1, y_1^2, y_1^3 and y_1^4, which correlate with high current (HC), low current (LC) supply, constant current (CC) supply and no current (NC) supply. Table 13.5 and Figure 13.22 are provided to illustrate the division of various response combinations and the corresponding control states. According to the PanBoolean algebra, the following relationships exist

$$y_1^1 = x_1^1 \cdot x_2^1 + x_1^3 \cdot x_2^3 \tag{13.60}$$

$$y_1^3 = x_1^1 \cdot x_2^3 + x_1^3 \cdot x_2^1 \tag{13.61}$$

$$y_1^4 = x_1^2 \cdot x_2^2 \tag{13.62}$$

Table 13.5 Principle of multi-level logic control algorithm

$\dot{x}_r - x_r$	$\dot{x}_r < -\dot{x}_0'$	$|\dot{x}_r| \le \dot{x}_0'$	$\dot{x}_r > \dot{x}_0'$
$x_r < -x_0'$	I_2 (HC)	II_3 (LC)	III_2 (CC)
$|x_r| \le x_0'$	II_2 (LC)	O (NC)	II_4 (CC)
$x_r > x_0'$	III_1 (CC)	II_1 (LC)	I_1 (HC)

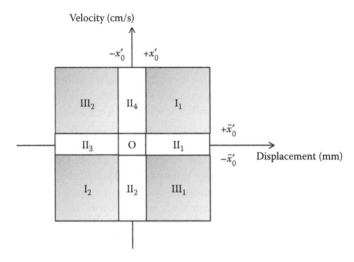

Figure 13.22 Diagram of multi-level logic control algorithm.

$$y_1^2 = 1 - \left(y_1^1 + y_1^3 + y_1^4 \right) \tag{13.63}$$

13.5.3 Seismic control of building complex with MR damper

The semi-active logic control presented above was programmed first by the MATLAB/Simulink and then converted to the executive program in the dSPACE via RTI, which consisted of three major parts: (1) converting the measured voltage signals of the two building responses at the third floor to the digital responses and calculating the relative velocity and displacement as feedbacks; (2) symbolising the feedback responses into the Boolean values, according to Equations 13.60 through 13.63, that are readable by the PowerPC controller for decision-making; and (3) sending a command voltage signal to the current controller to adjust the control force based on the semi-active logic control algorithm.

In the semi-active control, the index of equilibrium state was selected as $x'_0 = 0.3$ mm and $\dot{x}'_0 = 2$ mm/s. The states of applied control current were chosen as 0.6 A for HC, 0.45 A for MC, 0.35 A for LC and 0.0 A for NC. The passive-off mode in which the applied current was zero was also investigated in the study. The MR damper in the passive-off mode provided information on the minimum effectiveness of the damper if the external power was cut off.

Figure 13.23 shows the absolute peak and the RMS acceleration responses of the 12-storey building for the aforementioned four cases. Figure 13.24 gives the relative peak and RMS

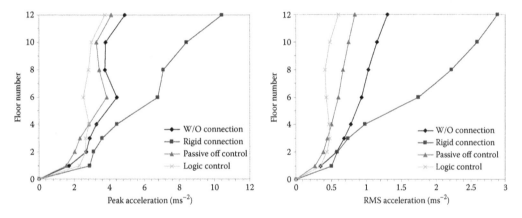

Figure 13.23 Comparison of acceleration responses of 12-story building.

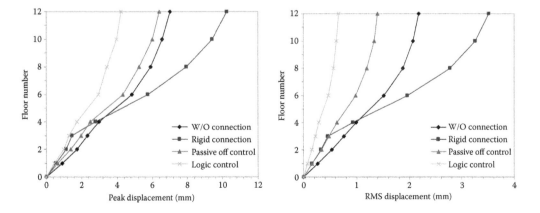

Figure 13.24 Comparison of displacement responses of 12-story building.

displacement responses of the 12-story building for the four cases. It can be seen from Figure 13.24 that both the peak and RMS displacement responses of the 12-story building in the passive-off mode are smaller than those of the 12-story building separated from the three-story building. Compared with the case of the two buildings rigidly connected, it is clearly seen that the MR damper in the passive-off mode totally eliminates the whipping effect that exists in the rigidly connected building. When the MR damper worked in the semi-active control mode with the logic control algorithm, both the peak and RMS displacement responses of the 12-story building were further reduced compared with those when the MR damper worked in the passive-off mode. Regarding the peak displacement response at the top of the 12-story building, the peak displacement response is 10.24 mm for the rigidly connected building, 7.02 mm for the totally separated building, 6.41 mm for the building with the MR damper in the passive-off mode and 4.24 mm for the building with the MR damper in the semi-active control mode. The same observations can be made for the absolute peak and RMS acceleration responses of the 12-story building above the fourth floor, as shown in Figure 13.23. However, the absolute peak and RMS acceleration responses of the first three floors of the building are slightly larger in the semi-active control mode than in the passive-off control model. The shear force in the fourth floor of the 12-story building is 1966 N for the rigidly connected building, 1126 N for the totally separated building, 1056 N for the building with the MR damper in the passive-off mode and 778 N for the building with the MR damper in the semi-active control mode.

Figure 13.25 displays the relative RMS displacement responses and the absolute RMS acceleration responses of the three-story building for the four cases. It can be seen that the RMS displacement responses of the three-story building in the passive-off mode are smaller than those of the three-story building either separated from the 12-story building or rigidly connected to the 12-story building. It can also be seen that the RMS displacement responses of the three-story building with the semi-active control are further reduced compared with those of the three-story building with the MR damper in the passive-off mode. With respect to the RMS acceleration responses, the performance of the semi-active control is much better than that of the passive-off control. With the semi-active control, the acceleration responses of the three-story building reach the same level as the three-story building rigidly connected to the 12-story building. Also by taking the peak displacement response at the top of the three-story building as an example, the peak displacement response is 1.48 mm for the rigidly connected building, 1.30 mm for the totally separated building, 1.11 mm

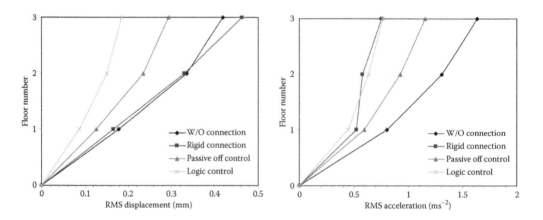

Figure 13.25 Comparison of RMS responses of 3-story building.

for the building with the MR damper in the passive-off mode and 0.83 mm for the building with the MR damper in the semi-active control mode. For the base shear force in the three-story building, it is 2096 N for the rigidly connected building, 2200 N for the totally separated building, 1912 N for the building with the MR damper in the passive-off mode and 1619 N for the building with the MR damper in the semi-active control mode.

The time-histories of displacement response at the top floor of the 12-story building for the four cases are depicted in Figure 13.26. It is clear that the use of the semi-active control with the MR damper can reduce the building response for the entire duration of the ground motion. Figure 13.27 displays the typical force-displacement hysteresis loops of the MR damper in the passive-off mode and in the semi-active control mode, respectively. The difference between the two loops indicates that the semi-active logic control algorithm works well.

In this study, the El Centro 1940 earthquake ground acceleration (N-S component) with the time scale of 1:1 and 1:2, the peak acceleration of 0.13 g and the harmonic ground excitation with a constant peak acceleration of 0.025 g but varying frequency from 3 to 5 Hz, was also generated by the simulator and used as other excitations to examine the performance of semi-active control. The maximum relative RMS displacement and absolute RMS acceleration response of both buildings, subjected to four types of ground excitations, are listed in Table 13.6 for the four connection cases. It can be seen that for the 12-story buildings, the positive results obtained by the semi-active control for the El-Centro ground excitation with the time scale of 1:3 remain at the same level for three other types of ground excitations. For the three-story building, the high performance of the semi-active control in reducing the maximum acceleration response is kept but it is not always true in reducing the maximum displacement response.

13.6 MULTI-OBJECTIVE HYBRID CONTROL OF HIGH-TECH EQUIPMENT IN HIGH-TECH FACILITY

The high-tech equipment engaged in the production of semiconductors and optical microscopes in high-tech facilities is extremely expensive. To ensure the high quality of ultra-precision products, high-tech equipment requires a normal working environment with extremely limited vibration. Some high-tech facilities are located in seismic zones where the safety of high-tech equipment during an earthquake becomes a critical issue. It is therefore imperative to find an effective way to ensure the functionality of high-tech equipment against microvibration under normal working conditions and to protect high-tech equipment from damage when an earthquake occurs.

Major sources of microvibration, affecting the normal operation of high-tech equipment, are due to traffic-induced ground motion, service machinery-induced floor vibration and direct disturbance caused by production-related activities in the form of a suddenly applied load (Ungar et al. 1990). Several well-known families of generic vibration criteria are in use for microvibration control of high-tech equipment in terms of velocity, such as the Bolt, Beranek and Newman (BBN) vibration criteria (Gordon 1991). Compared with traffic-induced ground motions, earthquake-induced ground motions are of relatively low frequency range but very high intensity. The building structures themselves in seismic zones are generally designed to meet seismic code requirements. However, fragile and vibration sensitive equipment is often mounted on the building floor without considering seismic provisions making it extremely vulnerable to a seismic-induced lateral load. Under a performance-based design concept, it is now recognised that high-tech equipment must function immediately after an earthquake and therefore seismic protection for high-tech equipment is

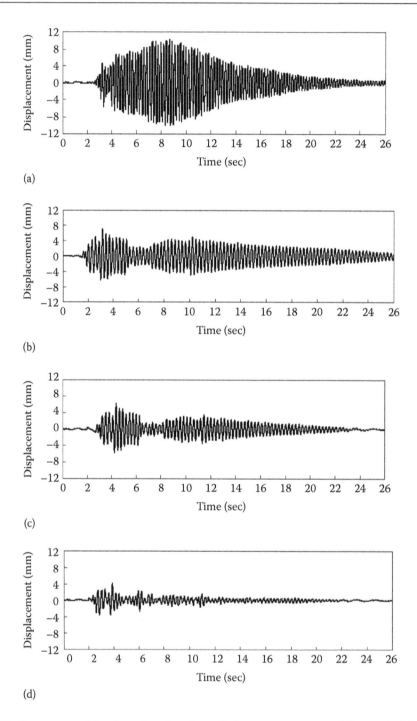

Figure 13.26 Comparison of time histories of top floor displacement responses of 12-story building: (a) with rigid connection, (b) without rigid connection, (c) with MR damper in passive control mode and (d) with MR damper in semi-active control mode.

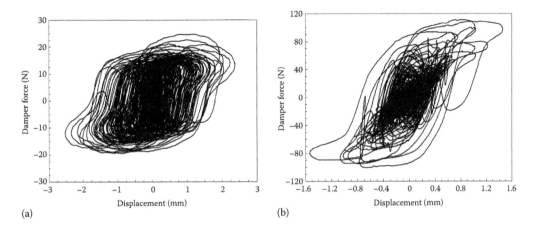

Figure 13.27 Displacement-force hysteresis loops of MR damper: (a) passive control mode and (b) semi-active control mode.

urgently needed (Amick and Bayat 1998). Although there are no consensual seismic design criteria for high-tech equipment, it is believed that excessive acceleration responses lead to damage to high-tech equipment.

Most of the previous studies for reducing microvibration of high tech equipment are concerned with vibration isolation systems. These vibration isolation systems include small passive mounts, hybrid tables and active tables, which are mainly used for isolating individual or a small quantity of high tech equipment (Serrand and Elliott 2000; Yang and Agrawal 2000; Yoshioka et al. 2001; Nakamura et al. 2006). The corresponding analytical model often takes building floor vibration as a direct base excitation to the vibration isolation system: the dynamic interaction between the building and the high tech equipment/vibration isolation system, however, is not considered. Recently, both experimental and numerical studies on microvibration control of a batch of high-tech equipment were carried out, in which a single-layer hybrid platform consisting of passive mounts and an actively controlled electromagnetic actuator was coupled with the building structure (Xu et al. 2003a,b; Yang et al. 2003). These studies, however, focused on microvibration control only without considering earthquake excitation.

This section presents a multi-objective hybrid control platform, which works as a passive platform aiming to mitigate the acceleration response of a batch of high-tech equipment during an earthquake and to function as an actively controlled platform, which intends to reduce the velocity response of a batch of high-tech equipment under normal working conditions (Xu and Li 2006). After the discussions on the single-layer passive isolation platform, the double-layer passive isolation platform and the hybrid platform with smart actuators, the analytical model of a coupled hybrid platform and building system incorporating smart actuators and LQG control algorithms was established. The analytical model was then applied to a high-tech facility to evaluate the performance of the proposed hybrid platform for both microvibration control and seismic response control. A three-story building model and a hybrid platform model were designed and manufactured in the laboratory as well. A series of shaking table tests, traffic vibration tests and impact tests were performed on the building to examine the performance of the two-layer hybrid platform for the functionality of high-tech equipment against microvibration and for the safety of high-tech equipment against earthquake (Xu et al. 2008).

428 Smart civil structures

Table 13.6 Maximum RMS response of building models subject to various ground motions

Earthquake	Separated		Rigidly connected		Passive-off control		Semi-active control	
	x_{RMS} (mm)	$\ddot{x}_{RMS}$ (m/s²)	x_{RMS} (mm)	$\ddot{x}_{RMS}$ (m/s²)	x_{RMS} (mm)	$\ddot{x}_{RMS}$ (m/s²)	x_{RMS} (mm)	$\ddot{x}_{RMS}$ (m/s²)
(a) 12-story building								
El-Centro (0.13 g, 1/1)	3.02	1.63	1.41 (53.3%)	1.05 (35.6%)	2.80(7.3%)	1.48 (9.2%)	1.38 (54.3%)	0.70 (57.1%)
El-Centro (0.13 g, 1/2)	4.11	2.30	1.74 (57.7%)	1.50 (34.8%)	3.66(10.9%)	2.00 (13.0%)	1.61 (60.8%)	0.94 (59.1%)
El-Centro (0.13 g, 1/3)	2.17	1.30	3.51 (−61.8)	2.88 (−122%)	1.39(35.9%)	0.84 (35.4%)	0.66 (69.6%)	0.60 (53.8%)
Harmonic (0.025 g, 3–5 Hz)	5.37	3.10	4.86 (9.5%)	3.84 (−23.9)	4.91(8.6%)	2.66 (14.2%)	2.33 (56.6%)	1.23 (60.3%)
(b) 3-Story Building								
El-Centro (0.13 g, 1/1)	0.20	0.53	0.23 (−15.0)	0.34 (−35.8%)	0.20(0%)	0.46 (13.2%)	0.19 (5.0%)	0.35 (34.0%)
El-Centro (0.13 g, 1/2)	0.24	0.95	0.28 (−16.7)	0.52 (45.3%)	0.22(8.3%)	0.84 (11.6%)	0.16 (33.3%)	0.56 (41.1%)
El-Centro (0.13 g, 1/3)	0.42	1.64	0.46 (−9.5%)	0.75 (54.3%)	0.29(31.0%)	1.16 (29.3%)	0.19 (54.8%)	0.77 (53.0%)
Harmonic (0.025 g, 3–5 Hz)	0.06	0.30	0.61 (−916%)	0.54 (−80.0)	0.06(0%)	0.30 (0%)	0.08 (−33.3)	0.27 (10.0%)

Note: Value in parenthesis represents the decrease/increase percentage in response with respect to separated buildings (+ sign denotes reduction, - sign denotes amplification).

13.6.1 Multi-objective hybrid control platform

As mentioned before, the multi-objective hybrid control platform is designed to work as a passive isolation platform to reduce the acceleration response of high-tech equipment during an earthquake and to function as an actively controlled platform to reduce mainly the velocity response of high-tech equipment under normal working conditions. In this regard, the stiffness and damping coefficients of a hybrid platform are determined according to the safety requirement. The active control of the hybrid platform is then designed according to the functionality requirement. This subsection mainly discusses important issues involved in the design of hybrid platform without considering the interaction between the platform and building.

13.6.1.1 Single-layer and double-layer passive isolation platform

A common practice for the microvibration control of high-tech equipment is to adopt a single-layer passive platform to isolate high-tech equipment from building floor vibration. The single-layer passive platform consists of a platform supported by several pneumatic isolators, as shown in Figure 13.28a. The determination of optimal stiffness and the damping coefficient of the pneumatic isolator is based on the possible maximum reduction of either velocity or the displacement response of the platform. However, the possible maximum reduction of the acceleration response of the platform without excessive drift is targeted in this study to resist the earthquake-induced building floor vibration. Research has found that, since the passive platform is targeted to reduce the absolute acceleration of high-tech equipment to the level below its allowable value, it is difficult to control the platform drift to the level below the allowable value for the safety of pneumatic isolators (Xu and Li 2006). Therefore, to satisfy both acceleration and drift requirements, a double-layer passive isolation platform is proposed.

A double-layer passive isolation platform is formed by adding an auxiliary isolation layer between the single-layer passive isolation platform and the building floor, as shown in Figure 13.28b. The auxiliary isolation layer consists of a light mounting plate supported by rubber isolators of high energy dissipating capacity. Comparative studies show that, with the given system parameters, the acceleration transmissibility of the double-layer platform is very close to that of the single-layer platform within the frequency ratio range from 0.1 to 10. For the range above 10, the acceleration attenuation performance of the double-layer platform is better than that of the single layer platform. It is also found that the drift transmissibility of either the platform or the mounting plate of the double-layer platform is much smaller than that of the single-layer platform as targeted. Furthermore, in the concerned excitation frequency range, the drift transmissibility of the platform is larger than that of the mounting plate in the double-layer platform as designed. A more detailed description of the single-layer and double-layer passive isolation platform can be found in the references (Xu and Li 2006).

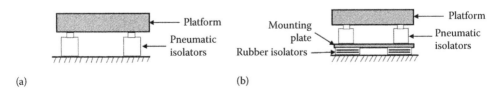

Figure 13.28 Schematic diagram of the passive isolation platform: (a) a single-layer passive isolation platform and (b) a double-layer passive isolation platform.

13.6.1.2 Hybrid platform with active actuator

If single- and double-layer isolation platforms can be used to satisfy the safety requirement of high-tech equipment during earthquakes, the next task is to satisfy the serviceability requirement for microvibration control of high-tech equipment installed on the platform against traffic-induced ground motion, machinery-induced floor vibration and direct disturbance to the platform. Though the passive platform is also beneficial for reducing microvibration of high-tech equipment against traffic-induced ground motion and machinery-induced floor vibration, the microvibration control performance of the passive platform is always limited because of its passive nature and the more stringent microvibration criteria. The passive platform also hardly mitigates direct disturbances such as occasional impact forces to the platform because the natural frequency of the platform is low and the impact forces will induce the platform response around its natural frequency, which can hardly be mitigated by the passive platform. Therefore, a hybrid control by coupling active actuators with a passive platform for microvibration control of high-tech equipment under normal working conditions is investigated in this section.

The basic components of a hybrid platform that incorporates active actuators to a double-layer passive isolation platform are shown in Figure 13.29. Conceptually, the actuator is installed on the mounting plate with a friction switch mechanism that can release the actuator during an earthquake event to prevent the actuator from damage and to switch the platform to a passive mode based on the signals from sensors. Under normal working conditions, the actuator is active and applies control force to the platform through its exciting rod connected to the platform according to a given control algorithm.

Consider two ideal cases: the first case is that only direct disturbance exists and the platform response is always zero because of the function of active control; and the second case is that only floor motion exists and the platform response is always zero as result of active control. Xu and Li (2006) found that when only floor motion is considered, the ideal active control force required by the double-layer platform is larger than that required by the single-layer platform within a relatively low frequency range, but the situation is reversed within a relatively large frequency range. Moreover, as compared with the case of floor motion, the ideal active control force required by the double-layer hybrid platform against direct disturbance is larger within a wide frequency range. This indicates that more control energy may be needed for the case of direct disturbance.

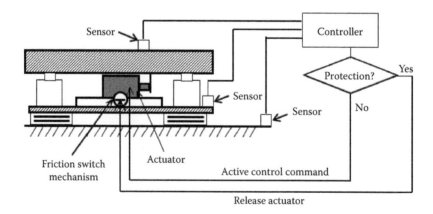

Figure 13.29 Configuration of a hybrid platform.

Since microvibration of high-tech equipment due to traffic-induced ground motion, machinery-induced floor motion and direct disturbance covers a wide range of frequency, actuators used in a double-layer hybrid platform must possess the capability of high control precision within a wide range of frequency. In this regard, the magnetostrictive material Terfenol-D, which has proven very effective at room temperature to deliver high control precision, is selected to make magnetostrictive actuators, which are then incorporated into a double-layer platform to form a double-layer hybrid platform. While magnetostrictive actuators can be utilised in linear regimes by maintaining low input levels, the microvibration control of high-tech equipment may require that they can be driven at high input levels where hysteresis and nonlinearities are inherent to actuator dynamics. Hence, it is necessary to design magnetostrictive actuators, which accommodate the material hysteresis and nonlinearities. The introduction of magnetostrictive actuators can be found in Chapter 4, and the details on the modelling of a magnetostrictive actuator in conjunction with the aforementioned double-layer platform to form a double-layer hybrid platform can be found in Xu and Li (2006).

13.6.2 Equation of motion of building with hybrid platform

Let us consider a three-story shear building with a double-layer platform installed on its first floor. Figure 13.30a shows that the friction switch mechanism is free and the platform works as a double-layer passive platform to protect high tech equipment against earthquake-induced ground motion. Figure 13.30b shows that the friction switch mechanism is locked and the platform works as a double-layer hybrid platform with magnetostrictive actuators for microvibration control of high-tech equipment against traffic-induced ground motion, machinery-induced floor vibration and direct table disturbance under normal working conditions.

13.6.2.1 Equation of motion of building with passive platform

In consideration that the performance evaluation of the system against earthquake-induced ground motion is based on the absolute acceleration response of the platform, the governing equation of the motion of the system is established in the absolute coordinate. For the building with a double-layer passive isolation platform, as shown in Figure 13.30a, the equation of motion of the coupled system under earthquake-induced ground motion can be derived as

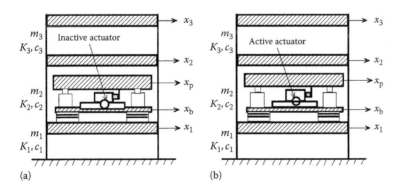

Figure 13.30 Three-story building with a double-layer platform: (a) passive platform and (b) hybrid platform.

$$
\begin{bmatrix} m_1 & & & & \\ & m_2 & & & \\ & & m_3 & & \\ & & & m_b & \\ & & & & m_p \end{bmatrix} \begin{Bmatrix} \ddot{x}_1 \\ \ddot{x}_2 \\ \ddot{x}_3 \\ \ddot{x}_b \\ \ddot{x}_p \end{Bmatrix} + \begin{bmatrix} c_1+c_2+c_b & -c_2 & 0 & -c_b & 0 \\ -c_2 & c_2+c_3 & -c_3 & 0 & 0 \\ 0 & -c_3 & c_3 & 0 & 0 \\ -c_b & 0 & 0 & c_b+c_s & -c_s \\ 0 & 0 & 0 & -c_s & c_s \end{bmatrix} \begin{Bmatrix} \dot{x}_1 \\ \dot{x}_2 \\ \dot{x}_3 \\ \dot{x}_b \\ \dot{x}_p \end{Bmatrix}
$$

$$
+ \begin{bmatrix} k_1+k_2+k_b & -k_2 & 0 & -k_b & 0 \\ -k_2 & k_2+k_3 & -k_3 & 0 & 0 \\ 0 & -k_3 & k_3 & 0 & 0 \\ -k_b & 0 & 0 & k_b+k_s & -k_s \\ 0 & 0 & 0 & -k_s & k_s \end{bmatrix} \begin{Bmatrix} x_1 \\ x_2 \\ x_3 \\ x_b \\ x_p \end{Bmatrix} = \begin{Bmatrix} c_1 \\ 0 \\ 0 \\ 0 \\ 0 \end{Bmatrix} \dot{x}_g + \begin{Bmatrix} k_1 \\ 0 \\ 0 \\ 0 \\ 0 \end{Bmatrix} x_g
$$

$$(13.64)$$

where:

m_i, k_i, c_i are the mass, stiffness coefficient and damping coefficient of the ith floor of the building, respectively

m_p, k_s and c_s stand for the mass, stiffness and damping coefficient of the platform, respectively

m_b, k_b and c_b are the mass, stiffness and damping coefficient of the mounting plate, respectively

x_p and x_b represent the displacement of platform and mounting plate, respectively

x_i is the absolute displacement of the ith floor of the building

x_g and $\dot{x}_g$ are the displacement and velocity of ground motion, respectively

13.6.2.2 Equation of motion of building with hybrid platform

For the building with a hybrid platform as shown in Figure 13.30b, the equation of the motion of the system under traffic-induced ground motion, service machinery-induced floor vibration and direct table disturbances can be expressed as

$$
\begin{bmatrix} m_1 & & & & \\ & m_2 & & & \\ & & m_3 & & \\ & & & m_b & \\ & & & & m_p \end{bmatrix} \begin{Bmatrix} \ddot{x}_1 \\ \ddot{x}_2 \\ \ddot{x}_3 \\ \ddot{x}_b \\ \ddot{x}_p \end{Bmatrix} + \begin{bmatrix} c_1+c_2+c_b & -c_2 & 0 & -c_b & 0 \\ -c_2 & c_2+c_3 & -c_3 & 0 & 0 \\ 0 & -c_3 & c_3 & 0 & 0 \\ -c_b & 0 & 0 & c_b+c_s & -c_s \\ 0 & 0 & 0 & -c_s & c_s \end{bmatrix} \begin{Bmatrix} \dot{x}_1 \\ \dot{x}_2 \\ \dot{x}_3 \\ \dot{x}_b \\ \dot{x}_p \end{Bmatrix}
$$

$$
+ \begin{bmatrix} k_1+k_2+k_b & -k_2 & 0 & -k_b & 0 \\ -k_2 & k_2+k_3 & -k_3 & 0 & 0 \\ 0 & -k_3 & k_3 & 0 & 0 \\ -k_b & 0 & 0 & k_b+k_s & -k_s \\ 0 & 0 & 0 & -k_s & k_s \end{bmatrix} \begin{Bmatrix} x_1 \\ x_2 \\ x_3 \\ x_b \\ x_p \end{Bmatrix} = \begin{Bmatrix} 0 \\ 0 \\ 0 \\ -f_a \\ f_a \end{Bmatrix} + \begin{Bmatrix} f_{\text{floor}} \\ 0 \\ 0 \\ 0 \\ 0 \end{Bmatrix} + \begin{Bmatrix} c_1\dot{x}_g+k_1x_g \\ 0 \\ 0 \\ 0 \\ 0 \end{Bmatrix} + \begin{Bmatrix} 0 \\ 0 \\ 0 \\ 0 \\ f_p \end{Bmatrix}
$$

$$(13.65)$$

where:

f_a denotes the active control force from magnetostrictive actuators, which is determined according to a given active control algorithm as will be discussed in the next section

f_{floor} denotes the exciting force induced by service machinery and applied on the first floor

f_p denotes direct disturbance to the platform

13.6.3 Active control algorithm

The equation of motion of the coupled building-hybrid platform system (see Equation 13.65) can be expressed in the state space form:

$$\dot{z} = A'z + B'f_a + E_1 f_{\text{floor}} + E_2 f_p + E_3 f_g \tag{13.66}$$

where:

z   is the state vector of the coupled system that can be expressed as $\{x_1, x_2, x_3, x_b, x_P, \dot{x}_1, \dot{x}_2, \dot{x}_3, \dot{x}_b, \dot{x}_P\}^T$

A'   is the 10×10 state matrix

B'   is the 10×1 location matrices of control force

E_1, E_2 and E_3 are the location matrices of excitations with the dimension of 10×1, 10×1 and 10×2, respectively, and

f_g is a vector that can be expressed as $\{x_g, \dot{x}_g\}$

From a practical point of view, the full state vector can be rarely measured as a whole. It is thus often necessary to replace the full state vector by an incomplete state measurement vector. To evaluate microvibration control performance properly, the absolute velocity of the platform $\dot{x}_p$, the absolute velocity of the mounting plate $\dot{x}_b$, the absolute velocity of the first floor $\dot{x}_1$ and the relative displacement (drift) between the platform and mounting plate are selected to form the measurement vector. The relationship between the full state vector and the measurement vector $\mathbf{y}'$ is thus expressed as

$$\mathbf{y}' = \left\{x_p - x_b, x_p - x_1, \dot{x}_1, \dot{x}_b, \dot{x}_p\right\}^T = C'z + \varepsilon(t) \tag{13.67}$$

where:

C'   is called the measurement matrix

$\varepsilon(t)$ is the measurement noise

The movement of the actuator can be calculated as $x_p - x_b$. The inclusion of measurement noise is necessary because the response signals are small in the microvibration environment and the measurement signals usually have a low signal-to-noise ratio. It is imperative to filter the measurement noise properly for a proper active control. In this study, the measurement noise $\varepsilon(t)$ is assumed as a Gaussian white noise with a zero mean. The optimal LQG control algorithm is also used in this section to design a controller for the hybrid platform. Similar to the procedures mentioned in Section 13.3.2, the optimal feedback control force $\mathbf{f}_a$ can be expressed as

$$\mathbf{f}_a = -K_c\{\hat{z}\} \tag{13.68}$$

where:

K_c   is the optimal feedback gain matrix determined by minimising the performance function

$\hat{z}$ is the estimated state vector of the state vector z from the Kaman filter constructed as

$$\dot{\hat{z}} = A'\hat{z} + B'f_a + E_1 f_{floor} + E_2 f_d + E_3 f_g + K_f \left(y' - \hat{y}' \right) \tag{13.69}$$

$$\hat{y}' = C'\hat{z} \tag{13.70}$$

where K_f is the Kalman filter gain as mentioned before. After the desired optimal control force is obtained from the LQG control algorithm, the inverse compensator technique is employed in the controller is to determine the precise input current for the magnetostrictive actuator to produce this desired control force (Xu and Li 2006). Figure 13.31 displays a block diagram for the active control of coupled building-hybrid platform system.

13.6.4 Experimental investigation of multiobjective hybrid platform

The numerical investigation shows that the proposed hybrid platform can reduce not only the traffic-induced and machinery-induced vibration but also the vibration caused by the direct disturbance. For more details refer to Xu and Li (2006). This section focuses on the experimental studies to examine the performance of the two-layer hybrid platform for the functionality of high-tech equipment against microvibration and for the safety of high-tech equipment against earthquake. A three-story building model and a hybrid-platform model were designed and manufactured. The two-layer hybrid platform was installed on the first floor of the building to work as a passive platform aiming to abate the acceleration response of the equipment during an earthquake and to function as an actively controlled platform, which intends to reduce the velocity response of the equipment under normal working conditions. The hybrid platform working as a passive platform was designed in such a way that its stiffness and damping ratio could be changed, whereas for the hybrid

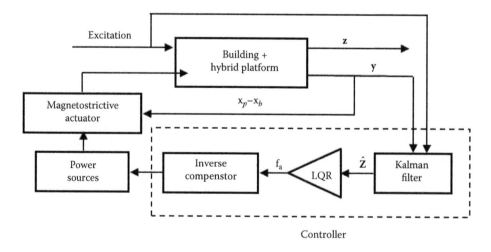

Figure 13.31 A block diagram of an active control system.

platform functioning as an active platform, a piezoelectric actuator with a sub-optimal velocity feedback control algorithm was used. A series of shaking table tests, traffic vibration tests and impact tests were performed on the building to examine the performance of the hybrid platform against both earthquake and microvibration.

13.6.4.1 Experimental arrangement

A three-story building model was designed and manufactured as shown in Figure 13.32. The building model consisted of three steel plates and four steel columns. Each plate had a rectangular shape of 850 × 500 mm in plane and a thickness of 25 mm with a mass of 82.9 kg. Each column of 75 × 16 mm rectangular cross section was embedded into, and welded to, the plates to ensure the rigid joints formed at the connections. The four columns of the building model were eventually welded to a thick steel plate that was in turn bolted to a shaking table. The cross section of the column was arranged in such a way that the stiffness of the building model in the y-direction was much higher than that in the x-direction. Each steel floor could be regarded as a rigid plate in horizontal, which leads to a shear-type building model in the x-direction. To properly simulate the inherent energy dissipation capacity of a real building, small dashpots were installed between every two floors. The geometric and time scales of the building model were assumed to be 0.125 and 0.353, respectively.

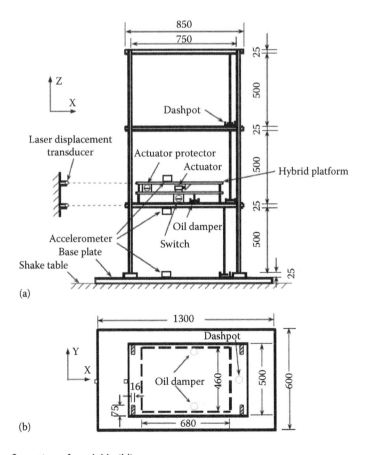

Figure 13.32 Configuration of model building.

A two-layer hybrid platform was manufactured and installed on the first floor of the building as shown in Figure 13.32. High-tech equipment is supposed to be installed on the top surface of hybrid platform rather than the first floor of the building. Accordingly, the control performance of the hybrid platform was assessed by comparing the response of its top surface with that of the first floor of the building without control. The two-layer platform was formed mainly by three rectangular steel plates with four columns of 0.7×16 mm rectangular cross section, which were embedded into and fastened by bolts to the three plates (see Figure 13.33a). Such an arrangement leads to only shear deformation of the two layers in the x-direction.

The bottom layer between the base plate and the middle plate was designed as a passive layer, in which two shear type oil dampers and one u-shape leaf spring were installed. Each oil damper consisted of a disk as a piston moving in a container filled with silicone oil (see Figure 13.33e). The stiffness of the passive layer was regulated by the u-shape leaf spring (see Figure 13.33d). One end of the u-shape leaf spring was fixed to the middle plate and the other was fixed to the base plate. By adjusting the effective length of the leaf spring, the stiffness of the passive layer could be changed within a wide range.

The top layer between the middle plate and the top plate was designed as an active layer. A flexitensional piezoelectric actuator (FPA) (model FPA-850) developed by the Dynamic Structures and Materials, LLC (DSM) was installed in the active layer, as shown in Figure 13.33b. With the input voltage from -30 to $+150$ V, the FPA provides a bi-directional stroke of 850 μm and an ultimate blocking force of 330 N. A specialised linear piezoelectric actuator (LPA) amplifier, which was part of an actuating system developed by DSM, was used with the FPA. The LPA accepted a command voltage signal from -1.5 to 7.5 V, which is an input via a Bayonet Neill–Concelman (BNC) connector on its front panel, and subsequently the LPA output an amplified voltage of -30 to 150 V to the FPA.

Here, the two-layer hybrid platform was designed to work as a passive platform aiming to abate acceleration of high-tech equipment during an earthquake and to function as an actively controlled platform, which intends to reduce velocity of high-tech equipment under normal working conditions. To fulfil this task, an electromagnetic mechanism was installed in the passive layer as a switch to lock the middle plate to the base plate under normal working conditions so that only the active layer functions against microvibration. There was another electromagnetic mechanism installed in the active layer as an actuator protector to lock the middle plate to the top plate when an earthquake occurs. Since the electromagnetic mechanism in the passive layer does not function and the active layer will be locked for an earthquake event, the hybrid platform becomes a purely passive platform against earthquake.

The details of the switch in the passive layer can be seen in Figure 13.33b. There are two blocks made of ferromagnetic alloy, each of which is appended to one free end of a beam. The beam is made of a silicon steel sheet of 0.5×40 mm rectangular cross section, and the middle part of the beam is bolted to the middle plate. Two electromagnets are mounted to the base plate just underneath the two ferromagnetic blocks with a gap of 1 mm between the ferromagnetic block and the electromagnet. Under normal working conditions, the electromagnet will be electrified and the ferromagnetic block will then be attached to the electromagnet without any gap. Since the stiffness of the beam in the x-direction is very large, the passive layer can be considered to be locked completely in the horizontal direction. Under an earthquake event, the electric supply to the electromagnets will be turned off and the ferromagnetic blocks will be disconnected from the electromagnets. The hybrid platform then becomes a passive platform.

The details of the actuator protector in the active layer can be seen in Figure 13.33c. Two electromagnets are fixed to the top plate. Two ferromagnetic blocks are appended to the

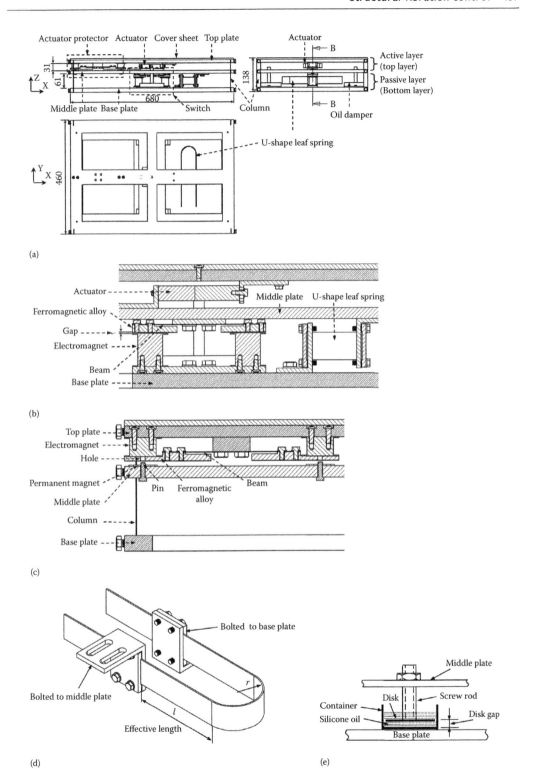

Figure 13.33 Configuration of hybrid platform: (a) schematic diagram (unit: mm), (b) details of actuator and switch (Section B-B), (c) details of actuator protector (Section B-B) and (d) u-shape-leaf spring unit.

free ends of a beam made of a silicon steel sheet. The middle part of the beam is bolted to the top plate. There are two permanent magnets fixed on the middle plate just underneath the two ferromagnetic blocks. Under normal working conditions, the ferromagnetic blocks are attached to the electromagnets when electrified. The top plate and the middle plate are thus connected by the four columns and the actuator only. The hybrid platform becomes an active platform against microvibration. When an earthquake occurs, the electric supply to the electromagnets is turned off. The ferromagnetic blocks are pulled down by the permanent magnets, and consequently the ferromagnetic blocks are locked to the middle plate by the pins: the top plate and the middle plate are firmly connected to each other and the hybrid platform becomes a passive platform. In this way, the actuator can be protected against overload due to seismic excitation.

In this experimental investigation, if the measured acceleration of the first floor of the building is less than 0.1 g, where g is the acceleration due to gravity in metres per second squared (m/s²), the computer will send a command to switch on the electromagnets in the two layers to turn the hybrid platform into the active platform. Otherwise, the electromagnets will be switched off and the hybrid platform becomes a passive platform. The complete experimental setup can be seen in Figure 13.32. A flowchart of hybrid control can be seen in Figure 13.34.

13.6.4.2 Instrumentation

Three KYOWA ASQ-1BL accelerometers (KYOWA Electronic Instruments Co. Ltd) were installed on the shaking table, the first floor of the building and the top surface of the platform, respectively (see Figure 13.32) to measure their accelerations. Two Keyence LK-503 laser displacement transducers of high sensitivity were used to measure the absolute displacement responses of the top plate of the platform and the first floor of the building to obtain the horizontal drift of the platform in the case of earthquake excitation. The velocity of the platform was obtained by integrating the measured acceleration time history of the platform with time on line. To eliminate the shift of velocity time history during the integration, a high-pass filter was applied to the acceleration time history before the integration. A dSPACE real-time system was used as the data acquisition and control signal generation system. The hybrid control of the platform in this experiment was realised by using the dSPACE system together with the MATLAB/Simulink program on the host computer via RTI.

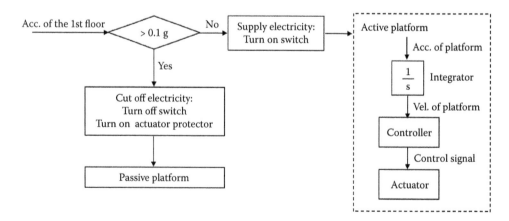

Figure 13.34 Flowchart of hybrid control system.

The experimental investigation on the performance of the hybrid platform against an earthquake was carried out on a shaking table housed in the Structural Dynamics Laboratory of The Hong Kong Polytechnic University. Three well-known earthquake records were chosen: the El Centro NS (1940, peak ground acceleration [PGA] = 3.417 m/s²), the Northridge NS (1994, PGA = 8.267 m/s²) and the Hachinohe NS (1968, PGA = 2.250 m/s²). They were compressed in time by a factor of 0.353 with the duration of 23 s for El-Centro, 30 s for Northridge, and 14 s for Hachinohe. The peak accelerations of the three records all were scaled to the same value of 1.5 m/s².

The experimental investigation on the performance of the hybrid platform against microvibration was carried out under traffic-induced ground motion, without the operation of the shaking table. The ground motion was mainly caused by vehicles running on the Hong Chong Road, which is very close to the laboratory. This experimental investigation also considered the direct disturbance caused by production-related activities in the form of a suddenly applied load on the active platform. An instrumental hammer was used to impact the active platform with and without active control to assess the performance of the hybrid platform for microvibration control of high-tech equipment.

13.6.4.3 System identification

To have a better understanding of the performance of both the building and hybrid platform under various excitations and to establish analytical models for simulation in parallel to the experiment, the system identification was carried out on the building and the hybrid platform.

The weights of the building components were directly measured as 100.7, 100.7 and 91.8 kg for the first, second and third floor, respectively. The natural frequencies, mode shapes and modal damping ratios of the building model without the platform were experimentally identified first by hammer test. Then, the stiffness and damping matrices of the building were identified and given as

$$\mathbf{K} = \begin{bmatrix} 3.53 & -1.94 & 0.21 \\ -1.94 & 3.49 & -1.81 \\ 0.21 & -1.81 & 1.68 \end{bmatrix} \times 10^6 \ (\text{N/m}); \mathbf{C} = \begin{bmatrix} 93.80 & -37.70 & 2.93 \\ -37.70 & 92.10 & -33.22 \\ 2.93 & -33.22 & 57.11 \end{bmatrix} \ (\text{N·s/m})$$

(13.71)

When the hybrid platform was used as a passive platform, the mass of the fixed part of the passive platform, which was composed of the mass of the base plate of the platform, the masses of the electromagnets and the accessories in the passive layer, was measured as 21.1 kg and added to the mass of the first building floor. The mass of the kinetic part of the passive platform, containing the masses of the cover sheet, top plate and middle plate of the platform and the accessories, was measured as 52.1 kg. The stiffness of the passive platform without the u-shape leaf spring was found to be 8501 N/m. Through the calibration, it was found that the total stiffness of the passive platform could vary from 8501 N/m without the u-shape leaf spring to 51912 N/m with the spring effective length of 25 mm. It was also found that the natural frequency of the passive platform could vary from 2.03 to 5.02 Hz, correspondingly. By selecting the proper viscosity of the silicon oil and adjusting the gap of the damper, a series of damping coefficient and damping ratio of the passive platform was obtained through free vibration tests of the platform. It was found that the damping ratio of the passive platform could vary from 0.7% to 78.2%, approximately.

When the hybrid platform was used as an active platform against microvibration, the mass of the fixed part of the active platform, including the masses of the base plate, the middle plate and the accessories of the passive layer, was measured as 40.2 kg and added to the first floor of the building. The mass of the kinetic part of the active platform that consisted of the masses of the cover sheet and top plate of the platform and the accessories (e.g. accelerometer and electromagnets), was measured as 33.1 kg. The stiffness of the active platform comprising the stiffness of the four columns and the actuator was measured as 5.53×10^5 N/m. The damping ratio of the platform was measured as 1.1% through free vibration test. The damping coefficient of the active platform was determined as 90.1 N·s/m.

13.6.4.4 Experimental results: Seismic response control

In the case of earthquake excitation, the hybrid platform works as a passive platform. It was found that the platform damping does not affect the peak acceleration of the first building floor (Xu et al. 2008). The peak drift of the platform decreases with increasing platform damping. However, there is an optimal platform damping, by which the peak acceleration of the platform is the smallest. The optimal platform damping is about 306.8 N·s/m in this study.

For a given optimal damping of 306.8 N.s/m of the platform, the acceleration responses of both the first building floor and the platform, as well as the drift of the platform, were then measured for different platform stiffness and earthquake excitations. It was found that the peak acceleration of the platform increases with the increase of platform stiffness. However, the variations of the peak acceleration of the first building floor and the peak drift of the platform with the platform stiffness are not monotonic. They depend on the frequency components of the earthquake and the natural frequency of the platform. Because the seismic response control concerns the peak acceleration of the platform, the smallest stiffness of the platform of 8501 N/m was used as the optimal value to obtain the smallest peak acceleration of the platform.

Based on the time history of the acceleration response of the platform with the optimal damping (306.8 N·s/m) and stiffness (8501 N/m) of the platform for the building under the El-Centro earthquake, the experimental and simulated peak accelerations of the platform were found to be 1.07 and 0.91 m/s², respectively. The experimental and simulated peak accelerations of the first floor of the building under the El-Centro earthquake without passive control were obtained as 5.89 and 5.47 m/s², respectively. In view of fact that without the platform, high-tech equipment will be installed on the first floor of the building, the experimental and simulated peak acceleration reductions of high-tech equipment subject to the El-Centro earthquake with the passive platform were 82% and 83%, respectively. Similar observations were also made for the building subject to the Northridge and Hachinohe earthquakes, which can be seen from the results listed in Table 13.7.

13.6.4.5 Experimental results: Microvibration control

In the case of microvibration, the hybrid platform is designed to function as an actively controlled platform to reduce mainly the velocity response of high-tech equipment under normal working condition. To have a good understanding of the control performance of

Table 13.7 Comparison of platform peak acceleration (m/s²)

	El-Centro	Hechinohe	Northridge
With inactive platform	5.89 (5.47)	5.90 (5.65)	4.68 (4.89)
Passive platform with optimal stiffness and damping	1.07 (0.91)	1.20 (1.31)	0.80 (0.61)

Note: The number in brackets is from simulation.

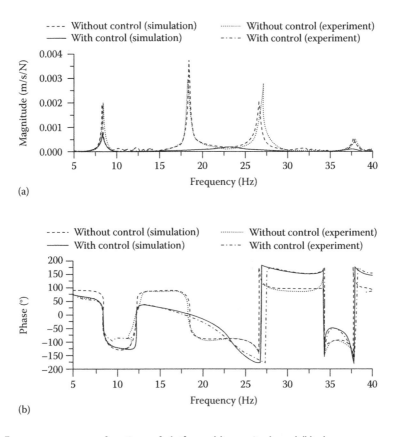

Figure 13.35 Frequency response functions of platform: (a) magnitude and (b) phase.

the active platform against microvibration, the frequency response functions (FRFs) of the absolute velocity of the platform to the force acting on the first building floor were measured and computed with and without active control. The Fourier transform was performed over both the time history of the velocity response of the platform and the measured force time history to obtain the magnitude and phase of the FRF. Both the experimental and simulation results with and without active control are plotted in Figure 13.35. It can be seen that without active control there are four peaks in the FRF, which correspond to the four natural frequencies of the building–platform system. The magnitude of the FRF at the second natural frequency is the largest among the four peaks. With active control, the magnitudes of the FRF at the second, third and fourth natural frequencies are almost reduced to zero. The magnitude of the FRF at the first natural frequency is also reduced by 70%, approximately. The magnitudes of the FRF with and without active control shown in Figure 13.35 indicate the effectiveness of the active platform.

Both the laboratory measurements and the computer simulations were carried out on the building with the active platform under the traffic-induced ground motion. The active control system was not actuated until the first 20 s. The recorded and simulated time histories of the velocity response of the platform are depicted in Figures 13.36a,b. The simulation results match the measurement results well. It can be seen that with the active control, the platform velocity is reduced significantly. The measured peak velocity of the platform without active control is about 516 μm/s. With the active control, the measured peak velocity of the platform is only 101 μm/s, a reduction of 80%. The recorded time history of velocity

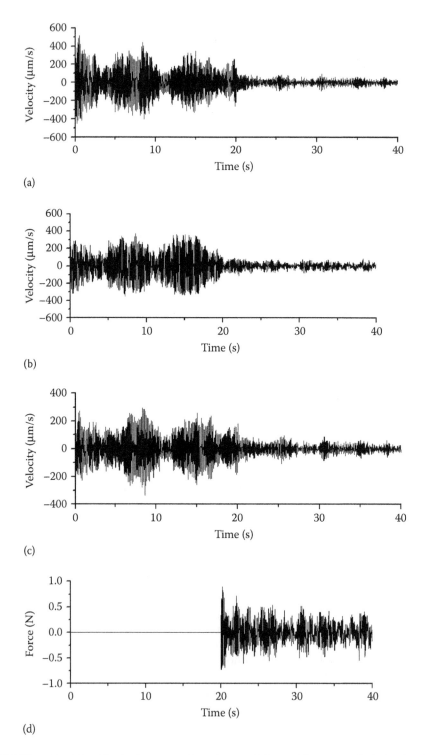

Figure 13.36 Time histories of velocity response and control force: (a) measured platform velocity, (b) simulated platform velocity, (c) measured first floor velocity and (d) simulated control force.

response of the first building floor is plotted in Figure 13.36c. The peak velocity of the first building floor is 330 μm/s without active control and 106 μm/s with active control. Clearly, the active platform reduces not only the platform velocity but also the velocity response of the first building floor. The time history of active control force is plotted in Figure 13.36d. The maximum control force is less than 0.9 N. The requirement of only small control force against microvibration enhances the feasibility of hybrid platform in practical use.

The time histories of velocity responses as already discussed, are also converted to a one-third octave band velocity spectra in dB referenced to 1 μm/s to investigate vibration energy distribution over the frequency. No attempt is made to compare spectra with the currently used microvibration criteria because of scaling issues. The velocity spectra of the platform from experiment and simulation are shown in Figure 13.37a for the case without control

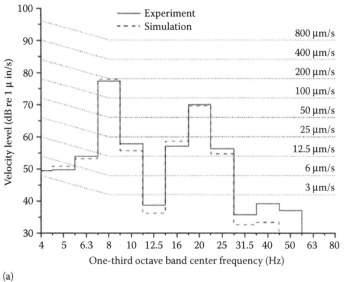

(a)

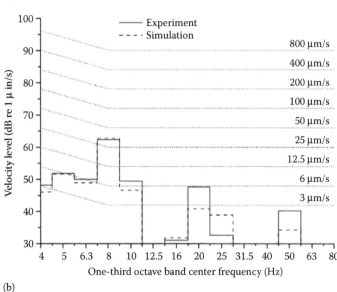

(b)

Figure 13.37 Comparison of platform velocity spectra: (a) without control and (b) with control.

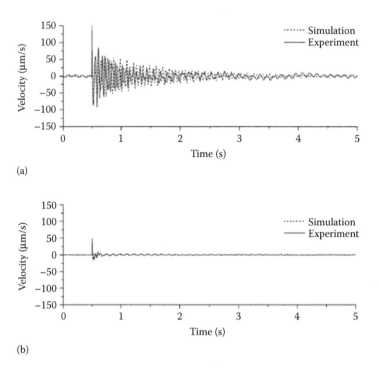

Figure 13.38 Velocity response of platform under impact force: (a) without control and (b) with control.

and in Figure 13.37b for the case with control. It can be seen that the traffic-induced vibration energy distributes over a wide frequency range. With active control, the velocity level of the platform is reduced significantly. The velocity level in the frequency band around 8 Hz is reduced by approximately 86% and the velocity level in the frequency band around 20 Hz is reduced by approximately 93%. The velocity spectra of the first building floor with and without active control were also obtained from the experiment. It was found that the active platform also reduced the velocity level of the first building floor over a wide frequency range. Further comparison was made in terms of the velocity spectrum of the platform with active control and the velocity spectrum of the first building floor with the inactive platform. It was found that the velocity level in the frequency band around 8 Hz was reduced by 80%, and the velocity level in the frequency band around 20 Hz was reduced by 76%.

To investigate the control performance of the active platform against the direct disturbance caused by production-related activities in the form of a suddenly applied load, an instrumental hammer was used to impact the active platform with and without control. The time histories of the measured and simulated velocity response of the platform under the unit impact force are shown in Figure 13.38a without active control and in Figure 13.38b with active control. By comparing Figure 13.38a with Figure 13.38b, one may see that the active platform is very effective in reducing velocity response of high-tech equipment due to direct disturbance on the platform.

NOTATION

e_k	The slip deformation of the kth friction damper
$\bar{e}_k$	The previously cumulated slip deformation of the kth damper

E	The index vector with all its elements equal to 1
$f_{dk}(t)$	The semi-active friction force provided by the kth variable friction damper in its slipping state
g_k^1, g_k^2	The positive gain coefficients in VRF controller for the slip $x_k^{\text{rel}}(t)$ and the slip rate $\dot{x}_k^{\text{rel}}(t)$, respectively, of the kth variable friction damper
H	The matrix denoting the location of control devices
$H_j(\omega)$	Frequency response function for the jth mode
k_{dk}	The spring stiffness of the kth variable friction damper if it is in sticking state
$\mathbf{K}_c$	The optimal control gain matrix
$\mathbf{K}_f$	The estimator gain matrix in KF algorithm
m_b, k_b, c_b	The mass, stiffness and damping coefficient of the mounting plate, respectively
m_p, k_s, c_s	The mass, stiffness and damping coefficient of the platform, respectively
μ	A proportional coefficient in the weighting matrix $\mathbf{Q}$
M, C, K	The mass, damping and stiffness matrices of structure, respectively
N	A total of DOFs of two adjacent buildings
$N_k(t)$	The clamping force
N_m	The number of sensors
$N_k^{\max}$	The maximum clamping force
N_1, N_2	The number of DOFs of Building 1 and Building 2, respectively
p_{km}	The pseudo response
$\mathbf{q}(t)$	The system state
$\hat{\mathbf{q}}(t)$	The system state estimate generated by KF technique
Q, R	The weighting matrices in the LQG algorithms associated with system states and control forces, respectively
$\mathbf{S}_{wv}$	The spectral density matrix of both the ground excitation and the measurement noise
$S_{\ddot{x}_g \ddot{x}_g}$	The ground acceleration spectrum
S_v	The intensity of noise
$\mathbf{u}(t)$	Control force vector
$\mathbf{v}(t)$	The measurement noise vector of N_m dimension
$w(t)$	Zero-mean Gaussian white noise with intensity S_0
$\dot{x}_k^{\text{rel}}$	The relative velocity between the two buildings
$x_k^{\text{rel}}(t)$	The relative displacement between the two buildings at the floor where the kth damper is mounted
$\mathbf{x}(t)$	The vector of relative displacement response with respect to the ground
$x_g(t), \dot{x}_g(t), \ddot{x}_g(t)$	The ground displacement, velocity and acceleration, respectively
x_p, x_b	The displacement of platform and mounting plate, respectively
$\mathbf{y}(t)$	The observation vectors
$\hat{\mathbf{z}}$	The estimated state vector of the state vector $\mathbf{z}$ from KF technique
Z	The state vector of the coupled system
$\alpha_{kmj}, \beta_{kmj}$	The coefficients for the determination of pseudo displacement, velocity or acceleration response
β_k^1	The positive control gain in MHF controller
β_k^2	The velocity control gain in MHF controller
γ_k^2	The velocity control gain in NSF controller

$\Gamma(t)$	The state of the seismic excitation model
λ	Eigenvalue
μ'	Dynamic friction coefficient
$\sigma^2_{p_{km}}$	The variance response of p_{ki} under the kth pseudo-excitation
φ	Eigenvector
Φ	Modal matrix
ω_g, ξ_g, S_0	The characteristics and the intensity of an earthquake in a particular geological location
$\omega_j, \omega_{dj}, \xi_j$	The modal frequency, the damped modal frequency and the modal damping ratio, associated with mode j, respectively

REFERENCES

Aiken, I.D. and J.M. Kelly. 1990. Earthquake simulator testing and analytical studies of two energy-absorbing systems for multistory structures. Research Report, UCB/EERC-90-03, University of California, Berkeley, CA.

Amick. H. and A. Bayat. 1998. Seismic isolation of semiconductor production facilities. In *Proceedings of Seminar on Seismic Design, Retrofit, and Performance of Nonstructural Components*, ATC-20-1, Applied Technology Council, San Francisco, CA, 297–312.

Cao, H., A.M. Reinhorn, and T.T. Soong. 1998. Design of an active mass damper for a tall TV tower in Nanjing. China. *Eng. Struct.*, 20(3): 134–43.

Caughey, T.K. and A. Vijayaraghavan. 1970. Free and forced oscillations of a dynamic system with linear hysteretic damping (non-linear theory). *Int. J. Non-Linear Mech.*, 5:533–55.

Chen, C.C. and G.D. Chen. 2002. Nonlinear control of a 20-story steel building with active piezo-electric friction dampers. *Struct. Eng. Mech.*, 14(1): 21–38.

Chen, C.C. and G.D. Chen. 2004a. Shake table tests of a quarter-scale three story building model with piezoelectric friction dampers. *Struct. Control Health Monit.*, 11(4): 239–57.

Chen, G.D. and C.C. Chen. 2004b. Semiactive control of the 20-story benchmark building with piezoelectric friction dampers. *J. Eng. Mech.*, 130(4): 393–400.

Chen, J., W.L. Qu, and N.L. Zhang. 2002. Structural control based on PanBoolean algebra. In *Proceedings of the 6th World Multi-Conference on Systematics, Cybernetics, and Informatics*, ed. N. Callaos, Orlando, FL: IIIS. 199–202.

Dyke, S.J., B.F. Spencer Jr., M.K. Sain, and J.D. Carlson. 1996. Modeling and control of magneto-rheological dampers for seismic response reduction. *Smart Mater. Struct.*, 5: 565–75.

Gordon, C.G. 1991. Generic criteria for vibration sensitive equipment. In *Proceedings of SPIE 1619, Vibration Control in Microelectronics, Optics, and Metrology*, San Jose, CA, 1619: 71–85.

He, W.L., A.K. Agrawal, and J.N. Yang. 2003. Novel semiactive friction controller for linear structures against earthquakes. *J. Struct. Eng.*, 129(7): 941–50.

Heredia-Zavoni, E. and E.H. Vanmarcke. 1994. Seismic random-vibration analysis of multisupport-structural systems. *J. Eng. Mech.*, 120(5): 1107–28.

Housner, G.W., L.A. Bergman, T.K. Caughey, et al. 1997. Structural control: Past, present, and future. *J. Eng. Mech.*, 123(9): 897–971.

Inaudi, J.A. 1997. Modulated homogeneous friction: A semi-active damping strategy. *Earthq. Eng. Struct. Dyn.*, 26(3): 361–76.

Kobori, T. 1994. Future direction on research and development of seismic-response-controlled structure. In *Proceedings 1st World Conference on Structural Control*, Los Angeles, CA, FA2: 19–31.

Kobori, T. 1998. Mission and perspective towards future structural control research. In *Proceedings of the 2nd World Conference on Structural Control*, Kyoto, Japan, 1: 25–34.

Kobori, T., M. Takahashi, T. Nasu, N. Niwa, and K. Ogasawara. 1993. Seismic response controlled structure with active variable stiffness system. *Earthq. Eng. Struct. Dyn.*, 22: 925–41.

Kurata, N., T. Kobori, M. Takahashi, N. Niwa, and H. Midorikawa. 1999. Actual seismic response controlled building with semi-active damper system. *Earthq. Eng. Struct. Dyn.*, 28(11): 1427–47.

Mahmoodi, P. 1969. Structural dampers. *J. Struct. Div.*, 95(8): 1661–72.

Nakamura, Y., M. Nakayama, M. Yasuda, and T. Fujita. 2006. Development of active six-degrees-of-freedom micro-vibration control system using hybrid actuators comprising air actuators and giant magnetostrictive actuators. *Smart Mater. Struct.*, 15: 1133–42.

Ng, C.L. and Y.L. Xu. 2007. Semi-active control of a building complex with variable friction dampers. *Eng. Struct.*, 29(6): 1209–25.

Pall, A.S. and R. Pall. 1993. Friction-dampers used for seismic control of new and existing buildings in Canada. In *Proceedings of First Conference on Structural Control*, Los Angeles, CA, FP5: 29–38.

Patten, W.N. 1998. The I-35 Walnut Creek Bridge: An intelligent highway bridge via semi-active structural control. In *Proceedings of the 2nd Conference on Structural Control*, Kyoto, Japan, 1: 427–36.

Patten, W.N., J. Sun, G. Li, J. Kuehn, and G. Song. 1999. Field test of an intelligent stiffener for bridges at the I-35 Walnut Creek bridge. *Earthq. Eng. Struct. Dyn.*, 28(2): 109–26.

Reinhorn, A.M., T.T. Soong, R.J. Helgeson, M.A. Riley, and H. Cao. 1998. Analysis, design and implementation of an active mass damper for a communication tower. In *Proceedings of the 2nd Conference on Structural Control*, Kyoto, Japan, 3: 1727–36.

Serrand, M. and S.J. Elliott. 2000. Multichannel feedback control for the isolation of base-excited vibration. *J. Sound Vibr.*, 234(40): 681–704.

Shahrooz, B.M. and J.P. Moehle. 1990. Seismic response and design of setback buildings. *J. Struct. Eng.*, 116(5): 1423–39.

Skinner, R.I., W.H. Robinson, and G.H. McVerry. 1993. *An Introduction to Seismic Isolation.* Chichester, UK: Wiley.

Skinner, R.I., R.G. Tyler, A.J. Heine, and W.H. Robinson. 1980. Hysteretic dampers for the protection of structures from earthquakes. *Bull. N. Z. National Soc. Earthq. Eng.*, 13(1): 22–36.

Soong, T.T. and G.F. Dargush. 1997. *Passive Energy Dissipation Systems in Structural Engineering.* Chichester: Wiley.

Soong, T.T. and B.F. Spencer Jr. 2002. Supplemental energy dissipation: State-of-the-art and state-of-the-practice. *Eng. Struct.*, 24(3): 243–59.

Spencer Jr., B.F. and S. Nagarajaiah. 2003. State of the art of structural control. *J. Struct. Eng.*, 129(7): 845–56.

Ungar, E.E., D.H. Sturz, and C.H. Amick. 1990. Vibration control design of high-technology facilities. *Sound Vib.*, 24(7): 20–27.

Weber, F., J. Distl, and M. Maślanka. 2013. Semi-active TMD concept for Volgograd bridge. In *Proceedings of the 31st IMAC, A Conference on Structural Dynamics, 2013, Topics in Dynamics of Civil Structures*, eds. F.N. Catbas, S. Pakzad, V. Racic, A. Pavic, and P. Reynolds, New York: Springer-Verlag, 4: 79–88.

Xu, H.B., C.W. Zhang, H. Li, et al. 2014. Active mass driver control system for suppressing wind-induced vibration of the Canton Tower. *Smart Struct. Syst.*, 13(2): 281–303.

Xu, Y.L., J. Chen, C.L. Ng, and W.L. Qu. 2005. Semiactive seismic response control of buildings with podium structure. *J. Struct. Eng.*, 131(6): 890–99.

Xu, Y.L., Q. He, and J.M. Ko. 1999. Dynamic response of damper-connected adjacent buildings under earthquake excitation. *Eng. Struct.*, 21(2): 135–48.

Xu, Y.L. and B. Li. 2006. Hybrid platform for high-tech equipment protection against earthquake and microvibration. *Earthq. Eng. Struct. Dyn.*, 35(8): 943–67.

Xu, Y.L., H.J. Liu, and Z.C. Yang. 2003a. Hybrid platform for vibration control of high-tech equipment in building subject to ground motion Part 1: Experiment. *Earthq. Eng. Struct. Dyn.*, 32: 1185–200.

Xu, Y.L. and C.L. Ng. 2008. Seismic protection of a building complex using variable friction damper: Experimental investigation. *J. Eng. Mech.*, 134(8): 637–49.

Xu, Y.L., Z.C. Yang, J. Chen, and H.J. Liu. 2003b. Microvibration control platform for high technology facilities subject to traffic-induced ground motion. *Eng. Struct.*, 25(8): 1069–82.

Xu, Y.L., Z.F. Yu, and S. Zhan. 2008. Experimental study of a hybrid platform for high-tech equipment protection against earthquake and microvibration. *Earthq. Eng. Struct. Dyn.*, 37(5): 747–67.

Xu, Y.L. and W.S. Zhang. 2002. Closed-form solution for seismic response of adjacent buildings with linear quadratic Gaussian controllers. *Earthq. Eng. Struct. Dyn.*, 31(2): 235–59.

Yang, J.N. and A.K. Agrawal. 2000. Protective systems for high-technology facilities against microvibration and earthquake. *Struct. Eng. Mech.*, 10(6): 561–67.

Yang, Z.C., Y.L. Xu, J. Chen, and H.J. Liu. 2003. Hybrid platform for vibration control of high-tech equipment in building subject to ground motion Part 2: Analysis. *Earthq. Eng. Struct. Dyn.*, 32: 1201–15.

Yoshioka, H., N. Murai, T. Abe, and Y. Hashimoto. 2001. Active microvibration isolation system for hi-tech manufacturing facilities. *J. Vib. Acoust.*, 123: 269–75.

Zhang, W.S. and Y.L. Xu. 1999. Dynamic characteristics and seismic response of adjacent buildings linked by discrete dampers. *Earthq. Eng. Struct. Dyn.*, 28(10): 1163–85.

Zhou, F.L. 2015. Earthquake tragedy and application of seismic isolation, energy dissipation and other seismic control systems to protect structures in China. *Energia, Ambiente e Innovazione (EAI)*, 5: 23–30.

Chapter 14

Synthesis of structural health monitoring and vibration control in the frequency domain

14.1 PREVIEW

Structural vibration control technologies have been developed for civil structures to reduce excessive vibrations caused by strong winds, severe earthquakes or other disturbances, as introduced in Chapter 13. Structural health monitoring technologies have been developed for civil structures to identify their dynamic characteristics and parameters (*system identification*) and to detect their possible damage (*damage detection*). System identification can be defined as the process of developing or improving the mathematical model of a physical system using measurement data, as discussed in Chapter 7. Structural damage can be defined as changes in structural parameters which adversely affect the current or future performance of the structure, whereas structural damage detection aims to find such changes in the structure using measurement data, as shown in Chapter 12. Although vibration control systems and health monitoring systems both require the use of sensors, data acquisition and signal transmission for their implementation, the areas of structural vibration control and health monitoring have generally been treated separately according to their respective primary objectives. This separate approach is neither practical nor cost-effective if structures require both a vibration control system and a health monitoring system. This approach is also unsuitable for creating smart civil structures with their own sensors (nervous systems), processors (brain systems) and actuators (muscular systems), thus mimicking biological systems. In this regard, this chapter gives a brief review of current research on the synthesis of structural health monitoring and vibration control. The concept of an integrated system using semi-active friction dampers is first introduced. This chapter then presents an integrated procedure for the health monitoring and vibration control of building structures using semi-active friction dampers in the frequency domain. In the integrated procedure, a model updating scheme is first presented based on adding known stiffness using semi-active friction dampers to obtain the variations of the frequency response functions (FRFs) of a building between the two states and to identify its structural parameters. By using updated system matrices, the chapter then investigates the control performance of semi-active friction dampers using local feedback control with a Kalman filter for a building subjected to earthquake excitation. A damage detection scheme based on adding known stiffness by semi-active friction dampers is proposed and used for damage detection by assuming that the building suffers certain damage after an extreme event or long-term service and by using the previously identified original structural parameters. The feasibility and accuracy of the proposed integrated procedure are finally demonstrated through detailed numerical and experimental studies.

449

14.2 CURRENT RESEARCH IN THE SYNTHESIS OF STRUCTURAL HEALTH MONITORING AND VIBRATION CONTROL

Substantial efforts have been made in the development of smart structures that synthesise both vibration control and health monitoring. A number of comprehensive reviews are available; however, they are mainly in the fields of mechanical and aerospace engineering (e.g. Rao and Sunar 1994; Chopra 2002; Hurlebausa and Gaul 2006). For civil structures, many challenges exist for the implementation of such an integrated system in reality, such as the large size and complexity of structural systems, uncertainties in modelling, measurement errors and the large control forces required. Therefore, although the concept of an integrated system is compelling, the research and development of this kind of system is limited in civil engineering. This section provides a brief introduction to current research on the synthesis of structural health monitoring and vibration control of civil structures in both frequency domain and time domain.

In the frequency domain, the whole or partial time histories of structural responses are generally required for the extraction of modal properties for the purpose of system identification and damage detection. Thus, the implementation of the integrated system in the frequency domain is basically not in real time, which can be viewed as a major shortcoming of the frequency domain approach. However, it is worth noting that the identified structural parameters can provide a better foundation for the follow-up control action and that the control performance will be more effective if the updated parameters are used. Based on variations in the natural frequencies and mode shapes, Chen and Xu (2008) and Xu and Chen (2008) proposed and numerically investigated an integrated vibration control and health monitoring system using semi-active friction dampers to fulfil the model updating, seismic response control and damage detection of building structures. More recently, on the basis of FRFs, an integrated method for the system identification and damage detection of controlled buildings equipped with semi-active friction dampers was proposed by Huang et al. (2012) and experimentally validated via a complex building structure with a 12-story main building and a 3-story podium structure (Xu et al. 2014). The details of the frequency domain integrated system will be given in the following sections of this chapter.

Compared with the frequency domain integrated approach, more attention has been paid to the development of the time domain integrated approach mainly due to the integrated system in the time domain being able to act on-line. During an event, in an ideal time domain integrated approach, the structural parameters should be accurately estimated and immediately employed for the determination of optimal control force on one hand, and the measured control force should be used to improve the identification accuracy on the other hand. In this regard, the health monitoring system and vibration control system are tightly interconnected and interact to make the whole system self-diagnose and self-adapt in an effective manner. Some attempts have been made to realise this kind of smart civil structure. For example, on the basis of a direct adaptive control algorithm, Gattulli and Romeo (2000) proposed an integrated procedure for both the vibration suppression and health monitoring of multi-degree-of-freedom (MDOF) shear-type building structures. Chen et al. (2008) proposed a general time domain approach to the integration of the vibration control and health monitoring of building structures accommodating various types of control devices and on-line damage detection. More recently, Ding and Law (2011) proposed a method for integrating structural control and evaluation into large-scale structural systems. The control system was implemented with linear quadratic Gaussian (LQG) and pseudo-negative stiffness (PNS) controls for the vibration mitigation of building structures, whereas in the structural evaluation system, a modified adaptive regularisation method was developed based on model updating techniques with the aim of providing updated structural parameters for the control system. Lin et al. (2012) proposed a hybrid health monitoring system linked to an adaptive structural control algorithm to improve control performance; this was verified via a three-story steel frame structure with a damaged column and a damaged joint. Karami and Amini (2012) proposed an algorithm including integrated on-line health monitoring and a semi-active control strategy for reducing both the damage to and seismic response of the main structure caused by strong seismic disturbance. Lei et al. (2013) proposed an integrated algorithm for the decentralised structural control of shear-type tall buildings and for the identification of unknown earthquake-induced ground motion. Based on the extended Kalman filter and error tracking techniques, an on-line integration technique was also proposed by Lei et al. (2014) for the health monitoring and active optimal vibration control of the undamaged/damaged structures. He et al. (2014) proposed a time domain integrated vibration control and health monitoring approach for the simultaneous vibration mitigation and damage detection of building structures without prior knowledge of external excitations. An integrated semi-active control and damage detection system was considered by Amini et al. (2015) to locate and characterise the damage in base-isolated structures and to mitigate base displacements under the effects of seismic excitation. On the basis of recursive least-square estimation, Xu et al. (2015) proposed a real-time integrated procedure for accurately identifying time-varying structural parameters and unknown excitations, as well as optimally mitigating excessive vibration in a building structure. Although the effectiveness of most integrated methodologies has been examined through numerical examples, several experimental investigations exploring the possibility of establishing such a smart civil structure can also be found. For example, a new real-time tuning algorithm that is able to identify the instantaneous frequency of linear time-varying systems and tune smart mass dampers for vibration attenuation were developed and experimentally investigated by Nagarajaiah (2009). By using a model reference adaptive control algorithm, a hybrid real-time health monitoring and control system for building structures during earthquakes was presented by Yang et al. (2014) and experimentally validated using a three-story aluminium frame structure. Moreover, by employing magnetorheological (MR) dampers, an experimental investigation of an integrated smart building structure subject to seismic excitations was conducted by He et al. (2016), details of which can be found in Chapter 15.

In this chapter, the synthesis method for the structural health monitoring and vibration control of building structures in the frequency domain in terms of FRFs (Huang et al. 2012; Xu et al. 2014) will be presented in detail. Integrated smart civil structures with synthesised health monitoring and vibration control in the time domain will be given in Chapter 15.

14.3 INTEGRATED PROCEDURE USING SEMI-ACTIVE FRICTION DAMPERS

This section presents an integrated procedure for the health monitoring and vibration control of building structures equipped with semi-active friction dampers to accomplish system identification and model updating, seismic response mitigation and damage detection systematically.

The first task is to update the stiffness matrix of a building and identify the stiffness parameters of its structural members based on the variations of FRFs between the two states of the building. The two states are created by adding known stiffness using semi-active friction dampers: (1) the original building without any additional stiffness (the clamping force is set at zero) and (2) the original building with additional stiffness (the damper is in a sticking state). This model updating scheme can avoid some of the shortcomings of current methods that require the measured modal matrix to be properly normalised with respect to the actual mass and stiffness matrices, which are not known a priori and fail to preserve the physical connectivity of the system. The updated system matrices and structural parameters facilitate the implementation of structural vibration control and provide a reference state for subsequent damage detection.

The second task is to present a local feedback control algorithm with a steady-state Kalman filter for the building with semi-active friction dampers subjected to an earthquake excitation. The primary purpose of implementing local feedback control is to involve minimal feedback signals to achieve a reliable and economical control design. The use of a steady-state Kalman filter makes it possible to use the accelerometers as sensors for both health monitoring and vibration control, leading to a common sensory system and a common data acquisition and transmission system. However, the localised feedback signal limits the capability of providing a full picture of the vibration level of the entire building. Thus, there is a need to investigate the effectiveness of local feedback control in comparison with the global feedback control counterpart. It is noted that the vibration control here is performed on a building with updated stiffness matrix and structural parameters, which are imperative for achieving the desirable control performance.

The final task is to apply the proposed model updating scheme to a damaged building to identify its structural parameters based on the variations of the FRFs between the two states of the building. The two states of the damaged building are created by adding known stiffness using semi-active friction dampers: (1) the damaged building without any additional stiffness (the clamping force is set at zero) and (2) the damaged building with additional stiffness (the damper is in a sticking state). By comparing these parameters with those of the undamaged structure, the location and severity of the structural damage can be determined. Figure 14.1 shows a schematic diagram for the proposed integrated health monitoring and vibration control system for a shear building. The details of each task are introduced in the subsequent sections.

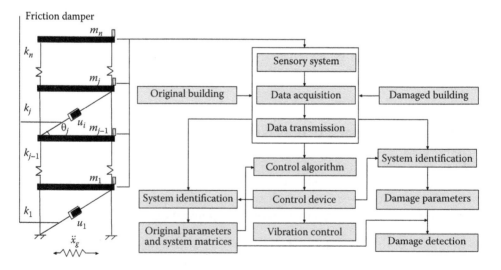

Figure 14.1 Integrated health monitoring and vibration control system.

14.4 SYSTEM IDENTIFICATION AND MODEL UPDATING

14.4.1 Equation of motion

The second-order differential equation of the motion of a building structure with n DOFs can be given as

$$M\ddot{X}(t) + C\dot{X}(t) + KX(t) = f(t) \tag{14.1}$$

where:

M, C and K	are the mass, damping and stiffness matrices of the building structure, respectively
$\ddot{X}(t), \dot{X}(t)$ and $X(t)$	are the structural acceleration, velocity and displacement response vectors, respectively
$f(t)$	is the applied force vector

If the applied force is harmonic with frequency ω, one may define $f(t) = F(\omega)e^{j\omega t}$, where $F(\omega)$ is the force amplitude vector in the frequency domain and j is the imaginary unit. The displacement response of the structure can then be expressed as $X(t) = X(\omega)e^{j\omega t}$, where $X(\omega)$ is the displacement response amplitude vector of the building structure in the frequency domain. Given that the commonly used sensors in practice to measure structural responses are accelerometers, the FRF of acceleration rather than the FRF of displacement is considered. In this regard, Equation 14.1 becomes

$$\frac{-\omega^2 M + j\omega C + K}{-\omega^2}\ddot{X}(\omega)e^{j\omega t} = F(\omega)e^{j\omega t} \tag{14.2}$$

or

$$\ddot{X}(\omega) = H(\omega)F(\omega) \tag{14.3}$$

where $\ddot{X}(\omega)$ is the acceleration response amplitude vector in the frequency domain. $H(\omega)$ is the FRF matrix of acceleration, defined as

$$H(\omega) = \left(\frac{-\omega^2 M + j\omega C + K}{-\omega^2}\right)^{-1} \tag{14.4}$$

Assume that the structural damping follows the Rayleigh damping, which can be expressed as $C = \alpha M + \beta K$, where α and β are the coefficients of the first-order and the second-order damping ratios, respectively. Equation 14.4 can then be written as

$$(1 + j\omega\beta)KH(\omega) = -\omega^2 I + (\omega^2 - j\omega\alpha)MH(\omega) \tag{14.5}$$

where I denotes the identity matrix.

14.4.2 FRF-based method with full excitation

For the sake of explanation, let us consider a shear-type building with external excitations applied at all the floors (full excitation), as shown in Figure 14.2. The additional stiffness

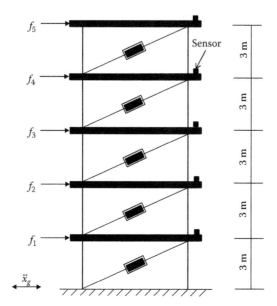

Figure 14.2 Example shear-building with semi-active friction dampers.

matrix contributed by the semi-active friction dampers that are installed in the building together with the brace systems is denoted as $\mathbf{K}_a$. To obtain the elements in the ith column of the FRF matrix $\mathbf{H}(\omega)$ in Equation 14.5 from a series of measurements, simply apply a harmonic excitation horizontally to the ith floor, record the acceleration responses at all the floors and then use Equation 14.3 to obtain the elements. The first state we consider is that all the clamping forces in the dampers are set at zero so that there is no additional stiffness in the original building. Denote the FRF matrix of the original building as $\mathbf{H}_o$, the frequency of the applied excitations as ω_o, and the coefficients of the Rayleigh damping as α_o and β_o, where the subscript o refers to the original building. Equation 14.5 can then be written as

$$(1 + j\omega_o\beta_o)\mathbf{K}\mathbf{H}_o(\omega_o) = -\omega_o{}^2\mathbf{I} + (\omega_o{}^2 - j\omega_o\alpha_o)\mathbf{M}\mathbf{H}_o(\omega_o) \tag{14.6}$$

The second state to be considered is that all semi-active friction dampers are set in a sticking state so as to produce an additional stiffness matrix $\mathbf{K}_a$. Denote the FRF matrix of the added-stiffness building as $\mathbf{H}_a$, the frequency of the applied excitations as ω_a and the coefficients of the Rayleigh damping as α_a and β_a, where the subscript a refers to the added-stiffness building. Applying Equation 14.5 yields

$$(1 + j\omega_a\beta_a)(\mathbf{K} + \mathbf{K}_a)\mathbf{H}_a(\omega_a) = -\omega_a{}^2\mathbf{I} + (\omega_a{}^2 - j\omega_a\alpha_a)\mathbf{M}\mathbf{H}_a(\omega_a) \tag{14.7}$$

Eliminating the unknown mass matrix $\mathbf{M}$ in Equations 14.6 and 14.7 yields

$$\mathbf{H}_o^T(\omega_o)\mathbf{K}\mathbf{H}_a(\omega_a) = \mathbf{S} \tag{14.8}$$

where:

$$S = \left\{ 1 \Big/ \Big[(\omega_a^2 - j\omega_a\alpha_a)(1 + j\omega_o\beta_o) - (\omega_o^2 - j\omega_o\alpha_o)(1 + j\omega_a\beta_a) \Big] \right\}$$

$$\times \Big[\omega_a^2(\omega_o^2 - j\omega_o\alpha_o)\mathbf{H}_o^T(\omega_o) - \omega_o^2(\omega_a^2 - j\omega_a\alpha_a)\mathbf{H}_a(\omega_a)$$

$$+ (1 + j\omega_a\beta_a)(\omega_o^2 - j\omega_o\alpha_o)\mathbf{H}_o^T(\omega_o)\mathbf{K}_a\mathbf{H}_a(\omega_a) \Big] \tag{14.9}$$

According to Equation 14.9, the matrix S can be determined by the known additional stiffness matrix $\mathbf{K}_a$, the FRF matrix $\mathbf{H}_o$ of the original building and the FRF matrix $\mathbf{H}_a$ of the added-stiffness building. Then, in order to find the stiffness matrix $\mathbf{K}$ of the original building in Equation 14.8 without inverting the FRF matrices, which may cause ill-posed problems, Equation 14.8 is rewritten so that $\mathbf{K}$ appears as an unknown column vector $\mathbf{k}$ as follows:

$$\mathbf{Ak} = \mathbf{b} \tag{14.10}$$

where:

$$\mathbf{A} = \mathbf{H}_a^T(\omega_a) \otimes \mathbf{H}_o^T(\omega_o) \tag{14.11}$$

$$\mathbf{b} = \text{vec}(\mathbf{S}) \tag{14.12}$$

$$\mathbf{k} = \text{vec}(\mathbf{K}) = [k_{11}\ k_{21}\ \cdots\ k_{n1}\ k_{12}\ k_{22}\ \cdots\ k_{n2}\ \cdots\ k_{1n}\ k_{2n}\ \cdots\ k_{nn}]^T \tag{14.13}$$

where the symbol $\otimes$ denotes the Kronecker product. The symbol 'vec' means an operation to arrange the matrix to a column vector, which can be realised conveniently by the function RESHAPE in MATLAB. For the shear building considered in this study, the stiffness matrix is tri-diagonal ($k_{ij} = 0$ (abs $(i - j) > 1$)); it is also a symmetric matrix with many zero elements due to the connectivity of structural members. The sparsity of information shall be taken as a constraint condition of the updated stiffness matrix. Mathematically, this can be achieved by eliminating all of the known zero elements from the vector $\mathbf{k}$ and by deleting all of the corresponding columns in the matrix $\mathbf{A}$. By doing this, Equation 14.10 can be simplified as

$$\mathbf{A}^e\mathbf{k}^e = \mathbf{b} \tag{14.14}$$

where:

$$\mathbf{k}^e = [k_{11}\ k_{21}\ \cdots\ k_{ij}\ \cdots\ k_{nn}]^T\ (\text{abs}(i - j) \leq 1) \tag{14.15}$$

$\mathbf{A}^e$ is obtained from $\mathbf{A}$ by deleting all the columns that multiply by $k_{ij} = 0$ (abs$(i - j) > 1$). In this way, the vector $\mathbf{k}$ of size $n^2 \times 1$ is reduced to $\mathbf{k}^e$ of size $(3n - 2) \times 1$. Note that the identified results from Equation 14.14 are the elements of the stiffness matrix rather than the structural parameters. Thus, the identified results from Equation 14.14 cannot be directly used for damage detection. In this regard, a transformation matrix is introduced to overcome the problem. For an n-story shear building, the number of unknown horizontal story stiffness coefficients is n, whereas the number of nonzero elements in the stiffness matrix is $(3n - 2)$. The relationship between the horizontal story stiffness coefficients k_i ($i = 1, 2,...,$ n) and the elements of the stiffness matrix can be established as

$$\mathbf{k}^e = \mathbf{T}\mathbf{k}^s \tag{14.16}$$

where the vector $\mathbf{k}^s = [k_1, k_2,..., k_n]^T$ is the horizontal story stiffness coefficient vector of the original building. The matrix $\mathbf{T}$ is the transformation matrix of size $(3n - 2) \times n$. The substitution of Equation 14.16 with Equation 14.14 yields the identification equation as

$$(\mathbf{A}^e\mathbf{T})\mathbf{k}^s = \mathbf{b} \tag{14.17}$$

Note that the elements of the FRF matrix $\mathbf{H}(\omega_o)$ and $\mathbf{H}(\omega_a)$ are the function of frequencies ω_o and ω_a, respectively. To identify the horizontal story stiffness coefficients using Equation 14.17, particular frequency points shall be chosen, but there are infinite frequency points ω_o and ω_a to be selected. This allows for great flexibility and high-quality identification. Nevertheless, if the excitation frequency is far away from the natural frequency of the building, the values of the elements in the FRF matrix will be too small to be properly used for parameter identification. It is also difficult in practice to excite the building with very high frequencies. Therefore, it is better to select excitation frequencies close to the first few natural frequencies of the building; however, they should not be the same as the natural frequencies because the resonance is sensitive to structural damping, which is of high uncertainty in practice. Let us assume that d excitation frequencies are selected to identify the stiffness coefficients. We then have d equations in the form of Equation 14.17. The combination of these d equations yields

$$\mathbf{A}_f\mathbf{k}^s = \mathbf{b}_f \tag{14.18}$$

where:

$$\mathbf{A}_f = \{\mathbf{A}_1^e\mathbf{T}, \mathbf{A}_2^e\mathbf{T}, ...,\mathbf{A}_d^e\mathbf{T} \} \tag{14.19}$$

$$\mathbf{b}_f = \{\mathbf{b}_1, \mathbf{b}_2,...,\mathbf{b}_d\} \tag{14.20}$$

Equation 14.18 yields an overdetermined problem, which can be solved by the least-square optimisation method to find the vector $\mathbf{k}^s = [k_1, k_2, ..., k_n]^T$. Figure 14.3 shows a flowchart of the proposed system identification and model updating method using the variations of the FRFs with full excitation.

14.4.3 FRF-based method with single excitation

The system identification and model updating methods using Equation 14.18 require all the FRFs of the building to be measured, which is usually difficult to be implemented

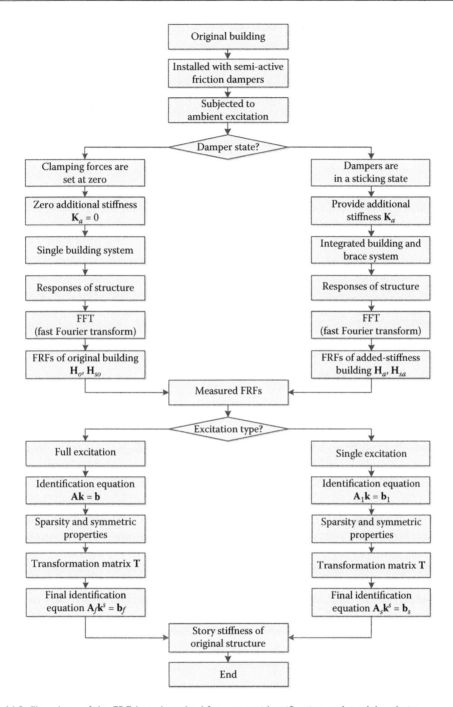

Figure 14.3 Flowchart of the FRF-based method for system identification and model updating.

in practice. To overcome this problem, model updating using single excitation is proposed herein. Let us consider only one excitation acting on the jth floor with a varying excitation frequency. Recording the responses at all the floors and using Equation 14.3 yield the jth column of the FRF $\mathbf{H}(\omega)$. Extracting the jth column from both sides of Equation 14.5 yields

$$(1 + j\omega\beta)\mathbf{K}\mathbf{H}_j(\omega) = -\omega^2\mathbf{I}_j^T + (\omega^2 - j\omega\alpha)\mathbf{M}\mathbf{H}_j(\omega) \tag{14.21}$$

where $\mathbf{H}_j$ denotes the jth column of the FRF matrix and $\mathbf{I}_j = [\overbrace{0\,0\cdots0}^{j-1}\,1\,\overbrace{0\,0\cdots0}^{n-j}]$.

Transposing Equation 14.21 and considering that both $\mathbf{K}$ and $\mathbf{M}$ are symmetric matrices, one can obtain

$$(1 + j\omega\beta)\mathbf{H}_j^T(\omega)\mathbf{K} = -\omega^2\mathbf{I}_j + (\omega^2 - j\omega\alpha)\mathbf{H}_j^T(\omega)\mathbf{M} \tag{14.22}$$

Given that $\mathbf{H}_j(\omega)$ is the function of the excitation frequency, one can use s different excitation frequencies. As a result, one may obtain the following equation (ω_i for $i = 1, 2, \ldots, s$):

$$\begin{bmatrix} \ddots & & \\ & 1 + j\omega_i\beta & \\ & & \ddots \end{bmatrix} \begin{bmatrix} \mathbf{H}_j^T(\omega_1) \\ \mathbf{H}_j^T(\omega_2) \\ \cdots \\ \mathbf{H}_j^T(\omega_s) \end{bmatrix} \mathbf{K} = -\begin{bmatrix} \omega_1^2 \\ \omega_2^2 \\ \cdots \\ \omega_s^2 \end{bmatrix} \mathbf{I}_j + \begin{bmatrix} \ddots & & \\ & \omega_i^2 - j\omega_i\alpha & \\ & & \ddots \end{bmatrix} \begin{bmatrix} \mathbf{H}_j^T(\omega_1) \\ \mathbf{H}_j^T(\omega_2) \\ \cdots \\ \mathbf{H}_j^T(\omega_s) \end{bmatrix} \mathbf{M} \tag{14.23}$$

This equation can also be written in the following form:

$$\mathbf{W}_\beta\mathbf{H}_s\mathbf{K} = -\mathbf{W}\mathbf{I}_j + \mathbf{W}_\alpha\mathbf{H}_s\mathbf{M} \tag{14.24}$$

where $\mathbf{H}_s = \begin{bmatrix} \mathbf{H}_j^T(\omega_1) \\ \mathbf{H}_j^T(\omega_2) \\ \cdots \\ \mathbf{H}_j^T(\omega_s) \end{bmatrix}$, $\mathbf{W} = \begin{bmatrix} \omega_1^2 \\ \omega_2^2 \\ \cdots \\ \omega_s^2 \end{bmatrix}$, $\mathbf{W}_\beta = \begin{bmatrix} \ddots & & \\ & 1 + j\omega_i\beta & \\ & & \ddots \end{bmatrix}$, $\mathbf{W}_\alpha = \begin{bmatrix} \ddots & & \\ & \omega_i^2 - j\omega_i\alpha & \\ & & \ddots \end{bmatrix}$.

Let us use a swept-sine force to excite the building structure at the jth floor to obtain two sets of FRFs: (1) all clamping forces in semi-active friction dampers are set to zero so that there is no any added stiffness to the building; (2) all semi-active friction dampers are set to be in a sticking state so that there is an added stiffness matrix $\mathbf{K}_a$ to the building. Select s different excitation frequencies and denote them as ω_{oi} and ω_{ai} for the original building and the added-stiffness building, respectively, where $i = 1, 2, \ldots, s$. The application of Equation 14.24 then leads to

$$\mathbf{W}_{\beta o}\mathbf{H}_{so}\mathbf{K} = -\mathbf{W}_o\mathbf{I}_j + \mathbf{W}_{\alpha o}\mathbf{H}_{so}\mathbf{M} \tag{14.25}$$

$$\mathbf{W}_{\beta a}\mathbf{H}_{sa}(\mathbf{K} + \mathbf{K}_a) = -\mathbf{W}_a\mathbf{I}_j + \mathbf{W}_{\alpha a}\mathbf{H}_{sa}\mathbf{M} \tag{14.26}$$

Transpose Equation 14.25 and pre-multiply the two sides of the equation by $\mathbf{W}_{\alpha a}\mathbf{H}_{sa}$. Then post-multiply Equation 14.26 by $\mathbf{H}_{so}^T\mathbf{W}_{\alpha o}$ and subtract the two equations, resulting in the following equation:

$$\mathbf{W}_{\alpha a}\mathbf{U}\mathbf{W}_{\beta o} - \mathbf{W}_{\beta a}\mathbf{U}\mathbf{W}_{\alpha o} = \mathbf{R} \tag{14.27}$$

where:

$$U = H_{sa}KH_{so}^T \tag{14.28}$$

$$R = -W_{\alpha a}H_{sa}I_j^T W_o^T + W_a I_j H_{so}^T W_{\alpha o} + W_{\beta a}H_{sa}K_a H_{so}^T W_{\alpha o} \tag{14.29}$$

Given that $W_{\alpha o}$, $W_{\alpha a}$, $W_{\beta o}$ and $W_{\beta a}$ all are diagonal matrices, Equation 14.27 can be expanded so that its (i, j)th element can be given by

$$[\omega_{\alpha a ii}\omega_{\beta o jj} - \omega_{\beta a ii}\omega_{\alpha o jj}]U_{ij} = R_{ij} \tag{14.30}$$

All the elements of R can be determined by Equation 14.29 using the known additional stiffness matrix and the measured FRFs of the building structure without additional stiffness and with additional stiffness. Therefore, all the unknown elements U_{ij} can be found using Equation 14.30, and as a result, the matrix U can be determined. Rewrite Equation 14.28 as follows:

$$A_1k = b_1 \tag{14.31}$$

where:

$$A_1 = H_{so} \otimes H_{sa} \tag{14.32}$$

$$b_1 = \text{vec}(R_{ij} / (\omega_{\alpha a ii}\omega_{\beta o jj} - \omega_{\beta a ii}\omega_{\alpha o jj})) \tag{14.33}$$

$$k = \text{vec}(K) = [k_{11} \ k_{21} \ \cdots \ k_{n1} \ k_{12} \ k_{22} \ \cdots \ k_{n2} \ \cdots \ k_{1n} \ k_{2n} \cdots \ k_{nn}]^T \tag{14.34}$$

Similar to the procedure for Equation 14.10, Equation 14.31 can finally be rewritten as

$$A_sk^s = b_s \tag{14.35}$$

where:

$$A_s = A_1^e T \tag{14.36}$$

$$b_s = b_1 \tag{14.37}$$

in which A_1^e is obtained from A_1 by deleting all the columns that multiply by $k_{ij} = 0$ $(\text{abs}(i - j) > 1)$, and the matrix T is the transformation matrix of size $(3n - 2) \times n$. As a result, the horizontal story stiffness can be identified from Equation 14.35 by exciting just one story of the building. Figure 14.3 shows a flowchart of the proposed system identification and model updating method using the variations of the FRFs with single excitation.

14.5 VIBRATION CONTROL USING SEMI-ACTIVE FRICTION DAMPERS

The development of semi-active friction dampers for vibration mitigation has attracted more and more attention in recent years. As mentioned in Section 13.4 of Chapter 13, the semi-active friction damper can be modelled with the components of a linear spring and a variable friction damper connected in series. A number of control strategies for semi-active friction dampers have been proposed, and they can be classified into two major types: (1) global feedback control strategies that use clipped strategies to allow semi-active friction dampers to work effectively with an LQG controller, and (2) local feedback control strategies that use local motions as feedback signals to change damper clamping forces in such a way that the friction dampers can slip as much as possible to achieve the maximum energy dissipation. The global feedback control strategy and the three local feedback controllers – viscous and Reid friction controllers, modulated homogeneous friction (MHF) controllers and non-sticking friction (NSF) controllers – have been investigated in Chapter 13. The LQG global control strategy and the NSF local control strategy are considered in this section for the structural vibration control of a building structure using semi-active friction dampers based on the updated system matrix and structural parameters of the building. The primary purpose of implementing the NSF local feedback control with a steady-state Kalman filter is to involve minimal feedback signals to achieve a reliable and economical control design. The use of a steady-state Kalman filter makes it possible to use the accelerometers as sensors for both health monitoring and vibration control, leading to a common sensory system and a common data acquisition and transmission system.

As mentioned in Chapter 13, the semi-active control force $u(t)$ depends on either the sticking or the slipping state of the damper, and it can be written as

$$u(t) = \begin{cases} f^k(t) & \text{if } |f^k(t)| < |f^d(t)| \text{ (sticking)} \\ f^d(t) & \text{if } |f^k(t)| \geq |f^d(t)| \text{ (slipping)} \end{cases} \tag{14.38}$$

$$f^k(t) = k^d[d(t) - e(t)] \tag{14.39}$$

where:
   k^d is the spring stiffness of the semi-active friction damper
   $f^d(t)$ is the friction force of the damper
   $f^k(t)$ is the axial force in the semi-active friction damper
   $d(t)$ is the axial displacement between the two ends of the friction damper
   $e(t)$ is the slip deformation of the friction damper that is given by

$$e(t) = \bar{e}(t) + \left| |d(t) - \bar{e}(t)| - \frac{|u(t)|}{k^d} \right| \text{sgn}\left(\dot{d}(t)\right) \tag{14.40}$$

where $\bar{e}(t)$ is the previously cumulated slip deformation of the friction damper and $\dot{d}(t)$ is the relative velocity between the two ends of the friction damper. The friction force of the semi-active friction damper is given by

$$f^d(t) = \mu^d N^d(t) \text{sgn}\left(\dot{d}(t)\right) \tag{14.41}$$

where μ^d is the dynamic friction coefficient and $N^d(t)$ is the clamping force of the semi-active friction damper, which is time dependent and determined by the feedback controller. The semi-active control force $u(t)$ is exactly the same as the semi-active friction force $f^d(t)$ if the friction damper slips continuously without sticking. This, however, depends on the control strategy and parameters.

14.5.1 Local feedback control strategy

As mentioned previously, only the NSF local control strategy is considered herein. In this strategy, the controllable clamping force $N^d(t)$ is given as

$$N^d(t) = N^d_{max}\left|\tanh\left[\gamma^d \dot{d}(t)\right]\right|$$
(14.42)

where N^d_{max} is the maximum clamping force and γ^d is the velocity control gain.

With reference to Figure 14.2 and by assuming one semi-active friction damper for every building story, the relationship between the jth damper displacement (velocity) and the relative displacement (velocity) of the jth building story at time instant t can be expressed as

$$d^j(t) = \left(x^j(t) - x^{j-1}(t)\right)\cos\theta_j$$
(14.43)

$$\dot{d}^j(t) = \left(\dot{x}^j(t) - \dot{x}^{j-1}(t)\right)\cos\theta_j$$
(14.44)

where:

$x^j(t)$ and $\dot{x}^j(t)$ are the displacement and velocity of the jth building floor, respectively

$x^{j-1}(t)$ and $\dot{x}^{j-1}(t)$ are the displacement and velocity of the $(j-1)$th building floor, respectively

θ_j is the angle between the jth semi-active friction damper and the $(j-1)$th building floor

Using Equations 14.43 and 14.44, the semi-active friction force of the damper, $f^d(t)$, can be determined using Equation 14.41 in terms of the relative displacement and velocity of the building story.

As previously mentioned, the accelerometer is the most widely used sensor in practice for system identification and health monitoring because of its high sensitivity and reliability. It is beneficial to install the same sensors for both the health monitoring and vibration control of a building. Therefore, a local feedback control strategy with a Kalman filter, by which the acceleration responses of the building are used as feedback signals rather than the displacement and velocity responses, is presented in this chapter for the vibration control of a building subjected to earthquake excitation using semi-active friction dampers. In this regard, the state space equation of the building can be expressed as

$$\dot{z} = \bar{A}z + \bar{B}u + \bar{E}\ddot{x}_g$$
(14.45)

$$\bar{A} = \begin{bmatrix} 0 & I \\ -M^{-1}K & -M^{-1}C \end{bmatrix}; \bar{B} = \begin{bmatrix} 0 \\ M^{-1}\bar{H} \end{bmatrix}; \bar{E} = \begin{bmatrix} 0 \\ -1 \end{bmatrix}; \bar{B} = \begin{bmatrix} 0 \\ M^{-1}\bar{H} \end{bmatrix}; \bar{E} = \begin{bmatrix} 0 \\ -1 \end{bmatrix}$$
(14.46)

where:

$\mathbf{z}$ is the state vector of the controlled building $[\mathbf{x} \ \dot{\mathbf{x}}]^T$

$\mathbf{u}$ is the semi-active control force vector

$\bar{\mathbf{H}}$ is the influence matrix reflecting the location of the semi-active friction dampers

$\mathbf{I}$ is the unit diagonal matrix

The measured responses $\mathbf{y}_m$ are selected as the absolute acceleration outputs of the building floors.

$$\mathbf{y}_m = \bar{\mathbf{C}}_m \mathbf{z} + \bar{\mathbf{D}}_m \mathbf{u} + \bar{\mathbf{F}}_m \ddot{x}_g + \mathbf{v} \tag{14.47}$$

where $\mathbf{v}$ is the measurement noise vector and $\bar{\mathbf{C}}_m$, $\bar{\mathbf{D}}_m$ and $\bar{\mathbf{F}}_m$ are the reduced-order coefficient matrices of $\bar{\mathbf{A}}$, $\bar{\mathbf{B}}$ and $\bar{\mathbf{E}}$, respectively.

The estimated state vector $\hat{\mathbf{z}}$ is described by the steady-state Kalman filter optimal estimator (Stengel 1986; Skelton 1988) in the following form:

$$\dot{\hat{\mathbf{z}}} = \bar{\mathbf{A}}\hat{\mathbf{z}} + \bar{\mathbf{B}}\mathbf{u} + \bar{\mathbf{L}}(\mathbf{y}_m - \bar{\mathbf{C}}_m\hat{\mathbf{z}} - \bar{\mathbf{D}}_m\mathbf{u}) \tag{14.48}$$

$$\bar{\mathbf{L}} = \left[\bar{\mathbf{R}}^{-1}(\bar{\gamma}_g \bar{\mathbf{F}}_m \bar{\mathbf{E}}^T + \bar{\mathbf{C}}_m \bar{\mathbf{S}}) \right]^T \tag{14.49}$$

where $\bar{\mathbf{S}}$ is the solution of the algebraic Riccati equation given by

$$\bar{\mathbf{S}}\tilde{\mathbf{A}} + \tilde{\mathbf{A}}^T\bar{\mathbf{S}} - \bar{\mathbf{S}}\tilde{\mathbf{G}}\bar{\mathbf{S}} + \tilde{\mathbf{H}} = 0 \tag{14.50}$$

and

$$\tilde{\mathbf{A}} = \bar{\mathbf{A}}^T - \bar{\mathbf{C}}_m^T\tilde{\mathbf{R}}^{-1}(\bar{\gamma}_g \bar{\mathbf{F}}_m \bar{\mathbf{E}}^T) \tag{14.51}$$

$$\tilde{\mathbf{G}} = \bar{\mathbf{C}}_m^T\tilde{\mathbf{R}}^{-1}\bar{\mathbf{C}}_m \tag{14.52}$$

$$\tilde{\mathbf{H}} = \bar{\gamma}_g \bar{\mathbf{E}}\bar{\mathbf{E}}^T - \bar{\gamma}_g^2 \bar{\mathbf{E}}\bar{\mathbf{F}}_m^T\tilde{\mathbf{R}}^{-1}\bar{\mathbf{F}}_m\bar{\mathbf{E}}^T \tag{14.53}$$

$$\tilde{\mathbf{R}} = \mathbf{I} + \bar{\gamma}_g \bar{\mathbf{F}}_m \bar{\mathbf{F}}_m^T \tag{14.54}$$

The measurement noises in all of the building floor responses are assumed to be the same as a stationary Gaussian white noise process. The power spectral density ratio in Equation 14.49 is defined as $\bar{\gamma}_g = \bar{S}_{\ddot{x}_g \ddot{x}_g} / \bar{S}_{v_i v_i}$, where $\bar{S}_{\ddot{x}_g \ddot{x}_g}$ and $\bar{S}_{v_i v_i}$ are power spectral density functions of the stationary white noise of $\ddot{x}_g$ and v_i, respectively. Once the estimated state vector $\hat{\mathbf{z}}$ is obtained from Equation 14.48, the corresponding estimates for the displacement and velocity responses, $\hat{\mathbf{x}}$ and $\dot{\hat{\mathbf{x}}}$, of the building can be obtained. The estimates of damper slip $\hat{d}^i(t)$ and slip rate $\dot{\hat{d}}^i(t)$ at time instant t can then be obtained from

$$\hat{d}^i(t) = (\hat{x}^i(t) - \hat{x}^{i-1}(t))\cos\theta_i \tag{14.55}$$

$$\hat{\dot{d}}^j(t) = (\hat{\dot{x}}^j(t) - \hat{\dot{x}}^{j-1}(t))\cos\theta_j \qquad (14.56)$$

where $\hat{x}^j(t)$ and $\hat{\dot{x}}^j(t)$ are the estimated displacement and velocity of the jth floor of the building and $\hat{x}^{j-1}(t)$ and $\hat{\dot{x}}^{j-1}(t)$ are the estimated displacement and velocity of the $(j-1)$th floor of the building. An estimation of clamping force and the friction force in the semi-active friction damper can be given by

$$\hat{N}^d(t) = N_{\max}^d \left| \tanh\left[\gamma^d \hat{d}(t)\right] \right| \qquad (14.57)$$

$$\hat{f}^d(t) = \mu^d \hat{N}^d(t) \operatorname{sgn}\left(\hat{\dot{d}}(t)\right) \qquad (14.58)$$

Equation 14.38 is then used to determine the semi-active control force, and finally, Equation 14.45 is used to determine the structural responses at the next time step. A flowchart of determining semi-active friction damper force using a local control strategy with a Kalman filter is shown in Figure 14.4.

14.5.2 Global feedback control strategy

To provide a comparative basis for the local feedback control strategy, an LQG controller (Stengel 1986; Skelton 1988) with a modified clipped strategy is also applied to a building with semi-active friction dampers. The same state space equations as Equations 14.45 and 14.46 are used. An additional equation for the regulated response is

$$y_{ed} = C'_{ed}z + D'_{ed}u + F'_{ed}\ddot{x}_g \qquad (14.59)$$

where y_{ed} is the regulated response vector and C'_{ed}, D'_{ed} and F'_{ed} are the reduced-order coefficient matrices of $\bar{A}$, $\bar{B}$ and $\bar{E}$, respectively. The acceleration feedback LQG controller in this study is basically designed to minimise a quadratic objective function by weighting the absolute acceleration responses of the building and the control forces. The objective function is given by

$$J = \lim_{\tau \to \infty} \frac{1}{\tau} E\left[\int_0^\tau \left\{ \left(C'_{ed}z + D'_{ed}u\right)^T Q'\left(C'_{ed}z + D'_{ed}u\right) + u^T R'u \right\} dt \right] \qquad (14.60)$$

where Q' and R' are the weighting matrices for the acceleration responses and the semi-active control forces, respectively. Based on the separation principle that allows feedback gain and Kalman gain to be determined separately (Stengel 1986; Skelton 1988), the optimal control force vector is obtained as

$$u = -G'\hat{z}. \qquad (14.61)$$

where G' is the full-state feedback gain matrix given by

$$G' = \tilde{R}'^{-1}(\tilde{N}'^T + \bar{B}^T\bar{P}') \qquad (14.62)$$

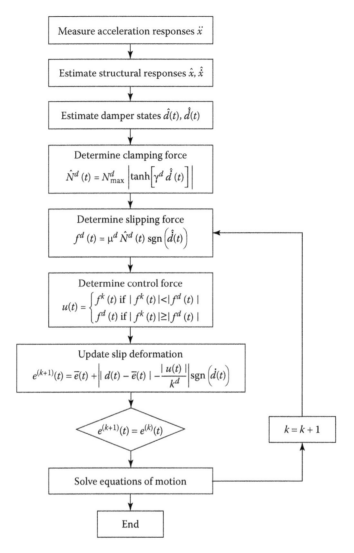

Figure 14.4 Flowchart determining control force using a local control strategy with a Kalman filter.

and $\overline{\mathbf{P}}'$ is the solution of the following algebraic Riccati equation:

$$\overline{\mathbf{P}}'\tilde{\mathbf{A}}' + \tilde{\mathbf{A}}'^T\overline{\mathbf{P}}' - \overline{\mathbf{P}}'\overline{\mathbf{B}}\tilde{\mathbf{R}}'^{-1}\overline{\mathbf{B}}^T\overline{\mathbf{P}}' + \tilde{\mathbf{Q}}' = 0 \tag{14.63}$$

in which

$$\tilde{\mathbf{Q}}' = \mathbf{C}_{ed}'^T\mathbf{Q}'\mathbf{C}_{ed}' - \tilde{\mathbf{N}}'\tilde{\mathbf{R}}'^{-1}\tilde{\mathbf{N}}'^T \tag{14.64}$$

$$\tilde{\mathbf{N}}' = \mathbf{C}_{ed}'^T\mathbf{Q}'\mathbf{D}_{ed}' \tag{14.65}$$

$$\tilde{\mathbf{R}}' = \mathbf{R}' + \mathbf{D}_{ed}'^T\mathbf{Q}'\mathbf{D}_{ed}' \tag{14.66}$$

$$\tilde{\mathbf{A}}' = \overline{\mathbf{A}} - \overline{\mathbf{B}}\tilde{\mathbf{R}}'^{-1}\tilde{\mathbf{N}}'^T \tag{14.67}$$

Equations 14.45, 14.48 and 14.61 form the basic equations for active control of the building using an LQG controller. Nevertheless, the control force of a semi-active friction damper depends on its motion status: when the damper is in its sticking stage, the control force is equal to the axial force in the brace; when the damper is its slipping stage, the magnitude of the control force depends on the controllable clamping force and its direction depends on the velocity of the damper. Therefore, the closed-loop feedback force determined by Equation 14.61 cannot always be achieved by semi-active friction dampers. Here, the following modified clipped control strategy is used for friction-based semi-active devices.

$$
N^d(t) = \begin{cases}
u^{LQG}(t)/\mu & \dot{d}(t) \cdot u^{LQG}(t) < 0 \text{ and } |u^{LQG}(t)| < \mu^d N_{max}^d \\
N_{max}^d \text{sgn}(u^{LQG}(t)) & \dot{d}(t) \cdot u^{LQG}(t) < 0 \text{ and } |u^{LQG}(t)| \geq \mu^d N_{max}^d \\
0 & \dot{d}(t) \cdot u^{LQG}(t) \geq 0
\end{cases}
\tag{14.68}
$$

where $u^{LQG}(t)$ is the optimal active control force determined by the LQG controller and N_{max}^d is the maximum clamping force which can be provided by the semi-active friction damper. It can be seen from Equation 14.68 that when the direction of the desired active control force is opposite to the velocity of the semi-active friction damper, and the magnitude of the desired active control does not exceed the maximum control force which the damper can tolerate, the semi-active friction damper is able to generate the desired control force in the same direction as required. If the semi-active friction damper cannot generate the desired active control force in the opposite direction to the velocity of the damper, a zero clamping force is then commanded to produce zero friction force at a given time instant. A flowchart of the vibration control process using either a local or a global control strategy with a Kalman filter is shown in Figure 14.5.

14.6 FRF-BASED STRUCTURAL DAMAGE DETECTION

Even with control devices installed, building structures may still suffer some damage after extreme events or long-term service. A rational approach is necessary to assess the damage of a controlled building. By referring to the system identification and model updating method introduced in Section 14.4 of this chapter, Equations 14.18 and 14.35 can both be used not only to update the stiffness of the original building but also to identify the stiffness of the damaged building. By comparing the identified stiffness coefficients of both undamaged and damaged buildings, the location and severity of the structural damage can be determined. Figure 14.6 displays a flowchart of the proposed damage detection method.

14.7 NUMERICAL INVESTIGATION

14.7.1 Description of a numerical example building

The example building is a simple five-story shear building which has an identical story height of 3 m (see Figure 14.2). The original building has uniform mass $m = 5.1 \times 10^3$ kg and uniform horizontal story (shear) stiffness $k = 1.334 \times 10^7$ N/m. The five natural frequencies of the original building without added stiffness are 2.317, 6.762, 10.661, 13.695

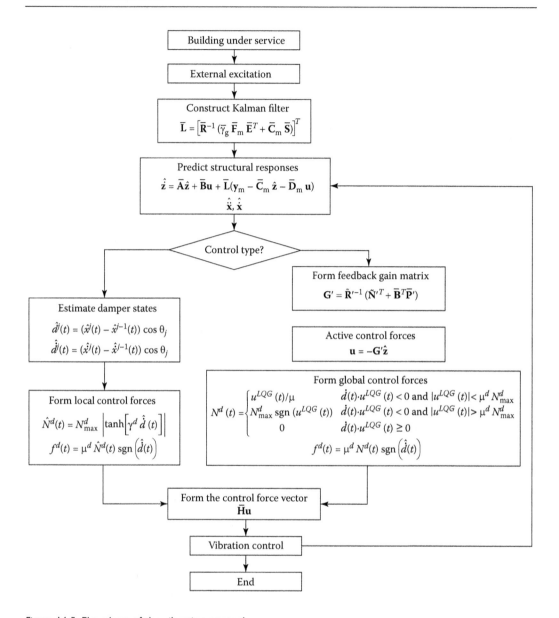

Figure 14.5 Flowchart of the vibration control process.

and 15.620 Hz. The structural damping matrix is assumed in the Rayleigh damping form, and the damping ratios for the first- and second-order modes of vibration of the building are assumed to be 2%. On each story of the building, a semi-active friction damper is installed with a diagonal brace that connects two neighbouring floors. The building is subjected to harmonic excitations and the acceleration responses are measured by five accelerometers, with one on each floor of the building, for parameter identification and damage detection.

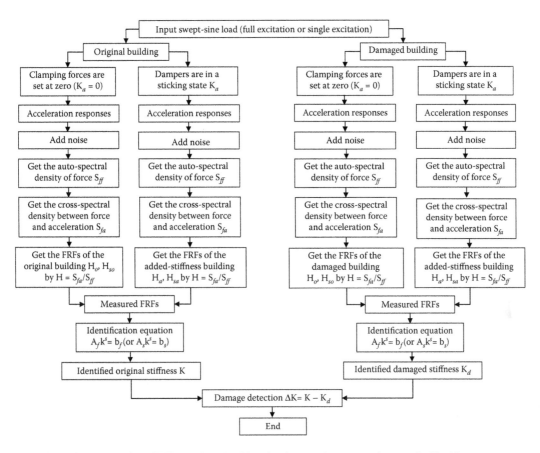

Figure 14.6 Flowchart of the FRF-based method for the damage detection of controlled building structures.

The transformation matrix $\mathbf{T}$ between $\mathbf{k}^e$ and $\mathbf{k}^s$ is formed as follows:

$$\mathbf{T} = \begin{bmatrix} 0 & 0 & 0 & 0 & 0 & 0 & 0 & 0 & 0 & 1 & -1 & -1 & 1 \\ 0 & 0 & 0 & 0 & 0 & 0 & 1 & -1 & -1 & 1 & 0 & 0 & 0 \\ 0 & 0 & 0 & 1 & -1 & -1 & 1 & 0 & 0 & 0 & 0 & 0 & 0 \\ 1 & -1 & -1 & 1 & 0 & 0 & 0 & 0 & 0 & 0 & 0 & 0 & 0 \\ 1 & 0 & 0 & 0 & 0 & 0 & 0 & 0 & 0 & 0 & 0 & 0 & 0 \end{bmatrix}^{T} \tag{14.69}$$

All the simulation cases consider six levels of noise, which are, 0%, 0.5%, 1%, 2%, 5% and 10%, respectively. All noises are uncorrelated and are added to the acceleration response data. The noise level is related to the noise intensity, which is defined as the ratio of the root mean square (RMS) of the noise to the RMS of the acceleration response.

14.7.2 Selection of additional stiffness

For either model updating or damage detection, the following two states must be created: setting the clamping forces in the semi-active friction dampers at zero, which creates the original building, and setting the clamping forces at maximum value, which creates the building with the additional stiffness provided by the damper brace systems. The stiffness ratio (SR) of the additional stiffness from the damper brace system to the horizontal story stiffness of the building is defined as

$$SR = \frac{K_d}{K_s} \tag{14.70}$$

where K_s is the horizontal story stiffness of the building and K_d is the additional stiffness from the damper brace system when the semi-active friction damper is in sticking state. The additional stiffness and the corresponding SR are assumed to be the same for each story of the building.

For the vibration control of the building with known stiffness K_s, the additional stiffness K_d from the damper brace system must be properly selected in order to achieve the best control performance. Similarly, to achieve the best results in system identification and damage detection, the additional stiffness K_d from the damper brace system must be optimised. The computed results of the average identification error against SR with full excitation and single excitation at the top floor are plotted in Figure 14.7, with the six levels of noise intensity considered. The average identification error is defined as the average value of the five identification errors in the five horizontal stiffness coefficients of the example building. For a given SR value and a given noise level, the proposed identification method is applied to the building. A total of 100 frequency points are selected from FRFs of the building, with 15 frequency points before and after each of the five natural frequencies at an interval of frequency resolution Δf of 0.12207 Hz. Equations 14.18 and 14.35 are then used to update the horizontal story stiffness coefficients of the building, from which the average identification error can be calculated. It can be seen that the average identification errors reduce rapidly with increasing SR for both the full-excitation and single-excitation cases when the SR values are smaller than 0.4. This is because small changes in FRFs are easily overlapped by measurement noise. As SR values further increase from 0.4, there are slight changes in the average identification errors. For the full-excitation case, the average identification errors are less than 0.5% when the SR value is greater than 0.4, and the measurement noise does not affect the identification results. For the single-excitation case, higher measurement noise leads to higher identification error, but the identification error is still smaller than 1.0% even for a 10% measurement noise level. The full excitations give better identification results than the single excitation, as expected. Based on the results shown in Figure 14.7, the SR value could be selected as 0.4, but a value of 0.9 is selected for the subsequent case studies in order to achieve the best control performance and to make comparisons with the results given by Chen and Xu (2008). Once the optimal value is found, the damper brace systems are then designed and manufactured based on the optimal value and, finally, installed in the building. As a result, the accuracy of the additional stiffness matrix K_d from the damper brace systems can be controlled, and in most cases no further identification in the field may be required. The five natural frequencies of the building after added known stiffness with an SR value of 0.9 are 3.193, 9.322, 14.695, 18.877 and 21.530 Hz, respectively.

14.7.3 Effects of the number of natural frequencies included

The proposed method is to identify structure parameters and detect damage using a few frequency points in the measured FRFs around the natural frequencies of the building through either Equations 14.18 or 14.35. One of the key problems is thus how many natural frequencies of the building shall be considered for parameter identification. Since it is difficult to excite the high natural frequencies of the building, it is better to select the lowest natural frequencies possible. Figure 14.8 shows the average identification errors in story

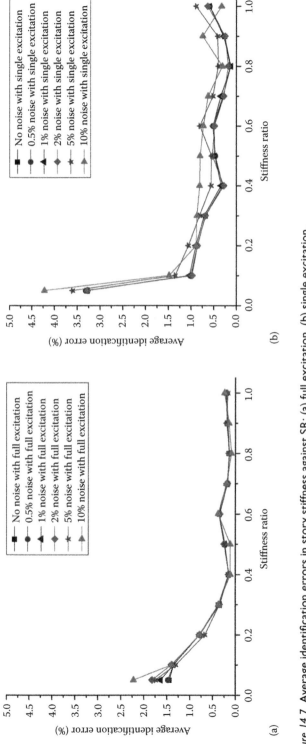

Figure 14.7 Average identification errors in story stiffness against SR: (a) full excitation, (b) single excitation.

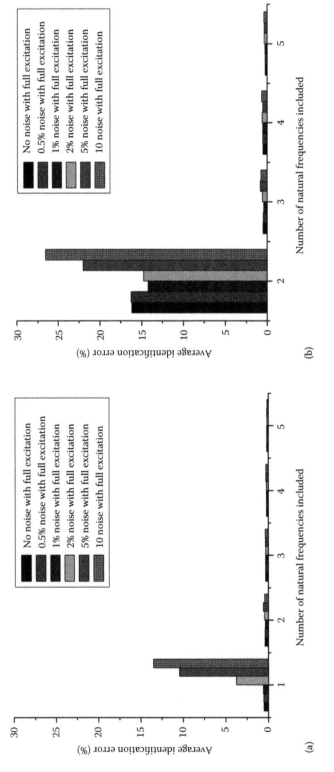

Figure 14.8 Average identification errors in story stiffness against the number of natural frequencies: (a) full excitation, (a) single excitation.

stiffness against the number of natural frequencies considered for the full-excitation case and single-excitation case, with six levels of noise intensity. In the computation, the SR value is set at 0.9 for all the stories of the building, and 15 frequency points are selected before and after each natural frequency at an interval of frequency resolution Δf of 0.12207 Hz. It can be seen from Figure 14.8a that if the full-excitation method is used, only the first two natural frequencies should be considered to keep the average identification error less than 1%. The effects of measurement noise on the identification results are very small. If only the first natural frequency is considered and the measurement noise level is less than 2%, the identification results are still acceptable, but as the measurement noise level increases the average identification error increases quickly. For the single-excitation method, at least the first three natural frequencies of the building should be considered to keep the average identification error less than 1% without significant influence from measurement noise, as shown in Figure 14.8b. As a result, the first three natural frequencies are included in the study of the example building.

14.7.4 Effects of the number of frequency points used

After the number of natural frequencies to be included in parameter identification is decided, one then decides how many frequency points need to be considered before and after each natural frequency of the building. Figure 14.9 displays the average identification error in story stiffness against the number of frequency points (NFP) for the full-excitation case and single-excitation case with the six levels of noise intensity. In the computation, the SR value is set at 0.9 for all the stories of the building and only the first three natural frequencies are taken into consideration. It can be seen from Figure 14.9a for the full-excitation case that the optimal NFP is about 15, for which the average identification error is minimal. The effects of measurement noise on the identification results are not obvious if the measurement noise level is not greater than 5%. For the single-excitation case, Figure 14.9b shows that the NFP will be greater than 9 if the average identification error is controlled below 2%. Thus, for the example building, an NFP of 15 is selected.

14.7.5 Comparison with a previous study

The parameter identification of the example building is also performed using the natural frequency and mode shape (NF&MS)-based method (Chen and Xu, 2008), and the identification quality is compared with the FRF-based method. The SR of the building varies from 0.05 to 1.0 in the computation. Since the NF&MS-based method is sensitive to measurement noise, three low noise levels of 0.5%, 1% and 2% are considered, but for the FRF-based method a 10% measurement noise level is considered, with the number of natural frequencies at three and the number of frequency points at 15.

Figure 14.10 shows the average identification error in the story stiffness of the two methods. It can be seen that the identification error in the FRF-based method is much smaller than in the NF&MS-based method, although a 10% noise level is considered in the FRF-based method but only a maximum 2% noise level is considered in the NF&MS-based method. This conclusion is particularly true when the SR value is less than 0.4. When the noise level is greater than 2%, the NF&MS-based method cannot identify the stiffness parameters properly. Therefore, one may conclude that the FRF-based method is more accurate in identifying the stiffness of the building than the NF&MS-based method, and that the FRF-based method is much less sensitive to measurement noise than the NF&MS-based method. Furthermore, the FRF-based method allows the use of a relatively small SR compared with the NF&MS-based method so as to facilitate practical application.

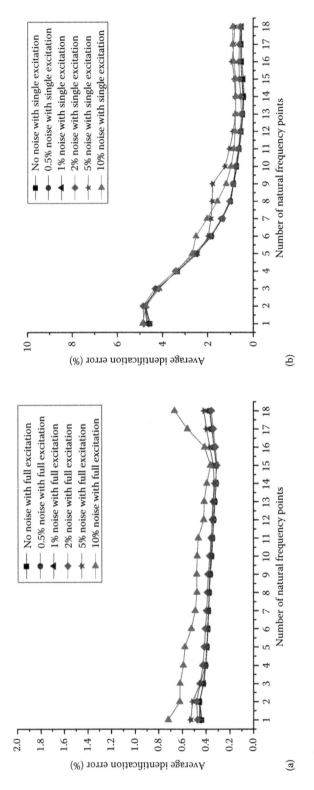

Figure 14.9 Average identification errors in story stiffness against the number of frequency points: (a) full excitation, (a) single excitation.

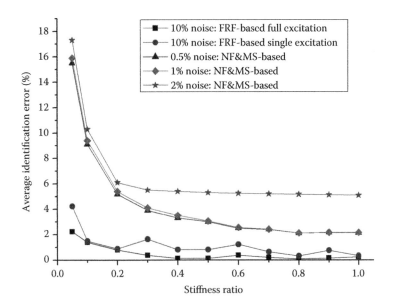

Figure 14.10 Comparison of the average identification errors between the two methods.

14.7.6 Damage detection using full excitation

Five cases of structural damage are considered in the example building: (1) no damage, (2) a 5% reduction of the horizontal stiffness coefficient in the second story of the building, (3) a 10% reduction of the horizontal stiffness coefficient in the top story of the building, (4) a 5% reduction of the horizontal stiffness coefficient in both the first and third stories of the building, and (5) a 10% reduction of the horizontal stiffness coefficient in both the fourth and fifth stories of the building.

For each damage case, the SR value is selected as 0.9, the number of natural frequencies is taken as three, and the number of frequency points opted for is 15. The frequency range of the FRF is chosen as 0.5 Hz to 100 Hz based on the acceleration response data with a frequency resolution Δf of 0.12207 Hz. Figure 14.11 shows the identification results of the building without added stiffness and without any damage. The measurement noise levels considered range from 0% to 10%. The identification results demonstrate that the building has no damage. Two single-damage cases, one with 5% damage at the second floor and the other with 10% damage at the fifth floor, are simulated, and the identification results are shown in Figure 14.12. Again, the two single-damage cases are clearly identified even though the measurement noise reaches 10%. Two double-damage cases, one with 5% damage at the first and third floors and the other with 10% damage at the fourth and fifth floors, are also examined. Figure 14.13 shows the identification results of the two cases, respectively. Clearly, the proposed full-excitation FRF-based damage detection method is capable of locating and quantifying multiple instances of damage in the building with a maximum error less than 1%, even though the noise level is as high as 10%.

14.7.7 Damage detection using single excitation

The investigation of the aforementioned five cases of structural damage is also considered using the single-excitation identification method. All the other conditions remain

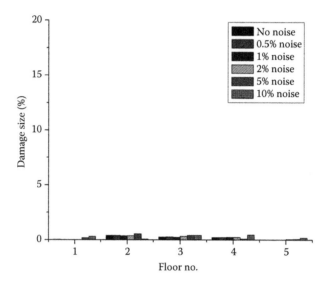

Figure 14.11 Damage detection results without damage (full excitation).

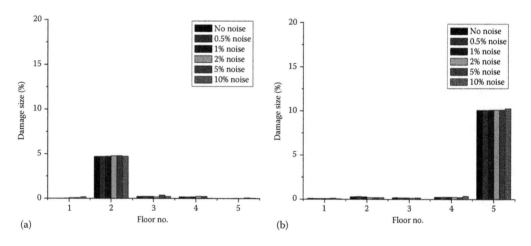

Figure 14.12 Single damage detection results (full excitation). (a) 5% damage at second floor, (b) 10% damage at fifth floor.

unchanged from the full-excitation case. Figures 14.14, 14.15 and 14.16 display the damage detection results for the five cases: without damage, with 5% damage at the second floor, with 10% damage at the fifth floor, with 5% damage at the first and third floors, and with 10% at the fourth and fifth floors, respectively. The results show that the single-excitation FRF-based damage detection method can also locate and quantify the damage even though the noise level is 10%. The maximum error is less than 2%. Comparatively speaking, the full-excitation FRF-based damage detection method provides more accurate results than the single-excitation FRF-based damage detection method, but the single-excitation FRF-based damage detection method is good enough for practical application.

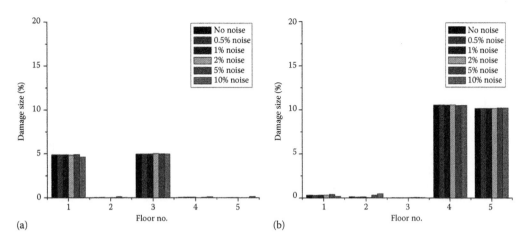

Figure 14.13 Multiple damage detection results (full excitation). (a) 5% damage at first and third floors, (b) 10% damage at fourth and fifth floors.

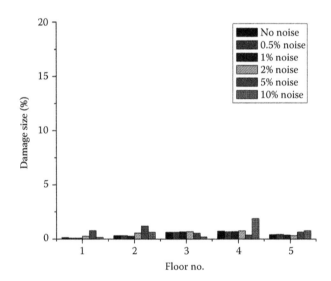

Figure 14.14 Damage detection results without damage (single excitation).

14.7.8 Seismic response control of the building structure

To evaluate the control performance and robustness of the semi-active friction dampers, four seismic records are selected as inputs to the example building: (1) El-Centro NS (1940), (2) Hachinohe NS (1968), (3) Northridge NS (1994) and (4) Kobe NS (1995). The original peak ground accelerations (PGAs) of the four seismic records are 3.417, 2.250, 8.2676 and 8.1782 m/s², respectively. The original time histories of the four seismic records are scaled to have the same PGA of 4.0 m/s² to facilitate the comparison. In consideration of the problem from a practical viewpoint, the previously identified stiffness parameters in the case of a 2% noise level are adopted and used for the construction of the damping matrix based on the Rayleigh damping assumption. The SRs of all five semi-active friction dampers are kept at the same value of 0.9 unless otherwise specified.

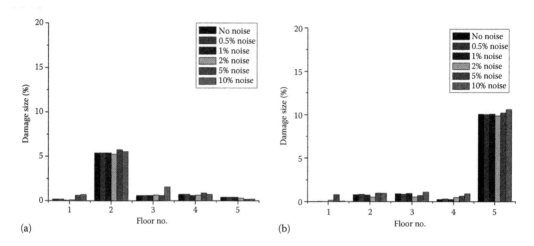

Figure 14.15 Single damage detection results (single excitation). (a) 5% damage at second floor, (b) 10% damage at fifth floor.

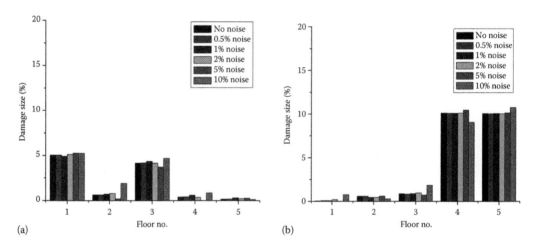

Figure 14.16 Multiple damage detection results (single excitation). (a) 5% damage at first and third floors, (b) 10% damage at fourth and fifth floors.

As mentioned before, two control strategies are used for performance evaluation: (1) local feedback control and (2) LQG global feedback control. To implement the first control strategy in practice, only five accelerometers, one for each building floor, are needed to obtain the feedback. For the implementation of the second control strategy in practice, five accelerometers and five force transducers, one accelerometer and one force transducer for each building story, are necessary to realise the feedback control. Clearly, the first control strategy is most attractive in terms of the type and number of sensors required if the control performance using this strategy is compatible with the other control strategies.

The control performance is evaluated in terms of a vibration reduction factor (VRF) defined as follows:

$$VRF = \frac{Z_{nc} - Z_{co}}{Z_{nc}} \tag{14.71}$$

where Z_{nc} is the maximum response (either displacement, velocity or acceleration) of a given building floor without control and Z_{co} is the maximum response of the same quantity of the same floor with control.

In the implementation of the global feedback control strategy, the two weighting matrices $\mathbf{Q}'$ and $\mathbf{R}'$ are selected as the unit diagonal matrix multiplied by a factor. The optimum factor is found to be 2.1×10^5 for $\mathbf{Q}'$ and 0.016 for $\mathbf{R}'$. For the implementation of a NSF controller, the velocity control gain γ^d defined in Equation 14.42 is set to be 2.1×10^6. To determine the maximum clamping force N_{max}^d, which can be provided by the semi-active friction damper, the maximum axial force in the brace of the example building without semi-active control is computed with the El-Centro NS seismic input. The maximum axial force is then taken as the maximum slipping force.

Figures 14.17a–c depict the variations of the peak displacement, velocity and acceleration responses of the building structure under the El-Centro NS seismic ground motion for three cases: (1) the original building without any control, (2) local feedback control and (3) global feedback control. It can be seen that with the aid of control strategies, the maximum responses of displacement, velocity and acceleration of the building are all reduced compared with the original building. The semi-active friction dampers with the global control strategy demonstrate the best control performance in the sense that they mostly reduce all three kinds of seismic responses on all building floors. Nevertheless, given the relatively simple sensor system in the local control strategy with a Kalman filter, the performance of this control strategy is still acceptable. Displayed in Figures 14.18a–c are the time histories of the displacement, velocity and acceleration responses at the top floor of the building without any control and with the local control strategy together with a Kalman filter. Clearly, the local control strategy can effectively suppress the seismic responses of the building when the building experiences large vibrations.

The feasibility of the proposed integrated procedure also depends on the brace (damper) stiffness for both vibration control and model updating. Figure 14.19 displays the variations in the

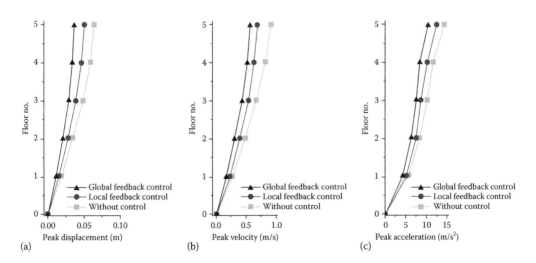

Figure 14.17 Comparison of control performance for various control strategies. (a) Peak displacement, (b) peak velocity, (c) peak acceleration.

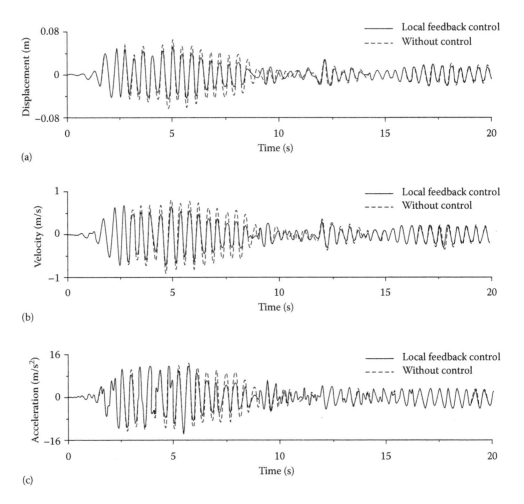

Figure 14.18 Performance of local feedback control strategy: (a) time history of displacement, (b) time history of velocity, (c) time history of acceleration.

VRFs of the displacement, velocity and acceleration responses of the top floor with SR during the El-Centro NS earthquake by using the local feedback control strategy. It can be seen that the VRFs for all three responses increase rapidly when the SR increases from 0.0 to about 0.1. Afterwards, the VRFs increase only slightly with increasing SR. Above 0.9, the SR has almost no effect on the VRFs. Similar results are found for other building floors. As a result, the optimum SR for the seismic response control of the example building should be 0.9, which is consistent with the optimum SR determined for the system identification of the same building.

To investigate the robustness of the control performance of the local feedback controller on the integrated building structure, similar investigations are performed on the example building with the other three seismic inputs. It is found that the optimum SR from the case of the El-Centro NS earthquake remains almost unchanged for the other three seismic inputs. The optimum gain coefficient obtained from the El-Centro NS earthquake can also be applied to the other three seismic inputs, although there are slightly different values for different seismic inputs. To have a reasonable comparison of the control performance of the semi-active friction dampers manipulated by the local and global feedback control strategy for the example building under different seismic inputs, two sets of normalised performance

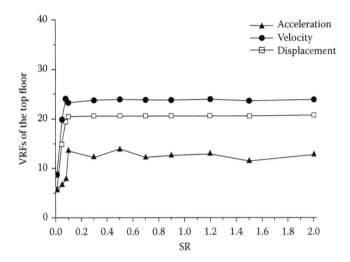

Figure 14.19 Variations of VRFs with SR using local feedback control strategy.

indices are used. The first set of the performance indices is related to the building responses (Spencer et al. 1998; Ohtori et al. 2004). They include peak- and RMS-based inter-story drift ratios (J_1 and J_3) and peak- and RMS-based absolute acceleration responses (J_2 and J_4).

$$J_1 = \left\{ \frac{\max_{t,i} |dx_i^c(t)|/h_i}{\max_{t,i} |dx_i^n(t)|/h_i} \right\}; \quad J_2 = \left\{ \frac{\max_{t,i} |\ddot{x}_i^c(t)|}{\max_{t,i} |\ddot{x}_i^n(t)|} \right\}; \quad J_3 = \left\{ \frac{\max_{t,i} \|dx_i^c(t)\|/h_i}{\max_{t,i} \|dx_i^n(t)\|/h_i} \right\};$$

$$J_4 = \left\{ \frac{\max_{t,i} \|\ddot{x}_i^c(t)\|}{\max_{t,i} \|\ddot{x}_i^n(t)\|} \right\} \tag{14.72}$$

where:

$dx_i^c(t)$ and $dx_i^n(t)$ are the inter-story drift of the ith story of the building, with control and without control, respectively

h_i is the height of the ith story

$\ddot{x}_i^c$ and $\ddot{x}_i^n$ are the absolute acceleration response of the ith floor of the building with control and without control, respectively

The RMS response quantities within the time duration t_f under each earthquake are calculated by $\|\cdot\| = \sqrt{1/t_f \int_0^{t_f} [\cdot]^2 dt}$. The sign max means to find the maximum value within the given time duration first and among all the building stories/floors afterwards.

The second set of performance indices are related to the capacity of control devices. Only the peak-based control force (J_5) is used in this example.

$$J_5 = \left\{ \frac{\max_{t,l} |u_l(t)|}{W} \right\} \tag{14.73}$$

where $u_l(t)$ is the control force generated by the lth control device and W is the seismic weight of the building – that is, the total weight of all the building floors in this example. A comparison of the performance indices of the controlled building using local and global feedback control strategies under different earthquakes are listed in Table 14.1 and Table 14.2, respectively. It can be seen that both control strategies can effectively reduce the RMS responses of the building, with the global control strategy being more effective. The peak responses of the building structure can also be reduced, but they are not as effective as the RMS responses, in particular, in the cases of Hachinole and Northridge earthquakes. The control forces when using the global control strategy are, however, much larger than those when using the local control strategy. These observations together with the sensory system required for each control strategy clearly demonstrate that the local control strategy with a Kalman filter is preferred for the integrated health monitoring and vibration control system.

14.8 EXPERIMENTAL INVESTIGATION

Since an experimental investigation of the vibration control of a building complex using semi-active friction dampers under earthquake excitation has been reported in Chapter 13, this section concerns only an experimental investigation of the system identification and damage detection of the same building complex based on variations of the FRFs between the two states of the building complex. To create damage scenarios, and given that the links between the main building and the podium structure are most susceptible to damage in practice, a circular steel ring was designed and used to connect the podium structure to the main building, and damage to the connection between the main building and the podium structure was simulated by varying the stiffness of the steel ring. To simulate the added stiffness provided by the friction dampers, another circular steel ring was used to connect the podium structure to the main building so that the additional stiffness could be selected relatively easily compared with using the available semi-active friction damper. Details of the experimental arrangements and test results are described in the following subsections.

14.8.1 Experimental setup

The model of the building complex has been described in detail in Chapter 13. This section focuses on a description of the self-designed circular steel rings, as shown in Figure 14.20. The beam/plate connections between the main building and the podium structure are most susceptible to damage in real situations. A circular steel ring, as shown in Figure 14.21, was designed and used to connect the top of the podium structure to the third floor of the main building along the middle line of the two models (see Figure 14.20). Changing the size of that connection ring varied the stiffness of the connection, which simulated damage to the building complex while the separate models remained unchanged.

Table 14.1 Performance indices for the local feedback control strategy

	Local feedback control strategy with Kalman filter			
Index	El-Centro	Hachinohe	Northridge	Kobe
J_1 (Peak drift ratio)	0.7342	0.8796	0.8839	0.7132
J_2 (Peak acc.)	0.8642	0.9453	0.9273	0.7720
J_3 (RMS drift ratio)	0.7046	0.7526	0.6276	0.6097
J_4 (RMS acc.)	0.7165	0.7642	0.6741	0.6244
J_5 (Control force)	0.0357	0.0407	0.0753	0.0457

Table 14.2 Performance indices for the global feedback control strategy

Index	Global feedback control strategy			
	El-Centro	Hachinohe	Northridge	Kobe
J_1 (Peak drift ratio)	0.6825	0.7651	0.8056	0.5260
J_2 (Peak acc.)	0.7398	0.8022	0.7627	0.4950
J_3 (RMS drift ratio)	0.4809	0.5684	0.4843	0.3946
J_4 (RMS acc.)	0.4677	0.5488	0.4822	0.3798
J_5 (Control force)	0.0982	0.1010	0.1425	0.1070

Four connection rings of different thicknesses were designed for damage simulation. The required stiffness of the connection ring was estimated as follows:

$$k_c = \frac{0.56Eb_ct_c^3}{R_c^3} \tag{14.74}$$

where:

E is the elastic modulus of the steel
b_c is the width of the ring
t_c is the thickness of the ring
R_c is the internal radius of the ring (see Figure 14.21)

Once made, the connection rings were first calibrated to find their actual stiffness, because uncertainties exist in the stiffness estimated by Equation 14.74 due to welding and other manufacturing errors. Table 14.3 lists the actual stiffnesses of the four connection rings along with their linear correlation coefficients. Using the number 2 ring listed in Table 14.3

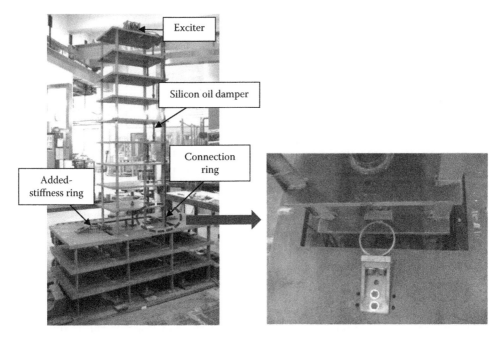

Figure 14.20 Experimental set-up for the scale model of the building complex.

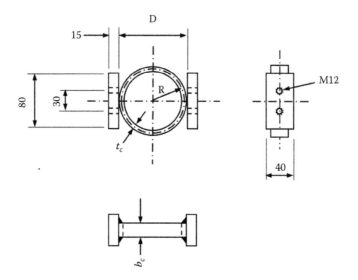

Figure 14.21 Configuration of circular steel rings (all dimensions in millimetres).

Table 14.3 Circular ring calibration results

Ring no.	k_c (N/mm)	Function	Damage ratio	Linear correlation
1	1062.8	Added-stiffness ring	—	0.99993
2	1398.4	Connection ring without damage	0	0.99992
3	1323.4	Connection ring with damage	5%	0.99986
4	1250.5	Connection ring with damage	10%	0.99988
5	1142.2	Connection ring with damage	18%	0.99996

as the connection ring without damage, rings 3, 4 and 5 can be used as connection rings with damage ratios of 5%, 10% and 18%, respectively. The installation details of the connection rings can be observed in Figure 14.20. Moreover, to simulate the added stiffness, another circular steel ring (ring number 1 in Table 14.3) was manufactured to connect the podium structure to the main building, as shown in Figure 14.20.

To accurately obtain the experimental FRFs of the building complex, an eccentric wheel excitation system was developed and installed on the top floor of the main building model, as shown in Figure 14.20. The sensing system consisted of 15 accelerometers, one current eddy sensor and 16 charge amplifiers. By installing or uninstalling the added-stiffness ring and by changing the connection ring, eight cases were considered in this experiment, as listed in Table 14.4. All of these cases were tested using the following procedure after the installation of all of the sensors and the set-up of the measurement system.

1. One connection ring was installed to connect the top floor of the podium structure to the third floor of the main building.
2. The added-stiffness ring was installed or demounted.
3. The top floor of the main building was excited using an electronic hammer to identify the natural frequencies of the building complex and to determine the frequency range for sweep-sine excitation.
4. The proper eccentric wheel was selected and the top floor of the main building excited within a proper frequency range around each of the first 10 modes of

Table 14.4 Test cases

Case no.	Added-stiffness ring	Connection ring no.
1	No	2
2	Yes	2
3	No	3
4	Yes	3
5	No	4
6	Yes	4
7	No	5
8	Yes	5

vibration for the building complex. For each mode of vibration, more than 20 frequency points were selected before and after the corresponding natural frequency. For each frequency point, the excitation lasted more than 2 min before sampling data for the FRFs.

5. The recorded data was analysed to obtain the FRFs of the building complex for each case and the structure parameters were identified using the proposed method expressed by Equation 14.35.

It may be useful to mention here that because the stiffness of the podium structure was much larger than that of the main building, the FRF data around the first 10 natural frequencies of the building complex were used to detect damage to the connection between the main building and the podium structure. For a general structure, FRF data for only a few lower modes of vibration are needed to detect structural damage, such as the first three natural frequencies in the numerical example.

14.8.2 Measured FRFs of the building complex

The hammer impact excitation was first used to obtain the initial natural frequency values of the building complex for each of the eight test cases. The eccentric wheel excitation was then used to precisely identify the natural frequencies, mode shapes and modal damping ratios. The natural frequencies and the first two modal damping ratios of the building complex were identified for all eight test cases; the results are listed in Table 14.5 and Table 14.6, respectively. Figure 14.22 shows examples of the FRFs around the first two natural frequencies for the building complex without damage and with or without added stiffness (cases 1 and 2). The first natural frequency of the building complex with added stiffness changed only 1.2% compared with the one without added stiffness, whereas the second natural frequency exhibited almost no change. Correspondingly, the FRF curve around the first natural frequency displayed a clear shift for the building complex with added stiffness, while the FRF curve around the second natural frequency remained almost unchanged.

The eccentric wheel excitation tests revealed that the FRF curves around each of the first 10 natural frequencies of the building complex could be obtained for each of the eight test cases. Figure 14.23 shows the FRF curves of the top story of the main building for cases 3 and 4. For clarity of expression, the FRF curves around the first and last three natural frequencies are shown in Figure 14.24. The FRF values changed more in some modes of vibration and less in others. The changes in the FRF around the last three natural frequencies were greater than those in the FRF around the first three natural frequencies, indicating that

Table 14.5 First 10 natural frequencies of the building complex (sweep-sine tests)

Case no.	Natural frequencies (Hz)									
	1st	2nd	3rd	4th	5th	6th	7th	8th	9th	10th
1	4.133	10.200	14.167	21.750	26.233	29.433	32.883	40.200	41.467	47.683
2	4.183	10.200	14.517	22.500	26.400	30.383	32.933	40.283	41.933	48.483
3	4.133	10.200	14.117	21.633	26.217	29.417	32.817	40.133	41.450	47.583
4	4.183	10.200	14.550	22.333	26.367	30.383	32.900	40.233	41.933	48.483
5	4.117	10.183	14.050	21.550	26.217	29.367	32.800	40.167	41.400	47.550
6	4.183	10.200	14.500	22.433	26.400	30.300	32.967	40.300	41.917	48.500
7	4.117	10.200	13.983	21.417	26.183	29.183	32.850	40.150	41.383	47.467
8	4.167	10.200	14.467	22.367	26.350	30.167	32.967	40.283	41.883	48.417

Table 14.6 First two modal damping ratios of the building complex (sweep-sine tests)

Case no.	1	2	3	4	5	6	7	8	Average
ζ_1 (%)	1.00	0.90	1.00	1.00	1.01	1.01	1.01	1.01	0.99
ζ_2 (%)	0.65	0.65	0.65	0.65	0.74	0.57	0.66	0.57	0.64

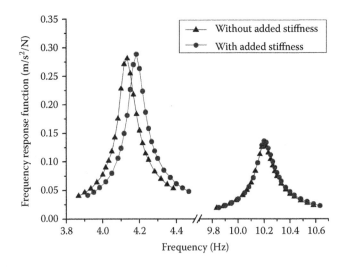

Figure 14.22 FRFs of the first two natural frequencies of the building complex with and without added stiffness (cases 1 and 2).

high-order information may enhance the accuracy of parameter identification and damage detection.

In summary, if the natural frequencies and mode shapes of a building are not changed, the FRF curves of the structure also remain unchanged (provided the structural damping stays the same). If the natural frequencies and mode shapes change slightly, the peaks in the FRF curves shift horizontally and vertically. The horizontal and vertical shifts of the peak then cause a detectable change in the FRF curve around the shifted peak compared with the original FRF curve. Furthermore, if the changes of the FRFs at all the frequency points

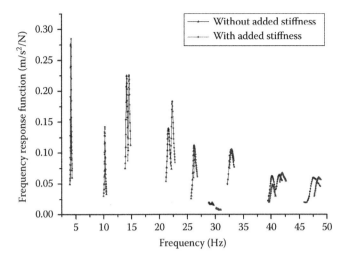

Figure 14.23 FRF curves of the top floor of the main building (cases 3 and 4).

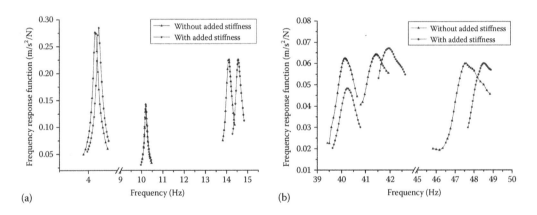

Figure 14.24 FRF curves of the top floor of the main building around the first and last three natural frequencies (cases 3 and 4): (a) the first three natural frequencies, (b) the last three natural frequencies.

around the shifted peak are used, the sensitivity of the FRF curve to the damage can be much higher than that based on either natural frequency or mode shape.

14.8.3 Stiffness identification of the building complex

For the building complex with the connection ring without damage (case 1: the original building complex), the horizontal story stiffness of the main building was estimated to be 1270 N/mm for all 12 stories by calculation, and the horizontal story stiffness of the podium structure was estimated to be 3572 N/mm for all three stories. The stiffness of the connection ring (number 2) was calibrated as 1398.4 N/mm. To identify the actual horizontal story stiffness for the original building complex, the number 2 ring could be used to connect the podium structure to the main building. The natural frequencies and FRFs of the original building complex without added stiffness were first obtained by exciting the top floor of the main building using the eccentric wheel excitation system.

Then, the number 2 ring was installed in the original building complex and excited in the same way as the original building complex without added stiffness, so that another set of the natural frequencies and FRFs of the original building complex with known added stiffness could be obtained. Since the external excitation is only applied on the top floor of the main building in this experiment, it could be considered as a case of single excitation, and the horizontal story stiffness of the model could be identified by using Equation 14.35.

Figure 14.25 shows the FRF curves around the first natural frequency of the original building complex without added stiffness. Figure 14.26 shows the first mode shapes of the original building complex with and without added stiffness. The average horizontal story

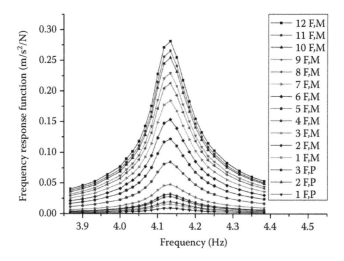

Figure 14.25 FRF curves around the first natural frequency of the original building complex without added stiffness (case 1).

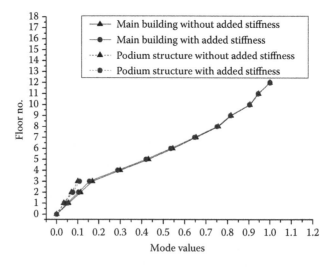

Figure 14.26 First mode shapes of the original building complex with and without added stiffness (cases 1 and 2).

stiffness of the main building was identified as 1247 N/mm compared with a theoretical value of 1270 N/mm, whereas the average horizontal story stiffness of the podium structure was identified as 3648 N/mm compared with a theoretical value of 3572 M/mm. The stiffness of the connection ring (number 2) was identified as 1392.5 N/mm with a 0.42% identification error compared with a calibration value of 1398.4 N/mm.

14.8.4 Structural damage detection

Connection rings 3, 4 and 5 were used to simulate the connection with 5%, 10% and 18% damage, respectively. These three connection rings were installed in the building complex one by one and the stiffness identification procedure described previously was followed. Equation 14.35 was used to identify the horizontal story stiffness of the damaged building, which was compared with that of the original building as given in the previous subsection. The damage to the building complex was thus located and quantified.

The identification results are shown in Figure 14.27 for the cases of 5%, 10% and 18% connection damage. For the 5%, 10% and 18% damage cases, the identified connection damage was 5.36%, 10.58% and 18.32%, respectively. The identified locations and sizes of connection damage were quite close to the real ones. Although less than 2.2% of the

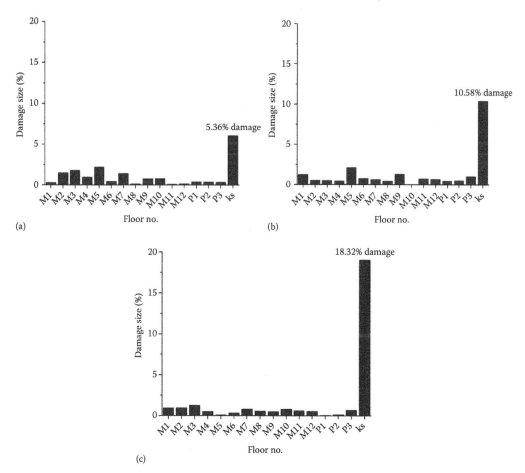

Figure 14.27 Structural damage identification results. (a) 5% connection damage case, (b) 10% connection damage case, (c) 18% connection damage case.

damage was also identified for some stories of the main building in the 5% damage case, the 5.36% connection damage that was identified was a clear indication that an inspection of the connection was required.

NOTATION

$\bar{\mathbf{A}}$	System matrix in the state space equation
b_c, t_c, R_c	Width, thickness and internal radius of the connection ring, respectively
$\mathbf{B}$	The matrix in the state space equation in relation to the control force
$d(t)$	The axial displacement between the two ends of the friction damper
$\dot{d}(t)$	The relative velocity between the two ends of the friction damper
$e(t)$	The slip deformation of the friction damper
$\bar{e}(t)$	The previously cumulated slip deformation of the friction damper
$f^d(t)$	Friction force of the damper
$f^k(t)$	Axial force in the semi-active friction damper
$\mathbf{f}(t)$	The applied force vector
$\mathbf{F}(\omega)$	Force amplitude vector in the frequency domain
$\mathbf{G}'$	Full-state feedback gain matrix
$\bar{\mathbf{H}}$	The influence matrix reflecting the location of the semi-active friction dampers
$\mathbf{H}_o, \mathbf{H}_a$	FRF matrix of the original building and added-stiffness building, respectively
$\mathbf{H}(\omega)$	FRF matrix of acceleration
j	Imaginary unit
J_1, J_3	Peak- and RMS-based inter-story drift ratios, respectively
J_2, J_4	Peak- and RMS-based absolute acceleration responses, respectively
J_5	Peak-based control force
k^d	Spring stiffness of the semi-active friction damper
$\mathbf{K}_a$	The additional stiffness matrix contributed by the semi-active friction dampers that are installed together with brace systems in the building
$\bar{\mathbf{L}}$	The optimal gain of the steady-state Kalman filter
$\mathbf{M}, \mathbf{C}, \mathbf{K}$	The mass, damping and stiffness matrices of the building structure, respectively
n	The number of DOFs
$N^d(t)$	The clamping force of semi-active friction dampers
$N^d_{\max}$	The maximum clamping force
$\mathbf{Q}', \mathbf{R}'$	Weighting matrices for the acceleration responses and the semi-active control forces, respectively
$\bar{S}_{\ddot{x}_g \ddot{x}_g}, \bar{S}_{v_i v_i}$	Power spectral density functions of the stationary white noise of $\ddot{x}_g$ and v_i, respectively
SR	Stiffness ratio of the additional stiffness from the damper brace system to the horizontal story stiffness of the building
$\ddot{\mathbf{X}}(t), \dot{\mathbf{X}}(t), \mathbf{X}(t)$	The structural acceleration, velocity and displacement response vectors, respectively
$\mathbf{T}$	Transformation matrix of size $(3n - 2) \times n$
$u^{\mathrm{LQG}}(t)$	The optimal active control force determined by the LQG controller
$u(t)$	Semi-active control force
$\mathbf{v}$	Measurement noise vector
VRF	Vibration reduction factor
$\ddot{x}_g$	Ground excitation

$X(\omega)$	Displacement response amplitude vector of the building structure in the frequency domain
$\ddot{X}(\omega)$	Acceleration response amplitude vector in the frequency domain
y_{ed}	The regulated response vector
$\hat{z}$	The estimated state vector
z	State vector of the controlled building
Z_{nc}, Z_{co}	Maximum response of a given building floor without control and with control, respectively
α, β	The coefficients of the first-order and the second-order damping ratios in Rayleigh damping assumption
α_o, α_a	Rayleigh damping coefficient α for the cases of the original building and added-stiffness building, respectively
β_o, β_a	Rayleigh damping coefficient β for the cases of the original building and added-stiffness building, respectively
γ^d	Velocity control gain
$\overline{\gamma}_g$	The power spectral density ratio
μ^d	Dynamic friction coefficient
θ_j	The angle between the jth semi-active friction damper and the $(j-1)$th building floor
ω	Frequency of the harmonic external force
ω_o, ω_a	The frequency of the external excitations applied to the original building and added-stiffness building, respectively

REFERENCES

Amini, F., S.A. Mohajeri, and M. Javanbakht. 2015. Semi-active control of isolated and damaged structures using online damage detection. *Smart Mater. Struct.*, 24: 105002.

Chen, B. and Y.L. Xu. 2008. Integrated vibration control and health monitoring of building structures using semi-active friction dampers: Part II. Numerical investigation. *Eng. Struct.*, 30(3), 573–87.

Chen, B., Y.L. Xu, and X. Zhao. 2008. Integrated vibration control and health monitoring of building structures: A time-domain approach. *Smart Struct. Syst.*, 6(7): 811–33.

Chopra, I. 2002. Review of state of art of smart structures and integrated systems. *AIAA J.*, 40(11): 2145–87.

Ding, Y. and S.S. Law. 2011. Integration of structural control and structural evaluation for large scale structural system. In *Proceedings of SPIE 7977: Active and Passive Smart Structures and Integrated Systems*, 2011, San Diego, CA, ed. M.N. Ghasemi-Nejhad, 797724, doi:10.1117/12.880690.

Gattulli, V. and F. Romeo. 2000. Integrated procedure for identification and control of MDOF structures. *J. Eng. Mech.*, 126(7), 730–37.

He, J., Q. Huang, and Y.L. Xu. 2014. Synthesis of vibration control and health monitoring of building structures under unknown excitation. *Smart Mater. Struct.*, 23: 105025.

He, J., Y.L. Xu, S. Zhan, and Q. Huang. 2016. Structural control and health monitoring of building structures with unknown ground excitations: Experimental investigation. *J. Sound Vibr.*, in press. http://dx.doi.org/10.1016/j.jsv.2016.11.035.

Huang, Q., Y.L. Xu, J.C. Li, Z.Q. Su, and H.J. Liu. 2012. Structural damage detection of controlled building structures using frequency response functions. *J. Sound Vibr.*, 331(15): 3476–92.

Hurlebausa, S. and L. Gaul. 2006. Smart structure dynamics. *Mech. Syst. Signal Proc.* 20(2): 255–81.

Karami, K. and F. Amini. 2012. Decreasing the damage in smart structures using integrated online DDA/ISMP and semi-active control. *Smart Mater. Struct.*, 21(10): 105017.

Lei, Y., D.T. Wu, and S.Z. Lin. 2013. Integration of decentralized structural control and the identifi-cation of unknown inputs for tall shear building models under unknown earthquake excitation. *Eng. Struct.*, 52: 306–16.

Lei, Y., H. Zhou, and L.J. Liu. 2014. An on-line integration technique for structural damage detection and active optimal vibration control. *Int. J. Struct. Stab. Dyn.*, 14(5): 1440003.

Lin, C.H., N. Sebastijanovic, H.T.Y. Yang, Q. He, and X.Y. Han. 2012. Adaptive structural control using global vibration sensing and model updating based on local infrared imaging. *Struct. Control Health Monit.*, 19(6): 609–26.

Nagarajaiah, S. (2009). Adaptive passive, semiactive, smart tuned mass dampers: Identification and control using empirical mode decomposition, Hilbert transform, and short-term Fourier trans-form. *Struct. Control Health Monit.*, 16(7–8): 800–841.

Ohtori, Y., R.E. Christenson, B.F. Spencer Jr., and S.J. Dyke. 2004. Benchmark control problems for seismically excited nonlinear buildings. *J. Eng. Mech.*, 130(4): 366–85.

Rao, S.S. and M. Sunar. 1994. Piezoelectricity and its use in disturbance sensing and control of flex-ible structures: A survey. *Appl. Mech. Rev.*, 47(4): 113–23.

Skelton, R.E. 1988. *Dynamic System Control: Linear Systems Analysis and Synthesis.* Wiley: New York.

Spencer Jr., B.F., S.J. Dyke, and H.S. Deoskar. 1998. Benchmark problems in structural control: I. Active mass driver system. *Earthq. Eng. Struct. Dyn.*, 27(11): 1127–47.

Stengel, R.F. 1986. *Stochastic Optimal Control: Theory and Application.* Wiley: New York.

Xu, Y.L. and B. Chen. 2008. Integrated vibration control and health monitoring of building structures using semi-active friction dampers: Part I. Methodology. *Eng. Struct.*, 30(7): 1789–801.

Xu, Y.L., Q. Huang, Y. Xia, and H.J. Liu. 2015. Integration of health monitoring and vibration control for smart building structures with time-varying structural parameters and unknown excitations. *Smart Struct. Syst.*, 15(3): 807–30.

Xu, Y.L., Q. Huang, S. Zhan, Z.Q. Su, and H.J. Liu. 2014. FRF-based structural damage detection of controlled buildings with podium structures: Experimental investigation. *J. Sound Vibr.*, 333(13): 2762–75.

Yang, H.T.Y., J.Z. Shan, C.J. Randall, P.K. Hansma, and W.X. Shi. 2014. Integration of health monitoring and control of building structures during earthquakes. *J. Eng. Mech.*, 140(5): 04014013.

Synthesis of structural health monitoring and vibration control in the time domain

15.1 PREVIEW

The current research on the synthesis of structural health monitoring and vibration control was introduced in Chapter 14. The synthesis of structural health monitoring and vibration control in the frequency domain, based on variations in the frequency response functions between the two states of a building that are created by using semi-active friction dampers, was also discussed in Chapter 14 in relation to system identification, vibration control and damage detection in a systematic way. However, the integrated procedure in the frequency domain introduced in Chapter 14 is not the real-time integration of health monitoring and vibration control of building structures; it requires the semi-active friction dampers to create the two states of a building structure and the external excitations and structural responses to be measured and transformed in the frequency domain. External excitations such as earthquake-induced ground motions are difficult, if not impossible, to measure directly and accurately on site and other semi-active control devices may not be able to create the two states of a building structure. The field measurements, including the input and output measurements, are always first recorded in the time domain. The transformation of data from the time domain to the frequency domain takes time, which may cause a time delay in the structural vibration control. For these reasons, it is necessary to develop a real-time integrated system of structural health monitoring and vibration control for building structures.

This chapter first presents an integrated health monitoring and vibration control system in the time domain in terms of the projection matrix and the extended Kalman filter (EKF) for simultaneous consideration of vibration mitigation and time-invariant parameter identification of building structures without the knowledge of the external excitations (He et al. 2014). The efficiency and accuracy of this approach is then numerically confirmed and experimentally validated via a five-story building structure equipped with magneto-rheological (MR) dampers (He et al. 2016). Taking into consideration that structural damage may occur during an extreme event and the structural parameters of damaged components actually vary with time during the event and to ensure that the control performance does not deteriorate with varying structural parameters, this chapter finally presents a real-time integrated system of health monitoring and vibration control to accurately identify time-varying structural parameters without knowing excitations on the one hand, and to effectively reduce excessive vibrations of the building structure with time-varying structural parameters on the other hand (Xu et al. 2015).

15.2 FORMULATION OF INTEGRATED SYSTEM WITH TIME-INVARIANT PARAMETERS

When a building structure is subjected to external excitations, such as a severe earthquake or strong wind, extensive vibration of the building structure may be induced. For the purpose of vibration mitigation, various control devices, as introduced in Chapter 4, can be employed and some control algorithms, as described in Chapter 9, can be utilised for the determination of control forces. The second-order differential equation of motion of the controlled building structure with n degrees of freedom (DOFs) can be given as follows:

$$\mathbf{M}\ddot{\mathbf{x}}(t) + \mathbf{C}\dot{\mathbf{x}}(t) + \mathbf{K}\mathbf{x}(t) = \boldsymbol{\varphi}^* \mathbf{f}^*(t) + \boldsymbol{\varphi}\mathbf{f}(t) \tag{15.1}$$

where:

M, C and **K**	represent the mass, damping and stiffness matrices of the building structure, respectively
$\ddot{\mathbf{x}}(t)$, $\dot{\mathbf{x}}(t)$ and $\mathbf{x}(t)$	denote structural acceleration, velocity and displacement response vectors, respectively
$\mathbf{f}^*(t)$	is the control force vector
$\mathbf{f}(t)$	is the external excitation vector
$\boldsymbol{\varphi}^*$ and $\boldsymbol{\varphi}$	denote the influence matrices associated with $\mathbf{f}^*(t)$ and $\mathbf{f}(t)$, respectively

Although the vibration control system is used for mitigating structural vibration and trying to keep the structure in a healthy state, it is still inevitable that damage may occur to some structural members due to a severe earthquake or strong wind. If damage does occur to some structural members, the control forces should be adjusted accordingly because the structural parameters of the damaged structural members will be employed in the determination of the control force so as to maintain control efficiency in active, semi-active or hybrid control. The structural parameters of the damaged structural members are assumed to be invariant (constant) in this section while the time-varying structural parameters will be considered in the last section of this chapter. Moreover, the structural state vectors, such as displacement and/or velocity, are often required for the determination of the control force. However, in many situations, it is almost impossible to obtain complete measurements of the structural responses because of the limitation of sensor numbers and the difficulty of sensor installation in some particular locations. The classic EKF method provides a possible way to identify the structural parameters and state vectors of a structure when the control forces and external excitations, shown on the right side of Equation 15.1, are both considered as *known information*. The control forces could be easily obtained by installing force transducers associated with the control devices and thus could be assumed to be known in this study. However, in consideration of practical limitations, the external excitations applied in the controlled building structure, such as earthquakes and winds, are assumed to be unknown in this study. Since the external excitations are not available, the classic EKF method cannot be directly employed. An integrated vibration control and health monitoring approach, based on the projection matrix and the EKF method, is proposed in this study for constant parameter and excitation identification as well as vibration mitigation. The formula of the proposed approach is presented in the following subsections.

15.2.1 Identification of constant structural parameters and excitations

Consider an extended state vector consisting of structural displacement, velocity and structural parameters to be identified as

$$\mathbf{Z}(t) = \begin{bmatrix} \mathbf{x}(t)^T & \dot{\mathbf{x}}(t)^T & \boldsymbol{\theta}^T \end{bmatrix}^T \tag{15.2}$$

in which $\boldsymbol{\theta}$ is an m-unknown parametric vector composed of structural stiffness and damping coefficients. As the unknown parameters are assumed to be constant during the event, that is, $\dot{\boldsymbol{\theta}} = 0$, one can obtain the following nonlinear state equation:

$$\frac{d\mathbf{Z}(t)}{dt} = \begin{bmatrix} \dot{\mathbf{x}}(t) \\ \mathbf{M}^{-1}\left(-\mathbf{C}\dot{\mathbf{x}}(t) - \mathbf{K}\mathbf{x}(t) + \boldsymbol{\varphi}^*\mathbf{f}^*(t) + \boldsymbol{\varphi}\mathbf{f}(t)\right) \\ 0 \end{bmatrix} = \mathbf{u}\left(\mathbf{Z}(t), \mathbf{f}(t), \mathbf{f}^*(t), t\right) \tag{15.3}$$

Let $\hat{\mathbf{Z}}_{k|k}$ and $\hat{\mathbf{f}}_{k|k}$, respectively, be the estimates of $\mathbf{Z}_k$ and $\mathbf{f}_k$ at time $t = k \cdot \Delta t$ with Δt the sampling interval. Equation 15.3 can be linearised with respect to $\hat{\mathbf{Z}}_{k|k}$ and $\hat{\mathbf{f}}_{k|k}$ as follows:

$$\mathbf{u}\left(\mathbf{Z}_k, \mathbf{f}_k, \mathbf{f}_k^*, k\Delta t\right) = \mathbf{u}\left(\hat{\mathbf{Z}}_{k|k}, \hat{\mathbf{f}}_{k|k}, \mathbf{f}_k^*, k\Delta t\right) + \mathbf{U}_{k|k}\left(\mathbf{Z}_k - \hat{\mathbf{Z}}_{k|k}\right) + \mathbf{B}_{k|k}\left(\mathbf{f}_k - \hat{\mathbf{f}}_{k|k}\right) \tag{15.4}$$

in which

$$\mathbf{U}_{k|k} = \left.\frac{\partial \mathbf{u}\left(\mathbf{Z}_k, \mathbf{f}_k, \mathbf{f}_k^*, k\Delta t\right)}{\partial \mathbf{Z}_k}\right|_{\mathbf{Z}=\hat{\mathbf{Z}}_{k|k}, \mathbf{f}_k=\hat{\mathbf{f}}_{k|k}} \tag{15.5}$$

$$\mathbf{B}_{k|k} = \left.\frac{\partial \mathbf{u}\left(\mathbf{Z}_k, \mathbf{f}_k, \mathbf{f}_k^*, k\Delta t\right)}{\partial \mathbf{f}_k}\right|_{\mathbf{Z}=\hat{\mathbf{Z}}_{k|k}, \mathbf{f}_k=\hat{\mathbf{f}}_{k|k}} = \begin{bmatrix} 0 \\ \mathbf{M}^{-1}\boldsymbol{\varphi} \\ 0 \end{bmatrix} \tag{15.6}$$

Substituting $\dfrac{\mathbf{Z}_{k+1} - \mathbf{Z}_k}{\Delta t} = \dfrac{d\mathbf{Z}_k}{dt} = \mathbf{u}\left(\mathbf{Z}_k, \mathbf{f}_k, \mathbf{f}_k^*, k\Delta t\right)$ into the left side of Equation 15.4, one obtains

$$\mathbf{Z}_{k+1} = \mathbf{Z}_k + \Delta t\left[\mathbf{u}\left(\hat{\mathbf{Z}}_{k|k}, \hat{\mathbf{f}}_{k|k}, \mathbf{f}_k^*, k\Delta t\right) + \mathbf{U}_{k|k}\left(\mathbf{Z}_k - \hat{\mathbf{Z}}_{k|k}\right) + \mathbf{B}_{k|k}\left(\mathbf{f}_k - \hat{\mathbf{f}}_{k|k}\right)\right] \tag{15.7}$$

The discretised observation equation associated with Equation 15.1 at time $t = k \cdot \Delta t$ can be described as

$$\mathbf{y}_k = \mathbf{K}\mathbf{x}_k + \mathbf{C}\dot{\mathbf{x}}_k - \boldsymbol{\varphi}\mathbf{f}_k + \mathbf{v}_k = \mathbf{h}(\mathbf{Z}_k) - \boldsymbol{\varphi}\mathbf{f}_k + \mathbf{v}_k \tag{15.8}$$

where:

$\mathbf{y}_k$ is the measurement vector, and is equal to $-\mathbf{M}\ddot{\mathbf{x}}_k + \boldsymbol{\varphi}^* \mathbf{f}_k^*$

$\mathbf{h}(\mathbf{Z}_k)$ is a combination of state vector and is equal to $\mathbf{K}\mathbf{x}_k + \mathbf{C}\dot{\mathbf{x}}_k$

$\mathbf{v}_k$ is the measurement noise vector assumed to be a Gaussian white noise vector with zero mean and a covariance matrix $E[\mathbf{v}_k \mathbf{v}_j^T] = \mathbf{R}_k \delta_{kj}$, where δ_{kj} is the Kroneker delta

It can be found from Equation 15.8 that there are two variable vectors $\mathbf{Z}_k$ and $\mathbf{f}_k$ to be predicted and estimated. Consequently, Equation 15.8 is a multiple regression equation, which means the classic EKF method cannot be directly applied. To transform the multiple regression problem described by Equation 15.8 to a single regression problem, a straightforward way in terms of the projection matrix is proposed here.

From Equation 15.8, one can obtain

$$\boldsymbol{\varphi}\mathbf{f}_k = -\mathbf{y}_k + \mathbf{h}(\mathbf{Z}_k) + \mathbf{v}_k \tag{15.9}$$

By assuming that the number of observed DOFs is larger than the number of excitations, the closest solution of $\mathbf{f}_k$ in Equation 15.9 can then be given by the following equation through least-square estimation (LSE)

$$\mathbf{f}_{k,LS} = \left(\boldsymbol{\varphi}^T \boldsymbol{\varphi}\right)^{-1} \boldsymbol{\varphi}^T \left[-\mathbf{y}_k + \mathbf{h}(\mathbf{Z}_k) + \mathbf{v}_k\right] \tag{15.10}$$

The error of the solution from Equation 15.10 is given by

$$\begin{aligned}
\text{err} &= \boldsymbol{\varphi}\mathbf{f}_k - \boldsymbol{\varphi}\mathbf{f}_{k,LS} \\
&= \left[-\mathbf{y}_k + \mathbf{h}(\mathbf{Z}_k) + \mathbf{v}_k\right] - \boldsymbol{\varphi}\left(\boldsymbol{\varphi}^T \boldsymbol{\varphi}\right)^{-1} \boldsymbol{\varphi}^T \left[-\mathbf{y}_k + \mathbf{h}(\mathbf{Z}_k) + \mathbf{v}_k\right] \\
&= \left(\mathbf{I}_n - \boldsymbol{\varphi}\left(\boldsymbol{\varphi}^T \boldsymbol{\varphi}\right)^{-1} \boldsymbol{\varphi}^T\right)\left[-\mathbf{y}_k + \mathbf{h}(\mathbf{Z}_k) + \mathbf{v}_k\right]
\end{aligned} \tag{15.11}$$

where:

$\mathbf{I}_n$ is the $n \times n$ identity matrix

The matrix $\boldsymbol{\varphi}(\boldsymbol{\varphi}^T\boldsymbol{\varphi})^{-1}\boldsymbol{\varphi}^T$ is the projection matrix that projects the vector $[-\mathbf{y}_k + \mathbf{h}(\mathbf{Z}_k) + \mathbf{v}_k]$ onto the space spanned by the columns of $\boldsymbol{\varphi}$.

As a limit, the error in Equation 15.11 tends to be zero, leading to

$$\boldsymbol{\Phi}\mathbf{y}_k = \boldsymbol{\Phi}\mathbf{h}(\mathbf{Z}_k) + \boldsymbol{\Phi}\mathbf{v}_k \tag{15.12}$$

where $\boldsymbol{\Phi} = \mathbf{I}_n - \boldsymbol{\varphi}(\boldsymbol{\varphi}^T\boldsymbol{\varphi})^{-1}\boldsymbol{\varphi}^T$ for simplicity of presentation.

It is noted that the projection matrix and naturally the matrix $\boldsymbol{\Phi}$ has two properties: (1) $\boldsymbol{\Phi}$ is a symmetric matrix, that is, $\boldsymbol{\Phi}^T = \boldsymbol{\Phi}$; and (2) $\boldsymbol{\Phi}^2 = \boldsymbol{\Phi}$.

It can be found from Equation 15.12 that the multiple regression equation expressed by Equation 15.8 is transformed into a simple regression equation.

It is obvious from Equation 15.10 that if the estimate vector $\hat{Z}_{k|k}$ of Z_k at the kth time step is obtained, the unknown excitation could be accordingly determined as

$$\hat{f}_{k|k} = \left(\varphi^T \varphi\right)^{-1} \varphi^T \left[-y_k + h\left(\hat{Z}_{k|k}\right)\right] \tag{15.13}$$

Let $\hat{Z}_{k+1|k}$ be the priori estimate of state Z_{k+1}, and linearise $h(Z_{k+1})$ with respect to $\hat{Z}_{k+1|k}$:

$$h\left(Z_{k+1}\right) = h\left(\hat{Z}_{k+1|k}\right) + H_{k+1|k}\left(Z_{k+1} - \hat{Z}_{k+1|k}\right) \tag{15.14}$$

in which

$$H_{k+1|k} = \left.\frac{\partial h\left(Z_{k+1}\right)}{\partial Z_{k+1}}\right|_{Z_{k+1} = \hat{Z}_{k+1|k}} \tag{15.15}$$

The priori estimation state $\hat{Z}_{k+1|k}$ could be given according to first-order Taylor expansions:

$$\hat{Z}_{k+1|k} = \hat{Z}_{k|k} + \Delta t \left[u\left(\hat{Z}_{k|k}, \hat{f}_{k|k}, f_k^*, k\Delta t\right)\right] \tag{15.16}$$

It should be noted that integrating the actual nonlinear equation, that is, Equation 15.3, forward at each sampling interval is usually implemented in the EKF approach to determine $\hat{Z}_{k+1|k}$ for the purpose of improving the accuracy of the estimate. The priori estimation error $\varepsilon_{k+1|k}$ of the unknown state vector Z_{k+1} at time $t = (k+1)\Delta t$ can then be obtained by the combination of Equation 15.7:

$$\varepsilon_{k+1|k} = Z_{k+1} - \hat{Z}_{k+1|k} = \left(I + \Delta t U_{k|k}\right)\left(Z_k - \hat{Z}_{k|k}\right) + \Delta t B_{k|k}\left(f_k - \hat{f}_{k|k}\right) \tag{15.17}$$

in which $U_{k|k}$ and $B_{k|k}$ are defined in Equations 15.5 and 15.6, respectively.

By combining Equations 15.6, 15.10 and 15.13, the second term on the right side of Equation 15.17 can be transformed as

$$\Delta t B_{k|k}\left(f_k - \hat{f}_{k|k}\right) = \Delta t \left[M^{-1}\varphi\left[\left(\varphi^T\varphi\right)^{-1}\varphi^T\left[-y_k + h\left(Z_k\right) + v_k\right] - \left(\varphi^T\varphi\right)^{-1}\varphi^T\left[-y_k + h\left(\hat{Z}_{k|k}\right)\right]\right]\right]$$

$$= \Delta t \left[M^{-1}\left(I - \Phi\right)\left[H_{k|k}\left(Z_k - \hat{Z}_{k|k}\right) + v_k\right]\right] \tag{15.18}$$

in which $H_{k|k}$ can be calculated according to Equation 15.15 when $Z_k = \hat{Z}_{k|k}$. Substituting Equation 15.18 into Equation 15.17, one obtains

$$\varepsilon_{k+1|k} = \left(I + \Delta t U_{k|k} + \Delta t \left[M^{-1}(I - \Phi)H_{k|k} \atop 0 \right]\right)\left(Z_k - \hat{Z}_{k|k}\right) + \left[0 \atop \Delta t M^{-1}(I - \Phi) \atop 0 \right] v_k$$

$$= \left(I + \Delta t U_{k|k} + \Delta t \left[0 \atop M^{-1}(I - \Phi)H_{k|k} \atop 0 \right]\right)\varepsilon_{k|k} + \left[0 \atop \Delta t M^{-1}(I - \Phi) \atop 0 \right] v_k \qquad (15.19)$$

$$= A_1 \varepsilon_{k|k} + A_2 v_k$$

where

$$A_1 = \left(I + \Delta t U_{k|k} + \Delta t \left[0 \atop M^{-1}(I - \Phi)H_{k|k} \atop 0 \right]\right); \quad A_2 = \left[0 \atop \Delta t M^{-1}(I - \Phi) \atop 0 \right] \qquad (15.20)$$

The priori estimation error covariance can then be calculated as

$$P_{k+1|k} = E\left(\varepsilon_{k+1|k}\varepsilon_{k+1|k}^T\right) = A_1 P_{k|k} A_1^T + A_2 R_k A_2^T \qquad (15.21)$$

Based on Equation 15.12, the recursive solution for the posteriori estimation state can be given as

$$\hat{Z}_{k+1|k+1} = \hat{Z}_{k+1|k} + G_{k+1}\left[\Phi y_{k+1} - \Phi h\left(\hat{Z}_{k+1|k}\right)\right] \qquad (15.22)$$

in which G_{k+1} denotes the EKF gain matrix at the $(k+1)$th time step. The posteriori estimation error $\varepsilon_{k+1|k+1}$ of the unknown state vector Z_{k+1} at time $t = (k+1)\Delta t$ can then be calculated as

$$\varepsilon_{k+1|k+1} = Z_{k+1} - \hat{Z}_{k+1|k+1}$$

$$= Z_{k+1} - \hat{Z}_{k+1|k} - G_{k+1}\left[\Phi y_{k+1} - \Phi h\left(\hat{Z}_{k+1|k}\right)\right]$$

$$= Z_{k+1} - \hat{Z}_{k+1|k} - G_{k+1}\left[\Phi h\left(Z_{k+1}\right) + \Phi v_{k+1} - \Phi h\left(\hat{Z}_{k+1|k}\right)\right]$$

$$= Z_{k+1} - \hat{Z}_{k+1|k} - G_{k+1}\left[\Phi\left(h\left(\hat{Z}_{k+1|k}\right) + H_{k+1|k}\left(Z_{k+1} - \hat{Z}_{k+1|k}\right)\right) + \Phi v_{k+1} - \Phi h\left(\hat{Z}_{k+1|k}\right)\right]$$

$$= \left(I - G_{k+1}\Phi H_{k+1|k}\right)\left(Z_{k+1} - \hat{Z}_{k+1|k}\right) - G_{k+1}\Phi v_{k+1}$$

$$= \left(I - G_{k+1}\Phi H_{k+1|k}\right)\varepsilon_{k+1|k} - G_{k+1}\Phi v_{k+1}$$

$$\qquad (15.23)$$

Similarly, the posteriori estimation error covariance can then be found as

$$P_{k+1|k+1} = E\left(\varepsilon_{k+1|k+1}\varepsilon_{k+1|k+1}^T\right)$$

$$= \left(I - G_{k+1}\Phi H_{k+1|k}\right)P_{k+1|k}\left(I - G_{k+1}\Phi H_{k+1|k}\right)^T + G_{k+1}\Phi R_{k+1}\Phi G_{k+1}^T$$

(15.24)

To obtain the optimal value of the gain matrix G_{k+1} that can minimise the estimation error covariance $P_{k+1|k+1}$ at time $t = (k+1)\Delta t$, the differentiation of $P_{k+1|k+1}$ in Equation 15.24 with respect to G_{k+1} produces

$$\partial P_{k+1|k+1}\Big/\partial G_{k+1} = 2G_{k+1}\Phi\left(H_{k+1|k}P_{k+1|k}H_{k+1|k}^T + R_{k+1}\right)\Phi - 2P_{k+1|k}H_{k+1|k}^T\Phi \qquad (15.25)$$

By setting the value of the partial derivative to zero, one can obtain

$$G_{k+1} = P_{k+1|k}H_{k+1|k}^T\Phi\left(\Phi\left(H_{k+1|k}P_{k+1|k}H_{k+1|k}^T + R_{k+1}\right)\Phi\right)^{-1} \qquad (15.26)$$

It can be found that the corresponding two sets of equations for EKF with unknown excitations, that is, time update equations, including Equations 15.16 and 15.21, as well as measurement update equations, including Equations 15.22, 15.24 and 15.26, are established. The unknown loadings at the current time step can then be identified according to Equation 15.13 by using the state vectors estimated through these two sets of equation. Notably, if the excitations are all assumed to be known, the matrix Φ would be equal to an identity matrix I and the aforementioned two sets of equations reduce to the classical EKF method.

The presented algorithm can identify the unknown external excitations and time-invariant structural parameters simultaneously, and the flowchart of the proposed algorithm for the identification is shown in Figure 15.1.

15.2.2 Structural vibration control with identified time-invariant parameters

In this study, vibration control is also simultaneously considered for a building structure. MR dampers rather than semi-active friction dampers are used in this instance because MR dampers are more appropriate for large civil structures. A variety of semi-active control strategies for adjusting the properties of MR dampers for the purpose of seismic control have been proposed and investigated (Dyke et al. 1996; Dyke and Spencer 1997; Jansen and Dyke 2000; Xu et al. 2000; Ou 2003). Here, a switching control algorithm is considered and can be expressed as follows (Ou, 2003):

$$u_d = \begin{cases} c_d\dot{e} + f_d^{\max}\,\mathrm{sgn}(\dot{e}) & (u\dot{e} < 0) \\ c_d\dot{e} + f_d^{\min}\,\mathrm{sgn}(\dot{e}) & (u\dot{e} \geq 0) \end{cases} \qquad (15.27)$$

where:

u_d is the MR damper force

c_d is the damping coefficient determined by the viscosity of MR fluid

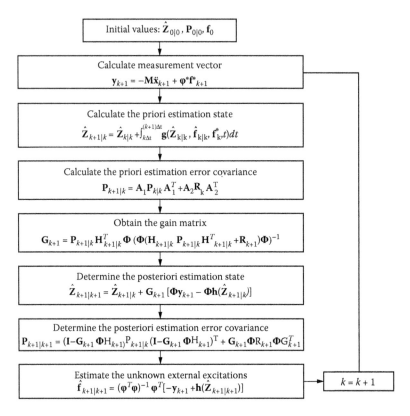

Figure 15.1 Flowchart of the proposed method for identifying time-invariant structural parameters and unknown excitations.

$f_d^{\max}$ and $f_d^{\min}$ denote the maximum value and minimum value of the frictional forces which are related to the yielding shear stress and can be achieved by adjusting the applied field respectively

$\dot{e}$ is the velocity of the MR damper

u is the optimal control force determined by the linear quadratic Gaussian (LQG) control algorithm

In case the optimal control force and the structural motion are in opposite directions, the maximum level of the current is applied to obtain the maximum frictional force $f_d^{\max}$ for the purpose of mitigating vibration as much as possible. Otherwise, the commanded current is set to zero and thus the minimum frictional force $f_d^{\min}$ is achieved.

Since the structural parameters and the state vectors of a structure are estimated in Section 15.2.1, they can be used for determining the desirable optimal control force u in the LQG control algorithm:

$$\mathbf{u} = -\bar{\mathbf{R}}^{-1}\bar{\mathbf{B}}^T\bar{\mathbf{P}}\{\bar{\mathbf{X}}\} \tag{15.28}$$

where $\bar{\mathbf{P}}$ is the solution of the classical Riccati equation:

$$\overline{P}\overline{B}^T\overline{R}^{-1}\overline{B}\overline{P} - \overline{A}^T\overline{P} - \overline{P}\overline{A} - \overline{Q} = 0 \tag{15.29}$$

in which $\overline{Q}$ and $\overline{R}$ are the weighting matrix for the structural response and the control force, respectively.

As mentioned before, the estimated state vectors are used for the determination of the control force, leading to the matrices $\overline{A}$ and $\overline{B}$ as well as the vector $\overline{X}$:

$$\overline{A} = \begin{bmatrix} 0 & I \\ -M^{-1}K_{id} & -M^{-1}C_{id} \end{bmatrix}; \quad \overline{B} = \begin{bmatrix} 0 \\ M^{-1}\varphi^* \end{bmatrix}; \quad \overline{X} = \begin{Bmatrix} x_{id} \\ \dot{x}_{id} \end{Bmatrix} \tag{15.30}$$

where:

K_{id} and C_{id} denote the identified stiffness and damping matrix

x_{id} and $\dot{x}_{id}$ denote the estimated structural displacement and velocity

The flowchart of the proposed semi-active vibration control using MR dampers is shown in Figure 15.2.

15.2.3 Implementation procedure of integrated system with time-invariant parameters

Based on the equations presented in the previous two subsections, the integrated procedure can be implemented for time-invariant structural parameter identification, excitation identification and vibration control of a building structure step-by-step as follows:

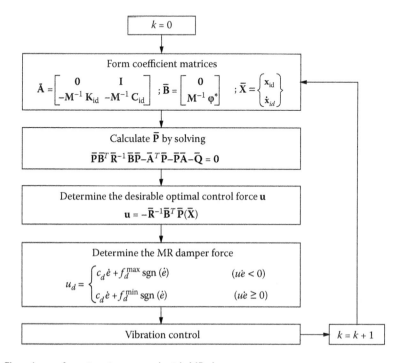

Figure 15.2 Flowchart of semi-active control with MR damper.

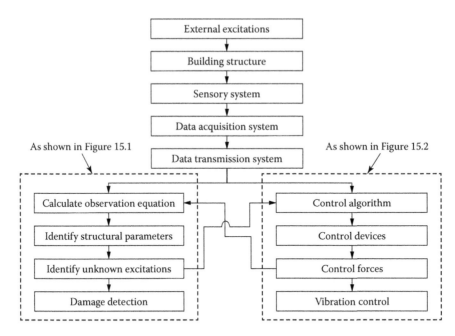

Figure 15.3 Schematic diagram of the implementation of the integrated system with time-invariant parameters.

Step 1: Calculate the observation equation based on the measured acceleration responses and MR damper forces.

Step 2: Determine the priori estimation state $\hat{Z}_{k+1|k}$ according to Equation 15.16.

Step 3: Determine the priori estimation error covariance $P_{k+1|k}$ using Equation 15.21.

Step 4: Obtain the gain matrix G_{k+1} according to Equation 15.26.

Step 5: Obtain the posteriori estimation state $\hat{Z}_{k+1|k+1}$ according to Equation 15.22.

Step 6: Update the posteriori estimation error covariance $P_{k+1|k+1}$ by Equation 15.24.

Step 7: Estimate the unknown excitations $\hat{f}_{k+1|k+1}$ according to Equation 15.13.

Step 8: Form the coefficient matrices according to Equation 15.30 with the aid of the estimated state $\hat{Z}_{k+1|k+1}$ determined in Step 5.

Step 9: Calculate $\bar{P}$ by solving Equation 15.29 and determine the optimal control force according to Equation 15.28.

Step 10: Determine the MR damper force according to Equation 15.27.

It is clear that in the proposed integrated approach, the identified structural parameters and responses are employed in the control devices and control algorithm on the one hand, and the measured control forces are used for parameter and excitation identification on the other hand. A schematic diagram for the implementation of such integrated health monitoring and vibration control system for a building structure is shown in Figure 15.3.

15.3 NUMERICAL INVESTIGATION OF INTEGRATED SYSTEM WITH TIME-INVARIANT PARAMETERS

To investigate the feasibility and reliability of the proposed integrated vibration control and health monitoring system with time-invariant parameters, a five-story shear building structure equipped with MR dampers is considered and shown in Figure 15.4. The mass

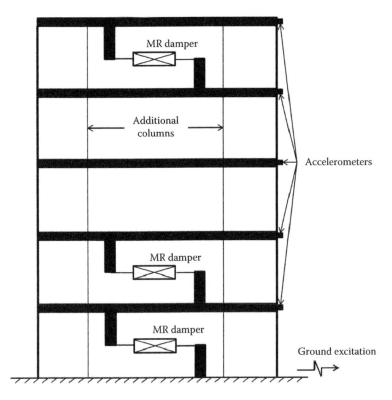

Figure 15.4 Five-story shear building structure with MR dampers.

and stiffness coefficients of each floor are, respectively, set as 60 kg and 2.6×10^5 N/m. The Rayleigh damping assumption with the proportional coefficient of $\alpha = 0.2$ s^{-1} for the mass matrix and $\beta = 2.5 \times 10^{-4}$ s for the stiffness matrix is used to construct the damping matrix. Three MR dampers are employed and, respectively, installed on the first, second and fifth floor. The parameters for the MR dampers are set as $c_d = 5.5$ N·s/m, $f_d^{max} = 1.29 \times 10^3$ N and $f_d^{min} = 251.5$ N. The structural damage is simulated by removing the additional columns from the original building model, as shown in Figure 15.4. In this numerical investigation, four damage scenarios are considered:

Case 1:   One additional column is removed from the first floor (10% damage occurs on the first floor)

Case 2:   Two additional columns are removed from the first floor (20% damage occurs on the first floor)

Case 3:   Two additional columns on the first floor and one additional column on the third floor are removed (i.e. 20% damage occurs on the first floor and 10% damage occurs on the third floor)

Case 4:   The additional columns on the first floor and third floor are all removed (i.e. 20% damage occurs on the first floor and third floor).

In this numerical investigation, the controlled building structure is subjected to the scaled Kobe earthquake with a peak acceleration of 0.84 m/s². The corresponding structural responses are determined by the state-space method with a time interval of 0.001 s. The acceleration of each floor and the three MR damper forces are measured and known for damage detection and vibration control. For practical consideration, all the structural response measurements are simulated by the theoretically computed responses superimposed with the white noise time history of a 2% noise-to-signal ratio in terms of root mean square (RMS).

The influence matrix $\boldsymbol{\varphi}$ associated with the unknown ground excitation is $\boldsymbol{\varphi} = -\mathbf{M} \cdot [1\ 1\ 1\ 1\ 1]^T$. The unknown quantities to be identified are (1) the extended state vector $\mathbf{Z} = \left[x_1, ..., x_5, \dot{x}_1, ..., \dot{x}_5, k_1, ..., k_5, \alpha, \beta \right]^T$ where x_i, $\dot{x}_i$ and k_i denote the displacement, velocity and stiffness of the ith floor, respectively; and (2) the unknown ground acceleration $\ddot{x}_g(t)$

The initial values of the stiffness, α and β are set as 1.2×10^5 N/m, 0.01 s^{-1} and 1×10^{-4} s, respectively. The covariance matrix of the measurement noise is chosen as $\mathbf{R} = 0.1 \times \mathbf{I}$.

By using the proposed integrated approach, the unknown structural parameters in the aforementioned four damage cases can be identified and the results are listed in Table 15.1. It can be seen from Table 15.1 that the values of the identified parameters are in excellent agreement with the corresponding actual values. Taking Case 4 as an example, the identified parameters during the earthquake are plotted in Figure 15.5 as solid lines, whereas the actual ones are shown as dashed lines. Only the identified results of k_1, k_3, k_5 and α are plotted in Figure 15.5 for demonstrative purposes. Similar results can also be obtained for the rest of the parameters. For ease of comparison, the irregular scale for the x-axis is selected in Figure 15.5. It is obvious that the identified structural parameters can be promptly and stably converged to the actual ones, which means that the proposed approach is capable of identifying structural parameters accurately and quickly.

The unknown ground excitation can also be identified by the proposed integrated approach. The time history of identified seismic input is plotted in Figure 15.6 as a solid curve, whereas the dashed curve is the actual one. Only the time segment from 4 to 5 s is shown in Figure 15.6 for clarity of comparison. It can be seen that the ground excitation can be identified accurately.

Furthermore, based on the proposed integrated approach and the identified structural parameters and excitation, the structural vibration can be simultaneously reduced with the aid of MR dampers. To demonstrate the control performance, the maximum acceleration and displacement responses of the building model in the cases with control and without control are given in Table 15.2 for comparison. As shown in Table 15.2, s_i and a_i ($i = 1, 2, ..., 5$) are the peak displacement and acceleration of the ith floor, respectively. It can be seen from Table 15.2 that the structural vibration is significantly reduced. Moreover, a comparison of the time histories of the displacement and acceleration responses on the top floor is given in Figure 15.7. Only the building model in Case 4 is plotted as an example. It can be seen in

Table 15.1 Comparison of the identified structural parameters

	Case 1		Case 2		Case 3		Case 4	
	Identified	Relative error (%)	Identified	Relative error (%)	Identified	Relative error (%)	Identified	Relative error (%)
k_1	2.339×10^5	0.04	2.079×10^5	0.05	2.082×10^5	0.10	2.081×10^5	0.05
k_2	2.599×10^5	0.03	2.598×10^5	0.07	2.602×10^5	0.07	2.597×10^5	0.12
k_3	2.598×10^5	0.07	2.601×10^5	0.03	2.342×10^5	0.08	2.079×10^5	0.05
k_4	2.602×10^5	0.07	2.597×10^5	0.12	2.603×10^5	0.12	2.602×10^5	0.07
k_5	2.603×10^5	0.12	2.602×10^5	0.07	2.601×10^5	0.03	2.598×10^5	0.07
α	0.201	0.5	0.201	0.5	0.199	0.5	0.201	0.5
β	2.499×10^{-4}	0.04	2.498×10^{-4}	0.08	2.499×10^{-4}	0.04	2.501×10^{-4}	0.04

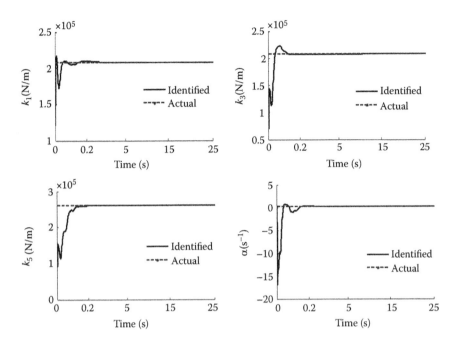

Figure 15.5 Comparison of the identified structural parameters.

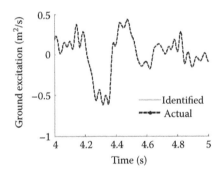

Figure 15.6 Comparison of identified ground excitation.

Table 15.2 Comparison of the peak responses of the building model

		s_i (mm)					a_i (m/s^2)				
		s_1	s_2	s_3	s_4	s_5	a_1	a_2	a_3	a_4	a_5
Case 1	Without control	4.01	7.26	9.96	11.91	12.92	1.48	2.44	3.31	3.94	4.48
	With control	2.13	3.87	5.28	6.29	6.82	0.91	1.52	1.92	2.26	2.55
Case 2	Without control	4.53	7.71	10.23	11.95	12.83	1.89	2.92	3.65	4.16	4.30
	With control	2.15	3.68	4.91	5.75	6.19	1.02	1.64	2.01	2.32	2.45
Case 3	Without control	4.41	7.51	10.26	11.96	12.79	1.92	2.96	3.58	4.04	4.16
	With control	2.05	3.51	4.81	5.60	6.01	1.01	1.63	2.05	2.35	2.49
Case 4	Without control	4.05	6.95	9.88	11.52	12.37	1.74	2.73	3.31	3.87	4.16
	With control	1.89	3.25	4.63	5.41	5.78	0.93	1.56	2.01	2.38	2.53

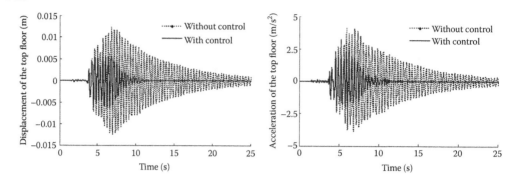

Figure 15.7 Comparison of the time histories of structural responses.

Figure 15.7 that the structural vibration is substantially attenuated. Similar results can be obtained from the other cases.

The results obtained from this numerical example demonstrate that the proposed integrated vibration control and health monitoring approach can not only satisfactorily identify the time-invariant structural parameters and unknown ground excitation but also simultaneously suppress the excessive vibration in an effective way.

15.4 EXPERIMENTAL INVESTIGATION OF INTEGRATED SYSTEM WITH TIME-INVARIANT PARAMETERS

15.4.1 Experimental setup

To experimentally investigate the performance of the proposed integrated system for damage detection and vibration control of a building structure, a five-story building model was designed and built in the Structural Dynamics Laboratory of The Hong Kong Polytechnic University, as shown in Figure 15.8. The building model consisted of five rigid plates of $850 \times 500 \times 16$ mm and four equal-sized rectangular columns of cross section 50×6 mm. The total height of this building model was 1750 mm, with the height of each story being identical. The cross section of the column was arranged in such a way that the stiffness of the building model in the y-direction was much larger than that in the x-direction. Each steel floor plate was highly rigid in the horizontal direction compared with the columns, which led to a shear-type deformation. Four additional columns of the cross section of 10×6 mm were added in the building model to simulate damage when they were symmetrically cut off in some stories. Each column was embedded into and welded to the plates to ensure that rigid joints formed at the connections. All of the columns on the first floor were eventually welded to a thick steel plate that was in turn bolted firmly to a shaking table. A series of silicon oil dampers were designed and installed between each of the adjacent floor plates to increase the structural damping. As can be seen from Figure 15.8, the lumped mass for each floor was composed of the masses of the plate, columns, oil damper, MR damper and the auxiliary for connecting the dampers to the plate. With the measurement of each component, the mass matrix for the building model can be determined as $\mathbf{M} = \mathrm{diag}\,[67.955, 61.918, 56.837, 63.079, 59.922]$ (unit: kg) in which diag[·] denotes a diagonal matrix. The Rayleigh damping assumption was used to construct the damping matrix in this study.

The building model was fixed on a shaking table of 3×3 m, which was built by MTS Corporation. During the tests, the structural acceleration response of each floor was

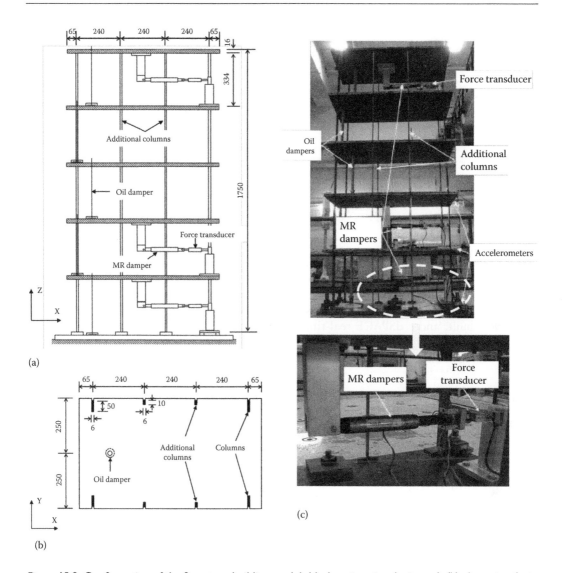

Figure 15.8 Configuration of the five-story building model: (a) elevation view (unit: mm), (b) plane view (unit: mm) and (c) photograph on the model.

measured by a piezoelectric accelerometer with a sampling frequency of 1000 Hz. The ground acceleration induced by the shaking table was also measured by the accelerometer and used for a comparison with the identified one. Moreover, the force transducers were placed in a series with the MR dampers to measure the control forces, which were then used for parameter and excitation identification. The displacement response of each floor was also obtained but only utilised for assessing the control performance.

Three MR dampers (RD-1097-01) manufactured by the Lord Corporation, United States, were, respectively, installed on the first, second and fifth floor, and used to provide the control forces to the building model. To achieve the best performance of the MR damper, the Rheonetic Wonder Box device controller kit (RD-3002-03) designed by the Lord Corporation was used along with the MR damper. The output current supplied to the MR damper by the Rheonetic Wonder Box device depended on the command voltage

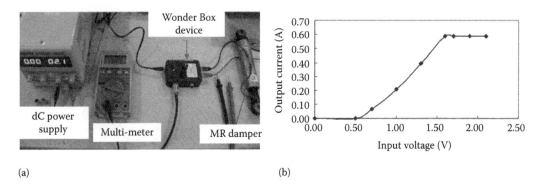

(a)   (b)

Figure 15.9 Calibration of Wonder Box device: (a) calibration setup of Wonder Box and (b) output voltages vs. input voltages.

(the input voltage) from the computer. Through the calibration of the Wonder Box device (see Figure 15.9), it was found that the output current of the device could reach its saturation value when the input voltages were larger than a certain value (about 1.8 V). Therefore, the maximum and minimum input voltages were set to be 1.8 and 0 V, respectively. A Simulink model was built, and a dSPACE real-time simulator and control system was employed to process and analyse the aforementioned measurements and provide the desired control signals according to Equation 15.27.

15.4.2 Damage scenarios

The static tests of the building model, as shown in Figure 5.10, were first conducted to determine the structural stiffness in the cases without and with additional columns. As the number of the calibrated mass block increased, the increased deformation of the building model in the x-direction was recorded by dial gauges and used for the determination of the structural stiffness. The measured results are given in Table 15.3.

Since structural damage usually results in a reduction in structural stiffness, the damage was simulated in this study by symmetrically cutting off the additional columns in certain

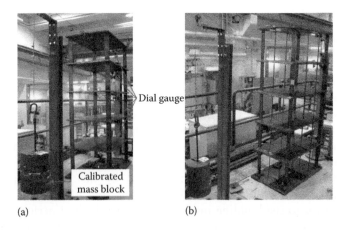

(a)   (b)

Figure 15.10 Static tests of the building model: (a) without additional columns and (b) with four additional columns.

Table 15.3 Measured stiffness of the building model

	k_1 (N/m)	k_2 (N/m)	k_3 (N/m)	k_4 (N/m)	k_5 (N/m)
Without additional columns	2.181×10^5	2.085×10^5	2.107×10^5	2.166×10^5	2.213×10^5
With two additional columns	2.418×10^5	2.323×10^5	2.363×10^5	2.412×10^5	2.483×10^5
With four additional columns	2.657×10^5	2.576×10^5	2.619×10^5	2.639×10^5	2.713×10^5

Figure 15.11 Damage pattern in the experiment.

stories as shown in Figure 15.11. One advantage of such a damage pattern is that the mass remains unchanged while the structural stiffness is reduced. In this experimental study, four damage scenarios were considered. In Case 1, two additional columns in the symmetrical position of the first floor were cut off; in Case 2, the remaining two additional columns (i.e. all of the four additional columns) in the first floor were cut off; in Case 3, two additional columns in the third floor were cut off; in Case 4, four additional columns in the third floor were cut off.

In each of the aforementioned four damage scenarios, hammer tests were carried out to capture the properties of the building model when the MR dampers were removed, as shown in Figure 15.12. The impact force was applied on the top floor of the building model and the acceleration response of each floor was measured. The modal parameters of the building, including its natural frequencies, damping ratios and modal shapes, were identified based on the analysis of the corresponding frequency response functions. Since the mass of the MR damper was 0.525 kg only, the loss of the total mass of each floor was relatively small as compared with the original building model. Moreover, the stiffness provided by the MR damper was quite limited. For these reasons, the variations in the natural frequencies caused by the removal of the MR dampers could be small. The first two natural frequencies and damping ratios of the building model in the four damage cases are shown in Table 15.4.

15.4.3 Implementation of identification and control algorithms

After the MR dampers were installed in the building model, the integrated system was established and the shaking table tests could be conducted for the aforementioned four damage cases. The dSPACE real-time simulator and control system was employed to realise the integration of damage detection and vibration control of the building model as schematically

Figure 15.12 Hammer tests of the building model without MR dampers.

Table 15.4 Properties of the model determined by the hammer test

	Natural frequencies		Damping ratios	
	f_1 (Hz)	f_2 (Hz)	ζ_1 (%)	ζ_2 (%)
Case 1	2.907	8.533	0.86	1.03
Case 2	2.845	8.408	0.89	1.08
Case 3	2.813	8.346	1.01	1.09
Case 4	2.782	8.283	1.09	1.15

shown in Figure 15.13. It can be seen that the dSPACE real-time simulator and control system consists of three processing blocks: (1) DS2003 (A/D) block that is used for receiving analogue signals from sensors and transforming them to digital signals; (2) DS2102 (D/A) block that is used for transforming digital signals to analogue signals and then sending them to the control devices; and (3) the algorithms written using the MATLAB/Simulink block program, which are used to identify the structural parameters and unknown excitations, as well as to create the corresponding control forces.

Use of the dSPACE central processing unit (CPU) and access to its memory can be realised by means of the main program ControlDesk of dSPACE, which offers an automatic implementation of the MATLAB/Simulink block program on the host computer via Real-Time Interface (RTI) and provides a real-time interactive data display and visualisation. Some details on how to build the Simulink model on the basis of the associated block programs are shown in Figure 15.14. It should be pointed out that each block marked by the bold line stands for the corresponding subsystem as shown in Figure 15.14b and c and it is built according to the equations given in Section 15.2.

After implementing the proposed identification and control algorithms, the shaking table tests were conducted. Two sets of earthquakes, that is, the Kobe earthquake and the Northridge earthquake, were considered in this experimental investigation. The structural parameter and earthquake identification results and the control performance under these two earthquakes are given in the following subsections.

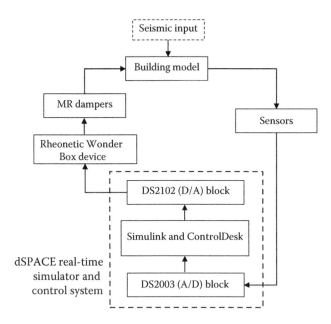

Figure 15.13 Schematic diagram of the integrated system.

15.4.4 Damage detection and vibration control under Kobe earthquake

The scaled Kobe earthquake with a peak acceleration of 0.84 m/s^2 was employed as the seismic input. The initial values for the identification of the structural stiffness and Rayleigh damping coefficients were assigned as 1.2×10^5 N/m, 0.01 s^{-1} and 1×10^{-4} s, respectively. The building model was static before the test and thus the initial values for the ground excitation and structural responses were all assumed to be zero. Since the seismic input was used and assumed to be unknown, the excitation influence matrix φ in Equation 15.1 was equal to $-M \cdot [1\ 1\ 1\ 1\ 1]^T$. As mentioned before, the structural acceleration responses and the MR damper forces were all measured and thus the observation equation could be calculated according to Equation 15.8 for the real-line parameter and excitation identification. The estimated structural parameters and the system states were then used for the vibration control based on the procedures described in Section 15.2.3. The aforementioned four damage cases were considered, and the identified structural stiffness under the Kobe earthquake are given in Table 15.5. By comparing the values of the identified stiffness with those obtained from the static tests, it can be found from Table 15.5 that the proposed approach is capable of identifying the structural stiffness satisfactorily. Taking Case 4 as an example, the identified stiffness during the earthquake is presented in Figure 15.15 as a solid line, whereas that determined by the static test is shown as a dashed line. It can be seen that the identified results from the proposed integrated approach can be stably converged to the measured ones.

By using the proposed approach, the Rayleigh damping coefficients can also be estimated. However, the values of Rayleigh damping coefficients are usually small and not easily assessed. Thus, the first two damping ratios, which are determined based on the identified Rayleigh damping coefficients and the first two natural frequencies as shown in Table 15.4, are used for ease of comparison. The identified results are given in Table 15.6. It can be seen that, as compared with the results obtained from the hammer tests (see Table 15.4), the identified results by means of the proposed approach are still acceptable.

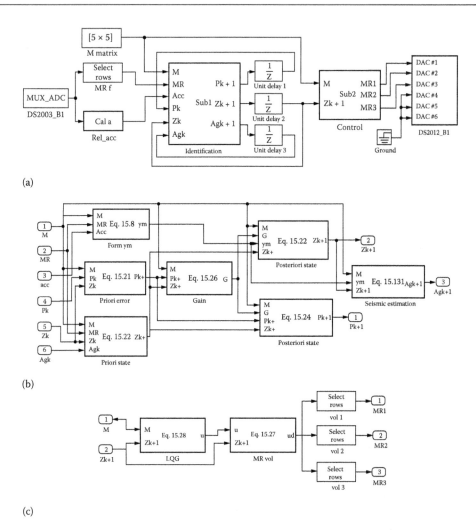

(a)

(b)

(c)

Figure 15.14 MATLAB/Simulink block diagrams of the implementation of the integrated system: (a) block diagram of identification and control integration, (b) block diagram for identification in Sub1 and (c) block diagram for control in Sub2.

Table 15.5 Identified structural stiffness (under the Kobe earthquake)

		k_1	k_2	k_3	k_4	k_5
Case 1	Identified (N/m)	2.389×10^5	2.601×10^5	2.533×10^5	2.604×10^5	2.742×10^5
	Relative error (%)	1.20	0.97	3.28	1.32	1.06
Case 2	Identified (N/m)	2.147×10^5	2.619×10^5	2.535×10^5	2.608×10^5	2.684×10^5
	Relative error (%)	1.51	1.67	3.21	1.17	1.07
Case 3	Identified (N/m)	2.111×10^5	2.638×10^5	2.388×10^5	2.662×10^5	2.765×10^5
	Relative error (%)	3.16	2.41	1.06	0.87	1.92
Case 4	Identified (N/m)	2.086×10^5	2.675×10^5	2.142×10^5	2.663×10^5	2.783×10^5
	Relative error (%)	4.31	3.84	1.64	0.91	2.58

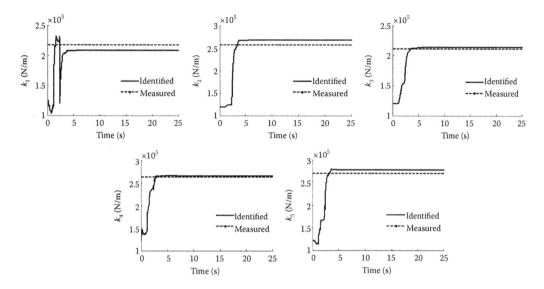

Figure 15.15 Identified structural stiffness under Kobe earthquake.

Table 15.6 Identified damping ratios (under the Kobe earthquake)

	Case 1	Case 2	Case 3	Case 4
ζ_1 (%)	0.67	0.86	0.82	1.06
ζ_2 (%)	0.28	0.45	0.88	0.89

Besides the identification of structural parameters, the unknown ground motion can also be simultaneously identified by the proposed approach. Also, taking Case 4 as an example, the time history of the identified ground excitation is plotted in Figure 15.16 as a dashed curve, whereas the solid curve is the corresponding measured ground excitation obtained directly from the accelerometer. Only the segment from 4 to 8 s is shown in Figure 15.16 for clarity of comparison. It can be found from Figure 15.16 that the identified ground excitation has a good agreement with the measured one.

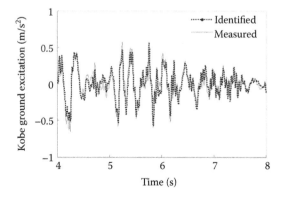

Figure 15.16 Comparison of the identified Kobe ground excitation.

Table 15.7 Peak responses of the building model under the Kobe earthquake

		$\bar{s}_i$ (mm)					$\bar{a}_i$ (m/s²)				
		$\bar{s}_1$	$\bar{s}_2$	$\bar{s}_3$	$\bar{s}_4$	$\bar{s}_5$	$\bar{a}_1$	$\bar{a}_2$	$\bar{a}_3$	$\bar{a}_4$	$\bar{a}_5$
Case 1	Without control	5.83	9.22	12.58	14.74	14.65	1.95	3.09	4.12	4.66	4.80
	With control	3.45	4.19	5.17	5.69	5.79	1.11	1.37	1.30	1.45	1.61
Case 2	Without control	6.22	9.40	12.45	14.31	15.14	1.97	2.93	3.67	4.17	4.35
	With control	3.53	4.24	5.19	5.70	5.77	1.12	1.35	1.31	1.46	1.67
Case 3	Without control	5.58	8.36	11.24	12.81	13.47	1.82	2.59	3.22	3.72	3.95
	With control	3.41	3.86	4.81	5.39	5.43	1.31	1.42	1.33	1.73	1.70
Case 4	Without control	5.19	7.86	10.88	12.38	13.04	1.59	2.28	2.97	3.53	4.03
	With control	3.36	3.81	4.78	5.33	5.36	1.15	1.29	1.21	1.56	1.53

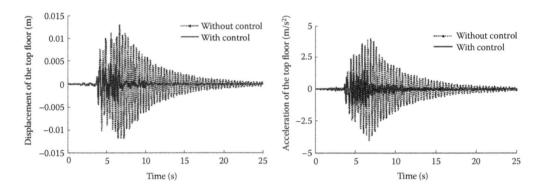

Figure 15.17 Comparison of the time histories of structural responses with and without control (Kobe earthquake).

Furthermore, based on the proposed integrated approach, the structural vibration can be significantly reduced with the aid of MR dampers. Two sets of shaking table tests, when the building model was equipped with and without MR dampers, were conducted. The aforementioned four damage scenarios were considered. The switching control law introduced in Section 15.2.2 was employed to adjust the properties of the MR dampers. The maximum acceleration and displacement responses of the building model under the Kobe earthquake are given in Table 15.7 for comparison. $\bar{s}_i$ and $\bar{a}_i$ ($i = 1, 2, ..., 5$) are the peak displacement and the acceleration of the ith floor obtained from the sensors, respectively. It can be seen from Table 15.7 that the structural vibration is significantly reduced. Moreover, the comparison of the time histories of the measured displacement and acceleration of the top floor are shown in Figure 15.17. Only the building model in Case 4 is plotted as an example. Similar results can be obtained from the other cases.

15.4.5 Damage detection and vibration control under Northridge earthquake

To investigate the robustness of the proposed integrated system, shaking table tests of the building model under the scaled Northridge earthquake with a peak acceleration of 2.28 m/s² were conducted. The aforementioned four damage scenarios were considered.

Based on the proposed integrated system, the structural stiffness can be identified and shown in Table 15.8. It can be found from Table 15.8 that the maximum relative error for all the cases is only 4.37%, indicating that the structural stiffness can be identified with acceptable accuracy. Similarly, taking Case 4 as an example, the identified stiffness during the earthquake is plotted in Figure 15.18 as a solid line whereas that determined by the static test is shown as a dashed line. It can be found that the identified results by means of the proposed integrated approach can be stably converged to the measured ones. The Rayleigh damping coefficients can be simultaneously identified as well, and the first two damping ratios determined on the basis of the identified damping coefficient and the first two natural frequencies are given in Table 15.9 for ease of comparison. It can be seen from Table 15.9 that the identification results of the structural damping are acceptable to some extent.

Furthermore, the unknown ground motion can also be identified by the proposed approach. By taking Case 4 as an example, the time history of the identified ground excitation is shown in Figure 15.19 as a dashed curve, whereas the solid curve is the

Table 15.8 Identified structural stiffness (under the Northridge earthquake)

		k_1	k_2	k_3	k_4	k_5
Case 1	Identified (N/m)	2.368×10^5	2.641×10^5	2.537×10^5	2.605×10^5	2.686×10^5
	Relative error (%)	2.07	2.52	3.13	1.29	1.01
Case 2	Identified (N/m)	2.143×10^5	2.616×10^5	2.537×10^5	2.612×10^5	2.668×10^5
	Relative error (%)	1.83	1.55	3.13	1.02	1.66
Case 3	Identified (N/m)	2.095×10^5	2.626×10^5	2.392×10^5	2.599×10^5	2.756×10^5
	Relative error (%)	3.89	1.94	1.23	1.52	1.58
Case 4	Identified (N/m)	2.121×10^5	2.634×10^5	2.199×10^5	2.598×10^5	2.751×10^5
	Relative error (%)	2.71	2.25	4.37	1.55	1.41

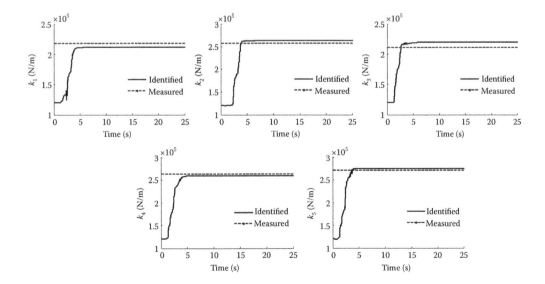

Figure 15.18 Identified structural stiffness under Northridge earthquake.

Table 15.9 Identified damping ratios (under the Northridge earthquake)

	Case 1	Case 2	Case 3	Case 4
ζ_1 (%)	0.68	0.65	0.9	0.75
ζ_2 (%)	0.67	0.54	0.95	0.85

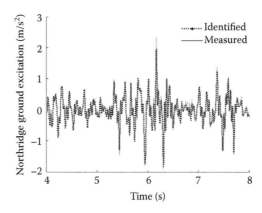

Figure 15.19   Comparison of the identified Northridge ground excitation.

corresponding measured one. Only the segment from 4 to 8 s is shown in Figure 15.19 for clarity of comparison. It is obvious that the identified ground excitation is close to the measured one.

A comparison of the structural responses in terms of the peak acceleration and displacement with and without control is shown in Table 15.10. Four damage scenarios are considered and the absolute acceleration and displacement measurements obtained from the sensors are employed for comparison. It can be found from Table 15.10 that the structural vibration can be reduced substantially. Moreover, Figure 15.20 gives a comparison of the time histories of the measured displacement and acceleration of the top floor. Only the building model in Case 4 is plotted as an example. Similar results can be obtained from the other cases.

Table 15.10 Peak responses of the building model (under the Northridge earthquake)

		$\bar{s}_i$ (mm)					$\bar{a}_i$ (m/s²)				
		$\bar{s}_1$	$\bar{s}_2$	$\bar{s}_3$	$\bar{s}_4$	$\bar{s}_5$	$\bar{a}_1$	$\bar{a}_2$	$\bar{a}_3$	$\bar{a}_4$	$\bar{a}_5$
Case 1	Without control	5.89	7.85	10.27	12.35	13.64	3.12	3.45	3.44	4.52	5.74
	With control	4.66	6.33	8.31	9.64	9.98	1.89	1.95	2.73	3.09	3.41
Case 2	Without control	5.51	7.48	9.84	11.73	12.86	2.85	2.86	3.16	4.18	4.88
	With control	4.76	6.36	8.30	9.59	9.92	1.67	1.93	2.62	2.76	3.03
Case 3	Without control	5.38	7.30	9.81	11.66	12.74	2.72	2.61	3.13	3.95	4.71
	With control	4.27	5.68	7.81	9.28	9.56	1.71	1.73	2.29	2.27	2.45
Case 4	Without control	5.36	7.18	9.91	11.72	12.78	2.74	2.60	3.02	3.76	4.56
	With control	4.12	5.37	7.56	8.94	9.20	1.77	1.69	2.13	2.11	2.26

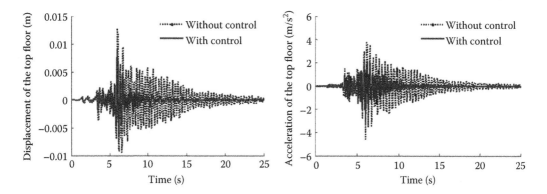

Figure 15.20 Comparison of the time histories of structural responses with and without control (Northridge earthquake).

15.5 FORMULATION OF INTEGRATED SYSTEM WITH TIME-VARYING PARAMETERS

It can be found that the integrated system described in Sections 15.2 through 15.4 is capable of effectively mitigating structural vibration and accurately identifying the parameters in the damage state without prior knowledge of the external excitations. A limitation of this integrated system is that the parameters to be identified should be time-invariant during the event. However, even for a building structure with control devices, structural damage may occur during an extreme event and the structural parameters of damaged components actually vary with time during the event. In these particular situations, the time-varying structural parameters shall be identified to ensure control performance, and the control parameters shall be adjusted accordingly. Thus, this section presents a real-time integrated system of health monitoring and vibration control to accurately identify time-varying structural parameters and unknown excitations on the one hand, and to optimally mitigate excessive vibration of the building structure on the other hand.

The basic equations for the identification of time-varying structural parameters and unknown excitations of a controlled building structure under earthquake excitation are presented based on the recursive LSE method as well as measured structural responses and control forces. The basic equations for the semi-active control of the building structure with MR dampers and clipped optimal displacement control algorithm are put forward based on the updated time-varying structural parameters and unknown excitations. The numerical algorithm is then followed to perform both identification and control simultaneously. The feasibility and accuracy of the proposed method are finally examined through a numerical investigation of an example building in the next section.

15.5.1 Identification of time-varying parameters and unknown excitation

In this study, the structural mass matrix **M** is assumed to be known and constant for simplicity of presentation. Other structural parameters, such as damping and stiffness coefficients, are time-varying as structural damage occurs with time. Excitations on a controlled building structure can be separated into two parts: control forces and unknown excitations. The second-order differential equation of motion of a controlled building structure can also be given by Equation 15.1.

Suppose that $\mathbf{Z}'$ is an $m \times 1$ unknown time-varying structural parameter vector, which can include both structural stiffness and damping parameters. The unknown time-varying structural parameter vector at time $t = k \times \Delta t$ is denoted as $\mathbf{Z}'_k$, in which Δt is the sampling interval.

The observation (measurement) equation associated with Equation 15.1 at time $t = k \times \Delta t$ can be described as

$$\mathbf{y}'_k = \mathbf{H}'_k \mathbf{Z}'_k - \boldsymbol{\varphi} \mathbf{f}_k + \mathbf{v}'_k \tag{15.31}$$

where:
  $\mathbf{y}'_k$ is an $n \times 1$ measurement vector that can be expressed as $-\mathbf{M}\ddot{\mathbf{x}}_k + \boldsymbol{\varphi}^* \mathbf{f}^*_k$
  $\mathbf{H}'_k$ is an $n \times m$ observation matrix composed of the measured structural velocity $\dot{\mathbf{x}}_k$ and displacement responses $\mathbf{x}_k$
  $\mathbf{v}'_k$ represents an $n \times 1$ noise vector, taking into consideration the model uncertainty of the structure and the measurement noise

$\mathbf{v}'_k$ represents an $n \times 1$ noise vector, taking into consideration the model uncertainty of the structure and the measurement noise

The noise vector can be assumed as a white noise with a normal probability distribution. Subscript k represents the values of matrices or vectors at time $t = k \times \Delta t$.

It can be found from Equation 15.31 that there are two variable vectors $\mathbf{Z}'_k$ and $\mathbf{f}_k$ to be predicted, which means Equation 15.31 is a multiple-linear regression equation. Furthermore, the elements in the predictor variable vectors are time-varying. Therefore, the traditional LSE method for solving the simple linear regression equation to find constant structural parameters cannot be directly applied.

Equation 15.31 can also be rearranged as follows:

$$\boldsymbol{\varphi} \mathbf{f}_k = -\mathbf{y}'_k + \mathbf{H}'_k \mathbf{Z}'_k + \mathbf{v}'_k \tag{15.32}$$

The closest solution of $\mathbf{f}_k$ in Equation 15.32 can be given as

$$\mathbf{f}_{k,LS} = \left(\boldsymbol{\varphi}^T \boldsymbol{\varphi} \right)^{-1} \boldsymbol{\varphi}^T \left(-\mathbf{y}'_k + \mathbf{H}'_k \mathbf{Z}'_k + \mathbf{v}'_k \right) \tag{15.33}$$

Similar to Equation 15.11, the error of the solution from Equation 15.33 is given by

$$\text{err} = \boldsymbol{\varphi} \mathbf{f}_k - \boldsymbol{\varphi} \mathbf{f}_{k,LS} = \left(\mathbf{I}_n - \boldsymbol{\varphi} \left(\boldsymbol{\varphi}^T \boldsymbol{\varphi} \right)^{-1} \boldsymbol{\varphi}^T \right) \left(-\mathbf{y}'_k + \mathbf{H}'_k \mathbf{Z}'_k + \mathbf{v}'_k \right) \tag{15.34}$$

in which $\mathbf{I}_n$ is the $n \times n$ identity matrix, and the matrix $\boldsymbol{\varphi}(\boldsymbol{\varphi}^T\boldsymbol{\varphi})^{-1}\boldsymbol{\varphi}^T$ is referred as the projection matrix. As a limit, the error in Equation 15.34 tends to be zero, leading to

$$\boldsymbol{\Phi} \mathbf{y}'_k = \boldsymbol{\Phi} \mathbf{H}'_k \mathbf{Z}'_k + \boldsymbol{\Phi} \mathbf{v}'_k \tag{15.35}$$

where the matrix $\boldsymbol{\Phi}$ is equal to $\mathbf{I}_n - \boldsymbol{\varphi}(\boldsymbol{\varphi}^T\boldsymbol{\varphi})^{-1}\boldsymbol{\varphi}^T$, and the definition and some properties of matrix $\boldsymbol{\Phi}$ are also mentioned in Equation 15.12. Obviously, with the proper aforementioned transformation, the multiple-linear regression equation expressed by Equation 15.31 is transformed into the simple-linear regression equation expressed by Equation 15.35.

When a structure is being damaged, the structural parameters vary with time. To track the parametric variation due to damage with unknown inputs, a simple time-varying correction factor is introduced. The recursive LSE of Equation 15.35 yields

$$\hat{\mathbf{Z}}'_k = \hat{\mathbf{Z}}'_{k-1} + \mathbf{K}'_k\left(\boldsymbol{\Phi}\mathbf{y}'_k - \boldsymbol{\Phi}\mathbf{H}'_k\hat{\mathbf{Z}}'_{k-1}\right) = \hat{\mathbf{Z}}'_{k-1} + \mathbf{K}'_k\boldsymbol{\Phi}\left(\mathbf{y}'_k - \mathbf{H}'_k\hat{\mathbf{Z}}'_{k-1}\right) \tag{15.36}$$

where:

$\hat{\mathbf{Z}}'_k$ and $\hat{\mathbf{Z}}'_{k-1}$	are the estimated values of $\mathbf{Z}'$ at time $t = k \cdot \Delta t$ and $t = (k-1)\cdot\Delta t$, respectively
$\mathbf{K}'_k$	is the LSE gain matrix for $\hat{\mathbf{Z}}'_k$ at time $t = k\cdot\Delta t$ with a size of $m \times n$

The term $\boldsymbol{\Phi}\mathbf{y}'_k - \boldsymbol{\Phi}\mathbf{H}'_k\hat{\mathbf{Z}}'_{k-1}$ is known as the correction term.

The current estimation error $\boldsymbol{\varepsilon}'_k$ of the unknown parameter vector $\mathbf{Z}'_k$ can be obtained as follows:

$$\begin{aligned}
\boldsymbol{\varepsilon}'_k &= \mathbf{Z}'_k - \hat{\mathbf{Z}}'_k \\
&= \mathbf{Z}'_k - \hat{\mathbf{Z}}'_{k-1} - \mathbf{K}'_k\left(\boldsymbol{\Phi}\mathbf{y}'_k - \boldsymbol{\Phi}\mathbf{H}'_k\hat{\mathbf{Z}}'_{k-1}\right) \\
&= \mathbf{Z}'_k - \hat{\mathbf{Z}}'_{k-1} - \mathbf{K}'_k\left(\boldsymbol{\Phi}\mathbf{H}'_k\mathbf{Z}'_k + \boldsymbol{\Phi}\mathbf{v}'_k - \boldsymbol{\Phi}\mathbf{H}'_k\hat{\mathbf{Z}}'_{k-1}\right) \\
&= \left(\mathbf{I}_m - \mathbf{K}'_k\boldsymbol{\Phi}\mathbf{H}'_k\right)\left(\mathbf{Z}'_k - \hat{\mathbf{Z}}'_{k-1}\right) - \mathbf{K}'_k\boldsymbol{\Phi}\mathbf{v}'_k
\end{aligned} \tag{15.37}$$

in which $\mathbf{I}_m$ is the $m \times m$ identity matrix. If the structural parameters are constants, that is, $\mathbf{Z}'_k = \mathbf{Z}'_{k-1}$, one can then have $\mathbf{Z}'_k - \hat{\mathbf{Z}}'_{k-1} = \mathbf{Z}'_{k-1} - \hat{\mathbf{Z}}'_{k-1} = \boldsymbol{\varepsilon}'_{k-1}$. However, the structural parameters vary with time, for example, a degradation of stiffness when structural damage occurs. To track the structural parametric variations and consequently detect structural damage on-line, a time-varying correction factor matrix $\boldsymbol{\lambda}'_k$ is introduced to reflect the structural parametric variations as follows:

$$\mathbf{Z}'_k - \hat{\mathbf{Z}}'_{k-1} = \boldsymbol{\lambda}'_k(\mathbf{Z}'_{k-1} - \hat{\mathbf{Z}}'_{k-1}) = \boldsymbol{\lambda}'_k\boldsymbol{\varepsilon}'_{k-1} \tag{15.38}$$

in which $\boldsymbol{\lambda}'_k$ is a diagonal matrix with size $m \times m$. By substituting Equation 15.38 into Equation 15.37, the current estimation error $\boldsymbol{\varepsilon}'_k$ could be calculated by

$$\boldsymbol{\varepsilon}'_k = \left(\mathbf{I}_m - \mathbf{K}'_k\boldsymbol{\Phi}\mathbf{H}'_k\right)\boldsymbol{\lambda}'_k\boldsymbol{\varepsilon}'_{k-1} - \mathbf{K}'_k\boldsymbol{\Phi}\mathbf{v}'_k \tag{15.39}$$

The estimation error covariance can then be obtained with the notice of $\boldsymbol{\Phi}$ being a symmetric matrix:

$$\mathbf{P}'_k = E(\mathbf{\varepsilon}'_k \mathbf{\varepsilon}'^T_k)$$

$$= E\left\{ \left[(\mathbf{I}_m - \mathbf{K}'_k \mathbf{\Phi} \mathbf{H}'_k) \mathbf{\lambda}'_k \mathbf{\varepsilon}'_{k-1} - \mathbf{K}'_k \mathbf{\Phi} \mathbf{v}'_k \right] \left[(\mathbf{I}_m - \mathbf{K}'_k \mathbf{\Phi} \mathbf{H}'_k) \mathbf{\lambda}'_k \mathbf{\varepsilon}'_{k-1} - \mathbf{K}'_k \mathbf{\Phi} \mathbf{v}'_k \right]^T \right\}$$

$$= (\mathbf{I}_m - \mathbf{K}'_k \mathbf{\Phi} \mathbf{H}'_k) \mathbf{\lambda}'_k E(\mathbf{\varepsilon}'_{k-1} \mathbf{\varepsilon}'^T_{k-1}) \mathbf{\lambda}'^T_k (\mathbf{I}_m - \mathbf{K}'_k \mathbf{\Phi} \mathbf{H}'_k)^T - \mathbf{K}'_k \mathbf{\Phi} E(\mathbf{v}'_k \mathbf{\varepsilon}'^T_{k-1}) \mathbf{\lambda}'^T_k (\mathbf{I}_m - \mathbf{K}'_k \mathbf{\Phi} \mathbf{H}'_k)^T$$

$$- (\mathbf{I}_m - \mathbf{K}'_k \mathbf{\Phi} \mathbf{H}'_k) \mathbf{\lambda}'_k E(\mathbf{\varepsilon}'_{k-1} \mathbf{v}'^T_k) (\mathbf{K}'_k \mathbf{\Phi})^T + \mathbf{K}'_k \mathbf{\Phi} E(\mathbf{v}'_k \mathbf{v}'^T_k) (\mathbf{K}'_k \mathbf{\Phi})^T$$

$$= (\mathbf{I}_m - \mathbf{K}'_k \mathbf{\Phi} \mathbf{H}'_k) \mathbf{\lambda}'_k \mathbf{P}'_{k-1} \mathbf{\lambda}'^T_k (\mathbf{I}_m - \mathbf{K}'_k \mathbf{\Phi} \mathbf{H}'_k)^T + \mathbf{K}'_k \mathbf{\Phi} \mathbf{R}'_k \mathbf{\Phi} \mathbf{K}'^T_k$$

$$(15.40)$$

where $\mathbf{R}'_k = E(\mathbf{v}'_k \mathbf{v}'^T_k)$ is the covariance of noise $\mathbf{v}'_k$. Moreover, the estimation error $\mathbf{\varepsilon}'_{k-1}$ at time $t = k \cdot \Delta t$ can be assumed to be independent of the noise vector $\mathbf{v}'_k$ at time $t = k \cdot \Delta t$, and accordingly $E(\mathbf{v}'_k \mathbf{\varepsilon}'^T_{k-1}) = E(\mathbf{\varepsilon}'_{k-1} \mathbf{v}'^T_k) = 0$ in Equation 15.40.

The time-varying correction factor matrix $\mathbf{\lambda}'_k$ can be calculated based on the current measurements. It is noted from Equation 15.36 that the current correction term at time $t = k \cdot \Delta t$ can be calculated based on the current measurements as follows:

$$\mathbf{r}'_k = \mathbf{\Phi} \mathbf{y}'_k - \mathbf{\Phi} \mathbf{H}'_k \hat{\mathbf{Z}}'_{k-1} = \mathbf{\Phi} \mathbf{H}'_k (\mathbf{Z}'_k - \hat{\mathbf{Z}}'_{k-1}) + \mathbf{\Phi} \mathbf{v}'_k = \mathbf{\Phi} \mathbf{H}'_k \mathbf{\lambda}'_k \mathbf{\varepsilon}'_{k-1} + \mathbf{\Phi} \mathbf{v}'_k \qquad (15.41)$$

Hence, the time-varying factor correction matrix $\mathbf{\lambda}'_k$ can be determined by

$$\mathbf{P}'_{r,k} = E(\mathbf{r}'_k \mathbf{r}'^T_k) = \mathbf{\Phi} \mathbf{H}'_k \mathbf{\lambda}'_k \mathbf{P}'_{k-1} \mathbf{\lambda}'_k \mathbf{H}'^T_k \mathbf{\Phi} + \mathbf{\Phi} \mathbf{R}'_k \mathbf{\Phi} \qquad (15.42)$$

To obtain the optimal value of the gain matrix $\mathbf{K}'_k$ that can minimise the estimation error covariance $\mathbf{P}'_k$ at time $t = k \cdot \Delta t$, the differentiation of $\mathbf{P}'_k$ in Equation 15.40 with respect to $\mathbf{K}'_k$ produces

$$\partial \mathbf{P}'_k / \partial \mathbf{K}'_k = 2(\mathbf{I}_m - \mathbf{K}'_k \mathbf{\Phi} \mathbf{H}'_k) \mathbf{\lambda}'_k \mathbf{P}'_{k-1} \mathbf{\lambda}'^T_k (-\mathbf{\Phi} \mathbf{H}'_k)^T + 2\mathbf{K}'_k \mathbf{\Phi} \mathbf{R}'_k \mathbf{\Phi}$$

$$= 2\mathbf{K}'_k \mathbf{\Phi} (\mathbf{H}'_k \mathbf{\lambda}'_k \mathbf{P}'_{k-1} \mathbf{\lambda}'^T_k \mathbf{H}'^T_k + \mathbf{R}'_k) \mathbf{\Phi} - 2\mathbf{\lambda}'_k \mathbf{P}'_{k-1} \mathbf{\lambda}'^T_k \mathbf{H}'^T_k \mathbf{\Phi} \qquad (15.43)$$

By setting the value of the partial derivative to zero, one can obtain

$$\mathbf{K}'_k = \mathbf{\lambda}'_k \mathbf{P}'_{k-1} \mathbf{\lambda}'^T_k \mathbf{H}'^T_k \mathbf{\Phi} / \left[\mathbf{\Phi} (\mathbf{H}'_k \mathbf{\lambda}'_k \mathbf{P}'_{k-1} \mathbf{\lambda}'^T_k \mathbf{H}'^T_k + \mathbf{R}'_k) \mathbf{\Phi} \right] \qquad (15.44)$$

It is noted that $\mathbf{P}'_{k-1}$, $\mathbf{R}'_k$ and $\mathbf{\Phi}$ are the symmetric matrices, the estimation error covariance expressed by Equation 15.40 could be simplified in terms of Equation 15.44 as

$$\mathbf{P}'_k = \left(\mathbf{I}_m - \mathbf{K}'_k\mathbf{\Phi}\mathbf{H}'_k\right)\lambda'_k\mathbf{P}'_{k-1}\lambda'^T_k \left(\mathbf{I}_m - \mathbf{K}'_k\mathbf{\Phi}\mathbf{H}'_k\right)^T + \mathbf{K}'_k\mathbf{\Phi}\mathbf{R}'_k\mathbf{\Phi}\mathbf{K}'^T_k$$

$$=\left(\mathbf{I}_m - \mathbf{K}'_k\mathbf{\Phi}\mathbf{H}'_k\right)\lambda'_k\mathbf{P}'_{k-1}\lambda'^T_k - \lambda'_k\mathbf{P}'_{k-1}\lambda'^T_k\mathbf{H}'^T_k\mathbf{\Phi}\mathbf{K}'^T_k + \mathbf{K}'_k\mathbf{\Phi}\left(\mathbf{H}'_k\lambda'_k\mathbf{P}'_{k-1}\lambda'^T_k\mathbf{H}'^T_k + \mathbf{R}'_k\right)\mathbf{\Phi}\mathbf{K}'^T_k$$

$$= \left(\mathbf{I}_m - \mathbf{K}'_k\mathbf{\Phi}\mathbf{H}'_k\right)\lambda'_k\mathbf{P}'_{k-1}\lambda'^T_k$$

$$(15.45)$$

Once the estimate value $\hat{\mathbf{Z}}'_k$ of the unknown parametric vector is calculated by Equation 15.36, the estimate value of the unknown excitation vector at time $t = k \cdot \Delta t$ can be estimated by

$$\hat{\mathbf{f}}'_k = -\left(\mathbf{\phi}^T\mathbf{\phi}\right)^{-1}\mathbf{\phi}^T\left(\mathbf{y}'_k - \mathbf{H}'_k\hat{\mathbf{Z}}'_k\right)$$

$$(15.46)$$

Equations 15.36, 15.42, 15.44, 15.45 and 15.4 form the recursive LSE for identifying both time-varying structural parameters and unknown excitations. If all the external excitations can be measured and the unknown parametric vector is constant (i.e. the influence matrix of the unknown excitation vector $\mathbf{\phi}$ in Equation 15.1 is a null matrix leading to $\mathbf{\Phi} = \mathbf{I}_n$ and the time-varying correction factor matrix $\lambda'_k = \mathbf{I}_m$), the proposed algorithm becomes the same as the traditional recursive LSE. A flowchart of the proposed recursive LSE algorithm for simultaneously identifying the unknown inputs and time-varying structural parameters is shown in Figure 15.21.

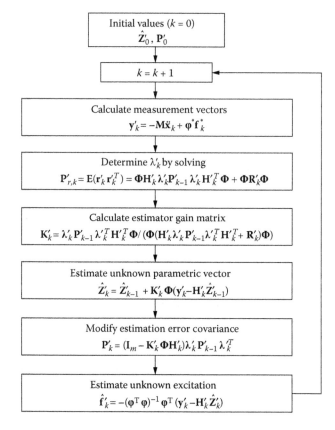

Figure 15.21 Flowchart of the proposed method for simultaneously identifying time-varying parameters and unknown inputs.

15.5.2 Vibration control with identified time-varying parameters

When a building structure is subjected to earthquake excitation, excessive vibration may occur and cause damage. Semi-active dampers can be installed in the building structure to reduce the excessive vibration and further damage. The vibration control system, including sensors, semi-active dampers, data transmission system, data acquisition system, control algorithm and data analysis system, should be installed in the building structure to provide essential feedback information and form a closed-loop control. The semi-active vibration control algorithm is then implemented based on the measured structural responses and the identified time-varying structural parameters to generate optimal control forces and achieve the maximum building response reduction. The basic equations for the semi-active control of the building structure with MR dampers and the clipped optimal displacement control algorithm are put forward here based on the updated time-varying structural parameters and unknown excitations.

As mentioned before, a number of mechanical models are available to describe the relationship between the force and motion of an MR damper. For the sake of simplicity, to illustrate the integration of structural control and health monitoring, the simple Bingham model is adopted here. Similar to Equation 15.27, the relationship between the force P_d and velocity $\dot{e}$ of the MR damper can be expressed in the following form (Gavin et al. 1996; Xu et al. 2000; Qu and Xu 2001):

$$P_d = C_d \dot{e} + F_d \, \text{sgn}(\dot{e}) \tag{15.47}$$

in which

$$C_d = C_1 \frac{12\eta L A_p}{bh^3} A_p \; ; \; F_d = C_2 \frac{L\tau_y}{h} A_p + P_y \tag{15.48}$$

For the fixed-plate-type damper:

$$C_1 = 1.0 \; ; \; C_2 = 2.07 + \frac{1.0}{1.0 + 0.4T} \; ; \; T = \frac{bh^2 \tau_y}{12 A_p \eta \dot{e}_i} \tag{15.49}$$

where:

η	is the viscosity coefficient
b	is the width of the rectangular plate
h	is the gap between two parallel plates
L	is the effective axial pole length
A_p	is the cross-sectional area of the piston
τ_y	represents the yielding shear stress controlled by the applied field
P_y	is the mechanical friction force in the damper

Clearly, P_d is a function of the yielding shear stress and it can be controlled through the applied field, but C_d is independent of the applied field.

Let us consider a multi-story shear building subjected to earthquake excitation as shown in Figure 15.22a; semi-active MR dampers can be positioned between the chevron braces and the rigid floor diaphragms to enhance its vibration energy dissipation capacity. In consideration of the stiffness of the chevron brace, the mechanical model for the MR

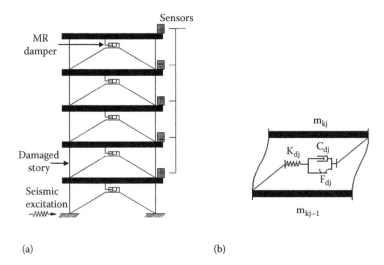

(a)   (b)

Figure 15.22 Schematic diagram of a building structure with MR damper-brace system: (a) sensors and MR dampers and (b) mechanical model of MR damper-brace system.

damper–chevron brace system can be seen as a damper and a spring connected in series, as shown in Figure 15.22b. When considering the MR damper and the chevron brace connected in series, the spring force in the brace is equal to the force on the piston of the damper. Equation 15.47 should thus be correspondingly changed to

$$C_d \dot{e}_i + F_d \, \mathrm{sgn}(\dot{e}_i) = K_d (u_s - e) \tag{15.50}$$

where:

u_s is the relative displacement between the two floors with the damper installed
K_d is the horizontal stiffness of the chevron brace

Notably, since the effect of chevron braces are considered, that is, the semi-active control system in this section is the MR damper–chevron brace system, the relative displacement of the MR damper on its two ends is not equal to the inter-story shift of the floor. Thus, a clipped optimal displacement control strategy (Xu et al. 2000), which is parallel to the clipped optimal force control strategy, is employed here to consider the influence of the brace system.

In terms of Equation 15.50, the equation of motion of an n-story frame structure with q dampers subject to earthquake excitation can be expressed as

$$\mathbf{M}\ddot{\mathbf{x}}(t) + \mathbf{C}(t)\dot{\mathbf{x}}(t) + \left(\mathbf{K}(t) + \mathbf{H}_c \mathbf{K}_d \mathbf{H}_c^T\right)\mathbf{x}(t) + \mathbf{H}_c \mathbf{K}_d \mathbf{e}(t) = \mathbf{H}_e \ddot{\mathbf{x}}_g(t) \tag{15.51}$$

$$\frac{C_{d_j}}{K_{d_j}} \dot{e}_j + e_j + \frac{F_{d_j}}{K_{d_j}} \, \mathrm{sgn}(\dot{e}_j) = \mathbf{H}_{cj}^T \mathbf{x} = x_{kj} - x_{kj-1} \quad (j = 1,2,...,m) \tag{15.52}$$

where:

$\mathbf{M}$ is the $n \times n$ mass matrix of the frame structure, which is assumed to be constant and known

$\mathbf{C}(t)$ and $\mathbf{K}(t)$ are the time-varying structural damping and stiffness matrices, which are updated by the health monitoring system with time

$\mathbf{K}_d$ is the $q \times q$ diagonal stiffness matrix, of which the element is the stiffness coefficient of the chevron brace

q is the number of stories with MR dampers installed

$\mathbf{H}_c$ is the $n \times q$ matrix converting the brace stiffness matrix into the global coordinate system

Superscript T means the transposition of a matrix

$\mathbf{x}$, $\dot{\mathbf{x}}$ and $\ddot{\mathbf{x}}$ are the $n \times 1$ displacement, velocity and acceleration vectors of the frame structure, respectively

$\mathbf{e}$ is the $q \times 1$ displacement vector of the MR dampers

$\mathbf{H}_{cj}$ is the jth column vector of the matrix $\mathbf{H}_c$

x_{kj} and x_{kj-1} are the displacements of the top and bottom floors of the kth story where the jth damper is installed

It can be easily found that Equation 15.51 can be written in the same form as Equation 15.1 by using the following substitutions:

$$\boldsymbol{\varphi}^* = \mathbf{H}_c, \quad \mathbf{f}^*(t) = -\mathbf{K}_d\left[\mathbf{H}_c^T\mathbf{x}(t) + \mathbf{e}(t)\right], \quad \boldsymbol{\varphi} = \mathbf{H}_e, \quad \mathbf{f}(t) = \ddot{x}_g(t) \tag{15.53}$$

Equation 15.51 can also be replaced by an equivalent first-order differential equation of the state-space form:

$$\dot{\mathbf{X}}_{MR}(t) = \mathbf{A}_{MR}(t)\mathbf{X}_{MR}(t) + \mathbf{B}_{MR}\mathbf{e}(t) + \mathbf{D}_{MR}\ddot{x}_g(t) \tag{15.54}$$

in which

$$\mathbf{A}_{MR}(t) = \begin{bmatrix} 0 & \mathbf{I} \\ -\mathbf{M}^{-1}\left(\mathbf{K}(t) + \mathbf{H}_c\mathbf{K}_d\mathbf{H}_c^T\right) & -\mathbf{M}^{-1}\mathbf{C}(t) \end{bmatrix},$$

$$\mathbf{B}_{MR} = \begin{bmatrix} 0 \\ \mathbf{M}^{-1}\mathbf{H}_c\mathbf{K}_d \end{bmatrix}, \quad \mathbf{D}_{MR} = \begin{bmatrix} 0 \\ \mathbf{M}^{-1}\mathbf{H}_e \end{bmatrix}, \quad \mathbf{X}_{MR}(t) = \begin{bmatrix} \mathbf{x}(t) \\ \dot{\mathbf{x}}(t) \end{bmatrix} \tag{15.55}$$

Similar to Equation 15.30, the stiffness matrix and damping matrix of the building structure are constructed using the stiffness coefficients and damping coefficients that are identified from the health monitoring system in real time. Accordingly, the matrix $\mathbf{A}_{MR}(t)$ in Equation 15.55 is reconstructed at each time step of the computation.

A clipped optimal displacement control approach is used in terms of the linear quadratic regular (LQR) control theory that minimises

$$J = \int_0^{t_f} \left[\mathbf{X}_{MR}^T(t)\mathbf{Q}_{MR}\mathbf{X}_{MR}(t) + \mathbf{e}^T(t)\mathbf{R}_{MR}\mathbf{e}(t)\right]dt \tag{15.56}$$

to control the displacement vector $\mathbf{e}_T(t)$ as

$$\mathbf{e}_T(t) = -\mathbf{R}_{MR}\mathbf{B}_{MR}\mathbf{P}_{MR}(t)\mathbf{X}_{MR}(t) \tag{15.57}$$

where:

$\mathbf{Q}_{MR}$ is the weighting matrix for the structure response in the optimal displacement control, which is an $n \times n$ positive-semi-definite matrix

$\mathbf{R}_{MR}$ is the weighting matrix for the damper displacement in the optimal displacement control, which is a $q \times q$ positive-definite matrix

$\mathbf{P}_{MR}(t)$ is the positive-definite solution of the following Riccati equation:

$$\mathbf{P}_{MR}(t)\mathbf{B}_{MR}^T\mathbf{R}_{MR}^{-1}\mathbf{B}_{MR}\mathbf{P}_{MR}(t) - \mathbf{A}_{MR}^T(t)\mathbf{P}_{MR}(t) - \mathbf{P}_{MR}(t)\mathbf{A}_{MR}(t) - \mathbf{Q}_{MR} = 0 \qquad (15.58)$$

The strategy in the clipped optimal displacement control approach (Xu et al. 2000) can be described as follows. When the jth damper displacement e_j is approaching the desired optimal damper displacement vector e_{Tj}, the friction force F_{dj} in the damper is set to its minimum value. When the jth damper moves in the opposite direction to the optimal damper displacement, the friction force F_{dj} in the damper should be set to a smaller value of the two quantities: F_{max} and the actual damper force $K_{dj}(x_{kj} - x_{kj-1} - e_j)$ minus a small quantity F_0. In this way, the damper is always in motion to dissipate vibration energy. This strategy can be stated as

$$F_{dj} = \begin{cases} F_{min} & \text{when } e_j(e_{Tj} - \dot{e}_j) > 0 \\ \min\{abs[K_{dj}(x_{kj} - x_{kj-1} - e_j)] - F_0, F_{max}\} & \text{when } e_j(e_{Tj} - \dot{e}_j) < 0 \end{cases} \quad (j = 1, 2, \ldots, q)$$

$$(15.59)$$

The flowchart of the presented semi-active control process with MR dampers is shown in Figure 15.23.

15.5.3 Implementation procedures of integrated system with time-varying parameters

Based on the equations presented in the previous two subsections, the integrated system can be implemented for real-time system identification and vibration control of the building structure with time-varying parameters step-by-step as follows:

Step 1:   Obtain the time-varying factor correction matrix λ'_k at time $t = k \cdot \Delta t$ by solving Equation 15.42 based on the current measurements.

Step 2:   Calculate the estimator gain matrix $\mathbf{K}'_k$ using Equation 15.44 with the time-varying factor correction matrix that is determined in Step 1.

Step 3:   Generate the unknown parametric vector $\hat{\mathbf{Z}}'_k$ using Equation 15.36 based on the estimator gain matrix that is calculated in Step 2 and the current correction term.

Step 4:   Update the estimation error covariance matrix $\mathbf{P}'_k$ using Equation 15.45.

Step 5:   Estimate the unknown excitation $\hat{\mathbf{f}}'_k$ using Equation 15.46 with the unknown parametric vector that is identified in Step 3.

Step 6:   Form the coefficient matrices $\mathbf{A}_{MR}$ and $\mathbf{B}_{MR}$ according to Equation 15.55 with the aid of the identified stiffness and damping coefficients in Step 3.

Step 7:   Calculate the matrix $\mathbf{P}_{MR}$ by solving Equation 15.58 and then find the optimal control displacement vector $\mathbf{e}_T$ according to Equation 15.57.

Step 8:   Determine the control forces based on the semi-active control strategy as shown in Equation 15.59.

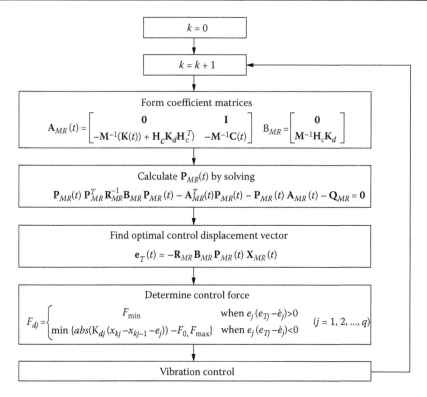

Figure 15.23 Flowchart of semi-active control process with MR dampers.

A flowchart of the integrated health monitoring and vibration control system for the building structure with time-varying parameters is shown in Figure 15.24. In this integrated system, the health monitoring system and vibration control system are combined for both system identification and vibration control. The control forces are first measured and transmitted to the health monitoring system in real time for structural parameter and excitation identification and damage detection. The time-varying structural parameters and ground excitation, identified from the health monitoring system in real time, are then transmitted to the vibration control system on-line to determine optimal control forces to mitigate the structural responses in the next step. The iteration of the above two steps of system identification and vibration control forms the on-line integrated structural health monitoring and vibration control.

15.6 NUMERICAL INVESTIGATION OF INTEGRATED SYSTEM WITH TIME-VARYING PARAMETERS

15.6.1 Description of example building structure

A simple five-story shear building is chosen as the example building structure (see Figure 15.22a). It is assumed to have the identical story height of 3 m. The building structure has a uniform mass of 5.1×10^3 kg and a uniform horizontal story (shear) stiffness of 1.334×10^7 N/m for all five stories. The Rayleigh damping assumption with the proportional coefficient of $\alpha = 0.4335$ s^{-1} for the mass matrix and $\beta = 7.015 \times 10^{-4}$ s for the stiffness matrix is used to construct the damping matrix. The scaled El-Centro earthquake with a peak ground acceleration

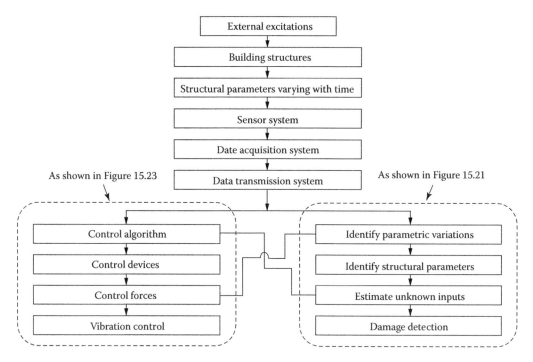

As shown in Figure 15.23

As shown in Figure 15.21

Figure 15.24 Schematic diagram of the implementation of the integrated system with time-varying parameters.

(PGA) of 4.0 m/s² is applied to the building model. The corresponding structural acceleration responses are measured by five accelerometers with one on each floor of the building structure. From a practical viewpoint, the white noise of 2% intensity is added to the calculated acceleration response as the measured acceleration response. The noise intensity is defined as the ratio of the RMS of the noise to the RMS of the acceleration response. The displacement and velocity responses of the building are obtained from the measured noise-polluted acceleration responses through numerical integrations. On each story of the building, a semi-active MR damper is installed with a chevron brace that connects two neighbouring floors. The ratios of the brace horizontal stiffness to the structure horizontal stiffness k_d/k are selected as one and the same for all the building stories. The properties of the MR dampers are listed in Table 15.11. The damper forces are assumed to be measured for the identification of the time-varying parameters and the unknown ground motion. Similarly, the RMS noise of 1% intensity is added to the calculated control force as the measured control force.

Damage occurs in one instance on the second story with a linear 5% degradation of the story stiffness from $t=8$ s to $t=9$ s. The structural damping matrix is assumed to be the Rayleigh damping matrix $C(t) = \alpha M + \beta K(t)$, where α and β are the two constant coefficients decided by the first and second modal damping ratios of 2% and the original mass and stiffness matrixes in this study. It is noted that the damping matrix is also time-varying because the stiffness matrix is time-varying.

Table 15.11 Basic parameters of an MR damper

Parameters	L (m)	h (m)	b (m)	A_p (m²)	P_y (kN)	η (kPa·s)	$\tau_{y\min}$ (kPa)	$\tau_{y\max}$ (kPa)
Value	0.5	0.002	0.75	0.04	0.05	0.0002	0.05	10

15.6.2 Accuracy of time-varying parameter and excitation identification

In this study, the building structure is subjected to earthquake-induced ground acceleration $\mathbf{f}(t) = \ddot{\mathbf{x}}_g(t)$. The unknown parameter vector at time $t = k \cdot \Delta t$ can be written as $\mathbf{Z}'_k = \begin{bmatrix} \mathbf{K}_\theta & \beta\mathbf{K}_\theta & \alpha \end{bmatrix}^T_k$, where $\mathbf{K}_\theta = [k_1 \, k_2 \cdots k_n]^T$ is the time-varying stiffness coefficients and $n = 5$.

The observation matrix can be worked out as

$$\mathbf{H}'_k = \begin{bmatrix} \mathbf{H}'_{1,k} & \mathbf{H}'_{2,k} & \mathbf{H}'_{3,k} \end{bmatrix} \tag{15.60}$$

in which

$$\mathbf{H}'_{1,k} = \begin{bmatrix} x_{1,k} & x_{1,k} - x_{2,k} & & & \\ & x_{2,k} - x_{1,k} & x_{2,k} - x_{3,k} & & \\ & & \cdots & & \cdots & \\ & & & x_{n-1,k} - x_{n-2,k} & x_{n-1,k} - x_{n,k} \\ & & & & x_{n,k} - x_{n-1,k} \end{bmatrix},$$

$$\mathbf{H}'_{2,k} = \begin{bmatrix} \dot{x}_{1,k} & \dot{x}_{1,k} - \dot{x}_{2,k} & & & \\ & \dot{x}_{2,k} - \dot{x}_{1,k} & \dot{x}_{2,k} - \dot{x}_{3,k} & & \\ & & \cdots & & \cdots & \\ & & & \dot{x}_{n-1,k} - \dot{x}_{n-2,k} & \dot{x}_{n-1,k} - \dot{x}_{n,k} \\ & & & & \dot{x}_{n,k} - \dot{x}_{n-1,k} \end{bmatrix}$$

and $\mathbf{H}'_{3,k} = \begin{bmatrix} \mathbf{M}\ddot{\mathbf{x}}_k \end{bmatrix}$.

The sampling interval Δt is set as 0.002 s, and accordingly the sampling frequency is 500 Hz for all the measurement responses. The initial value of time-varying story stiffness coefficients are taken as 1.25 times the original stiffness coefficients, and the two constant coefficients α and β are initially assumed as unit. As a result, the initial estimated unknown parametric vector can be written as $\hat{\mathbf{Z}}'_0 = \begin{bmatrix} 1.25\mathbf{K}_\theta & 1\times 1.25\mathbf{K}_\theta & 1 \end{bmatrix}^T$. For the controlled building structure with semi-active MR dampers subjected to earthquake excitation, the damper forces could be measured by force transducers as the measured excitation $\mathbf{f}^*(t)$, while the earthquake ground acceleration $\ddot{\mathbf{x}}_g(t)$ could be treated as the unknown excitation $\mathbf{f}(t)$. The influence matrix of the unknown excitations can be set as $\boldsymbol{\varphi} = \mathbf{H}_e = -\mathbf{M}\begin{bmatrix} 1 & 1 & 1 & 1 & 1 \end{bmatrix}^T$. The influence matrix of the known inputs $\boldsymbol{\varphi}^* = \mathbf{H}_c$ reflects the location of the semi-active MR dampers. The time-varying correction factor matrix is set as $\boldsymbol{\lambda}'_k = \mathbf{I}_m$ during the time period from $t = 0$ s to $t = 2$ s in order to obtain the covariance matrix of noise $\mathbf{R}' = 1/(1000-1) \times \sum_{i=1}^{1000}[(\boldsymbol{\Phi}\mathbf{y}'_k - \boldsymbol{\Phi}\mathbf{H}'_k\hat{\mathbf{Z}}'_k)(\boldsymbol{\Phi}\mathbf{y}'_k - \boldsymbol{\Phi}\mathbf{H}'_k\hat{\mathbf{Z}}'_k)^T]$. The noise covariance matrix calculated is then used for every subsequent time step.

Figure 15.25 presents the identified results of time-varying stiffness coefficients of the five stories of the building structure. Figure 15.26 shows the identified results of two coefficients, α and β. In Figures 15.25 and 15.26, the identified results are presented as dashed lines but the real values are presented as solid lines for comparison. It can be seen that after a very short time period (less than 1.25 s), the identified results converge to the actual ones. It can also be seen from Figure 15.25 that the proposed algorithm has a very good tracking ability for capturing

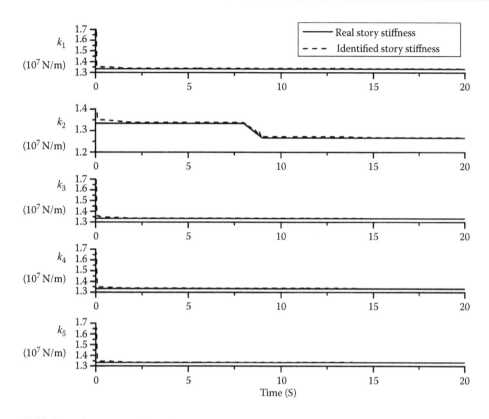

Figure 15.25 Identification results with time-varying stiffness.

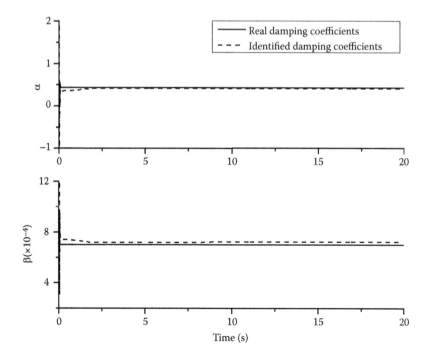

Figure 15.26 Identified results of Rayleigh damping coefficients.

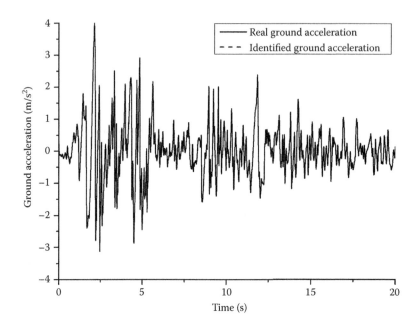

Figure 15.27 Identified results of unknown earthquake-induced ground acceleration.

slightly changed stiffness in a very short time period from 8 to 9 s. The results presented in Figure 15.26 also show that the proposed algorithm can identify the two damping coefficients accurately. In summary, the proposed algorithm can identify time-varying structural stiffness and damping coefficients accurately and therefore can detect structural damage precisely online. The identified results of the unknown earthquake-induced ground acceleration are presented in Figure 15.27 with dashed lines and compared with the actual ones with solid lines. Clearly, the proposed algorithm is capable of identifying the unknown excitation very well.

15.6.3 Performance of semi-active control with MR dampers

To evaluate the semi-active control performance, the seismic record El-Centro NS (1940) is selected as input to the example building. The PGA of the seismic record is scaled from 3.417 to 4.0 m/s². The stiffness matrix and damping matrix of the example building is constructed using the stiffness coefficients and damping coefficients that are identified from the health monitoring system accordingly. The matrix $\mathbf{A}_{MR}(t)$ in Equation 15.55 and the matrix $\mathbf{P}_{MR}(t)$ in Equation 15.58 are reconstructed at each time step. The ratios of the brace horizontal stiffness to the structure horizontal stiffness of all the five semi-active MR dampers are assigned the same value of 1. Five accelerometers and five force transducers with one accelerometer and one force transducer for each story are necessary to realise the feedback control. In the numerical investigation of the semi-active control performance, the corresponding computed building responses and damper forces are taken as the relevant feedback instead of the signals from the sensors in practice.

The variations of the peak displacement, velocity and acceleration responses of the example building without control and with semi-active control are demonstrated in Figure 15.28. It can be seen that the peak responses of displacement, velocity and acceleration of all the building floors under semi-active control are substantially reduced in comparison with those without control. Clearly, the semi-active control with the clipped optimal displacement control algorithm can effectively suppress the seismic responses of the building structure.

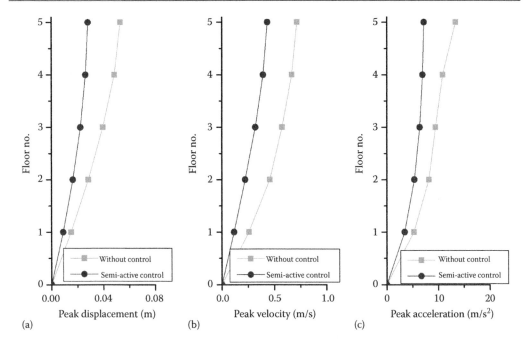

Figure 15.28 Comparison of control performance with and without semi-active control.

15.6.4 Comparisons

To further demonstrate the necessity and advantage of the proposed integrated procedure, the performance of semi-active control using on-line updated structural parameters is compared with that without updating structural parameters. Moreover, the accuracy of the parameter identification of the controlled building structure with semi-active control using on-line updated structural parameters is compared with that without updating structural parameters.

The semi-active control performance is evaluated in terms of two widely accepted sets of normalised performance indices. The first set of performance indices is related to the building responses, which include peak- and RMS-based inter-story drift ratios (J_1 and J_3) and peak- and RMS-based absolute acceleration responses (J_2 and J_4) expressed by

$$J_1 = \frac{\max\limits_{t,i} \left| dx_i^c(t) \right| / h_i}{\max\limits_{t,i} \left| dx_i^n(t) \right| / h_i} \tag{15.61}$$

$$J_2 = \frac{\max\limits_{t,i} \left| \ddot{x}_i^c(t) \right|}{\max\limits_{t,i} \left| \ddot{x}_i^n(t) \right|} \tag{15.62}$$

$$J_3 = \frac{\max\limits_{t,i} \left\| dx_i^c(t) \right\| / h_i}{\max\limits_{t,i} \left\| dx_i^n(t) \right\| / h_i} \tag{15.63}$$

$$J_4 = \frac{\max_{t,i} \left\| \ddot{x}_i^c(t) \right\|}{\max_{t,i} \left\| \ddot{x}_i^n(t) \right\|} \tag{15.64}$$

where:

$dx_i^c(t)$ and $dx_i^n(t)$	are the inter-story drifts of the ith story of the building with and without control, respectively
h_i	is the height of the ith story
$\left\| dx_i^c(t) \right\|/h_i$ and $\left\| dx_i^n(t) \right\|/h_i$	are the inter-story drift ratios of the ith story of the building with and without control, respectively
$\ddot{x}_i^c(t)$ and $\ddot{x}_i^n(t)$	are the absolute acceleration responses of the ith floor of the building with and without control, respectively

The RMS response quantities within the time duration t_f under earthquake excitation are calculated by $\|\cdot\| = \sqrt{1/t_f \int_0^{t_f} [\cdot]^2 dt}$. The sign $\max_{t_f}$ means to find the maximum value within the given time duration first and among all the building stories afterwards. The second set of performance indices are related to the capacity of control devices. The peak-based control force (J_5) is

$$J_5 = \frac{\max_{t,k} \left| u_k(t) \right|}{W} \tag{15.65}$$

where:

$u_k(t)$	is the control force generated by the kth control device (MR damper)
W	is the total weight of all the building floors

Table 15.12 shows the performance indices of the controlled building with and without parameter updating. It can be seen that the proposed semi-active control using MR dampers and considering on-line parameter updating, can effectively reduce both the peak and RMS responses of the example building under seismic excitation. The reduction of the RMS responses (J_3 and J_4) is even more than that of the peak responses (J_1 and J_2). It can also be seen that the semi-active control with on-line parameter updating has much higher control performance than that without on-line parameter updating. The control force index further shows that without on-line parameter updating, the control force required is also more than that with on-line parameter updating. Therefore, on-line parameter updating is necessary to ensure higher control performance and less control force. This observation can be further confirmed through a comparison of the time histories of the displacement, velocity and acceleration responses of the top floor of the building without control, with semi-active control but without on-line parameter updating and with semi-active control and with on-line parameter updating, as shown in Figure 15.29.

Table 15.12 Performance indices for semi-active vibration control using MR dampers

Index	J_1	J_2	J_3	J_4	J_5
With parameter updating	0.5766	0.5376	0.2439	0.3127	0.054
Without parameter updating	0.8117	0.5744	0.8301	0.6951	0.068

To investigate the necessity and advantage of the proposed integrated procedure, the accuracy of the parameter identification of the controlled building structure with semi-active control using on-line updated structural parameters is also compared with that without updating structural parameters. Figures 15.30 and 15.31 present the identified time-varying stiffness coefficients and the two damping coefficients α and β, respectively. In Figures 15.30 and 15.31, the identified results under control without parameter updating are presented as dashed lines, the results under control with parameter updating are presented as dotted lines but the real values are presented as solid lines for comparison. It can be seen that parameter identifications under control with and without parameter updating, have very good tracking ability for capturing slightly changing stiffness. It can also be seen from Figures 15.30 and 15.31 that parameter identification under control with parameter updating converges much faster for obtaining the real constant value than that without parameter updating, especially in the identification of damping coefficients. For the identification of unknown inputs, the identified results of the unknown ground acceleration in the two cases are not presented because they are almost the same.

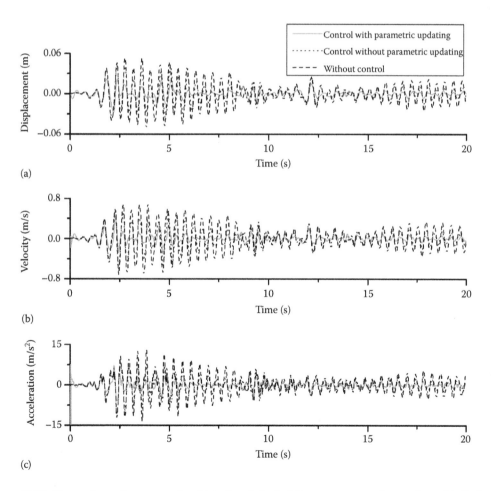

(a)

(b)

(c)

Figure 15.29 Comparison of response time histories of the building structure at the top floor: (a) displacement (m), (b) velocity (m/s) and (c) acceleration (m/s²).

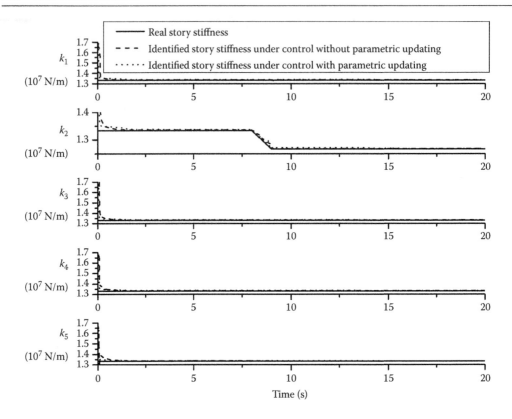

Figure 15.30 Comparison of identified results of story stiffness.

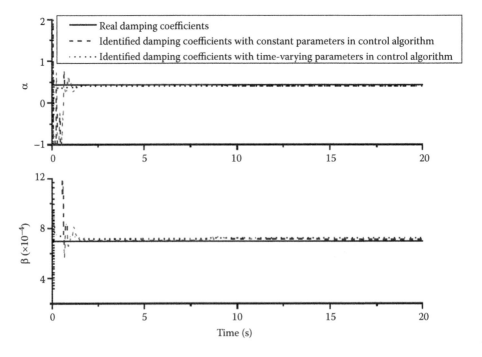

Figure 15.31 Comparison of identified results of Rayleigh damping coefficients.

NOTATION

A_p	The cross-sectional area of the piston
b	Width of the rectangular plate
c_d	Damping coefficient determined by the viscosity of MR fluid
$\underline{C}(t)$, $\underline{K}(t)$	The time-varying structural damping and stiffness matrices
$\dot{e}$	The velocity of the MR damper
$\mathbf{e}_T$	The desired optimal damper displacement vector
$f_d^{\max}$, $f_d^{\min}$	The maximum value and minimum value of the frictional forces, respectively
$\hat{\mathbf{f}}_k'$	The estimate value of the unknown excitation in the LSE-based algorithm
$\hat{\mathbf{f}}_{k|k}$	The estimates of $\mathbf{f}_k$ at the time instant $t = k \cdot \Delta t$
$\mathbf{f}(t)$	External excitation vector
$\mathbf{f}^*(t)$	Control force vector
$\mathbf{G}_{k+1}$	The EKF gain matrix at time $t = (k+1) \cdot \Delta t$
h	The gap between two parallel plates
$\mathbf{h}(\mathbf{Z}_k)$	A combination of state vector and equal to $\mathbf{K}\mathbf{x}_k + \mathbf{C}\dot{\mathbf{x}}_k$
$\mathbf{H}_c$	The matrix converting the brace stiffness matrix into the global coordinate system
$\mathbf{H}_k'$	The $n \times m$ observation matrix composed of the measured structural velocity $\dot{\mathbf{x}}_k$ and displacement responses $\mathbf{x}_k$
$\mathbf{I}$	Identity matrix
K_d	The horizontal stiffness of the chevron brace
$\mathbf{K}_{id}$, $\mathbf{C}_{id}$	The identified stiffness and damping matrix, respectively
$\mathbf{K}_k'$	The LSE gain matrix for $\hat{\mathbf{Z}}_k'$ at time $t = k \cdot \Delta t$
L	The effective axial pole length
$\mathbf{M}$, $\mathbf{C}$, $\mathbf{K}$	The mass, damping and stiffness matrices, respectively
$\mathbf{P}_k$	The estimation error covariance
$\mathbf{P}_{k+1|k}$	The priori estimation error covariance
$\mathbf{P}_{k+1|k+1}$	The posteriori estimation error covariance
P_y	The mechanical friction force in damper
$\bar{\mathbf{Q}}$, $\bar{\mathbf{R}}$	The weighting matrix for the structural response and control force, respectively
$\mathbf{Q}_{MR}$	The weighting matrix for the structure response in the optimal displacement control
$\mathbf{R}_k$, $\mathbf{R}_k'$	The covariance matrix of the measurement noise
$\mathbf{R}_{MR}$	The weighting matrix for the damper displacement in the optimal displacement control
$\mathbf{u}$	The optimal control force determined by LQG algorithm
u_d	The MR damper force
u_s	The relative displacement between the two floors with the damper installed
$\mathbf{v}_k$, $\mathbf{v}_k'$	Measurement noise vector
$\mathbf{x}_{id}$, $\dot{\mathbf{x}}_{id}$	The estimated structural displacement and velocity, respectively
$\ddot{\mathbf{x}}(t)$, $\dot{\mathbf{x}}(t)$, $\mathbf{x}(t)$	Structural acceleration, velocity and displacement response vectors, respectively
$\mathbf{y}_k$, $\mathbf{y}_k'$	The measurement vector obtained from the measured structural acceleration responses $\ddot{\mathbf{x}}_k$ and the measured control force $\mathbf{f}_k^*$
$\mathbf{Z}$	Extended state vector consisting of structural displacement, velocity and structural parameters to be identified

$\mathbf{Z}$	The $m \times 1$ unknown time-varying structural parameter vector including both structural stiffness and damping parameters
$\hat{\mathbf{Z}}_{k\|k}$	The estimates of $\mathbf{Z}_k$ at the time instant $t = k \cdot \Delta t$
$\hat{\mathbf{Z}}_{k+1\|k}$	The priori estimate of state $\mathbf{Z}_{k+1}$
$\hat{\mathbf{Z}}_{k+1\|k+1}$	The posteriori estimation state $\mathbf{Z}_{k+1}$
$\hat{\mathbf{Z}}'_k$	The estimated values of $\mathbf{Z}'$ at time $t = k \cdot \Delta t$
Δt	Time interval
ε'_k	The current estimation error of the unknown parameter vector $\mathbf{Z}'_k$
$\varepsilon_{k+1\|k}$	The priori estimation error of the unknown state vector $\mathbf{Z}_{k+1}$
$\varepsilon_{k+1\|k+1}$	The posteriori estimation error of the unknown state vector $\mathbf{Z}_{k+1}$
η	Viscosity coefficient of the MR fluid
Θ	m-unknown parametric vector composed of structural stiffness and damping coefficients
λ'_k	A time-varying correction factor matrix
τ_{y_*}	The yielding shear stress controlled by the applied field
$\boldsymbol{\varphi}^*, \varphi$	The influence matrices associated with $\mathbf{f}^*(t)$ and $\mathbf{f}(t)$, respectively

REFERENCES

Dyke, S.J. and B.F. Spencer Jr. 1997. A comparison of semi-active control strategies for the MR damper. In *Proceedings of the IASTED International Conference, Intelligent Information Systems*, The Bahamas, 580–84. IEEE Xplore digital library.

Dyke, S.J., B.F. Spencer Jr, M.K. Sain, and J.D. Carlson. 1996. Modeling and control of magnetorheological dampers for seismic response reduction. *Smart Mater. Struct.*, 5(5): 565–75.

Gavin, G.P., R.D. Hanson, and F.E. Filisko. 1996. Electrorheological dampers, Part I: Analysis and design. *J. Appl. Mech.*, 63(3): 669–75.

He, J., Q. Huang, and Y.L. Xu. 2014. Synthesis of vibration control and health monitoring of building structures under unknown excitation. *Smart Mater. Struct.*, 23: 105025.

He, J., Y.L. Xu, S. Zhan, and Q. Huang. 2016. Structural control and health monitoring of building structures with unknown ground excitations: Experimental investigation. *J. Sound Vibr.*, in press. http://dx.doi.org/10.1016/j.jsv.2016.11.035.

Jansen, L.M. and S.J. Dyke. 2000. Semiactive control strategies for MR dampers: Comparative study. *J. Eng. Mech.*, 126(8): 795–803.

Ou, J.P. 2003. *Structural Vibration Control: Active, Semi-Active, and Intelligent Control*. Beijing, China: Science Press.

Qu, W.L. and Y.L. Xu. 2001. Semi-active control of seismic response of tall buildings with podium structure using ER/MR dampers. *Struct. Des. Tall Buil.*, 10(3): 179–92.

Xu, Y.L., Q. Huang, Y. Xia, and H.J. Liu. 2015. Integration of health monitoring and vibration control for smart building structures with time-varying structural parameters and unknown excitations. *Smart Struct. Syst.*, 15(3): 807–30.

Xu, Y.L., W.L. Qu, and J.M. Ko. 2000. Seismic response control of frame structures using magnetorheological/electrorheological dampers. *Earthq. Eng. Struct. Dyn.*, 29(5): 557–75.

Chapter 16

Energy harvesting for structural health monitoring and vibration control

16.1 PREVIEW

Smart civil structures with integrated structural health monitoring (SHM) and vibration control functions have been discussed in Chapters 14 and 15 in the frequency domain and the time domain, respectively. The power supply for operating sensory systems and vibration control devices in such an integrated system is supposed to be externally provided. Nevertheless, it is hoped that a smart civil structure will be partially or fully self-powered. Therefore, the possibility of energy harvesting (EH) to provide electric power to small batteries used in wireless sensor networks has recently been actively investigated so that SHM systems in civil structures may operate autonomously and sustainably without the periodical replacement of batteries. In this regard, a variety of EH technologies using five typical energy sources are first introduced in this chapter. A brief review of these EH techniques for the health monitoring of civil structures is then given, followed by an introduction to several interesting applications, which include an EH system using the wind-induced vibration of a stay cable to power a wireless sensory system (Kim et al. 2013), the long-term testing of a vibration-based EH system for the health monitoring of bridges (McCullagh et al. 2014) and a self-powered strain sensor for damage detection (Elvin et al. 2003). Overviews of vibration control systems integrated with EH systems are given in the final part of this chapter, where much attention is paid to the investigation carried out by Shen and Zhu (2015) and Shen et al. (2016) using electromagnetic (EM) dampers to reduce structural vibration on one hand and to generate electric power for wireless sensor networks on the other.

16.2 ENERGY HARVESTING

16.2.1 Concept of energy harvesting

The process of extracting energy from the environment and surrounding system and converting it to usable electric energy is known as *energy harvesting* (Park et al. 2008). There are various kinds of energy harvesters and they can be categorised in two ways: what provides the energy for conversion and what type of energy is converted (Mateu and Moll 2005). Typical examples of widely used renewable energy harvesters include wind turbines and solar cells, which are able to covert renewable wind energy and solar energy, respectively, to electric energy. In the past few years, due to advances in wireless technology, there has been a surge of research into the development of EH technology for powering wireless sensor networks so as to create an autonomous and sustainable SHM system without the periodical replacement of batteries or the need for an external power supply. Moreover, considerable attention has also been paid to investigating how to simultaneously harvest power and mitigate vibration when the structural vibration control is needed for a civil structure. Some literature reviews are available on the development of energy harvesting for powering SHM sensing systems (e.g. Park et al. 2008, 2009; Gilbert and Balouchi 2008) and vibration control systems (e.g. Wang and Inman 2012), and details on the two topics will be given in Sections 16.3 and 16.4, respectively.

Notably, a number of scholars have proposed classification systems to categorise energy sources suitable for energy harvesting but, while broadly similar, they do exhibit some differences (e.g. Mateu and Moll 2005; Yerramilli and Tuluri 2012). Nevertheless, this chapter is not intended to provide an exhaustive literature survey of energy harvesting as this area is very broad and review articles with different emphases are already available in many publications (e.g. Sodano et al. 2004; Beeby et al. 2006; Harb 2011; Twiefel and Westermann 2013). Instead, this section will provide a brief introduction to wind, solar, vibration, radio frequency (RF) and thermal EH technologies.

16.2.2 Wind energy harvesting

The world has enormous potential for wind energy, which can be utilised for electricity generation in a sustainable and clean manner. As one of the world's fastest growing renewable energy resources, the installed wind generation capacity has increased from 25,000 MW to more than 200,000 MW in 10 years from 2001 to 2010 (Bhutta et al. 2012). Wind turbines are rotating machines that are able to capture and convert the kinetic energy in wind into either mechanical or, more significantly, electrical energy. They can be classified, firstly, in accordance with their aerodynamic function and, secondly, according to their constructional design (Hau 2013). The rotor's aerodynamic function is characterised by whether the wind energy harvester captures its power exclusively from the aerodynamic drag of the air stream acting on rotor surfaces or whether it is able to utilise the aerodynamic lift created by the flow against suitably shaped surfaces. Accordingly, there are so-called drag-type rotors and lift-type rotors. It is easy to show theoretically that it is much more efficient to use lift rather than drag when extracting power from wind (Hansen 2015). The classification according to constructional design is worth more attention: a turbine is called a *horizontal-axis wind turbine* (HAWT) if the blades are connected to a horizontal shaft for rotation, and the turbine is referred to as a *vertical-axis wind turbine* (VAWT) if the shaft is vertical. Both types of wind turbines have similar components (Paraschivoiu 2002), which include (1) a rotor that converts wind energy into mechanical power, (2) a tower to support the rotor, (3) a gearbox that is used to adjust the rotational speed of the rotor shaft of an electric generator or pump, (4) a control system that is connected to an anemometer and used to control the operation status of the wind turbine, including its starting and stopping, and (5) a foundation.

The majority of commercial wind turbines are HAWTs, which are designed and realised on the basis of a propeller-like concept. Unlike VAWTs, HAWTs are not omnidirectional. With changes of wind direction, in order to continue functioning, HAWTs have to change direction accordingly. A major advantage of HAWTs is that by pitching the rotor blades along their longitudinal axis, the rotor speed and power output can be effectively controlled on one hand and the wind turbines can be protected against extreme wind conditions on the other. Also, the rotor blade shape can be aerodynamically optimised, and it has been proven that the highest turbine efficiency can be achieved by exploiting aerodynamic lift to a maximum degree. However, a shortcoming of HAWTs is that their ability to generate power will be lost when the wind speed exceeds a certain value known as the *cut-off speed* (Rosen et al. 2010). For the purpose of safety and protecting the wind turbine structures, in particular the blades, a shutting down procedure is required at high wind speeds. Most HAWTs have a rotor cut-off speed ranging from 20 to 25 m/s. Thus, HAWTs are not suitable in storm-prone areas. Also, they are not suitable in urban areas due to their size and geometry and their sensitivity to wind turbulence. Many factors must be taken into consideration for the design, operation and maintenance of a HAWT, and some specifications and codes are available for guiding these procedures (e.g. IEC 2005; GL 2010).

Compared with HAWTs, VAWTs are still in the development stage, and their large-scale applications are relatively limited. There are two main types of VAWTs: *Darrieus rotor* and *Savonius rotor*. Some other types of VAWTs include combined Savonius and Darrieus rotors, two-leaf semi-rotary VAWTs, Sistan-type windmills and Zephyr turbines. In the Darrieus rotor, the blades are usually designed as an aerofoil in cross section, so the wind travels a longer distance on one side (convex) than on the other (concave), resulting in a higher wind speed on the convex side. It can be shown with the Bernoulli equation that a differential in wind speed over the aerofoil creates a differential pressure, which forces the rotor blade to turn around as the wind passes through the turbine (Rosen et al. 2010). Darrieus-type wind turbines have many variants and a typical Darrieus VAWT is characterised by its two or three C-shaped rotor blades. A Savonius rotor basically consists of hollow cup-shaped half-cylinders fixed with a central rotating shaft. Torque is generated due to the drag force acting on the half-cylinders. With an optimised aerodynamic design, a Savonius rotor can also make use of aerodynamic lift. The Savonius rotor is generally not employed for high-power electricity generation due to its low tip-speed ratio and its comparatively low power coefficient.

A merit of VAWTs is that they are suitable for electricity generation in conditions where traditional HAWTs are inefficient, such as areas of high wind velocity and turbulent wind flow (Bhutta et al. 2012). Moreover, VAWTs are omnidirectional, accepting wind from any direction without a yawing mechanism. This feature can lead to a simplified design as well as cost-effective construction and installation procedures. Also, the vertical axis of rotation provides the possibility of housing mechanical and electrical components, gearboxes and generators at ground level. However, some shortcomings still exist in VAWTs, such as their low tip-speed ratio, low self-starting torque and low power output.

Significant progress has been made in the research and development of both types of wind turbines. It is worth noting that VAWTs have been attracting increasing interest globally, primarily due to promising features absent from conventional HAWTs in some situations. However, whether the basic advantages of VAWTs can prevail over their disadvantages and whether they will become a serious rival to HAWTs cannot be foreseen (Hau 2013). In any case, this still requires a relatively long period of development.

16.2.3 Solar energy harvesting

Solar energy has existed for millions of years and represents the largest energy flow entering the terrestrial ecosystem. After reflection and absorption in the atmosphere, some 100,000 TW hit the surface of the earth. As solar energy is free, clean and extremely abundant, the usage of solar energy for electricity generation is compelling and offers both energy security and environmental benefits. To date, electricity production through solar energy is achieved mainly on the basis of two principles: the solar thermal effect and the solar photovoltaic (PV) effect.

Direct solar thermal power generation plants first produce high temperatures by gathering solar radiation on a concentrating solar collector and then directly convert the heat to electricity through a thermal electric conversion element. This direct conversion technology is able to convert the thermal energy into electricity power without the conventional intermediate mechanical conversion process, unlike traditional indirect solar thermal power technology, which uses a steam turbine generator. There are usually four types of direct thermoelectric conversion employed in the field of solar thermal power production technology (Yerramilli and Tuluri 2012): (1) thermoelectric conversion based on the Seebeck effect, (2) thermionic conversion based on the thermionic emission phenomenon, (3) alkali-metal thermoelectric conversion, which mainly generates current through the selective permeation of beta alumina solid electrolyte and (4) magneto-hydrodynamic (MHD) power generation based on Faraday's law of EM induction.

The PV effect is a physical phenomenon responsible for converting light to electricity and was discovered in 1839 by French physicist Edmund Becquerel. A *PV cell*, or *solar cell*, is a device composed of thin layers of semiconductor materials such as silicon which are able to free electrons and produce an electric current when exposed to sunlight. A single cell generates only a small amount of power and thus is usually not used individually. However, when groups of cells are hooked together to form a module, often named a *panel*, about 40–60 W of electricity can be produced. With the connection of two or more panels, an array can be formed and more electricity power can be generated. Moreover, since only a small portion of solar energy is converted to electric power by PV cells and a large amount of the remaining solar radiation produces heat energy, considerable attention has also been paid to integrating PV and thermoelectric technology together to improve the efficiency of the devices.

Since 2005, when the latest round of oil price hikes started, the solar cell industry has been experiencing phenomenal growth – about 50% annually on average (Tao 2014). However, even so, solar electricity makes a negligible contribution to our current energy mix, providing roughly 0.3% of the global electricity capacity. The high cost of solar technologies and their intermittent nature make them hardly competitive in an energy market dominated by cheap fossil fuels. Moreover, although the accumulated progress of solar cells in laboratories over the years is remarkable, the efficiency of commercial cells is typically about two-thirds of those in laboratories (Tao 2014). The potential for solar energy is enormous, and much more effort is required to remove the roadblocks and bottlenecks to making solar electricity a noticeable source of energy in our life.

16.2.4 Vibration energy harvesting

Due to their high availability in technical environments, mechanical vibrations are an attractive energy source, and thus there has been much interest in using micro-electromechanical systems (MEMS) to scavenge ambient vibration energy and transfer it to usable electrical power in the past few decades. The basic principle behind kinetic energy harvesting is the mechanical deformation of materials, such as piezoelectric materials, inside the EH device or the vibration of a moving part, such as a magnet or a proof mass. The most important conversion methods for ambient vibrations are based on piezoelectric, EM and electrostatic effects.

As introduced in Chapter 2, direct and converse piezoelectric effects basically reflect an interaction between the mechanical and electrical behaviour of piezoelectric materials. These materials, therefore, can be used to convert ambient vibration, which induces strain in the materials, into electric energy. Compared with the other two conversion mechanisms, piezoelectric vibration-to-electricity converters have received much more attention as they have high electromechanical coupling and no external voltage source requirement, and they are particularly attractive for use in MEMS (Shu 2009). In piezoelectric generators, the piezoelectric elements can be connected either in parallel or in series. In a parallel connection, the generated charge is added, whereas in a series connection, the produced charge corresponds to the strain of one of the piezoelectric elements connected, and the voltage of the piezoelectric elements is added. To adequately connect the piezoelectric elements in parallel or series, the orientation of the poling axis must be taken into consideration. Moreover, the optimum design and configuration of a piezoelectric generator also depends on the kind of surrounding kinetic energy to be exploited (amplitude and frequency) as well as on the electrical devices to be powered. Thus, to achieve higher efficiency, it is necessary to match the resonance frequency of the transducer with the most distinct frequency of the vibration source. Moreover, many researchers are devoted to increasing the working bandwidth of the energy harvester to ensure the most desirable energy generation performance is achieved (Twiefel and Westermann 2013). A number of literature reviews are available on the application of piezoelectric transducers as energy harvesters (e.g. Sodano et al. 2004; DuToit et al. 2005; Anton and Sodano 2007; Kim et al. 2011).

The basic principle behind EM generators is Faraday's law of EM induction, discovered by Michael Faraday in 1831, which says that when an electric conductor is moved through a magnetic field, a potential difference is induced between the ends of the conductor. EM systems are basically composed of a coil and a permanent magnet attached to a spring. The electrical power is generated by the variation in magnetic flux through an electrical circuit. This flux variation can be realised with a moving magnet whose flux is linked with a fixed coil or with a fixed magnet whose flux is linked with a moving coil. In most EM generators, the electric power is produced by the mechanical movement of the magnet, which is caused by structural vibration. EM energy conversion relies on relative velocity and variation in magnetic flux to generate electricity, and therefore, an EM device will not be limited in amplitude by the fatigue strength (Beeby and O'Donnell 2009). In general, bigger transducers with larger area coils will have better EH performance than smaller transducers, unless a larger time derivative is involved with the small-scale generators. Some detailed state-of-the-art reviews can be found on the micro- and macro-scale implementations of EM energy harvesters (e.g. Arnold 2007; Beeby and O'Donnell 2009).

The basic principle behind electrostatic generators is that the moving part of the transducer moves against an electrical field, thus generating energy (Mateu and Moll 2005). In general, electrostatic energy harvesters have an integrated mass forming one element of a variable capacitor, with power derived from an electrostatic force between the moving mass and the frame (Kiziroglou et al. 2009). The proof mass is coupled with the environmental vibrations through an elastic link, meaning the mass and the spring constitute a mechanical resonator. Thus, when the mechanical stimulus vibrates at that resonance frequency, maximum energy can be derived.

16.2.5 Radio frequency energy harvesting

Radio waves carry and transmit information by varying a combination of the amplitude, frequency and phase of the waves within a frequency band. On contact with a conductor such as an antenna, the EM wave or RF radiation can produce electrical energy on the conductor's surface, known as the *skin effect*. RF energy harvesting was investigated in the 1960s (Brown et al. 1965) using high-power microwave sources, as more of a proof-of-concept rather than a practical energy source due to the limitation of technologies at that time. However, with the modern development of low-power devices and smart materials, RF energy can be captured more effectively and becomes a viable alternative power supply to batteries in some applications, especially for wireless devices.

A variety of wireless sources of different frequencies radiating power have the potential to be exploited for RF energy harvesting applications. These sources might be, but are not limited to, TV and broadcast radio, mobile telephony, wireless networks and radar. A typical RF energy harvesting system contains an antenna that is used to capture the ambient RF sources, a matching EH circuit connected to the antenna and/or an associated power storage element. Much progress has been made in RF energy harvesting technologies, and this section provides a brief introduction to the two major developments: innovations in antennas and the optimisation of RF energy harvesting circuits.

Antennas play a key role in RF energy harvesting systems. A number of aspects should be considered for the design of the ideal antenna, including an appropriate physical size, an omnidirectional radiation pattern, a feeding point that can easily be attached to the rest of the harvesting circuits and good multi-band performance allowing electricity production for various frequency bands. Innovations involve explorations of new antenna variations, such as rectangular patch antennas (Rivière et al. 2010), slotted patch antennas (Georgiadis et al. 2010), microstrip patch antennas (Shrestha et al. 2013) and so forth. Typically, a single antenna is not sufficient to supply energy for reliable device operation. Consequently, significant efforts have also been made to investigate antenna array configurations with the aim of enhancing RF energy harvesting capabilities. For example, placing multiple EH antennas in one space or area was proposed and experimentally studied by Mi et al. (2005) as a means to increase the energy or power/area ratio. Four cooperating antennas were placed in a square area less than twice the area required for a single antenna. It was found that, in the chosen example, an increase of 83% in area gave a 300% increase in available power to a given device. More recently, Olgun et al. (2011) discussed the advantages and disadvantages of two antenna array configurations – that is, RF-combiner technology and DC-combiner technology. Based on a broadband 1×4 quasi-Yagi antenna array, Sun et al. (2013) presented a dual-band rectifier that can sufficiently enhance the RF-to-DC power conversion efficiency (PCE) at ambient RF power level. Measurement results showed that a PCE of 40% and an output DC voltage of 224 mV can be achieved over a 5 kΩ resistor when the dual-tone input power density is 455 µW/m^2.

Since the power efficiency of an RF energy harvesting system is also strongly dependent on its EH circuit, a great amount of attention has been paid to design and optimisation. For example, with the use of a P-type metal–oxide–semiconductor (PMOS) to reduce the body effect and the modification of the matching inductor and DC-blocking capacitor of a voltage doubler, Salter et al. (2007) found that a significant performance enhancement of 66% can be achieved. Based on an existing complementary metal–oxide–semiconductor (CMOS) voltage-doubler circuit, Jabbar et al. (2010) proposed a modified circuit to achieve a 160% increase in output power over traditional circuits at 0 dBm input power. Nintanavongsa et al. (2012) proposed a dual-stage EH circuit composed of a 7-stage and 10-stage design, and observed that the efficiency of the prototype could yield an almost 100% increase compared with that of a major commercially available EH circuit in the power range of –20 to 7 dBm.

16.2.6 Thermal energy harvesting

The temperature gradient and heat flow that naturally exist in the environment have the potential to produce electrical power by means of thermal-to-electric energy conversion. The temperature difference provides the potential for efficient energy conversion, while heat flow provides the power. However, even with a large heat flow, the converted power is typically small, mainly because of low Carnot and material efficiencies (Snyder 2009). In general, thermal energy can be directly converted into electricity by means of either the temperature gradient in space or the temperature variation in time, leading to two kinds of thermal EH techniques – that is, thermoelectric and pyroelectric energy harvesting.

Thermoelectric conversion is the most straightforward method of converting thermal energy into electric energy. Thermoelectric generators (TEGs) use the Seebeck effect, which describes the current generated when the junction of two dissimilar metals experiences a temperature difference. TEGs contain a number of thermoelectric couples consisting of *n*-type (containing free electrons) and *p*-type (containing free holes) thermoelectric elements wired electrically in series and thermally in parallel (Snyder 2009). Based on the Seebeck effect, the TEG is capable of producing an electrical current proportional to the temperature difference between the hot and cold junctions. However, the efficiency of this technology is low (<5%) if there is a small temperature gradient present. Thus, unless the temperature difference is large, the low efficiency of this type of conversion means that very little power can be extracted. In order to overcome the low energy conversion rates, efforts could be made to introduce new thermoelectric materials or to develop advanced thermoelectric devices able to operate at high temperature gradients. Good thermoelectric material is characterised by a large Seebeck coefficient: high electron conductivity to reduce Joule heating and low thermal conductivity to maintain the temperature difference across the material (Mouis et al. 2014). A major advantage of TEGs is that these devices only require a steady-state temperature gradient to operate, and thus no moving parts are needed, which can significantly improve the device's reliability and prolong its service life.

Compared with thermal energy harvesting via thermoelectric conversion, less attention has been paid to pyroelectric conversion techniques. This lack of interest stems from early studies which found that the energy conversion efficiency achieved by using pyroelectric materials was several times less than that potentially achievable with thermoelectric materials. However, recent modelling and experimental studies have shown that pyroelectric techniques can be cost-competitive with thermoelectrics and, on the basis of new temperature cycling techniques, has the potential to be several times more efficient than thermoelectrics under comparable operating conditions (Hunter et al. 2012). Moreover, many pyroelectric materials are stable up to temperatures approaching 1200°C, enabling energy harvesting from high-temperature sources with much improved thermodynamic efficiency. Thus, attention has gradually turned towards thermal energy harvesting with pyroelectric conversion techniques.

Pyroelectric energy generators rely on the properties of dielectric materials with polar point symmetry, which show a spontaneous electrical polarisation as a function of temperature. A temperature change in time (i.e. the temperature changing rate) determines the magnitude of the electrical current and energy conversion efficiency. In general, rapidly time-varying temperature gradients across a pyroelectric energy converter indicate large power generation and high conversion efficiency.

16.3 ENERGY HARVESTING FOR STRUCTURAL HEALTH MONITORING SENSORY SYSTEMS

16.3.1 Sensors and structural health monitoring sensory systems

As defined in Chapter 3, a sensor is a converter or transducer that measures a physical quantity and converts it into a signal which can be recorded by an observer or by an instrument, whereas a sensory system is a group of specialised sensors with a communication network intended to monitor the conditions at diverse locations of a structure. In general, an SHM sensory system is designed and established with the aim that the signals obtained from the sensors directly correlate with, and are sensitive to, the potential damage as much as possible. Meanwhile, one also strives to make the sensors as independent as possible from all other sources of environmental and operational variability, and, in fact, independent from each other (in an information sense), in order to provide maximal data with minimal sensor array outlay (Park et al. 2008). There are two basic types of sensory systems usually involved in the application of civil structures: wired sensory systems which utilise electrical cables, fibre-optic cables or another physical transmission medium between the sensing nodes for both power supply and communication, and wireless sensory systems which utilise wireless data connections and transmissions to connect network nodes.

A wired sensory system usually involves a sensor network directly connected to central processing hardware or a central database server, as mentioned in Chapter 5. The major advantages of this type of sensory system include the wide variety of commercially available off-the-shelf systems that can be used for monitoring, the reliable communication links provided by the cables and the wide variety of sensors, such as those introduced in Chapter 3, that can typically be interfaced with such a system. However, the installation and maintenance of the wired sensory system is expensive and labour-intensive, especially for civil structures due to their large sizes and complex structural systems. For example, the cost of installing over 350 sensing channels on the Tsing Ma suspension bridge in Hong Kong is estimated to have exceeded HK$8 million. Moreover, the direct wired connections to the processing unit make these systems one-point failure sensitive, as one wire may be as long as a few hundred metres. Also, these systems usually require AC power, making them vulnerable to malfunction during extreme events such as strong winds or severe earthquakes, because in these events the power supply would probably fail or be suspended.

To overcome the limitations of wired sensing networks, significant efforts have been made to integrate wireless communication technologies into SHM methods. Data transmission in wireless sensory systems is conducted without a physical link but based instead on RF communication. A core component of these systems is their 'smart' wireless sensors, which can be distinguished from standard sensors by their intelligence capabilities – that is, their on-board microprocessors. Microprocessors are typically used for digital processing, analogue-to-digital or frequency-to-code conversions, calculations and interfacing functions, which can facilitate self-diagnostics, self-identification and self-adaptation (decision-making) functions (Kirianaki 2002). It can also decide when to dump/store data and controls when and how long it will be fully awake so as to minimise power consumption (Spencer et al. 2004). Wireless communication provides a promising way to remedy the cabling problem of the traditional wire-based monitoring system and significantly reduces the maintenance cost. Nevertheless, for the large-scale implementation of SHM wireless sensory systems in practical situations, several serious issues arise with the current design and deployment scheme of the decentralised wireless sensing networks (Park et al. 2008). For example, since current wireless sensing design usually adopts ad hoc topologies, a network device has to receive several simultaneous requests to store or retrieve data from other devices on the network, resulting in a problem referred to as *data collision*. Thus, the nodes near the centralised base station are susceptible to data collision because most data flow through these nodes, and they will use up any battery power faster than the remote nodes. A number of excellent literature reviews are available on wireless sensors and their applications in the health monitoring of civil structures (e.g. Spencer et al. 2004; Lynch and Loh 2006; Farrar et al. 2006, 2011; Yun and Min 2011).

16.3.2 Power requirement and management

A major concern for SHM sensing networks is their long-term reliability and source of power. As mentioned before, the cables employed in wired sensory systems can be used for not only the transmission of data but also the supply of power to the sensors. Thus, the power supply of wired sensory systems is directly related to their connected external AC power source. The aim of this section is not to discuss external power supplies to wired sensory systems but to provide a brief introduction to their power requirements and the development of techniques to reduce power consumption in wireless sensors.

Given the wireless nature of some emerging sensors, it is necessary that they contain their own power supply, which is, in most cases, a conventional battery. In order to prolong the battery's lifetime, it is critically important to optimise the energy consumption in both computing and communication processes. At an elemental level, an appropriate low-power circuit design in the wireless sensors will be helpful for power saving. On the other hand, managing power dissipation at higher levels can considerably decrease energy requirements and thus increase battery lifetime and lower packaging and cooling costs (Park et al. 2008). Moreover, there are two different approaches to lowering power consumption at the system level: dynamic voltage scaling (DVS) and dynamic power management (DPM).

Lowering the power consumption by DVS algorithms is achieved by slowing down the execution and, when appropriate, lowering the component's voltage of operation. In DVS algorithms, instead of having longer idle periods, the central processing unit (CPU) is slowed down to the point where it completes the task in time for the arrival of the next processing request while at the same time saving quite a bit of energy. Unlike DVS, the reduction of energy consumption in DPM is realised by selectively placing idle components into lower-power states and then turning off (or reducing the performance of) system components when they are idle (or partially unexploited). DVS can only be used to decrease the energy consumption of CPUs, whereas DPM can be applied to wireless communication, CPUs and all other components that have low-power states. Thus, much more efficient power saving is usually achieved by applying DPM. In recent years, some attempts have been made to combine the two approaches to pursue the most optimal results. A more detailed review on the various DVS and DPM algorithms can be found in the literature (e.g. Benini et al. 2000; Kim and Simunic Rosing 2007).

Although much progress has been made in the development of low-power devices and the improvement of battery capacity, eventually the battery power will run out and a replacement or recharge will be required. However, because of the remote placement of wireless sensors in most civil structures, retrieving the sensors to replace or recharge their batteries can become a very expensive and tedious, or even impossible, task. Moreover, the concept of *embedded* wireless sensing, the autonomous monitoring features of which are extremely appealing, cannot be fully realised if the batteries have to be periodically replaced. As concluded by Roundy (2003), for a device whose desired lifetime is in the range of 1 year or less, battery technology alone is sufficient to provide enough energy. However, in many cases, a longer service life is required in an SHM sensing system. Recently developed EH technologies, which capture ambient energy in the surrounding medium to prolong the life of the power supply or ideally generate unlimited energy for the lifespan of the sensors, offer an attractive solution to the power requirement of wireless sensors. Thus, EH methods for wireless SHM sensory systems will be given in the following subsections.

16.3.3 Energy harvesting methods for structural health monitoring sensory systems

Although the technology for large-scale alternative energy generation using wind turbines and solar cells is mature, as mentioned before, the development of EH technology on a scale appropriate for small, low-power, embedded sensing systems is still in the developmental stages, particularly when applied to SHM sensing systems. This is mainly because the amount of harvested energy appears to fall significantly short of the level required (a conversion efficiency issue). Therefore, in order for harvested energy to become a practical source of power for wireless SHM sensing systems, the improvement of conversion efficiency to increase the amount of energy generated by the EH device or the development of new and innovative methods for accumulating ambient energy are highly desirable. The aim of this

section is to provide a brief introduction to some available EH methodologies that are potentially suitable for SHM sensing applications. As this area is quite broad and a number of
state-of-the-art reviews are available (e.g. Park et al. 2008; Gilbert and Balouchi 2008;
Bogue 2009; Ulukus et al. 2015; Shaikh and Zeadally 2016), this section is not intended to
provide an exhaustive literature survey but focuses on an overview of some applications of
SHM wireless sensing systems that integrate the aforementioned EH techniques in the field
of civil engineering. It is worth noting that every EH technique mentioned so far has its own
advantages and limitations. To improve the performance of wireless sensor nodes, some
applications introduced hereinafter are conducted on the basis of a hybrid or multi-source
EH solution, where energy is extracted from a combination of different sources.

As previously mentioned, though large-scale energy generation using wind turbines and
solar cells has been developed into mature and workable technology, significant efforts have
still been made in the development of small-scale SHM wireless sensors that synthesise these
EH techniques. For example, by using a small wind turbine as a power supply, an SHM test
bed using a smart wireless sensor system was constructed on the second cable-stayed Jindo
Bridge in Korea (Park et al. 2010a,b). It was demonstrated from the experimental results
that it took about 1.4 h to generate the electrical energy to power the Imote2 wireless sensor
for one-cycle sensing. Miller et al. (2010) developed a solar system to power Imote2 wireless
sensor networks and its effectiveness has been validated on a cable-stayed bridge. Spencer et
al. (2011) proposed solar or wind energy harvesting as the power supplies for all 113 wireless
sensors in the smart monitoring system for the Korean Jindo Bridge. A wireless sensing network for SHM purposes was designed, implemented and tested on the Beaufort #25 Bridge
on the east coast of North Carolina (Zhu et al. 2014). These wireless sensing nodes were
designed not only for the reliable communication of acceleration response measurements but
also for the self-powered capacity provided by solar panels and miniature wind turbines.

One of the most effective methods of implementing an EH system for powering wireless
sensor nodes is to convert mechanical vibrations into the desired electric power. The application of this kind of energy harvesting for wireless sensing networks can be found in many
fields, including electronic engineering (e.g. Mitcheson et al. 2008; Torah et al. 2008) as
well as mechanical and aerospace engineering (e.g. Roundy and Wrigh 2004; Pearson et al.
2012). Significant attention has also been paid in civil engineering structures, and much
progress has been made. For example, through the conversion of mechanical energy to
electrical energy, a self-powered wireless strain sensor was developed by Elvin et al. (2001)
and employed for damage detection (Elvin et al. 2003). A novel self-powered wireless sensing system was developed which was able to harvest the traffic-induced vibrations of the
bridge by means of a linear EM generator. The utilisation of this EM generator allowed up
to 12.5 mW of power to be generated in resonant mode with the frequency of excitation at
3.1 Hz (Sazonov et al. 2009). A segment-type piezoelectric energy harvester was designed
to efficiently generate electric energy from ambient vibrations with multi-modal frequencies,
and was used to power wireless sensors for real-time temperature monitoring to be used in
building automation (Lee et al. 2009). Based on the wind-induced vibrations of stay cables,
an EM energy harvesting system was proposed for powering the wireless sensor node on
the cable (Jung et al. 2011a). By replacing the EM induction part with a moving mass and
a rotational generator, the performance of this EH system was significantly improved by
Kim et al. (2013). From the field test, it was observed that the normalised output power
of the enhanced system was 35.67 mW/(m/s^2)^2, while that of the original device was only
5.47 mW/(m/s^2)^2. A self-powered sensing system with the aim of detecting the long-term
strain history of the pavement structure was introduced by Rhimi et al. (2012). Attempts
were made to integrate vibration-based and RF-based power harvesting elements within
the fully packaged Phoenix wireless sensor node with the aim of establishing a permanent

wireless monitoring system on the New Carquinez Bridge (NCB) in California (Kurata et al. 2010a,b). A novel parametric frequency-increased generator (PFIG) was developed by Galchev et al. (2011a,b,c) to harvest traffic-induced bridge vibrations for SHM purposes and also tested on the NCB. It was found that the fabricated device was able to generate a peak power of 57 µW and an average power of 2.3 µW from an input acceleration of 0.54 m/s^2 at only 2 Hz. Moreover, further short- and long-term tests of a vibration-based harvesting system based on the PFIG for the health monitoring of the NCB were conducted by McCullagh et al. (2012; 2014). An evolutionary computational approach for structural damage detection using the self-powered wireless sensor data was recently proposed and numerically investigated via a complicated case involving the gusset plate of a bridge (Alavi et al. 2016a,b).

Since the signal transmission in the wireless sensory system is basically in the form of RF waves, substantial attention has also been paid to the development of a sustainable SHM sensing system with integrated RF energy harvesting technologies. For example, based on the single-chip solution, a prototype impedance-based SHM wireless sensor node with a custom-designed RF wireless energy delivery system was presented by Mascarenas et al. (2007) and experimentally investigated in a portal frame structure. A new hybrid wireless sensor network paradigm, in which both power and data interrogation commands were provided by a mobile agent with the aid of RF signals, was proposed by Taylor et al. (2009) and validated via a field test on the Alamosa Canyon Bridge in southern New Mexico (Farinholt et al. 2009; Mascarenas et al. 2010). By extracting energy from microwave RF signals, an EH system was designed and tested by Liu et al. (2011) to power strain gauge sensors for sensing and data communication.

Compared with other EH methods, the application of the thermal energy harvesting technique for SHM wireless sensors in the field of civil engineering is relatively limited. This is primarily because the available temperature difference varies and may be very small in civil structures such as buildings and bridges, making reliable harvesting a challenge. Even so, some applications for this type of EH technique can still be found. For example, Inman and Grisso (2006) proposed an integrated autonomous sensory system including a vibration-based and temperature-based energy harvester, a battery charging circuit, local computing and memory, active sensors and wireless transmission. Farinholt et al. (2010) presented experimental investigations using EH and wireless energy transmission to power wireless structural health monitoring sensor nodes. Thermal harvesting from the temperature difference between the bridge surface and the outside air was investigated and showed promise.

It should be noted that it is impossible to present all of the available applications of the EH methods for the SHM sensory systems in this section. The aforementioned examples are only a part of the surge of research and development in this area. For the sake of better understanding, several interesting examples introduced earlier are discussed further in the following subsections.

16.3.4 Applications of energy harvesting systems to structural health monitoring

16.3.4.1 Energy harvesting system using the wind-induced vibrations of a stay cable to power a wireless sensory system

By using the translational vibration of a stay cable, an EM energy harvester was developed by Jung et al. (2011a) for the generation of sufficient electricity to power a wireless sensor node attached to a cable under gentle to moderate wind conditions. Nevertheless, from a practical perspective, several limitations exist for this energy harvester, such as its applicability in the case of an inclined cable. To address these limitations, Jung et al. (2012) developed a new EM energy harvester. However, the power generated by this energy harvester was quite small: a peak power of 0.16 mW on a measured RMS acceleration of 1 m/s² in the field test, insufficient for powering a wireless sensor, which consumes electrical energy of 26.14 mW for once-a-day measurement. Thus, by replacing the EM induction part with a moving mass and a rotational generator, an enhanced rotational energy harvester (EREH) was further developed by Kim et al. (2013) for the purpose of making the device applicable to a real cable. A brief introduction to this EREH follows.

Figure 16.1 shows the prototype EREH. Due to its low terminal resistance and high efficiency at nominal voltage, the 339150 Maxon electric motor was selected as a power generator. The position of the proof mass can be easily adjusted along the slit in the moment arm to tune the natural frequency of the harvester. The stiffness of the spiral spring is calculated as 0.65 N/m, and the weight of the proof mass is given as 139.5 g. The minimum and maximum values of the length of the moment arm are given respectively as 6 and 12 cm, and thus, the natural frequency of the device can be changed from 2.67 to 4.55 Hz.

The harvested power P_{EREH} can be calculated as

$$P_{EREH} = \frac{\eta_{EREH}}{4R_{EREH}} \left(\frac{60}{2\pi} \cdot \frac{N_{EREH}}{\upsilon_{EREH}} \cdot V_{EREH} \right)^2 \tag{16.1}$$

where:

R_{EREH} is the resistance of the generator
η_{EREH} is the power conversion efficiency
N_{EREH} is the gear ratio
υ_{EREH} is the speed constant
V_{EREH} is rotational velocity

For the purpose of improving the performance, EREH must be accurately tuned to the ambient vibration – that is, the vibration of an inclined cable. The cable, with a length of approximately 160 m, was installed in the middle of an in-service cable-stayed bridge (Cho et al. 2010). EREH was attached at 3% of the cable length from the bottom anchor, where a wireless sensor of the Imote2 was placed for recording the acceleration response. By analysing the measured acceleration, three natural frequencies of the cable (2.734, 3.369 and 4.053 Hz) were found and used to tune the EREH. In this study, the EREH was tuned to 2.734 Hz.

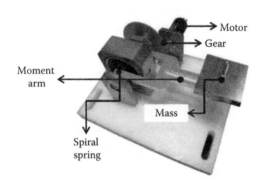

Figure 16.1 Prototype of EREH. (From Kim, I.H., et al., *Smart Mater. Struct.*, 22, 075004, 2013.)

The performance of the EREH is investigated via the field tests. In order to make the moment arm parallel with the cable, the rotational angle of the spiral spring is adjusted. To evaluate the performance of the EREH, the input accelerations and the output voltages obtained from the EREH are compared with those from the original rotational energy harvester (OREH) (Jung et al. 2012). The comparison results are given in Table 16.1. In the table, the electric powers are estimated based on the maximum power transfer theory. It can be seen from Table 16.1 that the average value of the output power increases 18.6 times (from 0.14 to 2.60 mW), while that of the input RMS acceleration rises 1.7 times (from 0.16 to 0.27 m/s^2).

The performance of the EREH under the different levels of cable vibration is also investigated (see Table 16.2). It can be found from Table 16.2 that as the RMS acceleration increases 4.7 times (from 0.34 to 1.60 m/s^2), the average power increases 9.6 times (from 2.1 to 20.3 mW). This shows that the EREH can exploit more vibrational energy as the input acceleration increases.

Finally, the possibility of using EREH to power a wireless sensor node for the SHM of the cable is investigated based on the field test results. When the amounts of energy consumption of a wireless sensor and the power production of an energy harvester are given, the required charging time can be calculated as follows:

$$\text{Required charging time (h)} = \frac{\text{Energy consumption (mWh)}}{\text{Generated power (mW)}} \times \text{Efficiency} \qquad (16.2)$$

In this equation, 'Efficiency' is the charging efficiency of the batteries of a wireless sensor node. If the measured performance of the EREH is applied to the Imote2, which consumes 26.14 mWh for once-a-day measurement and 41.48 mWh for twice-a-day measurement, the required charging time can be calculated according to Equation 16.2. In the case of once-a-day measurement, the required charging time at small, normal and large vibration levels is 13.8 h, 3.42 h and 1.43 h, respectively, whereas the required charging time in the case of twice-a-day measurement is 21.9 h, 5.42 h and 2.27 h, respectively. It can be found that it takes 3.42 and 5.42 h, in once-a-day and twice-a-day measurement mode, respectively,

Table 16.1 Performance comparison

	Acceleration (m/s^2)		Output voltage (V)		Estimated power (mW)	
	RMS	Max.	RMS	Max.	Average	Max.
OREH	0.16	1.00	0.129	0.327	0.14	1.10
EREH	0.27	1.13	0.073	0.323	2.60	50.52

Source: Kim, I.H., et al., Smart Mater. Struct., 22, 075004, 2013.

Table 16.2 Test results with different vibration levels

Vibration level	Small	Normal	Large
Maximum acceleration (m/s^2)	1.27	2.47	2.77
RMS acceleration (m/s^2)	0.34	1.02	1.60
Maximum voltage (V)	0.221	0.352	0.338
RMS voltage (V)	0.065	0.133	0.205
Maximum power (mW)	23.6	59.9	55.3
Average power (mW)	2.1	8.5	20.3

Source: Kim, I.H., et al., Smart Mater. Struct., 22, 075004, 2013.

to generate the electricity, under normal acceleration, to operate a wireless sensor for one day. It is not difficult to acquire this vibration level under gentle to moderate wind conditions (a wind speed of 5–6 m/s) at the bridge site considered in this study (Cho et al. 2010). Therefore, this demonstrates that the EREH can be used as a power supply for a wireless sensor node placed on an inclined stay cable under gentle to moderate wind conditions.

16.3.4.2 Long-term test of a vibration energy harvesting system for the health monitoring of bridges

As previously mentioned, a novel PFIG was developed by Galchev et al. (2011a,b,c) to generate electric energy from traffic-induced bridge vibrations for the purpose of powering wireless sensors to monitor the health of bridges. In order to increase the output power and improve the long-term reliability and robustness of the harvester, its technical characteristics were adjusted and an enhanced PFIG was created (McCullagh et al. 2014). For the sake of clarification, the original PFIG is referred to as *PFIG-B1*, whereas the enhanced one is called *PFIG-B2*. This section gives a brief introduction to the long-term testing of PFIG-B2 for the health monitoring of bridges.

To build PFIG-B2, the adjustments implemented to maximise power transfer to the circuit include decreasing the number of coil turns to lower the PFIG-B2's output impedance to 300 Ω, implementing a double magnet structure to better confine and route the flux, and changing the inertial mass suspension spring material to improve its reliability, in combination with new assembly techniques. These changes provide more flux linkage and achieve a better volumetric efficiency in the frequency-increased generator (FIG).

The goal of this development was to monitor the bridge harvesting system (BHS) over a long period of time and to monitor the FIG outputs and the storage capacitor voltage so that the BHS performance could be quantitatively assessed. For this goal, a special-purpose Narada wireless sensor node was used and deployed on the NCB. The Narada node samples the key BHS metrics once per hour for 90 s at a sampling rate of 100 Hz. The results are wirelessly transmitted to a base station and are then available through remote access (Kurata et al. 2010a). To facilitate long-term evaluation of the BHS, the Narada node automatically discharges the storage capacitor when it reaches 0.7 V. This allows for the average power to be estimated by the number of discharges in the 90-second sampling period. Solar cells on top of the bridge are wired and power the Narada node, allowing it to perform the described functions. It is important to note that the BHS is not used to power the Narada wireless sensor node. Figure 16.2 shows the long-term testing system. The PFIG-B2, harvesting circuit, buffer and level-shift circuits, transmission antenna, Narada node and rechargeable batteries (powered by solar panels on top of the bridge) are contained inside a commercially available water-tight box. The batteries are used to power the Narada system, the buffers and the level-shift circuits, which interface the BHS to the Narada node. The dimensions of the box are 30.5 × 20.3 × 13.2 cm.

Figure 16.3 summarises 1 week of data. Harvested power is greater during the daytime and on weekdays. Each discharge occurring at 0.7 V in the 10 μF capacitor, recorded during a 90 s period, corresponds to 27 nW of power delivered to a load. Once again, it should be noted that data recordings are only made once per hour and the number of discharges shown in Figure 16.3 is only during the several 90 s measurement windows.

The BHS operated continuously on the NCB for 13 months starting 30 April 2012 (see Figure 16.4). Starting in late October 2012, the wireless data collection system began to have outages. In many of these cases, measurements from the BHS locally written to non-volatile memory on the wireless sensor node were still received a few times a day. The root cause of this

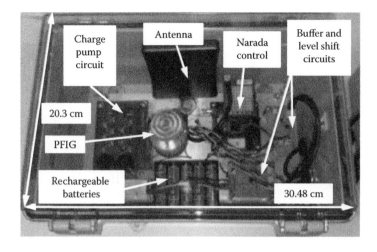

(a)

(b)

Figure 16.2 Long-term testing system on the NCB: (a) the water-tight box, (b) a schematic diagram of the long-term test system. (From McCullagh, J.J., et al., *Sens. Actuators, A: Phys.*, 217, 139–50, 2014.)

is suspected to be the solar harvesting used to power the Narada nodes. Likely, there was not enough light during this time of the year (winter season), when wet and foggy weather is common. This highlights a potential weakness of solar energy harvesting in unmanned systems. In the long-term test (see Figure 16.4), the same weekly pattern seen in Figure 16.3 continues to be observed, as well as sharp drops in harvested power on US national holidays. Over the first 6 weeks the power output remained relatively consistent. The highest power was recorded on the morning following Memorial Day on 29 May 2012 and is estimated to be 10.9 µW (22 discharges over 90 s; see Figure 16.4). The data clearly demonstrate that the harvested power correlates directly to traffic and thus could be a sensed variable itself. The PFIG exhibited a reduction in power at the end of June 2012, producing approximately half as much energy.

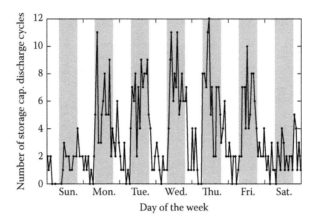

Figure 16.3 Circuit discharges for 1 week starting 13 May 2012. (From McCullagh, J.J., et al., *Sens. Actuators, A: Phys.*, 217, 139–50, 2014.)

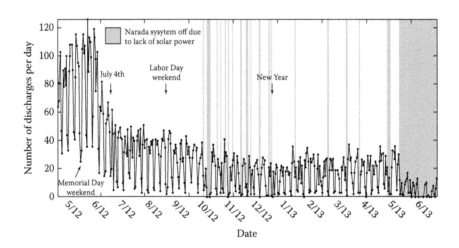

Figure 16.4 Entire transmitted data set since installation on 30 April 2012. (From McCullagh, J.J., et al., *Sens. Actuators, A: Phys.*, 217, 139–50, 2014.)

16.3.4.3 *Self-powered strain sensor for damage detection*

The performance of a self-powered strain energy sensor (SES) for measuring the load applied to a structure and transmitting the data wirelessly to a remote receiver was investigated by Elvin et al. (2001). The SES uses piezoelectric material to convert the applied mechanical strain to an electrical charge. This charge is then used to telemeter a signal to a remote receiver. The theoretical and experimental feasibility of using the SES for damage detection in a simple structure was further demonstrated by Elvin et al. (2003), and it is briefly introduced in the following.

One possible implementation of the SES for damage detection is schematically shown in Figure 16.5. In this case, a network of SESs (shown by SES1 and SES2) are embedded in the host structure during construction or attached to the structure at a later date. A loading rig (shown by a two-roller cart in this case) is used to apply the loads to the structure of interest with the aim of generating sufficient mechanical strain in the SESs to electrically power the transmitter. The transmitter signal is captured by the receiver and is analysed by

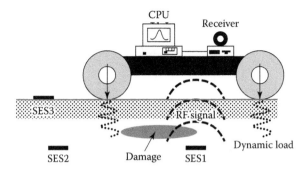

Figure 16.5 Possible implementation of an SES for damage detection. (From Elvin, N. et al., *J. Strain Anal. Eng. Des.*, 38(2), 115–24, 2003.)

the associated computer's CPU. By comparing the output of the SESs at periodic intervals throughout the life of the structure, the growth and appearance of damage zones can be ascertained.

A simply supported beam under four-point loading is considered to demonstrate the SES's ability to detect damage. The piezoelectric material for the SES is assumed to be 28 μm thick polyvinyl difluoride (PVDF) membrane. The structure is assumed to be linearly elastic and can be modelled using a plane stress finite element (FE) model. The damage of the structure is simulated by removing a rectangular section of material at the centre of the beam. The effect of increasing damage size can be studied by varying the fracture length, as shown in Figure 16.6. It should be noted that the results in Figure 16.6 are normalised so that the undamaged beam has a constant axial strain of 1. It can be found from Figure 16.6 that, firstly, substantial changes in strain occur only over a relatively localised area and, secondly, substantial strain changes only occur for relatively deep cracks. The voltage generated by surface-mounted piezoelectric material is proportional to the integrated axial strain $\int \varepsilon_{xx}$ applied over the entire sensing area. In the present illustrative model, the sensor is surface mounted and thus only in-plane strains produce an electric charge.

The performance of the SES in detecting damage in a real structure was experimentally conducted and compared with the numerical results. A simply supported beam structure was built, and a trapezoidal loading of 1 Hz frequency was applied to the beam and measured by a load cell. The 28 μm PVDF sensor was attached to the beam and electronic leads were then attached from the PVDF sensor to the transmitter circuit using CW2400 conductive epoxy. The damaged zone was introduced by sequentially cutting the beam in 2 mm increments using a standard band saw. The resulting cut was 1.5 mm wide.

A comparison of the experimental and numerical results at various crack depths is plotted in Figure 16.7. For consistency, Figure 16.7 is normalised so that the undamaged beam (crack length 0 mm) has a value of 1. It should be noted that the crack depths were made with a band saw with an accuracy of ±0.2 mm. Experimental data are only available for crack depths of less than 8 mm. When the crack depth was extended to 10 mm, and the same displacement of 2.2 mm used in the previous experiments was applied, the beam failed at the damaged zone. Data were collected three times per crack depth. The centre of each circle in Figure 16.7 corresponds to the average value of the three measurements. The deviation between the maximum and minimum measurements was small and it is thus not shown. It can be seen that the FE results broadly agree with those obtained from the experiment.

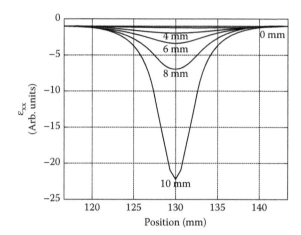

Figure 16.6 Axial strain distribution over the sensor length for various damage depths. (From Elvin, N., et al., *J. Strain Anal. Eng. Des.*, 38(2), 115–24, 2003.)

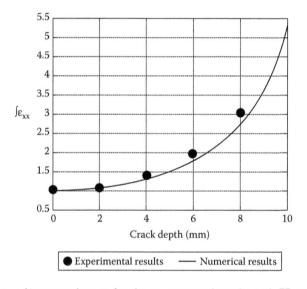

Figure 16.7 Comparison of integrated strain for the experimental results with FE results. (From Elvin, N., et al., *J. Strain Anal. Eng. Des.*, 38(2), 115–24, 2003.)

16.4 INTEGRATED VIBRATION CONTROL AND ENERGY HARVESTING SYSTEMS

The concept of using harvested energy as the main source of powering control systems to achieve a reasonable level of vibration mitigation has gradually become an important topic of research. A survey was conducted by Wang and Inman (2012) to review the existing methods of mitigating excessive vibration using harvested energy from ambient mechanical vibrations via piezoelectricity-based transduction. In the field of civil engineering, since the amount of power required in the active control system to mitigate civil structural responses is usually large, as mentioned in Chapter 4, it would be quite challenging to provide such large amounts for active control devices using EH techniques, and thus limited research can be found to date (e.g. Scruggs 2004; Scruggs and Iwan 2003, 2005). However, the integration of vibration control and EH techniques can still be found, mainly in passive and semi-active control systems, and a brief introduction to this topic is given in the follow paragraphs.

Since no external power is required for passive control devices, EH devices in passive control systems are designed to generate electric energy in an attempt to power wireless sensor nodes or other electrical devices. For example, Jung et al. (2010a,b) experimentally investigated the feasibility of an adaptive passive control system, consisting of a fluid damper and an EM induction part, for suppressing vibration in three-story shear building structures subjected to ground accelerations. Ni et al. (2011) conducted a 76-story benchmark simulation study considering along-wind and across-wind excitations. The energy harvesting potential was assessed assuming that the overall damping power dissipated by classical, parallel and series tuned mass dampers (TMDs) is the overall power output from the TMD. Shen et al. (2012) proposed a self-powered vibration control and monitoring system composed of a pendulum-type TMD, a rotary EM device, an EH circuit and a wireless smart sensor. The regenerative electromagnetic TMD (EMTMD) was able to convert the vibration energy of structures to electrical energy that can be further stored and used to power wireless sensors to monitor structural vibration responses. Tang and Zuo (2012) presented an approach to harvesting the vibration energy from tall buildings with TMDs by replacing the energy dissipating element with an EM harvester. The simultaneous energy harvesting and vibration control were demonstrated, for the first time, by an experiment based on a three-story building prototype with an electricity-generating TMD.

It is known that the power required to operate semi-active control devices is quite small and that many of them can even work at the battery power level. Thus, it is possible to use the energy generated by EH devices as the main source or at least an alternative source for powering semi-active control devices. Increasing attention has been paid in this area with the aim of establishing sustainable smart control systems. For example, Wang et al. (2009) proposed a self-powered semi-active control system based on an magnetorheological (MR) damper for the seismic protection of bridges. The proposed system used a linear permanent-magnet DC generator that served as a velocity sensor and a power generator. The harvested power was then used to power the MR damper through a control circuit. A self-powered smart damping system composed of an MR damper and an EM induction device was proposed for cable vibration reduction (Kim et al. 2010; Jung et al. 2011b). This EM device was able to convert vibration energy into electrical energy that was used as an alternative power source for the MR damper. The performance of this smart damping system was experimentally investigated via a full-scale, 44.7 m long, high-tension cable. Sapiński (2011) presented a self-powered, self-sensing MR damper–based vibration control system. The structural vibration energy was captured by the EM induction device and applied to control the damping characteristics of the MR damper. The energy balance of the self-powered vibration reduction system with the MR damper and EM generator was further investigated numerically and experimentally (Snamina and Sapiński 2011; Sapiński et al. 2011).

In the following subsections, an electromagnetic damping and energy harvesting (EMDEH) device proposed by Shen et al. (2012) and Shen (2014) is introduced. The two fundamental issues with this device – that is, how to extract the maximum power from the EM damper and how to achieve the optimal control for the structure – are discussed. The application of the EMDEH on a full-scale stay cable is given in the last part of this section (Shen and Zhu 2015; Shen et al. 2016).

16.4.1 Electromagnetic dampers for vibration damping and energy harvesting

An introduction to EM dampers for the purpose of structural vibration attenuation has been given in Chapter 4. In light of the energy conversion from mechanical energy to electrical energy, EM devices can also serve as energy harvesting or energy regeneration devices.

In general, an EM damper is composed of two major components: a permanent magnet and coils. According to Faraday's law, when the permanent magnet and coils move relative to each other, a back electromotive force (EMF) – that is, an open-circuit voltage – is generated in the coils. The open-circuit voltage is proportional to the velocity of the moving part in the EM damper but opposite in sign. The open-circuit voltage u_0 is given by (Okada et al. 1997; Palomera-Arias 2005)

$$u_0 = -e = K_e \dot{x}_d \tag{16.3}$$

where K_e is the back EMF constant (V·s/m), dependent on the geometric and magnetic properties of the EM damper, and $\dot{x}_d$ is the velocity time history of the EM damper.

According to Lorentz's law, the back EMF produces a current if the circuit is closed. Consequently, EM damping force f_{em} is exerted on the moving magnet, given by (Okada et al. 1997; Palomera-Arias 2005)

$$f_{em} = K_f i_0 \tag{16.4}$$

where i_0 is the instantaneous current in the coils and the proportional coefficient K_f is the force constant of the damper (N/A), which is equal to K_{em}. Hence, $K_{em} = K_f = K_e$, which is also known as the *machine constant* of EM damper.

The EM damping force is always against the relative movement and converts a portion of the vibration energy into electricity instead of heat. The electrical energy can thus be extracted if properly designed circuits are used.

The damping characteristics of EM dampers are established with respect to parasitic damping and EM damping. Here, the parasitic damping is modelled by a superposition of two components: viscous damping and Coulomb damping. These two damping forms are widely considered in the dynamic analyses of civil structures. The latter is a typical rate-independent damping form used to account for the friction effect. Under harmonic excitation, the average power of parasitic damping can be estimated by

$$P_p = 4F_c f d + 2c_m \pi^2 f^2 d^2 \tag{16.5}$$

where:

F_c represents the magnitude of the Coulomb friction force
c_m is the viscous damping coefficient
d is the displacement amplitude of harmonic motion
f is the oscillation frequency of harmonic motion

According to the equal energy dissipation rules, the parasitic damping coefficient C_p can then be given as

$$C_p = \frac{2F_c}{\pi^2 f d} + c_m \tag{16.6}$$

where F_c and c_m are the constants to be evaluated using P_p. It can be seen that the parasitic damping coefficient C_p varies with the frequency and amplitude of the harmonic oscillation because of the consideration of Coulomb damping.

EM damping can be determined as (Graves 2000; Stephen 2006)

$$C_{em} = \frac{K_{em}^2}{R_{coil} + R_{load}}$$

(16.7)

in which R_{load} is the external resistance and R_{coil} is the resistance of coils. It can be observed that the EM damping coefficient can be easily changed by adjusting the external resistance. Maximum EM damping can be achieved with a short circuit – that is,

$$C_{em,max} = K_{em}^2/R_{coil} \quad \text{when } R_{load} = 0$$

(16.8)

As mentioned before, the total damping coefficient of EM damper C_d is the superposition of the parasitic and EM damping, and thus can be expressed as

$$C_d = C_p + C_{em}$$

(16.9)

16.4.2 Energy harvesting circuits in electromagnetic dampers

The electrical energy converted by an EM damper can be harvested and stored when an EH circuit is connected to the EM damper. Then, the corresponding EM damping can be estimated by substituting the input resistance of the EH circuit into Equation 16.7. Therefore, the EH circuit features not only affect harvesting efficiency but also the damping characteristic of the designed EMDEH device. Achieving efficient energy harvesting and optimal damping performance should be addressed simultaneously in the design of an EH circuit.

For the design of an EH circuit, the selection of energy storage element should be considered first. A supercapacitor or rechargeable battery is usually employed as an energy storage element in energy harvesting, but a rechargeable battery is often preferable because of its higher power density and lower self-discharge rate (Casciati and Rossi 2007). Compared with a supercapacitor, the voltage of a rechargeable battery is more stable during the charge process. However, rechargeable batteries usually have stringent charge requirements to avoid potential overcharge, which may damage the batteries, while supercapacitors are able to withstand very high charge and discharge rates and require relatively simple charging methods. Besides, supercapacitors do not suffer from memory effects like some batteries and have virtually very long lives.

Given that the damping coefficient of an EMDEH device depends on the input resistance of the EH circuit, as shown in Equation 16.7, an EH circuit with nearly constant input resistance is favourable because the EMDEH device can maintain a stable damping coefficient close to the optimal value. The maximum harvesting efficiency can be maintained in a certain operating range if the input impedance is resistive and equals the optimal load resistance. A stand-alone simple EH circuit with low power consumption is also desirable, given the low output power in vibration-based energy harvesting. A theoretical and experimental study of linear EM dampers connected with four representative circuits was conducted by Zhu et al. (2012), where the dynamic characteristics of linear EM dampers, including parasitic damping, EM damping, energy conversion efficiency and output power, were modelled and discussed systematically for each circuit. In this section, a discontinuous conduction mode (DCM) buck-boost converter with a fixed duty cycle (Lefeuvre et al. 2007) is employed as the performance optimisation circuit to optimise both vibration damping and energy harvesting.

Figure 16.8 presents the architecture of the EH circuit, which is comprised of a full-wave bridge rectifier, a DCM buck-boost converter with a fixed duty cycle and a rechargeable battery. The bridge rectifier utilises Schottky diodes because of its low forward voltage drop (typically 0.25 V). An input capacitor connected to the bridge rectifier smoothens the DC voltage waveform input to the buck-boost converter. Another function of the input capacitor is to serve as a power supply for a crystal clock generating a control signal to drive the power switch SW (see Figure 16.8). The average input resistance of a DCM buck-boost converter can be estimated as follows (Erickson 1997; Lefeuvre et al. 2007):

$$R_{\mathrm{in,dcm}} = \frac{2Lf_{\mathrm{sw}}}{d_{\mathrm{c}}^2} \qquad (16.10)$$

where L and f_{sw} are the inductance and switching frequency of the buck-boost converter, respectively, and d_{c} denotes the duty cycle. Equation 16.10 indicates that the average input resistance of the DCM buck-boost converter is controlled by the duty cycle of the power switch, the inductance and the switching frequency. Given the prescribed circuit parameters, the DCM operation leads to a constant input resistance of the buck-boost converter.

16.4.3 Power flow in electromagnetic dampers

Power flow or energy flow is a fundamental issue in energy harvesting from a vibrating structure. According to the energy balance equation, the input energy to a structure subjected to dynamic external excitations is always equal to the summation of the kinetic energy of the structural mass, the elastic strain energy, the dissipative energy caused by inherent structural damping and the dissipative energy caused by passive dampers, if any (Soong and Dargush 1997). The last part of the energy would be absorbed by the EM dampers if the structure is equipped with an EMDEH system. On the other hand, this part of the power becomes input power P_{in} to the EMDEH system:

$$P_{\mathrm{in}} = f_{\mathrm{em}} \cdot \dot{x}_{\mathrm{m}} \qquad (16.11)$$

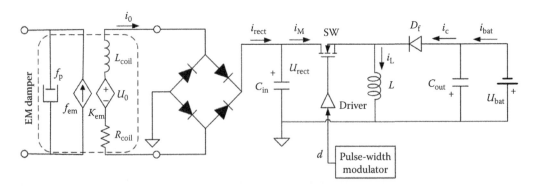

Figure 16.8 Schematic diagram of the performance optimisation circuit using a DCM buck-boost converter. (From Shen, W.A., and S.Y. Zhu., *J. Intell. Mater. Syst. Struct.*, 26(1), 3–19, 2015.)

where f_{em} and $\dot{x}_m$ are the instantaneous damper force and the velocity of the moving magnet, respectively. In the energy conversion process from the damping energy of an EMDEH device to the terminal electrical energy stored in an energy storage element, multiple power losses occur when the power flows through the EM damper and EH circuit. The input power is dissipated by two different damping effects of the EM dampers – that is, parasitic damping power P_p and EM damping power P_{em}. Parasitic damping arises due to various mechanical power losses when the EM dampers are in motion. In general, parasitic damping is independent with the current in circuit and can be evaluated in an open-circuit situation. EM damping arises only when the circuit is closed and the current flows in the circuit. Only a portion of electric power, output power P_{out}, can finally be stored in energy storage elements or utilised by end instruments. The other part will be dissipated by the copper loss of the coil, P_{coil}, and the power loss induced by the EH circuit, P_{ehc}. The energy balance equation can be written as

$$P_{in} = P_p + P_{em} = P_p + P_{coil} + P_g = P_p + P_{coil} + P_{ehc} + P_{out} \tag{16.12}$$

where P_g is the gross output power from the EM dampers. These average power terms can be calculated as follows:

$$P_p = \frac{1}{T_c} \int_{t_0}^{t_0+T_c} C_p \dot{x}_d^2(t) dt \; ; \quad P_{coil} = \frac{1}{T_c} \int_{t_0}^{t_0+T_c} i_0^2(t) R_{coil} dt \; ; \quad P_g = \frac{1}{T_c} \int_{t_0}^{t_0+T_c} u_i(t) i_0(t) dt$$

$$P_{out} = \frac{1}{T_c} \int_{t_0}^{t_0+T_c} u_c(t) i_c(t) dt \; ; \quad P_{ehc} = P_g - P_{out} \tag{16.13}$$

where:

T_c	is the calculation period
i_0	is the instantaneous current flowing in the damper coils; in particular, if a rectifier is used, $|i_0| \approx i_{rect}$
u_i	is the instantaneous output voltage from the EM damper coils
u_c and i_c	are the instantaneous voltage and charging current of the energy storage element, respectively

16.4.4 Energy harvesting efficiency of electromagnetic dampers

The EH efficiency of an EMDEH device is defined in this section to quantify the performance of EMDEH devices. This quantification enables the assessment of the design of an EMDEH device given different EM damper parameters and different circuits. The EH efficiency of an EMDEH is defined as follows:

$$\eta = \frac{P_{out}}{P_{in}} = \eta_1 \cdot \eta_2 \cdot \eta_3 \tag{16.14}$$

where:

η	is the overall EH efficiency of the EMDEH device (harvesting efficiency)
η_1	is the electromechanical coupling coefficient that describes the conversion efficiency from the mechanical power to electrical power
η_2	is the efficiency of the EM damper, which is affected by the power loss because of the coil resistance
η_3	is the efficiency of the EH circuit

These efficiency terms can be calculated as

$$\eta_1 = \frac{P_{em}}{P_{in}}; \quad \eta_2 = \frac{P_g}{P_{em}}; \quad \eta_3 = \frac{P_{out}}{P_g} \tag{16.15}$$

All of the efficiency terms expressed in Equations 16.14 and 16.15 should be less than one because of the power loss in each conversion process. Notably, all the efficiency terms in this section are average efficiency unless otherwise stated.

Minimising the total power loss in an EMDEH device, $P_p + P_{coil} + P_{ehc}$, is necessary to enhance the overall energy conversion efficiency. The power loss because of the parasitic damping should be minimised to enhance the efficiency η_1; a small resistance of the coil can maximise η_2; and a high-efficiency circuit design can maximise the ratio η_3. The optimisation of the EH performance of EMDEH devices aims to maximise output power, which claims for the maximum harvesting efficiency η_{max} and the maximum input power $P_{in,max}$. The harvesting efficiency η is dependent on the EMDEH device alone, including the EM damper used and the EH circuit, while the maximum input power needs to be optimised by the dynamic analysis of the structure-EMDEH system.

An optimal load resistance or optimal input resistance of the EH circuit is proposed by Shen (2014) for the purpose of achieving the maximum harvesting efficiency.

$$R_{opt} = R_{coil}\sqrt{1 + \frac{K_{em}^2}{C_p R_{coil}}} \tag{16.16}$$

16.4.5 Testing electromagnetic dampers

16.4.5.1 Design and fabrication

The performance optimisation circuit using a DCM buck-boost converter mentioned earlier was designed and fabricated for the device test in this subsection, and its experimental application to a scaled stay cable will be presented in Section 16.4.9. Figure 16.9 shows the experimental circuit diagram of the performance optimisation circuit using a DCM buck-boost converter. In this circuit, a low-power clock oscillator IC (OV-1564-C2, Micro Crystal, Switzerland) with a typical supply voltage in the 1.2 to 5.5 V range was used to generate a fixed-frequency ($f_{sw} = 32.768$ kHz) rectangular driving signal for the metal–oxide–semiconductor

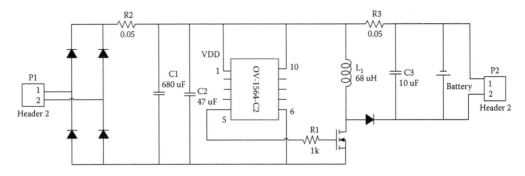

Figure 16.9 Experimental circuit diagram of the performance optimisation circuit using a DCM buck-boost converter. (From Shen, W.A., Electromagnetic damping and energy harvesting devices in civil structures, PhD diss., Department of Civil and Environmental Engineering, Hong Kong Polytechnic University, 2014.)

field-effect transistor (MOSFET). The duty cycle d_c of the clock oscillator IC ranged from 0.4 to 0.6 (typically, 0.5). Based on the typical duty cycle ($d_c = 0.5$) and switching frequency ($f_{sw} = 32.768$ kHz), an inductor ($L = 68$ μH) was used to emulate a 17.8 Ω resistor according to Equation 16.10. A rechargeable NiMH battery (nominal voltage 3.7 V) was used because of its low charging current, which ranges from 8.4 mA to 140 mA. An additional 10 μF tantalum capacitor (C3) was added parallel to the NiMH battery to reduce ripple current across the battery, thus enabling longer battery life. The printed circuit board was designed using industry-standard computer-aided design tools. A single-layer board, 3.0 × 9.5 cm, was designed using the Protel DXP software. The complete circuit placement design from Protel DXP is shown in Figure 16.10a. A picture of the prototype circuit board is shown in Figure 16.10b.

16.4.5.2 Experimental setup

The experimental study on an EMDEH device with a performance optimisation circuit was divided into two parts: a circuit board test alone with respect to input resistance and efficiency, and a test of the damping characteristics of the EMDEH device when connected to the performance optimisation circuit. The former is shown in Figure 16.11. A DC power supply was used to power the circuit board. The output signal of the clock oscillator IC and the driving signal for the MOSFET were measured by an Agilent oscilloscope. The rectifier output current and the charging current were measured through the voltage drop of two

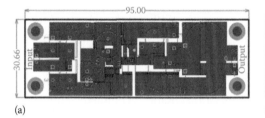 

(a)   (b)

Figure 16.10 The printed circuit board of the performance optimisation circuit: (a) printed circuit board layout, (b) photo of the prototype circuit board. (From Shen, W.A., Electromagnetic damping and energy harvesting devices in civil structures, PhD diss., Department of Civil and Environmental Engineering, Hong Kong Polytechnic University, 2014.)

Figure 16.11 Experimental setup of the performance optimisation circuit. (From Shen, W.A., Electromagnetic damping and energy harvesting devices in civil structures, PhD diss., Department of Civil and Environmental Engineering, Hong Kong Polytechnic University, 2014.)

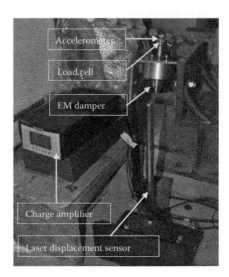

Figure 16.12 Experimental setup of the EM damper connected to the performance optimisation circuit. (From Shen, W.A., Electromagnetic damping and energy harvesting devices in civil structures, PhD diss., Department of Civil and Environmental Engineering, Hong Kong Polytechnic University, 2014.)

0.05 Ω high-precision resistors, where the small voltage drop signals were amplified by the voltage amplifier and filter with an amplification factor of 40dB. All the signals of interest, including the rectifier output voltage and NiHM battery voltage, were collected by a data acquisition system with a sampling frequency of 500 Hz. A low-pass analogue filter was set with a cut-off frequency of 100 Hz.

Figure 16.12 shows the experimental setup for the damping characteristics test of the EMDEH device connected to the performance optimisation circuit. A commercial linear voice coil motor (Moticont, model no. GVCM-095-051-01) was used as an EM damper. The diameter of the cylindroid linear EM damper was 95.3 mm and its length was 45.2 mm. A miniature load cell located between the EM damper and the cable was used to measure the damper force. In addition, the acceleration and displacement responses of the EM damper were measured by a miniature accelerometer and laser displacement sensor, respectively. The EM damper was driven by the motion of the scaled stay cable subjected to harmonic excitation. The damping characteristics test of the EM damper connected to the performance optimisation circuit was actually a part of the scaled cable dynamic test presented in Section 16.4.9.

16.4.5.3 Efficiency and input resistance of the circuit

This section presents the test results of the performance optimisation circuit with respect to input resistance and efficiency. Figure 16.13 shows the voltage–current relationship $(u_{rect}-i_{rect})$ of the rectifier output port (i.e. DCM buck-boost converter input port). The rectifier output current i_{rect} rises linearly with an increasing rectifier output voltage u_{rect} when the rectifier voltage is larger than 1.3 V, while a linear relationship with a smaller slope is observed when the rectifier voltage is smaller than 1.3 V, as shown in Figure 16.13. The linear $u_{rect}-i_{rect}$ relationship implies that the input impedance characteristic of the DCM buck-boost converter was nearly resistive, although the voltage–current curve does not pass through the origin. Thus, Figure 16.13 provides evidence that the performance optimisation circuit can approximately emulate a constant resistor. The average input resistance was estimated to be approximately 18.4 Ω. This is close to the theoretical prediction (17.8 Ω) given by Equation 16.10 where duty cycle $d_c = 0.5$.

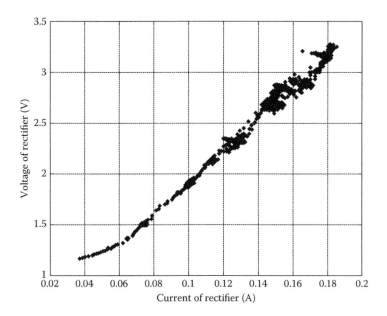

Figure 16.13 Voltage–current characteristics of the rectifier output port. (From Shen, W.A., Electromagnetic damping and energy harvesting devices in civil structures, PhD diss., Department of Civil and Environmental Engineering, Hong Kong Polytechnic University, 2014.)

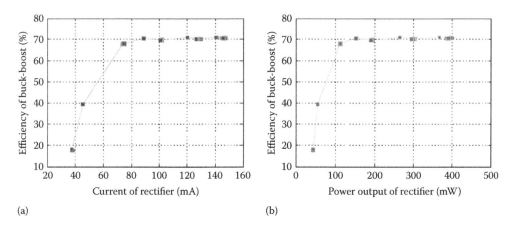

(a)   (b)

Figure 16.14 Measured efficiency–current and efficiency–power curve of the DCM buck-boost converter in the performance optimisation circuit: (a) efficiency–current, (b) efficiency–power. (From Shen, W.A., Electromagnetic damping and energy harvesting devices in civil structures, PhD diss., Department of Civil and Environmental Engineering, Hong Kong Polytechnic University, 2014.)

Figure 16.14 shows the efficiency varies with the rectifier current and rectifier output power. The DCM buck-boost converter efficiency increases rapidly from 18% to 68% when i_{rect} is in the range of 38–75 mA. The value then becomes approximately constant (71%). A similar relationship can be found in the efficiency–power curve presented in Figure 16.14b. The variation may be attributed to the change law of on-resistance of the MOSFET. The on-resistance in a DCM buck-boost converter may be up to several ohms high during low-power operation (<100 mW) but is closer to the typical value (0.33 Ω) during high-power operation (>100 mW). Therefore, a low on-resistance feature of the MOSFET is quite important in performance optimisation circuit design.

16.4.5.4 Damping characteristics

Harmonic and random vibration tests on the scaled stay cable with an EMDEH device (K_{em} = 38.0 V·s/m or N/A; R_{coil} = 9.3 Ω) connected to a performance optimisation circuit were conducted to calibrate the damping characteristics. Based on the measured displacement, acceleration and damper force, the displacement–force and velocity–force relationships can be obtained. Owing to the inertial effect of the moving mass of the EM damper, the true damping force is obtained by subtracting the inertial term.

Figure 16.15 exhibits a sample velocity–force plot obtained from the data collected. The damping force, which consists of the parasitic damping force and EM damping force, rises linearly with an increasing damper velocity. The total damping coefficient identified from the test results is 69.4 N·s/m, consisting of the EM damping coefficient C_{em} (33.5 N·s/m) and the parasitic damping coefficient C_p (30.9 N·s/m). The parasitic damping coefficient was identified in an open-circuit case.

Figure 16.16 shows a sample velocity–force plot obtained from the data collected in the random vibration test. The damping force rises almost linearly with increasing damper velocity. In this case, the measured total damping coefficient C_d is 69.6 N·s/m. When subtracting the parasitic damping coefficient C_p (32.4 N·s/m), the measured C_{em} is 37.2 N·s/m. The experimental results validate the modelling of the EMDEH device with a performance optimisation circuit. Both theoretical and experimental results confirm that the damping characteristics of the EM damper connected to a performance optimisation circuit are similar to those of linear viscous dampers. Because of the linear damping characteristics, the EMDEH device with a performance optimisation circuit can maintain a stable optimal damping setting for passive vibration control. This feature is desirable in EMDEH device design for a regenerative TMD or the vibration mitigation of bridge stay cables.

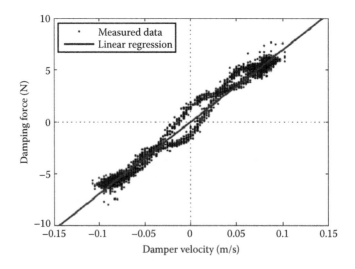

Figure 16.15 Measured velocity–force plot of the EMDEH device with the performance optimisation circuit subjected to sine sweep excitation (11.9 to 12.3 Hz). (From Shen, W.A., Electromagnetic damping and energy harvesting devices in civil structures, PhD diss., Department of Civil and Environmental Engineering, Hong Kong Polytechnic University, 2014.)

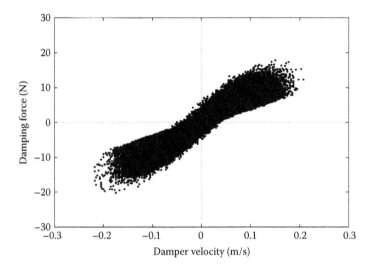

Figure 16.16 Measured velocity–force plot of the EMDEH device with the performance optimisation circuit under random vibration (0 to 25 Hz). (From Shen, W.A., Electromagnetic damping and energy harvesting devices in civil structures, PhD diss., Department of Civil and Environmental Engineering, Hong Kong Polytechnic University, Hong Kong, 2014.)

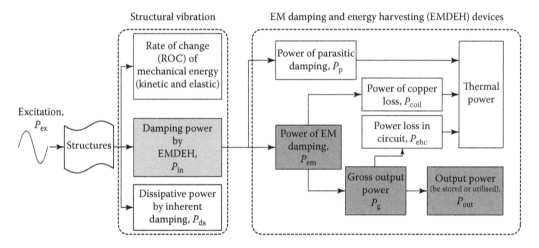

Figure 16.17 Power flow of a structure with EMDEH devices. (From Zhu, S.Y., W.A. Shen, and Y.L. Xu, *Eng. Struct.*, 34, 198–212, 2012.)

16.4.6 Power flow in integrated systems

As mentioned previously, the input energy to a structure subjected to dynamic external excitations is always equal to the summation of the kinetic energy of the structural mass, the structural strain energy, the dissipative energy caused by inherent structural damping and the dissipative energy caused by passive dampers, if any. This energy balance or energy conversion is represented in the power flow shown in Figure 16.17, which enables the direct assessment of the EH performance, in terms of output power and EH efficiency.

A general multi-degree-of-freedom (MDOF) structure equipped with EMDEH devices is considered. One or more EMDEH devices can be installed in an MDOF structure for vibration mitigation and energy harvesting simultaneously. N-dimensional vector $\mathbf{F_d}$ symbolises the damper forces provided by EMDEH devices, and the structure is assumed to respond within an entirely elastic range. As a result, the motion of the structure-EMDEH system is given by

$$\mathbf{M\ddot{x}} + \mathbf{C_s\dot{x}} + \mathbf{Kx} + \mathbf{F_d} = \mathbf{F} \qquad (16.17)$$

where:

$\mathbf{M}, \mathbf{C_s}$ and $\mathbf{K}$ are the $N \times N$ global mass matrix, damping matrix and stiffness matrix, respectively

$\mathbf{\ddot{x}}, \mathbf{\dot{x}}$ and $\mathbf{x}$ are the N-dimensional vectors that denote the acceleration, velocity and displacement response of the structure, respectively

$\mathbf{F}$ is the N-dimensional vector that denotes the external force vector imposed on the structure

The energy balance equation can be formed by integrating the individual force terms in Equation 16.17 over the entire displacement history (Soong and Dargush 1997). Based on the energy balance equation, the power flow can be straightforwardly obtained by calculating the average power of each energy term, given a specific calculation time T_c. The power flow of an MDOF structure is given by

$$P_{ex} = P_{ds} + P_{vs} + P_{in} \qquad (16.18)$$

where:

$$P_{ex} = \frac{1}{T_c} \int_{t_0}^{t_0+T_c} \mathbf{F}^{\mathbf{T}}\mathbf{\dot{x}}dt \; ; \quad P_{ds} = \frac{1}{T_c} \int_{t_0}^{t_0+T_c} \mathbf{\dot{x}}^{\mathbf{T}}\mathbf{C_s\dot{x}}dt \; ; \quad P_{vs} = \frac{E_{vs}(t_0 + T_c) - E_{vs}(t_0)}{T_c}$$

$$E_{vs} = \frac{1}{2}\mathbf{\dot{x}}^T\mathbf{M\dot{x}} + \frac{1}{2}\mathbf{x}^T\mathbf{Kx} \; ; \quad P_{in} = \frac{1}{T_c} \int_{t_0}^{t_0+T_c} \mathbf{F_d^T}\mathbf{\dot{x}}dt \qquad (16.19)$$

where:

t_0 is an arbitrary time

P_{ex} is the input power to the structure because of the external dynamic excitation

P_{ds} is the inherent damping power of the structure

P_{vs} is the rate of change (ROC) of the structural mechanical energy

E_{vs} is the mechanical energy of the structure, consisting of kinetic energy and elastic strain energy

P_{in} is the total damping power of the EMDEH devices – that is, the total input power from the structure to the devices

The input power to the structure caused by the external dynamic excitation P_{ex} is converted to three parts – namely, P_{ds}, P_{vs} and P_{in}, as shown in Figure 16.17. Assuming that the external excitation is stationary, the ROC of the structural mechanical energy P_{vs} tends to be zero. As a result, the input power caused by external disturbance is equal to the damping power contributed by the inherent damping mechanism and EMDEH devices. Thus,

$$P_{ex} = P_{ds} + P_{in} \qquad (16.20)$$

Equation 16.20 implies that the mechanical energy (the sum of the kinetic energy and the elastic strain energy) is stable, and all the power input to the primary structure is dissipated (or harvested) by the inherent damping mechanism and the EMDEH devices.

16.4.7 Analysis of structure-EMDEH systems

In this section, an analysis of structure-EMDEH systems is performed considering the stochastic forces acting on civil structures. The closed-form solutions of the damping power, as well as the output power of the EMDEH devices when attached to a single-degree-of-freedom (SDOF) or MDOF structure, are given by the analysis. The optimal damping coefficient for passive vibration control and for energy harvesting is discussed based on the analytical solutions.

To address the output power optimisation of EMDEH devices installed in a structure, an analysis of the coupled structure-EMDEH system is required. This analysis differs from the output power optimisation of energy harvesters in which the harvesters' reaction to primary structures is negligible, such that the problem is essentially uncoupled. In this analysis, the performance optimisation circuit as mentioned in Section 16.4.2 is adopted as the EH circuit attached to the EM device. The DCM buck-boost converter in a performance optimisation circuit can emulate resistance within a wide voltage range, which enables it to behave like a resistor. Under these conditions, an EMDEH device can be modelled as a linear viscous damper. Consequently, the analysis of the coupled structure-EMDEH system is simplified as a structural dynamic analysis. In this way, output power is optimised with respect to the damping of the EMDEH devices. In addition, forces acting on the civil structures such as wind loads, traffic loads and ocean wave loads are characterised as random excitations. For simplicity, the random force is assumed to be ideal white noise or band-limited white noise. Another assumption is that the structures installed with the EMDEH device(s) respond in a linearly elastic manner. In addition, the inherent damping of MDOF structures is modelled by the proportional damping matrix (also known as a classical damping matrix) in the presented formulation, and the damping contribution of the EMDEH devices is represented by the modal damping ratios. The resulting damping matrix, including inherent damping and the damping of EMDEH devices, is proportional. Consequently, the equations of motion of the structure-EMDEH system can be solved using the normal mode approach, also known as the *modal superposition method*. The output power analysis and optimisation can be performed using random vibration theory, which is presented in the subsequent subsections.

16.4.7.1 Integrated SDOF systems

To assess the optimal damping for vibration control and energy harvesting, an SDOF structure is first analysed. An SDOF structure subjected to an external force is shown in Figure 16.18, and its motion is given by

$$m\ddot{x} + c\dot{x} + kx = f(t) \tag{16.21}$$

where:

m, c and k	denote the mass, damping coefficient and stiffness of the structure, respectively
x	is the displacement response of the structure
$f(t)$	is the external force

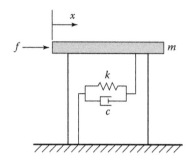

Figure 16.18 SDOF structure-EMDEH system.

Notably, the damping coefficient of the SDOF structure, c, is contributed by the inherent structural damping coefficient, C_s, and the total damping coefficient of the EMDEH device, C_d.

Equation 16.21 can be rewritten as

$$\ddot{x} + 2\zeta\omega_n\dot{x} + \omega_n^2 x = g(t) \tag{16.22}$$

where:

$$\omega_n = \sqrt{\frac{k}{m}}, \quad \zeta = \frac{c}{2\sqrt{km}}, \quad g(t) = \frac{f(t)}{m} \tag{16.23}$$

Here, ω_n, ζ and $g(t)$ are the natural frequency, damping ratio and the generalised force, respectively. The velocity complex frequency response function (FRF) of the SDOF structure can then be obtained using Fourier transforms:

$$H_v(j\omega) = \frac{j\omega}{\omega_n^2 - \omega^2 + j2\zeta\omega_n\omega} \tag{16.24}$$

where j is the imaginary unit and ω is the frequency of the external force.

Civil structures may undergo random vibration when wind and seismic loads are applied. Therefore, random excitation is considered here to investigate the EH performance of an EMDEH device attached to an SDOF structure. In this subsection, random excitation is modelled as white noise with a constant power spectral density (PSD) $S_f(\omega) = S_0$, which is generally employed to characterise random vibration. The PSD of velocity response is given by

$$S_{\dot{x}}(\omega) = \left|H_v(\omega)\right|^2 \frac{S_f(\omega)}{m^2} = \left|H_v(\omega)\right|^2 \frac{S_0}{m^2} \tag{16.25}$$

The average damping power of the entire SDOF system is then given by

$$P_d = E\left[p_d(t)\right] = E\left[c\dot{x}^2(t)\right] = c\int_{-\infty}^{+\infty} S_x(\omega)d\omega = \frac{cS_0}{m^2}\int_{-\infty}^{+\infty}\left|H_v(\omega)\right|^2 d\omega = \frac{\pi S_0}{m} \tag{16.26}$$

When the excitation is a band-limited white noise excitation, Equation 16.26 is an approximate estimation provided that the frequency band is sufficiently wide. In stationary vibration, the power transmitted to the SDOF structure due to random external excitation is equal to the damping power of the entire system, such that

$$P_{ex} = P_d = \frac{\pi S_0}{m} \tag{16.27}$$

The units of S_0 and m in Equations 16.26 and 16.27 are N²·s/rad and kg, respectively. If the unit of S_0 is N²/Hz, Equations 16.26 and 16.27 can be expressed as $P_{ex} = P_d = S_0/(2m)$. In addition, Equations 16.26 and 16.27 indicate that the average damping power of the entire SDOF structure, as well as the power transferred to the structure because of external excitation, is constant and independent of the damping coefficient c or the damping ratio ζ. The average damping power is proportional to the PSD of random excitation and is inversely proportional to the mass of the SDOF structure.

When the SDOF structure-EMDEH system is subjected to white noise ground motion excitation, the closed-form solution of the average damping power of the whole system is $P_{ex} = P_d = \pi m S_0$ (S_0: PSD of white noise ground motion acceleration; unit: [m/s²]²·s/rad) or $P_{ex} = P_d = 0.5 \, m S_0$ (S_0: PSD of white noise ground motion acceleration; unit: [m/s²]²/Hz). This analytical result indicates that the average damping power of the SDOF structure, when subjected to white noise ground motion excitation, is proportional to the mass. A more general conclusion was drawn by Mitcheson (2005) when considering the base motion input (seismic forces): 'Power is proportional to proof mass for any waveform.'

The preceding discussion indicates that a larger greater proof mass produces more output power from the EMDEH device when considering ground motions (base motions or seismic forces). In contrast, a larger proof mass results in less output power from the EMDEH device when considering external loads, such as wind loads.

Based on Equation 16.26, the input power of the EMDEH device is given by

$$P_{in} = P_d \cdot \frac{\zeta_d}{\zeta_s + \zeta_d} = \frac{\pi S_0}{m} \cdot \frac{\zeta_d}{\zeta_s + \zeta_d} \tag{16.28}$$

Equation 16.28 reveals that the input power to the EMDEH device P_{in} depends on the EMDEH device's contributed damping ratio ζ_d, although the overall damping power P_d is independent of damping and P_{in} increases with the increasing damping ratio contributed by the EMDEH device. It can be concluded that in the SDOF structure-EMDEH system, increasing the EMDEH device damping coefficient C_d – that is, increasing the damper size (or EMDEH device size) – results in better vibration control and EH performance.

Considering the harvesting efficiency of the EMDEH device (see Equation 16.14), the output power can be given by

$$P_{out} = \eta P_{in} = \frac{\pi S_0}{m} \cdot \frac{\zeta_d}{\zeta_s + \zeta_d} \cdot \eta \tag{16.29}$$

Equation 16.29 indicates that the maximum output power of the EMDEH device corresponds to the maximum harvesting efficiency if the damping coefficient of the EMDEH is determined. The condition for achieving maximum harvesting efficiency is expressed in Equation 16.16 for a specific EMDEH device in terms of the fixed K_{em}, R_{coil} and C_p.

16.4.7.2 Integrated MDOF systems

MDOF systems are more practical representations of real civil structures (such as high-rise buildings) that have potential applications to the proposed EMDEH device for simultaneous vibration control and energy harvesting. In this section, an N-DOF structure subjected to external random excitation is considered. It is known that, for a linearly elastic structure, the displacement vector x can be expressed as the sum of modal responses.

$$\mathbf{x} = \mathbf{\Phi y} \tag{16.30}$$

where $\mathbf{\Phi}$ is the mode shape matrix that consists of N independent modal vectors ($\mathbf{\Phi} = [\phi_1, \phi_2 \dots \phi_N]$), and y is the normal coordinate vector.

As mentioned earlier, suppose the damping matrix, including inherent damping and the damping of EMDEH devices, is proportional and satisfies the orthogonality condition. Consequently, the equation of motion of the MDOF structure can be easily decoupled to N independent SDOF differential equations in the frequency domain. The responses of the MDOF structure are the superposition of the modal responses of the N SDOF structures, which are given by Equation 16.30.

The instantaneous damping power of the MDOF structure can be given as

$$p_\mathrm{d}(t) = \dot{\mathbf{y}}^\mathrm{T} \begin{bmatrix} 2\zeta_1\omega_1\bar{M}_1 & & 0 \\ & \ddots & \\ 0 & & 2\zeta_N\omega_N\bar{M}_N \end{bmatrix} \dot{\mathbf{y}} = \sum_{j=1}^{N} 2\zeta_j\omega_j\bar{M}_j\dot{y}_j^2(t) \tag{16.31}$$

where ς_j, ω_j and $\bar{M}_j$ are the modal damping ratio, modal frequency and modal mass of the jth mode, respectively. Equation 16.31 indicates that the damping power of the MDOF structure is the sum of the damping power of each mode. The average damping power is given by

$$P_\mathrm{d} = E[p_\mathrm{d}(t)] = \sum_{j=1}^{N} 2\zeta_j\omega_j\bar{M}_j \, E\left[\dot{y}_j^2(t)\right] \tag{16.32}$$

With proper transformation, the average overall damping power can be rewritten as (Shen 2014)

$$P_\mathrm{d} = \sum_{j=1}^{N} \frac{\pi S_0}{\bar{M}_j} \sum_{p=1}^{N} \sum_{q=1}^{N} \phi_{pj}\phi_{qj}w_{pq} \tag{16.33}$$

where ϕ_{pj} and ϕ_{qj} are the components of jth mode shape vector at locations of p and q, respectively, and w_{pq} is the (p, q) entry in an $N \times N$ constant matrix that is used to represent the PSD matrix of the random force vector.

The power input to the MDOF structure by the external excitation (also called *excitation power*) is given by

$$P_\mathrm{ex} = P_\mathrm{d} = \sum_{j=1}^{N} \frac{\pi S_0}{\bar{M}_j} \sum_{p=1}^{N} \sum_{q=1}^{N} \phi_{pj}\phi_{qj}w_{pq} \tag{16.34}$$

Similar to the SDOF structure, Equations 16.33 and 16.34 indicate that both the average overall damping power of the entire MDOF structure P_d and the power input to the structure P_{ex} are independent of the damping ratio. The average overall damping power is proportional to the PSD of random excitation S_0 and is inversely proportional to the modal mass of the MDOF structure. The energy input to the MDOF structure by random excitations is dissipated by the inherent structural damping and the EMDEH devices attached to each DOF. Hence, the input power of the EMDEH devices is a portion of the overall damping power of the entire MDOF structure-EMDEH system, which is given by

$$P_{in} = \sum_{j=1}^{N} \frac{\zeta_{d,j}}{\zeta_{s,j} + \zeta_{d,j}} \cdot \frac{\pi S_0}{\bar{M}_j} \sum_{p=1}^{N} \sum_{q=1}^{N} \phi_{pj} \phi_{qj} w_{pq} \tag{16.35}$$

where $\zeta_{s,j}$ and $\zeta_{d,j}$ are the damping ratios contributed by the inherent structural damping and the EMDEH devices, respectively.

Assuming that the EMDEH devices attached to the MDOF structure have the same harvesting efficiency η, the output power of the EMDEH devices can be expressed as follows:

$$P_{out} = \eta P_{in} = \eta \sum_{j=1}^{N} \frac{\zeta_{d,j}}{\zeta_{s,j} + \zeta_{d,j}} \cdot \frac{\pi S_0}{\bar{M}_j} \sum_{p=1}^{N} \sum_{q=1}^{N} \phi_{pj} \phi_{qj} w_{pq} \tag{16.36}$$

Notably, the unit of the PSD of the white noise excitation is $N^2 \cdot s/rad$. If cyclic frequency is used, in which the unit of S_0 is N^2/Hz, the corresponding formulas in Equations 16.34 to 16.36 should be rewritten similarly to those in Section 16.4.7.1. Thus,

$$P_{ex} = P_d = \sum_{j=1}^{N} \frac{S_0}{2\bar{M}_j} \sum_{p=1}^{N} \sum_{q=1}^{N} \phi_{pj} \phi_{qj} w_{pq} \tag{16.37a}$$

$$P_{in} = \sum_{j=1}^{N} \frac{\zeta_{d,j}}{\zeta_{s,j} + \zeta_{d,j}} \cdot \frac{S_0}{2\bar{M}_j} \sum_{p=1}^{N} \sum_{q=1}^{N} \phi_{pj} \phi_{qj} w_{pq} \tag{16.37b}$$

$$P_{out} = \eta P_{in} = \eta \sum_{j=1}^{N} \frac{\zeta_{d,j}}{\zeta_{s,j} + \zeta_{d,j}} \cdot \frac{S_0}{2\bar{M}_j} \sum_{p=1}^{N} \sum_{q=1}^{N} \phi_{pj} \phi_{qj} w_{pq} \tag{16.37c}$$

It should be pointed out that the ideal white noise excitation with infinite bandwidth does not realistically exist. Therefore, the random excitation imposed on a structure is usually assumed as band-limited white noise in practical applications. In this case, the power items P_{ex}, P_d, P_{in} and P_{out} can be derived in the same manner. Each power item shares similar expressions in Equations 16.34 through 16.36 and is the sum of the powers of all the modes that are excited by the band-limited white noise.

16.4.8 Application to stay cables

Stay cables are critical load-carrying elements in cable-stayed bridges and are often vulnerable to excessive vibration under wind excitations because of their inherent low damping and high flexibility. Passive dampers are usually employed to enhance their damping to avoid excessive vibrations, such as wind- and rain-induced vibrations. When using EMDEH devices to replace conventional passive dampers, harmful vibrations can be suppressed and the vibration energy of the stay cable can be simultaneously harvested. The harvested electric power can then serve as the power supply of the wireless sensors which monitor the behaviours of the stay cable. This section presents the application of a novel EMDEH device to real stay cables through scaled cable model testing (Shen and Zhu 2015; Shen et al. 2016). The vibration control and EH performances of the EMDEH device with a performance optimisation circuit were investigated through both a sine sweep vibration test and a random vibration test.

16.4.8.1 Experimental setup

A scaled stay cable 5.85 m in length was employed in this experimental study. Table 16.3 shows the main parameters of the scaled stay cable model. Lumped masses at 90 mm intervals were arranged along the steel cable to simulate an appropriate mass density. A constant tension force (960 N) measured by a load cell (model no.: HONRE S314-2T) in the lower anchorage was applied to the scaled stay cable in this experimental study. The fundamental frequency of the scaled stay cable is 4.086 Hz, and the natural frequencies of the first six modes are shown in Table 16.4. A corresponding analytical model with 200 uniformly spaced segments using finite difference (FD) formulation (Mehrabi and Tabatabai 1998) was also established. All the parameters were set according to the values shown in Table 16.3. The calculated natural frequencies are consistent with the measured ones, as shown in Table 16.4.

Figure 16.19 schematically shows the test setup of the scaled stay cable vibration test with an EMDEH device. The test used an LDS V406 permanent magnet shaker, installed at the location $0.019l$ from the higher anchorage, to generate the exciting forces (see Figure 16.19).

Table 16.3 Main parameters of scaled stay cable and EM damper

Item of cable	Value	Item of EM damper	Value
Mass per unit length	0.442 kg/m	Machine constant	38.0 V·s/m or N/A
Cable length	5.85 m	Coil resistance	9.3 Ω
Inclination	15.5°	Diameter of damper	95.3 mm
Static tension force	980 N	Length of damper	45.2 mm
Diameter	4 mm	Moving mass of damper	0.443 kg
Cross-sectional area	7.28 mm²	Location of damper	0.05l
Young's modulus	8.242×10^4 Mpa		
Axial stiffness	6×10^5		

Source: Shen, W.A., et al., *Smart Mater. Struct.*, 25(6), 65011–27, 2016.

Table 16.4 Measured and computed natural frequencies of scaled stay cable

Measured frequency (Hz)	FD modelling (Hz)	Difference (%)
4.086	4.089	0.07
8.050	8.057	0.09
12.081	12.096	0.12
–	16.142	–
20.230	20.204	0.13
24.352	24.282	0.29

Source: Shen, W.A., et al., *Smart Mater. Struct.*, 25(6), 65011–27, 2016.

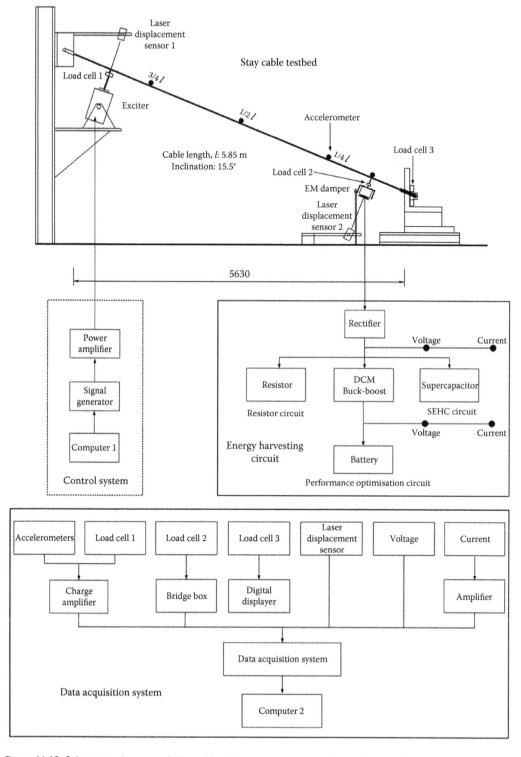

Figure 16.19 Schematic diagram of the cable vibration test setup. (From Shen, W.A., et al., *Smart Mater. Struct.*, 25(6), 65011–27, 2016.)

A digital signal generator generated sine sweep and random signals and fed them into a power amplifier that drove the shaker. Therefore, the desired excitations could be controlled by the signal generator. A miniature piezoelectric force transducer was arranged between the shaker and cable to measure the input force. The displacement of the shaker was also measured by the laser displacement sensor. The acceleration responses of the cable in the locations $0.25l$, $0.5l$ and $0.75l$ were measured by miniature accelerometers to evaluate the vibration control performance.

Figure 16.20 shows a picture of the cable vibration test setup. The EM damper ($K_{em} = 38$ V·s/m or N/A; $R_{coil} = 9.3$ Ω.) presented in Section 16.4.5 was installed at the location $0.05l$ (30 cm away from the lower anchorage) for simultaneous vibration control and energy harvesting (see Figure 16.20b). A performance optimisation circuit using the DCM buck-boost converter introduced in Section 16.4.2 was employed to achieve maximum output power and harvesting efficiency (see Figure 16.20c).

A series of test scenarios, as shown in Table 16.5, were considered in this experiment to validate the performance of the EMDEH device. Sine sweep and random excitations were applied to the scaled cable model at the location $0.019l$ from the higher anchorage. The performance of the EMDEH device, including its vibration control performance and EH performance, is discussed and given in the following subsections.

16.4.8.2 Vibration control performance

The EMDEH device with the performance optimisation circuit was designed to provide suboptimal damping for vibration mode 2 of the scaled cable model. Sine sweep excitations in the frequency range 7.7 to 8.1 Hz were imposed on the cable model with and without the EMDEH device to examine the vibration control performance.

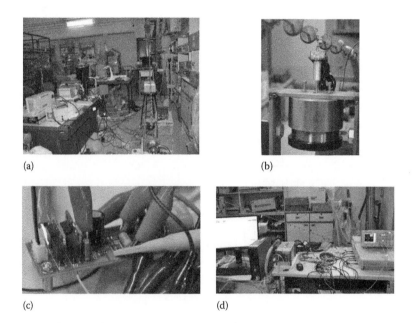

(a)

(b)

(c)

(d)

Figure 16.20 Pictures of the cable vibration test setup: (a) bridge stay cable, (b) EM damper, (c) performance optimisation circuit, (d) data acquisition system. (From Shen, W.A., et al., *Smart Mater. Struct.*, 25(6), 65011–27, 2016.)

Table 16.5 Scenarios of scaled stay cable vibration testing

Circuit	Excitation type	Freq. range (Hz)	Sweep rate (Hz/s)	Peak force (N)	RMS force (N)	Circuit parameter
Performance optimisation circuit	Sine sweep, mode 1	3.9–4.2	0.05	16.9	8.9	Refer to Figure 16.9
	Sine sweep, mode 2	7.7–8.1	0.05	18.1	11.5	
	Sine sweep, mode 3	11.9–12.3	0.05	19.8	12.1	
	Sine sweep, mode 4	15.9–16.3	0.05	17.2	11.1	
	Sine sweep, mode 5	19.9–20.2	0.05	20.6	10.6	
	Random	0.0–25.0	–	26.4	6.8	
	Random	0.0–25.0	–	35.7	9.0	
	Random	0.0–25.0	–	40.1	10.1	
	Random	0.0–25.0	–	44.5	11.3	
	Random	0.0–25.0	–	48.9	12.5	
Without control	Sine sweep, mode 2	7.7–8.	0.05	19.3	6.6	–
	Random	10.0–25.0	–	27.3	7.2	
	Random	0.0–25.0	–	31.6	8.4	

Source: Shen, W.A., et al., Smart Mater. Struct., 25(6), 65011–27, 2016.

Figures 16.21a and 16.21b show the acceleration responses with and without the EMDEH device at the locations 0.25*l* and 0.75*l* in two sweep cycles, respectively. The RMS acceleration at 0.25*l* can be reduced by 60%. A similar RMS response control effect can be observed at 0.75*l*, in which the RMS acceleration reduction was 57.9%. The peak acceleration response at 0.25*l* was reduced from 47.77 m/s² without control to 19.88 m/s² with the performance optimisation circuit, as shown in Figure 16.21a. A similar control effect can be found at the location 0.75*l* (see Figure 16.21b).

Moreover, in the case of random excitations, Figure 16.22 shows the acceleration time histories at the location 0.25*l* with and without the EMDEH device. The acceleration responses were markedly suppressed by the EMDEH device, in which the peak response was reduced from 100.3 m/s² to 38.1 m/s² and the RMS response was reduced from 21.6 m/s² to 8.2 m/s², a reduction of 62.1%.

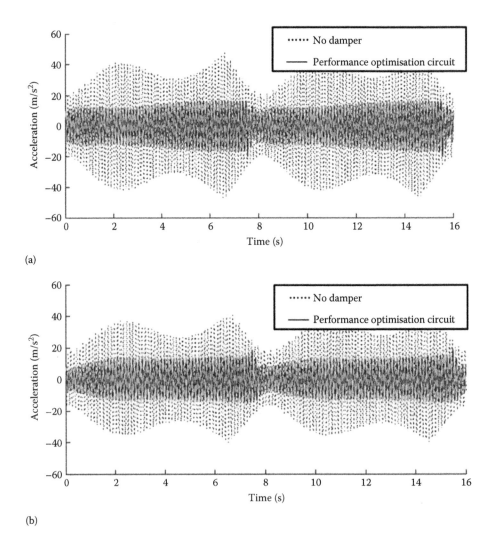

(a)

(b)

Figure 16.21 Acceleration time histories of the stay cable with and without the EMDEH device subjected to sine sweep excitation. (a) 0.25*l*, (b) 0.75*l*. (From Shen, W.A., et al., *Smart Mater. Struct.*, 25(6), 65011–27, 2016.)

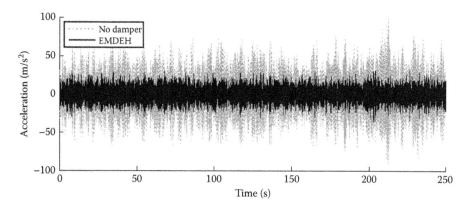

Figure 16.22 Acceleration time histories of the stay cable with and without the EMDEH device subjected to random excitation. (From Shen, W.A., et al., *Smart Mater. Struct.*, 25(6), 65011–27, 2016.)

16.4.8.3 Energy harvesting performance

Figure 16.23a shows a typical charging curve of the NiHM battery by the power output from the EMDEH device using a performance optimisation circuit in random vibration cases. The original voltage of 3.77 V was successfully charged up to 3.80 V over a duration of 600 s. The corresponding charging current is shown in Figure 16.23b, which shows that the peak charging current is 107 mA, while the average charging current is about 8.13 mA because of fluctuation. The charging current inevitably shares a random feature of the input force and does not stay constant during the charging process. Hence, there is a significant difference in charging between the power source using the ambient vibration-based EH technique and the routine power supplies with a fixed DC or AC voltage output.

Figure 16.24 shows the output power and corresponding harvesting efficiencies with different vibration levels in random vibration cases. The output power of the dual-function EMDEH device increases dramatically from 7.6 to 31.5 mW when increasing the vibration level from 9.86 to 18.32 m/s^2 (mid-span of the cable model). The corresponding harvesting efficiencies η are relatively stable, ranging from 13.2% to 14.9%. Detailed test data on power and efficiency in performance optimisation circuit cases are presented in Table 16.6. The intermediate efficiency η_1, representing electromechanical coupling, retains approximately 46%, except in the case of low random vibration (RMS force of 6.8 N). The results indicate that approximately 54% of the damping power of the EMDEH device was converted to heat by a parasitic damping dissipation mechanism (friction). Reducing the friction of the EM damper can dramatically improve the harvesting efficiency. The intermediate efficiency η_2, representing the proportion of the power input to the EH circuit of the total EM damping power, slightly varies from 64.0% to 72.4% in the vicinity of the optimal range (68%–76%). The intermediate efficiency η_2 is determined by the average input resistance of the performance optimisation circuit. The efficiency of the performance optimisation circuit itself (η_3) ranges from 45.0% to 49.5%, which is significantly lower than the theoretical upper limit.

The experimental results shown in this chapter demonstrate that an EMDEH device with the presented performance optimisation circuit is able to significantly mitigate the excessive vibration of stay cables (up to 60% reduction of RMS acceleration) on one hand, and can provide a certain amount of electric power (e.g. 31.5 mW of output power with a harvesting efficiency of 13.8% at a vibration level of 18.32 m/s^2) on the other. However, further

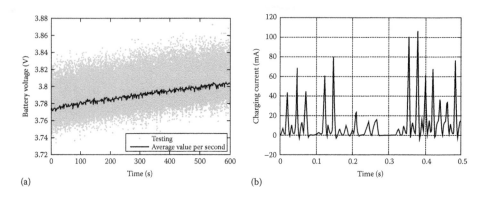

(a)   (b)

Figure 16.23 Typical charging curves in a random vibration case: (a) curve of charging voltage, (b) curve of charging current. (From Shen, W.A., et al., *Smart Mater. Struct.*, 25(6), 65011–27, 2016.)

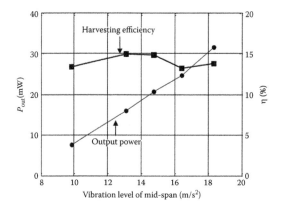

Figure 16.24 Output power and corresponding efficiency vary with cable vibration level. (From Shen, W.A., et al., *Smart Mater. Struct.*, 25(6), 65011–65027, 2016.)

Table 16.6 Average input resistance, power and efficiency in performance optimisation circuit cases

Force type	Frequency (Hz)	RMS force (N)	R_{in} (Ω)	Mean U_{rect} (V)	P_{out} (mW)	P_g (mW)	η(%)	$η_1$(%)	$η_2$(%)	$η_3$(%)
Sine sweep, mode 1	3.9–4.2	8.9	43.2	1.0	3.5	10.9	6.2	23.2	82.3	32.2
Sine sweep, mode 2	7.7–8.1	11.5	24.9	1.3	31.6	52.9	16.9	39.0	72.8	59.7
Sine sweep, mode 3	11.9–12.3	12.1	23.1	1.6	44.1	95.5	18.7	56.8	71.3	46.2
Sine sweep, mode 4	15.9–16.3	11.1	19.6	1.5	41.9	83.3	15.4	45.0	67.8	50.3
Sine sweep, mode 5	19.9–20.2	10.6	18.4	1.4	40.6	69.0	16.7	42.7	66.4	58.8
Random	0.0–25.0	6.8	25.8	1.0	7.6	15.7	13.4	37.7	73.5	48.3
Random	0.0–25.0	9.0	20.0	1.1	16.0	32.3	14.9	44.2	68.3	49.5
Random	0.0–25.0	10.1	18.4	1.2	20.7	43.1	14.9	46.6	66.5	48.0
Random	0.0–25.0	11.3	17.3	1.3	24.6	54.7	13.2	45.0	65.1	45.0
Random	0.0–25.0	12.5	16.6	1.4	31.5	68.0	13.8	46.5	64.0	46.3

Source: Shen, W.A., et al., *Smart Mater. Struct.*, 25(6), 65011–27, 2016.

investigation is still required to improve the harvesting efficiency and enhance the robustness of the EMDEH device. Moreover, the possibility of implementing the EMDEH device in real building structures may require further exploration as well.

NOTATION

c_m	Viscous damping coefficient
C_d	Total damping coefficient of the EM damper including parasitic and EM damping
C_{em}	EM damping
C_p	Parasitic damping coefficient
d_c	Duty cycle
D	Displacement amplitude of harmonic motion
E_{vs}	Mechanical energy of the structure, consisting of kinetic energy and elastic strain energy
f_{em}	EM damping force
F	Oscillation frequency of harmonic motion
$\mathbf{F}$	External force vector imposed on the structure
F_c	Magnitude of Coulomb friction force
H	Overall EH efficiency of an EMDEH
$H_v(j\omega)$	Velocity complex FRF of the SDOF structure
i_0	Instantaneous current in the coils
K_e	Back EMF constant
K_f	Machine constant of the EM damper
L, f_{sw}	Inductance and switching frequency of the buck-boost converter, respectively
$\mathbf{M}, \mathbf{C_s}, \mathbf{K}$	Global mass matrix, damping matrix and stiffness matrix, respectively
N_{EREH}	Gear ratio of the EREH generator
P_{coil}	Power dissipated by the copper loss of the coil
P_{ds}	Inherent damping power of the structure
P_{ehc}	Power loss induced by the EH circuit
P_{em}	EM damping power
P_{ex}	Input power to the structure because of the external dynamic excitation
P_g	Gross output power from the EM dampers
P_{in}	Total input power from the structure to EMDEH devices
P_{out}	Output power
P_p	Parasitic damping power
P_{vs}	The rate of change (ROC) of the structural mechanical energy
R_{coil}	Resistance of coils
R_{EREH}	Resistance of the EREH generator
R_{load}	External resistance
R_{opt}	Optimal load resistance or optimal input resistance of the EH circuit
$S_x(\omega)$	PSD of velocity response
S_0	Constant PSD of random excitation
T_c	Calculation period
u_c, i_c	Instantaneous voltage and charging current of the energy storage element, respectively
u_0	Open-circuit voltage
V_{EREH}	Rotational velocity of the EREH generator

$\ddot{x}, \dot{x}, x$	Acceleration, velocity and displacement response of the structure, respectively
$\dot{x}_d$	Velocity time history of the EM damper
$\dot{x}_m$	Velocity of the moving magnet
y	Normal coordinate vector
ζ_d	EMDEH device's contributed damping ratio
$\zeta_j, \omega_j, \bar{M}_j$	Modal damping ratio, modal frequency and modal mass of the jth mode, respectively
η_{EREH}	Power conversion efficiency of the EREH generator
η_1	Electromechanical coupling coefficient that describes the conversion efficiency from mechanical power to electrical power
η_2	Efficiency of the EM damper, which is affected by the power loss because of the coil resistance
η_3	Efficiency of the EH circuit
υ_{EREH}	Speed constant of the EREH generator
Φ	Mode shape matrix
ω	Frequency of external force
$\omega_n, \zeta, g(t)$	Natural frequency, damping ratio and the generalised force of the SDOF structure, respectively

REFERENCES

Alavi, A.H., H. Hasni, N. Lajnef, K. Chatti, and F. Faridazar. 2016a. Damage detection using self-powered wireless sensor data: An evolutionary approach. *Measurement*, 82: 254–83.

Alavi, A.H., H. Hasni, N. Lajnef, K. Chatti, and F. Faridazar. 2016b. An intelligent structural damage detection approach based on self-powered wireless sensor data. *Autom. Constr.*, 62: 24–44.

Anton, S.R. and H.A. Sodano. 2007. A review of power harvesting using piezoelectric materials (2003–2006). *Smart Mater. Struct.*, 16(3): R1–R21.

Arnold, D.P. 2007. Review of microscale magnetic power generation. *IEEE Trans. Magnetics*, 43(11): 3940–51

Beeby, S. and T. O'Donnell. 2009. Electromagnetic energy harvesting. In *Energy Harvesting Technologies*, ed. S. Priya and D.J. Inman, 129–61. New York: Springer Science+Business Media.

Beeby, S.P., M.J. Tudor, and N.M. White. 2006. Energy harvesting vibration sources for microsystems applications. *Meas. Sci. Technol.*, 17: R175.

Benini, L., A. Bogliolo, and G. De Micheli. 2000. A survey of design techniques for system-level dynamic power management. *IEEE Trans. Very Large Scale Integr. (VLSI) Syst.*, 8(3): 299–316.

Bhutta, M.M.A., N. Hayat, A.U. Farooq, et al. 2012. Vertical axis wind turbine: A review of various configurations and design techniques. *Renew. Sust. Energ. Rev.*, 16: 1926–39.

Bogue, R. 2009. Energy harvesting and wireless sensors: A review of recent developments. *Sens. Rev.*, 29(3): 194–9.

Brown, W., J. Mims, and N. Heenan. 1965. An experimental microwave-powered helicopter. Raytheon Company, Burlington, MA. *1965 IEEE International Record*, 13(5): 225–35.

Casciati, F. and R. Rossi. 2007. A power harvester for wireless sensing applications. *Struct. Control Health Monit.*, 14(4), 649–59.

Cho, S., H. Jo, S. Jang, et al. 2010. Structural health monitoring of a cable-stayed bridge using wireless smart sensor technology: Data analyses. *Smart Struct. Syst.*, 6: 461–80.

DuToit, N.E., B.L. Wardle, and S.G. Kim. 2005. Design considerations for MEMS-scale piezoelectric mechanical vibration energy harvesters. *Integr. Ferroelectr.*, 71(1), 121–60.

Elvin, N., A. Elvin, and D.H. Choi. 2003. A self-powered damage detection sensor. *J. Strain Anal. Eng. Des.*, 38(2): 115–24.

Elvin, N., A. Elvin, and M. Spector. 2001. A self-powered mechanical strain energy sensor. *Smart Mater. Struct.*, 10: 293–9.

Erickson, R.W. 1997. *Fundamentals of Power Electronics*. Boston, MA: Springer.

Farinholt, K.M., N. Miller, W. Sifuentes, et al. 2010. Energy harvesting and wireless energy transmission for embedded SHM sensor nodes. *Struct. Health Monit.*, 9(3): 269–80.

Farinholt, K.M., G. Park, and C.R. Farrar. 2009. RF energy transmission for a low-power wireless impedance sensor node. *IEEE Sens. J.*, 9(7): 793–800.

Farrar, C.R., D.W. Allen, G. Park, S. Ball, and M.P. Masquelier. 2006. Coupling sensing hardware with data interrogation software for structural health monitoring. *Shock Vib.*, 13(4): 519–30.

Farrar, C.R., G. Park, and M.D. Todd. 2011. Sensing network paradigms for structural health monitoring. In *New Developments in Sensing Technology for Structural Health Monitoring*, ed. S.C. Mukhopadhyay, 137–59. Berlin: Springer.

Galchev, T., H. Kim, and K. Najafi. 2011c. Micro power generator for harvesting low-frequency and nonperiodic vibrations. *J. Microelectromech. Syst.*, 20: 852–66.

Galchev, T., J. McCullagh, R.L. Peterson, and K. Najafi. 2011a. Harvesting traffic-induced vibrations for structural health monitoring of bridges. *J. Micromech. Microeng.*, 21(10): 104005.

Galchev, T., J. McCullagh, R.L. Peterson, K. Najafi, and A. Mortazawi. 2011b. Energy harvesting of radio frequency and vibration energy to enable wireless sensor monitoring of civil infrastructure. In *Proceedings of SPIE 7983, Nondestructive Characterization for Composite Materials, Aerospace Engineering, Civil Infrastructure, and Homeland Security 2011*, 798314. San Diego, CA. doi:10.1117/12.880174.

Georgiadis, A., G. Andia, and A. Collado. 2010. Rectenna design and optimization using reciprocity theory and harmonic balance analysis for electromagnetic (EM) energy harvesting. *IEEE Antennas Wirel. Propag. Lett.*, 9: 444–6.

Germanischer Lloyd (GL). 2010. *IV Rules and Guidelines Industrial Services, Part 1: Guideline for the Certification of Wind Turbines*. Hamburg, Germany: Germanischer Lloyd.

Gilbert, J.M. and F. Balouchi. 2008. Comparison of energy harvesting systems for wireless sensor networks. *Int. J. Autom. Comput.*, 5(4): 334–57.

Graves, K.E. 2000. Electromagnetic energy regenerative vibration damping. PhD diss., Swinburne University of Technology, Australia.

Hansen, M.O.L. 2015. *Aerodynamics of Wind Turbines*, third edition. New York: Routledge.

Harb, A. 2011. Energy harvesting: State-of-the-art. *Renew. Energy*, 36(10): 2641–54.

Hau, E. 2013. *Wind Turbines: Fundamentals, Technologies, Application, Economics*. New York: Springer.

Hunter, S.R., N.V. Lavrik, S. Mostafa, S. Rajic, and P.G. Datskos. 2012. Review of pyroelectric thermal energy harvesting and new MEMs-based resonant energy conversion techniques. In *Proceedings of SPIE 8377, Energy Harvesting and Storage: Materials, Devices, and Applications III*, 83770D, Baltimore, MA. doi:10.1117/12.920978.

IEC. 2005. Wind turbines, Part 1: Design requirements. International Electrotechnical Commission, International Standard IEC 61400-1, third edition.

Inman, D.J. and B.L. Grisso. 2006. Towards autonomous sensing. In *Proceedings of SPIE 6174, Smart Structures and Materials 2006: Sensors and Smart Structures Technologies for Civil, Mechanical, and Aerospace Systems*, ed. M. Tomizuka, C.B. Yun, and V. Giurgiutiu, 61740T. San Diego, CA.

Jabbar, H., Y.S. Song, and T.T. Jeong. 2010. RF energy harvesting system and circuits for charging of mobile devices. *IEEE Trans. Consum. Electron.*, 56(1): 247–53.

Jung, H.J., D.D. Jang, J.H. Koo, and S.W. Cho. 2010a. Experimental evaluation of a 'self-sensing' capability of an electromagnetic induction system designed for MR dampers. *J. Intell. Mater. Syst. Struct.*, 21(8): 827–35.

Jung, H.J., D.D. Jang, H.J. Lee, I.W. Lee, and S.W. Cho. 2010b. Feasibility test of adaptive passive control system using MR fluid damper with electromagnetic induction part. *J. Eng. Mech.*, 136(2): 254–9.

Jung, H.J., I.H. Kim, and S.J. Jang. 2011a. An energy harvesting system using the wind-induced vibration of a stay cable for powering a wireless sensor node. *Smart Mater. Struct.*, 20: 075001.

Jung, H.J., I.H. Kim, and J.H. Koo. 2011b. A multi-functional cable-damper system for vibration mitigation, tension estimation and energy harvesting. *Smart Struct. Syst.*, 7(5): 379–92.

Jung, H.J., J.S. Park, and I.H. Kim. 2012. Investigation of applicability of electromagnetic energy harvesting system to inclined stay cable under wind load. *IEEE Trans. Magn.*, 48: 3478–81.

Kim, H.S., J.H. Kim, and J. Kim. 2011. A review of piezoelectric energy harvesting based on vibration. *Int. J. Precis. Eng. Manuf.*, 12(6): 1129–41.

Kim, I.H., S.J. Jang, and H.J. Jung. 2013. Performance enhancement of a rotational energy harvester utilizing wind-induced vibration of an inclined stay cable. *Smart Mater. Struct.*, 22: 075004.

Kim, I.H., H.J. Jung, and J.H. Koo. 2010. Experimental evaluation of a self-powered smart damping system in reducing vibrations of a full-scale stay cable. *Smart Mater. Struct.*, 19: 115027.

Kim, J. and T. Simunic Rosing. 2007. Power-aware resource management techniques for low-power embedded systems. In *Handbook of Real-Time and Embedded Systems*, ed. I. Lee, J.Y.T. Leung, and S. Son. Boca Raton, FL: CRC Press.

Kirianaki, N.V., S.Y. Yurish, N.O. Shpak, and V.P. Deynega. 2002. *Data Acquisition and Signal Processing for Smart Sensors*, first edition. New York: Wiley.

Kiziroglou, M.E., C. He, and E.M. Yeatman. 2009. Rolling rod electrostatic microgenerator. *IEEE Trans. Ind. Electron.*, 56(4): 1101–8.

Kurata, M., J.P. Lynch, T. Galchev, et al. 2010b. A two-tiered self-powered wireless monitoring system architecture for bridge health management. In *Proceedings of SPIE 7649, Nondestructive Characterization for Composite Materials, Aerospace Engineering, Civil Infrastructure, and Homeland Security 2010*, ed. P.J. Shull, A.A. Diaz, and H.F. Wu. San Diego, CA. doi:10.1117/12.848212.

Kurata, M., J.P. Lynch, G. van der Linden, V. Jacob, and P. Hipley. 2010a. Preliminary study of a wireless structural monitoring system for the New Carquinez Suspension Bridge. In *Proceedings of the Fifth World Conference on Structural Control and Monitoring*, 5WCSCM-172: 1–14. Ann Arbor, MI.

Lee, S., B.D. Youn, and B. Jung. 2009. Segment-type energy harvester powering wireless sensor for building automation. In *ASME 2009 International Design Engineering Technical Conferences and Computers and Information in Engineering Conference*, paper no. DETC2009–87436, 495–504. San Diego, CA. doi:10.1115/DETC2009–87436.

Lefeuvre, E., D. Audigier, C. Richard, and D. Guyomar. 2007. Buck-boost converter for sensorless power optimization of piezoelectric energy harvester. *IEEE Trans. Power Electron.*, 22(5): 2018–25.

Liu, G., N. Mrad, G. Xiao, Z. Li, and D. Ban. 2011. RF-based power transmission for wireless sensors nodes. International Workshop on Smart Materials, Structures and NDT in Aerospace, Montreal, QC, Canada.

Lynch, J. P. and K.J. Loh. 2006. A summary review of wireless sensors and sensor networks for structural health monitoring. *Shock Vib. Dig.*, 38(2): 91–128.

Mascarenas, D.L., E.B. Flynna, M.D. Todd, et al. 2010. Experimental studies of using wireless energy transmission for powering embedded sensor nodes. *J. Sound Vibr.*, 329: 2421–33.

Mascarenas, D.L., M.D. Todd, G. Park, and C.R. Farrar. 2007. Development of an impedance-based wireless sensor node for structural health monitoring. *Smart Mater. Struct.*, 16(6): 2137.

Mateu, L. and F. Moll. 2005. Review of energy harvesting techniques and applications for microelectronics. In *Proceedings of SPIE 5837, VLSI Circuits and Systems II*, 359–73. doi:10.1117/12.613046.

McCullagh, J.J., T. Galchev, and R.L. Peterson, et al. 2014. Long-term testing of a vibration harvesting system for the structural health monitoring of bridges. *Sens. Actuators, A: Phys.*, 217: 139–50.

McCullagh, J.J, R.L. Peterson, T. Galchev, et al. 2012. Short-term and long-term testing of a vibration harvesting system for bridge health monitoring. In *The 12th International Workshop on Micro and Nanotechnology for Power Generation and Energy Conversion Application, PowerMEMS 2012*, 109–12. Atlanta, GA. Copyrighted by PowerMEMS 2012 and interteq.com.

Mehrabi, A.B. and H. Tabatabai. 1998. Unified finite difference formulation for free vibration of cables. *J. Struct. Eng.*, 124(11): 1313–22.

Mi, M., M.H. Mickle, C. Capelli, and H. Swift. 2005. RF energy harvesting with multiple antennas in the same space. *IEEE Antennas Propag. Mag.*, 47(5): 100–106.

Miller, T.I., B.F. Spencer Jr., J. Li, and H. Jo. 2010. Solar energy harvesting and software enhancements for autonomous wireless smart sensor network. NSEL report no. NSEL-022. University of Illinois at Urbana-Champaign, IL.

Mitcheson, P.D. 2005. Analysis and optimisation of energy-harvesting micro-generator systems. PhD diss., Imperial College London.

Mitcheson, P.D., E.M. Yeatman, G.K. Rao, A.S. Holmes, and T.C. Green. 2008. Energy harvesting from human and machine motion for wireless electronic devices. *Proc. IEEE*, 96(9): 1457–86.

Mouis, M., E. Chávez-Ángel, C. Sotomayor-Torres, et al. 2014. Thermal energy harvesting. In *Beyond-CMOS Nanodevices 1*, ed. F. Balestra, 135–219. Hoboken, NJ: John Wiley.

Ni, T., L. Zuo, and A. Kareem. 2011. Assessment of energy potential and vibration mitigation of regenerative tuned mass dampers on wind excited tall buildings. In *Proceedings of the ASME 2011 International Design Engineering Technical Conferences & Computers and Information in Engineering Conference*, IDETC/CIE 2011, Washington, DC.

Nintanavongsa, P., U. Muncuk, D.R. Lewis, and K.R. Chowdhury. 2012. Design optimization and implementation for RF energy harvesting circuits. *IEEE J. Emerging Sel. Top. Circuits Syst.*, 2(1): 24–33.

Okada, Y., H. Harada, and K. Suzuki. 1997. Active and regenerative control of an electrodynamic-type suspension. *J. Soc. Mech. Eng. Int. J. Ser. C: Mech. Syst., Mach. Elem. Manuf.*, 40(2): 272–8.

Olgun, U., C.C. Chen, J.L. Volakis. 2011. Investigation of rectenna array configurations for enhanced RF power harvesting. *IEEE Antennas Propag. Lett.*, 10: 262–5.

Palomera-Arias, R. 2005. Passive electromagnetic damping device for motion control of building structures. PhD diss., Massachusetts Institute of Technology, Cambridge.

Paraschivoiu I. 2002. *Wind Turbine Design: With Emphasis on Darrieus Concept*. Québec, Canada: Polytechnic International Press.

Park, G., K.M. Farinholt, C.R. Farrar, T. Rosing, and M.D. Todd. 2009. Powering wireless SHM sensor nodes through energy harvesting. In *Energy Harvesting Technologies*, eds. S. Priya and D.J. Inman, 493–506. New York: Springer Science+Business Media.

Park, G., T. Rosing, M.D. Todd, C.R. Farrar, and W. Hodgkiss. 2008. Energy harvesting for structural health monitoring sensor networks. *J. Infrast. Syst.*, 14(1): 64–79.

Park, J.W., S. Cho, H.J. Jung, et al. 2010a. Long-term structural health monitoring system of a cable-stayed bridge based on wireless smart sensor networks and energy harvesting techniques. In *Proceedings of 5th World Conference on Structural Control and Monitoring*, 5WCSCM-10436, 1–6. Shinjuku, Tokyo, Japan.

Park, J.W., H.J. Jung, H. Jo, S. Jang, and B.F. Spencer Jr. 2010b. Feasibility study of wind generator for smart wireless sensor node in cable-stayed bridge. In *Proceedings of SPIE 7647, Sensors and Smart Structures Technologies for Civil, Mechanical, and Aerospace Systems 2010*, ed. M. Tomizuka. San Diego, CA. doi:10.1117/12.853600.

Pearson, M.R., M.J. Eaton, R. Pullin, C.A. Featherston, and K.M. Holford. 2012. Energy harvesting for aerospace structural health monitoring systems. *J. Phys. Conf. Ser.*, 382(1): 012025.

Rhimi, M., N. Lajnef, K. Chatti, and F. Faridazar. 2012. A self-powered sensing system for continuous fatigue monitoring of in-service pavements. *Int. J. Pavement Res. Technol.*, 5(5): 303–10.

Rivière, S., A. Douyère, F. Alicalapa, and J.D. Lan Sun Luk. 2010. Study of complete WPT aystem for WSN applications at low power level. *Electron. Lett.*, 46(8): 597–8.

Rosen, M.A., J.O. Ajedegba, and G.F. Naterer. 2010. Predicting vertical axis wind turbine behavior: Effects of blade configuration on flow distribution and power output. In *Wind Turbines: Types, Economics and Development*, eds. G. Martin and J. Roux, Chapter 1, 1–63. New York: Nova Science Publishers.

Roundy, S. 2003. Energy scavenging for wireless sensor nodes with a focus on vibration to electricity conversion. PhD diss., Department of Mechanical Engineering, University of California, Berkeley.

Roundy, S. and P.K. Wrigh. 2004. A piezoelectric vibration based generator for wireless electronics. *Smart Mater. Struct.*, 13: 1131–42.

Salter Jr., T.S., G. Metze, N. Goldsman, 2007. Improved RF power harvesting circuit design. In *International Semiconductor Research Symposium Proceedings*, College Park, MD, 1–2. ISBN: 978-1-4244-1892-3.

Sapiński, B. 2011. Experimental study of a self-powered and sensing MR-damper-based vibration control system. *Smart Mater. Struct.*, 20(10): 105007.

Sapiński, B., J. Snamina, Ł. Jastrzębski, and A. Staśkiewicz. 2011. Laboratory stand for testing self-powered vibration reduction systems. *J. Theor. Appl. Mech.*, 49(4): 1169–81.

Sazonov, E., H. Li, D. Curry, and P. Pillay. 2009. Self-powered sensors for monitoring of highway bridges. *IEEE Sens. J.*, 9(11): 1422–9.

Scruggs, J.T. 2004. Structural control using regenerative force actuation networks. PhD diss., California Institute of Technology, Pasadena, CA.

Scruggs, J.T. and W.D. Iwan. 2003. Control of a civil structure using an electric machine with semiactive capability. *J. Struct. Eng.*, 129(7): 951–9.

Scruggs, J.T. and W.D. Iwan. 2005. Structural control with regenerative force actuation networks. *Struct. Control Health Monit.*, 12(1): 25–45.

Shaikh, F.K. and S. Zeadally. 2016. Energy harvesting in wireless sensor networks: A comprehensive review. *Renew. Sust. Energ. Rev.*, 55: 1041–54.

Shen, W.A. 2014. Electromagnetic damping and energy harvesting devices in civil structures. PhD diss., Department of Civil and Environmental Engineering, Hong Kong Polytechnic University.

Shen, W.A. and S.Y. Zhu. 2015. Harvesting energy via electromagnetic damper: Application to bridge stay cables. *J. Intell. Mater. Syst. Struct.*, 26(1): 3–19.

Shen, W.A., S. Y. Zhu, and Y.L. Xu. 2012. An experimental study on self-powered vibration control and monitoring system using electromagnetic TMD and wireless sensors. *Sens. Actuators, A: Phys.*, 180: 166–76.

Shen, W.A., S.Y. Zhu, and H.P. Zhu. 2016. Experimental study on using electromagnetic devices on bridge stay cables for simultaneous energy harvesting and vibration damping. *Smart Mater. Struct.*, 25(6): 65011–27.

Shrestha, S., S.K. Noh, and D.Y Choi. 2013. Comparative study of antenna designs for RF energy harvesting. *Int. J. Antennas Propag.*, 2013: 385260.

Shu, Y.C. 2009. Performance evaluation of vibration-based piezoelectric energy scavengers. In *Energy Harvesting Technologies*, eds. S. Priya and D.J. Inman, 79–105. London: Springer Science+Business Media.

Snamina, J. and B. Sapiński. 2011. Energy balance in self-powered MR damper-based vibration reduction system. *Bull. Pol. Acad. Sci. Tech. Sci.*, 59(1): 75–80.

Snyder, G.J. 2009. Thermoelectric energy harvesting. In *Energy Harvesting Technologies*, ed. S. Priya and D.J. Inman, 325–36. London: Springer Science+Business Media.

Sodano, H.A., D.J. Inman, and G. Park. (2004). A review of power harvesting from vibration using piezoelectric materials. *Shock Vibr. Dig.*, 36(3): 197–205.

Soong, T.T and G.F. Dargush. 1997. *Passive Energy Dissipation System in Structural Engineering*. Chichester, UK: John Wiley.

Spencer Jr., B.F., S. Cho, and S.H. Sim. (2011). Wireless monitoring of civil infrastructure comes of age. *Struct. Mag.*, 2011: 12–15.

Spencer Jr., B.F., M.E. Ruiz-Sandoval, and N. Kurata. 2004. Smart sensing technology: Opportunities and challenges. *Struct. Control Health Monit.*, 11(4): 349–68.

Stephen, N.G. 2006. On energy harvesting from ambient vibration. *J. Sound Vibr.*, 293(1–2): 409–25.

Sun, H.C., Y.X. Guo, M. He, and Z. Zhong. 2013. A dual-band rectenna using broadband Yagi antenna array for ambient RF power harvesting. *IEEE Antennas Wirel. Propag. Lett.*, 12: 918–21.

Tang, X. and L. Zuo. 2012. Simultaneous energy harvesting and vibration control of structures with tuned mass dampers. *J. Intell. Mater. Syst. Struct.* 23(18): 2117–27.

Tao, M. 2014. *Terawatt Solar Photovoltaics: Roadblocks and Opportunities*. Briefs in Applied Sciences and Technology. London: Springer.

Taylor, S.G., K.M. Farinholt, E.B. Flynn, et al. 2009. A mobile-agent-based wireless sensing network for structural monitoring applications. *Meas. Sci. Technol.*, 20(4): 045201.

Torah, R., P. Glynne-Jones, M. Tudor, et al. 2008. Self-powered autonomous wireless sensor node using vibration energy harvesting. *Meas. Sci. Technol.*, 19: 125202.

Twiefel, J. and H. Westermann. 2013. Survey on broadband techniques for vibration energy harvesting. *J. Intell. Mater. Syst. Struct.*, 24(11): 1291–1302.

Ulukus, S., A. Yener, E. Erkip, et al. 2015. Energy harvesting wireless communications: A review of recent advances. *IEEE J. Sel. Areas Commun.*, 33(3): 360–81.

Wang, Y. and D.J. Inman. 2012. A survey of control strategies for simultaneous vibration suppression and energy harvesting via piezoceramics. *J. Intell. Mater. Syst. Struct.*, 23(18): 2021–37.

Wang, Z., Z. Chen, and B.F. Spencer Jr. 2009. Self-powered and sensing control system based on MR damper: Presentation and application. In *Proceedings of SPIE Smart Structures/NDE*, San Diego, CA.,7292:729240.

Yerramilli, A. and F. Tuluri. 2012. *Energy Resources Utilization and Technologies*. Boca Raton, FL, CRC Press.

Yun, C.B. and J. Min. 2011. Smart sensing, monitoring, and damage detection for civil infrastructures. *KSCE J. Civil Eng.*, 15(1): 1–14.

Zhu, J.T., X.C. Zhang, C. Wan, et al. 2014. Development of a self-contained wireless-based SHM system for monitoring a swing bridge. *Int. J. Sustainable Mater. Struct. Syst.*, 1(4): 351–73.

Zhu, S.Y., W.A. Shen, and Y.L. Xu. 2012. Linear electromagnetic devices for vibration damping and energy harvesting: Modeling and testing. *Eng. Struct.*, 34: 198–212.

Chapter 17

Synthesis of energy harvesting, structural control and health monitoring

17.1 PREVIEW

Wind is one of the fastest growing renewable energy resources today, and the most efficient way to harvest wind energy is to use wind turbines. Modern wind turbines can be categorised mainly as horizontal axis wind turbines (HAWTs) rotating around a horizontal axis and vertical axis wind turbines (VAWTs) spinning around a vertical axis. VAWTs offer a number of advantages over modern HAWTs, but one of the major challenges facing VAWTs is their lower power efficiency, as the angles of attack (AOAs) of the blades vary rapidly in one revolution. Extensive research is therefore being carried out on the pitch control of the blades of a VAWT. Since VAWTs are getting bigger to generate more power, structural health monitoring (SHM) will be applied to large wind turbines to ensure their functionality, safety and integrity. This chapter focuses on VAWTs with particular interest in attempting to establish a smart VAWT through the synthesis of energy harvesting, pitch control and SHM functions. The current research status on energy harvesting of VAWTs is considered first in this chapter. The research on the control of the blade pitch angles of a VAWT for the improvement of its power efficiency, startup capability and shut-down performance is then introduced. Based on wind load simulation as well as the fatigue and ultimate strength analyses of laminated composite blades and other structural members in a VAWT, the SHM system is proposed for the VAWT. The concept of a smart VAWT comprising energy harvesting, pitch control and SHM functions is finally presented in the last part of this chapter.

17.2 CURRENT RESEARCH STATUS ON ENERGY HARVESTING OF VERTICAL AXIS WIND TURBINES

17.2.1 Introduction to VAWTs

Unlike burning fossil fuels for electricity, wind power for generating electricity produces almost no greenhouse gas emissions and therefore reduces the greenhouse effect. In fact, the world has an enormous resource of wind energy that can be utilised for electricity generation. The most efficient and widely used way of capturing and converting the kinetic energy in wind into electrical energy is to use wind turbines. The wind power generation capacity in the world has increased sharply from 25,000 MW to more than 200,000 MW in the 10-year period from 2001 to 2010 (Bhutta et al. 2012).

Wind turbines can be basically categorised into HAWTs and VAWTs, which were introduced in Chapter 16. HAWTs rotate around a horizontal axis while VAWTs spin around a vertical axis. In the early twentieth century, primarily due to the inventions of Savonius (1929) and Darrieus (1931), Savonius-type and Darrieus-type VAWTs were developed. Most

Savonius-type VAWTs, as shown in Figure 17.1a, use drag forces to generate power and operate at low tip-speed ratios, thus resulting in low power efficiency. Darrieus-type VAWTs use lift forces to generate power and improve the power coefficient of VAWTs to some extent. In terms of the shape of blades, Darrieus VAWTs can be further categorised as curve-bladed VAWTs (see Figure 17.1b) and straight-bladed VAWTs (see Figure 17.1c).

Compared with HAWTs, VAWTs possess some distinguishing features such as simplified design, easy installation and maintenance and the potential for larger size turbines for more power generation. However, one of the major challenges facing VAWTs is their lower power efficiency. This is because the AOAs of the blades of a VAWT vary rapidly in one revolution: the fixed pitched blades of a VAWT not only provide positive torques that can contribute to power generation only in some azimuth angles during one revolution, but they also generate negative torques that actually reduce power generation in other azimuth angles. Together with other outstanding challenges, VAWTs had a brief development in the 1980s but they were then gradually rescinded in the 1990s after a series of faults and accidents (Gipe 2009). HAWTs have therefore become mainstream today.

Nevertheless, research on the pitch control technology for improving the power efficiency of VAWTs never stops. Vandenbereghe and Dick (1986, 1987) found that there was some performance improvement when the pitch angles of the blades vary in a harmonic pattern. Accordingly, they proposed a gear mechanism to produce first- and second-order pitch control. Lazauskas (1992) compared three pitch control systems by using the double-disks multiple stream-tube (DMST) theory. Hwang et al. (2006) conducted numerical and experimental studies on a 1 KW VAWT to see if the pitch angle control could improve its performance. By comparing cycloidal pitch control with individual pitch control, they pointed out that both control mechanisms could improve the power efficiency while individual pitch control gained better results. By using a genetic algorithm, Paraschivoiu et al. (2009) proposed an approach for searching the best pitch variation. More recently, Kiwata et al. (2010) conducted studies on the performance of variable-pitch straight-bladed VAWTs with a four-bar linkage mechanism and achieved better power coefficients. By discussing the forces of the six-blades H-type VAWT under the stationary and rotating conditions, a variable pitch control method was proposed by Liu et al. (2015a). Their simulation results revealed that the variable pitch method could increase the utilisation efficiency of wind energy.

It is worthwhile to note that the aforementioned studies on pitch control technology mainly focused on the improvement of power generation of VAWTs. There are seldom

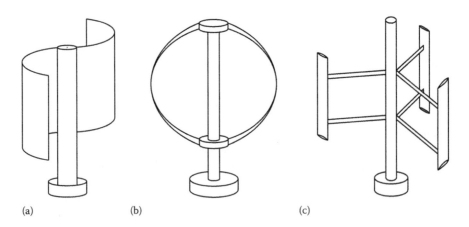

| (a) | (b) | (c) |

Figure 17.1 Schematic diagrams for types of VAWTs: (a) Savonius-type VAWT, (b) curve-bladed Darrieus-type VAWT and (c) straight-bladed Darrieus-type VAWT.

studies on the improvement of startup, rated power maintenance and shut-down performance of VAWTs using the pitch control technology. In fact, the startup performance of VAWTs under wind is another outstanding challenge, compared with that of HAWTs. The rated power maintenance and shut-down performance of VAWTs are expected to be better than those of HAWTs if the pitch control technology can be used appropriately. These topics in relation to the pitch control technology for VAWTs will be discussed in Section 17.3.

To pursue larger power generation, the size of wind turbines is getting bigger and bigger. The world's largest VAWT is the curved-blade Darrieus Éole turbine rated at 4 MW with a two-bladed rotor, 96 m in height and 64 m in diameter, as shown in Figure 17.2a. The tallest HAWT is the Vestas V164-8.0-MW wind turbine with a tip height of 220 m and a swept area of 21,000 m^2, as shown in Figure 17.2b. However, the further growth in tower height and rotor diameter of HAWTs meets many challenges, and very often the cost increases more rapidly than the power capacity. Some researchers have turned to VAWTs because VAWTs have a simpler structure than HAWTs and do not suffer from serious fatigue. A renewed interest in VAWTs has been seen in recent years, and there is a trend worldwide in building large-scale VAWTs.

Wind turbines are often established in an uninhabited area that is exposed to a harsh environment. During operation, large cyclic stresses in wind turbine blades are produced by time-varying wind loads, and the number of cycles is in the order of 10^8–10^9 over a 20–30 year life span. Besides fatigue loads, a wind turbine is also subjected to ultimate loads under extreme wind conditions. From 1989 to 2006, 64,000 incident reports from 1,500 onshore wind turbines were collected by the Institute for Solar Energy Technology (Pettersson 2010). The collapse of wind turbines due to failures of the blades are also reported in many references. Therefore, to ensure the functionality, safety and integrity of large-scale wind turbines, fatigue and ultimate strength analyses are required, the critical locations of fatigue and ultimate strength failure need to be figured out and SHM technology considered. In the past 10 years, researchers have made a great effort on fatigue and ultimate strength analyses of HAWTs (e.g. Kong et al. 2005; Shokrieh and Rafiee 2006; Nijssen 2006). The research on the SHM of wind turbines has also been actively conducted on HAWTs (e.g. Ciang et al. 2008; Hameed et al. 2009; Liu et al. 2015b). However, the studies on fatigue and ultimate strength analyses and the SHM of VAWTs are very limited. A framework for establishing an SHM system for VAWTs based on fatigue and ultimate strength analyses will be presented

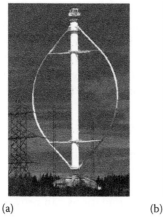

(a)   (b)

Figure 17.2 Large wind turbines: (a) Darrieus Éole wind turbine and (b) Vestas V164-8.0-MW wind turbine. (From (a) http://www.wind-works.org/cms/index.php?id=506; (b) http://www.shutterstock.com/pic-389971783.html.)

and discussed in Section 17.4. A concept of smart VAWTs by integrating pitch control, SHM and energy harvesting together will be presented and discussed in Section 17.5.

For ease of understanding, Section 17.3 on blade pitch control of VAWTs, Section 17.4 on SHM and Section 17.5 on smart VAWTs, the analytical method, computation simulation, wind tunnel test and field measurement currently used for investigating various aspects of VAWTs are briefly introduced in the following subsections.

17.2.2 Analytical methods for VAWTs

A number of analytical models are available for calculating wind forces on the blades and wind power generated by a VAWT. The two most frequently used models of VAWTs, the DMST model and the vortex model, are introduced in this subsection.

Based on the blade element momentum (BEM) theory, Templin (1974) proposed a single disk single stream-tube (SDST) model while Wilson and Lissaman (1974) proposed a multiple stream-tube (MST) model. On the basis of the MST concept, Paraschivoiu (1981) further developed the DMST model for HAWTs. BEM regards the rotor of HAWTs as an actuator disk. When air flows across the disk, the pressure drops suddenly and the forces acting on the blades are equal to the change in the momentum of the airflow. In the DMST model, the wind field is discretised into several stream-tubes. In VAWTs, in each stream-tube, the upwind and downwind sides are regarded as two actuator disks and the BEM theory is applied two times in these two actuator disks separately. Moreover, two assumptions are inherent in the DMST model: one is that the airflow has fully developed in the middle of the upwind and downwind sides so that the pressure at this point has resumed atmospheric pressure; and the other is that the stream-tube expansion can be ignored so that the area of the stream-tube is uniform. The SDST, MST and DMST models are featured by simple calculation but suffer deficiencies due to the lack of accuracy in the details of the flow field and sometimes the calculation will be divergent under situations of large tip-speed ratios. The fundamental theory of the DMST model will be given in Section 17.3 and the use of the DMST model to find pitch control algorithms for a VAWT will also be demonstrated in Section 17.3.

Vortex models, in which turbine blades are represented by bound vortices or lifting lines, were introduced and developed by Larsen (1975) and Strickland et al. (1979). The basic concept of vortex models is to calculate the velocity field of the wind turbine through the influence of vortices in the wake of blades while vortex strengths can be calculated from the aerodynamic coefficient dataset, the relative wind velocity and the AOA (Islam et al. 2008). The advantages of vortex models include their ability to determine blade-wake interactions and estimate the results in unsteady flow conditions and for finite aspect ratios of rotor blades (Bhutta et al. 2012). The main disadvantage of vortex models is that they require too much computational time and are difficult to converge under small tip-speed ratios.

17.2.3 Computational simulations of VAWTs

Computational fluid dynamics (CFD) simulations, with the increased availability of high-end computing capability and user-friendly commercial CFD codes, are gradually being adopted as an attractive tool to study VAWTs. Compared with the analytical methods, wind tunnel tests and field measurements, CFD simulations can provide detailed flow visualisation around, and wind pressure distribution over, the surfaces of the blades and other components of a VAWT. CFD simulations can also produce extremely large volumes of results at almost no added expense, and thus they provide an ideal tool for parametric studies. Nevertheless, there are some challenging issues remaining in CFD simulations and the accuracy of the results from CFD simulations needs to be validated against the results from other methods.

CFD simulations for a VAWT can be basically categorised as three-dimensional (3D), 2.5-dimensional (2.5D), and two-dimensional (2D) simulations. A sophisticated full 3D CFD simulation is always desirable for the performance assessment of a VAWT, which is particularly true in the case of high AOA. However, the extremely high computational cost of 3D CFD simulations prevents their wide applications by researchers or designers at present. Therefore, 2.5D CFD simulations are developed and investigated instead of 3D simulations. In a 2.5D simulation, the 2D VAWT model is extended in a span-wise direction for a considerable length in order to achieve a realistic reproduction of 3D separated vortices (Li et al. 2013). In other words, the 2.5D model differs from a full 3D model in that only a certain length of the blades is modelled with periodic boundaries at the two extremities of the domain. However, 2D CFD simulations need the least computational time, where the simulation of flow field is conducted in a planar area, but it should be used with caution for the accuracy of simulation results. A framework, which is based on the strip analysis method and 2D CFD simulation, will be developed and applied to a straight-bladed VAWT to determine its wind loads in Section 17.4 of this chapter. The influence of the existence of arms and a tower on the aerodynamic forces and the wind pressures on the blades, the variation of mean wind speed along the height of the VAWT and the effects of the turbulent wind, are all taken into consideration in this framework.

17.2.4 Wind tunnel tests of VAWTs

The main functions of wind tunnel tests of VAWTs, among others, are to validate the accuracy of analytical and CFD simulation results and to explore the performance of VAWTs with new shapes of blades. Many factors need to be considered for the design and implementation of a successful wind tunnel test of VAWTs. These factors include but are not limited to blockage effect, Reynolds number, tip-speed ratio, solidity ratio and so on. Biswas et al. (2007) conducted wind tunnel tests with the emphasis on the investigation of blockage effects on three-bucket Savonius rotors. Saha et al. (2008) performed wind tunnel tests to assess the aerodynamic performance of the Savonius rotor systems with different numbers of blades (two and three) and different geometries of the blades (semi-circular and twisted). Plourde et al. (2011) carried out wind tunnel tests to evaluate the performance of vented blades and blade capping for improving power generation. It was found that while venting provides only marginal improvement, capping greatly increases power generation. McLaren et al. (2012) conducted a series of wind tunnel tests to determine the aerodynamic behaviour of the aerofoils of an H-type VAWT of high solidity. Wind tunnel tests were also performed by Chong et al. (2013) to assess the performance of a five-bladed H-rotor VAWT with and without the integration of a novel omni-direction-guide-vane (ODGV). The experimental results showed an improvement in the VAWT's self-starting performance where the cut-in speed was reduced with the integration of the ODGV.

Most of the foregoing wind tunnel tests on VAWTs investigated the power efficiency of wind turbines. Limited attention was paid to the direct measurements of aerodynamic forces on the blades of a VAWT. This is probably because it is difficult to directly measure the aerodynamic forces on the blades of a rotating VAWT. However, it is important to know how the aerodynamic forces on a blade vary with azimuth angle and the AOA so that an appropriate pitch control algorithm can be developed. Recently, Peng et al. (2016) carried out a series of wind tunnel tests on a straight-bladed VAWT with different blade chord widths under different wind speeds at various tip-speed ratios. The experimental system was composed of a VAWT model with three straight blades, a speed control system used to

Figure 17.3 Vertical axis wind turbine in a wind tunnel.

control the rotational speed of the VAWT and a measuring system used to synchronously record the aerodynamic force time histories and the corresponding azimuth angle under high-speed rotating conditions (see Figure 17.3).

17.2.5 Field measurements of VAWTs

The investigation of VAWTs using wind tunnel tests is always conducted in a desirable and controllable environment. For a better understanding of the real performance of VAWTs or for validation of analytical, CFD simulation and wind tunnel test results, field measurements of VAWTs are often required.

A series of full-scale field tests were conducted on a group of 10-m VAWTs under natural wind conditions during summer 2010 (Dabiri 2010). The field test results showed that power densities of a greater order of magnitude could potentially be achieved by appropriately arranging VAWTs in layouts that enable them to extract energy from adjacent wakes and upwind farms. This improved performance did not require higher individual wind turbine efficiency, but only closer wind turbine spacing and a sufficient vertical flux of turbulence kinetic energy from the atmospheric surface layer. The National Research Council of Canada (NRC) also performed field tests of large-scale VAWTs and the test data were obtained from a 24-m VAWT operating at 29.4 rpm (Penna and Kuzina 1984). Sandia National Laboratories (SNL) designed and built a 17-m VAWT with and without struts. Worstell (1981) reported that a maximum rotor power coefficient of 0.467 could be achieved from this VAWT operating at 38.7 rpm. Moreover, the field measurements of a 34-m VAWT were conducted over three years by SNL to assess its structural dynamics, aerodynamics, fatigue and controls (Ashwill 1992). The field measurements of the largest VAWT, the curved-blade Darrieus Éole turbine as shown in Figure 17.2a, demonstrated that the maximum power output can exceed 1.3 MW at about 14.7 m/s (11.35 rpm) (Richards 1987). This turbine operated successfully for over 30,000 h during a five-year period from March 1988 and produced over 12 GW of electricity.

Figure 17.4 Hopewell VAWT in China.

A lift-type straight-bladed VAWT with a simple SHM system was constructed in 2011 at Yang-Jiang City of Guangdong Province, China, by the Hopewell Wind Power Limited of Hong Kong (see Figure 17.4). The reinforced concrete tower with a height of 24 m and a diameter of 5 m was used to support a vertical rotor, the rotating parts of the VAWT, of 26 m in height and 40 m in diameter. Three blades of NACA0018 type with a chord length of 2 m and a vertical height of 26 m were equally arranged at an interval angle of 120°. These blades were made of glass fibre-reinforced plastic (GFRP) laminate materials and supported by the Y-type steel arms connected to the tower via the main shaft at the top of the tower (the hub height). The ground clearance at the bottom end of the blades was 10.5 m. Each Y-type steel arm consisted of a main arm (10 m), an upper arm and a lower arm. The upper and lower arms supported the blade at one end and were connected to the main arm at the other end. Multi-type sensors were installed in this VAWT to monitor its performance, for example, the accelerometers used for the measurements of vertical and tangential acceleration of arms, strain gauges used for strain measurements of blades, load cells used for recording wind loads and ultrasonic anemometers used for wind velocity measurements. The first author of this book and his research team were involved in this project and part of the works will be reported in Section 17.4.

17.3 CONTROL OF BLADE PITCH ANGLES OF VERTICAL AXIS WIND TURBINES

Although VAWTs have attracted renewed attention in recent years, the research and development of VAWTs are still relatively insufficient and much more effort is required. Taking the research on pitch control of VAWTs as an example, most studies, as mentioned in Section 17.2, focus on pitch controlled VAWTs in a specific operational stage, that is, power generation below the rated wind speed. However, according to different wind conditions, there are different operational stages for a VAWT. The pitch control algorithm will be changed for different stages.

The operation of VAWTs can be divided into four stages. In the startup stage, defined as Stage 1, the aims of the pitch control algorithm are to maximise the aerodynamic torque and increase the rotational speed of the rotor to the designed rotational value as quickly as possible. In this stage, the pitch angle shall be controlled according to wind speed and rotational speed. When the rotational speed of the turbine rotor reaches the designed value, the wind turbine will be connected to the grid and the wind turbine then comes into the power generation stage, referred to as Stage 2. The goal of the pitch control algorithm will then be focused on maximising the aerodynamic torque for a given wind speed because the rotational speed remains unchanged in this stage. The pitch angle is thus controlled by wind speed only. When the wind speed is larger than the rated wind speed and lower than the cut-out wind speed, defined as Stage 3, the objective function of the control algorithm is to maintain the power generation at the rated value. The pitch angle is also controlled by wind speed only in this stage. When the wind speed surpasses the cut-out wind speed, defined as Stage 4, the pitch angle should be adjusted to increase AOA so that the stall can happen and the rotor can be stopped. It is obvious that in the different operational stages, the pitch control algorithms and their objective functions are different. In this section, a four-stage pitch control algorithm is developed to accomplish the following four objectives: assisting the startup in Stage 1, improving the power coefficient in Stage 2, maintaining the output power at the rated value in Stage 3 and assisting the shut-down in Stage 4 (Lin 2016).

Since a large number of pitch change cases are involved, the computational cost will be extremely high if using CFD simulations. To conduct this study in an effective way, the DMST model, which is first introduced, is adopted in this analysis. The four-stage control algorithm, including the startup control algorithm in Stage 1, the power maximisation control algorithm in Stage 2, the rated power control algorithm in Stage 3 and the parking control algorithm in Stage 4, is then discussed in detail. Based on these studies, a pitch control system for the Hopewell VAWT is given in the last part of this section.

17.3.1 Double disks multiple stream-tube model

The DMST model was developed on the basis of the BEM theory. In the DMST model, the wind field is discretised into several stream tubes (see Figure 17.5). In each stream-tube, the upwind and downwind sides are regarded as two actuator disks and the BEM theory is applied two times in these two actuator disks separately. Moreover, two assumptions are

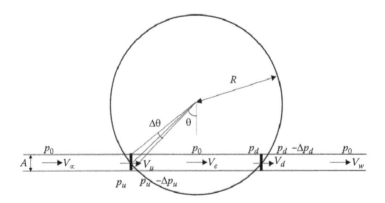

Figure 17.5 Wind speeds and pressures in a stream-tube.

adopted by the DMST model: one is that the airflow has fully developed in the middle of the upwind and downwind side so that the pressure at this point has resumed atmospheric pressure; and the other is that the stream tube expansion can be ignored so that the area of the stream tube is uniform.

In Figure 17.5, V_∞ is the inflow wind speed of the far upstream; V_u is the wind speed in the upwind rotor; V_d is the wind speed in the downwind rotor; V_e is the wind speed inside the rotor; V_w is the wind speed in the far wake; p_0 is the pressure of the far upstream; p_u is the pressure in front of the upstream actuator; Δp_u is the pressure drop due to the upwind actuator; p_d is the pressure in front of the downstream actuator; Δp_d is the pressure drop due to the downwind actuator; A is the area of the stream-tube; θ is the azimuth angle at which the stream-tube comes across the upwind side of the rotor; and $\Delta\theta$ is the angular width of the stream-tube.

By using the Bernoulli equation, the stream-wise force exerted by the upwind actuator disk and the downwind actuator disk can be derived and calculated as

$$\bar{F}_u = 2\rho R\Delta\theta a_u \left(1 - a_u\right) V_\infty^2 \sin\theta \tag{17.1a}$$

$$\bar{F}_d = 2\rho R\Delta\theta a_d \left(1 - a_d\right) V_e^2 \sin\theta \tag{17.1b}$$

where:

$\bar{F}_u$ and $\bar{F}_d$	are the stream-wise force exerted by the upwind actuator disk and the downwind actuator disk, respectively
ρ	is the air density
R	is the radius of the rotor
a_u	$= V_u/V_\infty$ is defined as the upwind induction factor
a_d	$= V_d/V_e$ is defined as the downwind induction factor

From the aerodynamics point of view, $\bar{F}_u$ and $\bar{F}_d$ are actually the average of aerodynamic drag force F_u and F_d, respectively, in one revolution of the VAWT. Based on the quasi-steady assumption, the aerodynamic drag force can also be calculated by the local relative velocity of the blade and the aerodynamic coefficients obtained from wind tunnel tests. The AOA of the blade at the azimuth angle of θ is shown in Figure 17.6a. The aerodynamic drag force F (equal to F_u or F_d), the normal force F_n and the tangential force F_t are shown in Figure 17.6b.

The relative wind speed V_r can be calculated by

$$V_r = \sqrt{\left(V\cos\theta\right)^2 + \left(V\sin\theta + \omega R\right)^2}, \ V = V_u \text{ or } V_d \tag{17.2}$$

where ω is the rotational speed.

The AOA for a zero pitch case can be calculated as follows:

$$a_r = \arctan\left(\frac{V\sin\theta}{\omega R + V\cos\theta}\right) = \arctan\left(\frac{\sin\theta}{\lambda + \cos\theta}\right), \ V = V_u \text{ or } V_d \tag{17.3}$$

where $\lambda = \omega R/V$ is defined as the tip-speed ratio.

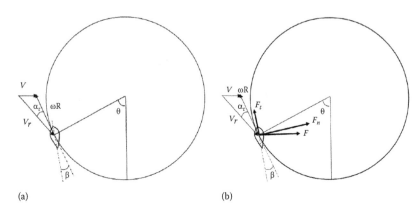

Figure 17.6 Relative wind speed, AOA and the corresponding aerodynamic forces: (a) relative wind speed and AOA and (b) aerodynamic forces.

From Equation 17.3, at a given azimuth angle θ, it can be seen that if λ is small, the AOA range is large and vice versa. When the pitch angle $\beta \neq 0$, the AOA can be estimated as

$$a = a_{\overline{r}} + \beta = \arctan\left(\frac{\sin\theta}{\lambda + \cos\theta}\right) + \beta, V = V_u \text{ or } V_d \tag{17.4}$$

For pitch control $\beta = \beta(\theta)$, it needs to use a different pitch angle in the corresponding stream tube.

The normal force and tangential force can be obtained by

$$F_n = \frac{1}{2}\rho V_r^2 c \Delta h C_n(\alpha) \tag{17.5a}$$

$$F_t = \frac{1}{2}\rho V_r^2 c \Delta h C_t(\alpha) \tag{17.5b}$$

where:

F_n and F_t	are the normal force and tangential force, respectively
C_n and C_t	are the normal force coefficient and tangential force coefficient obtained by wind tunnel tests, respectively
V_r	is the relative wind speed as shown in Figure 17.6a
c	is the chord length of the blade
Δh	is the height of the blade

In this study, the VAWT is regarded as 2D and Δh is equal to 1 m so that the aerodynamic per unit length is obtained. C_n and C_t are the function of AOA, which can be calculated by the lift coefficient C_l and the drag coefficient C_d as follows:

$$C_n(\alpha) = C_l(\alpha)\cos\alpha + C_d(\alpha)\sin\alpha \tag{17.6a}$$

$$C_t(\alpha) = C_l(\alpha)\sin\alpha - C_d(\alpha)\cos\alpha \tag{17.6b}$$

where the lift coefficient C_l and the drag coefficient C_d can be obtained by wind tunnel tests or numerical simulations. In this study, C_l and C_d of NACA0018 in the AOA region from $0°$ to $180°$ are used and shown in Figure 17.7.

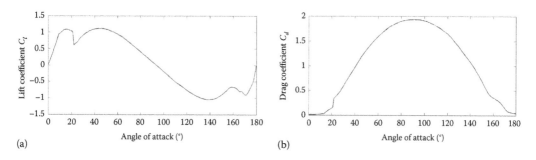

Figure 17.7 Lift and drag coefficients C_l and C_d of NACA0018: (a) lift coefficient C_l and (b) drag coefficient C_d.

Based on the relationship between the aerodynamic drag and tangential forces, F and F_t, and the normal force, F_n, as shown in Figure 17.6b, the following equation can be obtained via a proper transformation for a rotor with N' identical blades:

$$\bar{F} = \frac{N'\Delta\theta\rho V_r^2 c\Delta h}{4\pi}\Big[C_n \sin(\theta+\beta) - C_t \cos(\theta+\beta)\Big], \quad \bar{F} = \bar{F}_u \text{ or } \bar{F}_d \tag{17.7}$$

It can be found that Equations 17.1 and 17.7 describe the same forces from two aspects. By comparing these equations, the following equations can be obtained:

$$a_u = \frac{1}{1+G_u(a_u)} \tag{17.8a}$$

$$a_d = \frac{1}{1+G_d(a_d)} \tag{17.8b}$$

where

$$G_u(a_u) = \frac{N'c\Delta h}{8\pi R \sin\theta}\Big[C_n \sin(\theta+\beta) - C_t \cos(\theta+\beta)\Big]\left(\frac{V_r}{V_u}\right)^2 \tag{17.9a}$$

$$G_d(a_d) = \frac{N'c\Delta h}{8\pi R \sin\theta}\Big[C_n \sin(\theta+\beta) - C_t \cos(\theta+\beta)\Big]\left(\frac{V_r}{V_d}\right)^2 \tag{17.9b}$$

where V_r in Equation 17.9a is the relative wind speed at the upwind side while V_r in Equation 17.9b is the relative wind speed at the downwind side. Equations 17.8 and 17.9 give the iterative algorithm to calculate the induction factors a_u and a_d.

After calculating a_u and a_d, the aerodynamic forces can be obtained by Equation 17.5. The aerodynamic torque T and the power P can then be obtained by

$$T = \left(F_t \cos\beta - F_n \sin\beta\right)R \tag{17.10}$$

$$P = \left(F_t \cos\beta - F_n \sin\beta\right)\omega R \tag{17.11}$$

Substituting Equations 17.5 and 17.6 into Equation 17.10, one obtains

$$T = \frac{1}{2}\rho V_r^2 c\Delta h\big[C_l(\alpha)\sin(\alpha-\beta) - C_d(\alpha)\cos(\alpha-\beta)\big]R$$

$$= \frac{1}{2}\rho V_r^2 c\Delta h R\big[C_l(\alpha)\sin\alpha_r - C_d(\alpha)\cos\alpha_r\big] \quad\quad (17.12)$$

where α_r is defined in Equation 17.3.

In the following study, the blades of the NACA0018 with their force coefficients shown in Figure 17.7 are adopted in the Hopewell VAWT. Referring to Figure 17.7, the AOA region $|\alpha| < 21°$ can be seen as the stall-free region for the blade of NACA0018. The AOA region $|\alpha| \geq 21°$ can be seen as the post-stall region where the stall has already happened. From Equation 17.12, it can be seen that in the stall-free region, C_d remains at a small value as shown in Figure 17.7b and hence the torque T is mainly determined by the lift force. That is why this type of VAWT is called *lift-type*. It can also be seen that in this region, the torque T remains positive. When the AOA is in the post-stall region, C_l will drop and C_d will jump and hence the torque T is largely reduced. In the post-stall region, the larger is the absolute value of AOA $|\alpha|$ and the smaller is the torque T. When $|\alpha|$ increases to a certain extent, the torque T will turn to negative.

17.3.2 Startup control algorithm in Stage I

A drawback of VAWTs is their low self-starting torque, and a startup control algorithm is thus developed in this subsection to improve the self-starting capability of VAWTs. In the startup stage, the rotational speed of a VAWT will increase from zero to a specific value. As shown in Equations 17.4, 17.5 and 17.10, the aerodynamic torque is determined by the rotational speed, the wind speed and the pitch angle. The rotational speed and the wind speed cannot be controlled in this stage, and therefore the influence of the pitch angle on the torque is discussed. The aforementioned Hopewell VAWT (see Figure 17.4) with a 40 m diameter rotor and three 2 m-chord length NACA0018 blades is considered. The cut-in wind speed is 5 m/s and the considered rotational speed range is from 0.1 to 1.0 rad/s. The aerodynamic torque is calculated by Equation 17.10 and the influence of the pitch angle β on the aerodynamic torque is studied through the numerical simulation by using the DMST model. Two control algorithms are investigated: one is the fixed pitch control algorithm in one revolution and the other is the sinusoidal pitch control algorithm in one revolution. Nine cases are considered in fixed pitch control, that is, $\beta = -10°, -8°, -6°, -4°, -2°, 0°, 2°, 4°$ and $6°$. The pitch angle in the sinusoidal pitch control is set to be

$$\beta = \beta'\sin\theta \quad\quad (17.13)$$

where the amplitude β' is selected as the same as that of the fixed pitch. Detailed results of the simulated aerodynamic torques and the corresponding AOA for the aforementioned cases can be found in Lin (2016).

Because the kinetic energy of the rotor comes from the work done by wind loads, the work of wind loads can therefore be used as an index to evaluate the effect of pitch control in the startup process at a specific wind speed and rotational speed. By comparing the works done by the aerodynamic torques of the three blades in one revolution, the optimal pitch angle corresponding to the largest work at different wind speeds and rotational speeds can be found. For example, under the wind speed of 5 m/s and different rotational speeds, the works (calculated by the torque per length) in the cases of different pitch angles are shown in Figure 17.8. A positive work in Figure 17.8 stands for the work generated by the positive

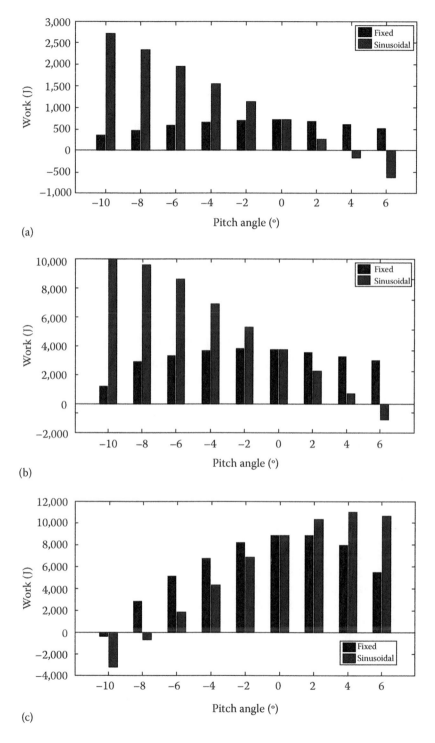

Figure 17.8 Works done by wind loads for different cases in Stage 1: (a) wind speed of 5 m/s, rotational speed of 0.1 rad/s, (b) wind speed of 5 m/s, rotational speed of 0.5 rad/s and (c) wind speed of 5 m/s, rotational speed of 1.0 rad/s.

torque, which is in the same direction of the rotation of the rotor, that is, clockwise in this study. It can be seen from Figures 17.8a and b that the sinusoidal pitch variation with an amplitude of -10° produces the largest positive work. However, under the wind speed of 5 m/s and the rotational speed of 1.0 rad/s, the sinusoidal pitch variation with an amplitude of 4° produces the largest positive work, as shown in Figure 17.8c. It can be found from these three cases that the sinusoidal pitch control can produce larger works than those produced by the fixed pitch control.

By considering the wind speed range from 5 to 10 m/s and the rotational speed range from 0.1 to 1.0 rad/s, the maximum powers of the two control algorithms at different wind speeds and rotational speeds are calculated and shown in Figure 17.9. It can be found from Figure 17.9 that the sinusoidal pitch control can obtain larger work than the fixed pitch control algorithm at any wind speed and rotational speed. To provide the reference pitch for the sinusoidal control, the optimal pitch angle amplitudes corresponding to the maximum power at each wind speed and rotational speed are given in Figure 17.10. It can be seen that at a low rotational speed, the sinusoidal pitch variation with an amplitude of −10° gets the largest work of the torque. As the rotational speed increases, the optimal pitch angle amplitude changes from negative value to positive value. This is because the range of the AOA is diminishing. When the stall does not happen, the optimal pitch angle amplitude becomes positive.

17.3.3 Power maximisation control algorithm in Stage 2

When a wind turbine is grid-connected, the rotational speed is basically locked. Therefore, the dominant factors for the determination of aerodynamic torque are the wind speed and the pitch angle in this stage. In this subsection, the influence of the pitch angle on the power maximisation is investigated with the rotational speed assumed at 2.1 rad/s. The considered wind speed region is from the cut-in wind speed (6 m/s) to the rated wind speed (14 m/s). Similar to the startup control, two control algorithms, the fixed pitch and the sinusoidal variation pitch, are considered. Nine cases are considered in the fixed pitch control, that is, $\beta = -10°, -8°, -6°, -4°, -2°, 0°,$ 2°, 4° and 6°. The pitch angle in the sinusoidal pitch control is determined according to Equation 17.13. Notably, the fixed pitch control in this stage means that the pitch is fixed at the same mean wind speed; when the wind speed changes, the pitch can be changed accordingly.

The optimal fixed pitch angles at different wind speeds are shown in Figure 17.11a and the optimal amplitudes of sinusoidal pitch are plotted in Figure 17.11b. For ease of comparison, the maximum powers obtained by using these two control algorithms are given in Figure 17.12. It can be observed that the powers generated by the sinusoidal pitch control algorithm are larger than those generated by the fixed pitch control algorithm at most wind speeds. Therefore, the sinusoidal pitch is preferred and the pitch control at different wind speeds can be found in Figure 17.11b. More details can be found in Lin (2016).

17.3.4 Rated power control algorithm in Stage 3

It is not the case of the higher the better with regard to produced power. There is rated power for every wind turbine, and the wind speed with respect to the rated power is called the *rated wind speed*. When the wind speed is above the rated wind speed, it is required to maintain the power in the rated value. In this study, the rated wind speed is selected as 14 m/s and the rotational speed is still assumed to be 2.1 rad/s. The rated

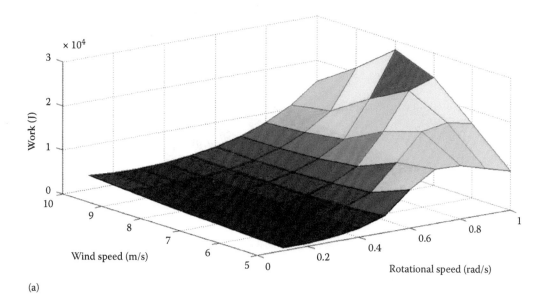

(a)

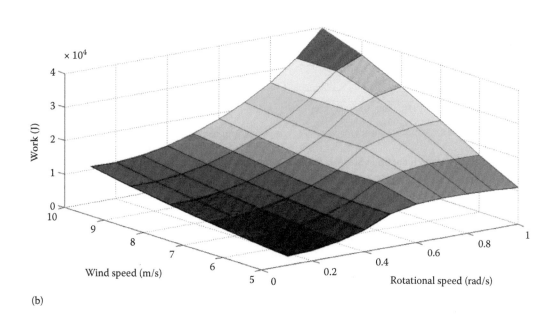

(b)

Figure 17.9 Maximum works of wind loads in different cases: (a) fixed pitch and (b) sinusoidal pitch.

power is about 800 kW. The power versus wind speed of the zero pitch case is shown in Figure 17.13a. It can be found that when wind speed increases, the power still increases and reaches a maximum value of about 1000 kW at a wind speed of 16 m/s and then the power decreases to about 600 kW. Hence, the zero pitch case cannot satisfy this requirement.

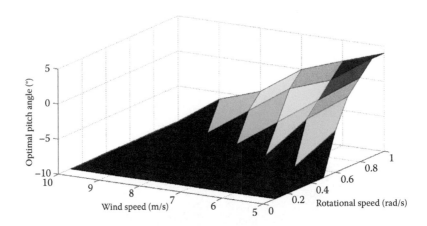

Figure 17.10 Optimal pitch angle amplitudes for sinusoidal pitch control in Stage 1.

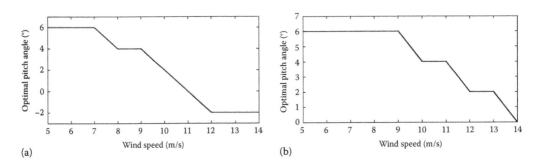

(a)

(b)

Figure 17.11 Optimal pitch angles in Stage 2: (a) fixed pitch control and (b) sinusoidal pitch angle control.

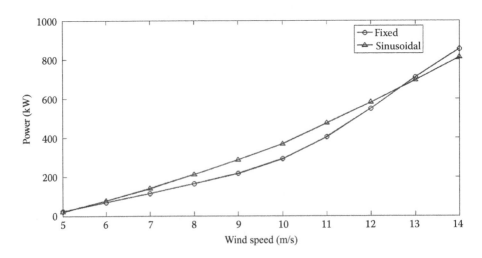

Figure 17.12 Maximum powers obtained by two control algorithms.

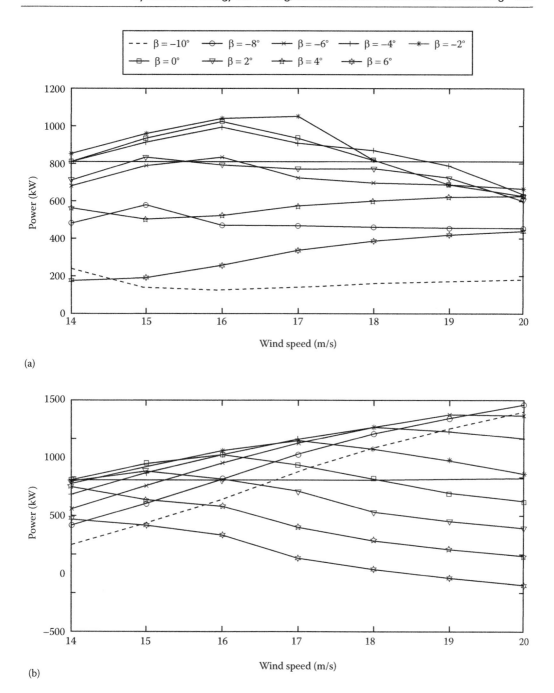

Figure 17.13 Output power vs. wind speed in Stage 3: (a) different fixed pitch control cases and (b) different sinusoidal pitch angle control cases.

The rated power is expected to be maintained in high wind speed through some techniques, such as pitch angle control. Similar to the previous subsections, two pitch control algorithms are considered: the fixed pitch in one revolution and the variable pitch in one revolution. The same fixed pitch angles as mentioned before are selected. The produced powers are shown in Figure 17.13a. The straight line is the target rated power. It is obvious that through adjusting the pitch angle, the rated power cannot be obtained by the fixed pitch control in the wind speed range from 19 to 20 m/s. The powers of the sinusoidal pitch control are also calculated and the results are shown in Figure 17.13b. It can be seen that the power can be properly maintained at the rated value from the rated wind speed (14 m/s) to the cut-out wind speed (20 m/s) by adjusting the amplitude of the sinusoidal pitch according to the mean wind speed. It means that the optimal pitch angle amplitudes in different wind speeds can be selected by comparing the corresponding produced power with the rated power (see Figure 17.14). A comparison of the powers generated in the cases with and without pitch controls are given in Figure 17.15. It can be observed that the sinusoidal pitch control can maintain the power at the rated value in the whole wind speed region, whereas the fixed pitch control cannot produce the rated power in the wind speed region from 19 to 20 m/s.

17.3.5 Parking control algorithm in Stage 4

When wind speed reaches a specific value, a wind turbine shall be delinked from the grid and parked for protection. This wind speed is called the *cut-out wind speed*. In this subsection, the parking of a VAWT with the aid of pitch control algorithms is studied. The cut-out wind speed is set to 20 m/s, and the rotational speeds from 2.1 to 0.1 rad/s are included. The fixed pitch control and the sinusoidal variation pitch control, as mentioned before, are considered.

Different from the previous subsections, the object of pitch control in this stage is not to prevent stall but keep the AOA in the stall region as far as possible. It is found that a fixed pitch cannot offer negative torque in a whole revolution to shut down the rotor while a sinusoidal pitch can offer negative torque in both upwind and downwind sides (Lin 2016). Since the turbine may be shut down in a wider range of wind speed, the aforementioned two control algorithms should be investigated in the wind speed region from 20 to 25 m/s and in the rotational speed region from 0.1 to 2.1 rad/s.

Similar to Stage 1, the work done by wind loads in one revolution is used as an index to evaluate the effect of a shut down. By comparing the work done by the aerodynamic torque in one revolution, the optimal pitch angle corresponding to the largest negative work at different wind speeds and rotational speeds can be determined. Taking the wind speed of 20 m/s as an example, the works done by wind loads in the fixed pitch and sinusoidal pitch controls at the rotational speeds of 2.1, 1.0 and 0.1 rad/s are shown in Figure 17.16. It can

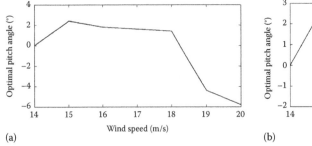

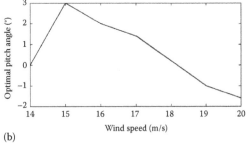

(a)   (b)

Figure 17.14 Optimal pitch angles in Stage 3: (a) optimal fixed pitch angle and (b) optimal amplitude for sinusoidal pitch angle control.

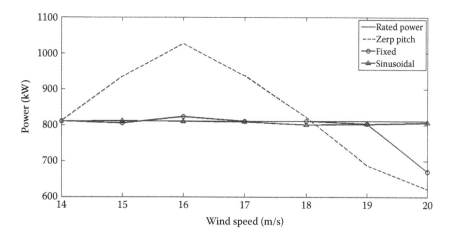

Figure 17.15 Power generations in the cases with and without pitch control.

be seen from Figure 17.16 that in the three rotational speeds, the work done by wind loads decreases as the amplitude of sinusoidal pitch increases. The largest minus work is obtained at $\beta' = 10°$. It can also be found that the negative torque can hardly be generated in the fixed pitch control, implying that it will not satisfy the requirement of a shut down. Therefore, the sinusoidal pitch control is recommended in the process of shut down. With the consideration of the whole wind speed region and the rotational speed region, the optimal amplitudes of sinusoidal pitch control within the range from –10° to 6° are calculated and shown in Figure 17.17. It can be observed that the largest positive amplitude of 6° can induce the largest minus torque at different wind speeds and rotational speeds.

17.3.6 Pitch control system

Based on the studies in Sections 17.3.2 through 17.3.5, a flowchart of the four-stage pitch control algorithm is presented and shown in Figure 17.18. The measured wind speed (V) and the rotational speed (ω) are continuously transmitted to the operational state indicator to determine the operational status of the VAWT. When the operational state is determined, the target pitch angle β can be selected from the control algorithms as shown in Figures 17.10, 17.11b, 17.14b and 17.17. The target value from the control algorithm is then sent to the controller to command the actuator to adjust the pitch. It should be noted that the measured pitch angles are also required. By comparing the measured pitch angle and the target one, the difference between these two values can be evaluated. If the difference is larger than the threshold, the controller should continue to give an order to the actuator to further regulate the pitch angle. If the difference is under the tolerance, the motion is finished and the control procedure should go back to the operational state indicator. The flowchart in Figure 17.18 assumes that the actuators can act quickly without too much time delay.

Besides the control algorithms introduced in Sections 17.3.2 through 17.3.5, the associated sensors and actuators are required to form the desired control system. In considering the Hopewell VAWT, the number and types of sensors and actuators should be determined. It can be found from the previous discussions that the wind speed and the rotational speed are required to determine the operational state, and the pitch angle of each blade needs to be changed to the target value at different azimuth angles. Therefore, an anemometer is needed to monitor the wind speed. The pitch angle is uniform along the blade while the mean wind speed is varying with height. Because the power of a wind turbine is usually

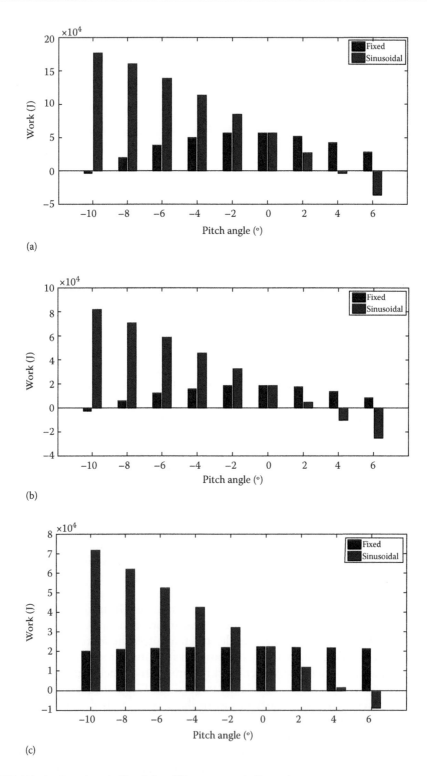

Figure 17.16 Works done by wind loads for different cases in Stage 4.

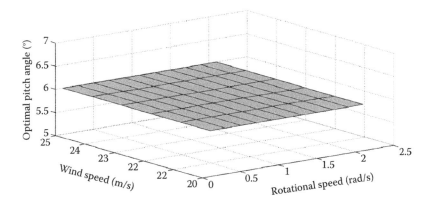

Figure 17.17 Optimal pitch angle amplitude for sinusoidal pitch control in Stage 4.

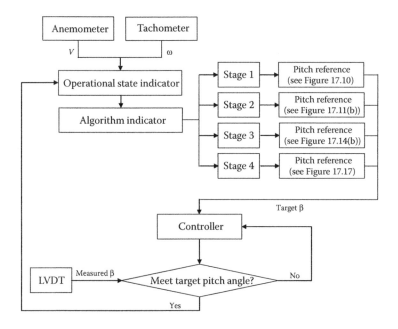

Figure 17.18 Flowchart of pitch control.

estimated using the mean wind speed at the hub height, the anemometer is installed at the hub height. A tachometer is needed to monitor the rotational speed and the azimuth angle. The tachometer is composed of a laser displacement sensor and a special gear wheel. The laser displacement sensor is fixed at the tower and the gear wheel is fixed at the shaft. The laser displacement sensor can record the distance variation and identify the azimuth angle of the arms and blades. Six actuators, which are installed on all the upper and lower arms to adjust the pitch of blade, are needed. To monitor the pitch of the blade, six linear variable differential transformers (LVDTs) with the same locations as the actuators are required to measure the pitch angle (see Figure 17.19). The schematic diagram of the tachometer is shown in Figure 17.20. A summary of the sensors and actuators is listed in Table 17.1.

A schematic diagram of the pitch control system is shown in Figure 17.21. It can be divided into two groups: one is the rotating parts and the other is the stationary parts. The

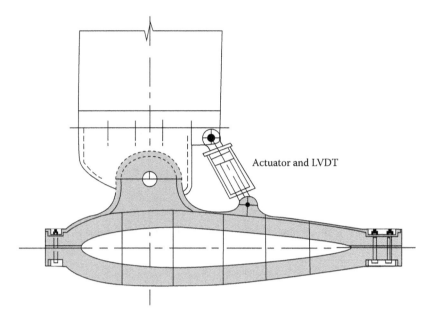

Figure 17.19 Location of actuator and LVDT.

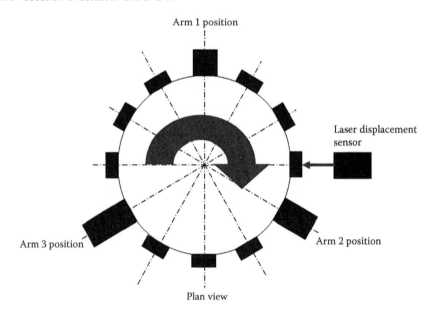

Figure 17.20 Schematic diagram of rotational speed measurement scheme.

power of the control system is obtained from the electrical grid. For the rotating parts, the transmission cables are connected to carbon brushes and supply the power to the rotating sensors, data recorder, computer and actuators. For the stationary parts, the power is directly obtained from the electrical grid. Data recorders are needed to record the measured data. Two channels are needed for the anemometer, one for wind speed and the other for wind direction; one channel is needed for the tachometer; six channels are needed for the LVDTs. The signals of the LVDTs are recorded by the data recorder of the rotating parts

Table 17.1 Summary of the sensors and actuators

	Anemometer	Tachometer	Actuator	LVDT
Wind mast	1	–	–	–
Shaft	–	1	–	–
Upper arm	–	–	3	3
Lower arm	–	–	3	3
Total	1	1	6	6

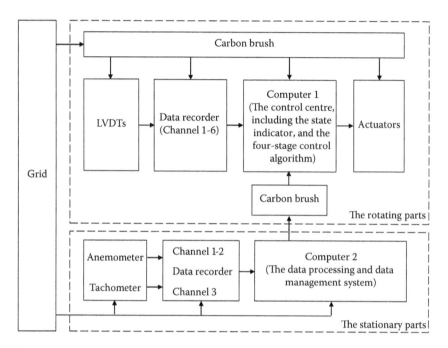

Figure 17.21 Schematic diagram of pitch control system.

with the sampling frequency of 25 Hz and transmitted to computer 1. These signals will be processed to obtain the corresponding pitch angles. The signals of the anemometer and tachometer can be recorded by the data recorder of the stationary parts with the same sampling frequency and transmitted to computer 2. After being processed by computer 2, the wind speed, rotational speed and azimuth angle are transmitted to computer 1 through carbon brushes. Then, the operational indicator in computer 1 can identify the present operational state and determine the appropriate control algorithm. Based on the referenced pitch angle proposed in Sections 17.3.2 through 17.3.5, control commands are generated and sent to the actuators to adjust the pitch angle.

17.4 STRUCTURAL HEALTH MONITORING OF VERTICAL AXIS WIND TURBINES

As mentioned before, the research on the SHM of wind turbines mainly focuses on HAWTs. Studies on the SHM of VAWTs are limited. Since VAWTs tends to become bigger when generating greater power, the SHM system will be important to ensure the functionality and safety of the VAWTs. To design an appropriate SHM system for a VAWT, fatigue and ultimate strength analyses must be conducted and the critical locations or potential failure locations must be identified. Since wind loads under different conditions are necessary for fatigue and ultimate strength analyses of the VAWTs, wind loading simulation for the Hopewell VAWT is discussed first in this section. The fatigue and ultimate strength analyses of laminated composite blades and other components of the Hopewell VAWT are then presented. A framework of the SHM system for the VAWT is presented in the last part of this section.

17.4.1 Wind loads on a VAWT

The Hopewell VAWT (see Figure 17.4) is taken as an example in this study. To determine wind loads on the VAWT during its operation, the mean wind speed and turbulent wind speed will be considered (see Figure 17.22). Since mean wind speed, turbulent wind speed and structural configuration vary along the height of the VAWT, it is insufficient and improper to determine wind loads by simplifying the VAWT as a planar structure. However, a truly 3D CFD simulation to determine wind loads on the VAWT is computationally difficult. Therefore, to take advantage of a 2D CFD simulation and at the same time to consider the varying wind load and a cross section of the VAWT, a strip analysis is employed. The strip method is widely used in the analysis of propellers (Chattot and Hafez 2015) and bridges (Davenport et al. 1992). The basic idea is that the 3D propeller or the bridge is first divided into several sections and the 2D simulation is then conducted on each of the sections. Based on this idea, the whole VAWT can be divided into several typical cross sections along its height and then the 2D simulation is applied to each of the sections with the simulated

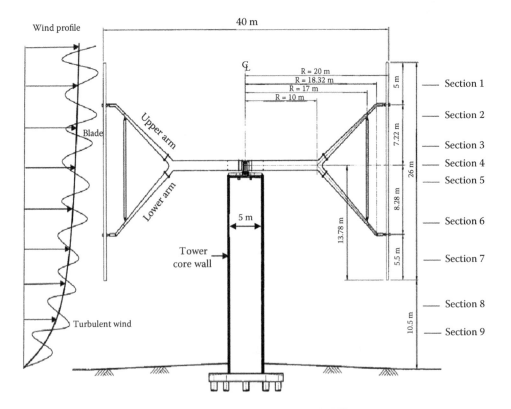

Figure 17.22 Wind field and structural configuration of Hopewell VAWT.

mean and turbulent wind field. For the VAWT in question, nine cross sections are chosen, as shown in Figure 17.22. After wind load simulations are completed for all nine sections under the simulated wind field, the simulated wind loads on all nine sections are used collectively and synchronously to yield the 3D wind loads on the VAWT.

CFD simulations on one blade, using the 2D SST k-ω method and the 2.5D LES method for the determination of wind loads, including wind pressures and aerodynamic forces, are first conducted and the results from the two methods are compared with each other. The comparative results show that the 2D SST k-ω method is an acceptable alternative to the 2.5D LES method in consideration of both the accuracy of the results and the efficiency of the computation (Lin 2016). To conduct the fatigue and ultimate strength analyses of the VAWT, the wind pressures on the blade and the aerodynamic forces on the whole VAWT under different wind conditions are required. The 2D CFD simulation is performed for the aforementioned nine cross sections together with the SST k-ω turbulence model. An interpolation technique is then used to obtain the spatial wind loads from these discrete sectional results.

Figure 17.23 shows the computed aerodynamic forces on a blade. It can be seen that the aerodynamic forces vary with time and height. The tangential force is much smaller than the normal force. Both the normal and tangential forces are reduced at the mid-span of the blade because of the existence of the main arm. The periodicity is obvious due to the rotation of the rotor. The aerodynamic forces and moment on other structural members, such as the upper and lower arms as well as the tower, are also simulated, and the details of such can be found in Lin (2016).

For the purpose of the ultimate strength analysis of the VAWT, wind pressures on the blade and the aerodynamic forces on the whole VAWT under extreme wind speed conditions are required. Different from the aforementioned operational stage of the VAWT, the wind turbine is standing still rather than rotating when it is under the extreme wind speed condition. Hence, wind loads vary with the inflow wind direction in the simulation, and the inflow wind direction is reflected by the azimuth angle. Figure 17.24 shows the normal and tangential forces and the moment on one blade. It can be seen that the largest positive and negative normal forces on the blade appear at $\theta \in [60°, 120°]$ and $\theta \in [240°, 300°]$, respectively. This is because in the range of these azimuth angles the drag force is large. The largest tangential force on the blade appears at $\theta = 90°$. The aerodynamic moment on the blade has the largest value at $\theta = 90°$ and $\theta = 300°$. It can also be found that similar to the operating condition, the aerodynamic forces in the higher section are generally larger than those in the lower section, which reflects the influence of the wind profile. Moreover, the forces near the fourth cross section are very different from those in other sections, which implies that the main arm has an obvious influence on the aerodynamic forces on the blade.

17.4.2 Fatigue and strength analyses of laminated composite blades

17.4.2.1 Framework of fatigue and ultimate strength analyses of blades

The framework of fatigue and ultimate strength analyses of laminated composite blades are shown in Figure 17.25. There are three parts in this framework. The first part is to determine the design loads, including wind loads and inertial forces. The second part is to establish a proper finite element (FE) model and conduct the model updating of the blade to obtain accurate laminar elastic constants. Finally, the fatigue and ultimate strength analyses are conducted in the third part.

Wind loading simulation was introduced in Section 17.4.1. An FE model of the laminated composite blade of the Hopewell VAWT is established by using ANSYS Shell91 elements, as

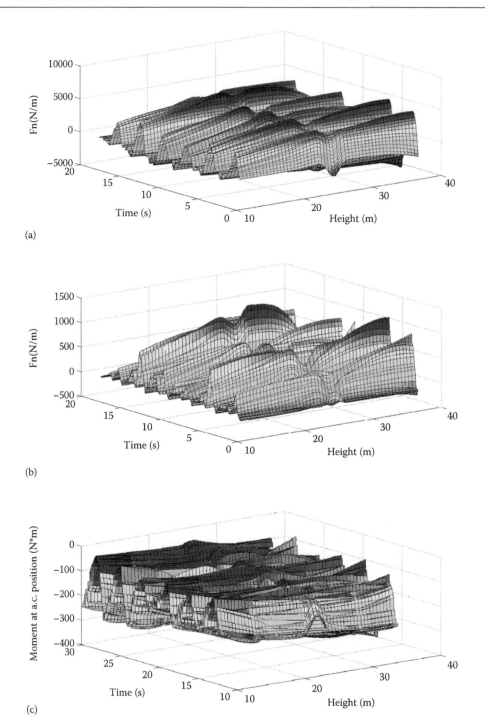

Figure 17.23 Aerodynamic forces and moment of a blade under operating wind speed condition: (a) normal force, (b) tangential force and (c) moment.

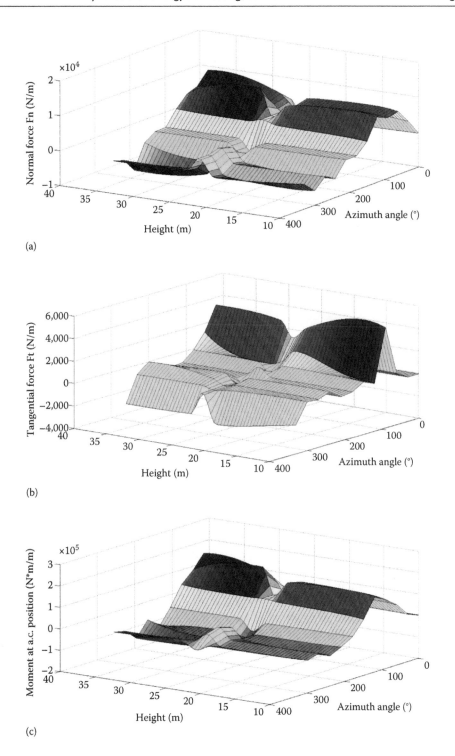

Figure 17.24 Aerodynamic forces and moment of a blade under extreme wind condition: (a) normal force, (b) tangential force and (c) moment.

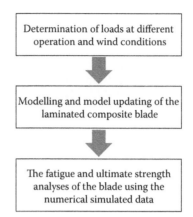

Figure 17.25 Framework of fatigue and ultimate strength analyses for a laminated composite blade.

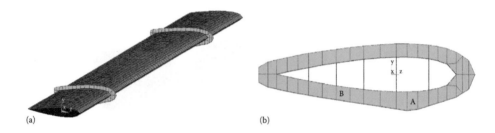

Figure 17.26 FE model of laminated composite blade: (a) overview and (b) side view.

shown in Figure 17.26. The blade has an NACA0018 cross section with 2 m chord length. The length of the blade is 26 m. Two identical steel brackets are used to fasten the blade near its two ends. The bracket is made of a steel stiffener and a steel flange. The connection between the element of stiffener and the element of flange is rigid. Rayleigh damping is used in this study and the damping ratio chosen is 0.5%. In this FE model, there are a total of 7,927 elements and 22,752 nodes in one blade, and the number of degrees of freedom (DOFs) is 136,512.

In order to analyse the laminated composite blade at the level of individual constituents and predict the properties of a composite material, a micromechanics model is employed. For the purpose of reducing modelling errors and producing a better FE model to represent the real one, the procedure of model updating is conducted on the basis of a micromechanics model. A more detailed description of the FE model of a laminated composite blade and its associated model updating can be found in Lin (2016). Based on the simulated wind loadings and the established FE model, the fatigue and strength analyses of the laminated composite blade can be conducted and discussed in subsections 17.4.2.2 and 17.4.2.3.

17.4.2.2 Fatigue analysis

The fatigue analysis for a laminated fibreglass reinforced plastic (FRP) blade is different from the fatigue analysis of steel components to some extent. Strain cycles, other than stress cycles, are used to evaluate the fatigue damage of an FRP. The cycle number to failure N of a laminated FRP plate is estimated by the $\varepsilon - N$ curve, which can be expressed in the form of Equation 17.14.

$$N = K(\varepsilon_r)^{-m} \tag{17.14}$$

where:

    N is the number of cycle to failure

    K is an empirical constant related to the material

    m is the slope parameter ($m = 9$ for laminates with polyester matrixes and $m = 10$ for laminates with epoxy matrixes)

    ε_r is the strain range

The service life of a wind turbine is more than 20 years and it is unrealistic to conduct a numerical simulation over such a long time. Hence, an alternative approach is presented as follows: first, the operational mean wind speed range is discretised into several wind speed bins; second, the fatigue damage in a short time period, for example 10 min, of the blade at its critical location under the wind speed at the centre of the wind bin is obtained; third, the proportion of the short-term damage in the total fatigue damage for the designated time period is evaluated by the mean wind speed distribution so that the final fatigue damage can be calculated.

For the VAWT under consideration, the cut-in and cut-out wind speeds of the VAWT are 5 and 21 m/s, respectively. The mean wind speed region is discretised into eight wind speed bins and the size of each bin is 2 m/s. The centres of these bins are 6, 8, 10, 12, 14, 16, 18 and 20 m/s, respectively. The probability of the mean wind speed in each bin is calculated using the Weibull distribution. The time length (in seconds) of the mean wind speed within one of the wind speed bins in n years can be calculated by $T_i = 365 \times 24 \times 3600 n P_i$, $i = 6, 8, 10, \ldots, 20$, where i represents the centre of the wind speed bin; P_i is the probability of the corresponding mean wind speed; and T_i is the total time length of the mean wind speed within the wind speed bin in n years in seconds.

The fatigue critical locations should be figured out. It is important to determine these locations, not only because the fatigue life of a wind turbine blade is determined by these locations but also because it can guide the sensor installation of the SHM system. To assess fatigue critical locations, the strain time history of any node of the blade should be simulated. In this regard, the strain time histories in 10 min of the blade at all nodes are simulated under the mean wind speeds at the centre of the wind speed bins. The fatigue critical locations are then found at the supports and the mid-span of the blade. More details can be found in Lin (2016).

The fatigue damage of the blade at the critical location is estimated using the ε–N curve. In this regard, the cycle numbers of different strain range and mean values of the concerned nodes are obtained by the rain-flow method. These cycle numbers are for 10 min. Divided by 600 s (10 min) and multiplying the time length of the mean wind speed for the corresponding wind speed bin, the cycle numbers for 10 min can be extended to the cycle numbers for 20 years. The fatigue damages of the blade at six critical nodes over different strain ranges and mean strains are shown in Figure 17.27 and the corresponding fatigue damage indices are listed in Table 17.2.

It can be seen from Table 17.2 that the fatigue life of this blade is determined by compressive fatigue damage at node 2. Although the fatigue damage at node 2 is only 0.1020, it cannot lead to the conclusion of safety because the fatigue mechanism of a composite material is complex and the uncertainty of the fatigue design of a composite material is very large. The fatigue damage due to the compressive stress state is different from metals in which the fatigue problem in the compressive stress state is not obvious.

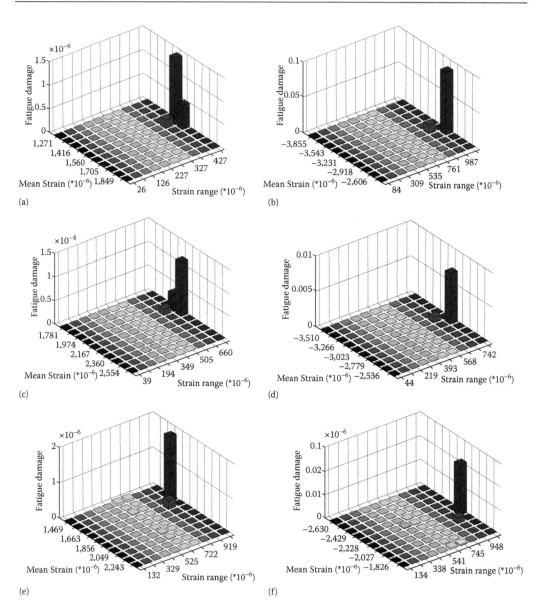

Figure 17.27 Matrix plot of fatigue damages at six nodes: (a) node 1, (b) node 2, (c) node 3, (d) node 4, (e) node 5 and (f) node 6.

Table 17.2 Fatigue damage indices of node s 1–6 in 20 years

	Node 1	Node 2	Node 3	Node 4	Node 5	Node 6
Fatigue damage	0.4581×10^{-6}	0.1020	0.1626×10^{-3}	0.0186	0.0186×10^{-3}	0.0122

17.4.2.3 Ultimate strength analysis

To evaluate the failure-critical locations of the blade, a failure criterion is required. A variety of failure criteria have been developed for laminated FRP material. The Tsai–Wu criterion (Gol'denblat and Kopnov 1965) is employed for the failure analysis of the concerned laminated blade. Under the extreme wind speed condition, the VAWT is parked and hence only wind pressures are applied on the blade.

The azimuth angles considered are 0°, 30°, 60°, 90°, 120°, 150°, 180°, 210°, 240°, 270°, 300° and 330°. At each azimuth angle, the stresses in each laminar of the blade can be obtained and the values of the Tsai–Wu strength index (P) can be calculated accordingly. It is then found that the failure critical locations (with large P value) are the same for each azimuth angle, that is, at the supports and the mid-span of the blades. The largest Tsai–Wu strength index appears at the upper support of the blade. By taking 90° as an example, the Tsai–Wu strength indices P in the three laminas are shown in Figure 17.28. The largest Tsai–Wu strength index in each lamina (P_{max}) is also given in the figure.

17.4.3 Fatigue and strength analyses of structural members

17.4.3.1 Framework of fatigue and ultimate strength analyses of structural members

The framework of the fatigue and ultimate strength analyses of the structural members of the Hopewell VAWT are shown in Figure 17.29. There are four parts in this framework. First, the design loads on all the structural components of the VAWT are determined, including the aerodynamic force, the inertial force, the gravity force and the Coriolis's force. Then, the FE model of the VAWT is established, followed by the procedure of model updating on the basis of the field measured data. Finally, the fatigue and ultimate strength analyses of the structural members of the VAWT are conducted.

To conduct the fatigue and ultimate strength analyses of the structural members of the VAWT, the FE model of the Hopewell VAWT is built and shown in Figure 17.30. The ANSYS beam188, a geometry-exact beam element, is used to model all components of the VAWT, including the tower, the shaft, the hub, the main arms, the upper and lower arms, the links and the blades. In this FE model, the tower is modelled by 26 beam elements; the hub by 8 beam elements; the shaft by 20 elements; each main arm by 20 elements; each upper arm by 28 elements; each lower arm by 28 elements; each link by 27 elements; and each blade by 50 elements. There are in total 519 elements, 673 nodes and 4020 DOFs in the FE model of the VAWT. Since various uncertainties exist in the FE model of VAWT, model updating on the basis of field measurements is necessary before using the established FE model for fatigue and ultimate strength analyses. More details on the modelling and model updating of the VAWT can be found in Lin (2016).

17.4.3.2 Fatigue analysis

Since the fatigue analysis of the blades has been performed in detail in Section 17.4.2, this subsection focuses on the fatigue analysis of other structural components, including the shaft, the main arms and the upper and lower arms, which are made of steel, and the tower, which is made of reinforced concrete.

The fatigue analysis is based on the stress responses of the structural members under the operational wind speeds for the designated time period of 20 years. First, the rain flow counting method is applied to the stress response time history of the structural member at its critical location to find out the number of stress cycles at different stress range levels. Then,

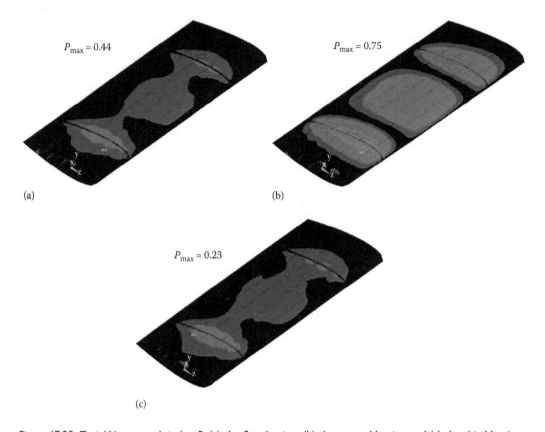

(a)   (b)

(c)

Figure 17.28 Tsai–Wu strength index P: (a) the first lamina, (b) the second lamina and (c) the third lamina.

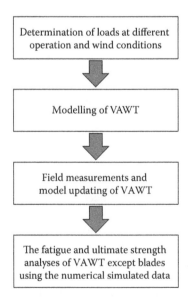

Figure 17.29 Framework of fatigue and ultimate strength analyses for the whole VAWT.

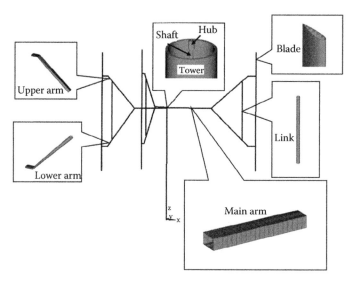

Figure 17.30 FE model of a straight-bladed VAWT.

the number of cycles to failure at each stress range level is estimated by the S–N curve and the Goodman diagram is used to transfer the stress range with nonzero mean to the zero-mean stress range. Finally, the Palmgren–Miner's rule is used to evaluate the fatigue damage of the structural members of the VAWT.

Due to the complexity of the VAWT, it is impossible to determine the fatigue-critical locations by calculating fatigue damage at all the nodes of the structural members. A method is presented to determine the fatigue-critical locations in consideration of the dominant mode shapes. It can be found that the concerned VAWT is dominated by the first four modes, and three critical sections A, B and C of the main arm and the shaft are selected via the modal analysis (see Figure 17.31a). The selection of section B is to give a comparison with the fatigue damage in section A. In sections A and B, the normal stresses due to vertical bending ($\sigma_{x,vA}$ and $\sigma_{x,vB}$) and the normal stresses due to horizontal bending ($\sigma_{x,hA}$ and $\sigma_{x,hB}$) are extracted at nodes A1, B1, A2 and B2, respectively, as shown in Figure 17.31b. Two normal stresses due to bending of the shaft ($\sigma_{x,C1}$ and $\sigma_{x,C1}$) in section C are also obtained at nodes C1 and C2, as shown in Figure 17.31c. In Figure 17.31, subscript 'x' means that the direction of the stress is normal to the cross section.

By using the S–N curve, the fatigue damage in 20 years at the 6 critical points in sections A, B and C is calculated and the results are listed in Table 17.3. It can be seen that the largest fatigue damage of 0.3476 occurs at the root of the main arm, caused by the vertical bending normal stress of the main arm. The second largest fatigue damage of 0.2687 occurs at the root of the main arm, which is caused by the horizontal bending normal stress of the main arm. The shaft has the least fatigue damage compared with other components of the VAWT. It is worth noting that although the fatigue damage of the shaft is small, the fatigue in this location should not be ignored because the influences of misalignment of the rotor and eccentric mass are not considered in this study.

Besides the fatigue analysis of the aforementioned steel components, a fatigue analysis of the tower is also conducted. The forces on the tower are the gravity force, the wind load and the reaction forces from the shaft. Under these forces, the largest stress and stress range appear at the bottom of the tower, which is the fatigue-critical location. Therefore, the cyclic

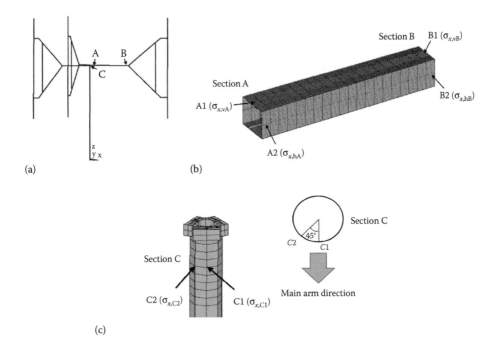

Figure 17.31 Fatigue critical locations of steel components and the positions of output stresses: (a) the positions of Sections A, B and C; (b) the output positions in Sections A and B; and (c) the output positions in Section C.

Table 17.3 Fatigue damage of fatigue-critical locations in 20 years

	$\sigma_{x,vA}$	$\sigma_{x,hA}$	$\sigma_{x,vB}$	$\sigma_{x,hB}$	$\sigma_{x,C1}$	$\sigma_{x,C2}$
Fatigue damage ($\bar{D}$)	0.3476	0.2687	0.0099	0.0088	0.0179	0.6567×10^{-4}

stresses at the bottom of the tower are calculated. Using the Miner's rule and the S-N curve, the maximum fatigue damage of the tower is calculated as 0.1017.

17.4.3.3 Ultimate strength analysis

Under the extreme wind speed condition, wind turbines must be parked because in such high wind speeds, the rotor is prone to the danger of overspeed. Although wind turbines will vibrate under these conditions, the occurrence probability of this happening is small and hence the fatigue damage caused under this condition is small as compared with that under the operation wind speed condition. Therefore, only an ultimate strength analysis is needed to be conducted under extreme wind speed conditions. In this case, the von Mises stress is used to evaluate the failure-critical locations. Since wind direction has influence on the magnitude of the von Mises stress at the failure-critical location, the influence of wind direction is considered in the analysis. Furthermore, because the ultimate strength analysis of the blades has been conducted and given in Section 17.4.2, this section only focuses on the analysis of the arms, the shaft and the tower.

Through a comparison of the von Mises stress at each azimuth angle, it is found that the locations of large von Mises stress in each component coincide with the fatigue-critical locations. Thus, the von Mises stresses at points A1, A2, B1, B2, C1 and C2 are considered. The von Mises stresses vary with the azimuth angle. For the main arm, the largest stress appears at 60° in the front wind side and 240° in the back wind side. The stresses at A1 and B1 caused by vertical bending are larger than those at A2 and at B2, caused by horizontal bending. For the von Mises stresses of the shaft, there are three peak stresses in a cycle due to the three-blade structure and the peaks occur at the azimuth angles 30°, 150° and 270°. For the von Mises stresses of the tower, there are also three peaks in a cycle. It has been concluded that for the VAWT of interest, the dangerous azimuth angles are 60° and 240° for the main arms, 30°, 150° and 270° for the shaft and 60°, 180° and 300° for the tower, under extreme wind speed conditions.

17.4.4 Structural health monitoring system for the VAWT

The SHM system used for a VAWT has five functions: to monitor the loading and responses of a structure, to assess its performance under various service loads, to verify or update the rules used in its design stage, to detect its damage or deterioration and to guide its inspection and maintenance.

Based on these objectives, the following sensor types are used. In order to monitor wind speed and wind profile, anemometers are required at different heights. Because rotational speed is another important parameter for a wind turbine, it has to be monitored. There are many methods to measure the rotational speed. One of the most widely used methods is a laser displacement sensor. To monitor the deformation of a VAWT, displacement sensors, accelerometers and strain gauges are required. To monitor global deformations, accelerometers need to be installed on the components, such as the tower, the main arms and the upper and lower arms. To monitor the swing of the shaft, two laser displacement sensors are used. To monitor local deformations, strain gauges are installed at the potential damage locations of the tower, shaft, main arms and blades. To assess wind power and monitor service loads, load cells are installed at the supports of the blade and the links connecting the upper and lower arms. A summary of the number, types and locations of sensors is given in Table 17.4. Details of the placements of sensors are shown in Figure 17.32. More descriptions on the number, location and function of these sensors can be found in Lin (2016).

After the number, location and function of the sensors have been determined, the SHM system is designed, as shown in Figure 17.33. The sensing system and the data acquisition and process system used in the SHM system can be divided into two groups: one is the rotating parts (on the blades, arms and shaft) and the other is the stationary parts (on the tower and the wind mast). The power for the sensors and other instruments is normally

Table 17.4 Summary of sensors used in SHM system

	Anemometer	Laser meter	Accelerometer	Strain gauges	Load cell	LVDT
Wind mast	2	–	–	–	–	–
Tower	–	–	2	4	–	–
Shaft	–	1	–	12	–	2
Main arm	–	–	6	3	–	–
Upper arm	–	–	6	–	3	3
Lower arm	–	–	6	–	3	3
Link	–	–	–	–	3	–
Blade	–	–	–	30	–	–
Total	2	1	20	49	9	8

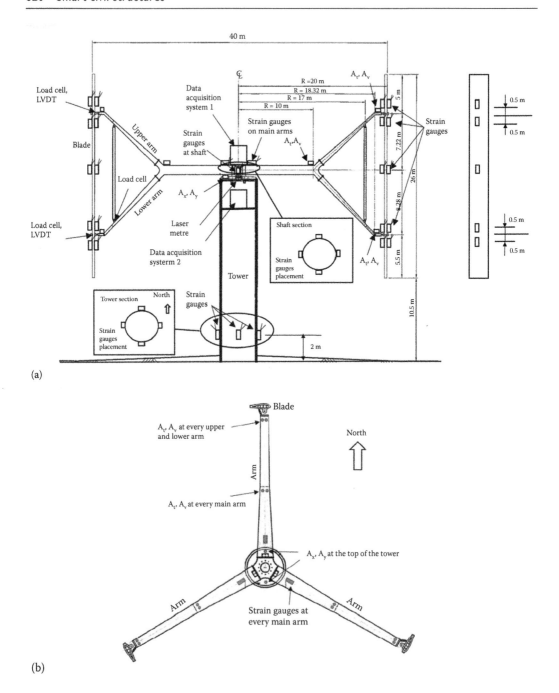

(a)

(b)

Figure 17.32 Schematic diagrams of sensor locations: (a) front view and (b) top view.

obtained from the power grid. For the rotating parts, carbon brushes are used to transmit the current from the stationary transmission cables to the rotating ones so that the equipment of the rotating parts can obtain the power supply, as shown in Figure 17.33a. For the stationary parts, the power is normally obtained from the electrical grid, as shown in Figure 17.33b.

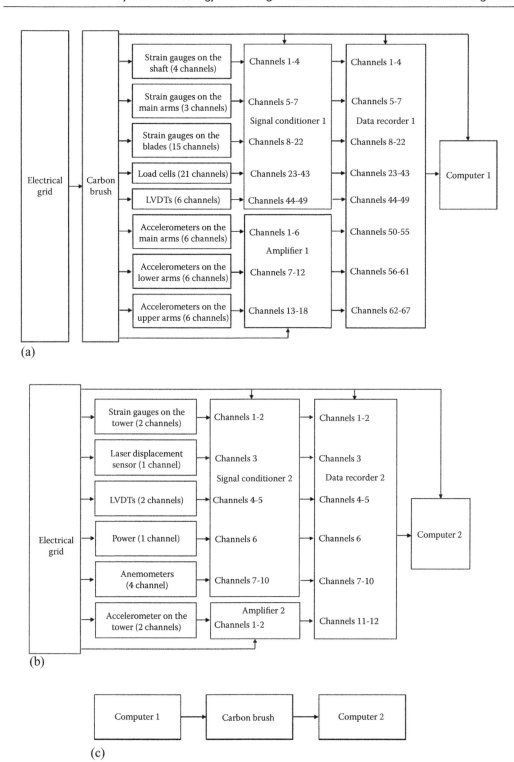

Figure 17.33 Proposed SHM system: (a) rotating part, (b) stationary part and (c) data transmission.

Two sets of signal conditioners are used to receive signals from load cells and strain gauges as shown in Figures 17.33a and b. One set is for the rotating parts and there are a total of 49 channels: 4 for the strain gauges on the shaft, 3 for the strain gauges on the main arms, 15 for the strain gauges on the blades, 21 for load cells and 6 for LVDTs. The other set is for the stationary parts and there are a total of eight channels: four for the anemometers, two for the strain gauges on the tower, one for the laser displacement sensor and one for the generated power.

Two sets of charge amplifiers are used to receive signals from accelerometers as shown in Figure 17.33a and b. One set is for the rotating parts and there are a total of 18 channels: 6 for accelerometers on the main arms, 6 for accelerometers on the upper arms and 6 for accelerometers on the lower arms. The other set is for the stationary parts and there are two channels for the accelerometers on the top of the tower. Two sets of data recorders are required. One set is to record the data from the rotating parts and there are 67 channels (49 for conditioners and 18 for amplifiers). The other set is to record the data from the stationary parts and there are a total of 10 channels. The sampling frequency of these two data recorders is 25 Hz.

Two sets of computers are used to process and store the data. One is installed on top of the centre of the rotor, rotating with the VAWT; the other is installed in the tower. These two computers act as the data processor systems, the data management systems and the structure evaluation systems. When there is a fault, an alarm will be given. Considering the convenience in getting the stored data, carbon brushes are used to transfer the data from the rotating computer to the stationary one, as shown in Figure 17.33c.

17.5 CONCEPT OF SMART VERTICAL AXIS WIND TURBINES

A pitch control system was proposed in Section 17.3.4. The aims of this control system are to improve the self-start capability of a VAWT when the wind speed reaches the cut-in wind speed, to maximise the power production when the wind speed is higher than the cut-in wind speed but lower than the rated wind speed, to maintain the power at the rated value when the wind speed is higher than the rated wind speed but lower than the cut-off wind speed, and to benefit the VAWT in the state of shut down when wind speed is larger than the cut-off wind speed. Moreover, an SHM system was presented in Section 17.4.4. This SHM system is designed to monitor wind condition, to ensure the functionality and safety of the VAWT and to guide the inspection and maintenance.

By comparing these two systems, it can be found that there are some common sensors (such as the anemometer, LVDT and the tachometer) and common data acquisition and processing systems. It will not be economic if these two systems are designed and used separately. Moreover, when a fault happens, it will be beneficial if a command can be sent from the SHM system to the control system to shut down the VAWT in time. Also, since a VAWT itself is an energy-harvesting machine, it will be quite promising if the generated power can be directly provided to the instruments of the SHM and pitch control systems.

Hence, it is highly desirable to develop an integrated SHM and control system for the VAWT to make it have self-sensing, self-diagnosis, self-control and self-power functions. A schematic diagram of such an integrated system is given in Figure 17.34. In this system, two sets of power supplies are used. In the operational condition, the partial power from the VAWT is offered to the sensors and other instruments of the integrated system, whereas the rest is transmitted to the electrical grid. When the VAWT is shut down, the power will not be produced by the generator. In this case, the power can be obtained from the grid. To deal with the transition of the power supply mode and the electrical grid outage, batteries,

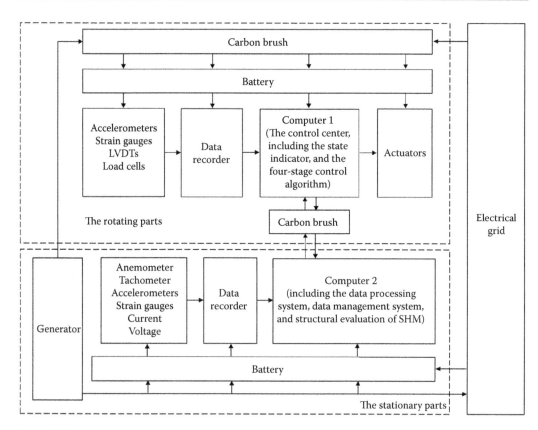

Figure 17.34 Concept of a smart VAWT.

such as uninterruptible power systems (UPS), can be used. The power from the generator or the grid is first connected to the batteries and the power is then supplied by the batteries.

The sensors, actuators and other instruments are also divided into two groups: one is the rotating parts and the other is the stationary parts. Data from the anemometers installed on the wind mast and the sensors installed on the tower are recorded by the data acquisition systems of the stationary parts, whereas the data from the sensors on the shaft, arms and blades are recorded by the data acquisition system of the rotating parts. Among the data of the stationary parts, the wind speed, the rotational speed and the azimuth angles are transferred to computer 1 of the rotating parts for the control purpose. Computer 1 is the control centre, which includes the operational state indicator and the aforementioned four-stage control algorithm. Computer 2 is the SHM centre, which includes the data processing system, the data management system and the structure evaluation system. Moreover, when a fault is found by the SHM system, a shut-down command will be sent to the control system to stop the rotation of the rotor. For ease of extracting the stored data, data in computer 1 are transferred to computer 2 for storage. In such a VAWT, the SHM system, the control system and the power supply system are synthesised to some extent. Hence, it is called the *smart VAWT*.

It is worthwhile to note that the smart VAWT is at the conceptual stage and does not include structural vibration control and self-repairing functions. A great amount of research and effort is still required in order to realise a truly smart VAWT in the near future.

NOTATION

a_u, a_d	Upwind and downwind induction factor, respectively
A	Area of the stream-tube
c	Chord length of the blade
C_l, C_d	Lift coefficient and drag coefficient, respectively
C_n, C_t	Normal force coefficient and tangential force coefficient, respectively
F	Aerodynamic drag force
F_n	Normal force
F_t	Tangential force
F_u, F_d	Aerodynamic drag force at the upwind and downside side, respectively
$\bar{F}_u$, $\bar{F}_d$	The stream-wise force exerted by the upwind actuator disk and the downwind actuator disk, respectively
K	An empirical constant related to the material
m	Slop parameter
N	The number of cycle to failure
N'	The number of the identical blades
p_d	Pressure in front of the downstream actuator
p_u	Pressure in front of the upstream actuator
p_0	Pressure of the far upstream
R	Radius of the rotor
V_d	Wind speed in the downwind rotor
V_e	Wind speed inside the rotor
V_r	Relative wind speed
V_u	Wind speed in the upwind rotor
V_w	Wind speed in the far wake
V_∞	Inflow wind speed of the far upstream
α	Angle of attack
α_r	Angle of attack for a zero pitch case
β	pitch angle
β'	The amplitude of pitch angle in the sinusoidal pitch control
Δh	Height of the blade
Δp_d	Pressure drop due to the downwind actuator
Δp_u	Pressure drop due to the upwind actuator
$\Delta \theta$	Angular width of the stream-tube
ε_r	Strain range
θ	Azimuth angle
λ	Tip-speed ratio
ρ	Air density
ω	Rotational speed

REFERENCES

Ashwill, T.D. 1992. Measured data for the Sandia 34-meter vertical axis wind turbine. Technical report SAND91-2228, Sandia National Laboratories, Albuquerque, NM.

Bhutta, M.M.A., N. Hayat, A.U. Farooq, et al. 2012. Vertical axis wind turbine: A review of various configurations and design techniques. *Renew. Sustainable Energy Rev.*, 16(4): 1926–39.

Biswas, A., R. Gupta, and K.K. Sharma. 2007. Experimental investigation of overlap and blockage effects on three-bucket Savonius rotors. *Wind Eng.*, 31(5): 363–68.

Chattot, J., and M. Hafez. 2015. Wind turbine and propeller aerodynamics: Analysis and design. In *Theoretical and Applied Aerodynamics*, eds. J. Chattot, and M. Hafez., 327–72. Dordrecht, the Netherlands: Springer Science+Business Media.

Chong, W.T., A. Fazlizan, S.C. Poh, et al. 2013. The design, simulation and testing of an urban vertical axis wind turbine with the omni-direction-guide-vane. *Appl. Energy*, 112: 601–9.

Ciang, C.C., J.R. Lee, and H.J. Bang. 2008. Structural health monitoring for a wind turbine system: A review of damage detection methods. *Meas. Sci. Technol.*, 19(12): 122001.

Dabiri, J.O. 2010. Potential order-of-magnitude enhancement of wind farm power density via counter-rotating vertical-axis wind turbine arrays. *J. Renewable Sustainable Energy*, 3: 043104.

Darrieus, G.J.M. 1931. Turbine having its rotating shaft transverse to the flow of the current. U.S. patent No. 1835018.

Davenport, A.G., J. King, and G.L. Larose. 1992. Taut strip model tests. In *Proceedings of the First International Symposium on Aerodynamics of Large Bridges*, Copenhagen, Denmark. ed. A. Larsen, 113–24. Rotterdam, the Netherlands: Balkema.

Gipe, P. 2009. *Wind Energy Basics: A Guide to Home and Community-Scale Wind Energy Systems*, 2nd Edition. Hartford, CT: Chelsea Green Publishing.

Gol'denblat, I., and V. Kopnov. 1965. Strength of glass-reinforced plastics in the complex stress state. *Polym. Mech.*, 1(2), 54–59.

Hameed, Z., Y.S. Hong, Y.M. Cho, S.H. Ahn, and C.K. Song. 2009. Condition monitoring and fault detection of wind turbines and related algorithms: A review. *Renewable Sustainable Energy Rev.*, 13(1): 1–39.

Hwang, I.S., S.Y. Min, I.O. Jeong, Y.H. Lee, and S.J. Kim. 2006. Efficiency improvement of a new vertical axis wind turbine by individual active control of blade motion. In *Proceedings of SPIE 6173, Smart Structures and Materials 2006: Smart Structures and Integrated Systems*, ed. Y. Matsuzaki, 617311. San Diego, CA.

Islam, M., D.S.K. Ting, and A. Fartaj. 2008. Aerodynamic models for Darrieus-type straight-bladed vertical axis wind turbines. *Renew. Sustainable Energy Rev.*, 12: 1087–109.

Kiwata, T., T. Yamada, T. Kita, et al. 2010. Performance of a vertical axis wind turbine with variable-pitch straight blades utilizing a linkage mechanism. *J. Environ. Eng.*, 5(1): 213–25.

Kong, C., J. Bang, and Y. Sugiyama. 2005. Structural investigation of composite wind turbine blade considering various load cases and fatigue life. *Energy*, 30(11–12): 2101–14.

Larsen, H. 1975. Summary of a vortex theory for the Cyclogiro. In *Proceedings of the Second US National Conference on Wind Engineering Research*. ed. J.E. Cermak, Fluid Mechanics and Wind Engineering Program. Fort Collins, CO: Colorado State University.

Lazauskas, L. 1992. Three pitch control systems for vertical axis wind turbines compared. *Wind Eng.*, 16: 269–82.

Li, C., S.Y. Zhu, Y.L. Xu, and Y.Q. Xiao. 2013. 2.5D large eddy simulation of vertical axis wind turbine in consideration of high angle of attack flow. *Renew. Energy*, 51: 317–30.

Lin, J.H. 2016. Structural analysis of large scale vertical axis wind turbines with monitoring and control. PhD diss., Department of Civil and Environmental Engineering, The Hong Kong Polytechnic University, Hong Kong.

Liu, L., C. Liu, and X. Zheng. 2015a. Modeling, simulation, hardware implementation of a novel variable pitch control for H-type vertical axis wind turbine. *J. Electr. Eng.*, 66(5): 264–69.

Liu, W.Y., B.P. Tang, J.G. Han, et al. 2015b. The structure healthy condition monitoring and fault diagnosis methods in wind turbines: A review. *Renewable Sustainable Energy Rev.*, 44: 466–72.

McLaren, K., S. Tullis, and S. Ziada. 2012. Measurement of high solidity vertical axis wind turbine aerodynamic loads under high vibration response conditions. *J. Fluids Struct.*, 32: 12–26.

Nijssen, R.P.L. 2006. Fatigue life prediction and strength degradation of wind turbine rotor blade composites. Wieringerwerf, the Netherlands: Knowledge Centre Wind Turbine Materials and Constructions. http://windpower.sandia.gov/other/067810P.pdf.

Paraschivoiu, I. 1981. Double-multiple streamtube model for Darrieus in turbines. In *NASA. Lewis Research Center Wind Turbine Dynamics*, 19–25. Document ID: 19820015811. https://ntrs.nasa.gov/archive/nasa/casi.ntrs.nasa.gov/19820015811.pdf.

Paraschivoiu, I., O. Trifu, and F. Saeed. 2009. H-Darrieus wind turbine with blade pitch control. *Int. J. Rotating Mach.*, 2009: 1–7. Article ID 505343.

Peng, Y.X., Y.L. Xu, and S. Zhan. 2016. Experimental and numerical study on aerodynamic forces on straight-bladed vertical axis wind turbine. In *8th International Colloquium on Bluff Body Aerodynamics and Applications*, Northeastern University, Boston, MA.

Penna, P.J., and J.C. Kuzina. 1984. Magdalen island VAWT summary and index of experimental data: 1980–1982. Technical Report: National Research Council of Canada, Ottawa, ON.

Pettersson, L., Andersson, J.O., Orbert, C., and Skagerman S. 2010. RAMS-database for wind turbines – pre-study. *Elforsk Report* 10:67.

Plourde, B., J. Abraham, G. Mowry, and W. Minkowycz. 2011. An experimental investigation of a large, vertical-axis wind turbine: Effects of venting and capping. *Wind Eng.*, 35(2): 213–20.

Richards, B. 1987. Initial operation of project Éole 4 MW vertical axis wind turbine generator. In *Proceedings of Windpower '87 Conference*, 22–27. American Wind Energy Association, Washington, DC.

Saha, U.K., S. Thotla, and D. Maity. 2008. Optimum design configuration of Savonius rotor through wind tunnel experiments. *J. Wind Eng. Ind. Aerodyn.*, 96(8–9): 1359–375.

Savonius, S.J. 1929. Rotor adapted to be driven by wind or flowing water. U.S. patent No. 1697574.

Shokrieh, M.M., and R. Rafiee. 2006. Simulation of fatigue failure in a full composite wind turbine blade. *Compos. Struct.*, 74(3): 332–42.

Strickland, J., B. Webster, and T. Nguyen. 1979. *A Vortex Model of the Darrieus Turbine: An Analytical and Experimental Study*. American Society of Mechanical Engineers, Winter Annual Meeting, New York, NY.

Templin, R. 1974. Aerodynamic performance theory for the NRC vertical-axis wind turbine. NASA STI/Recon Technical Report N, 76, 16618.

Vandenbereghe, D., and E. Dick. 1986. A theoretical and experimental investigation into the straight bladed vertical axis wind turbine with second order harmonic pitch control. *Wind Eng.*, 10(3): 122–38.

Vandenbereghe, D., and E. Dick. 1987. Optimum pitch control for vertical axis wind turbines. *Wind Eng.*, 11: 237–47.

Wilson, R.E. and P.B.S. Lissaman. 1974. Applied aerodynamics of wind power machines. Research report, Oregon State University, Corvallis, OR.

Worstell, M.H. 1981. Aerodynamic performance of the DOE/Sandia 17-m-diameter vertical-axis wind turbine. *J. Energy*, 5(1): 39–42.

Chapter 18

Synthesis of structural self-repairing and health monitoring

18.1 PREVIEW

Civil structures are often exposed to harsh environments and therefore in-service civil structures are under continuous deterioration. Civil structures can also be damaged to varying extents due to man-made and natural hazards such as typhoons, strong earthquakes, floods, fires and collisions. To keep the structures in service, maintenance and repair are inevitable although they are expensive and difficult. A considerable amount of effort has therefore been made to develop innovative smart materials and novel structural systems with a goal that rehabilitation can be effectively accomplished by the structures themselves. Structural self-rehabilitation can be roughly categorised as structural self-centring (SSC), structural self-healing (SSH) and structural self-repairing (SSR). With the onset of structural defects and damages, the SSR system can be activated by the structural health monitoring (SHM) system to prevent further development of structural damage and recover the structural capacity. This chapter first introduces the concept of structural self-rehabilitation and reviews the current research status of structural self-rehabilitation. It then elaborates the SSR systems, in which two studies on self-repairing (SR) concrete (Łukowski and Adamczewski 2013) and SR concrete beams (Kuang and Ou 2008) are concisely introduced. This chapter finally focuses on the synthesis of SSR and SHM systems, in which self-diagnosis and SR steel joints (Kim et al. 2009) and self-diagnosis and SR active tensegrity structures (Adam and Smith 2007) are presented.

18.2 CURRENT RESEARCH STATUS OF STRUCTURAL SELF-REHABILITATION

18.2.1 Concept of structural self-rehabilitation

Structural self-rehabilitation means that after suffering disturbance, the structure can restore itself to a state of health to continue performing its original functions. Such an inspiration comes from biological systems, which have the ability to heal or recover after being wounded or sick. Structural self-rehabilitation can be realised by SSC, SSH and SSR systems. The SSC systems are mainly used to enable the earthquake-excited structure to return to its initial configuration without much residual displacement after the earthquake is over. The SSH systems mainly refer to the SSH materials that have the structurally incorporated ability to repair the damage caused by mechanical usage over time. For a structural material to be strictly defined as autonomously SSH material, it is necessary that the healing process occurs without human intervention. The SSR systems are, however, embedded in a structure to enable the structure to repair damage if necessary. The SSR systems can be integrated with the SHM systems to form self-diagnosis and SR structures.

18.2.2 Structural self-centring systems

Observation of damaged structures in previous earthquakes has shown that large residual deformation may be induced in the structural or non-structural components of the structure although the collapse of the structure is prevented. The accumulation of such lateral deformation eventually causes partial or total loss of structural functionality. Repairing the damages and recovering the residual displacements are, in general, technically challenging and financially expensive. Therefore, these damaged structures are usually demolished, which in turn results in large economic losses. A recent investigation suggested that a residual inter-story drift ratio of 0.5% in Japan makes rebuilding a new structure more favourable than retrofitting or repairing the damaged structure (McCormick et al. 2008). In order to decrease the permanent deformation of structures after earthquakes and thereby minimise the possible economic losses, significant efforts have been made to develop innovative SSC systems.

SSC systems are a special class of structural systems that can undergo large lateral displacements with little or no residual drift (Li 2005). Compared with conventional structural systems, SSC systems are more attractive because the residual deformation is significantly reduced or even eliminated while the strength and stiffness degradation is small. In this section, two types of SSC systems, post-tension (PT)-based systems and shape memory alloy (SMA)-based systems, are introduced.

18.2.2.1 PT-based systems

PT steel tendons are popularly utilised in SSC systems to provide the desired self-centring (SC) forces, whereas energy dissipation devices are often used in parallel to offer damping capacity. Three broad classes of PT-based SC systems, including the SC rocking system, the SC moment-resisting frame (MRF) and the SC-braced frame, are briefly introduced in this section.

SC rocking systems allow a structure to form a rocking mechanism by permitting a gap between the structure and its foundation under earthquake excitation as schematically shown in Figure 18.1. Upon loading, concentrated deformation occurs at the gap location, and PT steel tendons then attempt to close the gap, bringing the structure back to its original

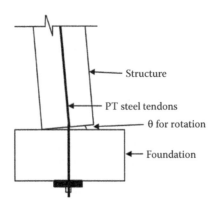

Figure 18.1 Schematic diagram of a simple SC rocking system.

position. The concept of SC rocking systems originated from the features incorporated in the design of the stepping railway bridge over the South Rangitikei River, New Zealand (Cormack 1988), where the SC rocking system is combined with a hysteretic energy dissipation device. Since then, the behaviour of SC rocking systems has been actively investigated analytically and experimentally. A number of analytical studies were carried out to consider potential applications of SC solutions to bridge columns (e.g. Kwan and Billington 2003a,b; Palermo et al. 2005). Significant effort has also been made on the experimental investigation of precast PT-based bridge columns. For example, an experimental investigation was carried out by Marriott et al. (2009) on three 1:3 scale un-bonded and post-tensioned cantilever bridge piers, in comparison with an equivalently reinforced monolithic benchmark. Minimal physical damage was observed and very stable energy dissipation and SC properties were exhibited in the PT rocking systems. Seven quasi-static cyclic tests of half-scale rocking frames were also conducted by Krawinkler et al. (2014). A precast PT composite steel-concrete hollow-core column with supplemental energy dissipaters was proposed by Guerrini et al. (2014) to minimise post-earthquake residual lateral displacements. Large inelastic rotations could be accommodated at the end joints of the columns with minimal structural damage, as gaps were allowed to open at these locations and to close upon load reversal. Besides the applications of SC rocking systems to columns, a similar SC solution was also employed by Stanton and Nakaki (2002) in the development of SC rocking wall systems. Additional experimental work on SC structural walls was also reported by Holden et al. (2003), Restrepo and Rahman (2007) and Toranzo et al. (2009).

SC-MRFs are constructed by post-tensioning beam-to-column connections using high-strength strands, as schematically shown in Figure 18.2. Top and seat angles are added to provide energy dissipation and redundancy under seismic loading. The advantages of this type of connection include that (1) field welding is not required; (2) the connection stiffness is similar to that of a welded connection; (3) the connection is SC; and (4) significant damage to the MRF is confined to the angles of the connection (Ricles et al. 2001). Further studies including cyclic load tests and full-scale application were conducted by Garlock et al. (2003, 2005, 2007). The seismic performance of a post-tensioned energy dissipating (PTED) connection for steel frames was investigated analytically and experimentally by Christopoulos et al. (2002). The results of the tests showed that the PTED test specimen was able to undergo large inelastic deformations without any damage in the beam or column and without residual drift. To avoid yielding in steel beam-to-column connections, energy dissipating elements using friction-damped connections on the outside of beam flanges (Rojas et al. 2005), a beam bottom flange friction device (Wolski et al. 2009) and a friction channel at the beam web (Lin et al. 2013) were also developed.

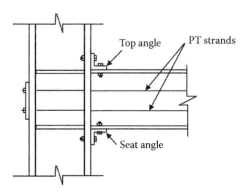

Figure 18.2 Schematic diagram of an SC-MRF system.

The SC braces that combine PT elements and energy dissipating components have an outlook similar to that of a conventional steel brace or a specialised damping device. This bracing member exhibits a repeatable flag-shaped hysteretic response with full SC capabilities, thereby eliminating residual deformations. The mechanics of this new type of brace was first explained by Christopoulos et al. (2008), and the equations governing its design and response were outlined as well. A comparison of the seismic responses of steel frame buildings with two types of bracing members, SC braces and buckling-restrained braces (BRBs), was numerically conducted by Tremblay et al. (2008). It was found that the SC bracing frames generally experienced smaller peak story drifts, less damage concentration over the building height and smaller residual lateral deformations compared with the BRB system. Shaking table tests on a three-story SC-braced frame were then conducted to validate the efficiency of the SC brace (Erochko et al. 2013). Moreover, a new enhanced-elongation telescoping SC brace was recently developed, which allows for an SC response over two times the range achieved with the original SC bracing system (Erochko et al. 2014).

18.2.2.2 SMA-based systems

In PT-based SC systems, PT strands remain elastic over the entire loading and unloading stage, and thus energy dissipation of this type of system is usually low. This can be viewed as a shortcoming of PT-based SC systems. As an alternative solution, SMA with its unique super-elastic ability to undergo large deformations and recover its original shape upon stress removal can be employed in SC systems.

SMAs are a remarkable class of metals that offer high strength, large energy dissipation through hysteretic behaviour, extraordinary strain capacity (up to 8%) with full shape recovery to zero residual strain and a high resistance to corrosion and fatigue. The basic characteristics of SMAs and their applications in civil structures were introduced in Chapter 2. A number of state-of-the-art reviews are also available (e.g. Wilson and Wesolowsky 2005; Song et al. 2006; Menna et al. 2015). It would be impossible to introduce the whole research work conducted for the development of SMA-based SC systems in this chapter, thus only a brief introduction to some SMA-based SC systems developed in recent years for columns, joints, braces and isolators is given in this section.

By placing the SMA rebar at the plastic hinge region, two quarter-scale spiral reinforced concrete (RC) columns representing bridge piers were designed, constructed and tested by Saiidi and Wang (2006). The results showed that the SMA-RC columns were capable of dissipating significant amounts of energy with negligible residual deformation and rotation during earthquakes. Alam et al. (2008) discussed the critical and essential design features of using SMA as reinforcement in concrete structures from an analytical point of view. Nikbakht et al. (2015) numerically investigated the performance of precast segmental bridge columns with SMA bars under lateral static seismic loading.

Feasibility studies on an SC beam-column connection using the super-elastic behaviour of SMAs were numerically investigated by Ma et al. (2007). It was found that the connection deformations were recoverable upon unloading. Two large-scale beam-column joints were designed and tested under reversed cyclic loading with the aim of assessing the seismic behaviour of the joints reinforced with super-elastic SMAs (Youssef et al. 2008). The results showed that the SMA-reinforced beam-column joint was able to recover most of its post-yield deformation. A 3/4 scale concrete beam-column joint reinforced with SMA bars at the plastic hinge region was designed and tested under reversed cyclic loading (Moncef et al. 2011). The results demonstrated that SMA-reinforced joints were able to recover nearly all of their post-yield deformation, requiring a minimum amount of repair.

Zhu and Zhang (2007) proposed a reusable hysteretic damping brace (RHDB) whose core component was made of stranded SMA wires. A comparative study of an RHDB frame and a BRB showed that the RHDB frame can achieve a seismic response level comparable with that of the BRB frame while having significantly reduced residual drifts. A similar study but with an SC friction damping brace was also conducted by Zhu and Zhang (2008). Miller et al. (2012) proposed an SC buckling-restrained brace composed of a typical BRB component, which was used for energy dissipation, and pre-tensioned super-elastic SMA rods, which were able to provide SC and additional energy dissipation ability.

Casciati et al. (2007) proposed a novel base-isolation device composed of two disks, one vertical cylinder with an upper enlargement sustained by three horizontal cantilevers and at least three inclined SMA bars. The role of the SMA bars is to limit the relative motion between the base and the superstructure, to dissipate energy by their super-elastic constitutive law and to guarantee the re-centring of the device. A displacement-based design approach for highway bridges with SMA isolators was proposed by Liu et al. (2011) on the basis of nonlinear dynamic analyses. Numerical simulation results indicated that a properly designed RC highway bridge with SMA isolators may achieve minor damage and minimal residual deformations under frequent and rare earthquakes. Nonlinear static analysis was also carried out to investigate the failure mechanism and the SC ability of the designed highway bridge. A novel SC isolator using super-elastic SMA was presented and installed between the piers and decks of highway bridges (Zhu and Qiu 2014). Based on the incremental dynamic analysis of a prototype highway bridge with SMA isolators, it was found that the SMA isolators can effectively protect the superstructure of the highway bridge and minimise the post-earthquake residual deformation.

18.2.3 Self-healing materials

Almost all civil structures are susceptible to natural or artificial degradation and deterioration with time. Taking concrete structures as an example, cracks can occur at any stage of the service life of the structures due to man-made factors such as overloading and improper maintenance, or volumetric changes caused by high temperatures, creep, plastic settlement or shrinkage. In the early stage, most cracks are in micro scale and are usually not visible. As the cracks expand and form a network, the permeation of aggressive substances becomes easier, which in turn leads to concrete corrosion and deterioration. To avoid the dangerous situations caused by such deterioration, proper inspection and maintenance techniques are required. In some cases, however, it is difficult for engineers to complete their repair work due to inaccessible damage location and/or environmental limitations. An ideal solution is that the structures can cure themselves, just like the natural process of blood clotting or repairing fractured bones. The motivation of conceiving such structural systems has led to increasing interest in the development of self-healing (SH) materials.

SH materials are a class of smart materials that have the ability to heal flaws or damages autogenously and autonomously without any external intervention (Ghosh 2009). Different types of materials such as polymers, coatings, alloys and concrete have their own SH mechanisms. Many excellent reviews on SH materials with different emphases can be found (e.g. Woo 2008; Mihashi and Nishiwaki 2012). This section, however, only gives a brief introduction to the development of SH materials for concrete structures. The possible mechanisms of the SH process in concrete are cited as follows (Ramm and Biscoping 1998): (1) further hydration of the concrete, (2) crystallisation (calcium carbonate), (3) expansion of the concrete in the crack flanks, (4) closing of the cracks by solid matter in the water and (5) closing of the cracks by spalling-off loose concrete particles resulting from cracking. Among these five mechanisms, the processes of hydration and carbonation are commonly accepted as the two main mechanisms, which can progressively fill the crack volume and, under certain specific conditions, can almost completely fill and heal the crack (Neville 2002). In general, the SH process of concrete can be categorised as natural SH, if the SH phenomenon of concrete happens naturally, and engineered SH, if extra engineered factors such as additional engineering materials or techniques are involved during the occurrence of SH.

18.2.3.1 Natural self-healing

It has been observed that some cracks in old concrete structures are lined with white crystalline material, suggesting the ability of concrete to self-seal the cracks with chemical products by itself, perhaps with the aid of rainwater and carbon dioxide in air (Li and Yang 2007). In the study of water flow through cracked concrete under a hydraulic gradient, many investigators also found a gradual reduction of permeability over time, again suggesting the ability of the cracked concrete to self-seal and slow the rate of water flow. Self-sealing is a commonly natural SH phenomenon and is important to prolong the service life of infrastructure, especially for waterproof structures.

Substantial attention has also been paid to investigating the influence of many other factors, such as mixture ratio, crack width, water pressure and ageing time, on the nature of SH action. For example, to investigate the SH and reinforcement corrosion of water-penetrated separation cracks in RC, experimental studies over a period of two years were carried out by Ramm and Biscoping (1998) considering the influence of crack width, structural thickness, water pressure and the degree of water acidity. More recently, SH of cracks in an ultra-high performance concrete was investigated by Granger et al. (2007). Mechanical tests demonstrated a recovery of the global stiffness for specimens initially cracked and then self-healed, and a slow improvement in structural strength. Gagné and Argouges (2012) investigated the natural SH of mortars using airflow measurements through a single crack of controlled geometry. They found that age at time of cracking only plays a minor role in mortar SH kinetics, and the final natural SH level is less than 20% after 5 months for the case of a crack opening greater than 300 μm.

18.2.3.2 Engineered self-healing

Different from natural SH of plain concrete, some engineering techniques are used to influence or stimulate the concrete SH capabilities in engineered SH. Studies on the two aspects of engineered SH, that is, engineered SH with fibre reinforcement or admixtures, are described briefly in this subsection.

Engineered cementitious composite (ECC) is a unique type of high performance fibre-reinforced cementitious composite, featuring high tensile ductility with moderate fibre volume fraction (2% volume or less) (Yang et al. 2011). Of special interest is that the tensile strain capacity of ECC is 2%–5%, several hundred times that of normal concrete, whereas the compressive strength of ECC ranges from 50 to 80 MPa, putting ECC in the class of high strength concrete materials but without the associated brittleness. With such attractive characteristics, ECC material is expected to have good potential to engage SH in a variety of environmental conditions, and much effort has been made in the investigation of SH with ECC material. For example, the performance of SH of ECC under two different cyclic wetting and drying regimes was investigated by Yang et al. (2009). They found that through SH, crack-damaged ECC recovered 76%–100% of its initial resonant frequency value and attained a distinct rebound in stiffness. Even for specimens deliberately pre-damaged with micro-cracks by loading up to 3% tensile strain, the tensile strain capacity after SH recovered close to 100% that of virgin specimens without any preloading. Moreover, Yang et al.

(2011) investigated the healing of early ages (3 days) ECC damaged by tensile preloading after exposure to different conditioning regimes: water/air cycles, water/high temperature air cycles, 90% RH/air cycles and submersion in water. Qian et al. (2010) investigated the SH behaviour of ECC with emphases on the influence of curing condition and pre-cracking time. It was found that for all curing conditions, deflection capacity after SH can recover or even exceed that of virgin samples with almost all pre-cracking ages. Some literature reviews can also be found that introduce the development of SH of ECC (e.g. Li 2003; Wu et al. 2012).

In order to stimulate the chemical reaction for the purpose of generating hydration products to fill cracks in concrete, some admixtures can be used, for example, mineral-producing bacteria. The principle mechanism of bacterial crack healing is that the bacteria themselves act largely as a catalyst, and transform a precursor compound to a suitable filler material (Jonkers 2011). Two basic points are required to take into account when applying bacteria-based SH methods (Jonkers 2007): (1) both bacteria and their produced compounds such as calcium carbonate–based mineral precipitates should effectively cure cracks without causing the loss of other concrete characteristics; and (2) the lifetime of bacteria should be long enough to perform long-term effective crack SH, preferably during the total service life of the concrete. In recent years, increasing attention has been paid to the development of bacteria-based SH. For example, the crack healing potential of bacteria and traditional repair techniques were compared by Tittelboom et al. (2010). Thermogravimetric analysis showed that bacteria were able to precipitate $CaCO_3$ crystals inside the cracks, and SH capabilities could be guaranteed when bacteria were protected in silica gel. The possibility to use silica gel or polyurethane as the carrier for protecting the bacteria was investigated by Wang et al. (2012). Experimental results indicated that polyurethane has more potential to be used as a bacterial carrier for SH of concrete cracks because a higher strength regain and lower water permeability coefficient can be achieved as compared with specimens healed by silica gel immobilised bacteria. More applications of bacteria as an SH agent for the development of sustainable concrete can be found in the literature (e.g. Jonkers 2011; Mihashi and Nishiwaki 2012).

18.2.4 Structural self-repairing materials and systems

The SSR materials or systems are embedded in a structure to enable the structure to repair damage if necessary. The SSR materials and systems can be integrated with the SHM systems to form self-diagnosis and SR structures. The SSR should proceed in the place and time expected by the designer, that is, where and when it is desirable. Smart materials with the following smart functions are usually involved in the SR process (Mihashi and Nishiwaki 2012): (1) sensing function for locating or detecting the presence of targeted changes such as cracks, (2) processing function for judging which action should be taken and/or when it should be taken and (3) actuation function for putting the planned repair operations into action. This is to say, a smart material can treat stimuli from the changing external environment as information to process the condition of the material itself. In this section, two types of SR process, i.e. passive SR and active SR, are introduced.

18.2.4.1 Passive self-repairing

In civil engineering, passive SR techniques are often used for concrete structures. A key feature of passive SR is that no external energy or SHM system is required. A common type of passive SSR system is based on the previously embedded agent, and the basic processes can be described as (1) mix brittle capsules or other types of carriers containing a healing agent or substance in concrete, (2) release the repair agent when the damage-induced triggering mechanism is activated and (3) repair agent penetrates into cracks and produces a chemical product to fill the cracks. According to this description of passive SR, some engineered SH, such as the aforementioned bacteria-based SH using silica gel or polyurethane as the carrier, can also be viewed as passive SR. This classification is also consistent with that defined by the Japan Concrete Institute (JCI) technical committee on autogenous healing in cementitious materials (Igarashi et al. 2009) to some extent.

The investigation of passive SR using a variety of adhesive agents has been actively carried out. Still focusing on the application in concrete, one of the earliest studies was conducted by Dry (1994a,b) using an adhesive agent contained in hollow brittle glass fibres. In this totally passive SR system, the tensile cracking of the matrix in brittle cementitious materials caused by overloading and the resulting breakage of the glass fibres stimulate the actuating mechanisms to release the adhesive agent for repairing the crack. The application of this idea to a one-story rigid portal frame was further considered by Dry (2001). More recently, Hilloulin et al. (2015) focused on the design of polymeric capsules which were able to resist the concrete mixing process and could break when cracks appeared. Łukowski and Adamczewski (2013) conducted experimental studies to evaluate the SR ability of a cement composite modified with epoxy resin without a hardener. More details of this study will be given in Section 18.3 as an example for ease of understanding passive SR.

Besides using an agent embedded in concrete for SR as previously mentioned, a number of researchers were devoted to developing the passive SR with the aid of the super-elastic SMA. For example, by using SMAs as the main reinforcing bars for concrete beams in order for large cracks under loading to be mechanically closed after unloading, a crack-closure system was proposed by Sakai et al. (2003). The comparison between a beam with SMAs and with steel wires indicated that, after maximum deflection, the mortar beam with SMAs could return to about one-tenth deflection compared with the maximum, and the range of deflection of the mortar beam with SMAs is more than seven times that of the beam with steels. Choi et al. (2010) examined the temperature hysteresis as well as the recovery and residual stress of NiTiNb and NiTi SMAs, and explained the possible applications for SR of concrete structures with the shape memory effect. The influence of un-bonded length on the SR capability of an SMA wire concrete beam was investigated theoretically and experimentally by Sun et al. (2013). The results clearly stated that a better crack-repairing situation could be achieved with the longer un-bonded length. Moreover, by taking advantage of the super-elastic effect of SMA and the cohering characteristic of repairing adhesive, a smart concrete beam with SR ability was developed by Kuang and Ou (2008), and the details will be described in Section 18.4.

18.2.4.2 Active self-repairing

A major difference between active SR and passive SR is that external power is required in active SR. Moreover, the SHM system can be incorporated into an active SR system making such SR system more self-controllable. In order to perform effectively, actuation and sensor elements are embedded or surface mounted properly to offer the measurement of physical quantities such as vibration, permeability, current flow, strain, acoustic emission, impedance and pH-changes, which are informative with respect to the state of structural health. An ideal active SR process in a smart system should be able to monitor the structural states online, to efficiently process the data or signals from sensors, to accurately perform the diagnosis and to effectively repair the damages if any. The health monitoring and repair cycle for an active SR system is sketched in Figure 18.3. From these points of view, the active SR could be considered as the highest level of SHM (Peairs et al. 2004; Kim et al. 2009). The

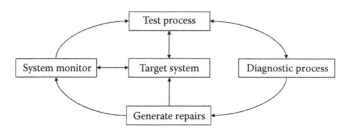

Figure 18.3 Health monitoring and repair cycle for an active self-repairing system. (From Coyle, E.A. et al., *Eng. Appl. Artif. Intell.*, 17(1), 1–9, 2004.)

concept of active SR has been investigated and applied in the areas of aerospace, mechanical engineering, civil engineering and life science. Focusing on the potential applications in civil structures, the majority of researches have been devoted to the development of active SR systems for concrete structures, steel structures and active tensegrity structures.

Concrete structures often suffer from cracking that leads to much earlier deterioration than the designed service life. To prevent such deterioration, the implementation of regular inspection and maintenance for concrete structures is usually required, although it is costly, time-consuming and difficult to execute. The recently emerging smart materials and the practical limitations of inspection and maintenance motivate civil engineers and researchers to investigate and develop active SR materials and systems for recovering the properties of concrete structure with the onset of defects such as cracks. Nishiwaki et al. (2006) developed a new active SR system which can automatically start in response to electrical signals triggered by cracking in concrete. The system was composed of pipes made with heat-plasticity film that contained a low viscosity epoxy resin as a repairing agent and a conductive composite simultaneously serving as a crack monitoring sensor and heating device for a specific location through electrification. When a crack in the concrete was detected by the sensor, the electrical resistance of the sensor was increased because part of the electrical conduction path was cut off around the crack. By means of electrification in this sensor, a partial increase in electrical resistance could provide selective heating around the crack, resulting in the release of the repair agent. An exhaustive experiment was carried out to investigate the quantitative relation between the strain for the diagnosis and the crack width, and to validate the effectiveness of such active SR system (Mihashi et al. 2008; Nishiwaki et al. 2009).

As one of the widely used smart materials, SMA possesses attractive sensing and actuation properties, and thereby shows the potential for its applications in active SR systems. Li et al. (2004) proposed an integrated self-diagnostic and self-repair system embedded with SMA bars. It was verified that embedding SMA bars could effectively decrease the moment of the RC beam. A finite-element analysis further indicated that the SMA bars could modify the tensile stress in the tensile zone or crack area and even close the cracks. A major roadblock for the active SR system using SMAs is that a large amount of energy is required for the recovery of cracks, thereby making such a concept almost impractical in civil engineering. Much attention has thus shifted towards using the shape memory effect of SMAs in active SR joints for steel structures.

The bolted joint is one of the most common mechanical components in steel structures, and these joints are often critical to the function of the structure. Unfortunately, bolted joints are subject to a variety of common modes of failure, such as self-loosening, shaking apart and breaking because of corrosion, stress cracking or fatigue. Self-loosening is the most frequent mode of failure for bolted joints. To reduce this mode of failure, the concept of self-diagnosis and SR bolted joints has been developed (Park and Inman

2001). This concept combines the impedance-based health monitoring technique with actuators which are usually included in the joint as SMA washers to restore tension in a loose bolt. Park et al. (2003) investigated the SR mechanism using SMAs along with a self-sensing mechanism with piezoelectric elements to measure the electrical impedance. Some practical issues of such self-diagnosis and SR bolted joints, especially for SMA actuators, were further investigated by Peairs et al. (2004). Antonios et al. (2006) also carried out an experimental study and numerical analysis of the SR bolted joint, trying to find the influence of initial clamping force, heating and cooling rate, heat loss for different insulation materials and analytical modelling. Faria et al. (2011) investigated the application of SMA washers as an actuator to increase the preload on loosened bolted joints. The application of SMA washers follows an SHM procedure to identify a damage occurrence. For a better understanding of the active SR bolted joints, an integrated digital impedance-based SHM and SSR system developed by Kim et al. (2009) will be described in Section 18.5.

Tensegrity structures are structures composed of tension elements (strings, tendons or cables) surrounding compression elements (bars or struts) in equilibrium (Motro and Raducanu 2003). Since only a small amount of energy is needed to change the shape of these structures, tensegrities are attractive solutions for controllable structures, and increasing effort has been made on the active control of tensegrity structures. The potential applications of tensegrities on footbridges or pedestrian bridges were actively investigated (e.g. Ali et al. 2010; Veuve et al. 2015). Adam and Smith (2007) further described how self-diagnosis, shape control and SR could be integrated into tensegrity structures to copy with unknown events. The identification of either loading or damage location was first involved in self-diagnosis, and then self-diagnosis results were used for control tasks such as shape control and/or SR. Self-repair involved stiffness increases and stress decreases with respect to the damage state. Further investigation of the SR of a damaged tensegrity pedestrian bridge with the aim of meeting safety and serviceability requirements was also conducted (Korkmaz et al. 2011). The introduction of self-diagnosis, SR and self-controlled tensegrity structures will be given in Section 18.6.

18.3 SELF-REPAIRING CONCRETE

This section presents a summary of the research work conducted by Łukowski and Adamczewski (2013) on SR polymer-cement concrete (PCC). The material model of the epoxy-cement composite without hardener is introduced. The material optimisation of the composite towards maximum SR ability is also given.

18.3.1 Epoxy-cement composites without hardener

PCC is a material in which the polymer forms a separate continuous phase as co-binder. The cross-linking of an epoxy resin in the environment of Portland cement paste can proceed, to the extent dependent on the accessibility of calcium hydroxide, without the presence of any hardener, which can make the production of epoxy-cement composites easier. This phenomenon can be used for the implementation of SR ability in concrete. At the polymer content of 20% in the whole binder, the degree of cross-linking of epoxy resin (when used without a hardener) is estimated to be about 50% (Butt et al. 1971). The excess of unhardened resin initially remains in the pores of the hardened cement paste. As the loadings occur, the resin is gradually released and fills the micro-cracks (see Figure 18.4).

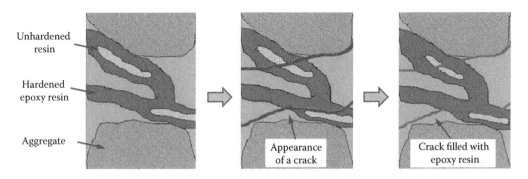

Figure 18.4 Self-repairing by modification of concrete using epoxy resin without hardener. (From Łukowski, P., and G. Adamczewski, *Bull. Pol. Ac.: Tech.*, 61(1), 195–200, 2013.)

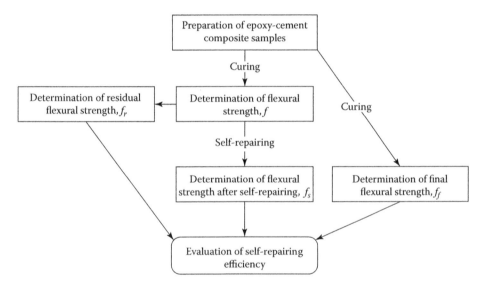

Figure 18.5 Scheme of investigation of self-repairing of epoxy-cement composites. (From Łukowski, P., and G. Adamczewski, *Bull. Pol. Ac.: Tech.*, 61(1), 195–200, 2013.)

18.3.2 Method of investigation

A flowchart of the investigation of the SR capacity of epoxy-cement composites is shown in Figure 18.5. The samples are divided into two groups as shown in Figure 18.5. For the first group, the residual flexural strength is determined immediately after weakening. The second group is left for a certain period of time to allow the SR process to take place and subsequently the flexural strength is determined. At the same time, the flexural strength of the reference samples (not weakened, but instead cured in normal conditions) is determined as well.

18.3.3 Method of evaluation of self-repairing ability of concrete

A self-repair degree (SRD) is employed as a measure of SR efficiency. SRD is defined as the ratio of the increase in strength caused by SR (i.e. excluding the effect of the natural curing of specimens, $C = B - D'$ as shown in Figure 18.6) to the decrease in strength caused by controlled weakening of the material (A):

$$\text{SRD} = \frac{C}{A} = \frac{B-D'}{A} = \frac{B-D}{A} = \frac{(f_s - f_r) - (f_f - f)}{f - f_r} \tag{18.1}$$

in which the definition of f, f_f, f_s and f_r can be found in Figure 18.5; A denotes the strength degradation without SR; B denotes the increase of strength after SR; C is the increase of strength connected directly to SR; and $D = D'$ denotes increase of strength connected to normal curing without SR. The definition of A, B, C, D and D' is clearly shown in Figure 18.6.

18.3.4 Results and discussions

In this study, an emulsion of epoxy resin was selected for SR tests. The introductory tests were first carried out using epoxy-cement mortars with constant binder (Portland cement + epoxy resin without a hardener) to sand ratio, equal to 1:3 by mass. Water to cement ratio was equal to 0.5. The water emulsion of epoxy resin was used, and therefore the water contained in the emulsion was counted as part of the mixing water. CEM I 42.5 R cement and standard sand according to PN-EN 196-1 were also used. The investigation was performed according to the procedure described in Section 18.3.2. The time of curing of the epoxy-cement mortars was 28 days in the conditions given in the European Standards. For SR, the specimens after weakening were left for another 28 days in laboratory conditions. According to Equation 18.1, the values of SRD can be calculated and are shown in Figure 18.7.

The flexural strength of the epoxy-cement composite without a hardener is worse than that of unmodified mortar. The reason is, most likely, the presence of unhardened resin inside the material as well as the presence of emulsifiers in the water emulsion of epoxy resin. The emulsifiers hinder the effective use of water from the emulsion to the cement hydration. They may also disturb the catalytic cross-linking of the resin without a hardener. Nevertheless, the results of the introductory tests made it possible to conclude that there is such content of the epoxy resin which gives the maximum effect of SR. The next step was, therefore, the material optimisation of the epoxy-cement mortar towards the maximum SR degree.

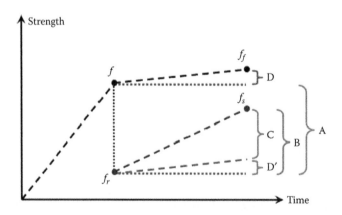

Figure 18.6 Evaluation of SR efficiency of epoxy-cement composite. (From Łukowski, P., and G. Adamczewski, *Bull. Pol. Ac.: Tech.*, 61(1), 195–200, 2013.)

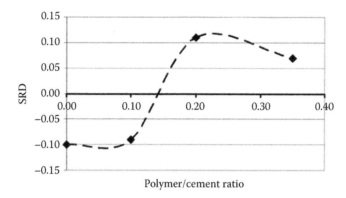

Figure 18.7 SRD of epoxy-cement mortars (without hardener). (From Łukowski, P., and G. Adamczewski, *Bull. Pol. Ac.: Tech.*, 61(1), 195–200, 2013.)

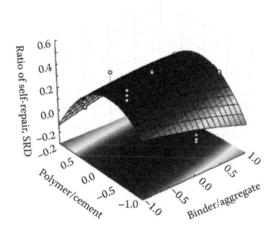

Figure 18.8 SRD as a function of p/c and b/a. (From Łukowski, P., and G. Adamczewski, *Bull. Pol. Ac.: Tech.*, 61(1), 195–200, 2013.)

Epoxy-cement mortars with various content of epoxy modifier (polymer/cement ratio, p/c, ranging from 0.1 to 0.35), applied without a hardener, and various binder/aggregate ratios, b/a (from 0.33 to 0.60), were considered. The results of the testing, performed according to the statistical design of the experiment, are plotted in Figure 18.8. Optimisation procedures using the genetic algorithm and the neural network method were then employed for determining the material composition enabling the maximum effectiveness of SR. The optimum values of the variables were finally determined as $p/c_{opt} = 0.19$ and $b/a_{opt} = 0.34$.

18.4 SELF-REPAIRING CONCRETE BEAMS

In this subsection, experimental studies carried out by Kuang and Ou (2008) on SR concrete beams using super-elastic SMA and fibres containing adhesives are briefly introduced.

18.4.1 SR system using super-elastic SMAs and adhesive-filled brittle fibres

Kuang and Ou (2008) built an SR concrete beam by embedding SMA wires and brittle fibres containing adhesives into concrete during fabrication. A schematic diagram of the SR concrete beam is shown in Figure 18.9. The brittle fibres were connected with a vessel by rubber pipes to ensure enough adhesive was provided for repairing cracks. As damage and cracks occurred, the fibres around the cracked areas ruptured. Once the mobile loads were removed, the deflections and deformations of the structural members were recovered with the aid of the super-elasticity of SMA wires. At the same time, the switch of the repairing vessel containing adhesives was turned on and the repairing adhesives flowed out from the broken-open fibres to fill/repair the cracks.

18.4.2 Experimental design

The specimens used in this experiment were reinforced normal concrete beams with a cross-section of 100×100 mm and a length of 400 mm. The main reinforcements were SMA wires (500 mm long and 2.0 mm in diameter in the tensile area). The diameter of the hoops was 3 mm and their pitch was 50 mm. There was no bond between the SMA wires and the concrete. Steel blocks were attached on both ends of the beam and the SMA wires were fixed in holes of the steel blocks through the frictional forces generated between the screws and the SMA wires. The following five specimens were tested: (L1): five main bars of SMA; (L2): two main bars of 4.0 mm diameter steel wires; (L3): seven main bars of SMA; (L4): five main bars of SMA and four adhesive-filled brittle fibres; (L5): seven main bars of SMA and four adhesive-filled brittle fibres. Four adhesive-filled glass fibres with a diameter of 6.0 mm and a thickness of 0.6 mm were distributed along the longitudinal axis below the neutral axis of the beam. A low viscosity epoxy adhesive served as the sealing/repairing chemical. The adhesive had a tensile strength of 25 MPa and a shear strength of 18 MPa. The concrete cover of the wires was 9 mm.

The specimens were tested at 28 days. A concentrated load was applied to the centre of the beam with a static deformation rate, in the direction orthogonal to the beam axis. The span of the beam was 350 mm by placing the centre of the bottom supports 25 mm away from the edges. The external load and the deflection at the centre of the specimen orthogonal to its axis were recorded. A crack scale was used to measure the crack widths at each peak loading and unloading point under loading cycles.

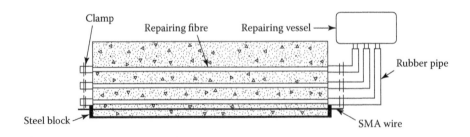

Figure 18.9 SR concrete beam. (From Kuang, Y.C. and J.P. Ou, *Smart Mater. Struct.*, 17(2), 025020, 2008.)

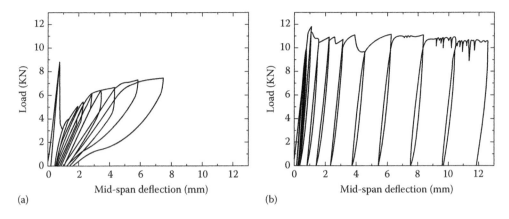

Figure 18.10 Load vs. mid-span deflection during the tests: (a) specimen L1 and (b) specimen L2. (From Kuang, Y.C. and J.P. Ou, *Smart Mater. Struct.*, 17(2), 025020, 2008.)

18.4.3 Experimental results and analysis

18.4.3.1 Performance of SC concrete beams with SMA wires

Figure 18.10a and b shows the relationship between the load and the mid-span deflection of specimen L1 and specimen L2, respectively. It can be seen from Figure 18.10a that on unloading after incurring an extremely large deflection in the beam with SMA, the deflection recovered almost completely through the super-elasticity of the SMA wires. However, for specimen L2 reinforced with steel wires, the deflection barely recovered (see Figure 18.10b), as compared with specimen L1 with SMA wires.

From the tests of all the SMA-reinforced beams, it was observed that the deformation and the width of the crack or gap of the beams increased during loading, but the deflection recovered almost completely and the crack almost closed at unloading. It was also observed that increasing the number or areas of SMA wires could effectively increase the bearing capacity and stiffness. These results clearly indicate that the concrete beams reinforced with SMA as main bars added a self-restoration capacity to concrete beams. Unfortunately, the cracked concrete itself was not repaired in the SMA SR concrete; therefore, the cracks experienced reopening during reloading.

18.4.3.2 Performance of SC concrete beams with SMA wires and adhesive-filled fibres

Specimens L4 and L5 were reinforced with main bars of SMA and adhesive-filled brittle fibres. Two specimens were first loaded to a certain deformation until an obvious crack appeared near the mid-span of the specimens and the repairing fibres in the cracked areas ruptured. Subsequently, the loading was removed and the super-elasticity of the SMA wires recovered the deflections of the specimens. Meanwhile, the clamp was opened up and the adhesive flowed out from the broken-open fibres to fill or repair the cracks. Then, the test was stopped and 10 days were allowed between the first and the second tests. During this period, the adhesive was allowed time to set. After the predetermined time, testing resumed on the specimens, and all information was recorded. The relationship between the load and the mid-span deflection of specimen L4 is shown in Figures 18.11 as an example. It can be seen from Figure 18.11 that the trend of the deflection in the second test is similar to the first test. Moreover, the cracking load in the second test for specimen L4 increases by 28.6%.

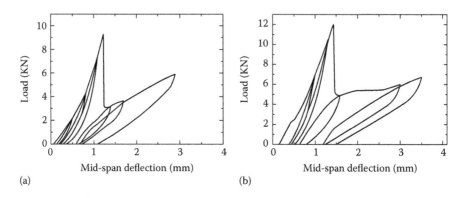

Figure 18.11 Mid-span deflection curve in specimen L4: (a) first testing and (b) second testing. (From Kuang, Y.C. and J.P. Ou, *Smart Mater. Struct.*, 17(2), 025020, 2008.)

Similar phenomenon can be found for specimen L5. These indicate that the strength of the cracked section is repaired and improved by the release of adhesive from the broken-open fibres into the crack.

18.5 SELF-REPAIRING STEEL JOINTS

This section gives a brief introduction of the research conducted by Kim et al. (2009) on SR bolted joints using SMA washers and the impedance-based SHM system.

18.5.1 Impedance-based SHM system

An impedance-based digital SHM system was developed by Kim et al. (2007). It excites a piezoelectric lead–zirconate–titanate (PZT) with a train of rectangular pulses instead of a sinusoidal signal and measures the phase of the response signal instead of the magnitude. The phase difference, specifically the time difference between the voltage and the current exerted on the structure, is measured by the digital SHM system.

The damage metric (DM) in this SHM system is defined as an absolute sum of difference (ASD) between the baseline profile and the phase profile of the structure under test (SUT) and is calculated by

$$DM = \sum_{w=w_1}^{w_2} \left| \varphi_{base}(w) - \varphi_{SUT}(w) \right| \tag{18.2}$$

where:

w	is the excitation frequency sweeping from the target frequency range from w_1 to w_2
$\varphi_{base}(w)$ and $\varphi_{SUT}(w)$	are the phase of baseline profile and the phase profile of the SUT, respectively

The DM is compared against a threshold value, which may be set based on field experience. If the DM is lower than the threshold value, the SUT is considered healthy. Otherwise, it is considered damaged.

Figure 18.12 A smart joint. (From Kim, J.K. et al., A structural health monitoring system for self-repairing. In *Proceedings of SPIE 7295, Health Monitoring of Structural and Biological Systems*, 729512, San Diego, CA, 2009.)

18.5.2 Self-repairing of loose bolted joints

Figure 18.12 shows a smart joint utilising an SMA washer as an actuator which is installed between a bolt and a nut. There are two possible heating methods for an SMA washer: resistive heating or the use of an external heater (Park et al. 2003; Peairs et al. 2004). The resistive heating treats an SMA ring as a solid wire and exploits the internal resistance of the SMA ring for heating. Using an external heater eliminates the need for large wires and unconventional power sources, but introduces new issues such as maintaining contact with a shrinking ring and increasing the possibility of uneven heating. These issues, however, can be addressed with an insulation silicon tape, which helps maintain contact with an SMA, preventing uneven heating. By taking these issues into account, an external heater with relatively small power, 12.5 W, was selected (Kim et al. 2009).

18.5.3 Integrated impedance-based SHM and SR system

Figure 18.13 shows an overall SHM system architecture for the SR of a loose bolted joint. It comprises three functional blocks: an impedance-based SHM system, a heater power controller with a battery and a structure equipped with an SR joint. The impedance-based SHM system is based on the TMS320F2812 Evaluation Module (EVM) from Texas Instruments. The heater power controller controls the delivery of power to the external heater attached to an SMA and consists of a buffer, a relay and a battery. The buffer implemented with two OP-amps turns on or off the relay, and the relay requires a minimum of 9 V and 18 mA to maintain the turn-on state. When the relay is turned on, the Li-ion battery supplies about 12.5 W (or 10 V and 1.25 A) to drive the heater.

18.5.4 Experimental investigation and results

A picture of the setup of the integrated impedance-based SHM and SR system is given in Figure 18.14a. The test specimen was two identical aluminium beams bolted together with a smart joint. Each beam was 298 mm long, 50 mm wide and 3 mm thick. The two beams were connected by a 16.5 mm bolt with an overlap of 194 mm. A PZT sensor with size of 27×22 mm was attached to an aluminium bar and separated by 5 cm from the bolt. The outer layer of the metal bolt was an SMA washer with an inner diameter of 24.4 mm, an

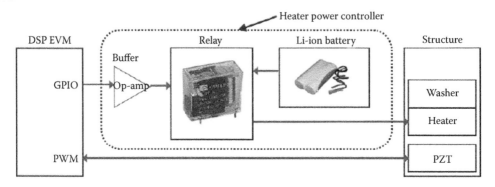

Figure 18.13 Overall system architecture. (From Kim, J.K. et al., A structural health monitoring system for self-repairing. In *Proceedings of SPIE 7295, Health Monitoring of Structural and Biological Systems,* 729512, San Diego, California, USA, 2009. doi: 10.1117/12.816398.)

outer diameter of 26.7 mm and 9.7 mm long. A ceramic washer with a diameter of 34 mm was attached at each end of the SMA washer to reduce the thermal loss, followed by a steel plate washer. Finally, an external heater was wrapped around the SMA washer (see Figure 18.14b). The size of the flexible heater was $1 \times 7.6 \times 79$ mm and was from Minco Products (model HR5208R6.4L12A).

Two different cases, that is, the bolt tightened with 25 N·m of torque and the bolt loosened with 10 N·m of torque, were first considered for the determination of the sweeping frequency range. It was found that the impedance difference between the two cases is most significant in the frequency range from 6 to 10 kHz, which was set as the sweeping frequency range for the subsequent experiments.

A baseline impedance profile in the frequency range of 6–8 kHz was obtained from the structure with the bolt tightened with 25 N·m of torque. Then, the bolt was loosened with 10 N·m of torque. The aforementioned impedance-based SHM system detected the loosened bolt defect, and correspondingly the heater of the smart joint was turned on. The SMA washer was heated to above the critical temperature of 165°C for several minutes, while the system repeatedly performed the SHM operations during the heating up. Upon reaching the DM below the threshold value through gradual tightening of the bolt, the system turned off the heater and the loosened bolt was tightened. The threshold value was set simply to multiple times the variation of the impedance profiles obtained by eight experiments. However, a more sophisticated method based on field testing could find a better threshold value.

Figure 18.15a shows the impedance profiles of the baseline (tightened bolt) and of the loosened bolt. Some peaks of the baseline profile shift to lower frequencies as the bolt is loosened and some peaks disappear. When the loosened bolt is tightened again by heating the SMA washer, the profile of the bolt is almost restored to its baseline profile, as shown in Figure 18.15b. It can be concluded that the SHM system can detect bolted-joint loosening defects, and such defects can be repaired effectively without human intervention.

18.6 SELF-DIAGNOSIS AND SELF-REPAIRING ACTIVE TENSEGRITY STRUCTURES

This section briefly introduces a framework of how to integrate self-diagnosis, shape control and SR of a tensegrity structure for unknown events (Adam and Smith 2007).

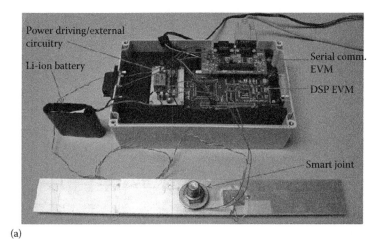

(a)

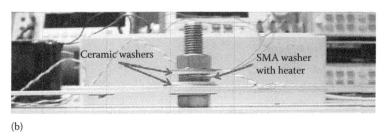

(b)

Figure 18.14 Picture of the experiment setup: (a) picture of configurations and (b) the smart joint with external heater. (From Kim, J.K. et al., A structural health monitoring system for self-repairing. In *Proceedings of SPIE 7295, Health Monitoring of Structural and Biological Systems*, 729512, San Diego, CA, 2009.)

18.6.1 Description of an active tensegrity structure

For the sake of easy understanding, the configuration of an active tensegrity structure is first introduced. The structure is composed of five modules and rests on three supports (see Figure 18.16). It covers a surface area of 15 m², has a static height of 1.20 m and withstands a distributed dead load of 300 N/m². It is composed of 30 struts and 120 tendons. Struts are fibre-reinforced polymer tubes of 60 mm diameter and 703 mm² cross section. Tendons are stainless steel cables of 6 mm in diameter. The structure is equipped with an active control system: 10 active struts allow for length adjustments and 3 displacement sensors measure vertical displacements at 3 nodes of the top surface edge. More detailed description can be found in Fest et al. (2004).

18.6.2 Self-diagnosis of active tensegrity structure

18.6.2.1 Three indicators for response changes

In this study, self-diagnosis involves identifying load positions and magnitudes in situations of partially defined applied loads, and damage location in situations of partially defined damage. Partially defined loading is a known type (e.g. single point load) but unknown magnitude and location. Partially defined damage is a known type (e.g. a broken cable) but unknown location. The response of the structure to a load and damage is measured and compared with the response of the structure to candidate solutions for load and damage.

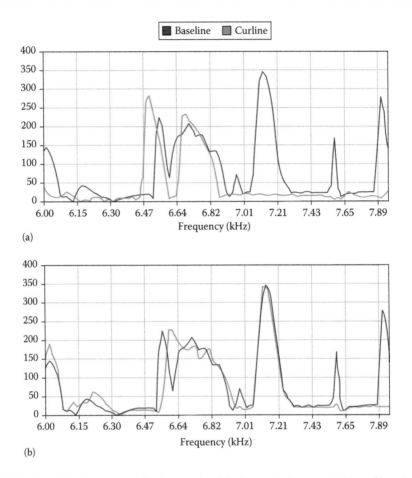

(a)

(b)

Figure 18.15 Profiles of the baseline and the loosened bolt before and after repair: (a) profiles of the baseline and the loosened bolt and (b) profiles after repairing the loosened bolt. (From Kim, J.K. et al., A structural health monitoring system for self-repairing. In *Proceedings of SPIE 7295, Health Monitoring of Structural and Biological Systems*, 729512, San Diego, CA, 2009.)

Figure 18.16 The active tensegrity structure used for tests. (From Adam, B. and I.F.C. Smith, *J. Struct. Eng.*, 133, 1752–61, 2007.)

Three indicators that reflect changes in structure response are used: (1) top surface slope deviation (TSSD), (2) transversal slope deviation (TSD) and (3) influence vector (v) obtained from the top surface slope variations that are induced by each of the 10 active struts. A detailed definition of these three indicators can be found in Adam and Smith (2007).

18.6.2.2 Loading identification

The load identification involves magnitude evaluation and load location by using the afore-mentioned three indicators. Loading is assumed to be single static vertical point loads in this study. They are applied one at a time on 1 of the 15 top surface nodes.

The following steps lead to load identification:

Step 1: TSSD is the first indicator. Load magnitudes are gradually increased until the relation shown in Equation 18.3 becomes true. Loads are incremented in steps of 50 N:

$$\text{TSSD}_c(Q_j) > \text{TSSD}_m \left(Q_j = 50, 100, 150, ...; \ j = 1, 2, ..., 15 \right) \tag{18.3}$$

where subscripts c and m stand for calculated and measured values, respectively.

Step 2: TSD is selected as the second indicator. Candidate solutions that do not satisfy Equation 18.4 are rejected.

$$\frac{\text{TSD}_c}{\text{abs}(\text{TSD}_c)} = \frac{\text{TSD}_m}{\text{abs}(\text{TSD}_m)} \tag{18.4}$$

Step 3: The influence vector (v) is the third indicator. The candidate that satisfies Equation 18.5 is taken to be the reference candidate:

$$\min \left| \mathbf{v}_c - \mathbf{v}_m \right| = \min \left(\sqrt{\sum_{j=1}^{10} \left(\Delta S_c(j) - \Delta S_m(j) \right)^2} \right) \tag{18.5}$$

where:
  j indicates the active strut
  ΔS is the top surface slope variation induced by active control perturbations

Practical applications of system identification include consideration of errors. An upper bound for the error on slope variations for one single active control perturbation has been observed to be $e_p = 0.11$ mm/100 m. Since 10 active control perturbations are applied by the 10 active struts, candidate solutions that satisfy Equation 18.6 are also taken to be load identification solutions.

$$\left| \mathbf{v}_{\text{ref}} - \mathbf{v}_c \right| \leq 10 \cdot e_p \tag{18.6}$$

where subscript 'ref' is introduced to define the reference candidate. This process results in a set of candidate solutions.

Step 4: For each of these solutions, load magnitudes are modified to approach more closely measured TSSD with 10 N increments. Improved candidates create the load identification solution set. In this set, the distance between candidate solution responses and measurements is less than or equal to the maximum error between measurements and numerical simulation. Load identification solutions are used as input to compute a control command for the shape control task.

18.6.2.3 Identification of damage location

In a way similar to the task of load identification, the TSSD and the influence vector are used as indicators for damage location. Damage is simulated by removing one cable from the structure. For damage location, a candidate is defined as the structure with one cable removed. The following steps are carried out:

Step 1: TSSD is the first indicator. Candidate solutions are retained in situations where Equation 18.7 is satisfied.

$$\left| \text{TSSD}_m - \text{TSSD}_c \right| \le e_s \tag{18.7}$$

where the maximal error $e_s = 96$ mm/100 m is related to model inaccuracies.

Step 2: The influence vector ($\mathbf{v}$) is the second indicator. Active control perturbations are applied to the damaged structure. The candidate with the minimum Euclidian distance between its influence vector $\mathbf{v}_c$ and the influence vector of the damaged structure $\mathbf{v}_m$ is taken to be the reference candidate, according to Equation 18.5. Since precision errors are considered, other candidate solutions are taken to be damage location solutions. In situations where Equation 18.6 is satisfied for other candidate solutions, they are also taken to be candidate solutions. These solutions are used as input to compute a control command for SR.

18.6.3 Self-repairing of active tensegrity structure

In the situation of damage, safety becomes more important than serviceability. SR measures have priority. The safety objective involves stiffness increases and stress decreases with respect to the damage state. Since stiffness increase and stress decrease are conflicting objectives, a multi-objective search method is attractive to compute control commands that maximise safety. The multi-objective task in this example is summarised as follows:

$$\text{Minimise} \left\{ F_{\text{stress}}(\mathbf{x}), F_{\text{stiffness}}(\mathbf{x}) \right\} \tag{18.8}$$

$$\text{Subject to } g_{\text{no\_rupture}}(\mathbf{x}, \mathbf{q}), g_{\text{no\_tension}}(\mathbf{x}, \mathbf{q}), g_{x,\text{min}}(\mathbf{x}), g_{x,\text{max}}(\mathbf{x}) \tag{18.9}$$

where:

$F_{\text{stress}}(\mathbf{x})$	is the objective function with respect to the force ratio of the most stressed element
$F_{\text{stiffness}}(\mathbf{x})$	is the objective function defined as the inverse of the global stiffness indicator
$\mathbf{x} = (x_1, x_2, \ldots, x_{10})$	are the position of the 10 active struts
$g_{\text{no\_rupture}}(\mathbf{x}, \mathbf{q})$	is the inequality constraint function for preventing strut buckling and cable rupture
$g_{\text{no\_tension}}(\mathbf{x}, \mathbf{q})$	is the constraint function for avoiding tension in struts
$g_{x,\text{min}}(\mathbf{x})$ and $g_{x,\text{max}}(\mathbf{x})$	are used to bound active strut positions

For a detailed description of the objective functions and constraints in this multi-objective process, refer to Adam and Smith (2007).

The presented methodology for computing SR control commands is based on Pareto filtering in order to avoid the use of weight factors. Sets of Pareto optimal solutions are built according to the stiffness and the stress objectives. Moreover, since multi-objective search supports control command computing, one single solution has to be automatically selected. The slope objective is now of tertiary importance. Among sets of Pareto optimal solutions for stress and stiffness, solutions that exhibit the highest slope compensation are taken to be the solution for the SR control command.

18.6.4 Experimental investigation and results

Adam and Smith (2007) carried out experimental work on the full-scale active tensegrity structure as shown Figure 18.16. The load identification, shape control, damage location identification and SR of the structure are briefly introduced in the following subsections.

18.6.4.1 Load identification and shape control

The load identification methodology is experimentally tested for 11 loading situations. Here, due to the length limitation of this section, only the first loading case is given, that is, load applied at node 26 with the magnitude of −625 N. The results of load identification and shape control in this case are given in Table 18.1. The detailed results of other loading situations can be found in Adam and Smith (2007). For loading identification, it can be found that (1) the exact load magnitude and location may not be identified; (2) the distance between load identification solutions and reality is less than or equal to the error between measurements and numerical simulation, according to the indicators; and (3) the reference candidate is not always the one that is located at the same node as the applied load situation. Load identification solutions are used as input for control command computation to improve the shape control performance. Considering the 11 loadings situations, it is found that the average error of slope compensation is equal to 16.6% without load identification, and it is improved down to 9.4% when using load identification solutions to compute control commands.

18.6.4.2 Identification of damage location

Cable 128 was removed from the structure to simulate a damage situation. As described before, the determination of damage location involves two steps.

Table 18.1 Load identification and shape control results

Load situation		Load identification solution		Shape control	
Loaded node	Load magnitude (N)	Loaded node	Load magnitude (N)	e_{sc} (%)	Sequence length (mm)
26	−625	51	−540	17	17.1
—	—	26	−470	11	18.1
—	—	39	−190	2	18.1
—	—	37	−290	5	20.3
—	—	48	−280	6	13.2

Source: Adam, B. and I.F.C. Smith, J. Struct. Eng., 133, 1752–61, 2007.

Table 18.2 Candidate solutions that induce TSSD$_c$ close to TSSD$_m$

Candidate	Broken cable	TSSD$_c$ (mm/100 m)
1	42	−122
2	121	−102
3	128	−102

Source: Adam, B. and I.F.C. Smith, *J. Struct. Eng.*, 133, 1752–61, 2007.

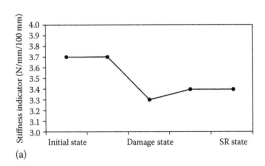
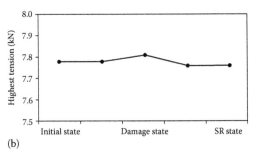

(a)   (b)

Figure 18.17 SR of damaged cable 128: (a) global stiffness indicator and (b) highest tension. (From Adam, B. and I.F.C. Smith, *J. Struct. Eng.*, 133, 1752–61, 2007.)

In Step 1, the top surface slope deviation, TSSD$_m$, is found to be –105 mm/100 m. According to Equation 18.7, three candidate solutions induce a TSSD$_c$ that is close to such value, as shown in Table 18.2.

In Step 2, the influence vector (**v**) serves as the second indicator. Slope variations, ΔS_c, due to active control perturbations are numerically simulated on the three remaining candidate solutions: cables 42, 121 and 128 broken.

By comparing the measured slope variations, ΔS_m, with the calculated ΔS_c obtained from Step 2, it is found that the Euclidian distance is least for Candidate 3, cable 128 broken with a value of 12.6 mm/100 m. It is taken to be the reference candidate. According to Equation 18.6, no other candidate solution is accepted since the Euclidian distance is equal to 20.3 mm/100 m and 14.9 mm/100 m for Candidate 1 (cable 43 broken) and Candidate 2 (cable 121 broken).

18.6.4.3 Self-repairing

As described earlier, SR involves increasing stiffness and decreasing stresses with respect to the damage state. Examine the self-repair process when cable 128 is removed from the structure. The damage location solution is exact in this situation. Figure 18.17 illustrates the global stiffness indicator and the force ratio evolution during damage and SR. The following observations are of interest: (1) the possibility of controlling objectives such as stiffness and stress by modifying the self-stress state of an active tensegrity structure is demonstrated and (2) the topology of the tensegrity structure in this study allows for redundant load-path behaviour for some types of damage.

NOTATION

A	Strength degradation without SR
B	The increase of strength after SR
C	The increase of strength connected directly to SR
D, D′	The increase of strength connected to normal curing without SR
DM	Damage metric
e_p	Upper bound for the error on slope variations for one single active control perturbation
e_s	The maximal error related to model inaccuracies
f_f	Final flexural strength
f_r	Residual flexural strength
f_s	Flexural strength after SR
F	Flexural strength
$F_{\text{stiffness}}(\mathbf{x})$	Objective function defined as the inverse of the global stiffness indicator
$F_{\text{stress}}(\mathbf{x})$	Objective function with respect to the force ratio of the most stressed element
$g_{\text{no\_rupture}}(\mathbf{x}, \mathbf{q})$	Inequality constraint function for preventing strut buckling and cable rupture
$g_{\text{no\_tension}}(\mathbf{x}, \mathbf{q})$	Constraint function for avoiding tension in struts
$g_{x,\min}(\mathbf{x}), g_{x,\max}(\mathbf{x})$	The minimum and maximum constraint functions used to bound active strut positions, respectively
SRD	Self-repair degree
$\text{TSD}_c, \text{TSD}_m$	The calculated and measured transversal slope deviation, respectively
$\text{TSSD}_c, \text{TSSD}_m$	The calculated and measured top surface slope deviation, respectively
$\mathbf{v}_c, \mathbf{v}_m$	The calculated and measured influence vector, respectively
W	Excitation frequency sweeping from the target frequency range from w_1 to w_2
X	The position vector of the 10 active struts
$\Delta S_c, \Delta S_m$	The calculated and measured top surface slope variation, respectively
$\varphi_{\text{base}}(w), \varphi_{\text{SUT}}(w)$	The phase of baseline profile and the phase profile of the structure under test, respectively

REFERENCES

Adam, B. and I.F.C. Smith. 2007. Self-diagnosis and self-repair of an active tensegrity structure. *J. Struct. Eng.*, 133: 1752–61.

Alam, M.S., M.A. Youssef, and M. Nehdi. 2008. Analytical prediction of the seismic behaviour of superelastic shape memory alloy reinforced concrete elements. *Eng. Struct.*, 30(12): 3399–411.

Ali, N.B.H., L. Rhode-Barbarigos, A.A.P. Albi, and I.F.C. Smith. 2010. Design optimization and dynamic analysis of a tensegrity-based footbridge. *Eng. Struct.*, 32(11): 3650–59.

Antonios, C., D.J. Inman, and A. Smaili, 2006. Experimental and theoretical behavior of self-healing bolted joints. *J. Intell. Mater. Syst. Struct.*, 17(6): 499–509.

Butt, J.M., G.V. Topilski, V.G. Mikulski, V.V. Kozłow, and A.K. Gorban. 1971. Investigation of interaction between epoxy resin and Portland cement. *Adv. Constr. Archit.*, 1: 75–80.

Casciati, F., L. Faravelli, and K. Hamdaoui. 2007. Performance of a base isolator with shape memory alloy bars. *Earthq. Eng. Eng. Vibr.*, 6(4): 401–8.

Choi, E., S.C. Cho, J.W. Hu, T. Park, and Y.S. Chung. 2010. Recovery and residual stress of SMA wires and applications for concrete structures. *Smart Mater. Struct.*, 19: 094013.

Christopoulos, C., A. Filiatrault, C. Uang, and B. Folz. 2002. Posttensioned energy dissipating connections for moment-resisting steel frames. *J. Struct. Eng.*, 128(9): 1111–20.

Christopoulos, C., R. Tremblay, H. Kim, and M. Lacerte. 2008. Self-centering energy dissipative bracing system for the seismic resistance of structures: Development and validation. *J. Struct. Eng.*, 134(1): 96–107.

Cormack, L.G. 1988. The design and construction of the major bridges on the Mangaweka rail deviation. *Trans. Inst. Prof. Eng. New Zealand*, 15(1): 6–23.

Coyle, E.A., L.P. Maguire, and T.M. McGinnity. 2004. Self-repair of embedded systems. *Eng. Appl. Artif. Intell.*, 17(1): 1–9.

Dry, C.M. 1994a. Matrix cracking repair and filling using active and passive modes for smart timed release of chemicals from fibers into cement matrices. *Smart Mater. Struct.*, 3: 118–23.

Dry, C.M. 1994b. Smart multiphase composite materials that repair themselves by a release of liquids that become solids. In *Proceedings of SPIE 2189, Smart Structures and Materials 1994: Smart Materials, 62*, ed. V.K. Varadan. Orlando, Florida. doi:10.1117/12.174085.

Dry, C.M. 2001. Design of self-growing, self-sensing, and self-repairing materials for engineering applications. In *Proceedings of SPIE 4234, Smart Materials, 23*, ed. A.R. Wilson, and H. Asanuma. Melbourne, Australia. doi:10.1117/12.424430.

Erochko, J., C. Christopoulos, and R. Tremblay. 2014. Design and testing of an enhanced-elongation telescoping self-centering energy-dissipative brace. *J. Struct. Eng.*, 141(6): 04014163.

Erochko, J., C. Christopoulos, R. Tremblay, and H.J. Kim. 2013. Shake table testing and numerical simulation of a self-centering energy dissipative braced frame. *Earthq. Eng. Struct. Dyn.*, 42(11): 1617–35.

Faria, C.T., V.L. Junior, and D.J. Inman. 2011. Modeling and experimental aspects of self-healing bolted joint through shape memory alloy actuators. *J. Intell. Mater. Syst. Struct.*, 22(14): 1581–94.

Fest, E., K. Shea, and I.F.C. Smith. 2004. Active tensegrity structure. *J. Struct. Eng.*, 130(10): 1454–65.

Gagné, R., and M. Argouges. 2012. A study of the natural self-healing of mortars using air-flow measurements. *Mater. Struct.*, 45(11): 1625–38.

Garlock, M.M., J.M. Ricles, and R. Sause. 2003. Cyclic load tests and analysis of bolted top-and-seat angle connections. *J. Struct. Eng.*, 129(12): 1615–25.

Garlock, M.M., J.M. Ricles, and R. Sause. 2005. Experimental studies of full-scale posttensioned steel connections. *J. Struct. Eng.*, 131(3): 438–48.

Garlock, M.M., R. Sause, and J.M. Ricles. 2007. Behavior and design of posttensioned steel frame systems. *J. Struct. Eng.*, 133(3): 389–99.

Ghosh, S.K. 2009. *Self-Healing Materials: Fundamentals, Design Strategies, and Applications*. Mörlenbach, Germany: Wiley-VCH Verlag GmbH & Co. KGaA.

Granger, S., A. Loukili, G. Pijaudier-Cabot, and G. Chanvillard. 2007. Experimental characterization of the self-healing of cracks in an ultra-high performance cementitious material: Mechanical tests and acoustic emission analysis. *Cem. Concr. Res.*, 37(4): 519–27.

Guerrini, G., J. Restrepo, M. Massari, and A. Vervelidis, 2014. Seismic behavior of posttensioned self-centering precast concrete dual-shell steel columns. *J. Struct. Eng.*, 141(4): 04014115.

Hilloulin, B., K.V. Tittelboom, E. Gruyaert, N.D. Belie, and A. Loukili. 2015. Design of polymeric capsules for self-healing concrete. *Cem. Concr. Compos.*, 55: 298–307.

Holden, T., J.I. Restrepo, and J.B. Mander. 2003. Seismic performance of precast reinforced and prestressed concrete walls. *J. Struct. Eng.*, 129(3): 286–96.

Igarashi, S., M. Kunieda, and T. Nishiwaki. 2009. Technical committee on autogenous healing in cementitious materials. Committee report: JCI-TC075B. http://www.jci-net.or.jp/j/jci/study/tcr/tcr_2009.html.

Jonkers, H.M. 2007. Self-healing concrete: A biological approach. In *Self-Healing Materials: An Alternative Approach to 20 Centuries of Materials Science*, ed. S. Zwaag, 195–204. Dordrecht, the Netherlands: Springer.

Jonkers, H.M. 2011. Bacteria-based self-healing concrete. *HERON*, 56(1/2): 1–12.

Kim, J., B.L. Grisso, D.S. Ha, and D.J. Inman. 2007. A system-on-board approach for impedance-based structural health monitoring, In *Proceedings of SPIE 6529, Sensors and Smart Structures Technologies for Civil, Mechanical, and Aerospace Systems*, 65290O, doi:10.1117/12.715791.

Kim, J.K., D. Zhou, D.S. Ha, and D.J. Inman. 2009. A structural health monitoring system for self-repairing. In *Proceedings of SPIE 7295, Health Monitoring of Structural and Biological Systems*, 729512, San Diego, California, doi:10.1117/12.816398.

Korkmaz, S., N.B.H. Ali, and I.F.C. Smith. 2011. Determining control strategies for damage tolerance of an active tensegrity structure. *Eng. Struct.*, 33(6): 1930–9.

Krawinkler, H., G. Deierlein, and J. Hajjar. 2014. Quasi-static cyclic behavior of controlled rocking steel frames. *J. Struct. Eng.*, 140(11): 04014083.

Kuang, Y.C. and J.P. Ou. 2008. Self-repairing performance of concrete beams strengthened using superelastic SMA wires in combination with adhesives released from hollow fibers. *Smart Mater. Struct.*, 17(2): 025020.

Kwan, W.P., and S. Billington. 2003a. Unbonded posttensioned concrete bridge piers. I: Monotonic and cyclic analyses. *J. Bridge Eng.*, 8(2): 92–101.

Kwan, W.P., and S. Billington. 2003b. Unbonded posttensioned concrete bridge piers. II: Seismic analyses. *J. Bridge Eng.*, 8(2): 102–11.

Li, H., Z.Q. Liu, Z.W. Li, and J.P. Ou. 2004. Study on damage emergency repair performance of a simple beam embedded with shape memory alloys. *Adv. Struct. Eng.*, 7(6): 495–502.

Li, P. 2005. Seismic response evaluation of self-centering structural systems. PhD diss., Department of Civil and Environmental Engineering, Stanford University, USA.

Li, V.C. 2003. On engineered cementitious composites (ECC): A review of the material and its applications. *J. Adv. Concr. Technol.*, 1(3): 215–30.

Li , V.C., and E.H. Yang. 2007. Self-healing in concrete materials. In *Self-Healing Materials: An Alternative Approach to 20 Centuries of Materials science*, ed. S. Zwaag, 169–93. Dordrecht, the Netherlands: Springer.

Lin, Y.C., R. Sause, and J.M. Ricles. 2013. Seismic performance of steel self-centering, moment-resisting frame: Hybrid simulations under design basis earthquake. *J. Struct. Eng.*, 139(11): 1823–32.

Liu, J.L., S.Y. Zhu, Y.L. Xu, and Y.F. Zhang. 2011. Displacement-based design approach for highway bridges with SMA isolators. *Smart Struct. Syst.*, 8(2): 173–90.

Łukowski, P., and G. Adamczewski. 2013. Self-repairing of polymer-cement concrete. *Bull. Pol. Ac.: Tech.*, 61(1): 195–200.

Ma, H.W., T. Wilkinson, and C. Cho. 2007. Feasibility study on a self-centering beam-to-column connection by using the superelastic behavior of SMAs. *Smart Mater. Struct.*, 16: 1555.

Marriott, D., S. Pampanin, and A. Palermo. 2009. Quasi-static and pseudo-dynamic testing of unbonded post-tensioned rocking bridge piers with external replaceable dissipaters. *Earthq. Eng. Struct. Dyn.*, 38(3): 331–54.

McCormick, J., H. Aburano, M. Ienaga, and M. Naashima. 2008. Permissible residual deformation levels for building structures considering both safety and human elements. In *Proceedings of the 14th World Conference on Earthquake Engineering*, Beijing, China. Open access online http://www.iitk.ac.in/nicee/wcee/article/14_05-06-0071.PDF.

Menna, C., F. Auricchio, and D. Asprone. 2015. Applications of shape memory alloys in structural engineering. In *Shape Memory Alloy Engineering: For Aerospace, Structural and Biomedical Applications*, ed. L. Lecce, and A. Concilio, 369–401. Butterworth-Heinemann, Elsevier.

Mihashi, H., and T. Nishiwaki. 2012. Development of engineered self-healing and self-repairing concrete: State-of-the-art report. *J. Adv. Concr. Technol.*, 10(5): 170–84.

Mihashi, H., T. Nishiwaki, K. Miura, and Y. Okuhara. 2008. Advanced monitoring sensor and self-repairing system for cracks in concrete structures. In *Proceedings of SACoMaTis*, Como Lake, Italy, 1: 401–9.

Miller, D.J., L.A. Fahnestock, and M.R. Eatherton. 2012. Development and experimental validation of a nickel-titanium shape memory alloy self-centering buckling-restrained brace. *Eng. Struct.*, 40: 288–98.

Moncef, N., M.S. Alam, and M.A. Youssef. 2011. Seismic behaviour of repaired superelastic shape memory alloy reinforced concrete beam-column joint. *Smart Struct. Syst.*, 7(5): 329–48.

Motro, R., and W. Raducanu. 2003. Tensegrity systems. *Int. J. Space Struct.*, 18(2): 77–84.

Neville, A. 2002. Autogenous healing-a concrete miracle. *Concr. Int.*, 24(11): 76–82.

Nikbakht, E., K. Rashid, F. Hejazi, and S.A. Osman. 2015. Application of shape memory alloy bars in self-centering precast segmental columns as seismic resistance. *Struct. Infrastruct. Eng.*, 11(3): 297–309.

Nishiwaki, T., H. Mihashi, B.K. Jang, and K. Miura. 2006. Development of self-healing system for concrete with selective heating around crack. *J. Adv. Concr. Technol.*, 4(2): 267–75.

Nishiwaki, T., H. Mihashi, and Y. Okuhara. 2009. Experimental study on a recovery assessment of effectiveness of a self-repairing concrete. *Proceedings of Japan Concrete Institute*, 31(1): 2167–72.

Palermo, A., S. Pampanin, and G.M. Calvi. 2005. Concept and development of hybrid solutions for seismic resistant bridge systems. *J. Earthq. Eng.*, 9(6): 899–921.

Park, G., and D.J. Inman. 2001. Smart bolts: An example of self-healing structures. *Smart Mater. Bull.*, 2001(7): 5–8.

Park, G., D.E. Muntges, and D.J. Inman. 2003. Self-repairing joints employing shape-memory alloy actuators. *JOM*, 55(12): 33–37.

Peairs, D.M., G. Park, and D.J. Inman. 2004. Practical issues of activating self-repairing bolted joints. *Smart Mater. Struct.*, 13(6): 1414.

Qian, S.Z., J. Zhou, and E. Schlangen. 2010. Influence of curing condition and precracking time on the self-healing behavior of engineered cementitious composites. *Cem. Concr. Compos.*, 32(9): 686–93.

Ramm, W., and M. Biscoping. 1998. Autogenous healing and reinforcement corrosion of water-penetrated separation cracks in reinforced concrete. *Nucl. Eng. Des.*, 179(2): 191–200.

Restrepo, J.I., and A. Rahman. 2007. Seismic performance of selfcentering structural walls incorporating energy dissipaters. *J. Struct. Eng.*, 133(11): 1560–70.

Ricles, J.M., R. Sause, M.M. Garlock, and C. Zhao. 2001. Post-tensioned seismic resistant connections for steel frames. *J. Struct. Eng.*, 127(2): 113–21.

Rojas, P., J.M. Ricles, and R. Sause. 2005. Seismic performance of post-tensioned steel moment resisting frames with friction devices. *J. Struct. Eng.*, 131(4): 529–40.

Saiidi, M.S., and H.Y. Wang. 2006. Exploratory study of seismic response of concrete columns with shape memory alloys reinforcements. *ACI J.*, 103: 435–42.

Sakai, Y., Y. Kitagawa, T. Fukuta, and M. Iiba. 2003. Experimental study on enhancement of self-restoration of concrete beams using SMA wire. In *Proceedings of SPIE 5057, Smart Structures and Materials 2003: Smart Systems and Nondestructive Evaluation for Civil Infrastructures*, ed. S.C. Liu, 178–86. San Diego, CA.

Song, G.B., N. Ma, and H.N. Li. 2006. Applications of shape memory alloys in civil structures. *Eng. Struct.*, 28: 1266–74.

Stanton, J.F., and S.D. Nakaki. 2002. Design guidelines for precast concrete structural systems. PREcast Seismic Structural Systems (PRESSS) Report No. 01/03-09.

Sun, L., D. Liang, Q. Gao, and J. Zhou. 2013. Analysis on factors affecting the self-repair capability of SMA wire concrete beam. *Math. Probl. Eng.*, 2013: 138162.

Tittelboom, K.V., N.D. Belie, W.D. Muynck, and W. Verstraete. 2010. Use of bacteria to repair cracks in concrete. *Cem. Concr. Res.*, 40(1): 157–66.

Toranzo, L.A., J.I. Restrepo, J.B. Mander, and A.J. Carr. 2009. Shake-table tests of confined-masonry rocking walls with supplementary hysteretic damping. *J. Earthq. Eng.*, 13(6): 882–98.

Tremblay, R., M. Lacerte, and C. Christopoulos. 2008. Seismic response of multistory buildings with self-centering energy dissipative steel braces. *J. Struct. Eng.*, 134(1): 108–20.

Veuve, N., S.D. Safaei, and I.F.C. Smith. 2015. Deployment of a tensegrity footbridge. *J. Struct. Eng.*, 141(11): 04015021.

Wang, J., K.V. Tittelboom, N.D. Belie, and W. Verstraete. 2012. Use of silica gel or polyurethane immobilized bacteria for self-healing concrete. *Constr. Build. Mater.*, 26(1): 532–40.

Wilson, J.C., and M.J. Wesolowsky. 2005. Shape memory alloys for seismic response modification: A state-of-the-art review. *Earthq. Spectra*, 21(2): 569–601.

Wolski, M., J.M. Ricles, and R. Sause. 2009. Experimental study of a self-centering beam-column connection with bottom flange friction device. *J. Struct. Eng.*, 135(5): 479–88.

Woo, P.R. 2008. Self-healing materials: A review. *Soft Matter*, 4: 400–18.

Wu, M., B. Johannesson, and M. Geiker. 2012. A review: Self-healing in cementitious materials and engineered cementitious composite as a self-healing material. *Constr. Build. Mater.*, 28(1): 571–83.

Yang, Y., M.D. Lepech, E.H. Yang, and V.C. Li. 2009. Show more autogenous healing of engineered cementitious composites under wet-dry cycles. *Cem. Concr. Res.*, 39(5): 382–90.

Yang, Y., E.H. Yang, and V.C. Li. 2011. Autogenous healing of engineered cementitious composites at early age. *Cem. Concr. Res.*, 41(2): 176–83.

Youssef, M.A., M.S. Alam, and M. Nehdi. 2008. Experimental investigation on the seismic behaviour of beam-column joints reinforced with super-elastic shape memory alloys. *J. Earthq. Eng.*, 12: 1205–22.

Zhu, S.Y., and C.X. Qiu. 2014. Incremental dynamic analysis of highway bridges with novel shape memory alloy isolators. *Adv. Struct. Eng.*, 17(3): 429–38.

Zhu, S.Y., and Y.F. Zhang. 2007. Seismic behaviour of self-centering braced frame buildings with reusable hysteretic damping brace. *Earthq. Eng. Struct. Dyn.*, 36(10): 1329–46.

Zhu, S.Y., and Y.F. Zhang. 2008. Seismic analysis of concentrically braced frame systems with self-centering friction damping braces. *J. Struct. Eng.*, 134(1): 121–31.

Chapter 19

Synthesis of structural life-cycle management and health monitoring

19.1 PREVIEW

An ideal smart civil structure is expected to possess self-sensing, self-adaptive, self-diagnostic, self-repairing and self-powered functions to confidently preserve the functionality, safety and integrity of the structures in various environmental conditions, as discussed in Chapter 1. Civil structures with one or more of the aforementioned functions have been introduced in Chapters 12–18, but none of them possesses all of these smart functions. The life-cycle management (LCM) of the ideal smart civil structures defined in Chapter 1 has therefore not been explored to date, although the LCM of ordinary civil structures has been actively investigated and implemented. However, the LCM of ordinary civil structures is subject to many uncertainties, and the reliability of existing LCM is difficult to quantify. This chapter therefore explores how to integrate structural health monitoring (SHM) technology with the LCM of civil structures. The concept of SHM-based LCM of civil structures is first introduced. Some preliminary works by the first author and his research team are then presented by taking the Tsing Ma suspension bridge in Hong Kong as an example.

19.2 CONCEPT OF SHM-BASED LIFE-CYCLE MANAGEMENT OF CIVIL STRUCTURES

Owing to the problem of a large stock of ageing infrastructure built 40 or 50 years ago in some countries and the limited availability of funding for its maintenance and repair, the LCM of civil structures has gained increasing attention over the last decade. The LCM of civil structures often requires a balance of both the structural performance and the total cost accrued over its entire life cycle. Although significant progress has been made in this field (e.g. Frangopol et al. 2004; Frangopol and Messervey 2008; Frangopol 2011), the current LCM of civil structures is subject to many uncertainties. These uncertainties can be classified into two broad categories: aleatory uncertainties, which describe the inherent randomness of phenomenon being observed, and epistemic uncertainties, which describe the errors associated with imperfect models of reality due to insufficient and inaccurate knowledge (Ang and Tang 2007). Because of these uncertainties, the reliability of current LCM is difficult to assess. SHM technology is based on a comprehensive sensory system and a sophisticated data processing system implemented with advanced information technology and supported by cultivated computer algorithms. This technology allows actual loading conditions to be monitored, various types of structural responses to be measured, and deterioration and damage to be identified, as demonstrated in previous chapters. SHM technology can thus provide a promising means and solid foundation for tackling the challenging issues in the LCM of civil structures.

The general components of an SHM-based LCM framework and their relationships are depicted in Figure 19.1. The framework involves seven major tasks: (1) to integrate multi-scale finite element (FE) modelling and model updating with stress analysis for predicting both global and local structural responses to external loadings, dynamic characteristics and responses measured by the SHM system; (2) to determine the optimal placement of multi-type sensors for the best global and local response reconstruction of civil structures, with the input of structural responses measured by the SHM system; (3) to assess the current health status of the structures based on previous loading histories and using the SHM-based damage detection method; (4) to perform proper inspection, maintenance and repair work on the basis of the evaluated structural health states; (5) to develop loading models based on incessant field measurement data from the SHM system so that previous loading histories can be analysed and future loadings can be forecast; (6) to conduct damage prognosis and predict the remaining service life of the civil structures; and (7) to develop LCM strategies for making optimal decisions under multiple objectives and various constraints.

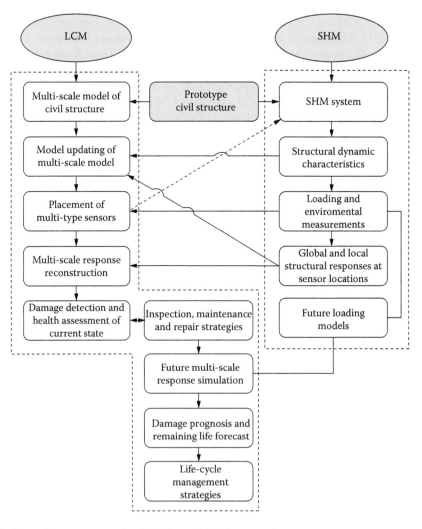

Figure 19.1 General components of an SHM-based LCM framework.

For the purpose of better understanding the proposed SHM-based LCM framework, the aforementioned seven major tasks are further discussed in the following subsections with reference to the Tsing Ma suspension bridge in Hong Kong. Some relevant works by the first author and his research team are then given in Sections 19.3 through 19.8.

19.2.1 Multi-scale modelling and model updating

Multi-scale modelling and model updating techniques have been introduced in Chapters 6 and 7 in detail. Taking a long-span bridge as an example, a finite element model (FEM) of the bridge at stress level is often required to conduct a fatigue damage prognosis (FDP) and evaluate the structural performance at the critical locations. However, modelling all of the structural components and joints of the bridge at the micro-scale level would be impractical due to the huge size and complexity of the structural system. Only critical areas vulnerable to stress concentration, yielding, fracture, fatigue, deterioration and other local damage should be modelled in detail. Therefore, the degree of multi-scale modelling of the bridge should be considered in conjunction with the critical areas identified and the computational demands of the stress analysis method. Moreover, the multi-scale FE model of a bridge should be based not only on modal data but also on stress data to ensure confidence in both global and local simulated responses.

In this regard, the basic steps that must be taken to integrate the multi-scale modelling and model updating with stress analysis of a long-span bridge are as follows: (1) the establishment of a consistent three-dimensional (3D) global FEM for the bridge – a linear elastic model; (2) the determination of critical components and locations where local damage may occur through a dynamic analysis of the global bridge model under combined traffic and wind loadings; (3) the development of micro-scale local models of the critical components, including material and geometric nonlinearities; (4) the incorporation of the local models into the global model using the sub-structuring technique and the interface coupling method with local geometric and material nonlinearities considered; (5) the model updating of the established multi-scale model using a hierarchic optimisation technique based on measured natural frequencies, displacement influence lines, stress influence lines and others; and (6) the performance of an efficient stress analysis based on the multi-scale model of the bridge under combined traffic and wind loadings.

19.2.2 Multi-type sensor placements for response reconstruction

The optimal placement of global sensors (e.g. accelerometers) and local sensors (e.g. strain gauges) for the SHM-based LCM of a long-span bridge is a challenging task. Economic factors that would limit the number of sensors must be considered. The limited number of global and local sensors must be judiciously placed on the global structure and local details of the bridge will provide adequate information for the LCM of the entire bridge. However, considering the bridge's huge size and the limited number of sensors, not all of the critical locations can be directly monitored and the use of measurement data alone may not provide sufficient information for the LCM of the entire bridge. An effective and practical approach should be developed for the optimal placement of multi-type sensors that can integrate global and local responses. In this regard, the first step is to perform an eigenvalue analysis of the multi-scale model of the bridge to determine its 3D global and local dynamic characteristics. The next step is to select the global and local modes that significantly contribute to the relevant global and local responses of the bridge. Finally, the number and location of the global and local sensors should be determined to minimise the estimation errors of both global and local responses at all critical locations. Unlike the conventional placement technique, in which the spatial configurations of the global and local sensors are usually designed in two separate and distinct processes, the approach here shall simultaneously optimise the number and location of both global and local sensors to fuse their limited measurements and give the best estimation of the global and local responses at all of the bridge's critical locations. As a result, the current state of the bridge damage can be estimated using the previously measured structural response time histories and the response reconstruction method. The existing SHM may need to be updated to accommodate the number and locations of multi-type sensors selected from this exercise. A detailed description of multi-type sensor placement and its application to the scaled Tsing Ma Bridge model can be found in Chapter 8.

19.2.3 Structural damage detection and health assessment

In a successful SHM-based LCM system, knowledge of the current status of the bridge structure is required for the managers to make optimal decisions on inspection, maintenance or repair schedules with the aim of ensuring its serviceability and safety. With information from the SHM system and a variety of the structural damage detection methods introduced in Chapter 12, the current health status of the bridge structure can be assessed appropriately.

Let us take fatigue damage as an example. Given the loading and structural response time histories, the current state of the bridge can be estimated based on the established multi-scale model of the bridge and the response reconstruction method, with reference to the initial state of the bridge when it is open to the public. A fatigue analysis of the bridge may be performed at this stage. There are two models which can be used for fatigue analysis: the linear damage accumulation model and the continuum damage mechanics model. The linear damage accumulation model is widely used to analyse the fatigue of steel bridges under traffic loading. Because wind loading on a bridge varies in an essentially random manner, fatigue analysis under wind loading can be performed either by a fully probabilistic approach with a closed-form solution or by a partial probabilistic approach involving deterministic cycle counting algorithms in the time domain.

If any symptom of potential damage is tracked through fatigue analysis, multi-scale damage detection methods incorporating multi-type sensors and the response reconstruction method can be further used to identify the locations and extent of structural damage (see Chapter 12). For example, a radial basis function (RBF) network can be adopted to estimate the strain and displacement mode shapes of the structure at the unmeasured locations based on the structural responses measured by the multi-type sensors. The natural frequencies and mode shapes given by the RBF network can then be used for response reconstruction. The reconstructed multi-type responses are finally fused and utilised in the response sensitivity-based FE model updating to identify the actual damage location and extent. Sparsity regulation, or specifically ℓ_1 norm regularisation, may be used to handle the *ill-posedness* problem in the response sensitivity-based FE model updating.

19.2.4 Inspection, maintenance and repair

To ensure the serviceability and safety of long-span suspension bridges, bridge rating systems are often adopted by bridge management authorities as guidance in determining the time intervals for inspection and the actions to be taken in the event of defects or damage being identified. In terms of finding desirable solutions, regular bridge inspection is one of the most important programs for the maintenance of satisfactory infrastructure performance from a long-term economic point of view. Nevertheless, most of the currently used rating methods for long-span suspension bridges are based on practical experience with some engineering analyses. There is an insufficient link between the bridge rating method and SHM technology to fulfil common goals. It is therefore necessary to develop the SHM-based bridge rating method for the inspection of long-span suspension bridges.

The SHM-based bridge rating method will be introduced in Section 19.8 of this chapter. The fuzzy-based analytic hierarchy approach (F-AHP) is employed, and a hierarchical structure for the synthetic rating of each structural component of a bridge is proposed. Criticality and vulnerability analyses are performed largely based on field measurement data from SHM systems so as to offer a relatively accurate condition evaluation of the bridges and to reduce uncertainties involved in the existing rating method. Procedures for determining relative weights and fuzzy synthetic ratings for both criticality and vulnerability are then suggested. The fuzzy synthetic decisions for inspection are made in consideration of the synthetic ratings of all structural components. Finally, the SHM-based bridge rating method is applied to the Tsing Ma suspension bridge as a case study.

After an inspection following SHM-based rating results, bridge management authorities may take certain maintenance measures, including preventative and corrective procedures (NYDOT 1997). Planned preventative procedures at appropriate regular intervals can significantly reduce the rate of deterioration of critical bridge elements. Cyclical preventative maintenance procedures include cleaning, sealing cracks, sealing the concrete deck and concrete substructures, replacing the asphalt wearing surface, lubricating bearings and painting steel. Corrective procedures are performed to remedy existing problems. These mainly include repairing the asphalt wearing surface, repairing the concrete deck, repairing or replacing joints, repairing or replacing concrete substructures and repairing erosion or scour.

19.2.5 Estimation of future loadings

When the current health status of the bridge is clear after inspection or maintenance works, the future loadings which will act on the bridge should be predicted so that the potential damage accumulation due to future loadings can be predicted using multi-scale model-based stress analysis methods. In principle, future loadings can be forecast based on an analysis of previous loading histories using various data-driven time-series prediction modelling techniques (Box et al. 2008). However, because uncertainties exist in different types of loadings (e.g. railway, highway and wind), probabilistic models shall be established for each individual loading based on data acquired by the SHM system. Then, using probabilistic loading models, the dominant loading parameters can be generated using Monte Carlo simulations (MCS), and stochastic stress responses induced by traffic and wind loadings can be calculated in terms of the multi-scale model and the stress analysis method. For example, the gross train weight (GTW), gross vehicle weight (GVW) and mean wind speed and direction can be selected as the dominant loading parameters. For wind loading, probabilistic models of typhoon and monsoon winds should be established separately. As uncertainties also exist in random combinations of multiple loadings, it is necessary to calculate multiple load-induced stochastic stress responses. Furthermore, in the prediction of future traffic loading, traffic growth plans shall be taken into consideration in future traffic loading simulations.

19.2.6 Damage prognosis and remaining life

Damage prognosis aims to forecast system performance by assessing the current damage state of a system, estimating its future loading environments and predicting through simulation and past experience its remaining useful life (Farrar and Lieven 2007).

Let us take one typical damage prognosis problem, fatigue damage prognosis (FDP) for long-span bridges, as an example. Having estimated the current state of the bridge's fatigue damage under combined traffic and wind loadings, stress characteristics (e.g. the number of cycles and the stress range) can be identified from the stress response time histories obtained by either direct measurements or the response reconstruction method. Hot spots of fatigue damage can then be located based on the stress characteristics identified. From the stress characteristics of the hot spots, fatigue damage accumulation can be evaluated using either the linear damage accumulation model or the continuum damage mechanics (CDM) model. In the CDM-based fatigue damage model, fatigue crack initiation and growth can be estimated in micro-scale, and the fatigue damage rate at each hot spot generated by one block of stress cycles can be naturally extended to estimate fatigue damage accumulation over any period. The following step is to predict the bridge's fatigue damage (remaining life) due to future fatigue loadings. Future loading within a given period can be simulated using future fatigue loading models. The stress time histories at the hot spots due to future fatigue loading can be computed using the multi-scale model and the stress analysis method with the input of the simulated future fatigue loadings. Based on the computed stress time histories, the stress characteristics at the hot spots due to future fatigue loading can be calculated and a fatigue damage prediction can be given in terms of the known current state of fatigue damage and the fatigue damage accumulation model. This evolutionary procedure can be repeated to update the fatigue damage in different periods. Because the loading models can be updated continuously using real-time measurement data, this asymptotic framework makes the fatigue life and reliability analyses increasingly accurate and reliable when additional measurement data from the SHM system are included in the analysis.

19.2.7 Life-cycle management strategies

Life-cycle cost and structural performance are two basic but conflicting aspects required for serious consideration in the LCM of bridges. A life-cycle cost consists of not only the initial design and construction costs, but also those due to operation, inspection, maintenance, repair and damage during a specified lifetime (Frangopol and Messervey 2007). Structural performance can be reflected mainly in terms of functionality, safety, durability and reliability. LCM strategies are urgently needed for structure managers with limited budgets to cost-effectively find desirable solutions to maintaining satisfactory infrastructure performance by optimally balancing long-term structural performance and life-cycle cost. Multi-objective optimisation techniques along with supplementary information, such as the cost of interventions, the status of structures and the effect of maintenance on structural performance, are thus employed to find the optimum inspection/maintenance types and application times.

It should be noted that uncertainties inevitably exist in structural models, response measurements, loading phenomena, deterioration mechanisms and surrounding environments. Probability-based concepts and methods provide a rational and more scientific basis for treating uncertainties and thus are usually employed in LCM analysis. A number of probabilistic models for maintaining and optimising the life-cycle performance of deteriorating structures have been developed, with each model being roughly divided into two parts: a deterioration model and a decision model (Frangopol et al. 2004). The deterioration model is used to approximate and predict the actual process of ageing in condition or in reliability. The decision model uses the deterioration model to determine the optimal times of inspection and maintenance with the ultimate goal of structural performance maximisation and life-cycle cost minimisation. Recently, Boller et al. (2015) introduced a damage tolerance approach used in aviation and attempted to explore the possibility of applying this concept to ageing civil infrastructures. Examples of the LCM of aged steel infrastructures were given in their studies, but significant challenges were encountered in their application to non-metallic structures; for example, neither a quantifiable degree of damage nor a description of damage accumulation so far exists for concrete structures. Further effort is required for the development of this concept in civil engineering together with a deep investigation of damage accumulation mechanisms in concrete structures.

It can be seen from the preceding discussion that the successful implementation of an entire SHM-based LCM framework is quite complicated and challenging, and many techniques and disciplines are involved in this framework. Although not all the aforementioned seven tasks of the framework have been completed, in a systematic way, some relevant works by the first author and his research team are presented in the following sections with reference to the Tsing Ma suspension bridge in Hong Kong.

19.3 FINITE ELEMENT MODEL AND MODEL UPDATING OF TSING MA BRIDGE

19.3.1 Tsing Ma Bridge

The Tsing Ma Bridge in Hong Kong is the longest suspension bridge in the world carrying both highway and railway. The Tsing Ma Bridge is also located in one of the most active typhoon-prone regions in the world. It has an overall length of 2132 m and a main span of 1377 m between the Tsing Yi tower in the east and the Ma Wan tower in the west (see Figure 19.2). The height of the two reinforced concrete towers is 203 m. The two main cables are 1.1 m in diameter and 36 m apart in the north and south and are accommodated by four saddles located at the top of the tower legs. The bridge deck is a hybrid steel structure consisting of Vierendeel cross frames supported on two longitudinal trusses acting compositely with stiffened steel plates. The bridge deck carries a dual three-lane highway on the upper level of the deck and two railway tracks and two carriageways on the lower level within the bridge deck. Further information on the bridge can be found in Xu et al. (1997).

19.3.2 Finite element model

Modelling work is executed using the commercial software packages MSC/PATRAN as a model builder and MSC/NASTRAN as an FE solver. The work is based on the previous model developed by Wong (2002), with the following principles: (1) model geometry should accurately represent actual geometry; (2) one analytical member should represent one real member; (3) stiffness and mass should be simulated and quantified properly; (4) boundary and continuity conditions should accurately represent reality; and (5) the model should be detailed enough at both the global and local levels to facilitate subsequent model updating methods. It may be worthwhile to mention that the FEM presented here was built some time ago. It is not a truly multi-scale model, as discussed in Chapter 6, but it can serve a purpose, except for the joints of structural components.

The deck is a hybrid steel structure consisting of Vierendeel cross frames supported on two longitudinal trusses acting compositely with stiffened steel plates that carry the upper and lower highways. The bridge deck at the main span is a suspended deck and the structural configuration is typical for every 18 m segment. Figure 19.3a illustrates a typical 18 m suspended deck module consisting of mainly longitudinal trusses, cross frames, highway decks, railway tracks and bracings. The upper and lower chords of the longitudinal trusses

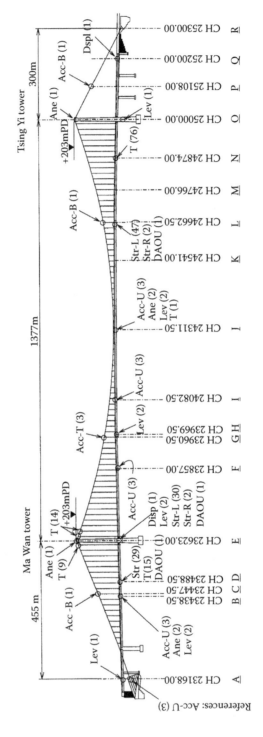

Figure 19.2 Tsing Ma Bridge and the layout of the sensory system: (a) an isometric view of a typical deck section, (b) FEM of an 18 m deck section.

are of box section, while the vertical and diagonal members of the longitudinal trusses are of I section. They are all modelled by 12 degrees of freedom (DOFs) beam elements (named CBAR in the software) based on the principle of one element for one member. The upper and lower chords of the cross frames are predominantly of T section, except for some segments with I section for the cross-bracing systems. The inner struts, outer struts and upper and lower inclined edge members of the cross frames all are of I section. All members in the cross frames are modelled as CBAR elements with actual section properties, except for the edge members, which are assigned a large elastic modulus and significantly small density to reflect the real situation, where the joint is heavily stiffened for the connection with the suspender. All the members in the cross bracings are of box section while all the members in the sway bracings are of circular hollow section. These members all are modelled as CBAR elements with actual section properties. Each railway track is modelled as an equivalent beam modelled by special 14-DOF beam elements (named CBEAM in the software), which are similar to the CBAR elements but with additional properties such as a variable cross section, shear centre offset from the neutral axis, wrap coefficient and others. The railway tracks are meshed every 4.5 m according to the interval of the adjacent cross frames. The modulus of elasticity, the density and Poisson's ratio for all members, except for the edge members, are taken as 2.05×10^{11} N/m^2, 8500 kg/m^3 and 0.3, respectively. Deck plates and deck troughs comprise orthotropic decks, and the accurate modelling of stiffened deck plates is complicated. To keep the problem manageable, two-dimensional anisotropic quadrilateral plate-bending elements (named CQUAD4 in the software) are employed to model the stiffened deck plates. The equivalent section properties of the elements are estimated roughly by a static analysis and the material properties of steel are used first but updated subsequently. The connections between the deck plates and the chords of the cross frames and the longitudinal trusses involve the use of a multi-point connection (MPC). Neutral axes for the connections between the components are properly offset to maintain the original configuration. In the modelling of the typical 18 m deck module in question, a total of 130 nodes with 172 CBAR elements, 16 CBEAM elements, 24 CQUAD4 elements and 50 MPCs are used. A skeleton view of the 3D FEM of the 18 m deck module is shown in Figure 19.3b. The deck modules at the Ma Wan tower, at the Ma Wan approach span, at the Tsing Yi tower and at the Tsing Yi approach span are constructed using the same principle as the deck module at the main span while taking into consideration the differences in the shape and size of cross frames, longitudinal trusses and other members.

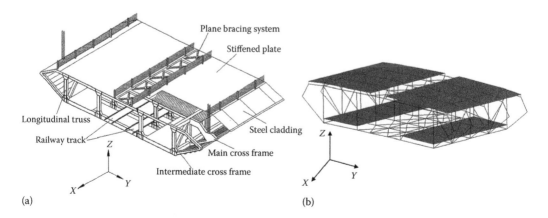

Figure 19.3 A typical 18 m deck section at the main span: (a) an isometric view, (b) FEM.

The Ma Wan tower and the Tsing Yi tower are reinforced concrete structures, and each tower consists of two reinforced concrete legs which are linked by four reinforced-concrete portal beams. The bridge towers are represented by multi-level portal frames in this study. The tower legs are modelled using CBAR elements. The tower leg from its foundation to the deck level is meshed with the element for a length of 5 m. At the deck level, the tower leg is meshed according to the positions of the lateral bearings. Though the dimension of the cross section of the tower leg varies from its bottom to its top, the geometric properties of the beam element are assumed to be constant along its axis, with an average value based on the design drawings. The four portal beams of either tower are also modelled using CBAR elements but with different geometric properties for each section. The deck-level portal beam of each tower is divided at the four particular positions, which correspond to the four vertical bearings between the bridge deck and the tower. The mass density, the Poisson's ratio and the modulus of elasticity of reinforced concrete for the towers are estimated to be 2500 kg/m^3, 0.2 and 3.4×10^{10} N/m^2, respectively.

The two side spans on the Ma Wan side and Tsing Yi side are supported by two and three piers, respectively. All supporting piers in the side spans are reinforced concrete structures. Piers M1, T2 and T3 are similarly modelled as a portal frame using CBAR elements. Pier M2 is also modelled as a portal frame using 12 CBAR elements, in which the upper portal beam is meshed according to the four vertical bearing positions. The wall panel of pier T1 is represented by an equivalent portal frame with 25 CBAR elements. The mass density, the Poisson's ratio and the modulus of elasticity of reinforced concrete for the piers are taken as 2500 kg/m^3, 0.2 and 3.4×10^{10} N/m^2, respectively.

The cable system is the major system supporting the bridge deck. The cable system consists of two main cables, 95 pairs of suspender units and 95 pairs of cable bands. CBEAM elements are used to model the main cables. The cable between the adjacent suspender units is modelled by one beam element of a circular cross section. The DOFs for the rotational displacements of each beam element are released at both ends because the cable is considered to be capable of resisting tensile force only. The cables in the main span are modelled by 77 beam elements, while 26 and 8 elements are used to model one cable on the Ma Wan side span and on the Tsing Yi side span, respectively. Each suspender unit is modelled by one CBEAM element to represent the four strands. A total of 190 elements are used to model all the suspender units. Since the stress distribution around the connection between the main cable and suspenders is not concerned in this chapter, the modelling of cable bands is ignored in the global bridge model. The connections between the main cables and suspenders are achieved by simply sharing their common nodes. To model the cable system, the geometry of the cable profile should be determined. The geometric modelling of the two parallel main cables follows the profiles of the cables under the dead load at a temperature of 23°C based on information from the design drawings. The horizontal tension in the main cable from pier M2 to the Ma Wan anchorage is 400,013 kN, but it is 405,838 kN in the other parts of the main cable. The tension forces in the suspenders on the Ma Wan side span are taken as 2610 kN, but they are 4060 kN in the other suspenders. The mass densities for both cables and suspenders are taken as 8200 kg/m^3. The area of cross section is 0.759 m^2 for the main cables and 0.018 m^2 for the suspenders. The modulus of elasticity is greatly influenced by the tension in the main cables and suspenders, which is estimated as 1.95×10^{11} N/m^2 and 1.34×10^{11} N/m^2, respectively, at a design temperature of 23°C, and will be updated subsequently.

By integrating the bridge components with the proper modelling of the connections and boundary conditions, the entire global bridge model is established as shown in Figure 19.4. The establishment of this global bridge model involves 12,898 nodes, 21,946 elements (2,906 plate elements and 19,040 beam elements) and 4,788 MPCs.

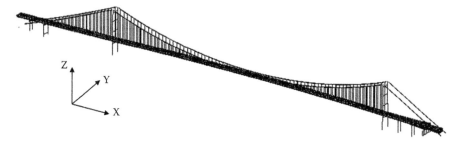

Figure 19.4 Three-dimensional FEM of the Tsing Ma Bridge.

19.3.3 Model updating

Although the geometric features and supports of the bridge deck are modelled in great detail in the established 3D FEM of the Tsing Ma Bridge as mentioned before, modelling discrepancies in the as-built bridge still exist. The discrepancies mainly come from four sources: (1) the simplified modelling of stiffened plates, (2) uncertainties in pavement mass and others, (3) uncertainties in the stiffness of bearings and (4) the assumption of rigid connections. Model updating is therefore necessary. From a theoretical viewpoint, both natural frequencies and mode shapes can be used for their objective functions in model updating. However, including mode shapes in model updating requires a lot of computational resources, because calculating the sensitivity of mode shapes with respect to structural parameters is very time-consuming. Moreover, the measured mode shapes are often less accurate than the measured natural frequencies in practice. This is particularly true for the Tsing Ma Bridge because there are only are 25 measurement points in the mode shape. Therefore, only the first 18 natural frequencies are used here, and the mode shapes are used for matching modes only via the modal assurance criterion (MAC).

It is assumed that the discrepancies due to the preceding sources can be minimised by updating the material properties of the relevant components of the bridge model. Table 19.1 lists the material properties selected for updating. They include mass densities M_{p1} and M_{p2}, elastic moduli E_{x1}, E_{y1}, E_{x2} and E_{y2}, and shear moduli G_{xy1} and G_{xy2} for the upper and bottom stiffened plates, respectively. By updating mass density M_d for all the beam elements of the bridge deck and mass density M_c for all the cable elements, the masses of pavement and other accessories could be included in the bridge model. Finally, elastic moduli E_h for horizontal bearings and E_v for vertical bearings are updated to overcome uncertainties in the stiffness of bearings. Furthermore, the lower and upper bounds for the mass densities are set to avoid physically meaningless updated results and impossible updated parameter values. The variations of mass densities are set as 20% of the initial values.

To reflect the relative importance of the measured data, the weight coefficients are set as 10 for the first two lateral modes (nos. 1 and 4), the first two vertical modes (nos. 2 and 3), the first two torsional modes (nos. 11 and 14) and as 1 for all other modes.

The initial and updated values of the selected parameters are listed in Table 19.1. It can be seen that the elastic moduli of plate-bending elements in the y direction and their shear moduli have significant changes. The major reason for the change is that the initial values of the elastic moduli of the plate elements are selected based on an isometric assumption, but they are not in the actual stiffened plates. It can also be seen that given the pavement mass and others, the mass densities of both cables and deck elements increase. Furthermore, the initial values for the stiffness of bearings are too small and the updated values increase significantly.

Table 19.1 Selected parameters for model updating

Component	Description	Parameter	Initial value	Updated value
Anisometric upper stiffened plates	Mass density (kg/m³)	M_{p1}	8500	9304
	Elastic modulus in x (Pa)	E_{x1}	2.05×10^{11}	1.85×10^{11}
	Elastic modulus in y (Pa)	E_{y1}	2.05×10^{11}	1.81×10^{12}
	Shear modulus (Pa)	G_{xy1}	7.885×10^{10}	1.2×10^{11}
Anisometric bottom stiffened plates	Mass density (kg/m³)	M_{p2}	8500	9107
	Elastic modulus in x (Pa)	E_{x2}	2.05×10^{11}	1.98×10^{11}
	Elastic modulus in y (Pa)	E_{y2}	2.05×10^{11}	1.807×10^{12}
	Shear modulus (Pa)	G_{xy2}	7.885×10^{10}	9.017×10^{11}
Beam elements in deck	Mass density (kg/m³)	M_d	8500	9817
Cable elements	Mass density (kg/m³)	M_c	8200	9173
Horizontal bearings	Elastic modulus (Pa)	E_h	3.4×10^{10}	3.078×10^{11}
Vertical bearings	Elastic modulus (Pa)	E_v	3.4×10^{10}	1.879×10^{11}

Table 19.2 Comparison of the measured and calculated modal parameters after updating

Mode no.	Measurement (Hz)	Calculation (Hz)	Difference (%)	MAC
1 (L1)	0.069	0.069	0.0	0.84
2 (V1)	0.113	0.122	8.0	0.82
3 (V2)	0.139	0.147	5.8	0.90
4 (L2)	0.164	0.160	−2.4	0.89
5 (V3)	0.184	0.198	7.6	0.87
6 (L3)	0.214	0.222	3.7	0.87
7 (L4)	0.226	0.231	2.2	0.87
8 (L5)	0.236	0.233	−1.3	0.56
9 (L6)	0.240	0.243	1.3	0.57
10 (V4)	0.241	0.250	3.7	0.87
11 (T1)	0.267	0.258	−3.4	0.93
12 (V5)	0.284	0.283	−0.4	0.84
13 (L7)	0.297	0.300	1.0	0.77
14 (T2)	0.320	0.282	−11.9	0.78
15 (V6)	0.327	0.340	4.0	0.78
16 (L8)	0.336	0.336	0.0	0.85
17 (L9)	0.352	0.327	−7.1	0.98
18 (L10)	0.381	0.350	−8.1	0.90

The natural frequencies and MAC values computed from the updated FEM are listed in Table 19.2. The discrepancies between the computed natural frequencies and the measured ones after model updating are reduced significantly. The largest difference in the measured and calculated natural frequency is 11.9% compared with 23.8% in the initial FEM. The average difference in the natural frequencies is 4% for the updated model compared with 9% in the initial model. However, the MAC values have not been improved because the mode shapes are not included in the objective function. More details of health monitoring–oriented FEM and model updating can be found in Zhang et al. (2007) and Liu et al. (2009).

19.4 SHM SYSTEMS OF TSING MA BRIDGE

19.4.1 WASHMS in Tsing Ma Bridge

The Hong Kong Highways Department installed a comprehensive wind and structural health monitoring system (WASHMS) and a global positioning system-on-structure instrumentation system (GPS-OSIS) in the Tsing Ma Bridge in 1997 and 2000, respectively. The WASHMS in the Tsing Ma Bridge is composed of five subsystems – namely, sensory system, data acquisition system, data processing and analysis system, computers for system operation and control, and fibre-optic cabling network system. The sensory system consists of about 300 sensors and associated interfacing units installed at different locations of the bridge (see Figure 19.2). They include anemometers, temperature sensors, accelerometers, strain gauges, level sensing stations, displacement transducers, weigh-in-motion (WIM) sensors, signal amplifiers and interfacing equipment. The data acquisition system refers to the computer-controlled data acquisition outstation units with appropriate data acquisition interfaces and the software for parameter configuration. The data acquisition outstation units are installed on/inside the bridge, and their major functions are to collect and digitise signals received from the sensory system and to deliver them to the data processing and analysis system through the fibre-optic cabling network system installed along the bridge alignment. The data processing and analysis system is located at the Tsing Yi administrative building; it is a workstation for overall data collection, transmission, storage, control and post-processing. The computer workstation for system operation and control is also located at the Tsing Yi administrative building and is equipped with appropriate software for the graphical inputs and outputs of structural modelling and analysis.

19.4.2 GPS-OSIS in Tsing Ma Bridge

The WASHMS in the Tsing Ma Bridge, however, has some weak points in bridge displacement response monitoring. The level-sensing stations provide real-time monitoring of displacements at typical stiffening deck sections but in a vertical direction only. To improve the efficiency and accuracy of the WASHMS in displacement monitoring, the Hong Kong Highways Department further installed a GPS-OSIS in December 2000 with ongoing updates until September 2002 to monitor the absolute displacements of the cables, the stiffening deck and the bridge towers. The WASHMS and GPS-OSIS are simply called the *SHM system for the Tsing Ma Bridge*. More details can be found in Xu and Xia (2012).

19.5 SHM-BASED LOADING ASSESSMENT AND MODELS

The Tsing Ma Bridge is subjected to four major types of loads: highway, railway, wind and temperature loads. The SHM system installed in the bridge makes it possible to assess the loading conditions and loading effects on the bridge. The four major types of loads applied to the Tsing Ma Bridge are further introduced in this section.

19.5.1 Highway Loading

The road traffic conditions and highway loading on the Tsing Ma Bridge are monitored by dynamic WIM stations for two carriageways (the airport-bound way and the Kowloon-bound way) at the approach to the Lantau Toll Plaza near the Lantau administration building. The WIM stations can measure axle numbers, axle spacing, vehicle speed and other measurements of road vehicles. The 8-class vehicle classification system is used to classify types of vehicles using the bridge according to their main features.

The WIM stations have recorded traffic information since August 1998. Owing to the need to conduct major updates for the entire system, both the airport- and Kowloon-bound stations were in suspension from July 2004 to March 2005. Furthermore, traffic information detailed to each lane was not available until April 2005. Therefore, for a detailed analysis of traffic conditions for each lane, the WIM data for the whole year 2006 are analysed. Vehicle amounts and the composition of different vehicle types are two important indices to represent vehicle traffic conditions. The traffic volume and traffic composition in each lane of the Tsing Ma Bridge in 2006 are shown in Figure 19.5 for the 8-class vehicle classification. It is observed that in 2006, a total of 8.5 million vehicles ran through the Tsing Ma Bridge in the airport-bound direction and 8.8 million vehicles in the Kowloon-bound direction. Cars, vans and taxis took up the biggest percentage among all types of vehicles: about 69.8% airport bound and 69.0% Kowloon bound. The percentage of heavy goods vehicles, including rigid and articulated heavy goods vehicles, is about 5%–6% of the total vehicles. More vehicles run on the middle lane than other lanes both ways and account for 49.1% and 55.5% of the total vehicles, respectively. Most heavy goods vehicles use the slow lane both ways, accounting for 95.1% and 89.1% of the total heavy vehicles, respectively. Most medium goods vehicles, including buses, use the slow lane both ways, accounting for 85.7% and 64.4% of the total medium vehicles, respectively.

Axle load is an important loading parameter for the bridge pavement and bridge structure. The percentage of axle load in different loading ranges with 1 ton intervals is determined according to the aforementioned vehicle classification system. Based on the statistical WIM data for the year 2006, the distributions of axle load are displayed in terms of vehicle category in Figure 19.6 and the bridge lane in Figure 19.7. It is observed that the axle number decreases with increasing axle load: only 12.5% of the total vehicle axles have an axle load more than 5 tons. Most vehicle axles have an axle load less than 1 ton, which accounts

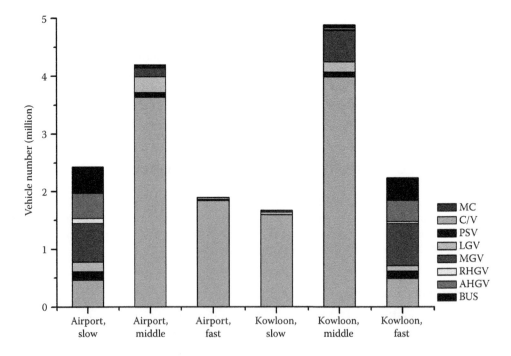

Figure 19.5 Vehicle count by 8-class vehicle classification.

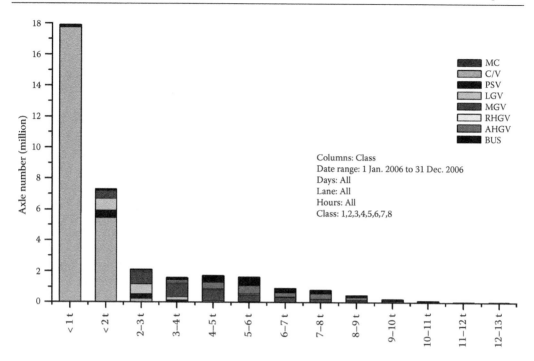

Figure 19.6 Axle loading distribution against axle number and vehicle category.

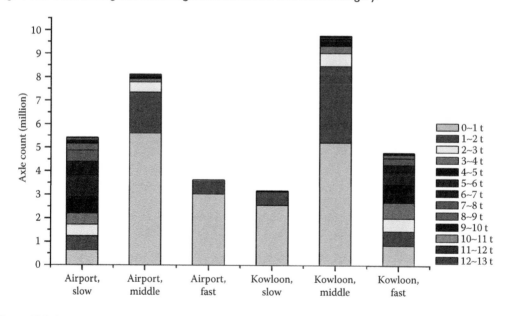

Figure 19.7 Axe loading distribution against axle number and bridge lane.

for 54.1% and 48.6% of the total axles, respectively, for both the airport- and Kowloon-bound ways. The second highest axle number corresponds to the axle load between 1 and 2 tons. More vehicle axles run in the middle lane than in the slow lane or fast lane, which accounts for about 47.2% and 55.1% of the total axles, respectively, both ways. Most of the axles with loads more than 5 tones run in the slow lane, accounting for about 95.5% for the airport-bound way and 84.2% for the Kowloon-bound way.

Focusing on the SHM-based FDP of the Tsing Ma Bridge, only road vehicles with a GVW of more than 4 tons are considered to influence fatigue damage. If a road vehicle is regarded as a concentrated force perpendicular to the bridge highway surface, the GVW, which describes the loading intensity of highway vehicles, and the time interval between adjacent vehicles, which is related to the occurrence frequency of highway vehicles, can be treated as two random variables.

Furthermore, road traffic conditions differ between the slow, middle and fast lane. The probabilistic distributions of GVW in different lanes should be respectively fitted. From the measurement data, it was found that some random variables could not be described by a single conventional probability distribution function. Mixture model distributions (McLachlan and Peel 2000) are thus used to describe these random variables based on measurement data and are used for railway loadings and stress range spectra as well. A mixture of Weibull density functions is utilised to fit the measured GVW density function for each lane. Figure 19.8 shows both measured and fitted GVW density functions for the slow, middle and fast lanes in the airport-bound direction. For probabilistic distributions of time intervals between adjacent vehicles, they may also differ between day and night, and accordingly, they should be fitted separately. In this chapter, the time period from 11:00 p.m. to 8:00 a.m. is defined as *normal hours*, in which less road vehicles pass over the bridge. The period from 8:00 a.m. to 11:00 p.m. is called *rush hours*, in which more road vehicles run on the bridge. Figure 19.9 shows the measured and fitted density functions of time intervals of

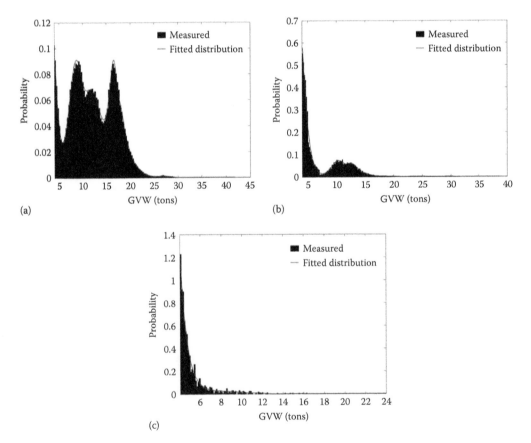

Figure 19.8 GVW and mixture Weibull distribution in the airport-bound direction: (a) slow lane, (b) middle lane, (c) fast lane.

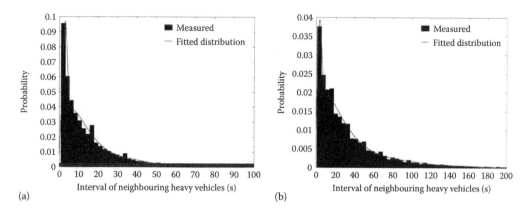

Figure 19.9 Time interval of adjacent vehicles and mixture Weibull distribution in the airport-bound slow lane: (a) rush hour, (b) normal hour.

adjacent vehicles on the slow lane in the airport-bound direction in rush hours and normal hours, respectively. The fitted density functions are a mixture of Weibull density functions.

19.5.2 Railway loading

The railway via the Tsing Ma Bridge is managed by the Mass Transportation Railway (MTR) Corporation, Hong Kong. The railway operation began in June 1998. Railway tracks supported on the bottom chords of the cross frames of the bridge deck consist of track plates, rail waybeams and tee diaphragms. To monitor train traffic flow and identify bogie load distribution, a set of strain gauges were installed on the inner waybeam of each pair of waybeams under the two rail tracks at chainage 24,662.5. Through proper calibration, the signals from the strain gauges can be converted to the bogie load data, by which the requested information on train traffic flow and bogie load distribution can be obtained. This special measurement system was put into operation in 2000.

Traffic volume and the composition of trains are two important indices of train traffic conditions. Based on the bogie load records, the annual number of trains passing through the bridge is calculated in terms of the number of bogies, shown in Figure 19.10 for 6 years from 2000 to 2005. It is observed that there is a significant increase in annual train count in 2003, and afterwards the annual train count becomes stable around 150,000. Before 2003, 14-bogie trains (seven-car trains) are dominant, at about 96% of all trains passing through the bridge annually. Since 2003, the number of 16-bogie trains (eight-car trains) has become more than that of 14-bogie trains. After the opening of the Hong Kong Disneyland in September 2005, 16-bogie trains became dominant. In November 2005, the percentage of 16-bogie trains was already 90%. November and December of 2005 can be considered as representing the current train traffic conditions. The monthly train count is about 12,000 and more than 90% of the trains are 16-bogie trains.

The bogie load data recorded in November 2005 are chosen as the database to investigate bogie load distribution. Because about 90% of the trains – that is, a total of 10,705 trains – passing through the Tsing Ma Bridge in November 2005 are 16-bogie trains, the investigation of bogie load distribution focuses on 16-bogie trains. The bogie load distribution in a 16-bogie train is actually the distribution of 16 bogie loads along the train. For each 16-bogie train passing through the bridge either airport bound or Kowloon bound, one bogie load distribution can be obtained. The distributions with the maximum bogie load

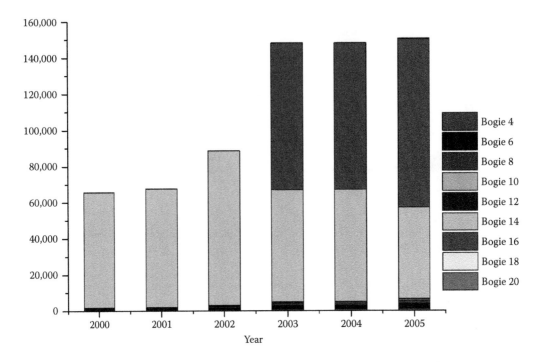

Figure 19.10 Annual train counts (2000–2005).

and the minimum bogie load in November 2005 are shown in Figure 19.11 for the airport-bound direction. The mean of the bogie load distributions is also plotted together with the design bogie load distribution provided by the MTR Corporation. It can be seen that the pattern of the measured bogie load distributions follows the design pattern. The bogie load distribution with the maximum bogie load is below the design bogie load distribution both airport and Kowloon bound.

For the SHM-based FDP of the Tsing Ma Bridge, the GTW and train arrival time are treated as two random variables. The random nature of the GTW is mainly due to uncertainties in passengers, while the random nature of the train arrival time is due to many reasons. However, as all the trains run following a scheduled timetable, the scheduled arrival time of each train is assumed constant and the random variable is used to represent only the minor difference between the actual arriving time and the scheduled arrival time. As a result, the actual arrival time of a train is the sum of the scheduled arrival instant and a random deviation. Figure 19.12 shows the measured and fitted probability density functions of the GTW of trains passing through the bridge based on the one-month measurement data in November 2005. The fitted probability density function is actually a weighted sum of the four Gaussian density functions. The random variable, representing the deviation of actual arrival time from the scheduled arriving time, could be described by a single Gaussian distribution as shown in Figure 19.13.

19.5.3 WIND CHARACTERISTICS AND MODELS

Hong Kong is situated at latitude N22.2° and longitude E114.1° and it is just on the southeastern coast of China facing the South China Sea. Two types of wind conditions dominate Hong Kong: monsoon wind prevailing in the months November through April and typhoon

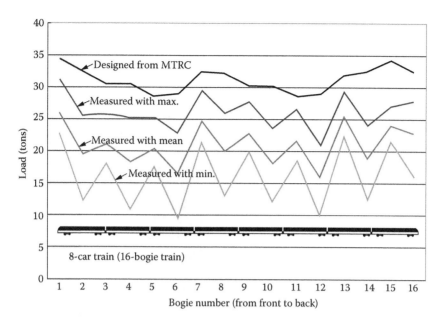

Figure 19.11 Comparison of measured and designed bogie load distributions in November 2005 (airport bound).

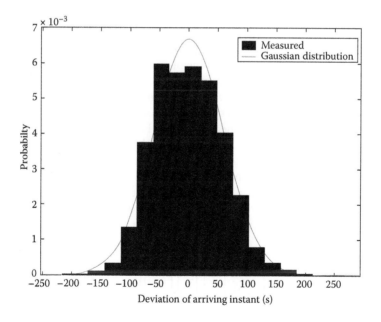

Figure 19.12 GTW and Gaussian distribution.

wind predominating in the summer. The local topography surrounding the bridge is quite unique and complex, and includes sea, islands and mountains 69–500 m high. The complex topography makes the wind characteristics at the bridge site very complicated. The SHM system installed in the bridge makes it possible to investigate wind characteristics and to gain a better understanding of wind loading. The SHM system of the bridge includes a total of six anemometers, with two at the middle of the main span, two at the middle of the Ma

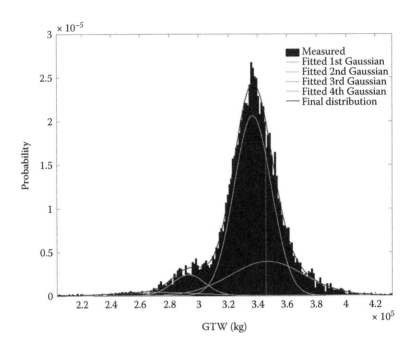

Figure 19.13 Deviations of train arrival time.

Wan side span and one in each of the Tsing Yi tower and Ma Wan tower. The sampling frequency of wind speeds was set at 2.56 Hz. Wind data recorded by the SHM system were analysed, and wind characteristics such as mean wind speed, mean wind direction, turbulence components, turbulence intensities, integral scales and wind spectra were obtained for both monsoon winds and typhoon winds. It is impossible to cover all the results in this section. Only the joint probability density function for monsoons at the bridge site is briefly introduced because it will be used in the subsequent FDP.

The joint probability density function of wind speed and wind direction is essential when assessing wind-induced fatigue damage to the bridge. A practical joint probability distribution function was adopted for a complete population of wind speed and direction based on two assumptions: (1) the distribution of the component of wind speed for any given wind direction follows the Weibull distribution; (2) the interdependence of wind distribution in different wind directions can be reflected by the relative frequency of occurrence of wind.

$$P_{u,\theta}(U,\theta) = P_\theta(\theta)\left(1 - \exp\left[-\left(\frac{U}{c(\theta)}\right)^{k(\theta)}\right]\right) = \iint f_\theta(\theta)f_{u,\theta}(U,k(\theta),c(\theta))\,du\,d\theta \qquad (19.1)$$

$$f_{u,\theta}(U,k(\theta),c(\theta)) = \frac{k(\theta)}{c(\theta)}\left(\frac{U}{c(\theta)}\right)^{k(\theta)-1}\exp\left[-\left(\frac{U}{c(\theta)}\right)^{k(\theta)}\right] \qquad (19.2)$$

$$P_\theta(\theta) = \int_0^\theta f_\theta(\theta)\,d\theta \qquad (19.3)$$

in which $0 \leq \theta \leq 2\pi$, and $P_\theta(\theta)$ is the relative frequency of occurrence of wind in wind direction θ. The occurrence frequency $P_\theta(\theta)$ as well as the distribution parameters $k(\theta)$ and $c(\theta)$ can be estimated using wind data recorded at the bridge site.

Monsoon wind records of hourly mean wind speed and direction within the period 1 January 2000 to 31 December 2005 from the anemometer installed at the top of the Ma Wan tower were used to ascertain the joint probability density function of hourly mean wind speed and direction. The height of the anemometer is 214 m above sea level. Wind records with an hourly mean wind speed lower than 1 m/s were removed in order to avoid any adverse effect on the statistics. As a result, 19,775 hourly monsoon records were available for calculation of the joint probability density function of wind speed and direction. All the monsoon records were classified into 16 sectors of the compass, with an interval of $\Delta\theta = 22.5°$ according to the hourly mean wind direction. In each sector, mean wind speed was further divided into 16 ranges from 0 to 32 m/s with an interval of $\Delta U = 2$ m/s. This leads to a total of 256 cells, and the relative frequency of the hourly mean wind speed and wind direction in each cell was calculated. Based on the calculated relative frequencies of wind speed and wind direction, the theoretical expression of joint probability density function was deduced based on Equation 19.1. The Weibull function was used to fit the histogram of hourly mean wind speed for each wind direction (see Figure 19.14). The relative frequency of wind direction and the scale and shape parameters of the Weibull function were finally obtained.

19.5.4 Temperature loading

Bridges are subjected to daily and seasonal environmental thermal effects induced by solar radiation and ambient air temperature. The variation of temperatures in bridge components significantly influences the overall deflection and deformation of the bridge.

By taking the Tsing Ma Bridge as an example, studies on the variation pattern, variation ranges and peak/trough occurrence period of the ambient air temperature were carried out (Xu et al. 2010). A total of six sensors were employed for measurement of the ambient air temperature. Figure 19.15 displays the monthly statistics of ambient air temperature, which include the mean, minimum and maximum values recorded by channel 82 from 1997 to 2005. The ambient air temperatures shown in Figure 19.15 indicate that there is a clear and fairly stable cycle of temperature variation. The temperature normally reaches its lowest value in January every year, while the highest level is in July or August. The minimum and maximum temperatures in the almost 9-year period are well above 0°C and below 40°C. This range of ambient temperature is within the design temperature limits of –2°C and 46°C for the maximum contraction and expansion movement of the bridge, respectively. It is also found that there are no significant differences in monthly statistical values among the six sensors.

It is understood that bridge deformation due to thermal effects is more directly related to the temperature of the structural members compared with the ambient air temperature. Therefore, the temperature range and variation of the bridge members were studied and compared with those of the ambient air temperature. To study the temperatures of the bridge deck section and the main cable section, the effective temperature was considered instead of the temperature measured from any single sensor in the cross section concerned. The effective temperature is theoretically calculated by weighting and adding temperatures measured at various locations within the cross section. It was found that the effective deck temperature has similar variation patterns to that of the ambient air temperature. The effective deck temperature of minimum value is almost the same as the ambient temperature. The magnitude differences regarding the mean values between the two types of temperature are

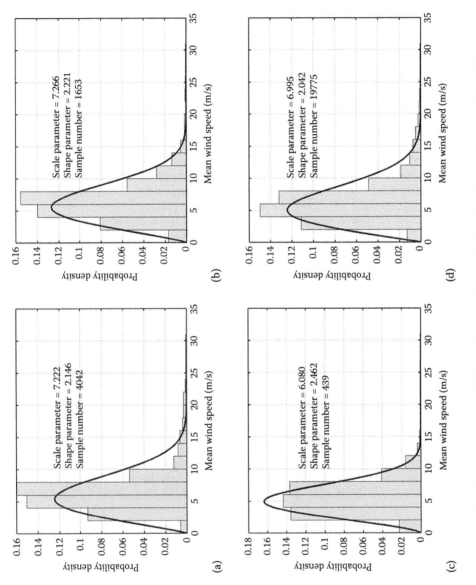

Figure 19.14 Weibull distribution of hourly mean wind speed: (a) east direction, (b) south direction, (c) west direction, (d) all directions.

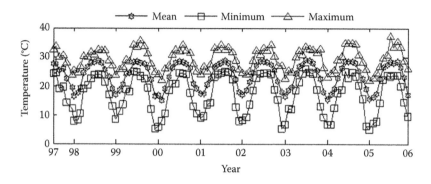

Figure 19.15 Monthly ambient air temperatures, 1997–2005 (Channel 82).

slight. For example, the mean temperature of the bridge deck is normally 2°C higher than the ambient air temperature. In view of the maximum temperature, the effective deck temperature is significantly higher than the ambient air temperature.

Given that temperature tends to cause static rather than dynamic deformation of the bridge, attention was primarily placed on the mean displacement in the evaluation of the predominant effects of temperature on bridge deformation. Furthermore, to facilitate the monitoring of temperature effects on-site, only the effective deck temperature nearby the Tsing Yi tower was selected as a reference temperature. It was noted that the effective mean deck temperature is almost the same as the effective mean cable temperature. Moreover, it was also found that the longitudinal displacement responses of the bridge towers, deck sections and cables show strong linear relationships with the effective deck temperature. Only the vertical displacement of the deck sections and cable section at the main span was closely correlated with the effective deck temperature. The lateral displacement responses of the Tsing Ma Bridge were observed not to be dominantly affected by the temperature.

As far as the bridge deck is concerned, the magnitude of the displacement variation rate gradually increases along the bridge span towards Tsing Yi Island. The variation rate of the vertical displacement of the deck section in the middle of the main span is almost the same as that of the cable section. Furthermore, based on the linear temperature displacement relationships determined through data fitting, the variation ranges of displacement responses can be effectively predicted. These relationships facilitate the monitoring of temperature effects on the Tsing Ma Bridge on-site.

It should be pointed out that the preceding observations are made for the Tsing Ma Bridge only. They may not suit other types of bridges, but they do have some implications for the monitoring of temperature effects on long-span suspension bridges.

19.6 SHM-BASED STRESS ANALYSIS DUE TO MULTIPLE DYNAMIC LOADINGS

19.6.1 Stress analysis framework

To undertake a fatigue analysis of a multi-loading long-span suspension bridge, a dynamic stress analysis of the bridge under multiple types of dynamic loads should first be conducted. A comprehensive framework based on the FE method is therefore developed to fulfil this task (Chen et al. 2011a). In the framework, a long-span suspension bridge, trains and road vehicles are regarded as three subsystems. An SHM-oriented FEM is established, as introduced earlier, for long-span suspension bridges such that the dynamic stresses of major bridge components can be predicted directly. The trains and road vehicles are also modelled using the FE method. Given that a large number of DOFs are involved in the FEMs of the three subsystems, the mode superposition method is adopted to make the dynamic stress analysis of the bridge under multiple types of dynamic loads manageable. The three subsystems are coupled through the contacts between the bridge and trains and between the bridge and road vehicles in terms of wheels–rails and tires–road surface. The nonlinear restoring forces and damping forces in the suspension units of the trains and road vehicles are treated as pseudo forces in the train and road vehicle subsystems. Wind forces may act on all three subsystems. The spatial distributions of both buffeting forces and self-excited forces over the bridge deck surface are considered for the dynamic stress analysis of the bridge. The aerodynamic wind forces acting on the car body of a train or road vehicle are determined using a quasi-steady approach. A stepwise explicit integration method is adopted to solve the coupled equations of motion for the wind–vehicle–bridge system. A set of computer programs coded in Fortran language and integrated with a commercial FE software package are developed to implement the framework for the dynamic stress analysis of a long-span suspension bridge under combined railway, highway and wind loading.

19.6.2 Verification of the framework

The accuracy of the proposed framework must be validated before it can be applied in practice. The data recorded by the SHM system in the Tsing Ma Bridge provide an excellent opportunity for validation. The SHM system provides not only the measurement data of different types of dynamic loads as input for a computer simulation but also the measured local stress responses for comparison with the computed ones. Due to the required stress analysis of local bridge components, the SHM-oriented FEM of the Tsing Ma Bridge established earlier (see Figure 19.4) is used.

To validate the framework for stress analysis in detail, three particular load cases are examined: (1) under strong wind only, (2) under strong wind and running trains and (3) under strong wind, running trains and running road vehicles. The selection and preprocessing of measurement data for wind, trains, road vehicles and structural responses are performed. The selected data corresponding to the three load cases are then analysed, respectively, to extract both input data for the computer simulation and output data for comparisons. Comparisons between the computed and measured dynamic stress responses are finally made for each load case in terms of time histories and amplitude spectra. The comparative results show that the time histories of the computed stress responses are close to those measured by the corresponding strain gauges. Displayed in Figure 19.16 are the computed and measured 140 s stress response time histories at the location of a strain gauge (SS-TLS-12) installed under a rail waybeam. During the time period concerned, one train, several heavy road vehicles and strong wind act on the bridge. The normal hourly mean wind speed at the bridge deck level is 11.91 m/s. Clearly, the proposed framework can accurately predict the dynamic stress responses of the local components of a suspension bridge under combined railway, highway and wind loading.

19.6.3 An engineering approach

In addition to computational accuracy, computational efficiency is very important in calculating the dynamic stress responses of a long-span suspension bridge because a great number of time histories at various locations of major bridge components should be computed for fatigue assessment. However, the level of computational efficiency of the preceding framework is very low, as several hours are required to compute the 140 s stress response time histories. It is thus necessary to develop an engineering approach based on the framework

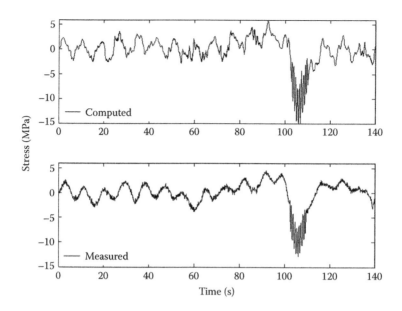

Figure 19.16 Comparison of stress responses under combined railway, highway and wind loadings.

mentioned previously for computational efficiency while maintaining computational accuracy to a certain extent. Two major assumptions are adopted to simplify the coupled dynamic stress analysis framework described previously for long-span suspension bridges (Chen et al. 2011a). As the dynamic response of a long-span suspension bridge rather than the safety of running vehicles is of interest herein, and also because a suspension bridge deck is massive and its fundamental frequencies are very low, trains and road vehicles running on the bridge deck can be simplified as moving loads, and the stress responses induced by trains and road vehicles are calculated based on stress influence lines. Furthermore, given the low level of wind-induced stress response in normal wind conditions and the relatively low level of traffic-induced stress response in extreme wind conditions, it is reasonable to assume that the coupled effects of dynamic stresses induced by railway, highway and wind loading can be neglected. Based on this assumption, the bridge stress responses at given points induced by railway, highway and wind loading can be computed separately. The three stress responses to the three individual types of dynamic loads are finally superposed to obtain the combined response to the multiple types of dynamic loads.

The feasibility of the proposed engineering approach is verified. The computational accuracy is validated by comparing the stress responses computed by the engineering approach with those by the coupled dynamic analysis framework. The comparative results show that the differences between the stress responses computed by the two methods are small and the engineering approach is applicable to long-span suspension bridges. The computational accuracy and efficiency of the engineering approach are further validated by comparing the computed daily stress time histories with the measured ones. Figure 19.17 shows the multiple load-induced stress time histories at the location of strain gauge SS-TLS-12 on 19 November 2005. During this particular day, 440 trains and 16,848 heavy road vehicles (weighing over 30 kN) ran across the bridge, and the hourly mean wind speed perpendicular to the bridge axis ranged from 2 to 13 m/s. The measurement data of the trains, road vehicles and wind recorded by the SHM system are utilised for the simulation. The results show that only several minutes are required for the engineering approach to compute the required stress time histories. The relative differences between the computed and measured stress time histories

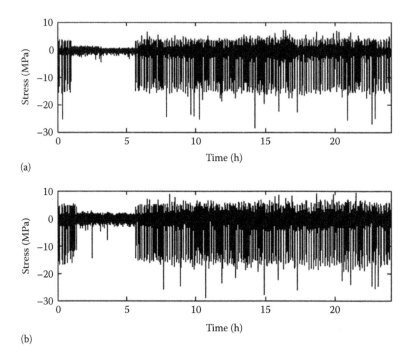

Figure 19.17 Daily stress time histories under multiple types of dynamic loads at the location of strain gauge SS-TLS-12: (a) calculated time histories, (b) measured time histories.

are small, which indicates that the engineering approach has a high level of computational efficiency and an acceptable level of computational accuracy. The engineering approach can be used for fatigue and reliability assessments of multi-loading suspension bridges.

19.7 SHM-BASED FATIGUE DAMAGE PROGNOSIS AND RELIABILITY

19.7.1 Fatigue damage prognosis

The deterministic approach based on Miner's rule (i.e. the linear damage accumulation model) is widely applied in the fatigue assessment of bridge structures in practice. Here, a general computational procedure based on this rule is presented for the fatigue prognosis of the Tsing Ma suspension bridge over its design life. There are thousands of structural members in the Tsing Ma Bridge. The fatigue-critical locations for fatigue assessment are first determined based on the maximum stress ranges in the stress time histories induced by a standard train running over the bridge. The results show that the fatigue-critical sections of the bridge deck are around the bridge towers, the pier on the Ma Wan side and the quarter span of the main span on the Tsing Yi side (Chen et al. 2011b). The fatigue-critical locations of the Tsing Ma Bridge which are most sensitive to wind loading are around the cross sections at the bridge towers (Xu et al. 2009). Therefore, the components around the bridge towers are fatigue-critical locations with respect to both traffic and wind loading, and six of them are chosen for fatigue analysis: the elements E32123, E34415, E40056, E40906, E55406 and E39417.

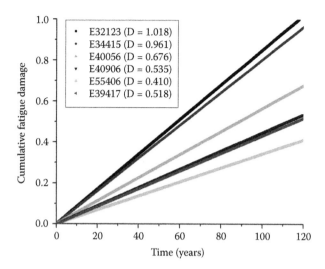

Figure 19.18 Cumulative fatigue damage curves at the fatigue-critical locations.

Databases for the dynamic stress responses at the critical locations induced by wind, railway and highway loadings are established. The time histories of dynamic stresses over 120 years induced by railway, highway and wind loading are then computed using the databases. The multiple load-induced stress time histories are finally generated. Fatigue analysis based on the stress time histories is performed to assess the cumulative fatigue damage over the bridge design life. Figure 19.18 shows the cumulative fatigue damage curves at the fatigue-critical locations within a design life of 120 years. The cumulative fatigue damage of 120 years at the most critical locations of the Tsing Ma Bridge, except for E32123, is lower than one.

The cumulative fatigue damage induced by each type of dynamic load and the damage magnification due to multiple types of dynamic loads are also investigated. It is found that railway loading plays the dominant role in bridge fatigue. The fatigue damage induced by highway loading is greater than that by wind loading for some structural components, but there is a reversal for other components. A multiple load magnification factor is defined herein as the ratio of the fatigue damage due to the combined effect of the three loadings and the sum of the damage due to each individual loading. The computed factors at the six fatigue-critical locations range from 1.06 to 1.35. The results indicate that it is necessary to consider the combined effect of multi-loads in the fatigue analysis of suspension bridges.

19.7.2 Fatigue reliability

Deterministic fatigue analysis is unable to consider the effects of uncertainties arising from load and structural properties. Another framework for fatigue reliability analysis is therefore proposed and applied to the Tsing Ma Bridge. Based on Miner's law and the empirical relationship between stress range and the number of cycles to failure, a limit state function is defined to describe the relationship between fatigue resistance and fatigue loading (Chung 2004; Chen et al. 2012).

$$g(X) = \left(\frac{K_0 \Delta}{S_{ef}^m} \right) - \left(\sum_{j=1}^{n} N_j \right) = \left(\frac{K_0 \Delta}{S_{ef}^m} \right) - (n\mu_N) \tag{19.4}$$

where:

K_0 and m	are the constants related to material
Δ	is Miner's damage accumulation index
S_{ef}^m	is calculated based on the stress range spectrum

The total number N of accumulated stress cycles is the sum of the number N_j of accumulated stress cycles on the jth day for all days, and it is also equal to the mean daily cycle number μ_N multiplied by the number of days n. Here, the parameter K_0 is assumed to follow a log-normal distribution. Miner's damage accumulation index Δ of fatigue failure for steel structures is also assumed to follow a log-normal distribution with a mean value μ_Δ of 1.0 and a coefficient of variation δ_Δ of 0.3. Employing the log-normal distribution models for K_0 and Δ in the limit state function Equation 19.4 and applying random variable transformations, the fatigue reliability index β can be found as follows:

$$\beta = \frac{(\lambda_{K_0} + \lambda_\Delta) - \ln\left(S_{ef}^m \right) - \ln(\mu_N) - \ln(n)}{\sqrt{\zeta_\Delta^2 + \zeta_{K_0}^2}} \tag{19.5}$$

Based on loading data acquired from the SHM system, the probabilistic models of railway, highway and wind loading are established to describe the uncertainties inherent in different loads, as discussed in Section 19.5. Using the probabilistic loading models, the dominant loading parameters are then generated using Monte Carlo Simulation (MCS), and the daily stochastic stress responses induced by railway, highway and wind loading are simulated at the fatigue-critical locations. The probability distribution of the daily sum of m-power stress ranges is estimated based on the daily stochastic stress responses. The probability distribution of the sum of m-power stress ranges over the period concerned is then estimated based on assumptions of future loading and traffic growth patterns. The future traffic loading is assumed to be a 30% increase in both railway and highway loading compared with the current traffic loading. Different traffic growth patterns are assumed to estimate the probability distribution of the sum of m-power stress ranges within the period concerned, including no traffic growth pattern (*Constant*) and growth in a linear pattern (*Linear*) and in two exponential patterns (*Exp-1* and *Exp-2*). Finally, the fatigue failure probabilities for different time epochs are solved at the fatigue-critical locations using the first-order reliability method (FORM). The evolution of fatigue failure reliability over time at the element E32123 is shown in Figure 19.19. The figure indicates that the fatigue failure probabilities increase with time, and that the failure probability without traffic growth is smaller than that of the three patterns with traffic growth. Among the three growth patterns, the failure probability is the largest for the Exp-2 pattern, followed by the linear pattern and then the Exp-1 pattern. The results demonstrate that the health condition of the bridge at the end of its design life is satisfactory under current traffic conditions without growth, but attention should be paid to future traffic growth because it may lead to a greater failure probability.

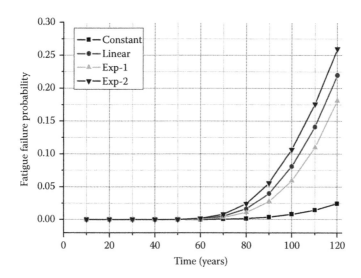

Figure 19.19 Evolution of fatigue failure probability over time.

19.8 SHM-BASED BRIDGE RATING SYSTEM AND INSPECTION

19.8.1 SHM-based long-span suspension bridge rating system

This section presents an SHM-based bridge rating method for the bridge inspection of long-span cable-supported bridges (Li et al. 2011).

The analytic hierarchy process (AHP) attracts the interest of many researchers because of its interesting mathematical properties and easy applicability. The main steps in the application of the AHP to the current problem are as follows: (1) to decompose a general decision problem into hierarchical sub-problems that can be easily comprehended and evaluated; (2) to determine the priorities of the items at each level of the decision hierarchy; and (3) to synthesise the priorities to determine the overall priorities of the decision alternatives. Since a long-span suspension bridge is a very complex system and the decision-making takes place in a situation in which the pertinent data and the sequences of possible actions are not precisely known, it is important to adopt fuzzy data to express such situations in the decision-making of inspection, leading to the so-called F-AHP bridge rating method, as shown in Figure 19.20.

The fuzzy synthetic rating R (see Figure 19.20) at the objective level can be calculated based on the calculated criticality rating T_c, the vulnerability rating T_v and the their relative weights ω_{cv} at the criterion level.

$$R = \omega_{cv}^{T} \begin{Bmatrix} T_c \\ T_v \end{Bmatrix} \tag{19.6}$$

The determination of weights ω_{cv} is related to the importance of criticality and vulnerability. For example, if the importance of criticality is regarded to be the same as that of vulnerability, the relative weight vector for the criterion level can be taken as $\omega_{cv} = \{0.5, 0.5\}^T$ for both cases. For different structural components at other levels of the hierarchical structure – that is, index level 1 and level 2 as shown in Figure 19.20 – the AHP often uses the

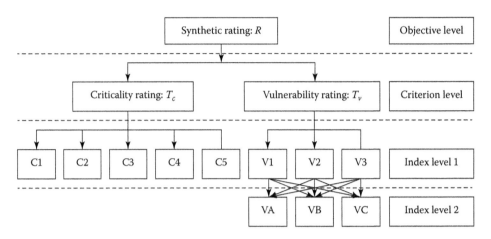

Figure 19.20 Analytic hierarchical structure for each bridge component.

eigenvalue solution of comparison matrices to find the best relative weights (relative importance). The details can refer to Li et al. (2011).

The criticality and vulnerability ratings (T_c and T_v) for each structural component are based on the criticality and vulnerability factors in relation to the criticality and vulnerability criteria. A definition and analysis of criticality and vulnerability factors are briefly discussed in the following subsections. With the determined criticality and vulnerability factors (see Figure 19.20) as well as their associated relative weights, the ratings T_c and T_v can be calculated accordingly. Formulas for the calculation of T_c and T_v can be found in Li et al. (2011) but are not listed in this section for the sake of easy understanding.

After the fuzzy synthetic ratings of all the structural components are obtained, the prioritisation or optimum for inspection frequency (fuzzy synthetic decision for inspection) can be determined. The larger the value of R, the smaller the inspection time interval. Notably, the numerical numbers for each of the factors are assumed to range from 0 to 100 to facilitate decision-making using the F-AHP-based bridge rating method.

19.8.2 Criticality and vulnerability factors

19.8.2.1 Criticality factors

Five criticality factors for a long-span cable-supported bridge follow in this subsection. Table 19.3 shows the definitions, range and points for each factor. As mentioned earlier, the numerical values of the five factors range from 0 to 100.

(a) Criticality factor C1: It is clear that a structural component with an alternative load path or redundancy is robust; that is, no serious failure or consequence will result from limited damage to this structural component. Therefore, the more the redundancy for a given structural component, the smaller the numerical value of C1 for this component. This criticality factor is represented in this chapter by three numerical values of 100, 67 and 0.

(b) Criticality factor C2: This factor represents the strength reliability of a structural component, which is determined based on the strength utilisation factor (SUF). The SUF is calculated using Equation 19.7. If the SUF of a certain structural component reaches the extreme value, the numerical value of the criticality factor C2 for this component should be 100. Otherwise, the numerical value should be the ratio of the strength utilisation factor divided by the extreme value.

Table 19.3 Criticality factor (CF): Definitions and values

CF	Definition	Range	Points
C1	Any alternative load path?	No	100
		Yes, will affect global structural performance	67
		Yes, will not affect global structural performance	0
C2	Design normal combined loads (based on strength utilisation factor)	0%–100%	0–100
C3	Design fatigue loads(based on fatigue life)	High: <200 years	100
		Normal: 200–300 years	67
		Low: >300 years or N/A	0
C4	Known or discovered imperfections but not serious enough to warrantimmediate repair	Any, non-repairable	100
		Any, repairable	67
		None	0
C5	Failure mechanisms	Catastrophic collapse	100
		Partial collapse	67
		Structural damage	33

$$\text{SUF} = \frac{\gamma_L \sigma_L (1.0 + I)}{\phi R_N - \gamma_D \sigma_D} \qquad (19.7)$$

where:
- R_N is the as-built nominal resistance
- ϕ is the strength reduction factor
- γ_L is the partial load factor for live loads
- γ_D is the partial load factor for dead loads
- σ_L is the stresses due to live loads
- σ_D is the stresses due to dead loads
- I is the impact factor for dynamic live loads

(c) Criticality factor C3: This item represents the fatigue reliability of a structural component. In a similar way to C1, the relative fatigue criticality of different structural components can be represented by three numerical values of 100, 67 and 0.

(d) Criticality factor C4: This item emphasises the imperfections (deterioration/damage) of structural components detected by the previous inspections. In this regard, the severity of imperfections and the urgency of repair is respectively considered and represented by the three numerical values of 100, 67 and 0 herein. Besides the visual inspection, the SHM system can now be used as a tool to detect imperfections by analysing the measurement data directly recorded by strain sensors, fatigue sensors, accelerometers and others.

(e) Criticality factor C5: This item represents the ultimate load-carrying capacity of structural components under an extreme loading event. The structural component of foremost failure will be the most critical component, and the corresponding energy demand for failure will be the least. The relative ultimate load-carrying capacity of different structural components could be represented by three numerical values of 100, 67 and 33 herein. Nonlinear pushover analysis or progressive collapse analysis as used in redundancy analysis can also be applied to determine C5.

19.8.2.2 Vulnerability factors

Three vulnerability factors follow in this subsection. Table 19.4 shows the definitions, range and points for each factor.

1. Vulnerability factor V1: This item represents damage due to extra-slowly varying effects, such as the carbonation of concrete and the corrosion of steel. The item V1 consists of three subitems, VA1, VB1 and VC1. Each subitem can be represented by the three numerical values of 100, 50 and 0, according to three different ranges.
2. Vulnerability factor V2: This item represents damage due to rapidly varying effects – for instance, accidental damage caused by vehicle or ship collision. The item V2 consists of three subitems, VA2, VB2 and VC2, each of which can be represented by the three numerical values of 100, 50 and 0, respectively.
3. Vulnerability factor V3: This item represents damage due to slowly varying effects – for example, the movement of bearings and joints because of daily temperature changes. The item V3 also has three subitems, and each subitem has three difference ranges represented by three corresponding numerical values.

The item VA1 can be quantified by the corrosion rate, which is defined as reciprocal of time demand to reach critical corrosion value once the exposure occurred. The corrosion rate can be calculated based on either data measured directly by corrosion sensors in the SHM system or the numerical model updated by the SHM system. The numerical values

Table 19.4 Vulnerability factor (VF): Definitions and values

VF	Definition	Range	Points
V1: Corrosion	Exposure or degree of protection (VA1)	Internal or adequate	0
		Partial or average	50
		Extreme or none	100
	Likelihood of detection in superficial inspection (VB1)	Likely	0
		Possible	50
		Unlikely	100
	Likely influence on structural integrity (VC1)	Likely	0
		Possible	50
		Unlikely	100
V2: Damage	Exposure to damage(VA2)	None	0
		Medium	50
		High	100
	Likelihood of detection in superficial inspection (VB2)	Likely	0
		Possible	50
		None	100
	Likely influence on structural integrity (VC2)	Low	0
		Medium	50
		High	100
V3: Wear	Relative wear rate per annum (VA3)	Low	0
		Medium	50
		High	100
	Likelihood of detection in routine maintenance (VB3)	Likely	0
		Medium	50
		Unlikely	100
	Likely influence on structural integrity (VC3)	Low	0
		Medium	50
		High	100

for the item VA2 can be allocated by experience just after any extreme accidental event has occurred and damage on components is detected. The item VA3 can be quantified either by visual inspection or by analysing the measurement data recorded by displacement sensors in the SHM system. The numerical values for the items VB1, VB2 and VB3 can be allocated to various structural components based on the experiences of the inspectors. The numerical values for the items VC1, VC2 and VC3 can be the same as those for criticality factor C1.

19.8.3 Criticality and vulnerability analyses

The Tsing Ma Bridge in Hong Kong is taken as an example to demonstrate the feasibility of the SHM-based F-AHP bridge rating method as guidance in determining the time intervals for inspection. The key structural components of the Tsing Ma Bridge are classified into 15 groups and 55 components for criticality and vulnerability analyses. The 15 groups, which are basically the key components of the Tsing Ma Bridge for direct and indirect load transfer, are (1) suspension cables, (2) suspenders, (3) towers, (4) anchorages, (5) piers, (6) outer-longitudinal trusses, (7) inner-longitudinal trusses, (8) main cross frames, (9) intermediate cross frames, (10) plan bracings, (11) the deck, (12) rail waybeams, (13) bearings, (14) movement joints and (15) the Tsing Yi approach deck. Details of classification in each group can refer to Li et al. (2011). The following determine the criticality and vulnerability factors for each structural component of the Tsing Ma Bridge.

19.8.3.1 Criticality analysis

For criticality factor C1, a load path analysis must be carried out based on the SHM-oriented FEM. However, since the computational effort for the load path analysis is beyond the current computer capacity because of the huge size of the model (21,946 elements in total), criticality factor C1 is adopted from the currently used values (Wong 2006), which are based on practical experience with some engineering analysis.

To determine criticality factor C2 for each of the structural components of the Tsing Ma Bridge, criticality analysis is performed on the strength of the bridge under design-normal combined loads in terms of the strength utilisation factor. To fulfil this task, the SHM-oriented FEM of the Tsing Ma Bridge is established (see Figure 19.4) and used. Seven types of loads (dead loads, superimposed dead loads, temperature loads, highway loads, railway loads, wind loads and seismic loads) and three load combinations have been considered in the stress analysis of the bridge (HKPU 2009a). Except for the dead loads, the superimposed dead loads and the seismic loads, there are 2, 24, 8 and 3 load cases for the temperature loads, the highway loads, the railways loads and the wind loads, respectively. In the three load combinations, there are also a total of 52 load cases. For each load case, the stresses in the major structural components are determined, and the stress distributions are obtained for each of the major structural components. Based on the obtained stress distribution results, the stresses in the structural components at five key bridge deck sections are provided. The strength utilisation factors of the major structural components are calculated, from which the critical location of each major structural component is identified. For the bridge towers made of reinforced concrete, strength analyses are carried out using the load moment strength interaction method, and the strength utilisation factors of the two tower legs are determined. Finally, strength utilisation factors for the key structural components are used together with other factors for rating the bridge components.

To determine criticality factor C3 for each of the structural components of the Tsing Ma Bridge, a criticality analysis of the bridge is performed focusing on the fatigue life of the structural components under traffic loads (HKPU 2009b). Railway loading and highway loading are considered to be major contributors to the fatigue damage of the bridge (Chen et al. 2012). The railway and highway loadings measured by the SHM system are then used to derive the actual train spectrum and the actual road vehicle spectrum for fatigue assessment. A traffic-induced stress analysis method is proposed based on the SHM-oriented FEM of the bridge and the influence line method for the determination of stress time histories. The fatigue-critical locations are identified for different bridge components through rigorous stress analysis. Finally, the fatigue lives due to both train and road vehicles at the fatigue-critical components are estimated using the vehicle spectrum method recommended by the British Standards Institution (BSI 1982). Finally, the estimated fatigue lives of the key structural components are used together with other factors for rating the bridge components.

Any damage and imperfections detected by the SHM system should be included in the determination of criticality factor C4. As the bridge has been completed and opened to public traffic for less than 15 years and the bridge has been carefully maintained, most C4 values are set at zero.

A collapse analysis, considering both geometrical nonlinearities and material nonlinearities, should be carried out to determine criticality factor C5. However, due to the huge size of FEMs of bridges, the computational effort for the collapse analysis is beyond the current computer capacity. As a result, criticality factor C5 is adopted from the currently used empirical values (Wong 2006).

19.8.3.2 Vulnerability analysis

V1 represents damage due to extra-slowly varying effects. Since there are no corrosion sensors installed in the Tsing Ma Bridge, the current vulnerability factor V1 (Wong 2006) is adopted without change in this case study.

V2 represents damage due to rapidly varying effects. VA2 can be evaluated by an impact analysis using the SHM-oriented FEM of the bridge, the dynamic loads predicted by the measurement data from the SHM system and the material properties confirmed by the SHM system. However, due to the huge size of the FEM, the computational effort for the dynamic impact analysis is beyond the current computer capacity. The currently used vulnerability factor V2 (Wong 2006) is adopted without change in this case study.

V3 represents damage due to slowing varying effects. From observations over the past 10 years, the movement joints and bearings are subjected to serious wear. The numerical values of VA3 allocated to all the joints and bearings could be set to 100, while other components could be set to 0. VB3 can be evaluated based on the experiences of the inspectors. The numerical value of criticality factor C1 can be used for VC3.

19.8.4 Inspection

To apply the proposed SHM-based F-AHP bridge rating method to the Tsing Ma Bridge, the relative weights shall be determined for each level. According to AHP procedures, the comparison matrix is calculated first, followed by the relative weights for the criticality index and vulnerability index. Since the comparison matrix is not unique, its effect on the final results shall be investigated. Two cases (Case 1 and Case 2) are considered, each of which has a different comparison matrix. More details can be found in Li et al. (2011). As previously mentioned, if the importance of criticality is regarded to be the same as that of vulnerability, the relative weight vector for the criterion level can be taken as $\omega_{cv} = \{0.5, 0.5\}^T$ for both cases.

Based on the determined relative weights and according to the presented SHM-based F-AHP bridge rating method, a decision on the time intervals for inspection can be reached, and the results are listed in Table 19.5 for the two cases. It can be seen that the time intervals

Table 19.5 Decision on time intervals for inspection

Group no.	Serial no.	Case 1		Case 2		
		Score of fuzzy rating (C1)	Time interval for inspection (year)	Score of fuzzy rating (C2)	Time interval for inspection (year)	(C2–C1)/C1 × 100%
1	1	53.3	2	54.4	2	2.06
	2	51.9	2	53.5	2	3.08
	3	51.9	2	53.5	2	3.08
	4	51.9	2	53.5	2	3.08
	5	45.3	2	46.2	2	1.99
2	6	44.3	2	43.9	2	−0.9
	7	57.6	1	60.2	1	4.51
	8	57.6	1	60.2	1	4.51
3	9	50.5	2	51.6	2	2.18
	10	50.1	2	51.5	2	2.79
	11	50.1	2	51.5	2	2.79
4	12	51.1	2	54	2	5.68
	13	55.1	2	57.4	2	4.17
	14	52.2	2	53.2	2	1.92
5	15	51.1	2	52.3	2	2.35
	16	50.1	2	51.5	2	2.79
6	17	50.6	2	51.3	2	1.38
	18	60.5	1	62.3	1	2.98
	19	50.1	2	50.5	2	0.8
	20	60.6	1	62.9	1	3.8
7	21	56.8	2	59.5	1	4.75
	22	45.1	2	45.8	2	1.55
	23	47.3	2	47.8	2	1.06
	24	56.8	2	59.5	1	4.75
8	25	51.4	2	53	2	3.11
	26	54.9	2	56.3	2	2.55
	27	59	1	61.3	1	3.9
	28	59	1	61.3	1	3.9
9	29	43	2	43.2	2	0.47
	30	52.8	2	54.6	2	3.41
	31	57	2	59.5	1	4.39
	32	57	2	59.5	1	4.39
10	33	54.4	2	56.2	2	3.31
	34	47.4	2	48.2	2	1.69
11	35	56.8	2	59.5	1	4.75
	36	55.1	2	57.5	1	4.36
12	37	48	2	48.5	2	1.04
	38	50.2	2	51	2	1.59
	39	50.2	2	51	2	1.59
13	40	59.6	1	62.2	1	4.36
	41	59.2	1	61.8	1	4.39
	42	59.2	1	61.8	1	4.39
	43	59.2	1	61.8	1	4.39

Table 19.5 (Continued) Decision on time intervals for inspection

		Case 1		Case 2		
Group no.	Serial no.	Score of fuzzy rating (C1)	Time interval for inspection (year)	Score of fuzzy rating (C2)	Time interval for inspection (year)	(C2–C1)/C1 × 100%
	44	59.2	1	61.8	1	4.39
	45	59.2	1	61.8	1	4.39
	46	59.6	1	62.2	1	4.36
	47	59.2	1	61.8	1	4.39
	48	59.6	1	62.2	1	4.36
14	49	62.3	1	64.1	1	2.89
	50	62.3	1	64.1	1	2.89
15	51	44.7	2	45.2	2	1.12
	52	44.2	2	44.6	2	0.9
	53	46.4	2	46.8	2	0.86
	54	53.8	2	55.8	2	3.72
	55	47.3	2	47.9	2	1.27

for inspection are almost the same for the two cases concerned, which indicates that the effect of the relative weights from different comparison matrices is small. It can also be seen that for the bridge components concerned, the time intervals for inspection are either 1 year or 2 years.

NOTATION

B	Fatigue reliability index
$g(X)$	Limit state function defined to describe the relationship between the fatigue resistance and the fatigue loading
$k(\theta)$, $c(\theta)$	Distribution parameters estimated using wind data recorded at the bridge site
K_0, m	Constants related to material
N	Number of days
$P_{u,\theta}(U,\theta)$	Joint probability density function of wind speed and wind direction
$P_\theta(\theta)$	Relative frequency of occurrence of wind in wind direction θ
R	Fuzzy synthetic rating
S_{ef}^m	Coefficient calculated based on the stress range spectrum
T_c, T_v	Criticality rating and vulnerability rating, respectively
Δ	Miner's damage accumulation index
θ	Wind direction
μ_N	Mean daily cycle number
μ_Δ, δ_Δ	Mean value and variation of the log-normal distribution of Δ, respectively
ω_{cv}	Weights related to the importance of criticality and vulnerability

REFERENCES

Ang, A.H.S. and W.H. Tang. 2007. *Probability Concepts in Engineering: Emphasis on Applications to Civil and Environmental Engineering*, second edition. New York: Wiley.

Boller, C., P. Starke, G. Dobmann, C.M. Kuo, and C.H. Kuo. 2015. Approaching the assessment of ageing bridge infrastructure. *Smart Struct. Syst.*, 15(3): 593–608.

Box, G.E., G.M. Jenkins, and G.C. Reinsel. 2008. *Time Series Analysis: Forecasting and Control*, fourth edition. Hoboken, NJ: Wiley.

BSI. 1982. BS5400, Part 10: Code of Practice for Fatigue.

Chen, Z.W., Y.L. Xu, Q. Li, and D.J. Wu. 2011a. Dynamic stress analysis of long suspension bridges under wind, railway and highway loadings. *J. Bridge Eng.*, 16(3): 383–91.

Chen, Z.W, Y.L. Xu, and X.M. Wang. 2012. SHMS-based fatigue reliability analysis of multi-loading suspension bridges. *J. Struct. Eng.*, 138(3): 299–307.

Chen Z.W., Y.L. Xu, Y. Xia, Q. Li, and K.Y. Wong. 2011b. Fatigue analysis of long-span suspension bridges under multiple loading: Case study. *Eng. Struct.*, 33:2, 3246–56.

Chung, HY. 2004. Fatigue reliability and optimal inspection strategies for steel bridges. PhD Diss., University of Texas at Austin.

Farrar, C.R. and N.A.J. Lieven. 2007. Damage prognosis: The future of structural health monitoring. *Phil. Trans. R. Soc. A*, 365: 623–32.

Frangopol, D.M. 2011. Life-cycle performance, management, and optimisation of structural systems under uncertainty: Accomplishments and challenges. *Struct. Infrastruct. Eng.*, 7(6): 389–413.

Frangopol, D.M., M.J. Kallen, and J.M. Noortwijk. 2004. Probabilistic models for life-cycle performance of deteriorating structures: Review and future directions. *Prog. Struct. Engng Mater.*, 6(4): 197–212.

Frangopol, D.M. and T.B. Messervey. 2007. Integrated life-cycle health monitoring, maintenance, management, and cost of civil infrastructure. In *Proceedings of 2007 International Symposium on Integrated Life-Cycle Design and Management of Infrastructure*, ed. L.C. Fan, L.M. Sun, and Z. Sun, 3–20. Shanghai, China: Tongji University Press.

Frangopol, D.M. and T.B. Messervey. 2008. Lifecycle cost and performance prediction: Role of structural health monitoring. In *Frontier Technologies for Infrastructures Engineering*, ed. S.S. Chen and A.H.S. Ang, 361–81. Boca Raton, FL: CRC Press.

HKPU. 2009a. Establishment of bridge rating system for Tsing Ma Bridge: Criticality and vulnerability analysis; Strength. Report no. 8. Hong Kong: Department of Civil and Structural Engineering, Hong Kong Polytechnic University.

HKPU. 2009b. Establishment of bridge rating system for Tsing Ma Bridge: Criticality and vulnerability analysis; Fatigue. Report no. 9. Hong Kong: Department of Civil and Structural Engineering, Hong Kong Polytechnic University.

Li, Q., Y.L. Xu, and Y. Zheng. 2011. SHM-based F-AHP bridge rating system with application to Tsing Ma Bridge. *Front. Archit. Civil Eng. China*, 5(4): 465–78.

Liu, T.T., Y.L. Xu, W. S. Zhang, et al. 2009. Buffeting-induced stresses in a long suspension bridge: Structural health monitoring orientated stress analysis. *Wind Struct.*, 12(6): 479–504.

McLachlan, G. and D. Peel. 2000. *Finite Mixture Models*. New York: Wiley.

NYDOT. 1997. *Fundamentals of Bridge Maintenance and Inspection*. New York: New York State Department of Transportation.

Wong, K.Y. 2002. Structural identification of Tsing Ma Bridge. *HKIE Transactions*, 10(1): 38–47.

Wong, K.Y. 2006. Criticality and vulnerability analyses of Tsing Ma Bridge. In *Proceedings of the International Conference on Bridge Engineering: Challenges in the 21st Century, Hong Kong*, ed. K.P. Yim. 209. Hong Kong: Hong Kong Institution Engineers.

Xu, Y.L., B. Chen, C.L. Ng, K.Y. Wong, and W.Y. Chan. 2010. Monitoring temperature effect on a long suspension bridge. *Struct. Control Health Monit.*, 17(6): 632–53.

Xu, Y.L., J.M. Ko, and W.S. Zhang. 1997 .Vibration studies of Tsing Ma long suspension bridge. *J. Bridge Eng.*, 2: 149–56.

Xu, Y.L., T.T. Liu, W.S. Zhang, et al. 2009. Buffeting-induced fatigue damage assessment of a long suspension bridge. *Int. J. Fatigue.* 31(3): 575–86.

Xu, Y.L. and Y. Xia. 2012. *Structural Health Monitoring of Long-Span Suspension Bridge*. New York: Spon Press.

Zhang, W.S., K.Y. Wong, Y.L. Xu, et al. 2007. Buffeting-induced stresses in a long suspension bridge: Structural health monitoring oriented finite element model. Research report, Department of Civil and Structural Engineering, Hong Kong Polytechnic University.

Chapter 20

Epilogue

Challenges and prospects

20.1 CHALLENGES

Smart civil structures are defined as civil structures that can mimic biological systems with smart self-sensing, self-adaptive, self-diagnostic, self-repair and self-powered functions, so as to perform any targeted functions under various environments and to preserve the safety and integrity of the structures during strong winds, severe earthquakes and other extreme events. A smart civil structure has the ability to sense, measure, process and diagnose, at critical locations, any change in selected variables, and to command appropriate action using its own power. A smart civil structure has a few key elements: smart materials, sensors, actuators, signal processors, communication networks, power sources, diagonal strategies, control strategies, repair strategies and life-cycle management (LCM) strategies. These key elements can be further integrated into four systems: the structural health monitoring (SHM) system, the structural vibration control (SVC) system, the structural self-repairing (SSR) system and the structural energy harvesting (SEH) system with optimised sensor and actuator placements and various optimised strategies. These four systems, together with an updated multi-scale finite element model (FEM) of civil structures, can then execute their tasks collectively and intelligently.

Although smart civil structures have not been completely realised yet, this book has presented a number of partially smart civil structures that possess two or more smart functions, such as the structures of self-sensing-diagnostic functions, self-sensing-adaptive functions, self-sensing–diagnostic-adaptive functions, self-sensing–diagnostic-repairing functions or self-sensing–diagnostic-adaptive-power functions. To make these partially smart civil structures even smarter and to realise truly smart civil structures, some important issues that have not been fully discussed in this book and some challenging issues to be solved in the near future are presented as future research topics in the area of smart civil structures.

20.1.1 Multi-scale modelling of smart civil structures

20.1.1.1 Nonlinear mix-dimensional finite element coupling for damaged structures at joints

A multi-scale analysis of a complex civil structure always encounters the problem of mix-dimensional finite element coupling, which is referred to as the connection of the interfaces between one-dimensional (beam), two-dimensional (plate), three-dimensional (solid) elements and others, as discussed in Chapter 6. This is particularly true if we want to know the details of the damage evolution of the joints of a civil structure, as both the geometric and material nonlinearities of the joints and interfaces should be considered and, accordingly, high quality nonlinear mix-dimensional finite element coupling methods will be developed. The high quality of the nonlinear mix-dimensional finite element coupling methods refers to the location automation of interfaces, the prediction accuracy of strains and stresses at the joints and interfaces, the displacement compatibility at the joints and interfaces, the computational convergence and the development or enhancement of commercial software for implementation.

20.1.1.2 Evolutionary multi-scale modelling

The simulation of damage evolution and structural failure of a complex civil structure is a multi-scale process, from micro-scale to meso-scale to macro-scale, which are parallel to the material, component and structural levels in the structural system. A structural failure could be very sensitive to microscopic and mesoscopic heterogenesis, and therefore the multi-scale damage evolution processes play an important role in the failure mechanism of a civil structure (Bai et al. 2005). However, it is challenging to incorporate microscopic damage models into mesoscopic damage models and macroscopic failure models in multi-scale modelling. Recently, an adaptive concurrent image-based multi-scale modelling method was developed and applied to simulate the damage evolution and failure process of concrete and steel structures (Sun and Li 2015; Sun et al. 2016). The adaptive modelling capability enables the mesoscopic damage models to be automatically implemented into the potential damage region at the structural level, to demonstrate continuous changes as a consequence of evolutionary microscopic damage without user intervention. Nevertheless, the structures that were investigated were small structures, not complex civil structures.

20.1.1.3 Multi-scale modelling of joints together with control devices

In a smart civil structure, control devices are often installed at structural joints, which make the damage evolution and failure of the joints more complicated. It is therefore necessary to incorporate the modelling of control devices into the multi-scale model of the joints. The multi-scale model of both the joints and the control devices can then be used together with hybrid tests in SVC. In such hybrid tests, a civil structure is often divided into two substructures: a numerical subsystem and a physical subsystem (Zapateiro et al. 2010). The numerical subsystem usually corresponds to that of a structure whose dynamics are well known, whereas the physical subsystem is, on the other hand, the critical component of the system, such as a joint with a nonlinear control device. However, in most of the current hybrid tests (Phillips and Spencer 2011), the FEM of the main structure is, in general, assumed to be linear and the structural control performance is of major concern. Little attention has been paid to either the control performance or the damage evolution of the joints. To solve this problem, multi-scale modelling, both of joints and control devices, must be developed together with hybrid test techniques.

20.1.2 Multi-scale damage detection with substructure techniques

A variety of damage detection methods were introduced in Chapter 12. Although most methods have been validated via numerical examples or experiments, it is quite difficult to apply these damage detection methods directly to a complex civil structure. With many degrees of freedom (DOFs) and uncertainties involved in a complex civil structure, the computation is rather time-consuming, sometimes even beyond current computer capacity. Particularly when multi-scale modelling of some critical joints or structural components is required, the number of DOFs in the FEM of the entire structure will be significantly increased. Since damage detection is basically an inverse problem, it would be difficult to converge if too many unknowns and uncertainties are involved at one time. The substructure techniques, which divide the entire structure into a few substructures of interest, may provide a promising way to address these issues, together with the multi-scale modelling technique.

The concept of multi-scale damage detection methods incorporated into substructure techniques can be expressed as follows. Firstly, the entire structure is divided into a few substructures by using a substructure technique with consideration of boundary conditions at structural level. Secondly, potential damage locations in terms of the substructures are identified using the damage detection methods introduced in Chapter 12. Thirdly, the multi-scale models and the microscopic and mesoscopic damage models are built with the substructures of potential damage identified in the first step. Finally, multi-scale damage detection methods, together with the response reconstruction method introduced in Chapter 12, may be used to identify the exact damage locations and intensities in the substructure.

Besides using substructure techniques to reduce the DOFs involved, a more direct method is to improve computation capacity, such as by using cloud computing. There seem to be many definitions of cloud computing; a more commonly used definition describes it as clusters of distributed computers that provide on-demand resources and services over a networked medium (usually the Internet) (Sultan 2010). Cloud computing is a promising paradigm for the delivery of information technology (IT) services as computing utilities. There are many obstacles required to be overcome for the adoption and growth of cloud computing (Armbrust et al. 2009), such as availability of utility computing services, customer data lock-in, data confidentiality and auditability, and so forth. In civil engineering, attempts to use cloud computing for SHM and damage detection of a smart civil structure are relatively limited. Cloud computing may provide a potential way for storing and processing the big data from the SHM system for the purpose of assessing structural performance effectively.

20.1.3 Distributed sensor systems and networks

20.1.3.1 Distributed fibre-optic sensor system measuring distributed strains at joints

Large amounts of stress often concentrate at joints where damage is likely to occur. To prevent substantial degradation or collapse of the structure, health monitoring of critical joints is significantly important. Fibre-optic sensors have been actively investigated and widely used in civil structures for SHM purposes, as introduced in Chapter 3. They possess some distinct advantages, such as being lightweight and small in size and having electromagnetic immunity, and can be used as distributed sensors for damage detection in critical joints with high spatial resolution (Murayama et al. 2011). There are a number of factors that influence measurement results. Although distributed fibre-optic sensor systems have been successfully used in damage detection of adhesive steel joints (Murayama et al. 2012) and welded steel joints (Murayama et al. 2013), these studies were conducted on a single joint without consideration of the interaction between the joint and other structural components. Moreover, research and application of distributed fibre-optic sensor systems for damage detection in reinforced concrete joints in civil engineering are still limited.

On the one hand, distributed fibre-optic sensors can be placed to obtain the strain fields at the joint concerned. On the other hand, a detailed multi-scale model of the joint can be built and corresponding updates to the model can be carried out on the basis of measurements from the distributed fibre-optic sensor system. The combination of the distributed fibre-optic sensor system with the multi-scale modelling method does provide a promising method for the performance monitoring of structural joints. Based on such a combination, joint damage can be estimated more accurately, even though the entire structural system is considered in multi-scale analysis.

20.1.3.2 Distributed fibre-optic sensor system measuring distributed loads over surfaces

The monitoring or identification of external loads applied to a structural system is one of the major targets of the SHM system. In many cases, the loads actually applied cannot be directly measured, but they can be estimated by using the information obtained from the sensors with the aid of some identification methods. Taking a fibre Bragg grating (FBG) sensor as an example, it has been shown that the principal strain/stress in the fibre cross section due to applied lateral loads induces birefringence effects that can cause the unique Bragg condition to break down the spectrum and produce two distinct Bragg wavelengths in accordance with the two polarisation modes (Wagreich et al. 1996). Since FBGs are frequently subjected to lateral loads during the service time of a civil structure, such loads can be estimated based on the understanding of the spectral response of birefringence in the structure (Murukeshan et al. 2000; Guemes and Menendez 2002; Botis et al. 2005; Wada et al. 2012). Nevertheless, research and application of distributed fibre-optic sensor systems for the distributed loading identification of a smart civil structure are very limited.

20.1.3.3 Distributed sensor networks

A distributed sensor network (DSN) has many autonomous sensor nodes spread out over a large area, in which each node is equipped with a processor, mission-specific sensors and short-range communications. The sensors can be deployed in various environments and the data gathered by sensors must be integrated to synthesise new information. Local interactions between sensor nodes allow them to reach global conclusions from the data provided by the sensor nodes. From a system perspective, once deployed, a DSN must organise itself, adapt to changes in the environment and nodes and continue to function reliably (Sastry and Iyengar 2005). It works in complementary, competitive or collaborative modes, using different data fusion and data mining techniques.

Since each DSN node possesses the ability to process and analyse measured information via the processor with which it is equipped, it has the potential to identify local damage. It is thus appealing to integrate the DSN with the multi-scale model updating and damage detection methods. The DSN can be placed at the critical joints or structural components of interest. The structural parameters around the DSN can be updated or identified by the local processor with the predefined algorithm at the substructure level.

The wireless sensor network is a typical successful DSN, as introduced in Chapter 5. The microprocessor in wireless sensors can be used for digital processing, analogue-to-digital or frequency-to-code conversions, calculations and interfacing functions, which can facilitate self-diagnostic, self-identification and self-adaptation (decision-making) functions. The properties of wireless communication and the merits of decentralised processing with each sensor node make such a DSN more attractive than the traditional SHM monitoring system.

20.1.4 Development of truly smart wind turbines

According to the description of a smart civil structure, an ideal smart wind turbine may be defined as a wind turbine that effectively integrates the SHM system, SVC system, SSR system and SEH system together, so that it has the smart functions of self-sensing, self-adaptation, self-diagnosis, self-repair and self-power to perform energy harvesting work in various environments and to preserve the safety and integrity of the structural system of the wind turbine.

Vertical axis wind turbines (VAWTs) were introduced in Chapter 17, with emphasis on their energy harvesting, pitch angle control and health monitoring. The aim of pitch angle control systems is to improve power efficiency and functionality of VAWTs. This is different from SVC systems, which have the goal of suppressing excessive vibration. For large wind turbines, structural vibration reduction is another class of scientific problems that has been actively investigated recently but has not been addressed in Chapter 17. Moreover, investigations concerning the self-repair of wind turbines are quite limited as compared with those on SHM systems, SVC systems and SEH systems of wind turbines. It is also of interest if the embedded FBGs can be used to identify wind loadings acting on the surface of the blades and to monitor the fatigue and ultimate strength of laminated composite blades of wind turbines.

The structural system of a wind turbine is relatively simple as compared with other large-scale civil structures, such as tall buildings or long-span bridges. Moreover, wind turbines are typically designed for power generation, which means that the function of energy harvesting is inherently exhibited in all wind turbines. Although the integration of the aforementioned four systems (SHM, SVC, SSR and SEH) for the establishment of a truly smart wind turbine has not yet been completed, smart wind turbines, as defined here, would probably be the first prototype of a truly smart civil structure.

20.1.5 Life-cycle management of smart civil structures

The framework of SHM-based LCM of conventional civil structures is presented and discussed in Chapter 19. Since truly smart civil structures are still at the conceptual level, the LCM of a truly smart civil structure may be a long way off. However, based on the research on LCMs of conventional civil structures and the definition of a truly smart civil structure, the profile of the LCM of a truly smart civil structure may be imagined.

Basically, life-cycle cost and structural performance are two basic but conflicting objectives in decision-making concerning LCMs of conventional civil structures. The ultimate goal of the LCM is to allocate minimum budgets to achieve the most desirable structural performance while considering various kinds of uncertainties and constraints. The LCMs of smart civil structures may be conducted with a similar goal in mind, but in a much more complicated process. This is mainly because the LCM is to be applied not only to the structure itself, but also to the four systems (SHM, SVC, SSR and SEH) and associated elements. For example, sensors should be optimally placed to consider the following factors: (1) from the economic perspective, fewer sensors should be used to achieve a cost-effective budget; but (2) from the viewpoint of structural performance, sufficient sensors are required for reliable assessment of structural performance and provision of sufficient information for structural vibration reduction and repair. Multi-objective optimisation techniques should therefore be developed to handle such situations.

20.2 PROSPECTS

Although there are many challenging issues ahead in the development of truly smart civil structures, the implementation of SHM and SVC technologies for practical applications is being widely considered and accepted by the civil engineering community. Rapid development in smart and bio-inspired materials, sensing technology, control technology, robotics technology, IT, computation simulation and LCM will eventually overcome the remaining challenging issues for the realisation of prototype truly smart civil structures. Mutual respect and effectively coordinated collaboration are required among professional engineers, academic researchers, government agencies, contractors, managers and owners to make smart civil structures a reality and a success. This is because all sectors will benefit from smart civil structures that can perfectly perform any intended functions in the surrounding environments and confidently preserve the safety and integrity of the structures under extreme events, while the LCM and operational cost of civil structures can be reduced.

Another important aspect is the education of the younger generation, who will finally make smart civil structures successful. It is most likely that some viewpoints introduced in this book will be overruled, and some methodologies improved, by our younger generation in the near future. This was the main purpose that compelled the authors to write this book.

REFERENCES

Armbrust, M., A. Fox, R. Griffith, et al. 2009. Above the clouds: A Berkeley view of cloud computing. Technical Report No. UCB/EECS-2009-28, Electrical Engineering and Computer Sciences, University of California at Berkeley.

Bai, Y.L., H.Y. Wang, M.F. Xia, and F.J. Ke. 2005. Statistical mesomechanics of solid, liking coupled multiple space and time scales. *Appl. Mech. Rev.*, 58: 372–88.

Botis, J., L. Humbert, F. Colpo, and P. Giaccari. 2005. Embedded fiber Bragg grating sensor for internal strain measurements in polymeric materials. *Opt. Laser. Eng.*, 43: 491–510.

Guemes, J.A., and J.M. Menendez. 2002. Response of Bragg grating fiberoptic sensors when embedded in composite laminates. *Compos. Sci. Technol.*, 62(7–8): 959–66.

Murayama, H., K. Kageyama, K. Uzawa, K. Ohara, and H. Igawa. 2012. Strain monitoring of a single-lap joint with embedded fiber-optic distributed sensors. *Struct. Health Monit.*, 11(3): 322–41.

Murayama, H., K. Ohara, N. Kanata, K. Kageyama, and H. Igawa. 2011. Strain monitoring and defect detection in welded joints by using fiber-optic distributed sensors with high spatial resolution. *E-J. Adv. Maintenance*, 2: 191–99.

Murayama, H., D. Wada, and H. Igawa. 2013. Structural health monitoring by using fiber-optic distributed strain sensors with high spatial resolution. *Photonic Sens.*, 3(4): 355–76.

Murukeshan, V.M., P.Y. Chan, L.S. Ong, and L.K. Seah. 2000. Cure monitoring of smart composites using fiber Bragg grating based embedded sensors. *Sens. Actuators, A*, 79: 153–61.

Phillips, B.M., and B.F. Spencer Jr. 2011. Model-based feedforward-feedback tracking control for real-time hybrid simulation. NSEL Report Series, Report No. NSEL-028. Newmark Structural Engineering Laboratory, University of Illinois at Urbana-Champaign.

Sastry, S., and S.S. Iyengar. 2005. A taxonomy of distributed sensor networks. In *Distributed Sensor Network*, eds. S.S. Iyengar and R.R. Brooks, 29–45. Boca Raton, FL, Chapman & Hall/CRC Press.

Sultan, N. 2010. Cloud computing for education: A new dawn? *Int. J. Inf. Manage.*, 30(2): 109–16.

Sun, B., and Z.X. Li. 2015. Adaptive image-based method for integrated multi-scale modeling of damage evolution in heterogeneous concrete. *Comput. Struct.*, 152: 66–81.

Sun, B., Y.L. Xu, and Z.S. Li. 2016. Multi-scale fatigue model and image-based simulation of collective short cracks evolution process. *Comput. Mater. Sci.*, 117: 24–32.

Wada, D., H. Murayama, and H. Igawa. 2012. Lateral load measurements based on a distributed sensing system of optical frequency-domain reflectometry using long-length fiber Bragg gratings. *J. Lightwave Technol.*, 30(14): 2337–44.

Wagreich, R.B., W.A. Atia, H. Singh, and J.S. Sirkis. 1996. Effects of diametric load on fibre Bragg gratings fabricated in low birefringent fibre. *Electron. Lett.*, 32: 1223–24.

Zapateiro, M., H.R. Karimi, N. Luo, and B.F. Spencer Jr. 2010. Real-time hybrid testing of semiactive control strategies for vibration reduction in a structure with MR damper. *Struct. Control Health Monit.*, 17(4): 427–51.

Index

ΔE-effect, 41
Accelerometers, 74–76
Acoustic emission method, 337–338
Active actuator, and SVC, 430–431
Active brace devices, 104–105
Active control algorithm, and SVC, 433–434
Active control devices, 98–105
    active brace devices, 104–105
    active mass dampers (AMDs), 103
    active tendon devices, 103–104
    magnetostrictive (MS) actuators, 100–101
    piezoelectric (PZT) actuator, 102
    pneumatic actuators, 99
    pulse generation devices, 105
    servovalve-controlled hydraulic
        actuators, 99
    shape memory alloy actuators, 99–100
Active mass dampers (AMDs), 103, 137
Active self-repairing, 634–636
Active tendon devices, 103–104
Actuator placement, and control device
        placement, 296–302
A/D, *see* Analogue-to-digital (A/D) conversion
Adaptive control, 273–274
Adhesive-filled brittle fibres, 640
AI, *see* Artificial intelligent (AI) control
Aliasing error, 129
AMDs, *see* Active mass dampers (AMDs)
American Society for Testing and Materials
        (ASTM), 335
American Society of Civil Engineers (ASCE), 15
Analogue-to-digital (A/D) conversion, 128
Anemometers, 61–63
Artificial intelligent (AI) control, 275–276
ASCE, *see* American Society of Civil Engineers
        (ASCE)
ASTM, *see* American Society for Testing and
        Materials (ASTM)
Austenitic phase, 31
Autogenic healing, 53

Backward sequential sensor placement (BSSP),
    218
Base isolation devices, 87–92

elastomeric bearings, 88–89
    friction pendulum bearings, 90–91
    high-damping rubber bearings (HDRBs), 90
    lead-plug bearings, 89–90
    types of, 91–92
Bayesian approach, and damage detection,
        372–373
Bayonet Neill–Concelman (BNC) connector,
        436
BBN, *see* Bolt, Beranek and Newman (BBN)
        vibration criteria
BeiDou Navigation Satellite System, 73
Bending actuator, 102
Berkeley Mote, 144, 146
Bio-inspiration
    self-healing materials, 52–53
    for sensing systems, 53
Biomimetics/biomimicry, *see* Bio-inspiration
Blade pitch angles control
    double disks multiple stream-tube (DMST)
        model, 592–596
    parking control algorithm, 602–603
    pitch control system, 603–607
    power maximisation control algorithm, 598
    rated power control algorithm, 598–602
    startup control algorithm, 596–598
BNC, *see* Bayonet Neill–Concelman (BNC)
        connector
Bolt, Beranek and Newman (BBN) vibration
        criteria, 425
Bolted joints, 23
BSSP, *see* Backward sequential sensor placement
        (BSSP)
CCD, *see* Charge-coupled-device (CCD) camera
CFD, *see* Computational fluid dynamics (CFD)
Charge-coupled-device (CCD) camera, 74
Chemical sensors, 80–82
CIB, *see* International Council for Research
        and Innovation in Building and
        Construction (CIB)
Civil structures
    description, 3–6
    design, construction and maintenance,
        13–15

dynamic, 185–186
LCM of (*see* SHM-based life-cycle management)
loading conditions and environments
    corrosion, 12
    dead loads, 6–8
    earthquake loads, 9–10
    highway loads, 10–11
    impact loads, 11–12
    live loads, 8
    railway loads, 11
    temperature effects, 12
    wind loads, 8–9
materials used in, 12–13
multi-scale modelling of, 158–159
Clipped optimal control law, 278–279
Closed-form solutions, of SVC
    application of, 398–406
    for dynamic characteristics, 394–395
    modal properties, 401–403
    seismic response, 395–398, 403–406
    selection of weighting matrices, 399–401
Coefficient matrix, 162–163
Collective placement of control devices and sensors
    case study, 318–328
        determination of configurations, 319–322
        and El-Centro ground excitation, 323–325
        Kobe ground excitation, 325–328
    optimal placement
        increment-based approach for, 312–313
        response reconstruction-based approach for, 313–318
    overview, 309–311
COMAC (COordinate MAC), 341
Computational fluid dynamics (CFD), 588–589
Computed tomography, 338
Constitutive modelling
    of piezoelectric materials, 37–38
    of shape memory effect, 34–35
Constraint equations
    application of unit force or moment, 162
    construction of coefficient matrix, 162–163
    linear, 160–161
    nonlinear
        in global coordinate system, 165–166
        in local coordinate system, 164–165
        transformation of coordinate systems, 164
    substructure and nodal force model, 161–162
Continuous wavelet transform (CWT), 346–347
Control device placement
    equivalent optimal parameters of, 294–296
    increment algorithms for, 289–292
    numerical example
        for actuator placement, 296–302

for passive damper placement, 302–303
    overview, 285–289
    suboptimal control gain and response, 292–294
Control devices and control systems
    active control devices, 98–105
        active brace devices, 104–105
        active mass dampers (AMDs), 103
        active tendon devices, 103–104
        magnetostrictive (MS) actuators, 100–101
        piezoelectric (PZT) actuator, 102
        pneumatic actuators, 99
        pulse generation devices, 105
        servovalve-controlled hydraulic actuators, 99
        shape memory alloy actuators, 99–100
    base isolation devices, 87–92
        elastomeric bearings, 88–89
        friction pendulum bearings, 90–91
        high-damping rubber bearings (HDRBs), 90
        lead-plug bearings, 89–90
        types of, 91–92
    configuration of, 113
    hybrid, 110–113
        hybrid base isolation devices, 111–112
        hybrid bracing control devices, 112–113
        hybrid mass dampers (HMDs), 110–111
    overview, 82
    passive energy dissipation devices, 92–98
        friction dampers, 93–94
        metallic dampers, 92–93
        tuned liquid column dampers (TLCDs), 98
        tuned mass dampers (TMDs), 97–98
        viscous fluid (VF) dampers, 95–96
    semi-active control devices, 105–110
        electrorheological (ER) dampers, 108–109
        magnetorheological (MR) dampers, 109–110
        semi-active friction dampers, 105–106
        semi-active hydraulic dampers, 106–107
        semi-active stiffness control devices, 108
        semi-active tuned liquid dampers, 107–108
Control force generation systems, 135
Controllability, and observability, 256–259
Coordinate systems
    global, 165–166
    local, 164–165
    transformation of, 164
Corrosion, 12
    sensors, 80–82
Criticality analysis, and LCM, 689–690
Criticality factors, and LCM, 686–687
Cut-off speed, 536
Cut-out wind speed, 602

CWT, *see* Continuous wavelet transform (CWT)
D/A, *see* Digital-to-analogue (D/A) converter
Damage location identification, 648
Damage prognosis (DP), 143
  and remaining life, 661–662
Damage spike, 347
Data acquisition and transmission system (DATS)
  configuration of, 126–127
  hardware of DAUs, 127–128
  network and communication, 128–129
  operation of, 129
Data acquisition control, 130
Data acquisition units (DAUs), 123
  hardware of, 127–128
Data collision, 542
Data fusion, 134–135
Data management systems, 141–142
  components and functions of, 141–142
  maintenance of, 142
Data mining, 131–132
Data processing systems (DPS)
  data acquisition control, 130
  data fusion, 134–135
  data mining, 131–132
  frequency domain analysis, 132–133
  signal post-processing and analysis, 131–135
  signal preprocessing, 130
  time–frequency domain analysis, 133–134
DAUs, *see* Data acquisition units (DAUs)
Dead loads, 6–8
Decision fusion, 134
Degrees of freedom (DOFs), 22, 185, 341, 492, 612
Design of experiment (DOE), 203
Digital-to-analogue (D/A) converter, 135
Discrete optical fibre sensors, 51
Discrete wavelet transform (DWT), 346
Displacement sensors
  charge-coupled-device (CCD) camera, 74
  global navigation satellite system (GNSS), 73–74
  level sensing station, 72
  linear variable differential transformer (LVDT), 70–72
  tilt-beams, 72–73
Distributed fibre-optic sensor system, 697–698
Distributed optical fibre sensors, 51
Distributed sensor network (DSN), 698
DMST, *see* Double disks multiple stream-tube (DMST) model
DOE, *see* Design of experiment (DOE)
DOFs, *see* Degrees of freedom (DOFs)
Double disks multiple stream-tube (DMST) model, 592–596
Double-layer passive isolation platform, 429
DP, *see* Damage prognosis (DP)
DPS, *see* Data processing systems (DPS)

DSN, *see* Distributed sensor network (DSN)
DSPACE, 415, 438
Dual-type sensor placement, 229–230
  experimental validation of, 228–231
    dual-type sensor placement, 229–230
    overhanging beam and FE model, 228–229
    reconstructed responses, 230–231
  numerical example, 225–228
  strain–displacement relationship, 220–221
  theoretical formulations, 222–225
Duff gauge sensor, 80
DWT, *see* Discrete wavelet transform (DWT)
Dynamic characteristics-based damage detection
  challenges in, 345–346
  comparison studies, 344–345
  flexibility changes, 344
  FRF changes, 342–343
  modal strain energy changes, 343–344
  mode shape changes, 341–342
  mode shape curvature changes, 343
  natural frequency changes, 340–341
Dynamic civil structures, 185–186
Dynamic response-based damage detection
  experimental investigation using EMD, 347–354
  statistical moment-based, 354–360
  using WT and HHT, 346–347
Earthquake Engineering Research Center (EERC), 92
Earthquake loads, 9–10
Eddy current method, 338
EERC, *see* Earthquake Engineering Research Center (EERC)
Effective independence (EfI), 218
EfI, *see* Effective independence (EfI)
EH, *see* Energy harvesting (EH)
Eigensystem realisation algorithm (ERA), 190
EKF, *see* Extended Kalman filter (EKF)
Elastomeric bearings, 88–89
El-Centro ground excitation, 323–325
Electric dipole, 40
Electromagnetic dampers
  analysis of structure-EMDEH systems, 565–569
  application to stay cables, 569–576
  circuits, 555–556
  damping characteristics, 562–563
  design and fabrication, 558–559
  efficiency of, 557–558
    input resistance of circuit, 560–562
  energy harvesting performance, 575–576
  experimental setup, 559–560, 570–572
  integrated MDOF systems, 568–569
  integrated SDOF systems, 565–567
  power flow in, 556–557
    integrated systems, 563–565
  vibration control performance, 572–575

for vibration damping, 554–555
Electromagnetic (EM) energy, 22
Electromagnetic TMD (EMTMD), 553
Electromechanical impedance (EMI), 39
Electromotive force (EMF), 554
Electrorheological and magnetorheological
    fluids, 43–49
    applications in smart civil structures, 48–49
    basic characteristics of, 44–45
    constitutive modelling of, 45–47
Electrorheological (ER) dampers, 108–109
EM, see Electromagnetic (EM) energy
EMA, see Experimental modal analysis (EMA)
EMD, see Empirical mode decomposition
    (EMD)
EMF, see Electromotive force (EMF)
EMI, see Electromechanical impedance (EMI)
Empirical mode decomposition (EMD), 134,
    191, 346–354
EMTMD, see Electromagnetic TMD
    (EMTMD)
Energy harvesting (EH)
    concept of, 535–536
    electromagnetic dampers
        analysis of structure-EMDEH systems,
            565–569
        application to stay cables, 569–576
        circuits, 555–556
        damping characteristics, 562–563
        design and fabrication, 558–559
        efficiency and input resistance of circuit,
            560–562
        efficiency of, 557–558
        energy harvesting performance, 575–576
        experimental setup, 559–560, 570–572
        integrated MDOF systems, 568–569
        integrated SDOF systems, 565–567
        power flow in, 556–557
        power flow in integrated systems,
            563–565
        vibration control performance, 572–575
        for vibration damping, 554–555
    integrated vibration control and
        overview, 552–553
    overview, 535
    power requirement and management,
        542–543
    and power supply, 146–148
    radio frequency, 539–540
    sensors and SHM sensory systems, 541–545
        enhanced rotational energy harvester
            (EREH), 545–548
        parametric frequency-increased generator
            (PFIG), 545, 548–550
        strain energy sensor (SES), 550–552
    solar, 537–538
    thermal, 540–541
    vibrations, 538–539
    wind, 536–537

Engineered self-healing materials, 632–633
Epoxy-cement composites, 636–637
Equation of motion, 239–240
    and SVC, 431–433
    and synthesis, of SHM and SVC, 453
ER, see Electrorheological (ER) dampers
ERA, see Eigensystem realisation algorithm
    (ERA)
Evolutionary multi-scale modelling, 696
Excitation power, 568
Experimental modal analysis (EMA), 360
Extended Kalman filter (EKF), 491
Fast Fourier transform (FFT), 191
Fatigue analysis, 612–618
Fatigue damage prognosis, 682–683
Fatigue reliability, 683–685
FBG, see Fibre Bragg grating (FBG)
FE, see Finite element (FE) model
Feedback-feedforward control, 123
Fibre Bragg grating (FBG), 51, 77
Fibreglass reinforced plastic (FRP), 612
Fibre-optic sensors, 76–78
Fibre-reinforced polymer (FRP), 13
Field measurements, of VAWTs, 590–591
FIM, see Fisher information matrix (FIM)
Finite element (FE) model, 155
    frame structure, 169–170
    and LCM, 663–667
    linear beam-to-plate coupling, 166–169
    nonlinear analysis of beam–shell coupling,
        170–172
    overhanging beam and, 228–229
    overview, 155–157
    response sensitivity-based, 362–365
    of Tsing Ma Bridge model, 237–238
Fisher information matrix (FIM), 218
FLC, see Fuzzy logic control (FLC)
Flexibility, and damage detection, 344
Flexible-to-flexible contact problem, 175
Flexitensional piezoelectric actuator (FPA), 436
Foil strain gauges, 69
The Foothill Communities Law and Justice
    Centre, 389
Force identification, 192
Forward sequential sensor placement (FSSP),
    218
FPA, see Flexitensional piezoelectric actuator
    (FPA)
Frame structure, 169–170
Frequency domain analysis, 132–133
Frequency response function (FRF)
    and damage detection, 342–343
    modal analysis and, 186–188
    synthesis, of SHM and SVC
        with full excitation, 453–456
        with single excitation, 456–459
        structural damage detection, 465
Frequency-time domain, 190–192
FRF, see Frequency response function (FRF)

Friction dampers, 93–94
Friction pendulum bearings, 90–91
FRP, *see* Fibreglass reinforced plastic (FRP); Fibre-reinforced polymer (FRP)
FSSP, *see* Forward sequential sensor placement (FSSP)
Full-state feedback, 140
Fuzzy logic control (FLC), 275
GAC, *see* Genetic algorithm control (GAC)
Galileo system of European Union, 73
GAs, *see* Genetic algorithms (GAs)
Gaussian distribution, 354
Gauss–Newton method, 198
Genetic algorithm control (GAC), 275
Genetic algorithms (GAs), 198–199, 218, 275
GFRP, *see* Glass fibre-reinforced plastic (GFRP)
Glass fibre-reinforced plastic (GFRP), 591
Global coordinate system, 165–166
Global-feedback control strategy
    semi-active control by PFDs, 410–411
    synthesis, of SHM and SVC, 463–465
Global navigation satellite system (GNSS), 73–74
Global rational fraction polynomial method, 189
Global residuals, 200
GLONASS, 73
GNSS, *see* Global navigation satellite system (GNSS)
GPS-OSIS, 669
H₂ and H∞ control, 269–273
    H₂ control algorithm, 271–272
    H∞ control algorithm, 272–273
    and transfer functions, 270–271
Hachinohe earthquake, 319
HAWT, *see* Horizontalaxis wind turbine (HAWT)
HDRBs, *see* High-damping rubber bearings (HDRBs)
Health assessment, and structural damage detection, 660
HHS, *see* Hilbert–Huang spectrum (HHS)
HHT, *see* Hilbert–Huang transform (HHT)
High-damping rubber bearings (HDRBs), 88
High-order modes, 50
Highway loading, 10–11
    and LCM, 669–673
Hilbert–Huang spectrum (HHS), 191
Hilbert–Huang transform (HHT), 133–134, 190, 346–347
Hilbert spectral analysis (HSA), 346
HMDs, *see* Hybrid mass dampers (HMDs)
Horizontalaxis wind turbine (HAWT), 536
HSA, *see* Hilbert spectral analysis (HSA)
Hybrid base isolation devices, 111–112
Hybrid bracing control devices, 112–113
Hybrid control devices, 110–113
    hybrid base isolation devices, 111–112
    hybrid bracing control devices, 112–113

hybrid mass dampers (HMDs), 110–111
Hybrid mass dampers (HMDs), 110–111
Hydration, 13
IDPCS, *see* Integrated data processing and controller systems (IDPCS)
IEEE, *see* Institute of Electrical and Electronics Engineers (IEEE)
IMF, *see* Intrinsic mode functions (IMF)
IMFs, *see* Intrinsic mode functions (IMFs)
Impact-echo/impulse-response methods, 336–337
Impact loads, 11–12
Impedance-based SHM system, 642–643
Impedance testing method, 337
Impulse response functions (IRFs), 189
Impulse-response method, 337
Inclinometers, *see* Tilt-beams
Increment algorithms, for control device placement, 289–292
Increment-based approach, for optimal placement, 312–313
Independent modal space control, 264–266
Information-passing multiscale method, 173
Infrared thermographic method, 339
Input-output methods, 189
Inspection
    and LCM, 690–692
    maintenance and repair, 660–661
Institute of Electrical and Electronics Engineers (IEEE), 37
Integrated data processing and controller systems (IDPCS), 140–141
Integrated impedance-based SHM and SR system, 643
Integrated MDOF systems, 568–569
Integrated SDOF systems, 565–567
Intelligent control systems, 275
Intel Mote, 144, 146
International Council for Research and Innovation in Building and Construction (CIB), 88
Intrinsic mode functions (IMFs), 134, 191, 347
IRFs, *see* Impulse response functions (IRFs)
Jindo Bridge, 22, 144, 145
Joule effect, 40
Kalman filter, 231, 232, 316, 322
Knowledge discovery, 131
Kriging meta-model, 202–203
    and QPRS method, 210–211
KYOWA ASQ-1BL accelerometers, 438
Latin hypercube design, 203
LCM, *see* Life-cycle management (LCM); SHM-based life-cycle management
Lead-plug bearings, 89–90
Level sensing station, 72
Levenberg–Marquardt algorithm, 198
Life-cycle management (LCM), 17
Light propagation, in optical fibres, 50–51
Linear beam-to-plate coupling, 166–169

Linear constraint equations, 160–161
Linear optimal control, 261–264
Linear piezoelectric actuator (LPA), 438
Linear quadratic Gaussian (LQG) control,
    263–264
Linear quadratic regulator (LQR) control,
    262–263
Linear time-invariant (LTI) system, 255
Linear variable differential transformer (LVDT),
    70–72, 605
Linear variable displacement transducer
    (LVDT), 415
Live loads, 8
Load cells
    description, 65–66
    weigh-in-motion (WIM), 67–68
Load identification, 647–648
Loading conditions and environments
    corrosion, 12
    dead loads, 6–8
    earthquake loads, 9–10
    highway loads, 10–11
    impact loads, 11–12
    live loads, 8
    railway loads, 11
    temperature effects, 12
    wind loads, 8–9
Local coordinate system, 164–165
Local-feedback control strategy
    semi-active control by PFDs, 411–412
    synthesis, of SHM and SVC, 461–463
Long-span suspension bridge rating system,
    685–686
Loose bolted joints, 643
Low strain integrity testing, 337
LPA, see Linear piezoelectric actuator (LPA)
LQG, see Linear quadratic Gaussian (LQG)
    control
LQR, see Linear quadratic regulator (LQR)
    control
LTI, see Linear time-invariant (LTI) system
Lumped mass, 156
LVDT, see Linear variable differential
    transformer (LVDT); Linear variable
    displacement transducer (LVDT)
Lyapunov direct method, 268–269
MAC, see Modal assurance criterion (MAC)
Machine constant, 554
Magnetic induction dampers, 22
Magnetorheological (MR) dampers, 109–110,
    389
    and semi-active control
        experimental arrangement, 420–421
        multilevel logic control algorithm,
            421–423
        and seismic control, 423–425
Magnetostrictive (MS) actuators, 100–101
Martensitic phase, 31
MATLAB/Simulink program, 438

Matteuci effect, 41
MDOF, see Multi-degrees of freedom (MDOF)
Mean square error (MSE), 203
Measurement noises, 393
MEMS, see Micro-electrical mechanical
    systems (MEMS)
Metallic dampers, 92–93
Micro-electrical mechanical systems (MEMS),
    54
Microvibration control, and SVC, 440–444
MIMO, see Multiple-input-multiple-output
    (MIMO)
Mixed-dimensional finite element coupling,
    158–159
    and frame structure, 169–170
    linear beam-to-plate coupling, 166–169
    nonlinear analysis of beam-shell coupling,
        170–172
Mobility, 337
Modal analysis, and FRF, 186–188
Modal assurance criterion (MAC), 193, 341
Modal identification
    in frequency domain/time domain, 189–190
    in frequency-time domain, 190–192
Modal properties, and SVC, 401–403
Modal strain energy, and damage detection,
    343–344
Modal tests, and model updating, 239
Model updating
    and LCM, 667–668
    and modal tests, 239
    multi-objective optimisation for, 205–207
    and multi-scale modelling, 659
    objective functions and constraints,
        194–195
    optimisation algorithm, 197–199
    overview, 192–193
    parameters, 196–197
    results and discussions, 207–209
    and synthesis, of SHM and SVC, 453–459
Mode shape
    curvature, and damage detection, 343
    and damage detection, 341–342
Moment-resisting frame (MRF), 628
Monte Carlo simulation, 373–374
Movable loads, 8
Moving loads, 8
MPC, see Multi-point constraint (MPC) method
MR, see Magnetorheological (MR) dampers
MRF, see Moment-resisting frame (MRF)
MS, see Magnetostrictive (MS) actuators
MSE, see Mean square error (MSE)
MTLCDs, see Multiple tuned liquid column
    dampers (MTLCDs)
Multi-degrees of freedom (MDOF), 312, 354,
    450, 563
    integrated systems, 568–569
Multi-level logic control algorithm, 421–423
Multi-mode-fibre, 50

Multi-objective hybrid control, and SVC, 425–444
  and active actuator, 430–431
  active control algorithm, 433–434
  and equation of motion, 431–433
  experimental arrangement, 435–438
  experimental investigation of, 434–444
  instrumentation, 438–439
  microvibration control, 440–444
  overview, 425–428
  seismic response control, 440
  single-layer and double-layer passive isolation platform, 429
  system identification, 439–440
Multi-objective optimisation, 200–202
  Kriging meta-model for, 202–203
  for model updating, 205–207
Multiple-input-multiple-output (MIMO), 189
Multiple tuned liquid column dampers (MTLCDs), 98
Multi-point constraint (MPC) method, 158
Multi-scale damage detection, 696–697
  experimental studies, 365–369
  RBF network for response reconstruction, 360–362
  response sensitivity-based FE model, 362–365
Multi-scale modelling
  of large civil structures, 158–159
  and model updating, 659
  of smart civil structures, 695–696
  of transmission tower, 174–175
    overview, 172–173
    physical model of, 173–174
    validation of, 175–181
Multi-scale model updating
  implementation procedure of, 203–205
  multi-objective optimisation, 200–202
    Kriging meta-model for, 202–203
    for model updating, 205–207
  objective functions and parameters for, 199–200
  and transmission tower, 203–205
    Kriging *vs.* QPSR method, 210–211
    model updating results and discussions, 207–209
Multi-scale response reconstruction
  equation of motion, 239–240
  implementation procedure, 241–242
  mode selection, 240–241
Multi-type sensor placement method
  and multi-scale response reconstruction
    equation of motion, 239–240
    implementation procedure, 241–242
    mode selection, 240–241
  state-space equation, 232
  suspension bridge model
    experimental validation, 246–248

    FE model of Tsing Ma Bridge model, 237–238
    modal tests and model updating, 239
    numerical analysis and results, 242–246
    physical bridge model, 236–237
    theoretical formulations, 232–236
Nanoparticles, 54
National Center for Atmospheric Research (NCAR), 64
Natural frequency, and damage detection, 340–341
Natural self-healing materials, 632
Naval Ordnance Laboratory (NOL), 40
NDT, *see* Non-destructive testing (NDT)
Network, and communication, 128–129
Neural network control (NNC), 275
Neuro-fuzzy control (NFC), 276
NFC, *see* Neuro-fuzzy control (NFC)
NFP, *see* Number of frequency points (NFP)
NNC, *see* Neural network control (NNC)
NOL, *see* Naval Ordnance Laboratory (NOL)
Non-contact sensors, 78–80
Nondestructive evaluation (NDE), *see* Non-destructive testing (NDT)
Nondestructive inspection (NDI), *see* Non-destructive testing (NDT)
Non-destructive testing (NDT), 333, 335–339
  acoustic emission method, 337–338
  Eddy current method, 338
  impact-echo/impulse-response methods, 336–337
  infrared thermographic method, 339
  radiographic method, 338
  ultrasonic pulse velocity method, 336
Non-inferiority, 201
Nonlinear analysis, of beam-shell coupling, 170–172
Nonlinear constraint equations
  in global coordinate system, 165–166
  in local coordinate system, 164–165
  transformation of coordinate systems, 164
Nonlinear mix-dimensional finite element coupling, 695–696
Null position, 70
Number of frequency points (NFP), 471
Observability, and controllability, 256–259
ODGV, *see* Omni-direction-guide-vane (ODGV)
Omni-direction-guide-vane (ODGV), 589
Optical fibres
  basic characteristics of, 49–50
  light propagation in, 50–51
  sensors and applications in smart civil structures, 51–52
Optimal bang-bang control law, 277–278
Optimal control algorithms, 140
Optimisation algorithm, 197–199
OREH, *see* Original rotational energy harvester (OREH)

Original rotational energy harvester (OREH), 547
Output feedback, pole assignment by, 261
Output only methods, 189
Overhanging beam, and FE model, 228–229
Parametric frequency-increased generator (PFIG), 545, 548–550
Parking control algorithm, 602–603
Partially smart structures, 19
Particle swarm optimisation (PSO), 218
Passive controller system, 135
Passive damper placement, 302–303
Passive energy dissipation devices, 92–98
    friction dampers, 93–94
    metallic dampers, 92–93
    tuned liquid column dampers (TLCDs), 98
    tuned mass dampers (TMDs), 97–98
    viscous fluid (VF) dampers, 95–96
Passive friction dampers (PFDs), 406
Passive self-repairing, 633–634
PDFs, *see* Probability density functions (PDFs)
Perturbation approach, 370–372
PFDs, *see* Passive friction dampers (PFDs)
Physical bridge model, 236–237
Piezo-driven variable friction damper (PVFD), 414–416
Piezoelectric (PZT) actuator, 102
Piezoelectric materials, 36–40
    applications in smart civil structures, 39–40
    basic characteristics of, 36–37
    constitutive modelling of, 37–38
PIMS, *see* Portable inspection and maintenance system (PIMS)
Pitch control system, 603–607
Pneumatic actuators, 99
Pole assignment, 259–261
    by output feedback, 261
    by state feedback, 260–261
Poling, 36
Portable inspection and maintenance system (PIMS), 123
Post-tensioned energy dissipating (PTED), 629
Power maximisation control algorithm, 598
Power supply, and energy harvesting, 146–148
Pressure transducers, 63
Probability density functions (PDFs), 354
Processors and processing systems
    configuration of integrated SHM and SVC, 125–126
    configuration of SHM, 121–123
    configuration of SVC, 123–125
    controller systems for SVC, 135–140
    data acquisition and transmission system (DATS)
        configuration of, 126–127
        hardware of DAUs, 127–128
        network and communication, 128–129
        operation of, 129
    data management systems, 141–142

        components and functions of, 141–142
        maintenance of, 142
    data processing systems for SHM
        data acquisition control, 130
        data fusion, 134–135
        data mining, 131–132
        frequency domain analysis, 132–133
        signal post-processing and analysis, 131–135
        signal preprocessing, 130
        time–frequency domain analysis, 133–134
    integrated data processing and controller systems (IDPCS), 140–141
    overview, 121
    power supply and energy harvesting, 146–148
    and structural health evaluation system (SHES), 142–143
    wireless sensors and sensory systems
        architectures and features of, 144–146
        challenges in, 146
        overview, 144
PSO, *see* Particle swarm optimisation (PSO)
PT-based structural self-centring systems, 628–630
PTED, *see* Post-tensioned energy dissipating (PTED)
Pulse generation devices, 105
Pulse velocity method, 336
PVFD, *see* Piezo-driven variable friction damper (PVFD)
PZT, *see* Piezoelectric (PZT) actuator
QPRS method, and Kriging meta-model, 210–211
Radio-frequency (RF) energy, 147
Radiographic method, 338
Railway loading, 11
    and LCM, 673–674
Random decrement technique (RDT), 191
Rated power control algorithm, 598–602
Rated wind speed, 598
Rayleigh damping, 156, 187, 318, 342, 368, 378, 453–454, 466, 475, 501, 504, 509, 513, 524, 525, 612
RBF network, for response reconstruction, 360–362
RDT, *see* Random decrement technique (RDT)
Real-time interface (RTI), 415
Real-time kinematic (RTK), 73
Reciprocity, 188
Reconstructed responses, 230–231
Relative percentage error (RPE), 230
Response reconstruction-based approach, for optimal placement, 313–318
Response sensitivity-based FE model, 362–365
Reynolds number, 16, 589
RF, *see* Radio-frequency (RF) energy
Rigid-to-flexible contact problem, 175

RMS, *see* Root mean square (RMS)
Root mean square (RMS), 322, 502
RPE, *see* Relative percentage error (RPE)
RTI, *see* Real-time interface (RTI)
RTK, *see* Real-time kinematic (RTK)
SDOF, *see* Single-degree-of-freedom (SDOF)
Seebeck effect, 68, 148
SEH, *see* Structural energy harvesting (SEH)
Seismic forces, 9
Seismic response control
    and MR dampers, 423–425
    and SVC, 395–398, 440
    and synthesis, of SHM and SVC, 475–480
Seismic sensors, 64–65
Self-diagnosis and self-repairing active
       tensegrity structures, 648–649
    damage location identification, 648
    description, 645
    experimental investigation and results,
      649–650
    indicators for response changes, 645–647
    load identification, 647–648
Self-healing materials, 52–53
    engineered, 632–633
    natural, 632
Self-repair degree (SRD), 637–638
Self-repairing concrete
    beams
      experimental design, 640–641
      experimental results and analysis,
        641–642
      super-elastic SMAs and adhesive-filled
        brittle fibres, 640
    epoxy-cement composites without hardener,
      636–637
    method of investigation, 637
    results and discussions, 638–639
    self-repair degree (SRD), 637–638
Self-repairing steel joints
    experimental investigation and results,
      643–644
    impedance-based SHM system, 642–643
    integrated impedance-based SHM and SR
      system, 643
    loose bolted joints, 643
Semi-active control, 276–279
    clipped optimal control law, 278–279
    optimal bang-bang control law, 277–278
    by PFDs, 406–420
      global-feedback control strategy,
        410–411
      local-feedback control strategy, 411–412
      and piezo-driven variable friction damper
        (PVFD), 414–416
      and variable friction dampers, 408–410,
        412–420
    simple bang-bang control law, 277
    using MR dampers
      experimental arrangement, 420–421

multilevel logic control algorithm,
    421–423
    and seismic control, 423–425
    SHM and SVC synthesis, 528–529
Semi-active control devices, 105–110
    electrorheological (ER) dampers, 108–109
    magnetorheological (MR) dampers,
      109–110
    semi-active friction dampers, 105–106
    semi-active hydraulic dampers, 106–107
    semi-active stiffness control devices, 108
    semi-active tuned liquid dampers, 107–108
Semi-active friction dampers, 105–106
    synthesis, of SHM and SVC, 451–452,
      460–465
Semi-active hydraulic dampers, 106–107
Semi-active stiffness control devices, 108
Semi-active tuned liquid dampers, 107–108
Sensing/sensors/sensory systems
    accelerometers, 74–76
    bio-inspiration for, 53
    chemical and corrosion sensors, 80–82
    displacement sensors
      charge-coupled-device (CCD) camera, 74
      global navigation satellite system (GNSS),
        73–74
      level sensing station, 72
      linear variable differential transformer
        (LVDT), 70–72
      tilt-beams, 72–73
    fibre-optic sensors, 76–78
    load cells
      description, 65–66
      weigh-in-motion (WIM), 67–68
    non-contact sensors, 78–80
    overview, 61
    performance, 82–83
    seismic sensors, 64–65
    strain gauges
      foil strain gauges, 69
      vibrating wire strain gauge, 69–70
    thermometers, 68
    weather stations, 80
    wind sensors
      anemometers, 61–63
      pressure transducers, 63
      wind profile measurements, 63–64
Sensor placement methods
    dual-type
      experimental validation of, 228–231
      numerical example, 225–228
      strain–displacement relationship,
        220–221
      theoretical formulations, 222–225
    overview, 217–219
Servovalve-controlled hydraulic actuators, 99
SES, *see* Strain energy sensor (SES)
Shape memory alloys (SMAs), 23
    actuators, 99–100

applications in smart civil structures, 35–36
basic characteristics of, 31–34
constitutive modelling of shape memory effect, 34–35
Shape memory effect, 34–35
SHES, *see* Structural health evaluation system (SHES)
SHM, *see* Structural health monitoring (SHM)
SHM-based life-cycle management
criticality analysis, 689–690
criticality factors, 686–687
damage prognosis and remaining life, 661–662
engineering approach, 680–682
estimation of future loadings, 661
fatigue damage prognosis, 682–683
fatigue reliability, 683–685
finite element model, 663–667
highway loading, 669–673
inspection, 690–692
maintenance and repair, 660–661
long-span suspension bridge rating system, 685–686
model updating, 667–668
multi-scale modelling and model updating, 659
multi-type sensor placements, 659–660
overview, 657
railway loading, 673–674
strategies, 662–663
stress analysis framework, 679–680
structural damage detection and health assessment, 660
temperature loading, 677–679
and Tsing Ma Bridge, 663
GPS-OSIS, 669
wind and structural health monitoring system (WASHMS), 669
verification of framework, 680
vulnerability analysis, 690
vulnerability factors, 688–689
wind characteristics and models, 674–677
Short-time Fourier transform (STFT), 133–134, 190
Sifting process, 347
Signal post-processing and analysis, 131–135
Signal preprocessing, 130
SIMO, *see* Single-input-multiple-output (SIMO)
Simple bang-bang control law, 277
Single-degree-of-freedom (SDOF), 354
integrated systems, 565–567
Single-input-multiple-output (SIMO), 189
Single-input-single-output (SISO), 189
Single-layer passive isolation platform, 429
Single-mode-fibre, 50
SISO, *see* Single-input-single-output (SISO)
Skin effect, 539
Sliding mode control (SMC)

design of controllers using Lyapunov direct method, 268–269
design of sliding surface, 266–268
Sliding surface design, and SMC, 266–268
SMA-based structural self-centring systems, 630–631
Smart civil structures
definition of, 19–20
distributed fibre-optic sensor system, 697–698
distributed sensor network (DSN), 698
ER and MR fluids applications in, 48–49
evolutionary multi-scale modelling, 696
historical developments of, 20–23
life-cycle management of, 699
magnetostrictive materials applications in, 42–43
multi-scale damage detection with substructure techniques, 696–697
multi-scale modelling of, 695–696
nonlinear mix-dimensional finite element coupling, 695–696
optical fibre sensors and applications in, 51–52
piezoelectric materials applications in, 39–40
and SHM, 16–17
and smart wind turbines, 698–699
SMAs applications in, 35–36
and SVC, 17–19
Smart materials
and bio-inspiration
self-healing materials, 52–53
for sensing systems, 53
electrorheological and magnetorheological fluids, 43–49
applications in smart civil structures, 48–49
basic characteristics of, 44–45
constitutive modelling of, 45–47
magnetostrictive materials
applications in smart civil structures, 42–43
basic characteristics of, 40–41
constitutive modelling of, 41–42
nanomaterials, 53–54
optical fibres
basic characteristics of, 49–50
light propagation in, 50–51
sensors and applications in smart civil structures, 51–52
overview, 31
piezoelectric materials, 36–40
applications in smart civil structures, 39–40
basic characteristics of, 36–37
constitutive modelling of, 37–38
shape memory alloys (SMAs)
applications in smart civil structures, 35–36

basic characteristics of, 31–34
constitutive modelling of shape memory
    effect, 34–35
Smart vertical-axis wind turbines, 622–623
Smart wind turbines, 698–699
SMAs, *see* Shape memory alloys (SMAs)
SRD, *see* Self-repair degree (SRD)
SSR, *see* Structural self-repairing (SSR)
Stability, 255–256
Startup control algorithm, 596–598
State feedback, pole assignment by, 260–261
State-space equation, 232
Statistical moment-based damage detection,
    354–360
Statistical pattern recognition, and damage
    detection, 373
Stay cables, and electromagnetic dampers,
    569–576
STFT, *see* Short-time Fourier transform (STFT)
Stiffness identification, 485–487
Stochastic damage detection method, 374–380
Strain-displacement relationship, 220–221
Strain energy sensor (SES), 550–552
Strain gauges
    foil strain gauges, 69
    vibrating wire strain gauge, 69–70
Stress analysis, and LCM, 679–680
Structural control theory
    adaptive control, 273–274
    artificial intelligent (AI) control, 275–276
    controllability and observability, 256–259
    H_2 and $H\infty$ control, 269–273
        H_2 control algorithm, 271–272
        $H\infty$ control algorithm, 272–273
        and transfer functions, 270–271
    independent modal space control, 264–266
    linear optimal control, 261–264
        linear quadratic Gaussian (LQG) control,
            263–264
        linear quadratic regulator (LQR) control,
            262–263
    overview, 255
    pole assignment, 259–261
        by output feedback, 261
        by state feedback, 260–261
    semi-active control, 276–279
        clipped optimal control law, 278–279
        optimal bang-bang control law, 277–278
        simple bang-bang control law, 277
    sliding mode control (SMC)
        design of controllers using Lyapunov
            direct method, 268–269
        design of sliding surface, 266–268
    stability, 255–256
Structural damage detection
    Bayesian approach, 372–373
    dynamic characteristics-based
        challenges in, 345–346
        comparison studies, 344–345

flexibility changes, 344
FRF changes, 342–343
modal strain energy changes, 343–344
mode shape changes, 341–342
mode shape curvature changes, 343
natural frequency changes, 340–341
dynamic response-based
    experimental investigation using EMD,
        347–354
    statistical moment-based, 354–360
    using WT and HHT, 346–347
FRF-based method, 465
and health assessment, 660
under Kobe earthquake, 509–512
Monte Carlo simulation, 373–374
multi-scale method
    experimental studies, 365–369
    RBF network for response
        reconstruction, 360–362
    response sensitivity-based FE model,
        362–365
non-destructive testing (NDT), 333,
    335–339
    acoustic emission method, 337–338
    Eddy current method, 338
    impact-echo/impulse-response methods,
        336–337
    infrared thermographic method, 339
    radiographic method, 338
    ultrasonic pulse velocity method, 336
under Northridge earthquake, 512–515
overview, 333–335
perturbation approach, 370–372
statistical pattern recognition, 373
stochastic method, 374–380
synthesis, of SHM and SVC, 487–488
    using full excitation, 473
    using single excitation, 473–475
uncertainties in, 369–370
Structural energy harvesting (SEH), 19
Structural health evaluation system (SHES),
    123, 142–143
Structural health monitoring (SHM), 12
    configuration of, 121–123
    data processing systems for
        data acquisition control, 130
        data fusion, 134–135
        data mining, 131–132
        frequency domain analysis, 132–133
        signal post-processing and analysis,
            131–135
        signal preprocessing, 130
        time–frequency domain analysis,
            133–134
    and smart civil structures, 16–17
    synthesis of
        comparison with previous study,
            471–472
        current research in, 450–451

damage detection using full excitation,
473
damage detection using single excitation,
473–475
and equation of motion, 453
experimental setup, 480–483
FRF-based method with full excitation,
453–456
FRF-based method with single excitation,
456–459
FRF-based structural damage detection,
465
global feedback control strategy,
463–465
local feedback control strategy, 461–463
measured FRFs, 483–485
and model updating, 453–459
number of frequency points (NFP), 471
number of natural frequencies, 468–471
numerical example building, 465–467
overview, 449
and seismic response control, 475–480
selection of additional stiffness, 467–468
semi-active friction dampers, 451–452
stiffness identification, 485–487
structural damage detection, 487–488
and system identification, 453–459
using semi-active friction dampers,
460–465
of vertical-axis wind turbines (VAWTs),
619–622
fatigue analysis, 612–618
framework, 609–612
ultimate strength analysis, 615, 618–619
wind loads on, 608–609
Structural self-centring systems
PT-based systems, 628–630
SMA-based systems, 630–631
Structural self-rehabilitation
concept of, 627–628
overview, 627
self-diagnosis and self-repairing active
tensegrity structures, 648–649
damage location Identification, 648
description, 645
experimental investigation and results,
649–650
indicators for response changes, 645–647
load identification, 647–648
self-healing materials
engineered, 632–633
natural, 632
self-repairing concrete
epoxy-cement composites without
hardener, 636–637
method of investigation, 637
results and discussions, 638–639
self-repair degree (SRD), 637–638
self-repairing concrete beams

experimental design, 640–641
experimental results and analysis,
641–642
super-elastic SMAs and adhesive-filled
brittle fibres, 640
self-repairing steel joints
experimental investigation and results,
643–644
impedance-based SHM system, 642–643
integrated impedance-based SHM and SR
system, 643
loose bolted joints, 643
structural self-centring systems
PT-based systems, 628–630
SMA-based systems, 630–631
structural self-repairing materials and
systems
active self-repairing, 634–636
passive self-repairing, 633–634
Structural self-repairing (SSR), 19
active self-repairing, 634–636
passive self-repairing, 633–634
Structural vibration control (SVC), 17–19
closed-form solutions
application of, 398–406
for dynamic characteristics, 394–395
modal properties, 401–403
seismic response, 395–398, 403–406
selection of weighting matrices, 399–401
configuration of, 123–125
controller systems for, 135–140
equations of motion, 391–393
full-scale implementations, 389–391
LQG controller, 393–394
and multi-objective hybrid control of
hightech equipment, 425–444
and active actuator, 430–431
active control algorithm, 433–434
and equation of motion, 431–433
experimental arrangement, 435–438
experimental investigation of, 434–444
instrumentation, 438–439
microvibration control, 440–444
overview, 425–428
seismic response control, 440
single-layer and double-layer passive
isolation platform, 429
system identification, 439–440
semi-active control by PFDs, 406–420
global-feedback control strategy, 410–411
local-feedback control strategy, 411–412
and piezo-driven variable friction damper
(PVFD), 414–416
and variable friction dampers, 408–410,
412–420
semi-active control using MR dampers
experimental arrangement, 420–421
multilevel logic control algorithm,
421–423

and seismic control, 423–425
synthesis of
  comparison with previous study, 471–472
  current research in, 450–451
  damage detection using full excitation,
    473
  damage detection using single excitation,
    473–475
  and equation of motion, 453
  experimental setup, 480–483
  FRF-based method with full excitation,
    453–456
  FRF-based method with single excitation,
    456–459
  FRF-based structural damage detection,
    465
  global feedback control strategy,
    463–465
  local feedback control strategy, 461–463
  measured FRFs, 483–485
  and model updating, 453–459
  number of frequency points (NFP), 471
  number of natural frequencies, 468–471
  numerical example building, 465–467
  overview, 449
  and seismic response control, 475–480
  selection of additional stiffness, 467–468
  semi-active friction dampers, 451–452
  stiffness identification, 485–487
  structural damage detection, 487–488
  and system identification, 453–459
  using semi-active friction dampers,
    460–465
Structure-EMDEH systems, 565–569
Suboptimal control gain and response, 292–294
Suboptimal controllers, 272
Substructure and nodal force model, 161–162
Superelasticity, 32
Super-elastic SMAs, 640
Suspension bridge model
  experimental validation, 246–248
  FE model of Tsing Ma Bridge model,
    237–238
  modal tests and model updating, 239
  numerical analysis and results, 242–246
  physical bridge model, 236–237
SVC, see Structural vibration control (SVC)
Synthesis, of SHM and SVC (frequency domain)
  comparison with previous study, 471–472
  current research in, 450–451
  damage detection
    using full excitation, 473
    using single excitation, 473–475
  and equation of motion, 453
  experimental setup, 480–483
  FRF-based method
    with full excitation, 453–456
    with single excitation, 456–459
    structural damage detection, 465

global feedback control strategy, 463–465
local feedback control strategy, 461–463
measured FRFs, 483–485
and model updating, 453–459
number of frequency points (NFP), 471
number of natural frequencies, 468–471
numerical example building, 465–467
overview, 449
and seismic response control, 475–480
selection of additional stiffness, 467–468
semi-active friction dampers, 451–452
stiffness identification, 485–487
structural damage detection, 487–488
and system identification, 453–459
using semi-active friction dampers, 460–465
Synthesis, of SHM and SVC (time domain)
  overview, 491
  time-invariant parameters
    constant structural parameters and
      excitations, 493–497
    damage detection, 509–515
    damage scenarios, 506–507
    experimental setup, 504–506
    identification, 497–499
    identification and control algorithms,
      507–509
    implementation procedure of, 499–500
    numerical investigation of, 500–504
    overview, 492
  time-varying parameters
    accuracy and excitation identification,
      526–528
    comparisons, 529–532
    description of example building
      structure, 524–525
    identification, 520–523
    implementation procedures of, 523–524
    numerical investigation of, 524–532
    semi-active control with MR dampers,
      528–529
    and unknown excitation, 515–519
System identification
  and SVC, 439–440
  and synthesis, of SHM and SVC, 453–459
Temperature effects, on civil structures, 12
Temperature loading, and LCM, 677–679
Thermometers, 68
Tilt-beams, 72–73
Time-frequency domain analysis, 133–134
Time-invariant parameters, and synthesis
  constant structural parameters and
    excitations, 493–497
  damage detection, 509–515
    under Kobe earthquake, 509–512
    under Northridge earthquake, 512–515
  damage scenarios, 506–507
  experimental setup, 504–506
  identification, 497–499
    and control algorithms, 507–509

implementation procedure of, 499–500
numerical investigation of, 500–504
overview, 492
Time lags, 140
Time-varying parameters, and synthesis
accuracy and excitation identification,
526–528
comparisons, 529–532
description of example building structure,
524–525
identification, 520–523
implementation procedures of, 523–524
numerical investigation of, 524–532
semi-active control with MR dampers,
528–529
and unknown excitation, 515–519
TinyOS, 146
TLCDs, see Tuned liquid column dampers
(TLCDs)
TMDs, see Tuned mass dampers (TMDs)
Transfer functions, 270–271
Transient response method, 337
Transmission tower
multi-scale modelling of, 174–175
overview, 172–173
physical model of, 173–174
validation of, 175–181
and multi-scale model updating, 203–205
Kriging vs. QPSR method, 210–211
model updating results and discussions,
207–209
Tsing Ma Bridge, 237–238
and SHM-based life-cycle management, 663
GPS-OSIS, 669
wind and structural health monitoring
system (WASHMS), 669
Tuned liquid column dampers (TLCDs), 98
Tuned mass dampers (TMDs), 97–98, 389, 553
UCS, see Uncoupled structures (UCS)
UHF, see Ultra-high-frequency (UHF)
Ultimate strength analysis, 615, 618–619
Ultra-high-frequency (UHF), 63–64
Ultrasonic pulse velocity method, 336
Uncoupled structures (UCS), 417
Under-sampling, 129
Uninterruptible power systems (UPS), 623
UPS, see Uninterruptible power systems (UPS)
Variable friction dampers, 408–410, 412–420
VAWT, see Vertical-axis wind turbine (VAWT)
Vertical-axis wind turbines (VAWTs), 536
analytical methods for, 588
computational fluid dynamics (CFD),
588–589
control of blade pitch angles
double disks multiple stream-tube
(DMST) model, 592–596
parking control algorithm, 602–603
pitch control system, 603–607

power maximisation control algorithm,
598
rated power control algorithm, 598–602
startup control algorithm, 596–598
field measurements of, 590–591
overview, 585–588
SHM of, 619–622
fatigue analysis, 612–618
framework, 609–612
ultimate strength analysis, 615, 618–619
wind loads on, 608–609
smart, 622–623
wind tunnel tests of, 589–590
Very-high-frequency (VHF), 64
VF, see Viscous fluid (VF) dampers
Vibrating wire strain gauge, 69–70
Vibration-based damage detection methods, 339
Vibration control performance, 572–575
Vibration damping, electromagnetic dampers
for, 554–555
Villari effect, 41
Viscous fluid (VF) dampers, 95–96
Vulnerability analysis, and LCM, 690
Vulnerability factors, and LCM, 688–689
WASHMS, see Wind and structural health
monitoring system (WASHMS); Wind
and Structural Health Monitoring
System (WASHMS)
Wavelet transform (WT), 190–191, 346–347
Weather stations, 80
Weibull distribution, 613
Weigh-in-motion (WIM), 67–68, 192
Wiedemann effect, 41
Wigner–Ville distribution (WVD), 190–191
WIM, see Weigh-in-motion (WIM)
Wind and structural health monitoring system
(WASHMS), 123, 669
Wind characteristics, and LCM, 674–677
Wind loads, 8–9
on vertical-axis wind turbines (VAWTs),
608–609
Wind profile measurements, 63–64
Wind sensors
anemometers, 61–63
pressure transducers, 63
wind profile measurements, 63–64
Wind tunnel tests, of VAWTs, 589–590
Wireless sensors and sensory systems
architectures and features of, 144–146
challenges in, 146
overview, 144
Wireless smart sensors (WSSs), 22
WSSs, see Wireless smart sensors (WSSs)
WT, see Wavelet transform (WT)
WVD, see Wigner–Ville distribution (WVD)
Yield stress, 45
Young's modulus, 41, 101
Zhaozhou Stone Arch Bridge, 12

Printed and bound by CPI Group (UK) Ltd, Croydon, CR0 4YY

01/11/2024

01782601-0018